CELL DEATH TECHNIQUES

A Laboratory Manual

ALSO FROM COLD SPRING HARBOR LABORATORY PRESS

RELATED TITLES

Cell Survival and Cell Death

Endocytosis

Mammalian Development: Networks, Switches, and Morphogenetic Processes

Means to an End: Apoptosis and Other Cell Death Mechanisms

Mitochondria

NF-κB: A Network Hub Controlling Immunity, Inflammation, and Cancer

Signal Transduction: Principles, Pathways, and Processes

OTHER LABORATORY MANUALS

Antibodies: A Laboratory Manual, Second Edition

Calcium Techniques: A Laboratory Manual

Imaging: A Laboratory Manual

Imaging in Developmental Biology: A Laboratory Manual

Imaging in Neuroscience: A Laboratory Manual

Manipulating the Mouse Embryo: A Laboratory Manual, Fourth Edition

Molecular Cloning: A Laboratory Manual, Fourth Edition

Molecular Neuroscience: A Laboratory Manual

Mouse Models of Cancer: A Laboratory Manual

Purifying and Culturing Neural Cells: A Laboratory Manual

RNA: A Laboratory Manual

Subcellular Fractionation: A Laboratory Manual

HANDBOOKS

At the Bench: A Laboratory Navigator, Updated Edition

At the Helm: Leading Your Laboratory, Second Edition

Experimental Design for Biologists, Second Edition

Lab Math: A Handbook of Measurements, Calculations, and Other Quantitative Skills for Use at the Bench

Lab Ref: A Handbook of Recipes, Reagents, and Other Reference Tools for Use at the Bench, Volume 1 and Volume 2

Statistics at the Bench: A Step-by-Step Handbook for Biologists

WEBSITE

www.cshprotocols.org

CELL DEATH TECHNIQUES

A Laboratory Manual

EDITED BY

Ricky W. Johnstone
Peter MacCallum Cancer Centre

John Silke
The Walter and Eliza Hall Institute

COLD SPRING HARBOR LABORATORY PRESS
Cold Spring Harbor, New York • www.cshlpress.org

CELL DEATH TECHNIQUES
A LABORATORY MANUAL

Publisher	John Inglis
Acquisition Editor	Richard Sever
Director of Editorial Services	Jan Argentine
Managing Editor	Maria Smit
Developmental Editor	Kaaren Janssen
Project Manager	Maryliz Dickerson
Production Editors	Joanne McFadden and Kathleen Bubbeo
Production Manager	Denise Weiss
Director of Product Development & Marketing	Wayne Manos
Cover Designer	Mike Albano

Cover art: Time-lapse images of a tumor cell undergoing apoptosis after exposure to perforin and granzyme B (top left, 5 min after exposure; bottom right, 4 h after exposure). On initiation of apoptosis, morphological changes—including cell rounding and blebbing—are immediately apparent. Annexin V (green) and propidium iodide (red) indicate phosphatidylserine exposure and loss of plasma membrane integrity in the final stages of cell death. (Image courtesy of Nigel Waterhouse, Jane Oliaro, Phil Bird, and Joe Trapani.)

Library of Congress Cataloging-in-Publication Data

Cell death techniques: a laboratory manual/edited by Ricky W. Johnstone, John Silke,
Peter MacCallum Cancer Centre, The Walter and Eliza Hall Institute.
    pages cm
ISBN 978-1-62182-012-3 (cloth)– ISBN 978-1-62182-005-5 (pbk.)
1. Cell death–Laboratory manuals. I. Johnstone, Ricky W. II. Silke, John.

QH671.C429 2015
571.9'36–dc23

2015020016

Contents

CHAPTER 4   **Biochemical Analysis of Initiator Caspase-Activating Complexes: The Apoptosome and Death-Inducing Signaling Complex**

INTRODUCTION

PROTOCOLS

CHAPTER 5   **Biochemical Assays of the Proapoptotic Proteins Bak and Bax**

INTRODUCTION

PROTOCOLS

CHAPTER 9   Strategies for Assaying Lysosomal Membrane Permeabilization

CHAPTER 10   Techniques to Distinguish Apoptosis from Necroptosis

CHAPTER 14   Monitoring Autophagy in *Drosophila*

INTRODUCTION

CHAPTER 15   Analysis of Apoptosis in *Caenorhabditis elegans*

INTRODUCTION

Acknowledgments

Putting this manual together has been a long haul and we sincerely thank all of the authors for their fantastic contributions and for their time, effort, and patience. . .thanks guys! We are indebted to the people at Cold Spring Harbor Laboratory Press for their help and support. In particular, Dr. Richard Sever, the Assistant Director, who originally approached us with the idea of a manual and maintained his enthusiasm throughout the project; Maryliz Dickerson, our project manager, who kept us heading in the right direction, and Maria Smit, managing editor, for her technical guidance. It has been a pleasure working with this dedicated group of professionals who never wavered in their support.

General Safety and Hazardous Material Information

This manual should be used by laboratory personnel with experience in laboratory and chemical safety or students under the supervision of such trained personnel. The procedures, chemicals, and equipment referenced in this manual are hazardous and can cause serious injury unless performed, handled, and used with care and in a manner consistent with safe laboratory practices. Students and researchers using the procedures in this manual do so at their own risk. It is essential for your safety that you consult the appropriate Material Safety Data Sheets, the manufacturers' manuals accompanying equipment, and your institution's Environmental Health and Safety Office, as well as the General Safety and Disposal Cautions in the Appendix for proper handling of hazardous materials in this manual. Cold Spring Harbor Laboratory makes no representations or warranties with respect to the material set forth in this manual and has no liability in connection with the use of these materials.

All registered trademarks, trade names, and brand names mentioned in this book are the property of the respective owners. Readers should please consult individual manufacturers and other resources for current and specific product information.

Appropriate sources for obtaining safety information and general guidelines for laboratory safety are provided in the General Safety and Hazardous Material Information Appendix.

CHAPTER 1

In the Midst of Life—Cell Death: What Is It, What Is It Good for, and How to Study It

John Silke[1,2,5] and Ricky W. Johnstone[3,4,5]

[1]The Walter and Eliza Hall Institute of Medical Research, Melbourne, Victoria 3052, Australia; [2]Department
of Medical Biology, University of Melbourne, Melbourne, Parkville, Victoria 3050, Australia; [3]Peter MacCallum
Cancer Centre, Melbourne, Victoria 3002, Australia; [4]The Sir Peter MacCallum Department of Oncology,
University of Melbourne, Melbourne, Victoria 3052, Australia

Cell death, one of the most fundamental biological processes, has not made it into the public consciousness in the same way that genetic inheritance, cell division, or DNA replication has. Everyone knows they get their genes from their parents, but few would be aware that even before they were born a lot of essential cell death has shaped their development. The greater population, for the most part, is blissfully unaware that every day millions of their own cells die in a programmed way and that this is essential for normal human physiology—their well-being, in fact. Nowhere is the burial liturgy, "In the midst of life we are in death," more apt. Despite this public underappreciation, cell death research is a major industry. A search in PubMed for "apoptosis," a special form of cell death that is caused by caspases, returns approximately 280,000 hits. The intense research interest arises from the realization that abnormal cell death responses play an important role in two of the biggest killers in the western world: cancer and cardio/cerebrovascular disease. Furthermore, the manner in which cells die can also influence the development of autoimmune and autoinflammatory diseases. It is therefore of paramount importance to ensure that experiments accurately quantitate and correctly identify cell death in all its guises. That is the goal of this protocol collection.

INTRODUCTION

In this introduction we take a very broad overview of the research area. Rather than try and attempt anything too profound, we provide a brief taster of some of the main ideas and questions in the field. As it is such a vibrant area of research, it would be impossible to summarize every facet of the area, and we would not even try. Rather, because this is a collection of protocols designed to allow the reader to experimentally explore questions about cell death, we will describe some basic concepts and pose some of the outstanding questions in the area.

One basic concept is that there are many different forms of cell death that have been identified and we therefore begin with very brief explanations of different types of cell death.

TYPES OF CELL DEATH

Accidental Cell Death Induced by Physical or Chemical Trauma

All cells are mortal—they can be killed by trauma, toxins, excessive temperatures, and chemical concentrations and other noxious stimuli. This type of cell death does not feature strongly in this

[5]Correspondence: j.silke@latrobe.edu.au; ricky.johnstone@petermac.org

Cite this introduction as Cold Spring Harb Protoc; doi:10.1101/pdb.top070508

collection. Noxious stimuli can activate programmed cell death pathways that this protocol collection does deal with, but at sufficiently high concentrations or intensities, these pathways may well be irrelevant as far as the ultimate fate of the cell is concerned.

Apoptosis

Apoptosis does not describe a cell death pathway but rather a cell death phenotype. The phenotype that the word describes consists of cell shrinkage, plasma membrane blebbing, nuclear chromatin condensation, and genomic DNA fragmentation among other things. Because these processes depend on caspase activity, apoptosis is, by definition, a caspase-dependent process. Initiator caspases are usually considered to become activated by conformational changes induced by oligomerization enforced by an upstream adaptor molecule, such as APAF-1 (caspase-9) or FADD and RIPK1 (caspase-8) in mammals (Boatright et al. 2003). These so-called initiator caspases, caspase-8 and caspase-9, become activated and are then able to process downstream (or executioner) caspases (e.g., caspase-3 and caspase-7) to induce their activation. Because the apoptotic morphology is widely conserved throughout evolution, conserved caspase substrates exist (Crawford et al. 2012) as nicely detailed by Gavin McStay and Doug Green (see Introduction: Measuring Apoptosis: Caspase Inhibitors and Activity Assays [McStay and Green 2014]). Frequently these are "effector" caspase substrates and their cleavage leads to the typical apoptotic morphology. For example, cleavage of ICAD, the inhibitor of the DNase CAD, by caspase-3 liberates CAD and results in DNA fragmentation. Likewise cleavage and inactivation of the enzymes involved in asymmetric distribution of phosphatidylserine (PS) in the plasma membrane, as well as cleavage and activation of flipase enzymes results in exposure of PS on the extracellular leaflet of the plasma membrane. Methods to detect and accurately measure molecular, biochemical, and biological hallmarks of apoptosis in mammals, flies, and worms are extensively covered in this collection. Moreover, advanced imaging techniques to assess dynamic apoptotic processes are now available, as showed in elegant detail by Lisa Bouchier-Hayes and Markus Rehm (see Introduction: Imaging-Based Methods for Assessing Caspase Activity in Single Cells [Parsons et al. 2015]).

Intrinsic Apoptosis?

The term "intrinsic apoptosis" is very frequently used and just as frequently misused. The signal that leads to activation of APAF-1/caspase-9/caspase-3 and an apoptotic morphology in mammals is the release of cytochrome c from the intermembrane space of mitochondria. Typically cytochrome c is released in response to such inducers of cell death as DNA damaging drugs, toxins, and lack of growth factors. Much developmental cell death is likely to be of this type. The APAF-1/caspase-9/caspase-3 has been termed the intrinsic apoptosis pathway because the signals are generated from within cells. However, and it is an important point to understand, the "point of no return" for a cell to survive has usually preceded this activation.

Cytochrome c is released following activation of Bax/Bak, a process covered in a series of protocols by Grant Dewson (see, e.g., Protocol 2: Investigating Bak/Bax Activating Conformation Change by Immunoprecipitation [Dewson 2015]). These proteins, once activated via the activities of BH3 proteins, are able to damage mitochondrial membranes and cause release of cytochrome c. In this case, if caspase inhibitors were applied to cells, they might be able to stop apoptosis (the particular morphology often associated with Bax/Bak-mediated cell death). But they would be unable to prevent cell death because, even if caspases are inhibited, the loss of cytochrome c and mitochondrial membrane disruption will cause cells to die from lack of ATP. Thus Bax/Bak-mediated cell death is often associated with caspase activation and cell death but it is not synonymous with "intrinsic apoptosis."

Extrinsic Apoptosis?

The signals that lead to the activation of caspase-8 in mammals are usually considered to be transmitted by plasma membrane receptors, often classified as "Death Receptors" such as Fas, TNF, and TRAIL-R, that respond to ligands delivered by other cells. Thus, cell death consequent on activation of caspase-8 is often considered to be via an "extrinsic" pathway. More recently, it has become clear that activation of caspase-8 to induce cell death can be accomplished by intracellular platforms including the "Ripoptosome" (containing RIPK1) but also MAVS/RIG and TLRs (Feoktistova et al. 2011; Tenev et al. 2011). In "extrinsic" apoptosis pathways, it is possible that blocking caspase-8 is sufficient to prevent cell death; however, as we shall discuss below, blocking caspase-8 may nevertheless trigger an alternative form of cell death called "necroptosis." Marion MacFarlane and Pascal Meier provide insight into these molecularly linked, yet phenotypically diverse processes and tools to study them (see Introduction: Biochemical Analysis of Initiator Caspase-Activating Complexes: The Apoptosome and Death-Inducing Signaling Complex [Langlais et al. 2015] and Introduction: Techniques to Distinguish Apoptosis from Necroptosis [Feoktistova et al. 2015]).

Finally, in some cells, activation of caspase-8 by plasma membrane receptors is insufficient to cause adequate caspase-3 activation to mediate the demise of a cell, yet is able to cause cell death. In these cells, caspase-8 is able to process and activate the BH3 protein Bid, and this activated BH3-only protein leads to the activation of Bax/Bak and cell death. Thus, as with intrinsic apoptosis, the term "extrinsic apoptosis" is potentially ambiguous and should be used and interpreted with caution.

Perforin and Granzyme B–Induced Cell Death

Certain cells, such as cytotoxic T cells and natural killer cells, are able to induce cell death by secreting a pore-forming protein, perforin, and delivering a class of serine proteases, called granzymes, into cells. Granzymes can enter cells in the absence of perforin, but perforin increases the efficiency and potency of delivery. Granzyme B has a unique substrate specificity and recognizes and cleaves the same short peptide motif that caspases cleave. Thus Granzyme B is able to cleave and activate downstream or effector caspases such as caspase-3, as well as the BH3 protein Bid (Waterhouse and Trapani 2002).

Caspase Activation Is Not Synonymous with Cell Death

Although caspase cleavage is frequently used to assess activation of caspases, it is not a required event for their activation. This is obviously true where initiator caspases are involved because conformational change induced by oligomerization platforms is all that is required to activate them (Boatright et al. 2003). It appears that caspase processing is probably a mechanism to ensure complete and irreversible activation. However, in most normal developmental situations, it is not easy to detect activated caspases, even in situations like the developing brain where there are supposed to be millions of cell deaths occurring because even early-stage apoptotic cells are rapidly engulfed. This phenomenon is discussed by Michael Bots and colleagues (see Introduction: Measuring Apoptosis in Mammals In Vivo [Newbold et al. 2014]).

Second, although apoptotic stimuli induce the activation of caspases, it has again become clear that even so-called apoptotic caspases may perform roles in addition to dismantling cells (Shalini et al. 2015). One of the best examples of alternative activity is that of caspase-8, which is linked to, among other things, the processing of cytokines and the inhibition of inflammasome activation. However, apoptotic caspases have also been linked to regulating cell cycle and other important physiological functions such as immune responses and cell fate determination (Shalini et al. 2015). This should not be too surprising because caspases, such as caspase-1, whose primary role appears to be proteolytic activation of cytokines such as IL-1β and IL-18, can nevertheless also induce cell death.

Thus, outstanding questions in this area are as follows.

1. How can apoptotic caspases be regulated such that they do not induce apoptosis?
2. To what extent does the extensive processing of caspases during apoptosis represent a physiological and/or pathological scenario? After all, even limited caspase activation may be sufficient to induce PS exposure and engulfment of a cell in vivo. It is possible to observe activated caspases in vivo in certain disease or mutant scenarios (i.e., in situations where massive cell death occurs or dead cells are ineffectively cleared). But in tissues where substantial developmental cell death occurs, such as in the thymus and brain, it is actually very difficult to quantify the degree of apoptosis because the

dying cells are so rapidly cleared from the microenvironment following PS exposure. Aspects of this are covered by Michael Bots and colleagues (see Introduction: Measuring Apoptosis in Mammals In Vivo [Newbold et al. 2014]).

Pyroptosis

Pyroptosis has certain similarities with necroptosis and necrosis. Like necrosis and necroptosis, pyroptosis is believed to result in cell swelling and the release of cellular contents, but it also has other apoptotic features such as DNA fragmentation and chromatin condensation. And, like apoptosis, it is caused by caspases. Both caspase-1 and caspase-11, in mice, have been implicated in this type of cell death, and depending on the stimulus, either may be used. It has been claimed that pyroptosis leads to release of IL-1β, IL-1α, and IL-18; however, more recent kinetic studies show that these cytokines, two of which are processed by caspase-1, are released before "cell death" (Gross et al. 2012). Although there are no caspase-1- or caspase-11-specific protocols, the assays to monitor caspase activity described in this collection are broadly applicable to all caspases (see Introduction: Imaging-Based Methods for Assessing Caspase Activity in Single Cells [Parsons et al. 2015]) and Introduction: Measuring Apoptosis: Caspase Inhibitors and Activity Assays [McStay and Green 2014]).

Mice deficient in these so called "inflammatory" caspases are highly susceptible to infection by bacteria, particularly intracellular pathogens; however, it is difficult to segregate the effects due to loss of cell death from those due to the loss of secretion of inflammatory mediators. Where it has been attempted, by testing mice deficient in signaling by the main cytokines that are dependent on caspase activity, IL-1β, IL-18, and IL-1α, but which retain the ability to activate caspase-1 and cause pyroptosis, the loss of cytokine production is far more detrimental than loss of caspase-1 activity (Aachoui et al. 2013). Most recently, a link between inflammasomes and caspase-8/caspase-3 activation has been described; thus, activation of inflammasomes, such as those mediated by AIM2 and NLRP3, can induce apoptosis. One study showed that lower doses of the inflammasome-activating stimulus preferentially induced apoptosis (Sagulenko et al. 2013).

Normally it is considered that a primary purpose, for the host, of infected cells undergoing cell death is to prevent the pathogen replicating and to expose it to the immune system. But pyroptosis is most frequently induced in the laboratory by bacterial PAMPs, not by infection. Do high doses of PAMPs mimic intracellular infection? Once dead, a cell cannot secrete more than its stockpile of cytokine; therefore, pyroptosis might be considered an antiinflammatory response or a mechanism to limit chronic inflammation that may ultimately be detrimental to the host (Martin et al. 2012). But this seems at odds with the idea that the release of intracellular contents may be inflammatory. The regulation of this process and how it is finely tuned and controlled to ensure maximal value to the host organism remains an area of intense study. In broader terms, the physiological importance of pyroptosis has yet to be fully articulated. If cells do not require pyroptosis to release inflammatory cytokines, why do they undergo this type of death? Given the potential involvement of caspase-8 in inflammasome activation, do inflammasome-activated cells only undergo a pyroptotic death?

Necrosis and Necroptosis

Necrosis is a type of cell death that is morphologically almost opposite to that of apoptosis. Instead of cell shrinkage, there is cell and organelle swelling, and there is a loss of intracellular contents, rather than a careful dismantling. In contrast to apoptosis, it was considered an accidental type of cell death. Methods to delineate necrosis from apoptosis are discussed by Nigel Waterhouse and colleagues (see Introduction: Dead Cert: Measuring Cell Death [Crowley et al. 2015a]). However, it has recently become apparent that in some cases this cell death morphology can also occur in a programmed manner. To indicate this necrotic cell morphology and the genetically encoded nature, this cell death has been coined "necroptosis." This cell death is caused by MLKL, and this protein, in an as-yet unknown manner, permeabilizes plasma membranes. MLKL can be activated by phosphorylation by RIPK3, and RIPK3 can in turn be activated by RIPK1, although RIPK1 is not always required. Activation of MLKL is often experimentally achieved by stimulating the receptors that initiate the extrinsic apoptosis

Cite this introduction as *Cold Spring Harb Protoc*; doi:10.1101/pdb.top070508

pathway and by preventing caspase-8 activation. Thus, one of the common ways to induce necroptosis is to add a ligand such as TNF and inhibit caspase-8 with a caspase inhibitor. In this pathway, caspase-8 cleaves RIPK1 and possibly RIPK3 to prevent them from activating MLKL. Pascal Meier, Martin Leverkus, and colleagues provide detailed methods to identify and quantify necroptotic features (see Introduction: Techniques to Distinguish Apoptosis from Necroptosis [Feoktistova et al. 2015]).

Cells displaying a necroptotic morphology are not particularly easy to observe in vivo, in contrast to apoptotic cells, where it is possible to detect cleaved DNA (using a TUNEL assay) or activated caspases in tissues. Necroptosis has been implicated as an active mediator of inflammation, and there is circumstantial evidence for a role in Crohn's disease, ulcerative colitis, or allergic colitis (Pasparakis and Vandenabeele 2015). Further-proposed pathological roles for necroptosis include situations of acute tissue damage, such as stroke and myocardial infarction, and in ischemia–reperfusion injury. Exactly how necroptosis contributes to the initiation, amplification, and/or maintenance of an inflammatory state and how cell autonomous or extrinsic factors regulate this process remain to be determined. However, the evidence for an important pathophysiological role for necroptosis is mounting, and the methods described by Pascal Meier, Martin Leverkus, and colleagues will facilitate further experimental investigation in this area (see Introduction: Techniques to Distinguish Apoptosis from Necroptosis [Feoktistova et al. 2015]).

Autophagy-Dependent Cell Death

Autophagy-dependent cell death is, like apoptosis, morphologically defined. The term really describes cell death with autophagy. Autophagy is an essential cellular process that disposes of organelles, cytosolic bacteria, and cytosol via a membrane vesicle mechanism that delivers these components to lysosomes to be degraded and recycled. Of all the types of cell death that are discussed in this collection, autophagy-dependent cell death is perhaps the most controversial and most representative of the fact that cell death maybe a consequence of an adaptive cellular response. There are clear cases, such as the remodeling *Drosophila* larval gut during metamorphosis, where the loss of genes required for autophagy reduces cell death, indicating that this is truly cell death caused by autophagy. Involuting *Drosophila* salivary glands also use autophagy; however, unlike the gut, caspases are also required for complete involution. Autophagy-dependent cell death also plays a role in developmental cell death in plants, such as *Arabidopsis*, and stalk formation in the single-celled organism *Dictyostelium* (Nelson and Baehrecke 2014).

However, from yeast to mammals, it is clear that autophagy is primarily used as a cell homeostasis and survival mechanism. The essential requirement for autophagy to dispose of organelles and recycle cellular components has been shown by genetic experiments. For example, when tissue-specific knockout mice of essential autophagy genes are analyzed, they accumulate damaged mitochondria that in turn cause severe disruption to cellular function. The ability of this process to recycle components also makes it valuable/vital in times of nutrient starvation and stress. A series of protocols within this collection describe contemporary methods used in mammalian, *Drosophila*, and *Caenorhabditis elegans* systems to study and read out molecular and biochemical features of autophagy (see, e.g., Protocol 1: LysoTracker Staining to Aid in Monitoring Autophagy in *Drosophila* [DeVorkin and Groski 2015] and Introduction: Detection of Autophagy in *Caenorhabditis elegans* [Palmisano and Meléndez 2015]). The discussion by Katja Simon and colleagues (Introduction: Techniques for the Detection of Autophagy in Primary Mammalian Cells [Puleston et al. 2015]) summarizes the types of autophagy, the cellular role it plays, and the chief molecular components required for it. Furthermore, they make it clear that the most important consideration/question in this field is whether autophagy executes cell death in vivo.

NETosis

Neutrophils, and some other myeloid cell types, have the ability to expel net-like structures containing their nuclear DNA coated with histones and proteases from their granules, such as elastase. The structures have been named neutrophil extracellular traps (NETs) and their formation is usually linked to plasma membrane disruption and cell death, hence NETosis. The fundamental questions in this area are: What is the physiological function of NETs? Are they a neutrophil host defense mechanism to disable bacteria and fungi or at least hinder their dissemination? Or are they a consequence of cell death induced by pathogen toxins? Is there a molecular link between NETosis and other forms of cell death such as necroptosis? And, finally and remarkably, there is even some question as to whether expulsion of nuclear DNA necessarily leads to cell death (Yipp and Kubes 2013).

THE BIGGER PICTURE

In this section, we have highlighted points that often get overlooked when investigating cell death and therefore are pertinent to raise by way of introduction to the content of this collection.

A core question not restricted to cell death and very topical is to what extent we can rely on published research (Begley and Ellis 2012; Errington et al. 2014; Morrison 2014; Begley and Ioannidis 2015). The scientific approach has always been one of healthy skepticism, and a key tenet is that others should be able to reproduce a scientific discovery for it to be valid. Francis Bacon, considered by many to be the father of modern science, recognized that following academic dogma and not asking one's own questions about the world obstructed scientific reasoning and discovery (Bacon 1620). However, increasing complexity and specialization, among other things, often makes replication hard in practice. Therefore, Vaux (2015) provides some useful advice for interpreting published research and for communicating one's own research to enable reproducibility with reference to cell death research.

Another point that is often overlooked in the field is the "simple" question of when can we consider a cell to be "dead"? As Vaux (2015) states, "at first glance it seems trivial to decide whether a cell is alive or dead." And the absurdity of the idea that anything can be both alive and dead is captured in such diverse areas as particle physics, Schrodinger's cat, and popular culture—for example, the song Grandad from Jake Thackeray reads "Although they stuffed him in a coffin and read out the will, Although he's six foot deep in darkness he'll never lie still."

It has been proposed that a cell should be considered dead if the integrity of the plasma membrane is irretrievably lost; the cell, including its nucleus, has undergone fragmentation; or its "corpse" has been engulfed by an adjacent cell (Kroemer et al. 2009). These are perfectly fine criteria but, as discussed above and by Michael Bots and colleagues (see Introduction: Measuring Apoptosis in Mammals In Vivo [Newbold et al. 2014]), in vivo the fact that an apoptotic cell "looks a little pale" (i.e., exposes PS), and long before membrane permeabilization occurs, can be sufficient to promote its rapid engulfment and disposal in vivo. On the other hand, the fact that engulfed cells can sometimes escape suggests that really only the first two are true markers of cell death (Galluzzi et al. 2015). Whatever weight one wishes to give these findings, it is very clear that apoptotic cells, which are so easy to distinguish in vitro, are very difficult to detect in vivo unless there is massive co-ordinated cell death that overwhelms the disposal of dead cells. Indeed, even in *C. elegans*, where dead cells can persist for ~40 min and can be easily distinguished with the appropriate microscope setup (see, e.g., Protocol 2: Visualizing Apoptosis in Embryos and the Germline of *Caenorhabditis elegans* [Lant and Derry 2014]), it was the use of *ced-1* mutants that are defective in dead cell disposal that allowed the identification of other cell death mutants. Thus one has to ask in what circumstances the intricate downstream dismantling of cells by caspases that takes several hours post PS exposure in vitro is relevant in vivo. One very interesting addition to this discussion is the finding that Apaf-1/caspase-9/caspase-3-induced apoptosis prevents production of IFNs in vivo. Specifically, Apaf-1/caspase-9/caspase-3 activation was shown to be necessary to correctly dispose of mitochondrial DNA. In the absence of these effectors, mitochondria become leaky and mitochondrial DNA is able to activate cytosolic DNA sensors to stimulate IFN production (Rongvaux et al. 2014; White et al. 2014). Along with a host of other evidence (Yoshida et al. 2005; Hanayama et al. 2006; Suzuki et al. 2013; Segawa et al. 2014), this indicates that a primary function of apoptosis is to promote the clearance of dead cells and thereby prevent an inflammatory or immune reaction. Whatever the purpose, in multicellular organisms, a cell can be effectively dead long before it displays many of the classical biochemical signatures of cell death.

Cite this introduction as *Cold Spring Harb Protoc*; doi:10.1101/pdb.top070508

The flip side of this question is whether a cell is alive. This question turns out to be difficult to answer when we consider the context in which it is usually asked. This question is often posed in short-term cell death assays that cannot evaluate whether a particular intervention or insult has simply delayed death, which is unlikely to be of much interest, or prevented it. Long-term survival assays in vitro are not necessarily easy to perform and usually conflate the two parameters of function and survival. For example, in vitro colony assays are useful to determine if cells that have the capacity to divide and form clones have retained that capacity following a death stimulus (see Protocol 5: Measuring Survival of Adherent Cells with the Colony-Forming Assay [Crowley et al. 2015b] and Protocol 6: Measuring Survival of Hematopoietic Cancer Cells with the Colony-Forming Assay in Soft Agar [Crowley and Waterhouse 2015]). If they have, then they were undoubtedly alive when seeded onto the tissue culture plate, *quod erat demonstratum*. However, this type of assay is not applicable to many cells, such as postmitotic neurons, whose death we might be interested in and wish to prevent. And even the colony assay is open to the criticism that the cells might still be living and functioning as ascertained by every other measure but have lost the ability to divide.

Furthermore biologists interested in multicellular animals are rarely concerned whether cells have a vegetative existence and much more interested in whether they have a functional existence. Thus we often determine whether cells display the biochemical signature of cell death, which as this collection shows is relatively straightforward to determine, when what we really want to know is whether the cells are alive and functioning, which is often far more difficult to ascertain in vitro. From a pragmatic viewpoint, as biologists, what we are interested in, more often than not, is whether a cell is alive and functioning. If we consider a neuron in an ischemic brain that has been prevented from dying by a novel cell death inhibiting drug, what we as drug deliverers want to achieve is to prevent and not delay its death and to maintain its functional existence not a vegetative state. Thus, although this is conflating two different issues (i.e., death and function) from a pragmatic perspective, if we wish to clinically prevent ischemic reperfusion brain injury, it is probably not enough to prevent cell death. However, in other scenarios, such as pyroptotic or necroptotic deaths that release DAMPs that activate the immune system and might thereby contribute to auto-inflammatory diseases, it maybe simply sufficient to stop cell death.

Naturally in vivo experiments often allow us to address this live/dead function conundrum directly, but it is very worthwhile considering that these experiments are far more dependent on genetic background than is widely acknowledged. In mammals, the loss of the key effector molecules of the intrinsic apoptosis pathway, apaf-1, caspase-9, and caspase-3, results in exencephaly in murine embryos and consequent lethality. In C57BL/6 mice, the loss of Bax and Bak, which are required for mitochondrial permeabilization, likewise result in failure of embryonic development. However it is worth remarking that many of these developmental failures are not absolute: Occasional Bax/Bak double knockout mice are born (Lindsten et al. 2000), and the effect of the loss of caspase-9 and apaf-1 is very dependent on the background strain of the knockout mice. In yet another example, $casp3^{-/-}$ mice generated on a mixed 129/Sv C57BL/6 background showed exencephaly and consequent perinatal lethality. However, when backcrossed further onto the C57BL/6 background, these knockout mice survive through adulthood with no obvious defects, whereas pure 129/Sv $casp3^{-/-}$ mice are embryonic lethal (Zheng 2000).

Another point often overlooked is whether cell death pathways really occur in isolation from each other. A prime example of this is the field of lysosomal membrane permeabilization (see Introduction: Methods for Probing Lysosomal Membrane Permeabilization [Jäättelä and Nylandsted 2015] and Introduction: Strategies for Assaying Lysosomal Membrane Permeabilization [Repnik et al. 2015]). Thus, the permeabilization of lysosomal and late endosomal membranes can result in the release of their lytic cargo into the cytosol and the activation of caspases, as well as the release of lysosomal proteases that can act as executioner proteases in their own right. On the other hand, apoptosis can be associated with the destabilization of these membranes and release of the contents of these organelles. This intimate link between these two pathways highlights the difficulties in studying these events as well as another fallacy that Francis Bacon identified—the very human tendency to categorize and impose more order than it finds. However these links exist in other pathways. It seems clear that reduced caspase-8 activity can lead to necroptosis, because caspase-8 limits activation of RIPK1 and RIPK3. But there still might be sufficient caspase-8 activity to activate caspase-3 so that we have a type of chimeric cell death. The tendency is for experimenters to look at what they want to see, and thus studies that exclude chimeric cell death are not abundant.

WHAT REMAINS TO DO AND FIND OUT?

With more than 280,000 articles in the mainstream scientific literature on "apoptosis" alone, we have clearly learned a lot about programmed cell death in recent years. Nevertheless the fact that this is a collection of experimental techniques makes it plain that there is still much to learn that will not be merely incremental advances in knowledge. Technological advances often precede and drive discovery, but there are a number of such advances that simply require the appropriate molecular tools to be fully used.

Markers and Reporters for Cell Death

For more recently described types of cell death, such as necroptosis and pyroptosis, there is still a lack of good markers that can be used in vitro. The recent development of a human-specific phospho-MLKL antibody (Wang et al. 2014) has helped for necroptosis, but more such tools are needed.

There is also a dearth of good markers and reporters for autophagy, pyroptosis, and necroptosis occurring in vivo in mammals, even in fixed tissues. This is unsurprising when we consider that in the well-researched area of apoptosis, most mammalian and *Drosophila* in vivo assays require fixing of tissues and cannot be used in real time (see Introduction: Measuring Apoptosis in Mammals In Vivo [Newbold et al. 2014] and Introduction: Studying Apoptosis in *Drosophila* [Denton and Kumar 2015]). This is one of the attractions of transparent organisms such as *C. elegans* and zebrafish, where apoptotic and autophagic cells are far easier to detect in situ with conventional microscopy techniques (e.g., see Protocol 2: Visualizing Apoptosis in Embryos and the Germline of *Caenorhabditis elegans* [Lant and Derry 2014] and Introduction: Detection of Autophagy in *Caenorhabditis elegans* [Palmisano and Meléndez 2015]). The next technological frontier will be to develop reporters that are able to detect cell death in real time in vivo. With such reporters, questions about the physiological relevance of particular types of cell death will be quickly resolved. Advances in microscopic techniques and equipment will undoubtedly contribute to such a revolution but molecular tools are still sorely needed.

Genetic Manipulation

The impact of new technologies is illustrated beautifully by the emerging role played in research by the relatively new technologies of CRISPR and TALENs. Genetic experiments are still some of the most reliable evidence for the involvement of a particular gene/protein in a particular pathway and the ability to now not only rapidly delete but even tag endogenous molecules with these technologies is likely to immediately lead to an increase in the reporter constructs that will help transform research of endogenous and physiological cell death (Wang et al. 2013; Yang et al. 2013). The CRISPR approach is also highly suited to high-throughput death assays; the contribution by Kaylene Simpson and colleagues (see Introduction: High-Throughput Approaches to Measuring Cell Death [Saunders et al. 2014]) gives examples of such assays and provides important design considerations.

Therapeutics and Small Drug-Like Molecules

As outlined elsewhere in this collection, defects in cell death pathways result in diseases. Indeed in the cancer setting it has been appreciated for some time that chemotherapeutic drugs, used for decades in the clinic, work by inducing apoptosis and that drug resistance can be mediated by defects in the apoptotic machinery or signaling processes (Hanahan and Weinberg 2000; Johnstone et al. 2002). A particularly exciting research area is therefore to try to target these cell death pathways with new drugs. This has been a spectacular, if still underappreciated, success story. Thus small molecules that mimic protein-protein interaction surfaces are able to activate apoptosis selectively in cancer cells and are now performing well in clinical trials (Mason et al. 2007; Varfolomeev et al. 2007; Vince et al. 2007; Souers et al. 2013; Condon et al. 2014). These specific small molecule inhibitors have also had an important knock-on effect in that they allow researchers to investigate the function of their targets in acute situations rather than with still comparatively cumbersome genetic techniques (Condon et al. 2014; Wong et al. 2014).

However the converse situation of morbidities where there is too much cell death, such as stroke and ischemia-reperfusion injuries has proven less easy to inhibit. The recent advances in robotic screening of well curated chemical libraries does provide an opportunity to develop such inhibitors and these are increasingly within reach of smaller academic institutions and not just major pharmaceutical companies. With this in mind, Kaylene Simpson and colleagues have outlined how to set up such high-throughput death assays (see Introduction: High-Throughput Approaches to Measuring Cell Death [Saunders et al. 2014]).

Go Forth

In conclusion, there is still much to discover in how, when, and where cell death occurs and what the consequences of deregulated cell death may be for human development and health. There have been incredible advances in our ability to recognize and accurately measure the molecular, biological, and biochemical hallmarks of different forms of cell death and we hope that the protocols described in this collection help you make new discoveries and advance our knowledge of this fascinating field of research.

COMPETING INTEREST STATEMENT

The authors declare no conflicts of interest.

ACKNOWLEDGMENTS

Our work is made possible through Victorian State Government Operational Infrastructure Support and Australian Government National Health and Medical Research Council, Independent Research Institutes Infrastructure Support Scheme, and National Health and Medical Research Council research Fellowships to J.S. and R.W.J.

REFERENCES

Aachoui Y, Sagulenko V, Miao EA, Stacey KJ. 2013. Inflammasome-mediated pyroptotic and apoptotic cell death, and defense against infection. *Curr Opin Microbiol* **16**: 319–326.

Bacon F. 1620. *"Novum organum"* London.

Begley CG, Ellis LM. 2012. Drug development: Raise standards for preclinical cancer research. *Nature* **483**: 531–533.

Begley CG, Ioannidis JP. 2015. Reproducibility in science: Improving the standard for basic and preclinical research. *Circ Res* **116**: 116–126.

Boatright KM, Renatus M, Scott FL, Sperandio S, Shin H, Pedersen IM, Ricci JE, Edris WA, Sutherlin DP, Green DR, et al. 2003. A unified model for apical caspase activation. *Mol Cell* **11**: 529–541.

Condon SM, Mitsuuchi Y, Deng Y, Laporte MG, Rippin SR, Haimowitz T, Alexander MD, Kumar PT, Hendi MS, Lee YH, et al. 2014. Birinapant —A Smac-mimetic with improved tolerability for the treatment of solid tumors and hematological malignancies. *J Med Chem* **57**: 3666–3677.

Crowley LC, Waterhouse NJ. 2015. Measuring survival of hematopoietic cancer cells with the colony-forming assay in soft agar. *Cold Spring Harb Protoc* doi: 10.1101/pdb.prot087189.

Crowley LC, Marfell BJ, Scott AP, Boughaba JA, Chojnowski G, Christensen ME, Waterhouse NJ. 2015a. Dead cert: Measuring cell death. *Cold Spring Harb Protoc* doi: 10.1101/pdb.top070318.

Crowley LC, Christensen ME, Waterhouse NJ. 2015b. Measuring survival of adherent cells with the colony-forming assay. *Cold Spring Harb Protoc* doi: 10.1101/pdb.prot087171.

Crawford ED, Seaman JE, Barber AE, David DC, Babbitt PC, Burlingame AL, Wells JA. 2012. Conservation of caspase substrates across metazoans suggests hierarchical importance of signaling pathways over specific targets and cleavage site motifs in apoptosis. *Cell Death Differ* **19**: 2040–2048.

Denton D, Kumar S. 2015. Studying apoptosis in *Drosophila*. *Cold Spring Harb Protoc* doi: 10.1101/pdb.top070433.

DeVorkin L, Gorski S. 2015. LysoTracker staining to aid in monitoring autophagy in *Drosophila*. *Cold Spring Harb Protoc* doi: 10.1101/pdb .prot080325.

Dewson G. 2015. Investigating Bak/Bax activating conformation change by immunoprecipitation. *Cold Spring Harb Protoc* doi: 10.1101/pdb .prot086454.

Errington TM, Iorns E, Gunn W, Tan FE, Lomax J, Nosek BA. 2014. An open investigation of the reproducibility of cancer biology research. *Elife* 3: e04333. doi: 10.7554/eLife.04333.

Feoktistova M, Geserick P, Kellert B, Dimitrova DP, Langlais C, Hupe M, Cain K, Macfarlane M, Hacker G, Leverkus M. 2011. cIAPs Block ripoptosome formation, a RIP1/Caspase-8 containing intracellular cell death complex differentially regulated by cFLIP isoforms. *Mol Cell* 43: 449–463.

Feoktistova M, Wallberg F, Tenev T, Geserick P, Leverkus M, Meier P. 2015. Techniques to distinguish apoptosis from necroptosis. *Cold Spring Harb Protoc* doi: 10.1101/pdb.top070375.

Galluzzi L, Bravo-San Pedro JM, Vitale I, Aaronson SA, Abrams JM, Adam D, Alnemri ES, Altucci L, Andrews D, Annicchiarico-Petruzzelli M, et al. 2015. Essential versus accessory aspects of cell death: recommendations of the NCCD 2015. *Cell Death Differ* 22: 58–73.

Gross O, Yazdi AS, Thomas CJ, Masin M, Heinz LX, Guarda G, Quadroni M, Drexler SK, Tschopp J. 2012. Inflammasome activators induce interleukin-1α secretion via distinct pathways with differential requirement for the protease function of caspase-1. *Immunity* 36: 388–400.

Hanahan D, Weinberg RA. 2000. The hallmarks of cancer. *Cell* 100: 57–70.

Hanayama R, Miyasaka K, Nakaya M, Nagata S. 2006. MFG-E8-dependent clearance of apoptotic cells, and autoimmunity caused by its failure. *Curr Dir Autoimmun* 9: 162–172.

Jäättelä M, Nylandsted J. 2015. Methods for probing lysosomal membrane permeabilization. *Cold Spring Harb Protoc* doi: 10.1101/pdb.top070367.

Johnstone RW, Ruefli AA, Lowe SW. 2002. Apoptosis: A link between cancer genetics and chemotherapy. *Cell* 108: 153–164.

Kroemer G, Galluzzi L, Vandenabeele P, Abrams J, Alnemri ES, Baehrecke EH, Blagosklonny MV, El-Deiry WS, Golstein P, Green DR, et al. 2009. Classification of cell death: Recommendations of the Nomenclature Committee on Cell Death 2009. *Cell Death Differ* 16: 3–11.

Langlais C, Hughes MA, Cain K, MacFarlane M. 2015. Biochemical analysis of initiator caspase-activating complexes: The apoptosome and death-inducing signaling complex. *Cold Spring Harb Protoc* doi:10.1101/pdb.top070326.

Lant B, Derry WB. 2014. Visualizing apoptosis in embryos and the germline of *Caenorhabditis elegans*. *Cold Spring Harb Protoc* doi: 10.1101/pdb.prot080218.

Lindsten T, Ross AJ, King A, Zong WX, Rathmell JC, Shiels HA, Ulrich E, Waymire KG, Mahar P, Frauwirth K, et al. 2000. The combined functions of proapoptotic Bcl-2 family members bak and bax are essential for normal development of multiple tissues. *Mol Cell* 6: 1389–1399.

Martin SJ, Henry CM, Cullen SP. 2012. A perspective on mammalian caspases as positive and negative regulators of inflammation. *Mol Cell* 46: 387–397.

Mason KD, Carpinelli MR, Fletcher JI, Collinge JE, Hilton AA, Ellis S, Kelly PN, Ekert PG, Metcalf D, Roberts AW, et al. 2007. Programmed anuclear cell death delimits platelet life span. *Cell* 128: 1173–1186.

McStay GP, Green DR. 2014. Measuring apoptosis: Caspase inhibitors and activity assays. *Cold Spring Harb Protoc* doi:10.1101/pdb.top070359.

Morrison SJ. 2014. Time to do something about reproducibility. *Elife* 3 doi: 10.7554/eLife.03981.

Nelson C, Baehrecke EH. 2014. Eaten to death. *FEBS J* 281: 5411–5417.

Newbold A, Martin B, Cullinane C, Bots M. 2014. Measuring apoptosis in mammals in vivo. *Cold Spring Harb Protoc* doi:10.1101/pdb.top070417.

Palmisano NJ, Meléndez A. 2015. Detection of autophagy in *C. elegans*. *Cold Spring Harb Protoc* doi: 10.1101/pdb.top070466.

Parsons MJ, Rehm M, Bouchier-Hayes L. 2015. Imaging-based methods for assessing caspase activity in single cells. *Cold Spring Harb Protoc* doi: 10.1101/pdb.top070342.

Pasparakis M, Vandenabeele P. 2015. Necroptosis and its role in inflammation. *Nature* 517 311–320.

Puleston D, Phadwal K, Watson AS, Soilleux EJ, Chittaranjan S, Bortnik S, Gorski SM, Ktistakis N, Simon AK. 2015. Techniques for the detection of autophagy in primary mammalian cells. *Cold Spring Harb Protoc* doi: 10.1101/pdb.top070391.

Repnik U, Česen MH, Turk B. 2015. Strategies for assaying lysosomal membrane permeabilization. *Cold Spring Harb Protoc* doi: 10.1101/pdb.top077479.

Rongvaux A, Jackson R, Harman CC, Li T, West AP, de Zoete MR, Wu Y, Yordy B, Lakhani SA, Kuan CY, et al. 2014. Apoptotic caspases prevent the induction of type I interferons by mitochondrial DNA. *Cell* 159: 1563–1577.

Sagulenko V, Thygesen SJ, Sester DP, Idris A, Cridland JA, Vajjhala PR, Roberts TL, Schroder K, Vince JE, Hill JM, et al. 2013. AIM2 and NLRP3 inflammasomes activate both apoptotic and pyroptotic death pathways via ASC. *Cell Death Differ* 20: 1149–1160.

Saunders DN, Falkenberg KJ, Simpson KJ. 2014. High-throughput approaches to measuring cell death. *Cold Spring Harb Protoc* doi: 10.1101/pdb.top072561.

Segawa K, Kurata S, Yanagihashi Y, Brummelkamp TR, Matsuda F, Nagata S. 2014. Caspase-mediated cleavage of phospholipid flippase for apoptotic phosphatidylserine exposure. *Science* 344: 1164–1168.

Shalini S, Dorstyn L, Dawar S, Kumar S. 2015. Old, new and emerging functions of caspases. *Cell Death Differ* 22: 526–539.

Souers AJ, Leverson JD, Boghaert ER, Ackler SL, Catron ND, Chen J, Dayton BD, Ding H, Enschede SH, Fairbrother WJ, et al. 2013. ABT-199, a potent and selective BCL-2 inhibitor, achieves antitumor activity while sparing platelets. *Nat Med* 19: 202–208.

Suzuki J, Denning DP, Imanishi E, Horvitz HR, Nagata S. 2013. Xk-related protein 8 and CED-8 promote phosphatidylserine exposure in apoptotic cells. *Science* 341: 403–406.

Tenev T, Bianchi K, Darding M, Broemer M, Langlais C, Wallberg F, Zachariou A, Lopez J, MacFarlane M, Cain K, et al. 2011. The Ripoptosome, a signaling platform that assembles in response to genotoxic stress and loss of IAPs. *Mol Cell* 43: 432–448.

Varfolomeev E, Blankenship JW, Wayson SM, Fedorova AV, Kayagaki N, Garg P, Zobel K, Dynek JN, Elliott LO, Wallweber HJ, et al. 2007. IAP antagonists induce autoubiquitination of c-IAPs, NF-κB activation, and TNFα-dependent apoptosis. *Cell* 131: 669–681.

Vaux DL. 2015. Presentation and interpretation of figures in cell death research publications. In *Cell death techniques: A laboratory manual* (ed Johnstone, Silke) Cold Spring Harbor Laboratory Press, Cold Spring Harbor, NY.

Vince JE, Wong WW, Khan N, Feltham R, Chau D, Ahmed AU, Benetatos CA, Chunduru SK, Condon SM, McKinlay M, et al. 2007. IAP antagonists target cIAP1 to induce TNFα-dependent apoptosis. *Cell* 131: 682–693.

Wang H, Yang H, Shivalila CS, Dawlaty MM, Cheng AW, Zhang F, Jaenisch R. 2013. One-step generation of mice carrying mutations in multiple genes by CRISPR/Cas-mediated genome engineering. *Cell* 153: 910–918.

Wang H, Sun L, Su L, Rizo J, Liu L, Wang LF, Wang FS, Wang X. 2014. Mixed lineage kinase domain-like protein MLKL causes necrotic membrane disruption upon phosphorylation by RIP3. *Mol Cell* 54: 133–146.

Waterhouse NJ, Trapani JA. 2002. CTL: *Caspases* terminate life, but that's not the whole story. *Tissue Antigens* 59: 175–183.

White MJ, McArthur K, Metcalf D, Lane RM, Cambier JC, Herold MJ, van Delft MF, Bedoui S, Lessene G, Ritchie ME, et al. 2014. Apoptotic caspases suppress mtDNA-induced STING-mediated type I IFN production. *Cell* 159: 1549–1562.

Wong WW, Vince JE, Lalaoui N, Lawlor KE, Chau D, Bankovacki A, Anderton H, Metcalf D, O'Reilly L, Jost PJ, et al. 2014. cIAPs and XIAP regulate myelopoiesis through cytokine production in an RIPK1- and RIPK3-dependent manner. *Blood* 123: 2562–2572.

Yang H, Wang H, Shivalila CS, Cheng AW, Shi L, Jaenisch R. 2013. One-step generation of mice carrying reporter and conditional alleles by CRISPR/Cas-mediated genome engineering. *Cell* 154: 1370–1379.

Yipp BG, Kubes P. 2013. NETosis: How vital is it? *Blood* 122: 2784–2794.

Yoshida H, Okabe Y, Kawane K, Fukuyama H, Nagata S. 2005. Lethal anemia caused by interferon-β produced in mouse embryos carrying undigested DNA. *Nat Immunol* 6: 49–56.

Zheng TS. Learning from deficiency: Gene targeting of caspases. In: Madame Curie Bioscience Database [Internet]. Austin (TX): Landes Bioscience; 2000. Available from http://www.ncbi.nlm.nih.gov/books/NBK6153/.

CHAPTER 2

Presentation and Interpretation of Figures in Cell Death Research Publications

David L. Vaux[1]

Department of Medical Biology, The Walter and Eliza Hall Institute, University of Melbourne, Melbourne, Victoria 3052, Australia

By convention, articles in scientific journals not only contain the authors' conclusions, but also the evidence to back them up. Evidence is given in the form of tables of data and figures. To convince skeptical reviewers and readers that the authors' conclusions are correct, the figures need to be presented in a way in which they can be easily understood. Additionally, because more than 20,000 new cell death papers appear in the literature each year, it is useful to have skills to rapidly identify those that merit further consideration. Here I suggest how to present and interpret data in papers describing cell death research. The goal is to improve rigor when critically evaluating not only others' results, but also one's own.

INTRODUCTION

In general, science is performed when researchers perform experiments and discover something new, then repeat their experiments, or perform further experiments to confirm that the first results were valid. They then try to make sense of their data, write up, and publish their findings. Research into cell death is no different.

Researchers investigating cell death spend much of their time either conducting experiments or reading papers that describe other researchers' experiments. Both involve the presentation and interpretation of data in figures in publications. Usually, some of the experiments involve obtaining quantitative information on the number of cells that are alive or dead. First, there must be an assay that can distinguish between live and dead cells. Second, enough data must be obtained to provide strong evidence that the results are reproducible. Inferences can only be made from independent data, and, where there is doubt, tests can be used to show whether differences reach statistical significance. If differences are not statistically significant, one cannot exclude the possibility that the differences observed are just due to chance, but it is also not possible to conclude there is no effect. When the results are statistically significant, comments should be made on the size of the effect, and, ideally, the conclusions should be verified by another, biologically different, but related, experiment (Cumming et al. 2007; Vaux et al. 2012).

ALIVE OR DEAD?

Although at first glance it seems trivial to decide whether a cell is alive or dead, it is often not as simple as it appears. The "gold standard" test to show that a cell is alive is a clonogenic assay—if a cell is

[1]Correspondence: vaux@wehi.edu.au

Cite this introduction as *Cold Spring Harb Protoc*; doi:10.1101/pdb.top070300

capable of dividing it is unequivocally alive. However, the converse is not true—if a cell does not form a colony, it does not necessarily mean that it is dead. Similarly, tests of plasma membrane integrity, such as the inability to exclude dyes such as trypan blue, eosin, or propidium iodide, can reliably show that a cell is dead, but it does not prove that it is viable, or that it has not committed to die in the immediate future. The results of most other tests give less definitive results, and even these tests are only good in one direction—if a cell divides it must have been alive, but if a cell has not divided, it does not necessarily mean that it is dead. If a cell takes up propiduim iodide and becomes highly fluorescent it is dead, but if a cell does not fluoresce in propidium iodide it is not necessarily alive, and might, for example, have already activated caspases, cleaved its DNA, and exposed phosphatidylserine on its surface.

Beyond measuring how many cells are alive or dead, research on cell death is concerned with the mechanisms, regulation, and role of cell death in both normal and pathological situations. Therefore, in cell death publications, the figures will show many other types of experiments in addition to those enumerating live and dead cells, such as western blots, flow cytometry, and microscopy. Usually, these figures provide evidence that is not statistical. It is important to note that just as statistical inferences can only be made when there is independent data, inferences from nonstatistical experiments can only be made if the experiment is repeated or if the hypothesis is confirmed by some independent test.

STATISTICS

In some types of research publications, such as those describing clinical drug trials, epidemiology, or genome-wide association studies, there is typically only a single experiment, and the analysis is purely statistical, with biology being treated as a black box. In these cases, it is important to have rigorous statistics, large numbers, proper randomization, and blinding. In contrast, papers describing experimental biology, including research into cell death, usually describe a number of different experiments that seek to put forward and test a hypothesis in several distinct ways. For example, one figure might describe the ability of two proteins to bind to each other in a coimmunoprecipitation assay; the second figure might show a structure–function analysis of the ability of mutant proteins to interact; the third might show the amount of cell death that occurs in cells overexpressing the wild-type and mutant proteins; the fourth might show whether inhibitors of the protein can prevent the interaction, or reduce the amount of cell death caused. There might be additional figures describing the response of cells in which the genes for the proteins have been knocked out or knocked down.

Because research into the biology of cell death does not rely entirely on statistical evidence, it is seldom necessary to obtain the dozens or hundreds of independent data points that are needed for drug trials, epidemiology, or genome-wide association studies. Indeed, in the quantitative experiments shown in cell death papers, n is seldom >3, and is often only 1. Although statisticians argue (correctly) that it is impossible to draw statistical inferences from such a small data set, considerable progress has been made in experimental cell biology because, in good papers, each of the figures provide evidence that supports the conclusions in a way that independently addresses the mechanism.

For quantitative experiments studying cell death (e.g., where the number or percentage of dead cells is determined), statistics should only be shown if the differences are small, the number of independent observations is large ($n > 10$), and there is no other supporting experimental evidence, that is, in very few cases. Most often it is best to just plot the data points, and where the differences are not clear-cut, show the results of another experiment that gives supporting evidence using a mechanistically independent test. As Rutherford said, "If your experiment needs statistics, you ought to have done a better experiment." Ideally, results of independently repeated experiments should be shown, and the figure legend should explicitly state how many times this was done.

A corollary of this is that caution should be exercised when interpreting other people's papers—if the conclusions are based on p values or standard errors of the mean (s.e.m.s) where n is from 1 to 3,

Cite this introduction as *Cold Spring Harb Protoc*; doi:10.1101/pdb.top070300

they are resting on very thin ice. Rather than supporting the investigators' conclusions, showing statistics or p values where $n = 1$ (i.e., from replicates) can provide hints to the authors' competence.

IMAGES

When interpreting images of gels and blots, two important considerations are the resolution of the image and the quality of the controls. The resolution should be similar to what you would see if you were looking at the film on a light box. No compression artifacts or pixelation should be apparent. In cases where not only the presence of a band is important, but also its relative intensity, loading controls must be included, and both sets of bands should be exposed within the dynamic range of the film. Being able to see the background of the gel and showing more than a narrow gel slice provide reassurance that that the contrast and brightness have not been used to hide unwanted bands, and show whether the antibody is specific.

OTHER CONSIDERATIONS

For all figures, the reader should be able to easily understand what was done and what conclusions can be drawn from an experiment just by looking at the figure and reading the legend. Avoiding jargon terms, indicating molecular masses, labeling axes, describing error bars (where shown), and stating the number of independent data points will all help convince the reader of the authors' conclusions, but if they are lacking, can undermine the conclusions, or even raise suspicions that the data have been misrepresented, misinterpreted, exaggerated, or fabricated.

In one of the associated protocols, I give suggestions on preparing figures for publications on cell death (see Protocol 1: Preparing Figures from Cell Death Experimental Data for Publication [Vaux 2014a]). The other associated protocol gives hints on what to look for when determining if the figures in a paper give convincing evidence that the authors' conclusions are correct (see Protocol 2: Interpreting Cell Death Data from Figures in Published Papers [Vaux 2014b]).

REFERENCES

Cumming G, Fidler F, Vaux DL. 2007. Error bars in experimental biology. *J Cell Biol* **177:** 7–11.

Vaux DL. 2014a. Preparing figures from cell death experimental data for publication. *Cold Spring Harb Protoc* doi: 10.1101/pdb.prot078964.

Vaux DL. 2014b. Interpreting cell death data from figures in published papers. *Cold Spring Harb Protoc* doi: 10.1101/pdb.prot078972.

Vaux DL, Fidler F, Cumming G. 2012. Replicates and repeats—What is the difference and is it significant? A brief discussion of statistics and experimental design. *EMBO Rep* **13:** 291–296.

Preparing Figures from Cell Death Experimental Data for Publication

David L. Vaux[1]

Department of Medical Biology, The Walter and Eliza Hall Institute, University of Melbourne, Melbourne, Victoria 3052, Australia

When preparing results of cell death experiments for figures for publication, keep in mind the needs of the people who will be reading the paper and trying to understand the experiments and conclusions. From looking at the figure and associated legend, readers should not only be told the key experimental conclusion, but also what assay was performed, how many times it was performed, what the controls were, whether the differences were statistically significant, and what the size of the effect was. This protocol describes steps to be taken when preparing cell death experimental data for publication.

MATERIALS

Equipment

Computer with appropriate software (spreadsheet, statistical analysis, and/or graphics programs) and scanner
Experimental results (e.g., numerical data, blot, gel picture)

METHOD

Before preparing your figures, read the authors' guidelines for the journal to which you intend to submit. Those from the Journal of Cell Biology *are a particularly good example (http://jcb.rupress.org/site/misc/print.xhtml).*

Plotting Data on a Graph

1. Enter your numerical data into a spreadsheet.

2. Plot the data (or the means of the data) using either the spreadsheet program or a separate graphics program.

   If it is possible to plot the independent data points and easily interpret the graph, do so, rather than plotting the means. This is generally feasible when the number of independent data points is ≤ 10. For suggestions on normalizing data and handling replicates, see Discussion.

3. Determine the scales. With the exception of log scales, start scales at zero.

   For cell death graphs, there is rarely a need to show negative numbers or >100% survival.

[1]Correspondence: vaux@wehi.edu.au

Cite this protocol as *Cold Spring Harb Protoc*; doi:10.1101/pdb.prot078964

A

B

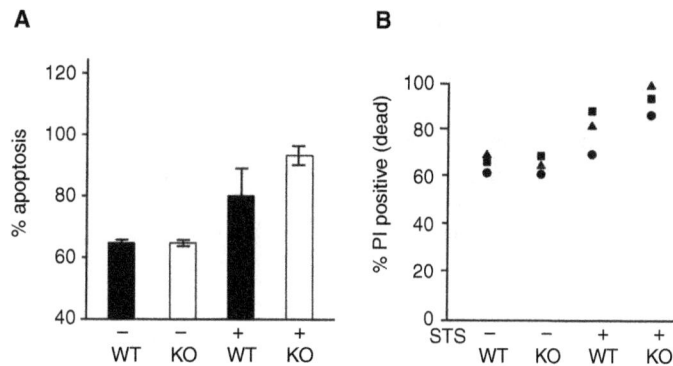

FIGURE 1. Figures and their legends should make it easy for the reader to understand what was done in the experiment. (A) An axis labeled "apoptosis" gives no indication of the assay performed. Scales should start at zero. Means ± (s.e.m.) of three independent experiments (each with quadruplicates) are shown, so $n = 3$. (B) The same data from A in a recommended presentation. The y-axis extends from 0% to 100%, and the label shows what assay was performed. The death stimulus (staurosporine [STS]) is shown on the figure, and the concentration used (1 μM) and time (24 h) is stated in the legend. The circle, triangle, and circle symbols depict three independent ($n = 3$) experiments, each using quadruplicate cultures.

4. Label the axes.

> *Each axis should reflect what was actually measured (e.g., red fluorescence, % trypan blue positive [dead], MTT intensity, % SYTOX-positive cells). Do not use jargon terms such as apoptosis, necroptosis, live, or dead. Figure 1 shows two presentations of the same data set, with Figure 1B showing the preferred presentation of the data.*

5. Present differences between control and experimental values appropriately.

- Include error bars only if the differences between the control and experimental values are not great, or if it is not clear by inspection whether you can exclude the null hypothesis (i.e., that differences of that size could arise by chance).

- If you wish to make statistical inferences from your data, show standard errors of the mean (s.e.m.) or confidence intervals rather than standard deviations (s.d.s).

- If it is obvious that the differences are highly statistically significant, or if you perform a statistical inferential test and show this to be the case, be sure to comment on the size and biological importance of the effect, rather than just saying "$p < 0.05$."

Presenting Blots and Gels

1. Scan blots and gels at a minimum of 300 dpi using image processing software. Keep a copy of the entire blot or gel as a backup.

> *Make sure resolution is not lost because of software compression.*

2. Minimize changes to the image.

- Adjust contrast and brightness only to make the data easier to see (no feature, spot, or band should be made less visible). Ensure that the background remains visible.

> *Most journals specify how images can be adjusted. For example, brightness and contrast may be changed, but only if it is done uniformly across the whole image (i.e., it should not be used to enhance or hide individual parts of an image). Check the images you intend to submit; if you can easily see pixilation or compression artifacts when the images are twice the size they would be when printed in the journal, start over (Rossner and Yamada, 2004). If there is concern about figure size, negotiate this with the editors after the paper has been seen and accepted by the reviewers.*

- Show as much of the gel or blot as possible; avoid cutting and pasting.

> *A single large blot or gel image that includes all of the experimental and control lanes is much more convincing than a cropped line of bands, or individual bands shown separately. If you wish to change the*

order of the lanes, or need to remove empty or irrelevant lanes from the image of a blot or gel, you may do so, but a clear gap or line must be left to indicate where the image was cut or joined. Consider running the samples again in the desired order. If you show black borders around the image, they should correspond to all places where the images were truncated.

3. Show loading controls.

Loading controls (such as actin, GAPDH, or tubulin) help show whether the loading of the lanes was even. The loading control should be generated from the same blot as the experimental bands. If instead the same lysates were run on identical gels to produce the loading control, this should be stated in the figure legend.

4. Label molecular masses.

This makes it easier to interpret the figure and to compare it with other figures examining the same protein.

Presenting Microscopy Images

1. Take an appropriate picture.

- If two types of cells are to be compared, consider showing an image where there is a 50:50 mix of cells, so that both types can be seen in the same field.

- For immunohistochemistry, consider including a full-length western blot.

  A single reactive band will give reassurance of the specificity of the antibody. If the antibody detects other, nonspecific epitopes, show these bands as well, and, if possible, show the specificity of the antibody by comparing wild-type cells to those lacking the gene for the relevant protein, or cells in which expression has been knocked down by gene silencing.

2. Include a scale bar and state what image processing software was used.

DISCUSSION

Data Normalization

Let us say you are testing the effect of an apoptosis inducing cancer drug on three mice. You do a white blood cell count before you give the drug, and then again 3 d after giving them the drug. You find that both before and after treatment the mice have very different blood counts, but for each of the three mice, their white blood cell count halves after they receive the drug. You could plot the mean white blood cell counts before and after treatment, but that would not show the consistent effect of the drug. You could normalize the data, and plot the mean percentage decrease in white blood cell count. Finally, you could plot the actual values, and use symbols or lines to indicate the individual mice. In most cases, it is best to manipulate the data as little as possible, and therefore I recommend the third option, because it is the easiest for the reader to understand. Figure 2 shows the multiple ways of presenting this hypothetical data.

Replicates

Replicates act as an internal check of the fidelity with which the experiment was performed. They are designed to be as closely linked as possible, that is, the exact opposite of being independent. Variation among replicates reflects the fidelity with which the replicates were created, such as the reproducibility of pipetting, and the consistency of measuring, and therefore has no bearing on the hypothesis being tested. Do not plot individual replicate data points. Plot the mean of replicates as $n = 1$ data point (Vaux et al. 2012) (Fig. 1). Similarly, never show statistics for replicates (never confuse independent experiments with replicate samples from one experiment). Replicates can give an indication of whether the experiment was performed well, but give no information that is relevant to the hypothesis being tested. Showing statistics for replicates can be misleading, as readers might wrongly assume that they are relevant to the hypothesis being tested.

Cite this protocol as *Cold Spring Harb Protoc*; doi:10.1101/pdb.prot078964

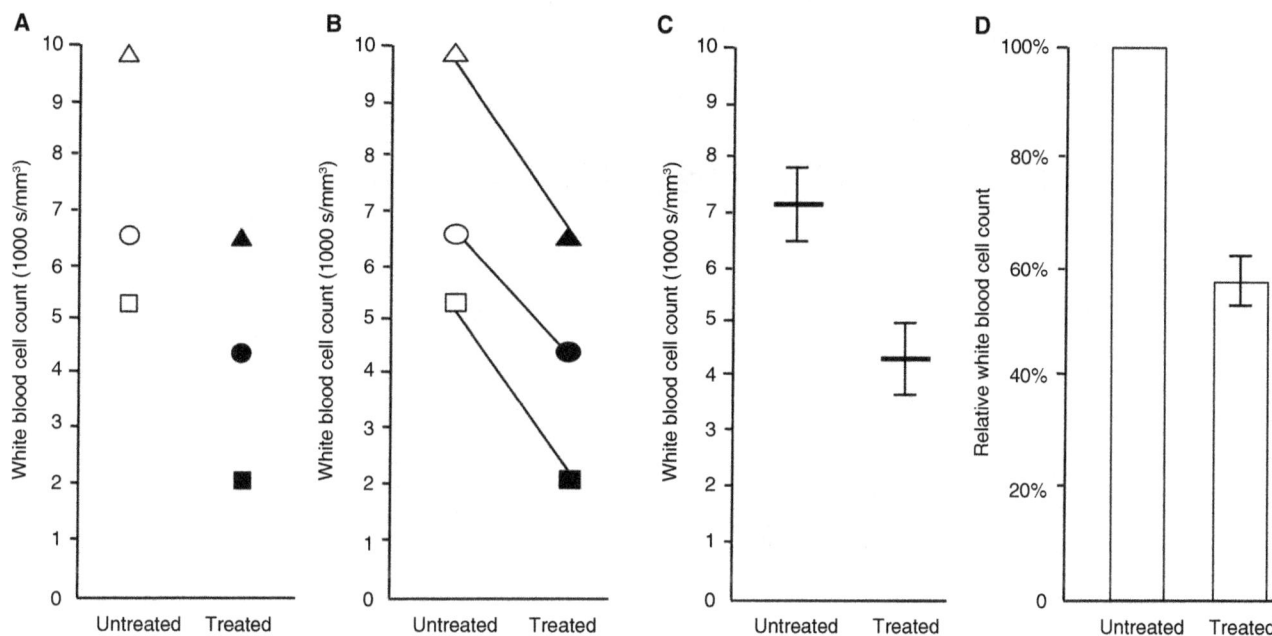

FIGURE 2. Avoid normalizing data. If an experimental stimulus has a consistent effect, but there is considerable variation between the values at the start of the experiment, it can be difficult choosing how to present the paired data. In this experiment, the three mice had differing peripheral white blood cell counts to begin with, and although the treatment halved the white blood cell counts in all three mice, the counts at the end of the experiment also showed variation. Rather than normalizing the results and presenting the mean ratios and SEM (*D*), or calculating means and SEMs with paired *t*-test statistics (*C*), it is better to show no statistics, but to present the results as in *A* or *B*, where the three individual mice are indicated by symbols (squares, triangles, and circles) or lines, and allow the readers to judge whether the treatment was having a consistent effect.

REFERENCES

Rossner M, Yamada KM. 2004. What's in a picture? The temptation of image manipulation. *J Cell Biol* **166:** 11–15.

Vaux DL, Fidler F, Cumming G. 2012. Replicates and repeats—What is the difference and is it significant? A brief discussion of statistics and experimental design. *EMBO Rep* **13:** 291–296.

Interpreting Cell Death Data from Figures in Published Papers

David L. Vaux[1]

Department of Medical Biology, The Walter and Eliza Hall Institute, University of Melbourne, Melbourne, Victoria 3052, Australia

About a million new papers are listed in PubMed each year, and >2% of them will be identified using the search terms "apoptosis" or "programmed cell death." This protocol provides guidance on interpreting the figures in these papers.

MATERIALS

Equipment

Computer with image analysis program
Paper(s) to be analyzed

METHOD

Analyzing Graphs

For a sample graph, see Figure 1B in Protocol 1: Preparing Figures from Cell Death Experimental Data for Publication (Vaux 2014).

1. Determine whether the figure legend and the y-axis label describe how cell death was measured, and whether the number of counts is stated either in the figure legend or in the Materials and Methods section.

   The figure legend should state how cell death was determined (e.g., by uptake of propidium iodide [measured by fluorescence], morphology [e.g., blebbing cells], counting TUNEL-stained cells with a microscope, trypan blue uptake using a hemocytometer, intensity of color in an MTT assay, or number of colonies on a plate). The number of counts should also be stated (e.g., "Cell death was determined by PI uptake using a flow cytometer; 10,000 events were counted in each sample." or "Cell death was determined by trypan blue exclusion using a hemocytometer; 300 cells were counted for each sample.")

2. Identify n (the number of independent experiments or data points). Determine whether replicates were performed. Check whether all data, or just "representative" results, are included.

   The figure legend should mention n. If the legend says "results are from one representative experiment; $n = 3$," then you are only being shown one-third of the results. Be extra suspicious.

   Replicates (duplicates, triplicates, etc.) are performed as an internal check of technical aspects of the experiment, such as accuracy of pipetting. Statistics (e.g., error bars and p values) should not be shown for replicates because they indicate only the fidelity with which the experiment was performed; they have no relevance to the hypothesis being tested. Small standard deviations (s.d.s) of replicates might indicate con-

---

[1]Correspondence: vaux@wehi.edu.au

Cite this protocol as *Cold Spring Harb Protoc*; doi:10.1101/pdb.prot078972

sistent pipetting, but provide no indication of whether differences between the control and experimental results are meaningful.

If error bars or p values are shown for replicates or a "representative" experiment, they are misleading, because n = 1. No conclusions can be drawn.

3. Analyze the error bars.

    i. Determine whether the figure legend defines the error bars (e.g., range, (s.d., standard error of the mean [s.e.m.], confidence interval [CI]).

    ii. Determine whether the statistical errors look plausible.

If an experimenter was sampling multiple times from the same suspension of cells, of which 50% were alive, and each time he was counting 100 cells, then ~95% of the time, his counts should be between 40% and 60% live, depending by chance on whether a dead or live cell entered his pipette or the counting chamber. If the results were from independent experiments, the variation should be even greater.

If the s.d.s (s.e.m. × √n) are consistently less than √ (cells counted × fraction live × fraction dead), there is a strong possibility the results are biased.

When a Poisson distribution is expected (such as the number of colonies on a dish, or the number of cells in a microscope field), the expected s.d. is √ mean. If the s.d.s are consistently smaller than this, there is a strong possibility the results are biased.

    iii. Determine whether the p values were calculated correctly.

If s.e.m. error bars overlap, p cannot be <0.05. If double the s.e.m., error bars do not overlap, p must be <0.05. (When n = 3, and double the s.e.m., error bars just touch, p ~ 0.05.)

Analyzing Blots and Gels

1. Carefully inspect the figures from a paper on a computer screen (printing reduces the resolution of the images). View them as PDFs, JPG images, or HTML files. Where available, use a "magnify" tool. Scan various parts of the image for potential problems with resolution, image splicing, or background (see Step 2).

2. Capture and import the images into an image analysis program.

    i. Check the resolution of the images.

Published images should be at high-enough resolution such that pixilation and compression artifacts are not obvious even if the figure is magnified to twice its printed size.

    ii. Adjust the contrast and levels to determine whether there are any discontinuities.

Sudden changes in contrast or linear discontinuities might indicate inappropriate splicing of images (Rossner and Yamada 2004).

    iii. Examine the background. Analyze the distribution of the brightness of the background pixels.

The background should be visible for gels and blots. If it cannot be seen, the contrast or brightness may have been adjusted too far. If the background pixels are all of the same intensity, the background may not be genuine, but may have been a grey box generated by software.

3. Examine the loading controls.

    i. Determine whether sufficient loading controls are present.

If inferences are made about the relative intensity of the bands, loading controls are necessary. There should be a set of loading controls for each of the experimental sets of lanes. It is reassuring if the legend states explicitly that the loading controls are from reprobing the same blot that was used for the experimental bands.

    ii. Check that the shapes of the loading control bands are consistent with those of the experimental bands.

For example, if the loading control bands are smiling upward, but the edges of test bands of similar molecular mass are drooping down, one or other sets of bands may have been pasted in upside down, or come from a different blot.

ACKNOWLEDGMENTS

This work was made possible through Victorian State Government Operational Infrastructure Support and Australian Government NHMRC IRIISS.

REFERENCES

Rossner M, Yamada KM. 2004. What's in a picture? The temptation of image manipulation. *J Cell Biol* **166:** 11–15.

Vaux DL. 2014. Preparing figures from cell death experimental data for publication. *Cold Spring Harb Protoc* doi: 10.1101/pdb.prot078964.

CHAPTER 3

Dead Cert: Measuring Cell Death

Lisa C. Crowley,[1] Brooke J. Marfell,[1] Adrian P. Scott,[1] Jeanne A. Boughaba,[1,2] Grace Chojnowski,[3] Melinda E. Christensen,[1,3,4] and Nigel J. Waterhouse[1,3,5,6]

[1]Apoptosis and Cytotoxicity Laboratory, Mater Research, Translational Research Institute, Woolloongabba, Brisbane, Queensland 4102, Australia; [2]Agroparistech, Paris Cedex 05, France; [3]Flow Cytometry and Imaging, QIMR Berghofer Medical Research Institute, Herston, Brisbane, Queensland 4006, Australia; [4]Division of Immunology, Mater Pathology, Mater Adult Hospital, South Brisbane, Queensland 4101, Australia; [5]School of Medicine, University of Queensland, St. Lucia, Brisbane, Queensland 4072, Australia

Many cells in the body die at specific times to facilitate healthy development or because they have become old, damaged, or infected. Defects in cells that result in their inappropriate survival or untimely death can negatively impact development or contribute to a variety of human pathologies, including cancer, AIDS, autoimmune disorders, and chronic infection. Cell death may also occur following exposure to environmental toxins or cytotoxic chemicals. Although this is often harmful, it can be beneficial in some cases, such as in the treatment of cancer. The ability to objectively measure cell death in a laboratory setting is therefore essential to understanding and investigating the causes and treatments of many human diseases and disorders. Often, it is sufficient to know the extent of cell death in a sample; however, the mechanism of death may also have implications for disease progression, treatment, and the outcomes of experimental investigations. There are a myriad of assays available for measuring the known forms of cell death, including apoptosis, necrosis, autophagy, necroptosis, anoikis, and pyroptosis. Here, we introduce a range of assays for measuring cell death in cultured cells, and we outline basic techniques for distinguishing healthy cells from apoptotic or necrotic cells—the two most common forms of cell death. We also provide personal insight into where these assays may be useful and how they may or may not be used to distinguish apoptotic cell death from other death modalities.

ASSAYS TO MEASURE CELL DEATH AND CELL SURVIVAL

Most cytotoxic stimuli trigger cell death by apoptosis; however, others may trigger alternative cell death mechanisms such as necrosis, autophagy, anoikis, or necroptosis. It is therefore wise to use a well-characterized model to evaluate and validate any protocol when embarking on cell death assays for the first time. Well-known inducers of apoptosis include death receptors such as TNF-related apoptosis-inducing ligand (TRAIL) (Corazza et al. 2009) and Fas ligand (Huang et al. 1999), serum withdrawal and other cell stressors (Schamberger et al. 2005), chemotherapeutic drugs including actinomycin D, staurosporine, or etoposide (Finucane et al. 1999a, 1999b; Waterhouse et al. 2001; Crowley et al. 2011), and ultraviolet (UV)- and γ-irradiation (Goldstein et al. 2000; Kulms and Schwarz 2000; Waterhouse et al. 1996; Lee et al. 2007). Although these stimuli normally trigger death by apoptosis, they can also result in different types of cell death, depending on the strength of the stimulus. For example, low levels of UV-irradiation have been shown to trigger apoptosis, while

[6]Correspondence: nigel.waterhouse@QIMRBerghofer.edu.au

Cite this introduction as Cold Spring Harb Protoc; doi:10.1101/pdb.top070318

higher levels can trigger necrosis. Of note, some cell types may be resistant to a variety of these stimuli because they contain specific defects in key signaling molecules. Thus, it is critical to select a cell type that is known to be sensitive to the planned stimulus. We currently use actinomycin D (1 µM) or low levels of UV-irradiation (80 mJ/m^2) to trigger apoptosis in U937 cells or HeLa cells, and higher levels of UV-irradiation (500 mJ/m^2) to trigger necrosis. Treatment of cells with these stimuli is described in our protocols for Protocol 1: Triggering Apoptosis in Hematopoietic Cells with Cytotoxic Drugs (Crowley et al. 2015a) and Protocol 2: Triggering Death of Adherent Cells with Ultraviolet Radiation (Crowley and Waterhouse 2015a). We also regularly use leucine zipper-TRAIL (230 ng/mL), staurosporine (1 µM), mitoxantrone (5 µM), doxetaxel (5 µM), and γ-irradiation (10 Gy) to trigger apoptosis in a variety of cell lines.

The above concentrations are provided as a guide only; a more extensive titration may be required in your model. Also, some stimuli may require training or clearance before use, depending on the occupational health and safety regulations of your country. Any restrictions should be discussed with your institution before proceeding.

Assays for Quantitating Cell Death

In many cases, it is sufficient to know the extent of cell death in a sample but not necessary to determine the mechanism of death. Cell death can be measured by staining a sample of cells with trypan blue, as we describe in Protocol 3: Measuring Cell Death by Trypan Blue Uptake and Light Microscopy (Crowley et al. 2015b). Because trypan blue is a colorimetric dye that only stains dead cells with ruptured plasma membranes but is excluded from live cells that have intact plasma membranes, dead cells (blue) can easily be distinguished from live cells (colorless) using a standard bright-field microscope. The extent of cell death can be determined by manually counting the number of colorless vs. blue cells in the sample. This assay is convenient, as trypan blue is inexpensive and loss of plasma membrane integrity is a terminal event in all cell death in culture. However, it cannot be used to quantitate the extent of death in fixed cells or in tissue sections where the plasma membrane of cells has been compromised by fixation or sectioning. Quantitating cell death using trypan blue can also be tedious when assaying large numbers of samples, because the cells must be counted manually.

Propidium iodide (PI), a small fluorescent dye that is similarly excluded from live cells but stains the nuclei of cells with ruptured plasma membranes, can also be used to measure the extent of death in a sample of cultured cells; see Protocol 4: Measuring Cell Death by Propidium Iodide Uptake and Flow Cytometry (Crowley et al. 2015c).

PI is maximally excited by light at 535 nm and emits at 617 nm. Dead cells can therefore be detected using a fluorescence microscope equipped with a green excitation filter (e.g., 495–570 nm) and a red emission filter (e.g., 610–750 nm). Counting can be performed manually or automated using various image analysis programs. Modern flow cytometers, which rapidly and accurately count large numbers of cells, are also equipped with lasers and filters that can detect PI fluorescence. Flow cytometry is therefore ideal for quantitating cell death in samples stained with PI (see Protocol 4: Measuring Cell Death by Propidium Iodide Uptake and Flow Cytometry [Crowley et al. 2015c]). PI is relatively inexpensive and flow cytometry can be automated. This assay is therefore ideal for screening cell death in large numbers of samples or for performing initial titrations for drug toxicity before proceeding to specific assays that are often more expensive.

Assays for Quantitating Cell Survival

Both trypan blue and PI will stain cells that have reached the terminal stage of death, but neither can be used to detect cells that are dying or marked for death but are not yet terminal. This is an important distinction, because the damage caused by treatment with some cytotoxic agents may not have an immediate measureable effect (Sedelies et al. 2008). If used in long-term assays, trypan blue staining and PI uptake may also underestimate the toxicity of some treatments. For example, in a culture where 50% of the cells died and 50% divided once, the percentage of cell death will only be measured as 33% using the trypan blue or PI assays. Distinguishing healthy from dying cells is particularly important in

Cite this introduction as *Cold Spring Harb Protoc*; doi:10.1101/pdb.top070318

applications such as cancer chemotherapy, where any surviving cells have the potential to cause relapse. One way to address this issue is by performing a colony-forming assay to define the number of cells in a population that are capable of proliferating and forming large groups of cells (Ludlow et al. 2008; Sedelies et al. 2008). Because each colony is generally derived from a single surviving cell, the results provide a measure of the number of healthy cells in a sample. Colony-forming assays are easily performed on adherent cells by adding a defined number of cells to a flat-bottomed dish and incubating the cells until visible colonies form, as we describe in Protocol 5: Measuring Survival of Adherent Cells with the Colony-Forming Assay (Crowley et al. 2015d). The colonies can be counted under a light microscope or fixed and stained with Crystal Violet for quantitation at a later stage. The percentage confluence can be determined using image analysis software. It is more difficult to perform colony-forming assays on suspension cells because the cells disperse in medium. However, it is possible to prevent dispersal by culturing the colonies in soft agar; see Protocol 6: Measuring Survival of Hematopoietic Cancer Cells with the Colony-Forming Assay in Soft Agar (Crowley and Waterhouse 2015b).

Using the above basic techniques, it is possible to begin investigating cell death in culture models. These assays are particularly useful when it is sufficient to simply determine whether a cell has died to evaluate whether a specific treatment triggers, enhances, or blocks cell death. They are also applicable to large screens or initial titrations for determining optimal treatment conditions (e.g., drug concentration or assay duration) for inducing cell death. More specific cell death assays are needed to quantitate a specific type of cell death or determine the mechanism of cell death (see below).

ASSAYS TO DISTINGUISH APOPTOSIS FROM NECROSIS AND OTHER DEATH MODALITIES

Most forms of cell death are characterized by distinct morphologies that can be visualized by light microscopy. For example, apoptosis is characterized by rounding, shrinking, plasma membrane blebbing, and formation of apoptotic bodies. This is in contrast to necrotic cells which swell and lyse, releasing their cellular contents. The usefulness of light microscopy in analysis of cell death is often overlooked and underestimated, because researchers have developed more robust definitions based on biochemical characteristics (Elmore 2007). However, looking down a microscope at your cells is one of the most convenient tools for distinguishing live cells from dead or dying cells, and for distinguishing apoptosis from necrosis or other death modalities. We strongly recommend visual inspection as the first step in any procedure for measuring cell death (see Box 1). This method is convenient, because cells can be easily observed at regular intervals to confirm that an experiment is proceeding as expected, or to predict the optimum timing for additional assays. Further, it is a useful tool for recording the health and quality of any cell culture before sampling for subsequent experimentation. It can be used in combination with other protocols including cytospinning, 4′,6-diamidino-2-phenylindole (DAPI) staining, or measuring immunofluorescence, or used as a visual reference for quantitative assays measured by flow cytometry (see below).

It is not always possible, however, to distinguish live cells from dead or dying cells using light microscopy. Some hematopoietic cells (e.g., T cells, B cells, or NK cells) and mitotic cells can have a similar appearance to apoptotic cells. Further, some cells also clump together, making it difficult to objectively measure cell death in individual cells. Finally, some treatments can induce morphological changes that appear similar to apoptosis. These issues can be addressed by staining and observing cellular structures such as the nucleus. The following techniques can be used to distinguish healthy cells from apoptotic or necrotic cells. Further assays will be required to distinguish these two most common forms of cell death from other modalities.

Detecting Nuclear Condensation

The nucleus is generally round in healthy cells but fragmented in apoptotic cells. Dyes such as Giemsa or hematoxylin, which are purple in color and therefore easily viewed using light microscopy, are

BOX 1. ANALYZING CELL DEATH BY LIGHT MICROSCOPY

The first and easiest question to ask when assaying cell death should always be "What do the cells look like?" Apoptotic and necrotic cells can easily be distinguished using standard light microscopes available in most laboratories. Observing cells by bright-field microscopy does not require a high level of expertise and can be performed by most laboratory staff. Images of the cells can be taken using a microscope camera attachment for further analysis. Cells that grow in suspension should be allowed to settle on the bottom of the dish before observation or imaging.

- Healthy cells generally have a round, well-defined nucleus surrounded by a cytoplasmic region enclosed by a continuous plasma membrane. In adherent lines, these cells remain attached as a monolayer on the bottom of the flask/dish (Fig. 1A).

- Apoptosis is characterized by nuclear condensation and fragmentation, cell rounding (in the case of adherent cells), shrinking, and blebbing of the plasma membrane (Fig. 1B.i and ii) (Hengartner 2000). Some cells eventually fragment into small, plasma membrane-bound apoptotic bodies, but other cells remain intact until they swell and lyse by a process known as secondary necrosis (Fig. 1B.iii). Cells may not progress to secondary necrosis in vivo because apoptotic cells are removed by phagocytosis (Tanaka et al. 2010), but this event provides a convenient end point for apoptosis in culture.

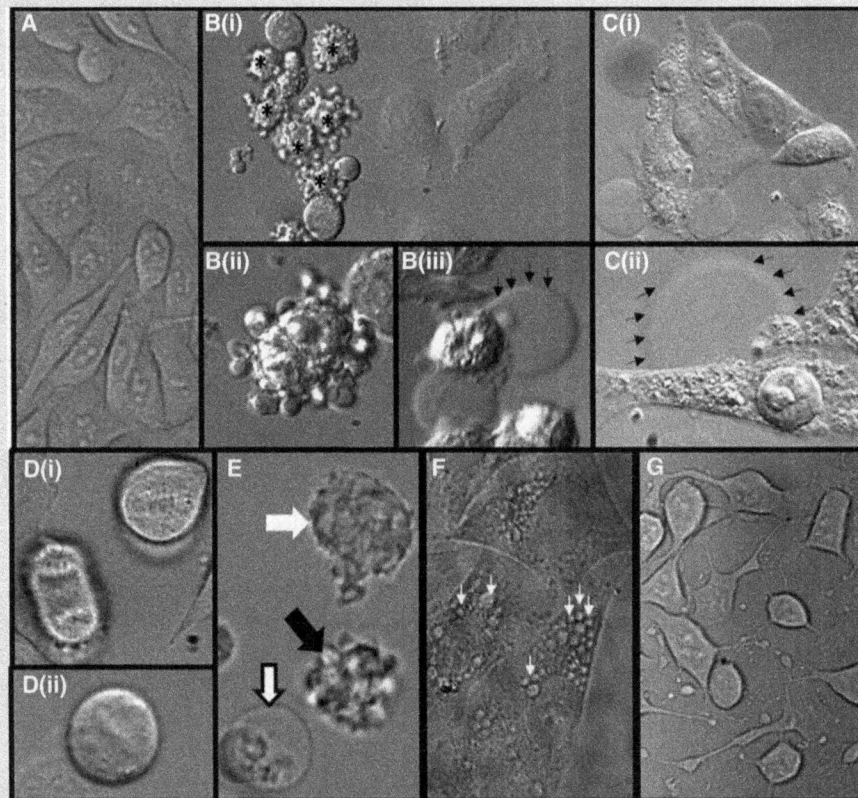

FIGURE 1. Images of cell survival and cell death. (A–C) Bright-field images of HeLa cells. (A) Untreated cells. (B) Cells treated with 230 ng/mL TRAIL for 4 h undergoing apoptosis. Apoptotic cells are indicated by asterisk (*). Features include plasma membrane blebbing, which appears as small bubble-like projections from the cell surface (shown at higher magnification in B(ii), and plasma membrane extrusions during secondary necrosis, which appear as large bubble-like projections of the plasma membrane (shown at higher magnification with arrows in B(iii). (C) Cells treated with UV 500 mJ/m² for 4 h undergoing necrosis. Features include large plasma membrane extrusions (shown at higher magnification with arrows in C(ii). (D(i), D(ii)) Mitotic cells from untreated cultures. Features include rounding and detachment, appearance of DNA midlines or chromatin segregation. (E) U937 cells treated with 1 μM actinomycin D for 16 h. Indicated are a live cell (white arrow), an apoptotic cell (black arrow), and a necrotic cell (white arrow with black outline). (F) Autophagic cells treated with 25 μM chloroquine for 24 h. Features include large vacuoles within the cytoplasmic region of the cells (white arrows). (G) Cells treated with 1 μM staurosporine for 6 h.

Cite this introduction as *Cold Spring Harb Protoc*; doi:10.1101/pdb.top070318

- Necrosis is characterized by overt swelling of the plasma membrane, which appears as large bubble-like projections that eventually rupture (Fig. 1C.i and ii, black arrows). Adherent cells do not generally detach during necrosis, making it easy to distinguish these cells from rounded apoptotic cells or cells that have progressed through apoptosis to secondary necrosis. The nuclei of necrotic cells do not fragment like apoptotic cells and this provides another distinguishing feature for this type of cell death.

- In some cases, it may be difficult to differentiate live cells from dead or dying cells. For example, healthy cells such as mitotic (Fig. 1D.i and ii) or some hematopoietic cells (Fig. 1E, white arrow) are round, and may therefore be difficult to distinguish from apoptotic (Fig. 1E, black arrow) or secondary necrotic (Fig. 1E, white arrow with black outline).

- Other forms of cell death such as autophagy can also be distinguished by morphology. Cell death by autophagy is less well-characterized, but one distinguishing morphological feature is the formation of large vacuole-like structures (Fig. 1F, white arrows). It is wise to confirm autophagy by assaying molecular events such as increased expression of microtubule-associated protein light chain 3 (LC3-II), a protein that is associated with formation of autophagosomes (Crowley et al. 2011).

- Some treatments can also induce morphological changes that appear similar to apoptosis. For example, staurosporine can induce changes in the plasma membrane that resemble the early stages of apoptosis (Fig. 1G). Apoptosis should therefore be verified by staining the cells for specific events associated with cell death.

commonly used to stain the nucleus. Other features of apoptosis and necrosis, such as plasma membrane blebbing or rupture, can be identified by staining the cytoplasm with eosin. Eosin is pinkish in color and can also be viewed using light microscopy. Hematoxylin and eosin are, therefore, commonly used together to stain cells. We recommend use of a rapid stain (e.g., Rapi-Diff or Diff Quick) that has been optomized for staining the nucleus and cytoplasm by light microscopy (Yang and Widmann 2002) as we describe in Protocol 7: Morphological Analysis of Cell Death by Cytospinning Followed by Rapid Staining (Crowley et al. 2015e).

Fixing and staining adherent cells poses few problems; however, apoptotic cells round up and may detach from the culture plate. These cells can therefore be washed away during the fixing and staining procedure. Similarly, it can prove difficult to fix and stain nonadherent cells for analysis by light microscopy. To overcome this, cells may be gently centrifuged onto glass slides by cytospinning and then fixed and stained, as described in the above protocol. Cytospinning is also ideal for use with other dyes that are commonly used to stain the nucleus. These include DAPI and Hoechst (33258 and 33342), which bind to the DNA of live or fixed cells and fluoresce at 461 nm when excited by UV light (Excitation maximum ~350 nm). Cells stained with these dyes can be visualized using standard fluorescence microscopy equipped with a UV light source and a blue/cyan filter; see Protocol 8: Analyzing Cell Death by Nuclear Staining with Hoechst 33342 (Crowley et al. 2015f).

Detection of DNA Fragmentation

The use of Hoechst 33342 and DAPI to identify and characterize apoptotic cells relies on a researcher's ability to distinguish the nuclear morphology of apoptotic cells from healthy cells or cells that have died by other mechanisms. Although this is generally easy, mitotic cells also have condensed chromatin that may be confused with apoptotic cells by inexperienced users. However, the condensed chromatin of mitotic cells differs from that of apoptotic cells because the DNA of latter is fragmented but the DNA of the former is not. The fragmented DNA in apoptotic cells can be labeled with fluorescently tagged UTP by terminal deoxynucleotidyl transferase, and terminal dUTP nick end labeling (TUNEL) can be used to specifically label apoptotic cells. UTP can be tagged with a variety of fluorochromes, including fluoresceinisothiocyanate (FITC), which is excited by light at 495 nm and emits at 520 nm. FITC-labeled UTP can be used in combination with Hoechst 33342 or DAPI to easily distinguish apoptotic cells from mitotic cells using fluorescence microscopy using the protocol for

Protocol 9: Detection of DNA Fragmentation in Apoptotic Cells by TUNEL (Crowley et al. 2015g). The number of apoptotic cells can be determined as a percentage of the total by counting both the TUNEL-positive and the Hoechst 33342- or DAPI-positive cells. Cell counting of TUNEL-stained cells can also be automated using flow cytometry, as only the apoptotic cells are fluorescently labeled (Trapani et al. 1998a, 1998b).

Apoptotic cells with fragmented DNA can also be identified and distinguished from live cells by staining with PI and measuring DNA content by flow cytometry. This assay, described in Protocol 10: Measuring the DNA Content of Cells in Apoptosis and at Different Cell-Cycle Stages by Propidium Iodide Staining and Flow Cytometry (Crowley et al. 2015h), is based on the fact that PI intercalates with DNA at a specific ratio. The level of PI staining is therefore directly proportional to the DNA content of the cell. PI staining is commonly used for cell cycle studies, as cells in G2/M have twice the DNA content as cells in G0/G1, and therefore bind twice the level of PI. Because apoptotic cells have fragmented DNA, and therefore show lower fluorescence than cells in G0/G1, PI staining is also ideal for quantitating the level of apoptosis in a sample. It should be noted that this assay, in which all cells are fixed and stained with PI, is distinct from the PI uptake assay in Protocol 4: Measuring Cell Death by Propidium Iodide Uptake and Flow Cytometry (Crowley et al. 2015c), in which live cells exclude the dye. It should also be noted that cells can die by apoptosis in the absence of nuclear fragmentation (e.g., because of defects in activation of caspase-activated DNAse [Kagawa et al. 2001; Zhang et al. 2001]). Although this is rare, lack of nuclear fragmentation should not be used in isolation to determine that cells are healthy or have not died by apoptosis. This issue can be addressed by performing assays for other apoptotic events that occur in the absence of DNA fragmentation.

Detecting Phosphatidylserine Exposure

Apoptosis is also characterized by exposure of phosphatidylserine (PS) on the outside of apoptotic cells, which acts as a signal that triggers removal of the dying cell by phagocytosis (Komoriya et al. 2000). Annexin V, a 36 kDa calcium-dependent protein, has been found to selectively bind to PS, and can be used to label apoptotic cells in which PS is exposed. Purified annexin V can be conjugated to various fluorochromes, which can then be visualized by fluorescence microscopy or detected by flow cytometry. Binding of fluorescently tagged annexin V is therefore ideal for detecting and quantitating the number of apoptotic cells in a sample. However, it should be noted that annexin V can also enter cells with ruptured membranes and bind to PS in necrotic cells. These cells can easily be distinguished from apoptotic cells by assaying PI uptake in combination with annexin V binding as we describe in Protocol 11: Quantitation of Apoptosis and Necrosis by Annexin V Binding, Propidium Iodide Uptake, and Flow Cytometry (Crowley et al. 2015i). Using this method, apoptotic cells are stained with annexin V but not PI (annexin V+/PI−), necrotic cells are stained with annexin V and PI (annexin V+/PI+) and healthy cells are not stained with either dye (annexin V−/PI−). It is wise to select an early time point to detect large numbers of apoptotic cells in this assay, because the annexin V+/PI− population is transient, and all apoptotic cells will eventually progress to secondary necrosis (annexin V+/PI+) (D'Amours et al. 1998). In fact, the majority of cells in long-term assays are either annexin V−/PI− (healthy) or annexin V+/PI+ (dead), regardless of how they died. Using annexin V binding in such assays is therefore unlikely to yield extra information compared to measuring PI uptake alone.

Detecting Caspase Activity

DNA fragmentation and PS exposure are both orchestrated by caspases during apoptosis. Caspases are a family of cysteine proteases that cleave their substrates at the carboxyl terminus of specific aspartic acid residues. This generally results in substrate proteolysis at only one or two sites (Martin et al. 1995, 1996; Waterhouse et al. 1996; Sedelies et al. 2008). The cleaved fragments can therefore be detected as distinct bands by western blotting using substrate-specific antibodies. Initially, the number of caspase substrates was believed to be limited, but the number of proteins known to be cleaved by caspases is now quite sizeable (Luthi and Martin 2007; Timmer and Salvesen 2007; Poreba et al. 2013). Caspases themselves are also activated by cleavage and undergo autolytic processing. Detection of cleaved caspases or proteins cleaved by caspases is therefore commonly used as a marker of apoptosis. Although western blotting is convenient, it can be difficult use this technique to quantitate the level of apoptosis in a sample. However, antibodies that specifically recognize the cleaved fragments of caspases and their substrates can be used to specifically detect caspase activity in apoptotic cells by immunocytochemistry. Flow cytometry (using primary antibodies conjugated to fluorescent molecules, or by counterstaining with fluorescently labeled antibodies against the primary antibody) can then be used to quantitate the number of apoptotic cells. This assay, described in Protocol 12: Detecting Cleaved Caspase-3 in Apoptotic Cells by Flow Cytometry (Crowley and Waterhouse 2015c), is particularly useful because the activation of specific caspases (e.g., caspase-3) is unique to apoptotic cells, and it is possible to detect cleaved caspases in cells that have died by apoptosis but progressed to secondary necrosis.

Detecting Mitochondrial Damage

Caspases are activated during apoptosis by proteolysis. This may occur via direct cleavage by granzyme B or by the caspases themselves (Martin et al. 1996; Sedelies et al. 2008). The latter can be facilitated by bringing several caspase zymogens together in sufficient proximity to allow autoactivation. This can be triggered by death-inducing signaling complexes formed following ligation of death receptors (the extrinsic pathway), or by apoptosome formation following the release of cytochrome c from mitochondria (the intrinsic pathway) (Waterhouse and Green 1999; Waterhouse et al. 2001). Cytochrome c release is a critical event in the intrinsic pathway to apoptosis and can be measured using several different methods, including western blotting, immunocytochemistry, and flow cytometry. The choice of assay will depend on a variety of criteria (Waterhouse and Trapani 2003). Protocol 13: Analysis of Cytochrome c Release by Immunocytochemistry (Crowley et al. 2015j) is an ideal starting point, because the difference between a cell with intact and permeabilized mitochondria is visually striking. Cytochrome c release can also be quantitated by flow cytometry using a modified immunocytochemistry protocol (Christensen et al. 2013). Alternatively, flow cytometry can be used to quantitate the number of cells that have reduced mitochondrial transmembrane potential, which is commonly associated with cytochrome c release during apoptosis; see Protocol 14: Measuring Mitochondrial Transmembrane Potential by TMRE Staining (Crowley et al. 2015k). Measuring loss of mitochondrial transmembrane potential by flow cytometry is easier than measuring cytochrome c release; however, it should be noted that there are cases where measurements of these events are not interchangeable. The relative benefits and pitfalls of measuring release of cytochrome c versus loss of mitochondrial transmembrane potential are discussed in the two protocols above and by Christensen et al. (2013).

CONCLUSION

Apoptosis was originally defined by distinct morphological features such as nuclear condensation and plasma membrane blebbing (Kerr et al. 1972; Wang et al. 2005). These features are still important in identifying and characterizing apoptosis, but new assays that measure key molecular events are more convenient for analyzing and quantitating apoptosis in a sample. Although the protocols introduced here comprise an effective overall strategy for measuring cell death, it is essential to understand the benefits and limitations of each assay. The importance of morphology should also not be underestimated for distinguishing apoptosis from other death modalities. Essentially, when it comes to assaying cell death: Keep it simple at the start, use "seeing is believing," and do not rely on the outcomes of one assay alone. These protocols should be used in combination to develop a basic understanding of cell death in your model.

ACKNOWLEDGMENTS

This work was supported by an Australian Research Council futures fellowship and project grants from the Leukaemia Foundation Queensland, the Prostate Cancer Foundation Australia, and the Australian Department of Defence to N.J.W.

REFERENCES

Christensen ME, Jansen ES, Sanchez W, Waterhouse NJ. 2013. Flow cytometry based assays for the measurement of apoptosis-associated mitochondrial membrane depolarisation and cytochrome c release. *Methods* **61:** 138–145.

Corazza N, Kassahn D, Jakob S, Badmann A, Brunner T. 2009. TRAIL-induced apoptosis. *Ann N Y Acad Sci* **1171:** 50–58.

Crowley LC, Waterhouse NJ. 2015a. Triggering death of adherent cells with ultraviolet radiation. *Cold Spring Harb Protoc* doi: 10.1101/pdb .prot087148.

Crowley LC, Waterhouse NJ. 2015b. Measuring survival of hematopoietic cancer cells with the colony-forming assay in soft agar. *Cold Spring Harb Protoc* doi: 10.1101/pdb.prot087189.

Crowley LC, Waterhouse NJ. 2015c. Detecting cleaved caspase-3 in apoptotic cells by flow cytometry. *Cold Spring Harb Protoc* doi: 10.1101/ pdb.prot087312.

Crowley LC, Elzinga BM, O'Sullivan GC, McKenna SL. 2011. Autophagy induction by Bcr-Abl-expressing cells facilitates their recovery from a targeted or nontargeted treatment. *Am J Hematol* **86:** 38–47.

Crowley LC, Marfell BJ, Scott AP, Waterhouse NJ. 2015a. Triggering apoptosis in hematopoietic cells with cytotoxic drugs. *Cold Spring Harb Protoc* doi: 10.1101/pdb.prot087130.

Crowley LC, Marfell BJ, Christensen ME, Waterhouse NJ. 2015b. Measuring cell death by trypan blue uptake and light microscopy. *Cold Spring Harb Protoc* doi: 10.1101/pdb.prot087155.

Crowley LC, Scott AP, Marfell BJ, Boughaba JA, Chojnowski G, Waterhouse NJ. 2015c. Measuring cell death by propidium iodide uptake and flow cytometry. *Cold Spring Harb Protoc* doi: 10.1101/pdb. prot087163.

Crowley LC, Christensen ME, Waterhouse NJ. 2015d. Measuring survival of adherent cells with the Colony-Forming Assay. *Cold Spring Harb Protoc* doi: 10.1101/pdb.prot087171.

Crowley LC, Marfell BJ, Waterhouse NJ. 2015e. Morphological analysis of cell death by cytospinning followed by rapid staining. *Cold Spring Harb Protoc* doi: 10.1101/pdb.prot087197.

Crowley LC, Marfell BJ, Waterhouse NJ. 2015f. Analyzing cell death by nuclear staining with Hoechst 33342. *Cold Spring Harb Protoc* doi: 10.1101/pdb.prot087205.

Crowley LC, Marfell BJ, Waterhouse NJ. 2015g. Detection of DNA fragmentation in apoptotic cells by TUNEL. *Cold Spring Harb Protoc* doi: 10.1101/pdb.prot087221.

Crowley LC, Chojnowski G, Waterhouse NJ. 2015h. Measuring the DNA content of cells in apoptosis and at different cell-cycle stages by propidium iodide staining and flow cytometry. *Cold Spring Harb Protoc* doi: 10.1101/pdb.prot087247.

Crowley LC, Marfell BJ, Scott AP, Waterhouse NJ. 2015i. Quantitation of apoptosis and necrosis by annexin V binding, propidium iodide uptake, and flow cytometry. *Cold Spring Harb Protoc* doi: 10.1101/pdb .prot087288.

Crowley LC, Marfell BJ, Scott AP, Waterhouse NJ. 2015j. Analysis of cytochrome c release by immunocytochemistry. *Cold Spring Harb Protoc* doi: 10.1101/pdb.prot087338.

Crowley LC, Christensen ME, Waterhouse NJ. 2015k. Measuring mitochondrial transmembrane potential by TMRE staining. *Cold Spring Harb Protoc* doi: 10.1101/pdb.prot087361.

D'Amours D, Germain M, Orth K, Dixit VM, Poirier GG. 1998. Proteolysis of poly(ADP-ribose) polymerase by caspase 3: Kinetics of cleavage of mono(ADP-ribosyl)ated and DNA-bound substrates. *Radiat Res* **150:** 3–10.

Elmore S. 2007. Apoptosis: A review of programmed cell death. *Toxicol Pathol* **35:** 495–516.

Finucane DM, Bossy-Wetzel E, Waterhouse NJ, Cotter TG, Green DR. 1999a. Bax-induced caspase activation and apoptosis via cytochrome c release from mitochondria is inhibitable by Bcl-xL. *J Biol Chem* **274:** 2225–2233.

Finucane DM, Waterhouse NJ, Amarante-Mendes GP, Cotter TG, Green DR. 1999b. Collapse of the inner mitochondrial transmembrane potential is not required for apoptosis of HL60 cells. *Exp Cell Res* **251:** 166–174.

Goldstein JC, Waterhouse NJ, Juin P, Evan GI, Green DR. 2000. The coordinate release of cytochrome c during apoptosis is rapid, complete and kinetically invariant. *Nat Cell Biol* **2:** 156–162.

Hengartner MO. 2000. The biochemistry of apoptosis. *Nature* **407:** 770–776.

Huang DCS, Hahne M, Schroeter M, Frei K, Fontana A, Villunger A, Newton K, Tschopp J, Strasser A. 1999. Activation of Fas by FasL induces apoptosis by a mechanism that cannot be blocked by Bcl-2 or Bcl-xL. *Proc Nat Acad Sci* **96:** 14871–14876.

Kagawa S, Gu J, Honda T, McDonnell TJ, Swisher SG, Roth JA, Fang B. 2001. Deficiency of caspase-3 in MCF7 cells blocks Bax-mediated nuclear fragmentation but not cell death. *Clin Cancer Res* **7:** 1474–1480.

Kerr JF, Wyllie AH, Currie AR. 1972. Apoptosis: A basic biological phenomenon with wide-ranging implications in tissue kinetics. *Br J Cancer* **26:** 239–257.

Komoriya A, Packard BZ, Brown MJ, Wu ML, Henkart PA. 2000. Assessment of caspase activities in intact apoptotic thymocytes using cell-permeable fluorogenic caspase substrates. *J Exp Med* **191:** 1819–1828.

Kulms D, Schwarz T. 2000. Molecular mechanisms of UV-induced apoptosis. *Photodermatol Photoimmunol Photomed* **16:** 195–201.

Lee JH, Kim SY, Kil IS, Park J-W. 2007. Regulation of ionizing radiation-induced apoptosis by mitochondrial NADP⁺-dependent isocitrate dehydrogenase. *J Biol Chem* **282:** 13385–13394.

Ludlow LE, Purton LE, Klarmann K, Gough DJ, Hii LL, Trapani JA, Keller JR, Clarke CJ, Johnstone RW. 2008. The role of p202 in regulating hematopoietic cell proliferation and differentiation. *J Interferon Cytokine Res* **28:** 5–11.

Luthi AU, Martin SJ. 2007. The CASBAH: A searchable database of caspase substrates. *Cell Death Differ* **14:** 641–650.

Martin SJ, O'Brien GA, Nishioka WK, McGahon AJ, Mahboubi A, Saido TC, Green DR. 1995. Proteolysis of fodrin (non-erythroid spectrin) during apoptosis. *J Biol Chem* **270:** 6425–6428.

Martin SJ, Amarante-Mendes GP, Shi L, Chuang TH, Casiano CA, O'Brien GA, Fitzgerald P, Tan EM, Bokoch GM, Greenberg AH, et al. 1996. The cytotoxic cell protease granzyme B initiates apoptosis in a cell-free system by proteolytic processing and activation of the ICE/CED-3 family protease, CPP32, via a novel two-step mechanism. *EMBO J* **15:** 2407–2416.

Poreba M, Strozyk A, Salvesen GS, Drag M. 2013. Caspase substrates and inhibitors. *Cold Spring Harb Perspect Biol*.

Schamberger CJ, Gerner C, Cerni C. 2005. Caspase-9 plays a marginal role in serum starvation-induced apoptosis. *Exp Cell Res* **302:** 115–128.

Sedelies KA, Ciccone A, Clarke CJ, Oliaro J, Sutton VR, Scott FL, Silke J, Susanto O, Green DR, Johnstone RW, et al. 2008. Blocking granule-mediated death by primary human NK cells requires both protection of mitochondria and inhibition of caspase activity. *Cell Death Differ* **15:** 708–717.

Tanaka M, Asano K, Qiu C-H. 2010. Immune regulation by apoptotic cell clearance. *Ann N Y Acad Sci* **1209:** 37–42.

Timmer JC, Salvesen GS. 2007. Caspase substrates. *Cell Death Differ* **14:** 66–72.

Trapani JA, Jans DA, Jans PJ, Smyth MJ, Browne KA, Sutton VR. 1998a. Efficient nuclear targeting of granzyme B and the nuclear consequences of apoptosis induced by granzyme B and perforin are caspase-dependent, but cell death is caspase-independent. *J Biol Chem* **273:** 27934–27938.

Trapani JA, Jans P, Smyth MJ, Froelich CJ, Williams EA, Sutton VR, Jans DA. 1998b. Perforin-dependent nuclear entry of granzyme B precedes apoptosis, and is not a consequence of nuclear membrane dysfunction. *Cell Death Differ* **5:** 488–496.

Wang ZQ, Liao J, Diwu Z. 2005. *N*-DEVD-*N'*-morpholinecarbonyl-rhodamine 110: Novel caspase-3 fluorogenic substrates for cell-based apoptosis assay. *Bioorg Med Chem Lett* **15:** 2335–2338.

Waterhouse NJ, Green DR. 1999. Mitochondria and apoptosis: HQ or high-security prison? *J Clin Immunol* **19:** 378–387.

Waterhouse NJ, Trapani JA. 2003. A new quantitative assay for cytochrome *c* release in apoptotic cells. *Cell Death Differ* **10:** 853–855.

Waterhouse N, Kumar S, Song Q, Strike P, Sparrow L, Dreyfuss G, Alnemri ES, Litwack G, Lavin M, Watters D. 1996. Heteronuclear ribonucleoproteins C1 and C2, components of the spliceosome, are specific targets of interleukin 1β-converting enzyme-like proteases in apoptosis. *J Biol Chem* **271:** 29335–29341.

Waterhouse NJ, Goldstein JC, von Ahsen O, Schuler M, Newmeyer DD, Green DR. 2001. Cytochrome *c* maintains mitochondrial transmembrane potential and ATP generation after outer mitochondrial membrane permeabilization during the apoptotic process. *Cell Biol* **153:** 319–328.

Yang JY, Widmann C. 2002. A subset of caspase substrates functions as the Jekyll and Hyde of apoptosis. Eur Cytokine Netw **13:** 404–406.

Zhang M, Li Y, Zhang H, Xue S. 2001. BAPTA blocks DNA fragmentation and chromatin condensation downstream of caspase-3 and DFF activation in HT-induced apoptosis in HL-60 cells. *Apoptosis* **6:** 291–297.

Triggering Apoptosis in Hematopoietic Cells with Cytotoxic Drugs

Lisa C. Crowley,[1] Brooke J. Marfell,[1] Adrian P. Scott,[1] and Nigel J. Waterhouse[1,2,3,4]

[1]Apoptosis and Cytotoxicity Laboratory, Mater Research, Translational Research Institute, Woolloongabba, Brisbane, Queensland 4102, Australia; [2]Flow Cytometry and Imaging, QIMR Berghofer Medical Research Institute, Herston, Brisbane, Queensland 4006, Australia; [3]School of Medicine, University of Queensland, St. Lucia, Brisbane, Queensland 4072, Australia

Cytotoxic agents are commonly added to cultured cells in the laboratory to investigate their efficacy, mechanism of action, and therapeutic potential. Most of these agents trigger cell death by apoptosis, which is also the most common form of cell death during development, aging, homeostasis, and eradication of disease. Treatment of cells with cytotoxic agents is therefore useful for investigating basic mechanisms of cell death in the human body. Actinomycin D, a cytotoxic agent isolated from *Streptomyces*, induces apoptosis in a variety of cell lines including the histiocytic lymphoma cell line U937. Treatment of U937 cells with actinomycin D provides an ideal model of drug-induced apoptosis that can also be used as a positive control for comparison with other treatments.

MATERIALS

It is essential that you consult the appropriate Material Safety Data Sheets and your institution's Environmental Health and Safety Office for proper handling of equipment and hazardous materials used in this protocol.

RECIPE: Please see the end of this protocol for recipes indicated by <R>. Additional recipes can be found online at http://cshprotocols.cshlp.org/site/recipes.

Reagents

Actinomycin D (1 mM in dimethyl sulfoxide [DMSO]) <R>

> *It is possible to add small volumes of concentrated actinomycin D (1 mM in DMSO) directly to cells; however, this may result in unequal distribution of the drug because DMSO is denser than medium. We therefore prefer to dilute the actinomycin D stock solution in fresh medium before use and add larger volumes of this working solution to our cells; see Step 2.*
> *Other cytotoxic agents can be substituted for actinomycin D in this protocol; see Table 1.*

U937 cells (2–3×10^5 cells/mL) with appropriate culture medium

> *U937 cells are derived from hematopoietic cells and grow in suspension. This protocol can also be used for treatment of other hematopoietic cells in suspension or for adherent cells that grow attached to culture flasks. Adherent cells should be seeded at subconfluent levels before treatment.*
> *It is essential to begin this protocol with a healthy population of cells. Do not use neglected or overconfluent cells. Ideally, U937 cells should be 95% viable before experimentation. (The standard level of viability may vary according to cell type.) Viability can be determined by trypan blue staining as described in Protocol 3: Measuring Cell Death by Trypan Blue Uptake and Light Microscopy (Crowley et al. 2015a).*

---

[4]Correspondence: nigel.waterhouse@qimrberghofer.edu.au

Cite this protocol as *Cold Spring Harb Protoc*; doi:10.1101/pdb.prot087130

TABLE 1. Cytotoxic agents for induction of apoptosis in cultured cells

Reagents	Suggested concentration	Suggested duration of treatment
Actinomycin D	100 nM–1 μM	0–24 h
Staurosporine	100 nM–1 μM	0–24 h
Etoposide	0.1–1 μg/mL	0–24 h
Mitoxantrone	500 nM–5 μM	0–24 h
Doxetaxel	500 nM–5 μM	0–48 h
γ-irradiation	1–20 Gy	0–24 h
TRAIL (Leucine Zipper)	20–250 ng/mL	0–8 h

Equipment

Bright-field microscope
Incubator at 37°C
Tissue-culture flasks and/or plates

METHOD

A graphic flowchart of the method is provided in Figure 1.

1. Seed cells at $2–3 \times 10^5$ cells/mL in a culture flask or plate and incubate overnight at 37°C.

   Ideally, cells should be treated in the log phase of growth, or $\sim 5 \times 10^5$ cells/mL for U937 cells. U937 cells seeded at $2–3 \times 10^5$ cells/mL will continue to proliferate and should reach $\sim 5 \times 10^5$ cells/mL overnight.

2. Add actinomycin D to a final concentration of 1 μM and mix gently by swirling the flask or plate.

   We recommend diluting 1 mM actinomycin D to 100 μM with growth medium and adding 10 μL of this solution to 990 μL of cells.

3. Incubate the cells at 37°C for 4–24 h before assaying cell death by morphological or biochemical assay(s).

   An early time point should be selected to observe high levels of apoptotic cells by morphology and phosphatidylserine exposure. Longer times may be required when measuring late events, such as plasma membrane rupture as detected by trypan blue staining (see Protocol 3: Measuring Cell Death by Trypan Blue Uptake and Light Microscopy [Crowley et al. 2015a]) or propidium iodide (PI) uptake (see Protocol 4: Measuring Cell Death by Propidium Iodide Uptake and Flow Cytometry [Crowley et al. 2015b]).

DISCUSSION

It is wise to begin any study of drug-induced apoptosis using a robust, well-defined model, as some cytotoxic agents can also trigger nonapoptotic death mechanisms such as necrosis, necroptosis, anoikis, and autophagy. This protocol describes a common method for inducing apoptosis in cultured cells using actinomycin D. It can serve as a primary stimulus to study the mechanism of apoptosis, or provide a positive control with which to compare apoptosis with the mechanism of cell death induced

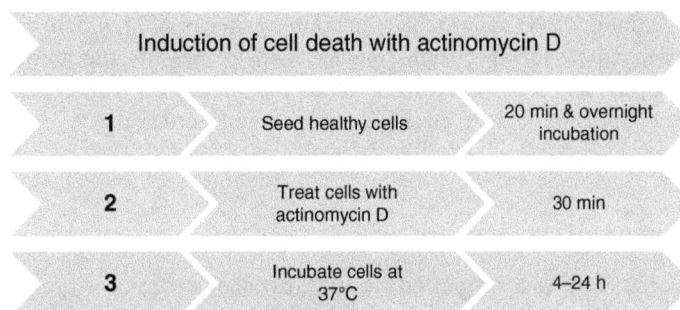

FIGURE 1. A graphic flowchart of treatment of cells with actinomycin D.

by other chemotherapeutic agents. Note that this protocol uses DMSO as a diluent for actinomycin D; care should be taken when using other chemotherapeutic agents to ensure they are soluble in culture medium, which is predominantly water. Cytotoxic agents which are insoluble in DMSO or water may require specialized delivery mechanisms.

Treatment of U937 cells with actinomycin D provides an ideal starting point for studying cell death by a variety of assays; however, various cytotoxic agents such as staurosporine, etoposide, doxetaxel, mitoxantrone, ultraviolet irradiation, γ-irradiation, and ligation of death receptors can be directly substituted for actinomycin D in this protocol (see Table 1). These stimuli all trigger cell death by apoptosis, but the mechanisms by which they initiate apoptosis may differ (Bellosillo et al. 1998; Kulms and Schwarz 2000; Schneider and Tschopp 2000; Kabir et al. 2002; Karpinich et al. 2006; Mhaidat et al. 2007; Kim et al. 2010). Treatment of cells with these agents may therefore result in discrete differences between individual apoptosis signaling pathways. For example, actinomycin D prevents transcription, which activates proapoptotic signaling pathways that are independent of p53 (Merkel et al. 2012). In contrast, etoposide blocks topoisomerase II, which activates proapoptotic signaling pathways that are p53-dependent (Grandela et al. 2008). This is significant because cancer cells often contain mutations in p53 that can block apoptosis when induced by cytotoxic agents that use similar proapoptotic signaling pathways to etoposide (Gimenez-Bonafe et al. 2009).

RELATED TECHNIQUES

The level of cell death induced by this protocol can be quantitated by trypan blue staining (see Protocol 3: Measuring Cell Death by Trypan Blue Uptake and Light Microscopy [Crowley et al. 2015a], PI uptake (see Protocol 4: Measuring Cell Death by Propidium Iodide Uptake and Flow Cytometry [Crowley et al. 2015b]), or a variety of other cell death assays. The choice of assay will depend on a variety of criteria. A combination of protocols may be performed by assaying multiple aliquots of the treated cells. For example, images can be taken of the cells at regular intervals to provide valuable records for analysis of morphology. In addition, aliquots can be preserved by cytospin for analysis by haematoxylin and eosin staining, stained with PI to quantitate loss of plasma membrane integrity, or lysed for analysis of protein content by western blotting.

RECIPE

Actinomycin D (1 mM in DMSO)

Prepare a solution of 1 mg actinomycin D (Sigma-Aldrich A1410) in 816 µL of DMSO (Sigma-Aldrich D2650). Determine the exact concentration of stock actinomycin D by measuring the absorbance at 441 nm and applying Beers law:

$$\text{Absorbance} = (\text{Extinction Coefficient}) \times \text{Concentration}\left(\frac{\text{mol}}{\text{L}}\right) \times \text{Light Path}.$$

Actinomycin D has a molar extinction coefficient of 21,900 (in water). The light path will depend on the spectrophotometer used to measure the absorbance at 441 nm. This is generally 1 cm when using older-style machines with cuvettes, but can be shorter (<1 mm) in modern machines such as the NanoDrop.

ACKNOWLEDGMENTS

This work was supported by an Australian Research Council futures fellowship and project grants from the Leukaemia Foundation Queensland, the Prostate Cancer Foundation Australia, and the Australian Department of Defence to N.J.W.

REFERENCES

Bellosillo B, Colomer D, Pons G, Gil J. 1998. Mitoxantrone, a topoisomerase II inhibitor, induces apoptosis of B-chronic lymphocytic leukaemia cells. *Br J Haematol* **100:** 142–146.

Crowley LC, Marfell BJ, Christensen ME, Waterhouse NJ. 2015a. Measuring cell death by trypan blue uptake and light microscopy. *Cold Spring Harb Protoc* doi:10.1101/pdb.prot087155.

Crowley LC, Scott AP, Marfell BJ, Boughaba JA, Chojnowski G, Waterhouse NJ. 2015b. Measuring cell death by propidium iodide uptake and flow cytometry. *Cold Spring Harb Protoc* doi:10.1101/pdb.prot087163.

Gimenez-Bonafe P, Tortosa A, Perez-Tomas R. 2009. Overcoming drug resistance by enhancing apoptosis of tumor cells. *Curr Cancer Drug Targets* **9:** 320–340.

Grandela C, Pera MF, Wolvetang EJ. 2008. p53 is required for etoposide-induced apoptosis of human embryonic stem cells. *Stem Cell Research* **1:** 116–128.

Kabir J, Lobo M, Zachary I. 2002. Staurosporine induces endothelial cell apoptosis via focal adhesion kinase dephosphorylation and focal adhesion disassembly independent of focal adhesion kinase proteolysis. *Biochem J* **367:** 145–155.

Karpinich NO, Tafani M, Schneider T, Russo MA, Farber JL. 2006. The course of etoposide-induced apoptosis in Jurkat cells lacking p53 and Bax. *J Cell Physiol* **208:** 55–63.

Kim C-H, Won M, Choi C-H, Ahn J, Kim B-K, Song K-B, Kang C-M, Chung K-S. 2010. Increase of RhoB in γ-radiation-induced apoptosis is regulated by c-Jun N-terminal kinase in Jurkat T cells. *Biochem Biophys Res Commun* **391:** 1182–1186.

Kulms D, Schwarz T. 2000. Molecular mechanisms of UV-induced apoptosis. *Photodermatol Photoimmunol Photomed* **16:** 195–201.

Merkel O, Wacht N, Sifft E, Melchardt T, Hamacher F, Kocher T, Denk U, Hofbauer J, Egle A, Scheideler M, et al. 2012. Actinomycin D induces p53–independent cell death and prolongs survival in high-risk chronic lymphocytic leukemia. *Leukemia* **26:** 2508–2516.

Mhaidat NM, Zhang XD, Jiang CC, Hersey P. 2007. Docetaxel-induced apoptosis of human melanoma is mediated by activation of c-Jun NH2-terminal kinase and inhibited by the mitogen-activated protein kinase extracellular signal-regulated kinase 1/2 Pathway. *Clin Cancer Res* **13:** 1308–1314.

Schneider P, Tschopp J. 2000. Apoptosis induced by death receptors. *Pharm Acta Helv* **74:** 281–286.

Triggering Death of Adherent Cells with Ultraviolet Radiation

Lisa C. Crowley[1] and Nigel J. Waterhouse[1,2,3,4]

[1]*Apoptosis and Cytotoxicity Laboratory, Mater Research, Translational Research Institute, Woolloongabba, Brisbane, Queensland 4102, Australia;* [2]*Flow Cytometry and Imaging, QIMR Berghofer Medical Research Institute, Herston, Brisbane, Queensland 4006, Australia;* [3]*School of Medicine, University of Queensland, St. Lucia, Brisbane, Queensland 4072, Australia*

Ultraviolet (UV) radiation is a convenient stimulus for triggering cell death that is available in most laboratories. We use a Stratalinker UV cross-linker because it is a safe, cheap, reliable, consistent, and easily controlled source of UV irradiation. This protocol describes using a Stratalinker to trigger UV-induced death of HeLa cells.

MATERIALS

It is essential that you consult the appropriate Material Safety Data Sheets and your institution's Environmental Health and Safety Office for proper handling of equipment and hazardous materials used in this protocol.

RECIPE: Please see the end of this protocol for recipes indicated by <R>. Additional recipes can be found online at http://cshprotocols.cshlp.org/site/recipes.

Reagents

Adherent cells (e.g., HeLa) (1×10^5 cells/mL) with appropriate culture medium

> *The protocol can also be used to treat other adherent cell lines, or suspension cells with modifications; see Discussion.*
>
> *It is essential to begin this protocol with a healthy population of cells. Ideally, HeLa cells should be 95% viable before experimentation. (The standard level of viability may vary according to cell type.) Viability can be determined by trypan blue staining as described in Protocol 3: Measuring Cell Death by Trypan Blue Uptake and Light Microscopy (Crowley et al. 2015a).*

Phosphate-buffered saline (PBS) <R>

Equipment

Bright-field microscope
Incubator at 37°C
Stratalinker ultraviolet (UV) cross-linker (Stratagene)

> *It is important to know the level of UV exposure administered to cells, as low levels may cause damage but not cell death, moderate levels can trigger apoptosis, and high levels can trigger necrosis. The Stratalinker is a fully enclosed source of UV and contains a built-in light detector which reports $\mu J/cm^2 \times 100$. This detector should be calibrated for each machine and when the bulbs are changed to ensure accurate and reproducible exposure.*

[4]Correspondence: nigel.waterhouse@qimrberghofer.edu.au

Cite this protocol as *Cold Spring Harb Protoc*; doi:10.1101/pdb.prot087148

The level of UV administered by other sources can be determined by using a light detector or using the equation: Joules (J) = Watts × time (sec).

Tissue-culture plates

METHOD

UV is significantly blocked by materials such as tissue-culture plastic and culture media. For best results, remove all materials in the direct light path between the UV source and the cells during irradiation. Removing the culture dish lid and replacing medium with PBS before irradiation (Steps 4 and 5) ensures cells are exposed to the full dose of UV.

The steps in this assay are depicted pictorially in Figure 1.

1. Seed the cells at 1×10^5 cells/mL in culture plates with removable lids. Carefully label all plates.

2. Incubate the cells overnight at 37°C.

   An overnight incubation allows HeLa cells to adhere to the culture plate.

3. Observe the cells by light microscopy to ensure they are healthy and evenly distributed.

   Cells should not be under- or overconfluent when treated.

4. Remove the culture medium and add sufficient PBS to cover the cells.

   Cells should not be irradiated in growth medium containing phenol red, as it partially blocks UV light and may become photoactivated.

5. Remove the plate lid and place the cells in the center of the Stratalinker.

   The lid of the culture plate must be removed before irradiation, as UV will not penetrate plastic efficiently. Control cells can be covered with cardboard to prevent UV exposure (Fig. 1).

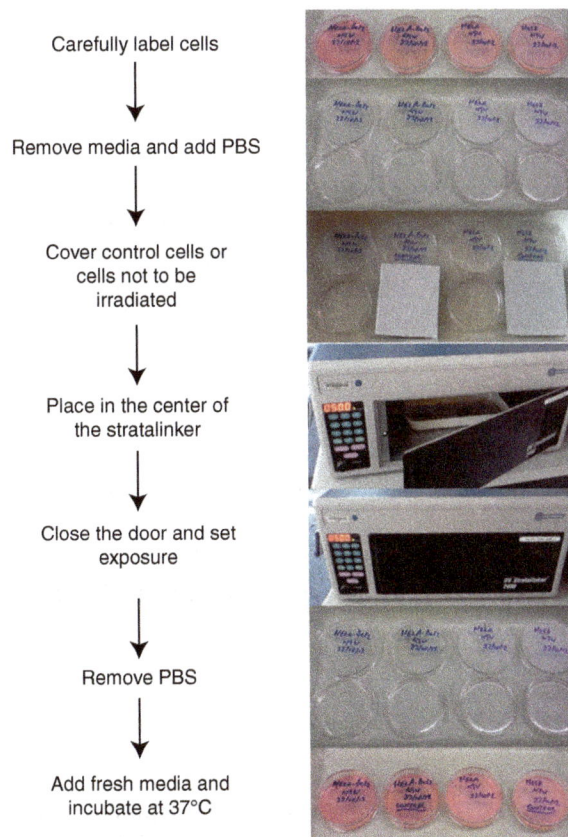

FIGURE 1. Images of work flow for efficient irradiation of cells with UV.

6. Program the Stratalinker with the desired settings and press "Start" (e.g., to expose the cells to 40 mJ/cm^2, press "Reset," press "Energy," insert "400," and press "Start").

   Exposure to different levels of UV can result in either apoptosis or necrosis. The level of UV exposure can be altered by changing the exposure time as determined by the equation: Joules (J) = Watts × time (sec).

7. Once the timer has counted down, remove the cells from the Stratalinker.

8. Remove the PBS from the cells and replace with fresh medium.

   It is important to use fresh medium at this step because the medium that was removed from the cells before irradiation will contain mitotic cells that have not been exposed to UV. These cells may survive and continue to proliferate.

9. Incubate the cells for 4–12 h at 37°C before assaying cell death by morphological or biochemical assay(s).

   The morphological features of apoptosis and necrosis are easy to observe in HeLa cells treated with UV. A time-point at which 40%–60% of cells are dying is optimal for observing high levels of apoptotic cells by morphology or annexin V staining. Longer incubation times may be required when measuring late events such as plasma membrane rupture as detected by trypan blue staining (see Protocol 3: Measuring Cell Death by Trypan Blue Uptake and Light Microscopy [Crowley et al. 2015a]) or propidium iodide (PI) uptake (see Protocol 4: Measuring Cell Death by Propidium Iodide Uptake and Flow Cytometry [Crowley et al. 2015b]).

DISCUSSION

Exposure to UV radiation has been shown to trigger cell death in many cell lines in the laboratory (Martin et al. 1995; Goldstein et al. 2000; Gentile et al. 2003). UV is also commonly used to sterilize laboratory equipment such as tissue culture hoods because of its cytotoxic and germicidal properties. It is used in transillumination boxes, cross-linkers and other equipment that employs lasers, such as microscopes and flow cytometers (Box et al. 1982). This protocol describes treatment of HeLa cells with UV irradiation to trigger cell death. It is an ideal method because HeLa cells grow adhered to culture plastic and are therefore easily manipulated for treatment. The protocol can also be used to treat other adherent cell lines with no modification, but care should be taken with lines that may not adhere to the plastic as strongly as HeLa cells when removing and adding solutions. Suspension cells can also be treated by simply resuspending the cells in a small amount of PBS and spreading the cells over the culture dish for irradiation. Use of UV light to trigger cell death is somewhat unique, but γ-irradiation can be used in place of UV. Sources of γ-irradiation are not commonly available and may require significant training and certification before use.

Although UV irradiation is a very convenient stimulus for triggering cell death in the laboratory that does not involve the use of cytotoxic drugs, it can also activate signaling pathways that directly alter the morphology of cells. It should be noted that UV irradiation is subdivided into three regions of the spectrum: UVA (315–400 nm), UVB (280–314 nm), and UVC (180–280 nm). The sun, which is our primary source of UV, emits light in all three regions; however, we are mostly exposed to UVA because the earth's atmosphere shields us from UVC and most of UVB (de Gruijl 2000; Korac and Khambholja 2011). The Stratalinker and most germicidal lamps generally emit UVC at ∼250 nm. Other laboratory equipment can emit UV in different regions. For example, lasers attached to microscopes can emit UV at various wavelengths. These may be applicable if you wish to investigate the cytotoxic effects of sunlight or a specific wavelength of UV irradiation.

RELATED TECHNIQUES

A variety of assays can be used to quantitate cell death, including counting live and dead cells based on their morphological characteristics. Other quantitative assays include Protocol 3: Measuring Cell Death by Trypan Blue Uptake and Light Microscopy (Crowley et al. 2015a) and Protocol 4: Measuring Cell Death by Propidium Iodide Uptake and Flow Cytometry (Crowley et al. 2015b).

RECIPE

Phosphate-Buffered Saline (PBS)

Reagent	Amount to add (for 1× solution)	Final concentration (1×)	Amount to add (for 10× stock)	Final concentration (10×)
NaCl	8 g	137 mM	80 g	1.37 M
KCl	0.2 g	2.7 mM	2 g	27 mM
Na_2HPO_4	1.44 g	10 mM	14.4 g	100 mM
KH_2PO_4	0.24 g	1.8 mM	2.4 g	18 mM

If necessary, PBS may be supplemented with the following:

$CaCl_2 \cdot 2H_2O$	0.133 g	1 mM	1.33 g	10 mM
$MgCl_2 \cdot 6H_2O$	0.10 g	0.5 mM	1.0 g	5 mM

PBS can be made as a 1× solution or as a 10× stock. To prepare 1 L of either 1× or 10× PBS, dissolve the reagents listed above in 800 mL of H_2O. Adjust the pH to 7.4 (or 7.2, if required) with HCl, and then add H_2O to 1 L. Dispense the solution into aliquots and sterilize them by autoclaving for 20 min at 15 psi (1.05 kg/cm^2) on liquid cycle or by filter sterilization. Store PBS at room temperature.

ACKNOWLEDGMENTS

This work was supported by an Australian Research Council futures fellowship and project grants from the Leukaemia Foundation Queensland, the Prostate Cancer Foundation Australia, and the Australian Department of Defence to N.J.W.

REFERENCES

Box JA, Sugden JK, Younis NM. 1982. The use of ultraviolet light to sterilize water. *Pharm Acta Helv* 57: 330–333.

Crowley LC, Marfell BJ, Christensen ME, Waterhouse NJ. 2015a. Measuring cell death by trypan blue uptake and light microscopy. *Cold Spring Harb Protoc* doi: 10.1101/pdb.prot087155.

Crowley LC, Scott AP, Marfell BJ, Boughaba JA, Chojnowski G, Waterhouse NJ. 2015b. Measuring cell death by propidium iodide uptake and flow cytometry. *Cold Spring Harb Protoc* doi:10.1101/pdb.prot087163.

de Gruijl FR. 2000. Photocarcinogenesis: UVA vs UVB. *Methods Enzymol* 319: 359–366.

Gentile M, Latonen L, Laiho M. 2003. Cell cycle arrest and apoptosis provoked by UV radiation-induced DNA damage are transcriptionally highly divergent responses. *Nucleic Acids Res* 31: 4779–4790.

Goldstein JC, Waterhouse NJ, Juin P, Evan GI, Green DR. 2000. The coordinate release of cytochrome *c* during apoptosis is rapid, complete and kinetically invariant. *Nat Cell Biol* 2: 156–162.

Korac RR, Khambholja KM. 2011. Potential of herbs in skin protection from ultraviolet radiation. *Pharmacogn Rev* 5: 164–173.

Martin SJ, Newmeyer DD, Mathias S, Farschon DM, Wang HG, Reed JC, Kolesnick RN, Green DR. 1995. Cell-free reconstitution of Fas-, UV radiation- and ceramide-induced apoptosis. *EMBO J* 14: 5191–5200.

Measuring Cell Death by Trypan Blue Uptake and Light Microscopy

Lisa C. Crowley,[1] Brooke J. Marfell,[1] Melinda E. Christensen,[1,2,3] and Nigel J. Waterhouse[1,2,4,5]

[1]Apoptosis and Cytotoxicity Laboratory, Mater Research, Translational Research Institute, Woolloongabba, Brisbane, Queensland 4102, Australia; [2]Flow Cytometry and Imaging, QIMR Berghofer Medical Research Institute, Herston, Brisbane, Queensland 4006, Australia; [3]Division of Immunology, Mater Pathology, Mater Adult Hospital, Raymond Terrace, South Brisbane, Queensland 4101, Australia; [4]School of Medicine, University of Queensland, St. Lucia, Brisbane, Queensland 4072, Australia

Trypan blue is a colorimetric dye that stains dead cells with a blue color easily observed using light microscopy at low resolution. The staining procedure is rapid and cells can be analyzed within minutes. The number of live (unstained) and dead (blue) cells can be counted using a hemocytometer on a basic upright microscope. Trypan blue staining is therefore a convenient assay for rapidly determining the overall viability of cells in a culture before commencing scientific experimentation, or for quantitating cell death following treatment with any cytotoxic stimuli.

MATERIALS

It is essential that you consult the appropriate Material Safety Data Sheets and your institution's Environmental Health and Safety Office for proper handling of equipment and hazardous materials used in this protocol.

RECIPE: Please see the end of this protocol for recipes indicated by <R>. Additional recipes can be found online at http://cshprotocols.cshlp.org/site/recipes.

Reagents

Cells, with appropriate culture medium
Cytotoxic agent of choice (see Step 1)
Phosphate-buffered saline (PBS) (as needed) <R>
Trypan blue staining solution (0.4%, w/v) (Sigma-Aldrich T8154)
Trypsin (0.025%)–EDTA (0.01%) in PBS (as needed, for adherent cells)

Equipment

Bright-field microscope
Centrifuge (as needed)
Centrifuge tubes
Hemocytometer and coverslips

[5]Correspondence: nigel.waterhouse@qimrberghofer.edu.au

Cite this protocol as Cold Spring Harb Protoc; doi:10.1101/pdb.prot087155

FIGURE 1. A graphic flowchart of trypan blue staining.

METHOD

It is relatively easy to distinguish stained from unstained cells; however, careful and complete mixing of the dye with the cells is essential to ensure accuracy.

A graphic flowchart of the method is provided in Figure 1.

1. Treat the cells with a cytotoxic stimulus as described in Protocol 1: Triggering Apoptosis in Hematopoietic Cells with Cytotoxic Drugs (Crowley et al. 2015) or Protocol 2: Triggering Death of Adherent Cells with Ultraviolet Radiation (Crowley and Waterhouse 2015).

2. Harvest all cells as follows:

   • For suspension cells: Collect cells directly in a centrifuge tube.

   • For adherent cells: Remove and save the medium, which contains dead and mitotic cells. Detach live cells using standard tissue culture techniques such as incubation with trypsin–EDTA, making sure to keep any washes (e.g., in PBS), which may contain dead or mitotic cells. Add the cells in medium plus the cells from any wash steps to the detached cells.

     Cells must be in suspension for analysis by trypan blue staining using a hemocytometer. Adherent cells may need to be concentrated by centrifugation (500g for 5 min) and resuspended at a convenient concentration (e.g., 5×10^5/mL) if they are plated at low density or if they are washed before detaching with trypsin.

3. Mix 50 µL of cells with an equal volume of trypan blue solution (0.4%).

   It is wise to mix at least 50 µL of cells with 50 µL of trypan blue to obtain efficient mixing and ensure an accurate cell count. If there are too many cells per milliliter to count on the hemocytometer, cell density can be reduced by diluting the sample in PBS or medium before adding trypan blue. This will affect the dilution factor used to calculate the total number of cells per milliliter and should be taken into account when calculating the final cell density in Step 9.

4. Set up the hemocytometer by placing a coverslip over the sample chamber. Gently pipette up and down to mix the sample thoroughly. Add 10 µL of the sample to the hemocytometer under the coverslip.

   Cells will sink to the bottom of the culture medium and should be resuspended directly before analysis.

A   **B**

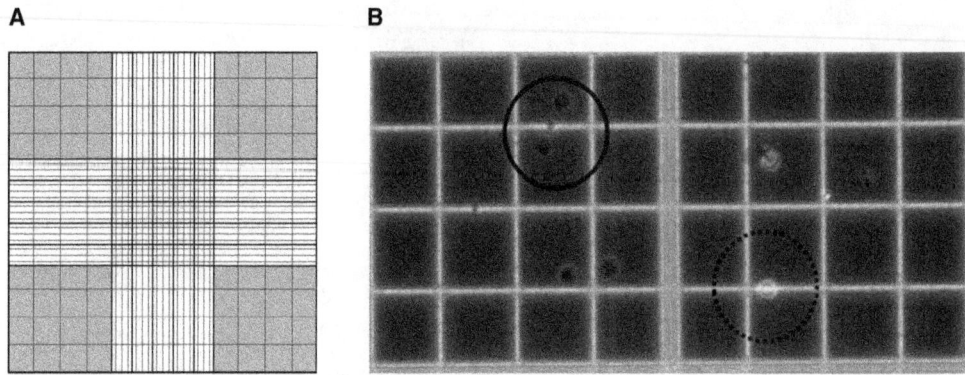

FIGURE 2. (A) Counting squares on a hemocytometer. Each field for cell counting consists of 16 small squares in each corner. The four counting fields are highlighted in gray. (B) Trypan blue-stained cells imaged on a hemocytometer as visualized by bright-field microscopy. Cells were mixed with an equal volume of trypan blue (0.4%). Dead cells (blue) are indicated by the solid circle. A live cell (colorless) is indicated by the dashed circle.

5. Observe trypan blue-stained cells using a standard bright-field microscope. Count at least 100 cells across all four fields of the hemocytometer (Fig. 2A) as follows:

   i. Perform a total cell count by counting all blue (dead) and colorless (healthy) cells (Fig. 2B).

   ii. Perform a live cell count by counting all colorless (healthy) cells.

   iii. Perform a dead cell count by counting all blue (dead) cells.

6. Calculate the total number of cells per milliliter using the following equations:

$$\text{Average number of cells per field} = \frac{\text{Sum of cells per field}}{\text{Number of fields}}.$$

$$\text{Total number of cells per mL} \left(x * 10^4/\text{mL}\right) = \text{Average number of cells per field} \times \text{dilution factor}.$$

   In this protocol, the dilution factor is 2 because the cells were originally diluted with equal volumes of trypan blue; adjust as needed (see Step 3).

7. Calculate the percentage viability using the following equation:

$$\%\text{Viability} = \frac{\text{Number of colorless cells}}{\text{Total number of cells}} \times 100.$$

8. Calculate the percentage of nonviable/dead cells using the following equation:

$$\%\text{Dead} = \frac{\text{Number of colorless cells}}{\text{Total number of cells}} \times 100.$$

DISCUSSION

The basic principle of trypan blue staining relies on the fact that live cells have intact plasma membranes that can exclude various chemicals, including trypan blue. Dead cells invariably have ruptured plasma membranes, regardless of the mechanism of death, and cannot exclude trypan blue (Strober 2001). Dead cells stained with trypan blue can be observed using low-resolution bright-field microscopes, which makes this assay accessible to most medical research scientists (Lee and Soh 2010). The staining procedure is rapid and cells can be analyzed within minutes;

Cite this protocol as *Cold Spring Harb Protoc*; doi:10.1101/pdb.prot087155

it is therefore convenient for use in combination with other cell death assays. It must be noted that trypan blue staining detects all forms of cell death and cannot be used to differentiate among types.

RELATED TECHNIQUES

Assays for loss of plasma membrane integrity can complement other assays for cell death, including cytospinning cells for analysis by light microscopy, DAPI staining to assay nuclear fragmentation, and/or immunocytochemistry to assay protein translocation during cell death.

RECIPE

Phosphate-Buffered Saline (PBS)

Reagent	Amount to add (for 1× solution)	Final concentration (1×)	Amount to add (for 10× stock)	Final concentration (10×)
NaCl	8 g	137 mM	80 g	1.37 M
KCl	0.2 g	2.7 mM	2 g	27 mM
Na_2HPO_4	1.44 g	10 mM	14.4 g	100 mM
KH_2PO_4	0.24 g	1.8 mM	2.4 g	18 mM
If necessary, PBS may be supplemented with the following:				
$CaCl_2 \cdot 2H_2O$	0.133 g	1 mM	1.33 g	10 mM
$MgCl_2 \cdot 6H_2O$	0.10 g	0.5 mM	1.0 g	5 mM

PBS can be made as a 1× solution or as a 10× stock. To prepare 1 L of either 1× or 10× PBS, dissolve the reagents listed above in 800 mL of H_2O. Adjust the pH to 7.4 (or 7.2, if required) with HCl, and then add H_2O to 1 L. Dispense the solution into aliquots and sterilize them by autoclaving for 20 min at 15 psi (1.05 kg/cm^2) on liquid cycle or by filter sterilization. Store PBS at room temperature.

ACKNOWLEDGMENTS

This work was supported by an Australian Research Council futures fellowship and project grants from the Leukaemia Foundation Queensland, the Prostate Cancer Foundation Australia, and the Australian Department of Defence to N.J.W.

REFERENCES

Crowley LC, Waterhouse NJ. 2015. Triggering death of adherent cells with ultraviolet radiation. *Cold Spring Harb Protoc* doi: 10.1101/pdb.prot087148.

Crowley LC, Marfell BJ, Scott AP, Waterhouse NJ. 2015. Triggering apoptosis in hematopoietic cells with cytotoxic drugs. *Cold Spring Harb Protoc* doi: 10.1101/pdb.prot087130.

Lee B-C, Soh K-S. 2010. Visualization of acupuncture meridians in the hypodermis of rat using trypan blue. *J Acupunct Meridian Stud* 3: 49–52.

Strober W. 2001. Trypan blue exclusion test of cell viability. *Curr Protoc Immunol* Appendix 3: Appendix 3B.

Measuring Cell Death by Propidium Iodide Uptake and Flow Cytometry

Lisa C. Crowley,[1] Adrian P. Scott,[1] Brooke J. Marfell,[1] Jeanne A. Boughaba,[1,2] Grace Chojnowski,[3] and Nigel J. Waterhouse[1,3,4,5]

[1]Apoptosis and Cytotoxicity Laboratory, Mater Research, Translational Research Institute, Woolloongabba, Brisbane, Queensland 4102, Australia; [2]Agroparistech, Paris Cedex 05, France; [3]Flow Cytometry and Imaging, QIMR Berghofer Medical Research Institute, Herston, Brisbane, Queensland 4006, Australia; [4]School of Medicine, University of Queensland, St. Lucia, Brisbane, Queensland 4072, Australia

Propidium iodide (PI) is a small fluorescent molecule that binds to DNA but cannot passively traverse into cells that possess an intact plasma membrane. PI uptake versus exclusion can be used to discriminate dead cells, in which plasma membranes become permeable regardless of the mechanism of death, from live cells with intact membranes. PI is excited by wavelengths between 400 and 600 nm and emits light between 600 and 700 nm, and is therefore compatible with lasers and photodetectors commonly available in flow cytometers. This protocol for PI staining can be used to quantitate cell death in most modern research facilities and universities.

MATERIALS

It is essential that you consult the appropriate Material Safety Data Sheets and your institution's Environmental Health and Safety Office for proper handling of equipment and hazardous materials used in this protocol.

RECIPE: Please see the end of this protocol for recipes indicated by <R>. Additional recipes can be found online at http://cshprotocols.cshlp.org/site/recipes.

Reagents

Cells, with appropriate culture medium

Cytotoxic agent of choice (see Step 1)

Phosphate-buffered saline (PBS) (as needed) <R>

Propidium iodide (PI) (Sigma-Aldrich 81845) (0.5 mg/mL in PBS)

PI is light-sensitive. Store PI stock solution (0.5 mg/mL) at 4°C in the dark. Immediately before use, prepare PI-FACS buffer containing 20 µL of PI stock solution per 1 mL of PBS.

Trypsin-EDTA (e.g., 0.25% with 1mM EDTA, Gibco 25200-056) or trypsin replacement (e.g., TrypLE Gibco 12604-013) (as needed, for adherent cells)

[5]Correspondence: nigel.waterhouse@qimrberghofer.edu.au

Cite this protocol as *Cold Spring Harb Protoc*; doi:10.1101/pdb.prot087163

Equipment

Bright-field microscope
Centrifuge
Centrifuge tubes
Flow cytometer

METHOD

A graphic flowchart of the method is provided in Figure 1.

1. Treat the cells with a cytotoxic stimulus as described in Protocol 1: Triggering Apoptosis in Hematopoietic Cells with Cytotoxic Drugs (Crowley et al. 2015) or Protocol 2: Triggering Death of Adherent Cells with Ultraviolet Radiation (Crowley and Waterhouse 2015).

2. Observe the cells with a bright-field microscope.

   It is wise to validate your experiment by bright-field microscopy before analysis by flow cytometry. For example, you may not expect your cells to have high fluorescence if they look morphologically intact. This will also highlight any issues that might affect flow cytometry (e.g., clumping or infection).

3. Harvest all cells as follows:

   • For suspension cells: Collect cells directly in a centrifuge tube and pellet by centrifugation at 500*g* for 5 min.

   • For adherent cells: Remove and save the medium, which contains dead and mitotic cells. Detach live cells using standard tissue culture techniques such as incubation with trypsin–EDTA, making sure to keep any washes (e.g., in PBS). Add the cells in medium and the cells from any washes to the detached cells and harvest all cells by centrifugation at 500*g* for 5 min.

4. Resuspend the harvested cells in PI-FACS buffer. Incubate the cells at room temperature for 15 min in the dark.

   Cells must be in suspension for analysis by flow cytometry. It is important to make sure cells are evenly distributed by flicking the tube or vortexing gently directly before analysis, because cells in suspension will eventually sink to the bottom of the flow cytometry tube. PI-stained cells should be protected from light whenever possible.

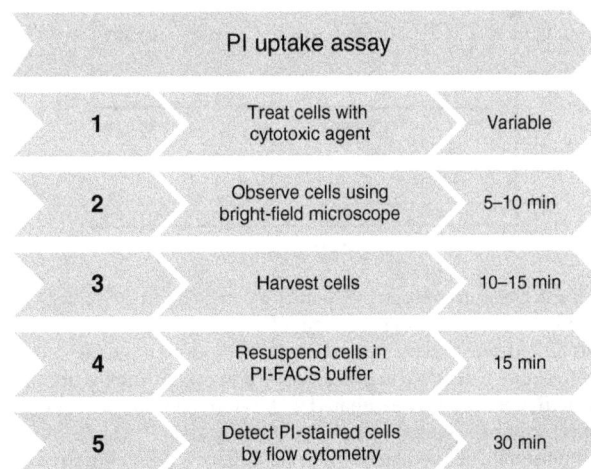

	PI uptake assay	
1	Treat cells with cytotoxic agent	Variable
2	Observe cells using bright-field microscope	5–10 min
3	Harvest cells	10–15 min
4	Resuspend cells in PI-FACS buffer	15 min
5	Detect PI-stained cells by flow cytometry	30 min

FIGURE 1. A graphic flowchart of the PI uptake assay.

Cite this protocol as *Cold Spring Harb Protoc*; doi:10.1101/pdb.prot087163

5. Measure cell death by flow cytometry as follows:

   i. Turn on the appropriate laser on the flow cytometer.

     Although PI has a maximum excitation of 535 nm, it is also excited by 488 nm, which is available on most flow cytometers. The choice of laser for your flow cytometer can be discussed with facility staff.

   ii. Set up a dot plot to detect size (forward scatter [FSC]) and granularity (side scatter [SSC]) using linear scale.

   iii. Set up a histogram plot to detect PI using log scale.

   iv. Run an untreated sample and set the voltage and gain for FSC and SSC to ensure live cells can be detected.

     Make sure to set the gates for FCS and SSC to include both live and dead cells (Fig. 2A). This is important to note, because many researchers who use flow cytometry to phenotype and characterize live cells routinely set their gates to exclude dead cells. It is therefore wise to discuss your experiment fully with the flow cytometry operator.

   v. Run a treated sample and adjust the voltage and gain for FSC and SSC to ensure all cells (live and dead) can be detected based on FSC and SSC.

   vi. Set exclusion gates based on FSC and SSC to exclude cellular debris.

     Dots that come up in the bottom left corner of the FSC versus SSC plot are unlikely to be cells because of their small size and should be gated out. However, it is essential to note this population because some cells fragment into apoptotic bodies that will appear as dots in this region. Although apoptotic bodies should not be counted as cells, observing this population may yield essential information on the level of cell death in the culture.

   vii. Run an untreated sample and adjust the voltage and gain for the PI detector so that all of the cells can be detected on the left side of a histogram plot.

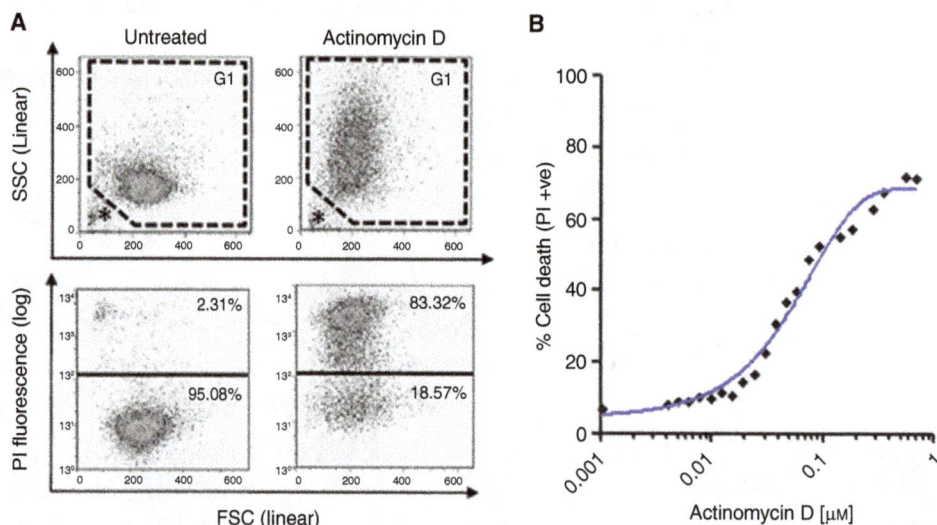

FIGURE 2. Measurement of cell death by PI uptake. (A) HeLa cells were treated with actinomycin D (1 µM) for 24 h and assayed by PI uptake. Cells were gated based on FSC and SSC using a linear scale (*upper* panels) to gate out cellular debris (G1; delineated by dashed line). The fluorescence intensity was then determined for PI and plotted using a log scale (*bottom* panels). The percentage of live cells (below the black line) and dead cells (above the black line) is shown. (B) U937 cells were treated with serial dilutions of actinomycin D for 24 h and assayed for PI uptake. Medium was prepared by first adding 5 µL of actinomycin D (1 mM) to 245 µL of medium (20 µM actinomycin D). This medium was then serially diluted by transferring 200 to 50 µL of fresh medium in the subsequent well. This step was repeated for the rest of the wells of a 96-well culture plate. The maximum concentration of actinomycin D was 10 µM. U937 cells (1×10^6/mL in fresh medium) were added to each well at 50/50 (v/v).

Cite this protocol as *Cold Spring Harb Protoc*; doi:10.1101/pdb.prot087163

It is important that the unstained and stained cells appear within the detection range of the plot. It is wise to first alter the voltage and gain so that unstained cells are as far to the left-hand side of the plot as possible. The PI-stained cells, which fluoresce brightly, will then be detectable on the right-hand side of the plot.

viii. Run a treated sample to ensure that dead cells appear to the right of the untreated cells in the PI channel.

ix. Run each sample and acquire data for at least 10,000 events.

Although it is not absolutely essential to collect 10,000 cells, collecting this many events will ensure smooth peaks when presenting data visually and will also provide statistical power when calculating the percentage of cells with high PI fluorescence.

See Discussion.

DISCUSSION

Advantages and Limitations

The PI exclusion/uptake assay is a relatively simple, inexpensive and convenient procedure for quantitating cell death. Further, detection of cells by flow cytometry can be automated, making this assay ideal for measuring large numbers of samples. PI is regarded as a universal cell death indicator because loss of plasma membrane integrity is a common event in all forms of cell death (Darzynkiewicz et al. 1997). However, PI uptake cannot be used to detect dying cells or to distinguish between different types of cell death (Silva 2010). Note that PI fluorescence can be masked by chemotherapeutic agents that have similar fluorescent properties, such as mitoxantrone (Homolya et al. 2011). In these cases, fluorescent dyes with different excitation/emission wavelengths, such as Live/Dead Aqua (Excitation 367 nm/Emission 526 nm; Molecular Probes L34957) can be used instead of PI.

Data Analysis

Data from each sample can be presented visually as a histogram plot or graphically as the percentage of cells with high PI fluorescence. First, gates should be set for FSC and SSC to ensure all cells are counted but all debris is excluded, as per the acquisition protocol. Histogram plots of cell number (*y*-axis) vs. PI fluorescence (*x*-axis) can then be generated using log scale. These data can be presented visually or used to determine the percentage of cells with high PI fluorescence. If the data are presented visually, the *x*-axis should be labeled "Log PI Fluorescence" and the *y*-axis should be labeled "Cell Count" or "Cell Number." To determine the percentage of cells with high PI fluorescence, a gate can be set around the cells with high or low fluorescence and the percentage of cells within each gate will automatically be determined by the analysis software.

The size and granularity of many cells change as they die. Live cells are larger (higher FSC), less granular (lower SSC), and have low PI fluorescence. Dead cells are smaller (lower FCS), more granular (higher SSC) and have high PI fluorescence. A dot plot of PI fluorescence in log scale vs. FSC or SSC in linear scale can therefore be useful for distinguishing live from dead cells (Fig. 2).

RELATED TECHNIQUES

Trypan blue also stains cells with permeabilized plasma membranes, and can be used in place of PI to quantitate cell death. Trypan blue staining is a colorimetric assay and therefore must be quantitated using a light microscope. More specialized techniques are required to detect dying cells or to distinguish between different types of cell death. PI uptake can be used in combination with fluorescently labeled annexin V to distinguish apoptotic cells from necrotic cells.

RECIPE

Phosphate-Buffered Saline (PBS)

Reagent	Amount to add (for 1× solution)	Final concentration (1×)	Amount to add (for 10× stock)	Final concentration (10×)
NaCl	8 g	137 mM	80 g	1.37 M
KCl	0.2 g	2.7 mM	2 g	27 mM
Na_2HPO_4	1.44 g	10 mM	14.4 g	100 mM
KH_2PO_4	0.24 g	1.8 mM	2.4 g	18 mM

If necessary, PBS may be supplemented with the following:

$CaCl_2 \cdot 2H_2O$	0.133 g	1 mM	1.33 g	10 mM
$MgCl_2 \cdot 6H_2O$	0.10 g	0.5 mM	1.0 g	5 mM

PBS can be made as a 1× solution or as a 10× stock. To prepare 1 L of either 1× or 10× PBS, dissolve the reagents listed above in 800 mL of H_2O. Adjust the pH to 7.4 (or 7.2, if required) with HCl, and then add H_2O to 1 L. Dispense the solution into aliquots and sterilize them by autoclaving for 20 min at 15 psi (1.05 kg/cm^2) on liquid cycle or by filter sterilization. Store PBS at room temperature.

ACKNOWLEDGMENTS

This work was supported by an Australian Research Council futures fellowship and project grants from the Leukaemia Foundation Queensland, the Prostate Cancer Foundation Australia, and the Australian Department of Defence to N.J.W.

REFERENCES

Crowley LC, Waterhouse NJ. 2015. Triggering death of adherent cells with ultraviolet radiation. *Cold Spring Harb Protoc* doi: 10.1101/pdb.prot087148.

Crowley LC, Marfell BJ, Scott AP, Waterhouse NJ. 2015. Triggering apoptosis in hematopoietic cells with cytotoxic drugs. *Cold Spring Harb Protoc* doi: 10.1101/pdb.prot087130.

Darzynkiewicz Z, Juan G, Li X, Gorczyca W, Murakami T, Traganos F. 1997. Cytometry in cell necrobiology: Analysis of apoptosis and accidental cell death (necrosis). *Cytometry* **27**: 1–20.

Homolya L, Orbán TI., Csanády L, Sarkadi B. 2011. Mitoxantrone is expelled by the ABCG2 multidrug transporter directly from the plasma membrane. *Biochimica et Biophysica Acta (BBA) - Biomembranes* **1808**: 154–163.

Silva MT. 2010. Secondary necrosis: The natural outcome of the complete apoptotic program. *FEBS Letters* **584**: 4491–4499.

Measuring Survival of Adherent Cells with the Colony-Forming Assay

Lisa C. Crowley,[1] Melinda E. Christensen,[1,2,3] and Nigel J. Waterhouse[1,2,4,5]

[1]*Apoptosis and Cytotoxicity Laboratory, Mater Research, Translational Research Institute, Woolloongabba, Brisbane, Queensland 4102, Australia;* [2]*Flow Cytometry and Imaging, QIMR Berghofer Medical Research Institute, Herston, Brisbane, Queensland 4006, Australia;* [3]*Division of Immunology, Mater Pathology, Mater Adult Hospital, South Brisbane, Queensland 4101, Australia;* [4]*School of Medicine, University of Queensland, St. Lucia, Brisbane, Queensland 4072, Australia*

Measuring cell death with colorimetric or fluorimetric dyes such as trypan blue and propidium iodide (PI) can provide an accurate measure of the number of dead cells in a population at a specific time; however, these assays cannot be used to distinguish cells that are dying or marked for future death. In many cases it is essential to measure the proliferative capacity of treated cells to provide an indirect measurement of cell death. This can be achieved using the colony-forming assay described here. This protocol specifically applies to measurement of HeLa cells but can be used for most adherent cell lines with limited motility.

MATERIALS

It is essential that you consult the appropriate Material Safety Data Sheets and your institution's Environmental Health and Safety Office for proper handling of equipment and hazardous materials used in this protocol.

RECIPE: Please see the end of this protocol for recipes indicated by <R>. Additional recipes can be found online at http://cshprotocols.cshlp.org/site/recipes.

Reagents

Adherent cells (e.g., HeLa), with appropriate culture medium

This protocol can be used for measuring the proliferative potential of HeLa or other adherent cells with the ability to form distinct colonies from a single cell. It cannot be used for suspension cells or adherent cells with high motility because these types of cells will not form distinct colonies; in these cases, see Protocol 6: Measuring Survival of Hematopoietic Cancer Cells with the Colony-Forming Assay in Soft Agar (Crowley and Waterhouse 2015a).

Crystal violet stain (Sigma-Aldrich C0775)

Prepare a staining solution of 0.5% crystal violet in 25% methanol.

Cytotoxic agent of choice (see Step 1)

Methanol (100%)

Phosphate-buffered saline (PBS) (as needed) <R>

Trypsin-EDTA (e.g., 0.25% with 1mM EDTA, Gibco 25200-056) or trypsin replacement (e.g., TrypLE Gibco 12604-013) (as needed, for adherent cells)

---

[5]Correspondence: nigel.waterhouse@QIMRBerghofer.edu.au

Cite this protocol as *Cold Spring Harb Protoc*; doi:10.1101/pdb.prot087171

Equipment

Bright-field microscope
Centrifuge
Centrifuge tubes
Colony counting software, or plastic sheet with grid for manual counting (see Step 14)
Incubator at 37°C
Tissue-culture plates (24-well, flat-bottomed)

METHOD

A graphic flowchart of the method is provided in Figure 1.

Plating Cells for Colony Formation

1. Treat the cells with a cytotoxic stimulus as described in Protocol 1: Triggering Apoptosis in Hematopoietic Cells with Cytotoxic Drugs (Crowley et al. 2015) or Protocol 2: Triggering Death of Adherent Cells with Ultraviolet Radiation (Crowley and Waterhouse 2015b).

2. Observe the cells with a bright-field microscope.

   Observing your cells by microscopy before harvest can highlight any issues, such as infection, that may negatively affect colony formation.

3. Harvest all cells as follows:

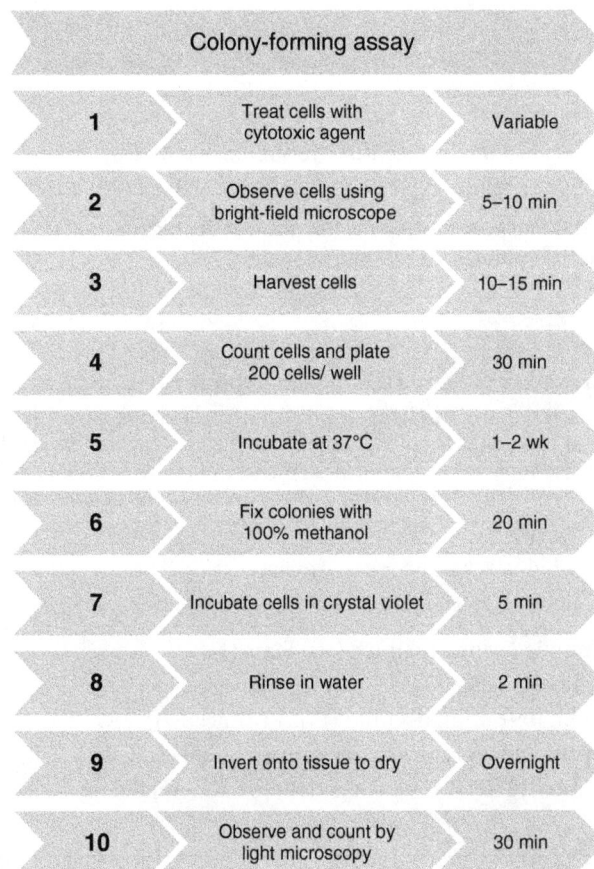

FIGURE 1. A graphic flowchart of the colony-forming assay for adherent cells.

Cite this protocol as *Cold Spring Harb Protoc*; doi:10.1101/pdb.prot087171

i. Remove the medium from the cells and save for Step 3.iii.

> *It is essential to harvest all cells for this assay. The culture medium may contain mitotic cells or dying cells that have detached from the culture plates. Cells may be washed in PBS at this stage, but it is important to keep all washes for Step 3.iii.*

ii. Cover the cells with trypsin–EDTA solution and incubate at 37°C until they become detached.

iii. Add the detached cells (Step 3.ii) to the medium (Step 3.i) and any PBS washes.

iv. Pellet the cells by centrifugation at 500g for 5 min.

v. Resuspend the cells in fresh medium at 200 cells/mL.

> *This cell density should include all cells (live and dead). If only live cells are counted, the number of colonies is likely to be similar for all treatments.*

4. Add 1 mL of cell suspension to each well of a 24-well culture plate.

> *Larger or smaller plates can also be used, and the number of cells in each well can be increased or decreased to obtain the best size and distribution of colonies. It is important to seed cells so they have sufficient space to proliferate without merging with neighboring colonies. The optimum number of cells per well can be determined by plating different numbers of untreated cells or performing a serial dilution.*

5. Incubate the cells for 7 d at 37°C. Check for the formation of colonies using a bright-field microscope. Proceed to fixation (Step 6) when the colonies are a sufficient size for counting.

> *Plates should be housed on their own shelf in the incubator and remain undisturbed. Excessive movement can result in dispersion of mitotic cells, which may result in formation of new colonies.*

> *One to two weeks are sufficient for HeLa cells to form colonies that are easy to count. The time required to grow colonies from other cell types will vary based on the proliferation rates of each individual cell type. Colonies of ~50 cells are sufficient for counting by light microscopy at low resolution. Cells can be returned to the incubator for an extended duration if required.*

Fixing and Staining Colonies for Counting

6. Gently remove all medium from the dish. Carefully add sufficient 100% methanol to cover the cells. Cover the dish and incubate at room temperature for 20 min.

> *Colonies can become detached from the dish if washed too vigorously. This can be avoided by tilting the dish when removing and adding liquid (medium and methanol). Less care is required after the cells are fixed because the colonies become more difficult to remove.*

7. Remove the methanol and rinse the cells with H_2O. Add sufficient crystal violet staining solution to cover the cells. Incubate the dish for 5 min at room temperature.

8. Wash the cells with H_2O until excess dye is removed.

> *Plates at this stage can be washed under running water or in a large water bath, as the colonies are now well fixed.*

9. Invert the plates on tissue paper to dry overnight.

10. View the cells by bright-field microscopy and count the colonies.

> *A colony is considered to be a defined, nonoverlapping group of at least 50 cells. Colonies can be counted manually by light microscopy, and a grid printed on a transparent plastic sheet and stuck to the bottom of the dish can be used to keep track of colonies counted. Alternatively, images of each plate can be scanned using colony counting software. An example of data obtained from this protocol is presented in Figure 2.*

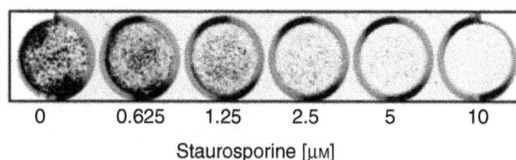

0   0.625   1.25   2.5   5   10
Staurosporine [µM]

FIGURE 2. HeLa cells were treated with 0–10 µM staurosporine as indicated for 8 h. Samples from each treatment were transferred to flat-bottomed 24-well plates and incubated for 7 d. Cells were fixed and stained with crystal violet.

DISCUSSION

This protocol can yield important information about the long-term proliferative potential of cells that cannot be determined by short-term assays. For example, it provides a good indication of whether cells in a population will die although they would not yet be counted as dead by trypan blue staining or propidium iodide (PI) uptake. However, the inability to proliferate in colony-forming assays does not specifically indicate that a cell has died or that it has lost its ability to function; therefore, this assay cannot be used to definitively indicate death or survival. Addressing this issue may require longer assays in which the cells are observed over many months. Cells that survive but do not proliferate may still provide essential functions in the body. Although such cells are unlikely to cause relapse in cancer, it is possible for them to eventually reenter the cell cycle (Kuilman et al. 2010). They may still be capable of producing cytokines and growth factors, or may act as adhesion points for neighboring cells that provide essential functions (Davalos et al. 2010). Combining the colony-forming assay with assays that directly measure cell death is likely to provide more complete information about the fate of cells in a population (Sedelies et al. 2008). Further, these issues provide an interesting topic for debate and another perspective on an important question: What is cell death and when does it matter?

RECIPE

Phosphate-Buffered Saline (PBS)

Reagent	Amount to add (for 1× solution)	Final concentration (1×)	Amount to add (for 10× stock)	Final concentration (10×)
NaCl	8 g	137 mM	80 g	1.37 M
KCl	0.2 g	2.7 mM	2 g	27 mM
Na_2HPO_4	1.44 g	10 mM	14.4 g	100 mM
KH_2PO_4	0.24 g	1.8 mM	2.4 g	18 mM

If necessary, PBS may be supplemented with the following:

$CaCl_2 \cdot 2H_2O$	0.133 g	1 mM	1.33 g	10 mM
$MgCl_2 \cdot 6H_2O$	0.10 g	0.5 mM	1.0 g	5 mM

PBS can be made as a 1× solution or as a 10× stock. To prepare 1 L of either 1× or 10× PBS, dissolve the reagents listed above in 800 mL of H_2O. Adjust the pH to 7.4 (or 7.2, if required) with HCl, and then add H_2O to 1 L. Dispense the solution into aliquots and sterilize them by autoclaving for 20 min at 15 psi (1.05 kg/cm^2) on liquid cycle or by filter sterilization. Store PBS at room temperature.

ACKNOWLEDGMENTS

This work was supported by an Australian Research Council futures fellowship and project grants from the Leukaemia Foundation Queensland, the Prostate Cancer Foundation Australia, and the Australian Department of Defence to N.J.W.

REFERENCES

Crowley LC, Waterhouse NJ. 2015a. Measuring survival of hematopoietic cancer cells with the colony-forming assay in soft agar. *Cold Spring Harb Protoc* doi: 10.1101/pdb.prot087189.

Crowley LC, Waterhouse NJ. 2015b. Triggering death of adherent cells with ultraviolet radiation. *Cold Spring Harb Protoc* doi: 10.1101/pdb.prot087148.

Crowley LC, Marfell BJ, Scott AP, Waterhouse NJ. 2015. Triggering apoptosis in hematopoietic cells with cytotoxic drugs. *Cold Spring Harb Protoc* doi: 10.1101/pdb.prot087130.

Davalos AR, Coppe JP, Campisi J, Desprez PY. 2010. Senescent cells as a source of inflammatory factors for tumor progression. *Cancer Metastasis Rev* 29: 273–283.

Kuilman T, Michaloglou C, Mooi WJ, Peeper DS. 2010. The essence of senescence. *Genes Dev* 24: 2463–2479.

Sedelies KA, Ciccone A, Clarke CJP, Oliaro J, Sutton VR, Scott FL, Silke J, Susanto O, Green DR, Johnstone RW, et al. 2008. Blocking granule-mediated death by primary human NK cells requires both protection of mitochondria and inhibition of caspase activity. *Cell Death Differ* 15: 708–717.

Cite this protocol as *Cold Spring Harb Protoc*; doi:10.1101/pdb.prot087171

Measuring Survival of Hematopoietic Cancer Cells with the Colony-Forming Assay in Soft Agar

Lisa C. Crowley[1] and Nigel J. Waterhouse[1,2,3,4]

[1]*Apoptosis and Cytotoxicity Laboratory, Mater Research, Translational Research Institute, Woolloongabba, Brisbane, Queensland 4102, Australia;* [2]*Flow Cytometry and Imaging, QIMR Berghofer Medical Research Institute, Herston, Brisbane, Queensland 4006, Australia;* [3]*School of Medicine, University of Queensland, St. Lucia, Brisbane, Queensland 4072, Australia*

Colony-forming assays measure the ability of cells in culture to grow and divide into groups. Any cell that has the potential to form a colony may also have the potential to cause cancer or relapse in vivo. Colony-forming assays also provide an indirect measurement of cell death because any cell that is dead or dying will not continue to proliferate. The proliferative capacity of adherent cells such as fibroblasts can be determined by growing cells at low density on culture dishes and counting the number of distinct groups that form over time. Cells that grow in suspension, such as hematopoietic cells, cannot be assayed this way because the cells move freely in the media. Assays to determine the colony-forming ability of hematopoietic cells must therefore be performed in solid matrices that restrict large-scale movement of the cells. One such matrix is soft agar. This protocol describes the use of soft agar to compare the colony-forming ability of untreated hematopoietic cells to the colony-forming ability of hematopoietic cells that have been treated with a cytotoxic agent.

MATERIALS

It is essential that you consult the appropriate Material Safety Data Sheets and your institution's Environmental Health and Safety Office for proper handling of equipment and hazardous material used in this protocol.

RECIPES: Please see the end of this protocol for recipes indicated by <R>. Additional recipes can be found online at http://cshprotocols.cshlp.org/site/recipes.

Reagents

Bottom and top layer media for colony-forming assay <R>

Cell culture medium (1×)

Prepare fresh from powdered media (e.g., Life Technologies 3180089) and filter-sterilize. The type of medium required will depend on the type of cells under investigation. Some cells may require addition of growth supplements.

Cell layer media for colony-forming assay <R>

Cell line of interest (e.g., hematopoietic cells)

This protocol is designed to assay the proliferative capacity of cells that normally grow in suspension. It can also be used with adherent cell lines that have been detached from the plate and resuspended in media. However,

[4]Correspondence: nigel.waterhouse@qimrberghofer.edu.au

Cite this protocol as *Cold Spring Harb Protoc*; doi:10.1101/pdb.prot087189

adherent cells will only proliferate in soft agar if they do not require attachment for survival. A colony-forming assay for adherent cells is described in Protocol 5: Measuring Survival of Adherent Cells with the Colony-Forming Assay (Crowley et al. 2015a).

Cytotoxic agent of choice (see Step 3)
Noble agar (1.3%)

Prepare fresh and place in a water bath at 55°C to prevent agar from solidifying.

Equipment

Cell culture incubator
Cell culture plates (6 cm)
Conical tube (10 or 15 mL)
Water bath

METHOD

A graphic flowchart of the method is provided in Figure 1.

Prepare Assay Plates

1. Thoroughly mix Noble agar (1.3%) with one half of the bottom and top layer media for colony-forming assay at a ratio of 55:45 and immediately pour into the assay plates (e.g., 2.5 mL per well of a 6 cm dish).

2. Place the media/agar plates in a cell culture incubator overnight at 37°C to solidify.

Prepare and Plate the Cell Layer

3. Treat the cells with a cytotoxic stimulus as described in Protocol 1: Triggering Apoptosis in Hematopoietic Cells with Cytotoxic Drugs (Crowley et al. 2015b). Count cells and resuspend at 2×10^4/mL in 1× cell culture medium.

4. Add 1 mL of cells to a clean 10- or 15-mL tube.

5. Mix the cell layer media for colony-forming assay with Noble agar (1.3%) at a ratio of 80:20.

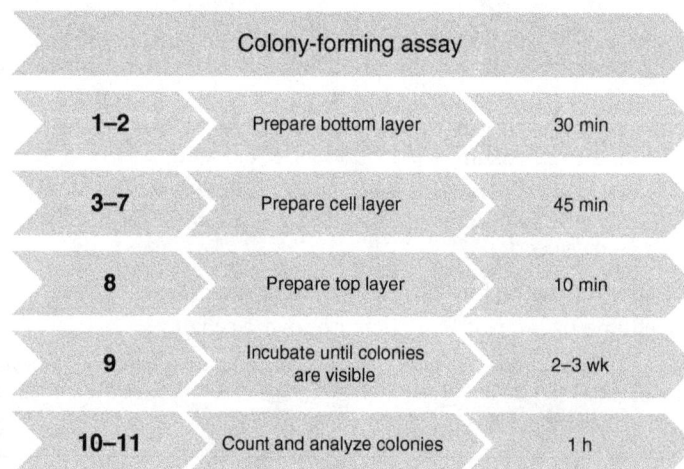

	Colony-forming assay	
1–2	Prepare bottom layer	30 min
3–7	Prepare cell layer	45 min
8	Prepare top layer	10 min
9	Incubate until colonies are visible	2–3 wk
10–11	Count and analyze colonies	1 h

FIGURE 1. Graphic flowchart of the colony-forming assay.

Cite this protocol as *Cold Spring Harb Protoc*; doi:10.1101/pdb.prot087189

6. Add 9 mL of the media/agar mix to the cells in Step 4 and immediately add the cell/media/ agar mixture to the top of the solid bottom layer of agar (e.g., 5 mL/6 cm dish, or 1×10^4 cells per dish).

   10 mL of cell/media/agar mix is sufficient for plating duplicate wells.

   Cells must be plated so they have sufficient space to form colonies without merging with neighboring colonies. Determine the optimum density by plating different numbers of untreated cells.

7. Incubate assay plates at room temperature for 30 min to solidify.

Prepare and Plate the Top Layer

8. Thoroughly mix Noble agar (1.3%) with the second half of the bottom and top layer media for colony-forming assay at a ratio of 55:45 and immediately pour into the assay plates (e.g., 2.5 mL per well of a 6 cm dish).

Grow and Count Colonies

9. Incubate the assay plates at 37°C in a cell culture incubator for ~2–3 wk until colonies are easily visible by the human eye.

   Grow colonies until they are easily seen by the naked eye but can still be distinguished from each other (i.e., nonoverlapping).

10. Count the number of colonies or make a photocopy of the plates for counting at a later stage.

11. Calculate the percentage survival relative to an untreated control.

$$\% \text{ survival} = \frac{\text{no. colonies in treated sample}}{\text{no. colonies in untreated sample}} \times 100 \tag{1}$$

These data can be represented graphically. For sample results, see Ludlow et al. (2008).

DISCUSSION

Colony-forming assays can yield important information about the long-term proliferative potential of cells, providing a good indication of whether the cells are healthy or whether they are sick and will eventually die (Ludlow et al. 2008; Sedelies et al. 2008). However, they cannot be used to definitively show that a cell has not died because some cells that are not dead may not continue to proliferate. These cells may still provide essential functions in the body such as production of cytokines and growth factors or provide support for other cells (Davalos et al. 2010). Combining colony-forming assays with assays that directly measure cell death is therefore likely to be most informative.

RECIPES

Bottom and Top Layer Media for Colony-Forming Assay

Cell culture medium (2×)	45 mL
Fetal bovine serum (FBS)	9 mL
Glutamine (200 mM)	1 mL

Prepare fresh 2× cell culture medium appropriate for the cell line of interest using a powdered medium (e.g., Life Technologies 3180089) and filter-sterilize. Then add FBS and glutamine. Prepare sufficient volume for the samples to be assayed. One-half of this recipe is to be used for the bottom layer and the other half for the top.

Cell Layer Media for Colony-Forming Assay

Cell culture medium (2×)	20 mL
Cell culture medium (1×)	50 mL
Fetal bovine serum (FBS)	9 mL
Glutamine (200 mM)	1 mL

Prepare fresh 1× and 2× cell culture media appropriate for the cell line of interest using a powdered medium (e.g., Life Technologies 3180089) and filter-sterilize. Then add FBS and glutamine. Prepare sufficient volume for the number of samples to be assayed.

ACKNOWLEDGMENTS

This work was supported by an Australian Research Council futures fellowship and project grants from the Leukaemia Foundation Queensland, the Prostate Cancer Foundation Australia, and the Australian Department of Defence to N.J.W.

REFERENCES

Crowley LC, Christensen ME, Waterhouse NJ. 2015a. Measuring survival of adherent cells with the colony-forming assay. *Cold Spring Harb Protoc* doi: 10.1101/pdb.prot087171.

Crowley LC, Marfell BJ, Scott AP, Waterhouse NJ. 2015b. Triggering apoptosis in hematopoietic cells with cytotoxic drugs. *Cold Spring Harb Protoc* doi: 10.1101/pdb.prot087130.

Davalos AR, Coppe JP, Campisi J, Desprez PY. 2010. Senescent cells as a source of inflammatory factors for tumor progression. *Cancer Metastasis Rev* 29: 273–283.

Ludlow LE, Purton LE, Klarmann K, Gough DJ, Hii LL, Trapani JA, Keller JR, Clarke CJ, Johnstone RW. 2008. The role of p202 in regulating hematopoietic cell proliferation and differentiation. *J Interferon Cytokine Res* 28: 5–11.

Sedelies KA, Ciccone A, Clarke CJ, Oliaro J, Sutton VR, Scott FL, Silke J, Susanto O, Green DR, Johnstone RW, et al. 2008. Blocking granule-mediated death by primary human NK cells requires both protection of mitochondria and inhibition of caspase activity. *Cell Death Differ* 15: 708–717.

Cite this protocol as *Cold Spring Harb Protoc*; doi:10.1101/pdb.prot087189

Morphological Analysis of Cell Death by Cytospinning Followed by Rapid Staining

Lisa C. Crowley,[1] Brooke J. Marfell,[1] and Nigel J. Waterhouse[1,2,3,4]

[1]Apoptosis and Cytotoxicity Laboratory, Mater Research, Translational Research Institute, Woolloongabba, Brisbane, Queensland 4102, Australia; [2]Flow Cytometry and Imaging, QIMR Berghofer Medical Research Institute, Herston, Brisbane, Queensland 4006, Australia; [3]School of Medicine, University of Queensland, St. Lucia, Brisbane, Queensland 4072, Australia

Identifying and characterizing different forms of cell death can be facilitated by staining internal cellular structures with dyes such as hematoxylin and eosin (H&E). These dyes stain the nucleus and cytoplasm, respectively, and optimized reagents (e.g., Rapi-Diff, Rapid Stain, or Quick Dip) are commonly used in pathology laboratories. Fixing and staining adherent cells with these optimized reagents is a straightforward procedure, but apoptotic cells may detach from the culture plate and be washed away during the fixing and staining procedure. To prevent the loss of apoptotic cells, cells can be gently centrifuged onto glass slides by cytospinning before fixing and staining. In addition to apoptotic cells, this procedure can be used on cells in suspension, or adherent cells that have been trypsinized and removed from the culture dish. This protocol describes cytospinning followed by Rapi-Diff staining for morphological analysis of cell death.

MATERIALS

It is essential that you consult the appropriate Material Safety Data Sheets and your institution's Environmental Health and Safety Office for proper handling of equipment and hazardous materials used in this protocol.

RECIPES: Please see the end of this protocol for recipes indicated by <R>. Additional recipes can be found online at http://cshprotocols.cshlp.org/site/recipes.

Reagents

Cell line of interest

This method can be used for adherent or suspension cell lines as well as primary cells.

Cytotoxic agent of choice (see Step 1)
Methanol
Phosphate-buffered saline (PBS) <R>
Rapid staining kit

Various hematology staining kits are available through pathology supply companies, which differ by country. Suitable products include Rapid Stain from Amber Scientific (Australia), Quick Dip from Fronine (Australia), Dip Quick from Jorgensen Labs (United States), and the Rapi-Diff II Stain Kit from Atom Scientific (United Kingdom, Ireland, and Europe). We recommend Rapi-Diff, and its use is described here.

[4]Correspondence: nigel.waterhouse@qimrberghofer.edu.au

Cite this protocol as *Cold Spring Harb Protoc*; doi:10.1101/pdb.prot087197

Slide mounting medium (e.g., DPX Mountant for histology, Sigma-Aldrich 06522)
Water

Equipment

Bright-field microscope and camera (optional; see Step 12)
Cytospin apparatus (e.g., Thermo Scientific Cytospin 4 cytocentrifuge), filter paper, and funnels
Glass slides and coverslips
Pencil
Standard staining dishes and slide racks

METHOD

See Figure 1 for a flowchart of this protocol. All steps should be performed at room temperature. Cold temperatures will distort the morphology of the cells.

Cytospinning

1. Treat cells with cytotoxic stimuli as described in Protocol 1: Triggering Apoptosis in Hematopoietic Cells with Cytotoxic Drugs (Crowley et al. 2015) or Protocol 2: Triggering Death of Adherent Cells with Ultraviolet Radiation (Crowley and Waterhouse 2015).

2. Harvest the cells using a method appropriate for the cell line of interest, and resuspend them in PBS at $2.5-3 \times 10^6$ cells/mL.

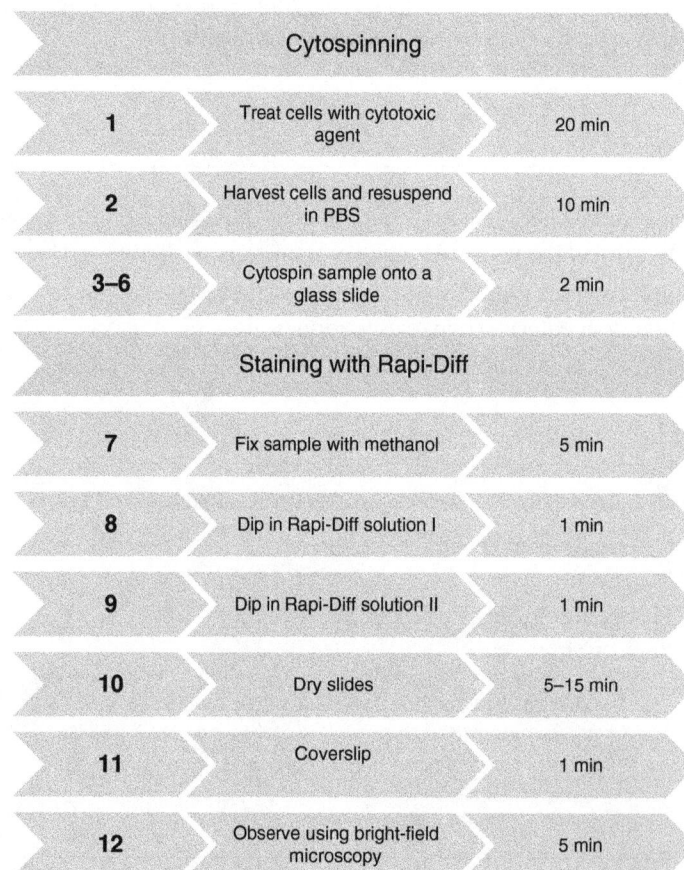

	Cytospinning	
1	Treat cells with cytotoxic agent	20 min
2	Harvest cells and resuspend in PBS	10 min
3–6	Cytospin sample onto a glass slide	2 min
	Staining with Rapi-Diff	
7	Fix sample with methanol	5 min
8	Dip in Rapi-Diff solution I	1 min
9	Dip in Rapi-Diff solution II	1 min
10	Dry slides	5–15 min
11	Coverslip	1 min
12	Observe using bright-field microscopy	5 min

FIGURE 1. Graphic flowchart of the cytospin and Rapi-Diff staining protocols.

Cite this protocol as *Cold Spring Harb Protoc*; doi:10.1101/pdb.prot087197

Mitotic, apoptotic, or dead cells become detached from the culture dish and therefore it is essential to harvest both the attached and nonattached cells before cytospinning. This can be achieved by harvesting the culture media containing the suspended cells followed by trypsinization of adherent cells. Combine both populations before cytospinning.

Optimize the number of cells for individual cell types. A total of $4–5 \times 10^5$ cells per microscope slide is ideal for cell lines such as U937 or other leukemic cell lines.

3. Assemble the cytospin apparatus by inserting the labeled microscope slides, filter paper, and cytospin funnel into the metal trap.

   It is essential to use cytospin filter paper between the funnel and the slide. The filter paper and funnel must be lined up correctly. If the hole in the filter paper and funnel are not aligned, the sample will not correctly spin on to the glass slide and the sample will be absorbed by the filter paper.

   Label the microscope slides with pencil as ink may be removed by the subsequent fixation and staining steps.

4. Add 150 µL of cells to the cytospin funnel.

   Volumes >150 µL can reduce the quality of the cytospin.

5. Centrifuge the samples in the cytospin apparatus at 400 rpm for 2 min at room temperature with medium acceleration.

6. Remove the slides from the cytospin apparatus.

Staining with Rapi-Diff

7. Fix the cells by placing the slides in methanol (provided in the kit) for 5 min.

8. Dip the slides in Solution 1 (eosin) 10 times.

   Staining intensity can be changed by incubating the slides in Solution 1 or Solution 2 for different lengths of time.

9. Dip the slides in Solution 2 (hematoxylin) 10 times.

10. Rinse the slides with water, and allow the slides to dry for 5–15 min.

11. Add a small drop of Slide mounting medium such as DPX on the stained cells and add a coverslip to protect the sample. Dry the slides overnight at room temperature.

    Wait for the coverslip fixative to dry before viewing the cells with the microscope. The fixative can coat and damage the lens, making it difficult to view the cells.

12. View the cells using a bright-field microscope.

    This technique is primarily qualitative; however, the cells can be counted to provide quantitative information. Images of the cells (e.g., Fig. 2) can also be taken for presentation or future analysis using a camera attached to the microscope view port. The level of information that can be obtained is dependent on the quality of the microscope and lens used to view the cells. It is important to ensure that the images taken are representative of the cells on the slide.

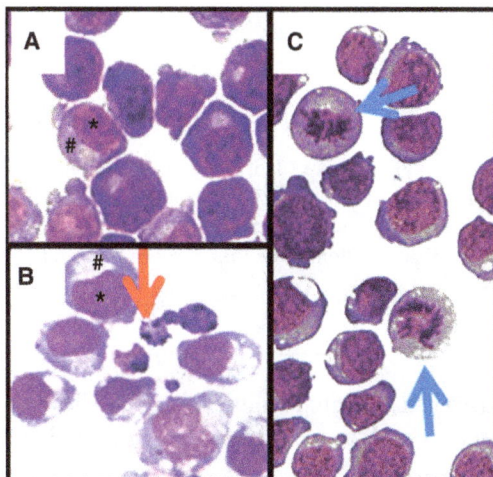

FIGURE 2. Cells stained with Rapi-Diff. Untreated K562 cells (A,C) and K562 cells treated with 5 µM Imatinib for 24 h (B) were pelleted onto glass slides by cytospinning and stained with Rapi-Diff. Features of healthy cells include clearly defined nuclei (*) that can easily be distinguished from the surrounding cytoplasm (#). Mitotic cells (blue arrows) may also be visible. Features of dead cells (red arrow) include shrinkage and nuclear condensation.

DISCUSSION

Cytospinning followed by fixation and staining with H&E using Rapid Stain (e.g., Rapi-Diff) is extremely useful for viewing the internal cell structure of treated cells with light microscopy (Fischer et al. 2008; Crowley et al. 2013). These dyes are commonly used by pathology laboratories to stain blood smears and tissues for further analysis (Fox 2000) and are therefore readily available. The clear differences in morphology between live and dead cells make this technique ideal for identifying and characterizing cell death in culture. Healthy cells generally have a round, well-defined nucleus surrounded by a cytoplasmic region enclosed by a continuous plasma membrane. Apoptosis is characterized by shrinking and blebbing of the plasma membrane and condensation and fragmention of the nucleus. Some cells can eventually fragment into small plasma membrane bound apoptotic bodies. The nucleus of necrotic cells is less well defined and the plasma membrane does not appear intact. It is important to note that the nuclei of healthy mitotic cells also appear condensed but these can be distinguished from apoptotic cells because the DNA of mitotic cells forms two parallel lines as the chromosomes are segregated. Examples of live and dead cells are presented in Figure 2.

RELATED TECHNIQUES

This technique allows the visualization of internal structures of cells by staining with colorimetric dyes. These structures can also be visualized by immunofluorescence using antibodies to proteins that are specific to those structures. Combining this technique with other protocols such as 4′, 6-diamidino-2-phenylindole (DAPI) or immunofluorescence staining can allow in-depth analysis of cell death. It can also be used as a visual reference for assays that include flow cytometry, which is useful for quantitating specific events but cannot be used to visualize cellular structures.

RECIPE

Phosphate-Buffered Saline (PBS)

Reagent	Amount to add (for 1× solution)	Final concentration (1×)	Amount to add (for 10× stock)	Final concentration (10×)
NaCl	8 g	137 mM	80 g	1.37 M
KCl	0.2 g	2.7 mM	2 g	27 mM
Na_2HPO_4	1.44 g	10 mM	14.4 g	100 mM
KH_2PO_4	0.24 g	1.8 mM	2.4 g	18 mM

If necessary, PBS may be supplemented with the following:

$CaCl_2 \cdot 2H_2O$	0.133 g	1 mM	1.33 g	10 mM
$MgCl_2 \cdot 6H_2O$	0.10 g	0.5 mM	1.0 g	5 mM

PBS can be made as a 1× solution or as a 10× stock. To prepare 1 L of either 1× or 10× PBS, dissolve the reagents listed above in 800 mL of H_2O. Adjust the pH to 7.4 (or 7.2, if required) with HCl, and then add H_2O to 1 L. Dispense the solution into aliquots and sterilize them by autoclaving for 20 min at 15 psi (1.05 kg/cm^2) on liquid cycle or by filter sterilization. Store PBS at room temperature.

ACKNOWLEDGMENTS

This work was supported by an Australian Research Council futures fellowship and project grants from the Leukaemia Foundation Queensland, the Prostrate Cancer Foundation Australia, and the Australian Department of Defence to N.J.W.

Cite this protocol as *Cold Spring Harb Protoc*; doi:10.1101/pdb.prot087197

REFERENCES

Crowley LC, Waterhouse NJ. 2015. Triggering death of adherent cells with ultraviolet radiation. *Cold Spring Harb Protoc* doi: 10.1101/pdb .prot087148.

Crowley LC, O'Donovan TR, Nyhan MJ, McKenna SL. 2013. Pharmacological agents with inherent anti-autophagic activity improve the cytotoxicity of imatinib. *Oncol Rep* **29:** 2261–2268.

Crowley LC, Marfell BJ, Scott AP, Waterhouse NJ. 2015. Triggering apoptosis in hematopoietic cells with cytotoxic drugs. *Cold Spring Harb Protoc* doi: 10.1101/pdb.prot087130.

Fischer AH, Jacobson KA, Rose J, Zeller R. 2008. Hematoxylin and eosin staining of tissue and cell sections. *Cold Spring Harbor Protocols* **2008:** pdb.prot4986.

Fox H. 2000. Is H&E morphology coming to an end? *J Clin Pathol* **53:** 38–40.

Analyzing Cell Death by Nuclear Staining with Hoechst 33342

Lisa C. Crowley,[1] Brooke J. Marfell,[1] and Nigel J. Waterhouse[1,2,3,4]

[1]Apoptosis and Cytotoxicity Laboratory, Mater Research, Translational Research Institute, Woolloongabba, Brisbane, Queensland 4102, Australia; [2]Flow Cytometry and Imaging, QIMR Berghofer Medical Research Institute, Herston, Brisbane, Queensland 4006, Australia; [3]School of Medicine, University of Queensland, St. Lucia, Brisbane, Queensland 4072, Australia

The nuclei of healthy cells are generally spherical, and the DNA is evenly distributed. During apoptosis the DNA becomes condensed, but this process does not occur during necrosis. Nuclear condensation can therefore be used to distinguish apoptotic cells from healthy cells or necrotic cells. Dyes that bind to DNA, such as Hoechst 33342 or 4′,6-diamidino-2-phenylindole (DAPI), can be used to observe nuclear condensation. These dyes fluoresce at 461 nm when excited by ultraviolet light and can therefore be visualized using conventional fluorescent microscopes equipped with light sources that emit light at ∼350 nm and filter sets that permit the transmission of light at ∼460 nm. This protocol describes staining and visualization of cells stained with Hoechst 33342, but it can be adapted for staining with DAPI or other dyes.

MATERIALS

It is essential that you consult the appropriate Material Safety Data Sheets and your institution's Environmental Health and Safety Office for proper handling of equipment and hazardous material used in this protocol.

RECIPES: Please see the end of this protocol for recipes indicated by <R>. Additional recipes can be found online at http://cshprotocols.cshlp.org/site/recipes.

Reagents

Cells of interest

Cytotoxic agent of choice

DPX mounting medium

Hoechst 33342 (Sigma-Aldrich)

> Hoechst 33342 binds to DNA and therefore has the potential to be mutagenic. Reagents containing this dye should be treated with caution. Hoechst 33342 can be substituted with Hoechst 33258, 4′,6-diamidino-2-phenylindole (DAPI), ProLong Gold Antifade with DAPI, or propidium iodide. DAPI is reportedly less toxic than Hoechst 33342, but it is still a DNA-binding agent and should therefore also be treated as mutagenic.

Paraformaldehyde (PFA) (4%) in PBS <R>

Phosphate-buffered saline (PBS) <R>

---

[4]Correspondence: nigel.waterhouse@QIMRBerghofer.edu.au

Cite this protocol as *Cold Spring Harb Protoc*; doi:10.1101/pdb.prot087205

Equipment

Coverslips

Fluorescence microscope

Hoechst 33342 binds to the DNA of live or fixed cells and fluoresces at 461 nm when excited by ultraviolet (UV) light (excitation maximum ∼350 nm). The dye can be visualized using a standard fluorescence microscope equipped with a UV light source and a blue/cyan filter (∼460 nm).

Light microscope

METHOD

For a flowchart summarizing this protocol, see Figure 1.

1. Treat cells of interest with a cytotoxic agent as described in Protocol 1: Triggering Apoptosis in Hematopoietic Cells with Cytotoxic Drugs (Crowley et al. 2015a) or in Protocol 2: Triggering Death of Adherent Cells with Ultraviolet Radiation (Crowley et al. 2015b). Observe the cells by bright-field microscopy to determine whether the cytotoxic agent has been effective. Harvest the cells (e.g., by cytospinning as described in Protocol 7: Morphological Analysis of Cell Death by Cytospinning Followed by Rapid Staining [Crowley et al. 2015c]) and fix them in 4% PFA for 20 min at room temperature.

   At this stage, the cells can be stored in PBS for up to 2 wk at 4°C.

2. Wash the cells with PBS for 5 min at room temperature. Pipette or aspirate off the PBS.

   To avoid damaging the cells, do not touch them with the pipette.

3. Incubate the cells with Hoechst 33342 (100 ng/mL in PBS) in the dark for 15 min at room temperature.

   Hoechst 33342 is light-sensitive, and samples should remain protected from light after this step.

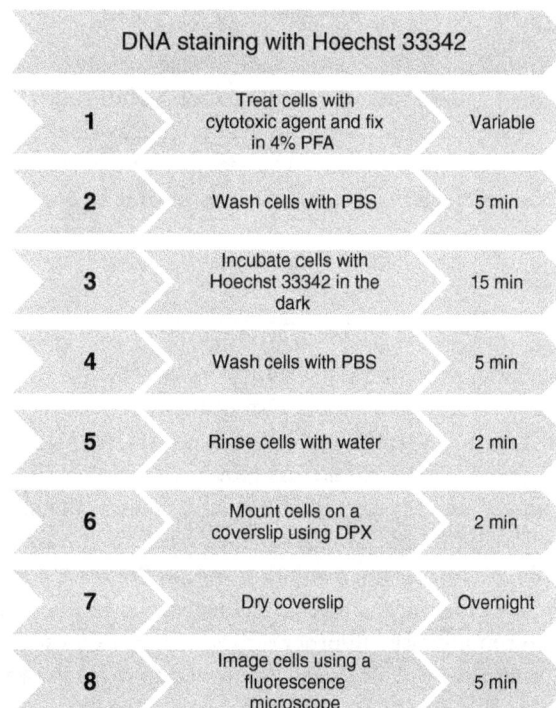

	DNA staining with Hoechst 33342	
1	Treat cells with cytotoxic agent and fix in 4% PFA	Variable
2	Wash cells with PBS	5 min
3	Incubate cells with Hoechst 33342 in the dark	15 min
4	Wash cells with PBS	5 min
5	Rinse cells with water	2 min
6	Mount cells on a coverslip using DPX	2 min
7	Dry coverslip	Overnight
8	Image cells using a fluorescence microscope	5 min

FIGURE 1. Flowchart depicting steps in the Hoechst 33342 staining protocol.

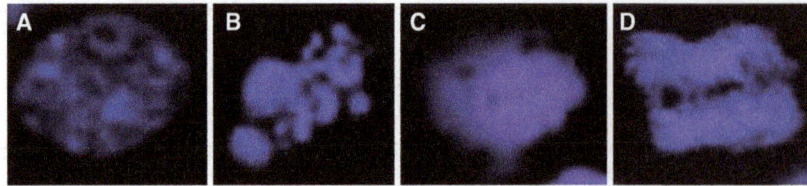

FIGURE 2. Images of K562 cells stained with Hoechst 33342. K562 cells were treated with imatinib (5 μM) for 24 h and then stained with Hoechst 33342. In healthy, untreated cells (A), the nuclei appear round and evenly stained. In apoptotic cells (B) the nuclei are generally fragmented and stained more intensely because of condensation of the DNA. In necrotic cells (C) the DNA is not condensed and the edges of the nucleus are less clearly defined. In healthy cells undergoing mitosis, the nuclei also appear condensed, but these nuclei can be distinguished from those of apoptotic cells because the DNA of mitotic cells forms two parallel lines as the chromosomes segregate (D). These two lines may have finger-like projections as the chromosomes pull apart during anaphase.

4. Wash the cells with PBS for 5 min at room temperature. Pipette or aspirate off the PBS.

   To avoid damaging the cells, do not touch them with the pipette.

5. Rinse the cells with water.

6. Mount the cells on a coverslip using DPX mounting medium.

7. Leave to dry in the dark overnight at room temperature.

   Make sure the DPX is dry before viewing your cells on the microscope. The mounting medium can coat the lens, damaging it and making it difficult to view the cells.

8. View the cells using a fluorescence microscope with emission ~350 nm and a blue/cyan emission filter (~460 nm).

   For an example of images obtained from K562 cells stained with Hoechst 33342, see Figure 2.

   See Troubleshooting.

TROUBLESHOOTING

Problem (Step 8): The fluorescence is weak.

Solution: The cells should be viewed soon after they are mounted on coverslips as the fluorescence intensity can fade over time. Fluorescence intensity can be maintained by mounting the slides in ProLong Gold Antifade. The staining intensity can also be adjusted by incubating the slides in stain for longer or shorter periods of time in Step 3.

DISCUSSION

Staining cells with Hoechst 33342 is an ideal assay for distinguishing apoptotic cells from healthy or necrotic cells (Zhivotosky and Orrenius 2001) because cells that have died by apoptosis will generally display condensed DNA and fragmented nuclei (Matassov et al. 2004; Errami et al. 2013), whereas healthy and necrotic cells do not. However, healthy cells undergoing mitosis may also have condensed DNA and some cells can still die by apoptosis in the absence of nuclear fragmentation (Zhang et al. 2001). This is the case, for example, with cells that have defects in the activation of caspase 3 (Kagawa et al. 2001). Nuclear fragmentation should therefore not be used in isolation to indicate the presence or absence of apoptosis. Because Hoechst 33342 fluoresces at 461 nm, it can be distinguished from other commonly used fluorochromes such as fluorescein isothiocyanate or Texas Red using fluorescence microscopy. Antibodies conjugated to these fluorochromes can therefore be used in combination with Hoechst 33342 to detect other cellular events that are characteristic of apoptosis such as caspase activation or cytochrome *c* release.

Cite this protocol as *Cold Spring Harb Protoc*; doi:10.1101/pdb.prot087205

RECIPES

Paraformaldehyde in PBS

Paraformaldehyde, EM grade

- PBS (10×)
- NaOH (1 M)
- HCl (1 M)

For a 4% paraformaldehyde solution, add 4 g of EM grade paraformaldehyde to 50 mL of H_2O. Add 1 mL of 1 M NaOH and stir gently on a heating block at ∼60°C until the paraformaldehyde is dissolved. Add 10 mL of 10× PBS and allow the mixture to cool to room temperature. Adjust the pH to 7.4 with 1 M HCl (∼1 mL), then adjust the final volume to 100 mL with H_2O. Filter the solution through a 0.45-μm membrane filter to remove any particulate matter. Make the paraformaldehyde solution fresh prior to use, or store in aliquots at −20°C for several months. Avoid repeated freeze/thawing.

Phosphate-Buffered Saline (PBS)

Reagent	Amount to add (for 1× solution)	Final concentration (1×)	Amount to add (for 10× stock)	Final concentration (10×)
NaCl	8 g	137 mM	80 g	1.37 M
KCl	0.2 g	2.7 mM	2 g	27 mM
Na_2HPO_4	1.44 g	10 mM	14.4 g	100 mM
KH_2PO_4	0.24 g	1.8 mM	2.4 g	18 mM

If necessary, PBS may be supplemented with the following:

$CaCl_2 \cdot 2H_2O$	0.133 g	1 mM	1.33 g	10 mM
$MgCl_2 \cdot 6H_2O$	0.10 g	0.5 mM	1.0 g	5 mM

PBS can be made as a 1× solution or as a 10× stock. To prepare 1 L of either 1× or 10× PBS, dissolve the reagents listed above in 800 mL of H_2O. Adjust the pH to 7.4 (or 7.2, if required) with HCl, and then add H_2O to 1 L. Dispense the solution into aliquots and sterilize them by autoclaving for 20 min at 15 psi (1.05 kg/cm^2) on liquid cycle or by filter sterilization. Store PBS at room temperature.

ACKNOWLEDGMENTS

This work was supported by an Australian Research Council Future Fellowship and project grants from the Leukaemia Foundation Queensland, the Prostate Cancer Foundation Australia, and the Australian Department of Defence.

REFERENCES

Crowley LC, Marfell BJ, Scott AP, Waterhouse NJ. 2015a. Triggering apoptosis in haematopoietic cells with cytotoxic drugs. *Cold Spring Harb Protoc* doi: 10.1101/pdb.prot087130.

Crowley LC, Marfell BJ, Scott AP, Waterhouse NJ. 2015b. Triggering death of adherent cells with ultraviolet radiation. *Cold Spring Harb Protoc* doi: 10.1101/pdb.prot087148.

Crowley LC, Marfell BJ, Scott AP, Waterhouse NJ. 2015c. Morphological analysis of cell death by cytospinning followed by Rapid staining. *Cold Spring Harb Protoc* doi: 10.1101/pdb.prot087197.

Errami Y, Naura AS, Kim H, Ju J, Suzuki Y, El-Bahrawy AH, Ghonim MA, Hemeida RA, Mansy MS, Zhang J, et al. 2013. Apoptotic DNA fragmentation may be a cooperative activity between caspase-activated deoxyribonuclease and the poly(ADP-ribose) polymerase-regulated DNAS1L3,

an endoplasmic reticulum-localized endonuclease that translocates to the nucleus during apoptosis. *J Biol Chem* **288**: 3460–3468.

Kagawa S, Gu J, Honda T, McDonnell TJ, Swisher SG, Roth JA, Fang B. 2001. Deficiency of caspase-3 in MCF7 cells blocks Bax-mediated nuclear fragmentation but not cell death. *Clin Cancer Res* **7**: 1474–1480.

Matassov D, Kagan T, Leblanc J, Sikorska M, Zakeri Z. 2004. Measurement of apoptosis by DNA fragmentation. *Methods Mol Biol* **282**: 1–17.

Zhang M, Li Y, Zhang H, Xue S. 2001. BAPTA blocks DNA fragmentation and chromatin condensation downstream of caspase-3 and DFF activation in HT-induced apoptosis in HL-60 cells. *Apoptosis* **6**: 291–297.

Zhivotosky B, Orrenius S. 2001. Assessment of apoptosis and necrosis by DNA fragmentation and morphological criteria. In *Current Protocols in Cell Biology*. John Wiley & Sons, Inc., New york.

Protocol 9

Detection of DNA Fragmentation in Apoptotic Cells by TUNEL

Lisa C. Crowley,[1] Brooke J. Marfell,[1] and Nigel J. Waterhouse[1,2,3,4]

[1]Apoptosis and Cytotoxicity Laboratory, Mater Research, Translational Research Institute, Woolloongabba, Brisbane, Queensland 4102, Australia; [2]Flow Cytometry and Imaging, QIMR Berghofer Medical Research Institute, Herston, Brisbane, Queensland 4006, Australia; [3]School of Medicine, University of Queensland, St. Lucia, Brisbane, Queensland 4072, Australia

Degradation of DNA into oligonucleosomal-sized fragments is a unique event in apoptosis that is orchestrated by caspase-activated DNase. Traditionally, this event is observed by resolving cellular DNA by gel electrophoresis, which results in a characteristic "ladder" pattern. However, this technique is time-consuming and cannot be used to quantitate the number of apoptotic cells in a sample. Terminal dUTP nick-end labeling (TUNEL) of fragmented DNA allows researchers to identify DNA fragmentation at the single-cell level. This method involves the specific addition of fluorescently labeled UTP to the 3'-end of the DNA fragments by terminal deoxynucleotidyl transferase. The TUNEL assay is both fast and sensitive. Here, we describe a protocol in which cells are treated with TUNEL reagent and counterstained with Hoechst 33342. In contrast to TUNEL, which only stains apoptotic cells, Hoechst 33342 stains the DNA of all cells.

MATERIALS

It is essential that you consult the appropriate Material Safety Data Sheets and your institution's Environmental Health and Safety Office for proper handling of equipment and hazardous material used in this protocol.

RECIPES: Please see the end of this protocol for recipes indicated by <R>. Additional recipes can be found online at http://cshprotocols.cshlp.org/site/recipes.

Reagents

Cells of interest

Citrate buffer (0.1% Triton X-100 in 0.1% sodium citrate)

Cytotoxic agent of choice

Hoechst 33342 (Sigma-Aldrich)

Hoechst 33342 can be substituted with DAPI, ProLong Gold Antifade with DAPI, or propidium iodide. These dyes bind to DNA and should be considered mutagenic.

In Situ Cell Death Detection Kit (Roche Diagnostics 11684795910)

The kit comprises a label solution (containing fluorescently labeled nucleotides) and an enzyme solution (terminal deoxynucleotidyl transferase [TdT]).

Paraformaldehyde (PFA) (4%) in PBS <R>

Phosphate-buffered saline (PBS) <R>

ProLong Gold Antifade (Life Technologies)

[4]Correspondence: nigel.waterhouse@QIMRBerghofer.edu.au

Cite this protocol as *Cold Spring Harb Protoc*; doi:10.1101/pdb.prot087221

Equipment

Fluorescence microscope

METHOD

For a flowchart summarizing this protocol, see Figure 1.

1. Treat cells of interest with a cytotoxic agent (see Protocol 1: Triggering Apoptosis in Hematopoietic Cells with Cytotoxic Drugs (Crowley et al. 2015c) or Protocol 2: Triggering Death of Adherent Cells with Ultraviolet Radiation ([Crowley et al. 2015d]) and fix the cells in 4% PFA for 20 min at room temperature following incubation with the cytotoxic agent for the required period of time.

   Adherent cells can be fixed in PFA directly. Nonadherent cells will need to be harvested and attached to the coverslip (see Protocol 7: Morphological Analysis of Cell Death by Cytospinning Followed by Rapid Staining [Crowley et al. 2015b]).

2. Wash the cells twice with PBS for 5 min each at room temperature. Pipette or aspirate off the PBS.

   To avoid damaging the cells, do not touch them with the pipette.

3. Permeabilize the cells by incubating them in citrate buffer for 2 min on ice.

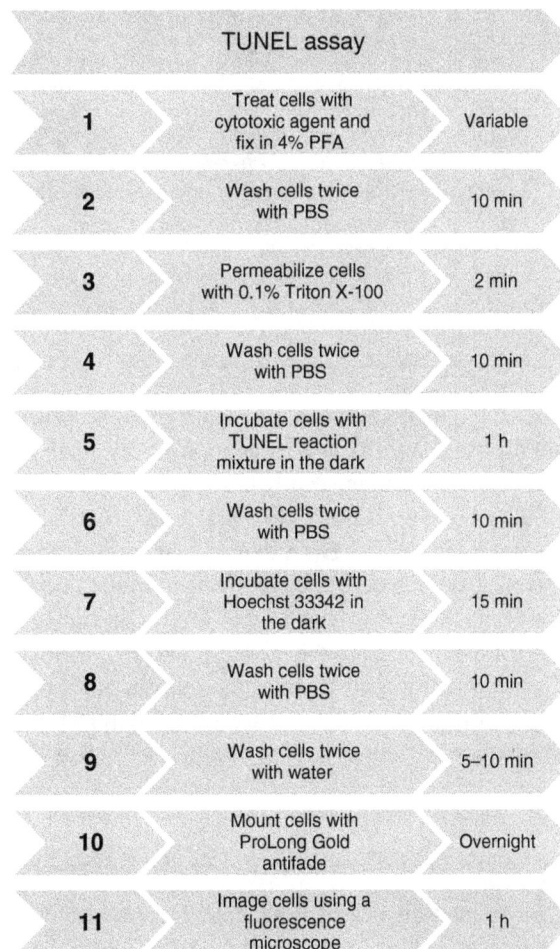

FIGURE 1. Flowchart depicting steps in the TUNEL assay.

4. Wash the cells twice with PBS for 5 min each at room temperature. Pipette or aspirate off the PBS.

    To avoid damaging the cells, do not touch them with the pipette.

5. Prepare the terminal dUTP nick-end labeling (TUNEL) reagent by mixing 5 µL of enzyme solution with 45 µL of label solution per slide. Add 50 µL of TUNEL reaction mixture per slide and cover with Parafilm or a coverslip to ensure even coating of the cells. Incubate in the dark in a humidified incubator for 1 h at 37°C.

    The TUNEL reagent must be freshly prepared before use. The reagent is light-sensitive and should be protected from light at all stages.

    A negative staining control should be performed on a control slide using 50 µL of label solution without enzyme solution.

6. Remove the TUNEL reagent by aspiration. Wash the cells twice with PBS for 5 min each at room temperature.

7. Incubate the cells with Hoechst 33342 (100 ng/mL in PBS) in the dark for 15 min at room temperature.

    Staining intensity can be changed by incubating the slides in stain for different periods of time.

8. Wash the cells twice with PBS for 5 min each at room temperature. Pipette or aspirate off the PBS.

    To avoid damaging the cells, do not touch them with the pipette.

9. Rinse the cells with water.

10. Mount the cells with ProLong Gold Antifade reagent and cover with a coverslip.

11. Leave to dry overnight at room temperature. View the slides immediately using a fluorescence microscope or store in the dark.

    TUNEL labeling indicates that DNA strand breaks have occurred in those cells. In contrast, cells stained only with Hoechst 33342 (and not TUNEL) do not have DNA strand breaks. The number of cells stained with both TUNEL and Hoechst 33342 (apoptotic) versus the number of cells stained with Hoechst 33342 alone (healthy) can therefore be used to determine the level of apoptosis in a sample. Images of cells can be captured using a fluorescence microscope with an attached camera and counted for graphical representation.

    TUNEL-stained cells may also be assayed by flow cytometry. It is advisable to seek advice from expert users if analysis by flow cytometry is desired.

DISCUSSION

Large-scale fragmentation of DNA into high-molecular-weight and nucleosome-sized DNA fragments is generally considered a marker of apoptotic cell death (Elmore 2007). This TUNEL assay measures the specific addition of fluorescently labeled UTP to the 3'-end of the DNA fragments by TdT (Gavrieli et al. 1992; Kressel and Groscurth 1994), and is therefore ideal for detecting these breaks. However, this assay is not fool-proof as DNA damage can also occur in the late stages of necrotic cell death (Grasl-Kraupp et al. 1995; Kelly et al. 2003) and apoptosis can occur in the absence of nuclear fragmentation (Kagawa et al. 2001; Zhang et al. 2001). A second assay should therefore be used in combination with the TUNEL assay to accurately discriminate apoptotic cells from cells that are in late-stage necrosis (e. g., annexin V binding, cytochrome *c* release, or caspase activation). The TUNEL assay can also be performed using nucleotides labeled with different fluorophores, allowing researchers to use TUNEL in combination with immunocytochemistry to measure other events in apoptosis.

RELATED TECHNIQUES

DNA fragmentation in apoptotic cells can also be detected with propidium iodide staining as described in Protocol 10: Measuring the DNA Content of Cells in Apoptosis and at Different Cell-Cycle Stages by Propidium Iodide Staining and Flow Cytometry (Crowley et al. 2015a).

Cite this protocol as *Cold Spring Harb Protoc*; doi:10.1101/pdb.prot087221

RECIPES

Paraformaldehyde in PBS

Paraformaldehyde, EM grade

• PBS (10×)
• NaOH (1 M)
• HCl (1 M)

For a 4% paraformaldehyde solution, add 4 g of EM grade paraformaldehyde to 50 mL of H_2O. Add 1 mL of 1 M NaOH and stir gently on a heating block at ~60°C until the paraformaldehyde is dissolved. Add 10 mL of 10× PBS and allow the mixture to cool to room temperature. Adjust the pH to 7.4 with 1 M HCl (~1 mL), then adjust the final volume to 100 mL with H_2O. Filter the solution through a 0.45-μm membrane filter to remove any particulate matter. Make the paraformaldehyde solution fresh prior to use, or store in aliquots at −20°C for several months. Avoid repeated freeze/thawing.

Phosphate-Buffered Saline (PBS)

Reagent	Amount to add (for 1× solution)	Final concentration (1×)	Amount to add (for 10× stock)	Final concentration (10×)
NaCl	8 g	137 mM	80 g	1.37 M
KCl	0.2 g	2.7 mM	2 g	27 mM
Na_2HPO_4	1.44 g	10 mM	14.4 g	100 mM
KH_2PO_4	0.24 g	1.8 mM	2.4 g	18 mM

If necessary, PBS may be supplemented with the following:

$CaCl_2 \cdot 2H_2O$	0.133 g	1 mM	1.33 g	10 mM
$MgCl_2 \cdot 6H_2O$	0.10 g	0.5 mM	1.0 g	5 mM

PBS can be made as a 1× solution or as a 10× stock. To prepare 1 L of either 1× or 10× PBS, dissolve the reagents listed above in 800 mL of H_2O. Adjust the pH to 7.4 (or 7.2, if required) with HCl, and then add H_2O to 1 L. Dispense the solution into aliquots and sterilize them by autoclaving for 20 min at 15 psi (1.05 kg/cm^2) on liquid cycle or by filter sterilization. Store PBS at room temperature.

ACKNOWLEDGMENTS

This work was supported by an Australian Research Council Future Fellowship and project grants from the Leukaemia Foundation Queensland, the Prostate Cancer Foundation Australia, and the Australian Department of Defence.

REFERENCES

Crowley LC, Chojnowski G, Waterhouse NJ. 2015a. Measuring the DNA content of cells in apoptosis and at different cell-cycle stages by propidium iodide staining and flow cytometry. *Cold Spring Harb Protoc* doi: 10.1101/pdb.prot087247.

Crowley LC, Marfell BJ, Waterhouse NJ. 2015b. Morphological analysis of cell death by cytospinning followed by rapid staining. *Cold spring Harb Protoc* doi: 10.1101/pdb.prot087197.

Crowley LC, Marfell BJ, Scott AP, Waterhouse NJ. 2015c. Triggering apoptosis in haematopoietic cells with cytotoxic drugs. *Cold Spring Harb Protoc* doi: 10.1101/pdb.prot087130.

Crowley LC, Marfell BJ, Scott AP, Waterhouse NJ. 2015d. Triggering death of adherent cells with ultraviolet radiation. *Cold Spring Harb Protoc* doi: 10.1101/pdb.prot087148.

Elmore S. 2007. Apoptosis: A review of programmed cell death. *Toxicol Pathol* 35: 495–516.

Gavrieli Y, Sherman Y, Ben-Sasson SA. 1992. Identification of programmed cell death in situ via specific labeling of nuclear DNA fragmentation. *J Cell Biol* 119: 493–501.

Grasl-Kraupp B, Ruttkay-Nedecky B, Koudelka H, Bukowska K, Bursch W, Schulte-Hermann R. 1995. In situ detection of fragmented DNA

(TUNEL assay) fails to discriminate among apoptosis, necrosis, and autolytic cell death: A cautionary note. *Hepatology* **21:** 1465–1468.

Kagawa S, Gu J, Honda T, McDonnell TJ, Swisher SG, Roth JA, Fang B. 2001. Deficiency of caspase-3 in MCF7 cells blocks Bax-mediated nuclear fragmentation but not cell death. *Clin Cancer Res* **7:** 1474–1480.

Kelly KJ, Sandoval RM, Dunn KW, Molitoris BA, Dagher PC. 2003. A novel method to determine specificity and sensitivity of the TUNEL reaction in the quantitation of apoptosis. *Am J Physiol Cell Physiol* **284:** C1309–C1318.

Kressel M, Groscurth P. 1994. Distinction of apoptotic and necrotic cell death by in situ labelling of fragmented DNA. *Cell Tissue Res* **278:** 549–556.

Zhang M, Li Y, Zhang H, Xue S. 2001. BAPTA blocks DNA fragmentation and chromatin condensation downstream of caspase-3 and DFF activation in HT-induced apoptosis in HL-60 cells. *Apoptosis* **6:** 291–297.

Cite this protocol as *Cold Spring Harb Protoc*; doi:10.1101/pdb.prot087221

Measuring the DNA Content of Cells in Apoptosis and at Different Cell-Cycle Stages by Propidium Iodide Staining and Flow Cytometry

Lisa C. Crowley,[1] Grace Chojnowski,[2] and Nigel J. Waterhouse[1,2,3,4]

[1]Apoptosis and Cytotoxicity Laboratory, Mater Research, Translational Research Institute, Woolloongabba, Brisbane, Queensland 4102, Australia; [2]Flow Cytometry and Imaging, QIMR Berghofer Medical Research Institute, Herston, Brisbane, Queensland 4006, Australia; [3]School of Medicine, University of Queensland, St. Lucia, Brisbane, Queensland 4072, Australia

All cells are created from preexisting cells. This involves complete duplication of the parent cell to create two daughter cells by a process known as the cell cycle. For this process to be successful, the DNA of the parent cell must be faithfully replicated so that each daughter cell receives a full copy of the genetic information. During the cell cycle, the DNA content of the parent cell increases as new DNA is synthesized (S phase). When there are two full copies of the DNA (G_2/M phase), the cell splits to form two new cells (G_0/G_1 phase). As such, cells in different stages of the cell cycle have different DNA contents. The cell cycle is tightly regulated to safeguard the integrity of the cell and any cell that is defective or unable to complete the cell cycle is programmed to die by apoptosis. When this occurs, the DNA is fragmented into oligonucleosomal-sized fragments that are disposed of when the dead cell is removed by phagocytosis. Consequently apoptotic cells have reduced DNA content compared with living cells. This can be measured by staining cells with propidium iodide (PI), a fluorescent molecule that intercalates with DNA at a specific ratio. The level of PI fluorescence in a cell is, therefore, directly proportional to the DNA content of that cell. This protocol describes the use of PI staining to determine the percentage of cells in each phase of the cell cycle and the percentage of apoptotic cells in a sample.

MATERIALS

It is essential that you consult the appropriate Material Safety Data Sheets and your institution's Environmental Health and Safety Office for proper handling of equipment and hazardous materials used in this protocol.

RECIPES: Please see the end of this protocol for recipes indicated by <R>. Additional recipes can be found online at http://cshprotocols.cshlp.org/site/recipes.

Reagents

Cell line or cell type of interest treated with cytotoxic agent of choice (see Step 1)
Ethanol (100%)
Phosphate-buffered saline (PBS) <R>
Propidium iodide (PI)-containing solution appropriate for chosen procedure (Steps 2–9 or Steps 10–12):

[4]Correspondence: nigel.waterhouse@qimrberghofer.edu.au

Cite this protocol as Cold Spring Harb Protoc; doi:10.1101/pdb.prot087247

Cell cycle buffer (for Steps 2–9) <R>

Nicoletti buffer (for Steps 10–12) <R>

Trypsin-EDTA (e.g., 0.25% with 1mM EDTA, Gibco 25200-056) or trypsin replacement (e.g., TrypLE Gibco 12604-013) (as needed, for adherent cells)

Equipment

Centrifuge

Eppendorf tubes

Flow cytometer

Ice bucket

Vortex mixer

METHOD

PI staining of DNA was one of the first techniques used in the flow cytometric analysis of cells and is still commonly used to analyze the cell cycle and measure apoptosis. It is important to be able to assess cell cycle status of a population because many cytotoxic agents trigger apoptosis by blocking various stages of the cell cycle (Campos and Dizon 2012), whereas other treatments only have antiproliferative effects (Fromberg et al. 2011). A workflow of the assay is presented in Figure 1.

Preparing Cells

1. Treat the cells with a cytotoxic stimulus as described in Protocol 1: Triggering Apoptosis in Hematopoietic Cells with Cytotoxic Drugs (Crowley et al. 2015a) or Protocol 2: Triggering Death of Adherent Cells with Ultraviolet Radiation (Crowley and Waterhouse 2015).

    Proceed to either Step 2 or Step 10 to stain the cells with PI.

Staining Cells with PI

This protocol includes two options for PI staining: PI staining after ethanol fixation (Steps 2–9) or PI staining of detergent-treated, permeabilized cells using the Nicoletti buffer (Steps 10–12). The rapid protocol developed by Nicoletti is more commonly used to measure apoptosis in cases where assessment of the cell cycle is not required. However, it should be noted that the permeabilized cells appear smaller when measured by flow cytometry using forward scatter (FSC) compared with cells fixed in ethanol. PI staining does not require a high level of expertise and can be performed by most laboratory staff. However, PI is mutagenic and care should be taken to avoid exposure even though the concentration of PI used in this assay is very low.

PI Staining of Ethanol-Fixed Cells

2. Harvest the cells, and resuspend 5×10^5 cells in 1 mL of ice-cold PBS in an Eppendorf tube.

    Cells must be in suspension for analysis by flow cytometry. For in vitro cultured adherent cells, remove and save medium that contains dead and mitotic cells. Detach live cells using standard tissue culture techniques such as incubation with trypsin–EDTA. Add the medium-suspended cells to the detached cells to ensure that mitotic and apoptotic cells are assayed.

    After harvesting, samples should be kept on ice until Step 9.

3. Slowly add 2.3 mL of ice-cold 100% ethanol while gently vortexing the sample.

    The cells will clump if the ethanol is added quickly. It is therefore important to add the ethanol very slowly in a drop-wise fashion while gently vortexing the sample.

4. Incubate for 20 min at 4°C.

5. Centrifuge at 300g for 5 min.

6. Discard supernatant and wash cells once with 500 µL of ice-cold PBS by carefully inverting or flicking the tube or by gently pipetting or vortexing.

7. Centrifuge at 300g for 5 min.

Cite this protocol as *Cold Spring Harb Protoc*; doi:10.1101/pdb.prot087247

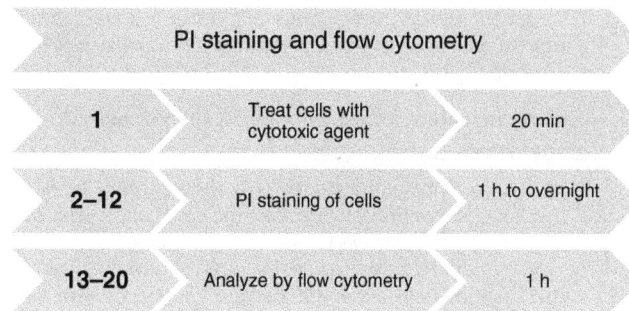

FIGURE 1. Workflow diagram for analyzing cell death and cell cycle using PI staining and flow cytometry.

8. Discard supernatant and resuspend cells in 500 μL of ice-cold cell cycle buffer.

   PI binds to DNA and RNA. It is therefore important to ensure the buffer contains active RNase and that all RNA is digested to ensure that only the DNA content is measured.

9. Incubate for 45 min at room temperature in the dark.

   Proceed to Step 13 to analyze cells by flow cytometry. Samples can be stored in PBS for up to 2 wk at 4°C before processing after they are fixed with 100% ethanol. However, PI is light sensitive and the samples should remain protected from light.

PI Staining of Detergent-Treated Cells

10. Resuspend 5×10^5 cells in PBS and centrifuge at 300*g* for 5 min.

11. Discard supernatant and resuspend cells in 1 mL of ice-cold Nicoletti buffer.

12. Vortex cells and incubate in the dark overnight at 4°C.

    Cells must be incubated for long enough to be fully permeabilized. Incubation for 1 h is generally sufficient but overnight incubation is convenient as the cells do not need to be analyzed by flow cytometry immediately. After incubation, proceed to Step 13 to analyze cells by flow cytometry.

Flow Cytometry

It is recommended to discuss experimental parameters and conditions for flow cytometry with experts in your facility.

13. Turn on the appropriate laser on the flow cytometer.

    Although the maximum excitation wavelength of PI is 535 nm, it is also excited by 488 nm, which is available on most flow cytometers. The choice of laser to use on your flow cytometer can be discussed with facility staff.

14. Set up a dot plot to detect size FSC and granularity SSC using linear scale.

15. Set up two histogram plots to detect PI using log scale and linear scale.

    PI emits maximally at 620 nm but can be detected between 400 and 700 nm. Histogram plots for detectors within this range should be sufficient to perform analysis of cell cycle and apoptosis. Collecting data in two detectors with one set on linear and the other set on log scale can facilitate analysis of cell cycle and apoptosis in the same sample.

16. Run an untreated sample and set the voltage and gain for FSC and SSC to ensure that live cells can be detected.

    For Steps 16 and 17, make sure to set the gates for FCS and SSC to include live and dead cells. This is important because many flow cytometric applications only gate for phenotyping of live cells.

17. Run a treated sample and adjust the voltage and gain for FSC and SSC to ensure all cells (live and dead) can be detected based on FSC and SSC.

18. Set exclusion gates based on FSC and SSC to exclude cellular debris.

    Dots that come up in the bottom left corner of the FSC versus SSC plot are unlikely to be cells due to their small size and should be gated out. However, it is essential to observe this population because some cells fragment into apoptotic bodies that will appear as dots in this region. Although apoptotic bodies should not be counted as cells, observing this population may yield essential information on the level of cell death in the culture.

19. Run an untreated sample and adjust the voltage and gain for the PI detector so that all of the cells can be detected.

It is important that the stained cells appear within the detection range of the plot. First alter the voltage and gain so that control cells are far to the right hand side of the histogram plot set on log scale. This is necessary because apoptotic cells will appear as a wide peak on the left of the plot. For the histogram plot in linear gate the gain should be set so that the G_0/G_1 peak has a mean fluorescence index (MFI) of ~200. The G_2/M peak will then be detected at 400 because these cells have twice the amount of DNA as cells in G_0/G_1 (Fig. 2). Apoptotic cells will then be detected below 200 MFI and polyploid cells (with multiple copies of DNA) will be detected at 800 MFI.

20. Run each sample and acquire data for at least 10,000 events.

It is not essential to collect data on 10,000 cells; however, collecting this number of events will ensure smooth peaks when presenting your data visually and will also provide statistical power when calculating the percentage cells in each phase of the cell cycle.

The data from each sample can be presented visually as a histogram plot (Fig. 2) or graphically as the percentage of cells in each phase of the cell cycle. The FSC and SSC (linear scales) should be set to include all cells but to exclude debris, and the histogram plots for PI fluorescence should be set in the linear and log

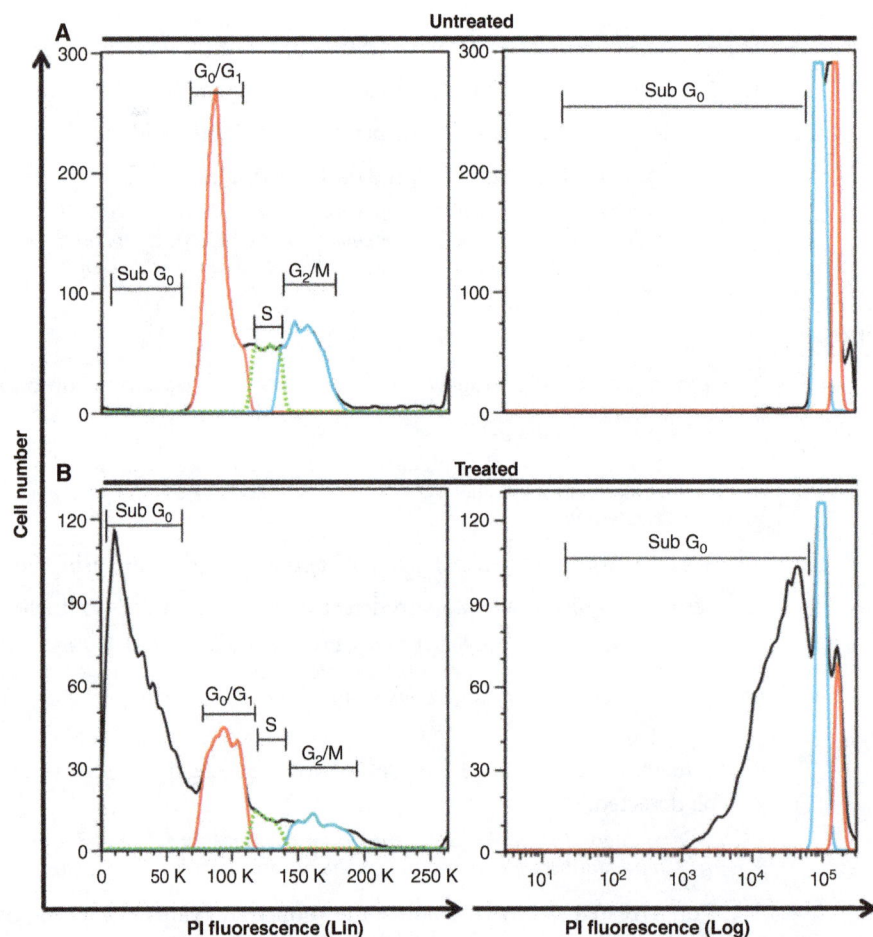

FIGURE 2. Cell cycle analysis of K562 cells stained with propidium iodide. (*A*) Untreated cells or (*B*) cells treated with 1 μM actinomycin D for 16 h were stained with PI and analyzed by flow cytometry. Histogram plots in linear scale show cells in G_0/G_1 (blue) and G_2/M (red) phase of the cell cycle. Cells in S phase appear between the red and blue peaks while apoptotic cells (Sub G_0) appear to the left of the G_0/G_1 population.

Cite this protocol as *Cold Spring Harb Protoc*; doi:10.1101/pdb.prot087247

scales. The linear scale plot should show two individual peaks. The first peak on the left-hand side shows cells in G_1/G_0. The second on the right-hand side shows cells in G_2/M. Apoptotic cells appear as an elongated peak to the left of the G_1/G_0 cells when viewed in linear scale. Apoptosis is often better viewed/measured in log scale. The difference between G_1/G_0 and G_2/M peaks may be indistinguishable but the protracted peak to the left of the G_1/G_0 and G_2/M peaks is indicative of cells with DNA that has been fragmented during apoptosis. The percentage of cells in each peak can be measured using a range-bar gate and represented graphically.

RELATED TECHNIQUES

Although DNA fragmentation is commonly observed in cells undergoing apoptosis, it should be noted that DNA fragmentation is not required for apoptosis per se and may not occur in all cells (Kagawa et al. 2001). Detection of apoptosis should, therefore, include other assays (e.g., Annexin V binding or caspase activation), especially in cases where DNA fragmentation is not observed.

It is also important to note that this assay is different than the PI uptake assay described in Protocol 4: Measuring Cell Death by Propidium Iodide Uptake and Flow Cytometry (Crowley et al. 2015b), in which only dead cells are permeabilized and stained with PI. This enables the level of cell death to be measured but cannot be used to analyze the cell cycle or to determine whether a cell has undergone apoptosis. However, in the assay described here, all cells whether alive, dead or dying, are permeabilized and stained with PI; therefore, cell cycle and apoptosis can be assessed.

RECIPES

Cell Cycle Buffer

30 µg/mL propidium iodide (Sigma-Aldrich 81845)
100 µg/mL DNase-free RNase A (Sigma-Aldrich R4875)

Prepare in PBS (pH 7.4).

Nicoletti Buffer

0.1% Sodium citrate
0.1% Triton X-100
50 µg/mL propidium iodide (Sigma-Aldrich 81845)

Prepare in PBS (pH 7.4).

Phosphate-Buffered Saline (PBS)

Reagent	Amount to add (for 1× solution)	Final concentration (1×)	Amount to add (for 10× stock)	Final concentration (10×)
NaCl	8 g	137 mM	80 g	1.37 M
KCl	0.2 g	2.7 mM	2 g	27 mM
Na_2HPO_4	1.44 g	10 mM	14.4 g	100 mM
KH_2PO_4	0.24 g	1.8 mM	2.4 g	18 mM

If necessary, PBS may be supplemented with the following:

$CaCl_2 \cdot 2H_2O$	0.133 g	1 mM	1.33 g	10 mM
$MgCl_2 \cdot 6H_2O$	0.10 g	0.5 mM	1.0 g	5 mM

PBS can be made as a 1× solution or as a 10× stock. To prepare 1 L of either 1× or 10× PBS, dissolve the reagents listed above in 800 mL of H_2O. Adjust the pH to 7.4 (or 7.2, if required) with HCl, and then add H_2O to 1 L. Dispense the solution into aliquots and sterilize them by autoclaving for 20 min at 15 psi (1.05 kg/cm^2) on liquid cycle or by filter sterilization. Store PBS at room temperature.

ACKNOWLEDGMENTS

This work was supported by an Australian Research Council futures fellowship and project grants from the Leukaemia Foundation Queensland, the Prostrate Cancer Foundation Australia, and the Australian Department of Defence to N.J.W.

REFERENCES

Campos SM, Dizon DS. 2012. Antimitotic inhibitors. *Hematolo Oncol Clin North Am* **26:** 607–628.

Crowley LC, Waterhouse NJ. 2015. Triggering death of adherent cells with ultraviolet radiation. *Cold Spring Harb Protoc* doi: 10.1101/pdb .prot087148.

Crowley LC, Marfell BJ, Scott AP, Boughaba JA, Chojnowski G, Christenson ME, Waterhouse NJ. 2015a. Triggering apoptosis in hematopoietic cells with cytotoxic drugs. *Cold Spring Harb Protoc* doi: 10.1101/pdb.prot087130.

Crowley LC, Scott AP, Marfell BJ, Boughaba JA, Chojnowski G, Waterhouse NJ. 2015b. Measuring cell death by propidium iodide uptake and flow cytometry. *Cold Spring Harb Protoc* doi: 10.1101/pdb .prot087163.

Fromberg A, Gutsch D, Schulze D, Vollbracht C, Weiss G, Czubayko F, Aigner A. 2011. Ascorbate exerts anti-proliferative effects through cell cycle inhibition and sensitizes tumor cells towards cytostatic drugs. *Cancer Chemother Pharmacol* **67:** 1157–1166.

Kagawa S, Gu J, Honda T, McDonnell TJ, Swisher SG, Roth JA, Fang B. 2001. Deficiency of caspase-3 in MCF7 cells blocks Bax-mediated nuclear fragmentation but not cell death. *Clinical Cancer Research* **7:** 1474–1480.

Quantitation of Apoptosis and Necrosis by Annexin V Binding, Propidium Iodide Uptake, and Flow Cytometry

Lisa C. Crowley,[1] Brooke J. Marfell,[1] Adrian P. Scott,[1] and Nigel J. Waterhouse[1,2,3,4]

[1]*Apoptosis and Cytotoxicity Laboratory, Mater Research, Translational Research Institute, Woolloongabba, Brisbane, Queensland 4102, Australia;* [2]*Flow Cytometry and Imaging, QIMR Berghofer Medical Research Institute, Herston, Brisbane, Queensland 4006, Australia;* [3]*School of Medicine, University of Queensland, St. Lucia, Brisbane, Queensland 4072, Australia*

The surface of healthy cells is composed of lipids that are asymmetrically distributed on the inner and outer leaflet of the plasma membrane. One of these lipids, phosphatidylserine (PS), is normally restricted to the inner leaflet of the plasma membrane and is, therefore, only exposed to the cell cytoplasm. However, during apoptosis lipid asymmetry is lost and PS becomes exposed on the outer leaflet of the plasma membrane. Annexin V, a 36-kDa calcium-binding protein, binds to PS; therefore, fluorescently labeled Annexin V can be used to detect PS that is exposed on the outside of apoptotic cells. Annexin V can also stain necrotic cells because these cells have ruptured membranes that permit Annexin V to access the entire plasma membrane. However, apoptotic cells can be distinguished from necrotic cells by co-staining with propidium iodide (PI) because PI enters necrotic cells but is excluded from apoptotic cells. This protocol describes Annexin V binding and PI uptake followed by flow cytometry to detect and quantify apoptotic and necrotic cells.

MATERIALS

It is essential that you consult the appropriate Material Safety Data Sheets and your institution's Environmental Health and Safety Office for proper handling of equipment and hazardous materials used in this protocol.

RECIPE: Please see the end of this protocol for recipes indicated by <R>. Additional recipes can be found online at http://cshprotocols.cshlp.org/site/recipes.

Reagents

Annexin V binding buffer <R>
Annexin V-FITC (fluorescein isothiocyanate) (BD Biosciences 556420)
Cell line or cell type of interest treated with cytotoxic agent of choice (see Step 1)
Propidium iodide (PI) (0.5 mg/mL; Sigma-Aldrich 81845)
Phosphate-buffered saline (PBS; pH 7.4; Sigma-Aldrich P5368)

Equipment

Bright-field microscope
Flow cytometer

[4]Correspondence: nigel.waterhouse@qimrberghofer.edu.au

Cite this protocol as *Cold Spring Harb Protoc*; doi:10.1101/pdb.prot087288

METHOD

A workflow diagram of the assay is presented in Figure 1.

Staining Cells with Annexin V/PI

Cells must be stained and assayed live because fixing will render all cells permeable to PI. Apoptosis is generally a rapid process that is complete within several hours. Using Annexin V is unlikely to yield extra information compared with using PI uptake alone in assays that have progressed for several days.

Annexin V/PI staining does not require a high level of expertise and can be performed by most laboratory staff. PI is mutagenic and care should be taken to avoid exposure even though the concentration of PI used in this assay is very low.

1. Treat the cells with a cytotoxic stimulus as described in Protocol 1: Triggering Apoptosis in Hematopoietic Cells with Cytotoxic Drugs (Crowley et al. 2015) or Protocol 2: Triggering Death of Adherent Cells with Ultraviolet Radiation (Crowley and Waterhouse 2015). Observe the cells using a bright-field microscope.

   It is wise to validate your experiment by bright-field microscopy before analysis by flow cytometry. For example, you may not expect your cells to have high fluorescence if they look morphologically intact. This will also higlight any issues that might affect flow cytometry (e.g., clumping or infection).

2. Harvest the cells and resuspend 1×10^5 cells in 200 µL of Annexin V binding buffer containing 4 µL of 0.5 mg/mL PI and 2 µL of Annexin V-FITC.

   When assaying adherent cells it is essential to retain the culture medium that is removed before cell harvesting because it may contain nonadherent apoptotic cells. Following trypsinization, combine the saved medium and trypsinized cells to ensure that mitotic and apoptotic cells are measured.

   Fluorescently conjugated Annexin V should be used at the concentration recommended by the manufacturer. If necessary, this can be titrated to achieve optimal Annexin V concentrations for individual cell lines.

   PI and fluorescently tagged Annexin V are light sensitive and should be protected from light in all following steps.

3. Incubate for 15 min at room temp in the dark.

   Cells should be assayed by flow cytometry as soon as possible after staining because they will not survive for long periods in the Annexin V binding buffer. Keep the cells in the dark until assayed by flow cytometry.

Flow Cytometry

It is recommended to discuss experimental parameters and conditions for flow cytometry with experts in your facility.

4. Turn on the 488-nm laser on the flow cytometer.

   Annexin V-FITC has maximum excitation at 488 nm and emission at 520 nm. PI has maximum excitation at ~535 nm and emission at 617 nm. However, both can be excited by a 488 nm laser.

5. Set up a dot plot to detect size (forward scatter; FSC) and granularity (side scatter; SSC) using linear scale.

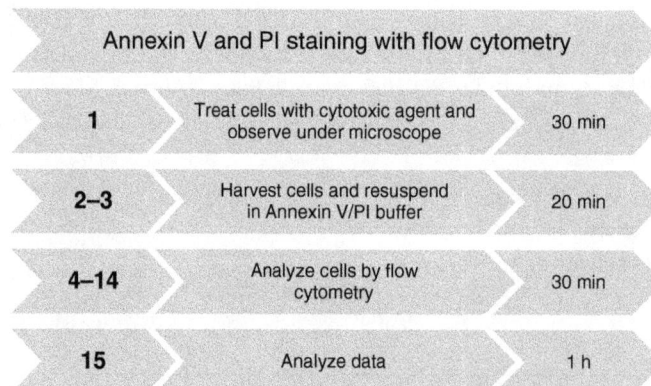

FIGURE 1. Workflow diagram for flow cytometric analysis of Annexin- and PI-stained cells.

Cite this protocol as *Cold Spring Harb Protoc*; doi:10.1101/pdb.prot087288

6. Set up dot plots to detect FITC (520 nm) and PI (620 nm) using log scale.

   FITC and PI have wide excitation spectra and can, therefore, be measured by a wide variety of detectors and filter sets. However, it is recommended to choose sets whose bandpass wavelengths are far apart, such as FL-1 and FL-3.

7. Run an untreated sample and set the voltage and gain for FSC and SSC to ensure your live cells can be detected.

   For Steps 7 and 8, it is essential to set large gates to encompass both live and dead cell populations. This is in contrast to many flow cytometric applications that only gate for phenotyping of live cells.

8. Run a treated sample and adjust the voltage and gain for FSC and SSC to ensure all cells (live and dead) can be detected based on FSC and SSC.

9. Set exclusion gates based on FSC and SSC to exclude cellular debris.

   Dots in the bottom left corner of the FSC versus SSC plot are unlikely to be cells because of their small size and should be gated out. However, it is essential to observe this population because some cells fragment into apoptotic bodies that will appear as dots in this region. Although apoptotic bodies should not be counted as cells, observing this population may yield essential information on the level of cell death in the culture.

10. Run an untreated sample and adjust the voltage and gain for the FITC and PI detectors so that all cells can be detected in the bottom left quadrant.

11. Run a treated sample stained with FITC alone and adjust the voltage and gain for the FITC detector so that the dead cells appear in the bottom right quadrant.

12. Run a treated sample stained with PI alone and adjust the voltage and gain for the PI detector so that the dead cells appear in the top left quadrant.

13. Run a treated sample stained with FITC and PI. Adjust the compensation so that the live cells appear in the bottom left, the apoptotic cells appear in the bottom right, and the necrotic cells appear in the top right quadrants (Fig. 2).

14. Run each sample and acquire data for at least 10,000 events.

    It is not essential to collect data on 10,000 cells; however, collecting this number of events will ensure smooth peaks when presenting your data visually and will also provide statistical power when calculating the percentage of cells in each phase of the cell cycle.

15. Analyze the data.

    i. Set a dot plot using FSC/SSC in linear scale and draw a gate to exclude debris from further analysis.

       It is, however, essential to observe this population as it represents the fragmentation of cells into apoptotic bodies.

    ii. Set a dot plot for FITC and PI in log scale.

    iii. Present the data visually or graphically.

        The individual populations can be defined using quadrant gates, which automatically quantify the number of cells in each quadrant as follows: quadrant 1 = live cells (PI^{-ve}/Annexin V^{-ve}); quadrant 2 = apoptotic cells (PI^{-ve}/Annexin V^{+ve}); quadrant 3 = necrotic cells = PI^{+ve}/Annexin V^{+ve}); quadrant 4 = nuclei without plasma membrane = (PI^{+ve}/Annexin V^{-ve}). The total number of dead or dying cells can be determined by adding the percentages of cells in each of the Q2, Q3, and Q4 quadrants.

        The assay is depicted in cartoon form and an example of data obtained using this protocol is presented in Figure 2.

        See Troubleshooting.

TROUBLESHOOTING

Problem (Step 15): Annexin V staining is weak.

Solution: Annexin V requires calcium to bind to phosphatidylserine (PS); therefore, ensure sufficient $CaCl_2$ is present in the Annexin V binding buffer used in Step 2. Fluorescently labeled Annexin V in Annexin V binding buffer can be purchased from many suppliers.

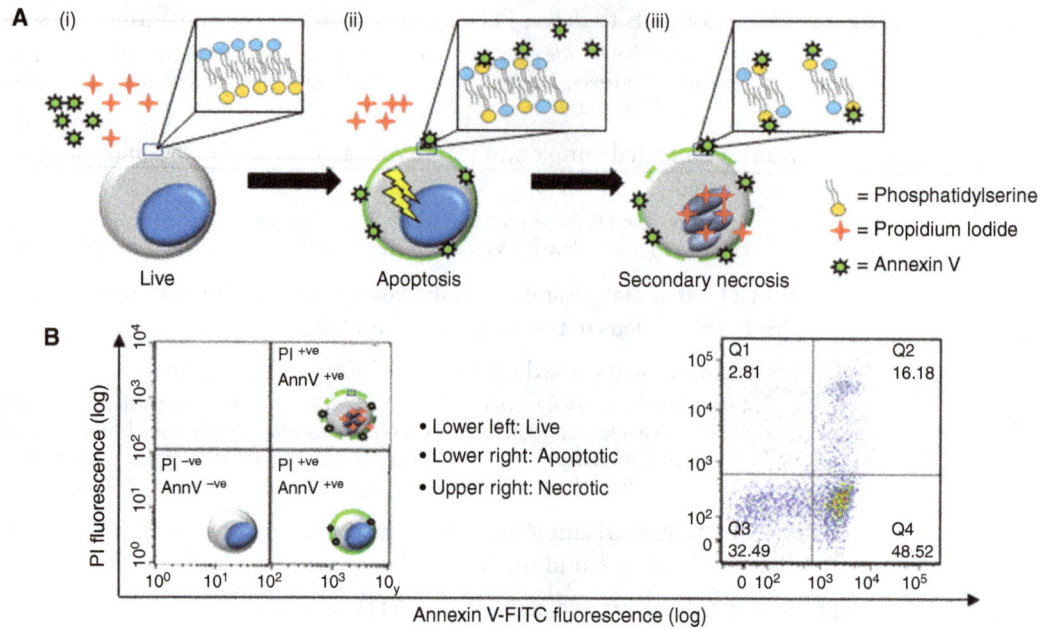

FIGURE 2. Measurement of cell death by Annexin V binding and PI uptake. (*A*) A cartoon depicting Annexin V binding and PI uptake by apoptotic cells. (i) Phosphatidylserine (PS) is on the inner leaflet of the intact plasma membrane of live cells and these cells do not stain with Annexin V or PI. (ii) PS flips to the outer leaflet of the plasma membrane of apoptotic cells and these cells bind Annexin V on the outside but still exclude PI. (iii) Cells that have undergone secondary necrosis have ruptured membranes. Annexin V binds to PS on the plasma membrane and PI is taken up and binds to the DNA. (*B*) Depicts where the populations in i, ii, and iii will be observed in dot plots of flow cytometry data (*left* side) and an example of U937 cells treated with actinomycin D (1 μM) for 16 h and stained with Annexin V-FITC and PI (*right* side).

DISCUSSION

Annexin V binding and PI uptake is one of the most commonly used assays to measure apoptosis and necrosis (Reutelingsperger and van Heerde 1997; Komoriya et al. 2000; Roy and Nicholson 2000; Elmore 2007). The number of apoptotic and necrotic cells in a population will depend on the overall percentage of cell death, and the duration and severity of a treatment. It is important to note that apoptosis is a rapid process that is complete in a few hours, after which cells undergo secondary necrosis (Annexin V^{+ve}/PI^{+ve}). It is, therefore, recommended to assay cell death at several times during an assay to observe the Annexin V^{+ve}/PI^{-ve} population. This assay measures the distinct event of PS exposure at the plasma membrane. However, care should be taken to ensure that Annexin V^{+ve} are indeed dying because some researchers have anecdotally suggested that Annexin V can bind healthy cells that have the potential to continue to proliferate. Similarly, others have suggested that some cells do not bind Annexin V when alive or dead. These potential problems can be overcome by following the assay to its completion (all cells PI^{+ve}) or by measuring other features of cell death, such as caspase 3 activity, DNA fragmentation and plasma membrane blebbing, which can confirm the presence of apoptotic cells in a sample. Annexin V can be conjugated to various fluorochromes and different combinations may be required depending on the cells used (e.g., cells that have been engineered to overexpress fluorescent proteins), the treatment used (e.g., some chemotherapeutic agents auto-fluoresce) and the availability of excitation wavelengths on the flow cytometer in your facility. Fluorescently tagged Annexin V is generally expensive but protocols are available to produce Annexin V by bacterial expression, and to purify and FITC-label recombinant Annexin V (Logue et al. 2009). Initial drug titrations can be performed using PI alone to enable optimal drug concentrations and assay times to be selected before Annexin V is used.

Cite this protocol as *Cold Spring Harb Protoc*; doi:10.1101/pdb.prot087288

RECIPE

Annexin V Binding Buffer

10 mM HEPES (pH 7.4)
150 mM NaCl
2.5 mM CaCl$_2$

Prepare in PBS (pH 7.4).

ACKNOWLEDGMENTS

This work was supported by an Australian Research Council futures fellowship and project grants from the Leukaemia Foundation Queensland, the Prostrate Cancer Foundation Australia, and the Australian Department of Defence to N.J.W.

REFERENCES

Crowley LC, Waterhouse NJ. 2015. Triggering death of adherent cells with ultraviolet radiation. *Cold Spring Harb Protoc* doi: 10.1101/pdb.prot087148.

Crowley LC, Marfell BJ, Scott AP, Waterhouse NJ. 2015. Triggering apoptosis in hematopoietic cells with cytotoxic drugs. *Cold Spring Harb Protoc* doi: 10.1101/pdb.prot087130.

Elmore S. 2007. Apoptosis: A review of programmed cell death. *Toxicol Pathol* 35: 495–516.

Komoriya A, Packard BZ, Brown MJ, Wu ML, Henkart PA. 2000. Assessment of caspase activities in intact apoptotic thymocytes using cell-permeable fluorogenic caspase substrates. *J Exp Med* 191: 1819–1828.

Logue SE, Elgendy M, Martin SJ. 2009. Expression, purification and use of recombinant Annexin V for the detection of apoptotic cells. *Nat Protoc* 4: 1383–1395.

Reutelingsperger CP, van Heerde WL. 1997. Annexin V, the regulator of phosphatidylserine-catalyzed inflammation and coagulation during apoptosis. *Cell Mol Life Sci* 53: 527–532.

Roy S, Nicholson DW. 2000. Criteria for identifying authentic caspase substrates during apoptosis. *Methods Enzymol* 322: 110–125.

Detecting Cleaved Caspase-3 in Apoptotic Cells by Flow Cytometry

Lisa C. Crowley[1] and Nigel J. Waterhouse[1,2,3,4]

[1]*Apoptosis and Cytotoxicity Laboratory, Mater Research, Translational Research Institute, Woolloongabba, Brisbane, Queensland 4102, Australia;* [2]*Flow Cytometry and Imaging, QIMR Berghofer Medical Research Institute, Herston, Brisbane, Queensland 4006, Australia;* [3]*School of Medicine, University of Queensland, St. Lucia, Brisbane, Queensland 4072, Australia*

Apoptosis is orchestrated by caspases, a family of cysteine proteases that cleave their substrates on the carboxy-terminal side of specific aspartic acid residues. These proteases are generally present in healthy cells as inactive zymogens, but when stimulated they undergo autolytic cleavage to become fully active. They subsequently cleave their substrates at one or two specific sites, which can result in activation, inactivation, relocalization, or remodeling of the substrate. Consequently, many of the cleaved fragments remain intact during apoptosis and can be detected using substrate-specific antibodies. These fragments are most commonly detected by western blotting, which resolves proteins and their fragments based on molecular mass. However, antibodies that only recognize cleaved fragments can be used to specifically label cells in which caspase cleavage has occurred. It is then possible to quantify these cells by flow cytometry. A number of antibodies that specifically recognize caspase-cleaved fragments have been generated, including antibodies that recognize the cleaved form of caspase-3. This caspase is responsible for the majority of proteolysis during apoptosis, and detection of cleaved caspase-3 is therefore considered a reliable marker for cells that are dying, or have died by apoptosis. This protocol outlines the quantification of apoptosis by flow cytometric detection of cleaved caspase-3.

MATERIALS

It is essential that you consult the appropriate Material Safety Data Sheets and your institution's Environmental Health and Safety Office for proper handling of equipment and hazardous materials used in this protocol.

RECIPE: Please see the end of this protocol for recipes indicated by <R>. Additional recipes can be found online at http://cshprotocols.cshlp.org/site/recipes.

Reagents

Alexa Fluor 488-conjugated donkey anti-rabbit IgG (Life Technologies A-21206)

Prepare secondary antibody master mix by diluting Alexa Fluor 488-conjugated donkey anti-rabbit IgG 1:200 in IFA-Tx.

Anticleaved caspase-3 (Cell Signaling 9661L)

Prepare primary antibody master mix by diluting the anticleaved caspase-3 1:500 in IFA-Tx buffer.

---

[4]Correspondence: nigel.waterhouse@qimrberghofer.edu.au

Cite this protocol as *Cold Spring Harb Protoc*; doi:10.1101/pdb.prot087312

Cell line or cell type of interest
Cytotoxic agent of choice
IFA-Tx buffer <R>
Paraformaldehyde (PFA) (4% in PBS) <R>
Phosphate-buffered saline (PBS) <R>

Equipment

Centrifuge
Flow cytometer
Vortex mixer

METHOD

Sample Preparation

1. Treat cells of interest with a cytotoxic agent. Harvest 5×10^5 cells per sample and pellet by centrifugation at 300*g* for 5 min.

   It is essential to harvest all cells, both alive and dead. For adherent cells, collect the culture medium (containing dead cells) and recover live cells from the plate using standard trypsin–EDTA treatment. Combine the trypsinized cells with the reserved medium before centrifugation.

2. Wash the cells in 500 µL of ice-cold PBS. Resuspend the cells in the PBS, and then centrifuge the cells at 300*g* for 5 min.

3. Discard the supernatant and resuspend the cells in 1 mL of ice-cold PBS. Carefully add 2 mL of ice-cold 4% PFA. Incubate for 20 min at 4°C.

   Add the PFA drop-wise while vortexing to prevent the cells forming clumps.

4. Wash cells twice in PBS as described in Step 2.

   At this stage, the samples can be stored in PBS for up to 2 wk at 4°C.

Primary Antibody Labeling

5. Pellet cells by centrifugation at 300*g* for 5 min and discard the supernatant. Wash the sample once with 1 mL of IFA-Tx buffer and centrifuging again at 300*g* for 5 min.

6. Discard the supernatant and resuspend each sample in 150 µL of primary antibody master mix.

   For Steps 6 and 9, it is important to remove all excess wash liquid to avoid excess dilution of the antibody in subsequent steps. Extra care should be taken to avoid disturbing the pellet when discarding the supernatant.

   The primary antibody dilution is important and should be titrated to ensure specificity.

7. Incubate for 1 h at 4°C.

8. Wash the cells three times in IFA-Tx buffer. For each wash, resuspend the cells in 1 mL of IFA-Tx buffer, centrifuge at 300*g* for 5 min, and discard the supernatant.

Secondary Antibody Labeling

Cells should be stained with secondary antibody alone as a negative control.

9. Discard the IFA-Tx buffer and resuspend each sample in 150 µL of secondary antibody master mix.

10. Incubate for 1 h at 4°C.

11. Add 850 µL IFA-Tx buffer and wash cells three times in 1 mL of PBS as in Step 8 to remove all of the IFA-Tx and any traces of antibodies.

12. Resuspend the sample in 200 µL of PBS.

Quantification by Flow Cytometry

Assaying caspase-3 cleavage by flow cytometry does not require a high level of expertise but it is recommended to discuss experimental parameters and conditions for flow cytometry with experts in your facility.

13. Turn on the 488 nm laser on the flow cytometer.

    This protocol uses an Alexa 488-conjugated secondary antibody but other fluorochromes may also be used. This may, however, necessitate the use of different lasers.

14. Set up a dot plot to detect size (forward scatter; FSC) and granularity (side scatter; SSC) using linear scale.

15. Run a treated and untreated sample and set the voltage and gain for FSC and SSC to ensure live and dead cells can be detected.

    Ensure the gates for FCS and SSC are set to include live and dead cells. This is important to note because gates are routinely set to exclude dead cells in many flow cytometry applications.

16. Set exclusion gates based on FSC and SSC to exclude cellular debris.

    Dots that come up in the bottom left corner of the FSC versus SSC plot are unlikely to be cells due to their small size and should be gated out. However, it is essential to observe this population because some cells fragment into apoptotic bodies that will appear as dots in this region. Although apoptotic bodies should not be counted as cells, observing this population may yield essential information on the level of cell death in the culture.

17. Set up a histogram to detect Alexa 488 emission (~520 nm) using log scale.

18. Run an untreated sample and adjust the voltage and gain for the Alexa 488 detector so that all of the cells can be detected on the left side of a histogram.

19. Run a treated sample to ensure that dead cells appear to the right of the untreated cells.

20. Run each sample and acquire data for at least 10,000 events

    It is not essential to collect data for 10,000 cells; however, collecting this number of events will ensure smooth peaks when presenting your data visually and will also provide statistical power when calculating the percentage of cells with high PI fluorescence.

    Data may be represented visually as a histogram or graphically as the percentage of cells that are positive for cleaved caspase-3 (the peak to the right side of the histogram plot). This can be calculated automatically by drawing a line gate on the histogram plot using the log scale on the x-axis.

DISCUSSION

Cleavage of caspase-3 is generally considered a universal marker of apoptosis because caspase-3 activity is required for the majority of morphological and biochemical events associated with apoptosis (Waterhouse et al. 1996; Chang and Yang 2000; Elmore 2007; Poreba et al. 2013). Caspase-3 is cleaved at two sites that result in three fragments; the prodomain, the large subunit, and the small subunit. The first cleavage only partially activates caspase-3 and the second autolytic cleavage results in full activation (Martin et al. 1996). It is, therefore, essential to ensure that caspase-3 is fully activated to indicate that apoptosis has occurred. This can be confirmed by adding a caspase inhibitor, such as zVAD-fmk or DEVD-fmk, before the initial caspase-3 cleavage, which will prevent the autolytic cleavage event and reduce staining by cleavage-specific antibodies (Stratos et al. 2012). Detection of caspase-3 activity may also be achieved by using antibodies that detect fragments of substrates that are cleaved by caspase-3, such as poly(adenosine ribose) polymerase PARP (Shi 2004). However, it is important to ensure that these antibodies are specific for the cleaved fragment as concomitant detection of a fragment and the uncleaved form cannot be used to quantify apoptotic cells by flow cytometry. These antibodies can also be used to show that caspase-3 is sufficiently activated to cause functional cleavage of key substrates in the dying cell, while specific cleavage by caspase-3 can be confirmed by the use of caspase-3 inhibitors.

Cite this protocol as *Cold Spring Harb Protoc*; doi:10.1101/pdb.prot087312

RECIPES

IFA-Tx Buffer

4 mL fetal calf serum
10 mL HEPES (100 mM, pH 7.4)
100 µL Triton X-100
0.1 µg sodium azide
86 mL sodium chloride (0.9%)

Paraformaldehyde in PBS

Paraformaldehyde, EM grade
PBS (10×)
NaOH (1 M)
HCl (1 M)

For a 4% paraformaldehyde solution, add 4 g of EM grade paraformaldehyde to 50 mL of H_2O. Add 1 mL of 1 M NaOH and stir gently on a heating block at ~60°C until the paraformaldehyde is dissolved. Add 10 mL of 10× PBS and allow the mixture to cool to room temperature. Adjust the pH to 7.4 with 1 M HCl (~1 mL), then adjust the final volume to 100 mL with H_2O. Filter the solution through a 0.45-µm membrane filter to remove any particulate matter. Make the paraformaldehyde solution fresh prior to use, or store in aliquots at −20°C for several months. Avoid repeated freeze/thawing.

Phosphate-Buffered Saline (PBS)

Reagent	Amount to add (for 1× solution)	Final concentration (1×)	Amount to add (for 10× stock)	Final concentration (10×)
NaCl	8 g	137 mM	80 g	1.37 M
KCl	0.2 g	2.7 mM	2 g	27 mM
Na_2HPO_4	1.44 g	10 mM	14.4 g	100 mM
KH_2PO_4	0.24 g	1.8 mM	2.4 g	18 mM
If necessary, PBS may be supplemented with the following:				
$CaCl_2 \cdot 2H_2O$	0.133 g	1 mM	1.33 g	10 mM
$MgCl_2 \cdot 6H_2O$	0.10 g	0.5 mM	1.0 g	5 mM

PBS can be made as a 1× solution or as a 10× stock. To prepare 1 L of either 1× or 10× PBS, dissolve the reagents listed above in 800 mL of H_2O. Adjust the pH to 7.4 (or 7.2, if required) with HCl, and then add H_2O to 1 L. Dispense the solution into aliquots and sterilize them by autoclaving for 20 min at 15 psi (1.05 kg/cm^2) on liquid cycle or by filter sterilization. Store PBS at room temperature.

ACKNOWLEDGMENTS

This work was supported by an Australian Research Council futures fellowship and project grants from the Leukaemia Foundation Queensland, the Prostrate Cancer Foundation Australia, and the Australian Department of Defence to N.J.W.

REFERENCES

Chang HY, Yang X. 2000. Proteases for cell suicide: Functions and regulation of caspases. *Microbiol Mol Biol Rev* **64:** 821–846.

Elmore S. 2007. Apoptosis: A review of programmed cell death. *Toxicol Pathol* **35:** 495–516.

Martin SJ, Amarante-Mendes GP, Shi L, Chuang TH, Casiano CA, O'Brien GA, Fitzgerald P, Tan EM, Bokoch GM, Greenberg AH, et al. 1996. The cytotoxic cell protease granzyme B initiates apoptosis in a cell-free system by proteolytic processing and activation of the ICE/CED-3 family protease, CPP32, via a novel two-step mechanism. *EMBO J* **15:** 2407–2416.

Poreba M, Strozyk A, Salvesen GS, Drag M. 2013. Caspase substrates and inhibitors. *Cold Spring Harb Perspect Biol.*

Shi Y. 2004. Caspase activation, inhibition, and reactivation: A mechanistic view. *Protein Sci* **13:** 1979–1987.

Stratos I, Li Z, Rotter R, Herlyn P, Mittlmeier T, Vollmar B. 2012. Inhibition of caspase mediated apoptosis restores muscle function after crush injury in rat skeletal muscle. *Apoptosis* **17:** 269–277.

Waterhouse N, Kumar S, Song Q, Strike P, Sparrow L, Dreyfuss G, Alnemri ES, Litwack G, Lavin M, Watters D. 1996. Heteronuclear ribonucleoproteins C1 and C2, components of the spliceosome, are specific targets of interleukin 1β-converting enzyme-like proteases in apoptosis. *J Biol Chem* **271:** 29335–29341.

Cite this protocol as *Cold Spring Harb Protoc*; doi:10.1101/pdb.prot087312

Protocol 13

Analysis of Cytochrome *c* Release by Immunocytochemistry

Lisa C. Crowley,[1,4] Brooke J. Marfell,[1] Adrian P. Scott,[1] and Nigel J. Waterhouse[1,2,3,4]

[1]*Apoptosis and Cytotoxicity Laboratory, Mater Research, Translational Research Institute, Woolloongabba, Brisbane, Queensland 4102, Australia;* [2]*Flow Cytometry and Imaging, QIMR Berghofer Medical Research Institute, Herston, Brisbane, Queensland 4006, Australia;* [3]*School of Medicine, University of Queensland, St. Lucia, Brisbane, Queensland 4072, Australia*

Cytochrome *c* is normally localized between the inner and outer membranes of mitochondria in healthy cells. However, during apoptosis, it is released into the cytoplasm, where it binds to apoptotic protease activating factor. Caspase-9 is then recruited and activated by this complex in a process known as the induced proximity model. Release of cytochrome *c* from mitochondria is therefore a critical event in apoptosis and various protocols are available for its measurement. Cytochrome *c* in mitochondria has a punctate localization pattern in the cell and its translocation to the cytoplasm results in a diffuse distribution. This is visually striking and easily observed by immunocytochemistry. This protocol describes the use of immunocytochemistry to assay cytochrome *c* release during apoptosis.

MATERIALS

It is essential that you consult the appropriate Material Safety Data Sheets and your institution's Environmental Health and Safety Office for proper handling of equipment and hazardous materials used in this protocol.

RECIPES: Please see the end of this protocol for recipes indicated by <R>. Additional recipes can be found online at http://cshprotocols.cshlp.org/site/recipes.

Reagents

AlexaFluor 568 Goat Anti-Mouse IgG (Life Technologies A11004)
Anti-cytochrome *c* (BD Pharmingen Clone 6H2B4 556432)
Anti-cytochrome *c* blocking buffer <R>
Cell line or cell type of interest
Clear nail varnish
Cytotoxic drug(s) appropriate for experimental goals
Paraformaldehyde (PFA) (4%) in PBS
Phosphate-buffered saline (PBS; pH 7.4; Sigma-Aldrich P5368)
ProLong Gold Antifade with DAPI (Life Technologies P-36931) or Hoechst stain and DPX mounting
    medium (see Step 10)
zVAD-fmk

Equipment

Fluorescence microscope
Microscope slides and coverslips

---

[4]Correspondence: nigel.waterhouse@qimrberghofer.edu.au

Cite this protocol as *Cold Spring Harb Protoc*; doi:10.1101/pdb.prot087338

METHOD

The steps in this procedure are outlined in Figure 1.

1. Treat cells with cytotoxic stimuli according to experimental goals in the presence of zVAD-fmk (100 μM) as described in Protocol 1: Triggering Apoptosis in Hematopoietic Cells with Cytotoxic Drugs (Crowley et al. 2015a) or Protocol 2: Triggering Death of Adherent Cells with Ultraviolet Radiation (Crowley and Waterhouse 2015).

   For adherent cells, it is easiest to perform this protocol on cells cultured in chamber slides with thin glass bottoms for viewing using an inverted fluorescence microscope. The cells can also be grown on chamber slides that can be removed and mounted with coverslips or grown directly on coverslips or slides and viewed using an upright microscope.

   For adherent cells it is essential not to lose any detached (apoptotic) cells before fixation. This may be achieved by adding a pan caspase inhibitor, such as zVAD-fmk, to the sample before treatment. zVAD-fmk prevents cells rounding and detaching but does not block cytochrome c release. However, zVAD-fmk cannot be used for assaying cytochrome c release in cells treated with death receptor agonists/antagonists because it blocks caspase-8, which is upstream of cytochrome c release in this pathway.

   Suspension cells should be attached to a slide by cytospinning before staining (see Protocol 7: Morphological Analysis of Cell Death by Cytospinning Followed by Rapid Staining [Crowley et al. 2015]). Treating suspension cells in the presence of zVAD-fmk will prevent these cells from rapid progression to secondary necrosis, which culminates in plasma membrane rupture and release of all cytochrome c from the cell.

2. Remove growth medium. Fix cells with 4% PFA for 15–20 min at room temperature.

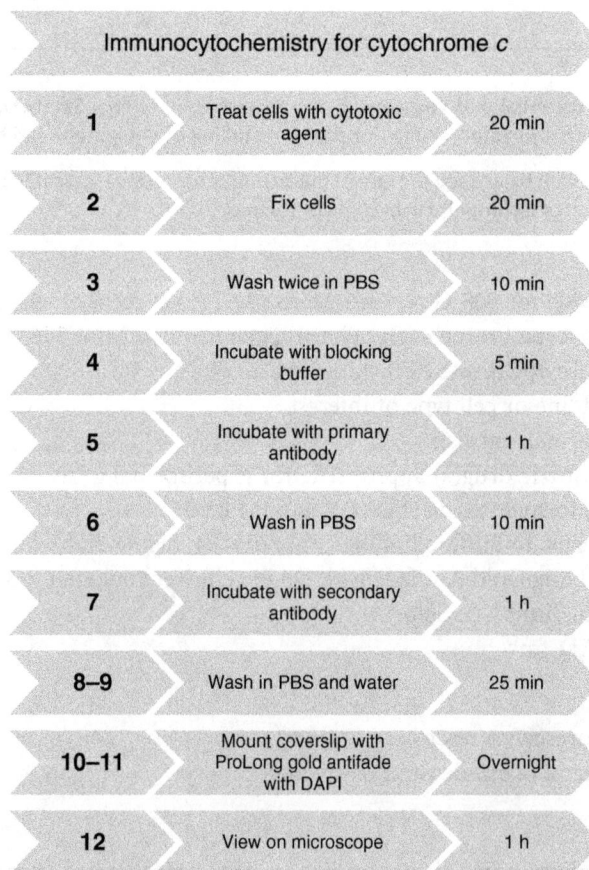

FIGURE 1. Workflow diagram for cytochrome *c* immunocytochemistry.

 Cite this protocol as *Cold Spring Harb Protoc*; doi:10.1101/pdb.prot087338

3. Wash cells twice with PBS for 5 min.

> *Coverslips can be washed by dipping and swirling in PBS. Cells grown in chamber slides can be washed by gently adding and removing PBS with a pipette. Slides can be stored in PBS for 1 mo at 4°C after this step.*

4. Incubate cells in sufficient blocking buffer for 5 min at room temperature.

5. Remove the blocking buffer. Add primary cytochrome *c* antibody diluted 1:200 in blocking buffer. Incubate for 1 h at room temperature.

> *For increased specificity, incubate overnight at 4°C.*

6. Wash twice with PBS for 5 min.

7. Add secondary AlexaFluor 568 antibody diluted 1 : 200 v/v in blocking buffer. Incubate for 1 h at room temperature in the dark.

> *The slides must be protected from light after this step. Staining using secondary antibody alone should be performed as a negative control.*

8. Wash twice with PBS and twice with water.

> *Cells stained in chamber slides with thin glass bottoms can be viewed directly on an inverted fluorescence microscope at this stage. Cells can also be stained with ProLong Antifade with DAPI or Hoeschst 33342 before viewing (see Protocol:* **Analyzing Cell Death by Nuclear Staining with Hoechst 33342** *[Crowley et al. 2015c]).*

9. Remove any remaining water by aspiration or pipetting.

10. Mount coverslips with ProLong Antifade with 4', 6-diamidino-2-phenylindole (DAPI). Leave slides to dry overnight in the dark at room temperature.

> *If ProLong Antifade with DAPI is unavailable, add Hoechst stain for the last 15 min of the secondary antibody incubation and then use DPX to mount the coverslip (see Protocol 8: Analyzing Cell Death by Nuclear Staining with Hoechst 33342 [Crowley et al 2015c]).*

11. Seal the edges of the coverslip using a clear nail varnish.

> *Be careful not to get the nail varnish on any part of the slide you want to view.*
>
> *Fluorescently labeled slides should be protected from light and may be stored for up to 3 mo at −20°C.*

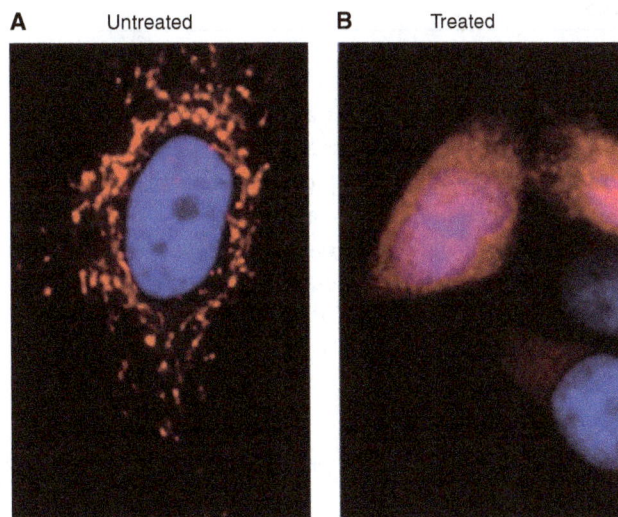

FIGURE 2. Immunofluorescence staining of cytochrome *c*. PC3 cells (a prostate cancer cell line) were treated with a cytotoxic agent (UV at 60 mJ/m^2) for 24 h in the presence of zVAD-fmk (100 μM). (*A*) Untreated cells or (*B*) treated cells were fixed and stained with anti-cytochrome *c* (red) and DAPI (blue). The red punctate pattern (*left*) shows a cell with intact mitochondria and the diffuse pattern (*right*) shows a cell in which cytochrome *c* has been released from mitochondria.

12. View the cells using fluorescence microscopy.

Mitochondria are distinct spaghetti-like organelles in healthy cells that can be easily identified by immunostaining for mitochondrial-specific proteins. In healthy cells, cytochrome c immunostaining results in a clear punctate pattern confined to mitochondria. During apoptosis, cytochrome c rapidly translocates from the mitochondria to the cytoplasm; therefore, immunostaining of apoptotic cells for cytochrome c results in a diffuse staining pattern. The distinct staining pattern of cytochrome c immunocytochemistry in healthy (punctate mitochondrial distribution) and apoptotic (diffuse cytoplasmic distribution) cells makes visual representation of cytochrome c staining particularly appealing (Fig. 2). Counting the number of cells with punctate and diffuse staining can be statistically analyzed and graphically presented. This can be contrasted with immunostaining for mitochondrial proteins that are not released during apoptosis (e.g., Cox IV, which resides in the mitochondrial inner membrane [Siskova et al. 2010]).

DISCUSSION

Cytochrome c release is a key event in the intrinsic apoptosis pathway that triggers activation of caspase proteases (Waterhouse and Green 1999). Blocking caspase activity downstream from cytochrome c release can prevent apoptosis but the cell will eventually die by a caspase-independent process (Amarante-Mendes et al. 1998). In contrast, blocking cytochrome c release can prevent apoptosis and caspase-independent cell death, and the cell may even continue to proliferate (Sedelies et al. 2008). Cytochrome c release is therefore considered a point of no return during cell death and modulating this event is generally believed to hold the key to understanding and treating several diseases. Cytochrome c release is believed to occur following selective permeabilization of the mitochondrial outer membrane but it is still not known how this occurs. Immunocytochemistry for cytochrome c is therefore an essential tool for understanding and characterizing the mitochondrial apoptosis pathway. It should be noted that cytochrome c can be released from any damaged mitochondria, which can occur during other forms of cell death. However, this generally occurs after rupture of the plasma membrane. To use cytochrome c release as a marker of the intrinsic apoptosis pathway it is essential to show that this event occurs before plasma membrane rupture. This can be achieved by using caspase inhibitors to block plasma membrane rupture but not cytochrome c release.

RELATED TECHNIQUES

Cytochrome c release can also be measured by western blotting or flow cytometry (Waterhouse and Trapani 2003; Christensen et al. 2013). Several other proteins are also released from mitochondria during apoptosis, including second mitochondrial activator of caspases (SMAC) and HTRA2. Release of these proteins during apoptosis can also be assayed by immunocytochemistry; however, SMAC is quickly degraded upon release and detection is difficult (MacFarlane et al. 2002). Cytochrome c release results in loss of mitochondrial membrane potential, which can be measured by staining cells with TMRE (see Protocol 14: Measuring Mitochondrial Transmembrane Potential by TMRE Staining [Crowley et al. 2015d] and Protocol 8: Analyzing Cell Death by Nuclear Staining with Hoechst 33342. [Crowley et al 2015c]).

RECIPE

Anti-Cytochrome c Blocking Buffer

0.05% Saponin
3% BSA

Prepare in PBS, fresh before use.

ACKNOWLEDGMENTS

This work was supported by an Australian Research Council futures fellowship and project grants from the Leukaemia Foundation Queensland, the Prostrate Cancer Foundation Australia, and the Australian Department of Defence to N.J.W.

REFERENCES

Amarante-Mendes GP, Finucane DM, Martin SJ, Cotter TG, Salvesen GS, Green DR. 1998. Anti-apoptotic oncogenes prevent caspase-dependent and independent commitment for cell death. *Cell Death Differ* 5: 298–306.

Christensen ME, Jansen ES, Sanchez W, Waterhouse NJ. 2013. Flow cytometry based assays for the measurement of apoptosis-associated mitochondrial membrane depolarisation and cytochrome *c* release. *Methods* 61: 138–145.

Crowley LC, Waterhouse NJ. 2015. Triggering death of adherent cells with ultraviolet radiation. *Cold Spring Harb Protoc* doi: 10.1101/pdb.prot087148.

Crowley LC, Marfell BJ, Scott AP, Waterhouse NJ. 2015a. Triggering apoptosis in hematopoietic cells with cytotoxic drugs. *Cold Spring Harb Protoc* doi:10.1101/pdb.prot087130.

Crowley LC, Marfell BJ, Waterhouse NJ. 2015b. Morphological analysis of cell death by cytospinning followed by rapid staining. *Cold Spring Harb Protoc* doi:10.1101/pdb.prot087197.

Crowley LC, Marfell BJ, Waterhouse NJ. 2015c. Analyzing cell death by nuclear staining with Hoechst 33342. *Cold Spring Harb Protoc* doi: 10.1101/pdb.prot087205.

Crowley LC, Christensen ME, Waterhouse NJ. 2015d. Measuring mitochondrial transmembrane potential by TMRE Staining. *Cold Spring Harb Protoc* doi: 10.1101/pdb.prot087361.

MacFarlane M, Merrison W, Bratton SB, Cohen GM. 2002. Proteasome-mediated degradation of Smac during apoptosis: XIAP promotes Smac ubiquitination in vitro. *J Biol Chem* 277: 36611–36616.

Sedelies KA, Ciccone A, Clarke CJ, Oliaro J, Sutton VR, Scott FL, Silke J, Susanto O, Green DR, Johnstone RW, et al. 2008. Blocking granule-mediated death by primary human NK cells requires both protection of mitochondria and inhibition of caspase activity. *Cell Death Differ* 15: 708–717.

Siskova Z, Mahad DJ, Pudney C, Campbell G, Cadogan M, Asuni A, O'Connor V, Perry VH. 2010. Morphological and functional abnormalities in mitochondria associated with synaptic degeneration in prion disease. *Am J Pathol* 177: 1411–1421.

Waterhouse NJ, Green DR. 1999. Mitochondria and apoptosis: HQ or high-security prison? *J Clin Immunol* 19: 378–387.

Waterhouse NJ, Trapani JA. 2003. A new quantitative assay for cytochrome *c* release in apoptotic cells. *Cell Death Differ* 10: 853–855.

Measuring Mitochondrial Transmembrane Potential by TMRE Staining

Lisa C. Crowley,[1] Melinda E. Christensen,[1,2] and Nigel J. Waterhouse[1,2,3,4]

[1]Apoptosis and Cytotoxicity Laboratory, Mater Research, Translational Research Institute, Woolloongabba, Brisbane, Queensland 4102, Australia; [2]Flow Cytometry and Imaging, QIMR Berghofer Medical Research Institute, Herston, Brisbane, Queensland 4006, Australia; [3]School of Medicine, University of Queensland, St. Lucia, Brisbane, Queensland 4072, Australia

Adenosine triphosphate (ATP) is the main source of energy for metabolism. Mitochondria provide the majority of this ATP by a process known as oxidative phosphorylation. This process involves active transfer of positively charged protons across the mitochondrial inner membrane resulting in a net internal negative charge, known as the mitochondrial transmembrane potential ($\Delta\Psi m$). The proton gradient is then used by ATP synthase to produce ATP by fusing adenosine diphosphate and free phosphate. The net negative charge across a healthy mitochondrion is maintained at approximately −180 mV, which can be detected by staining cells with positively charged dyes such as tetramethyl-rhodamine ethyl ester (TMRE). TMRE emits a red fluorescence that can be detected by flow cytometry or fluorescence microscopy and the level of TMRE fluorescence in stained cells can be used to determine whether mitochondria in a cell have high or low $\Delta\Psi m$. Cytochrome c is essential for producing $\Delta\Psi m$ because it promotes the pumping the protons into the mitochondrial intermembrane space as it shuttles electrons from Complex III to Complex IV along the electron transport chain. Cytochrome c is released from the mitochondrial intermembrane space into the cytosol during apoptosis. This impairs its ability to shuttle electrons between Complex III and Complex IV and results in rapid dissipation of $\Delta\Psi m$. Loss of $\Delta\Psi m$ is therefore closely associated with cytochrome c release during apoptosis and is often used as a surrogate marker for cytochrome c release in cells.

MATERIALS

It is essential that you consult the appropriate Material Safety Data Sheets and your institution's Environmental Health and Safety Office for proper handling of equipment and hazardous materials used in this protocol.

Reagents

Carbonyl cyanide 4-(trifluoromethoxy)phenylhydrazone (FCCP) (Sigma-Aldrich C2920)

Cell culture medium

Use the appropriate cell culture medium for the cell type/line being cultured. Mitochondria react quickly to stress. Cells should therefore be cultured and stained in media that support healthy mitochondria.

Cell line or cell type of interest

Cytotoxic drug(s) appropriate for experimental goals

Tetramethylrhodamine ethyl ester (TMRE) (Life Technologies T669)

Dissolve TMRE in ethanol or DMSO to make a 1 mM stock solution. Store at −20°C.

[4]Correspondence: nigel.waterhouse@qimrberghofer.edu.au

Cite this protocol as *Cold Spring Harb Protoc*; doi:10.1101/pdb.prot087361

Equipment

Bright-field microscope
Flow cytometer

METHOD

The entire experiment should be performed at room temperature because temperature will directly impact mitochondrial transmembrane potential and TMRE staining. Cells should never be placed, centrifuged, incubated, or washed at 4°C or have ice-cold buffers or media added. A workflow of the assay is presented in Figure 1.

Treating and Staining Cells with TMRE

1. Treat the cells with a cytotoxic stimulus as described in Protocol 1: Triggering Apoptosis in Hematopoietic Cells with Cytotoxic Drugs (Crowley et al. 2015) or Protocol 2: Triggering Death of Adherent Cells with Ultraviolet Radiation (Crowley and Waterhouse 2015). Observe the cells with a bright-field microscope.

   It is wise to validate your experiment by bright-field microscopy before analysis by flow cytometry. For example, you may not expect your cells to have high fluorescence if they look morphologically intact. This will also highlight any issues that might affect flow cytometry (e.g., clumping or infection).

2. Harvest cells and resuspend at 5×10^5 cells/mL in culture medium containing 150 nM TMRE. Incubate for 5 min at room temperature in the dark.

   This assay should be performed on live cells. Do not fix cells before or after staining because TMRE cannot be used to stain fixed cells and is not bound covalently so will be released from mitochondria that lose their charge. The cells can be analyzed any time after staining.

   TMRE will also bind to other negative charges in the cell. For example, the plasma membrane maintains a charge of −30 mV. This charge may be negated by adding 137 mM KCl to the staining medium because the charge across the plasma membrane is maintained by a potassium gradient.

3. Add FCCP (5 μM final concentration) to an aliquot of untreated cells and incubate for 5 min at room temperature in the dark.

   TMRE binds to mitochondria based on charge. This charge can be dissipated using a protonophore such as FCCP to determine the expected level of TMRE fluorescence in cells that have low ΔΨm.

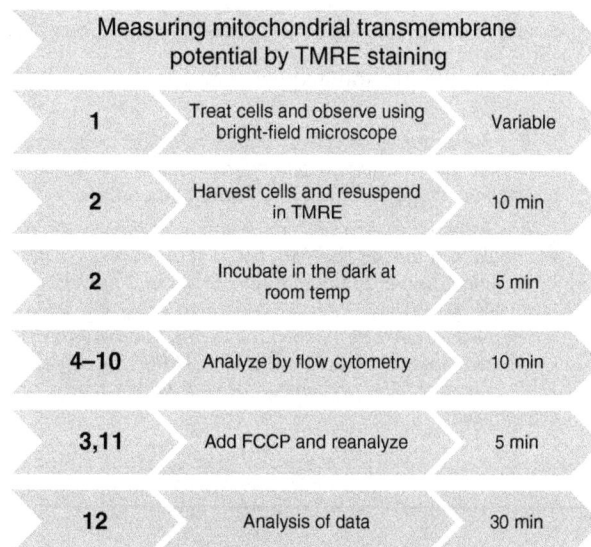

	Measuring mitochondrial transmembrane potential by TMRE staining	
1	Treat cells and observe using bright-field microscope	Variable
2	Harvest cells and resuspend in TMRE	10 min
2	Incubate in the dark at room temp	5 min
4–10	Analyze by flow cytometry	10 min
3,11	Add FCCP and reanalyze	5 min
12	Analysis of data	30 min

FIGURE 1. Workflow diagram for measuring mitochondrial transmembrane potential by TMRE staining.

Quantitation of TMRE Fluorescence Using Flow Cytometry

It is recommended to discuss experimental parameters and conditions for flow cytometry with experts in your facility.

4. Turn on the appropriate laser on the flow cytometer.

    TMRE is maximally excited at ~550 nm but can also be excited by the 488 laser line that most flow cytometers are equipped with. TMRE emits maximally at 575 nm.

5. Set up a dot plot to detect size (forward scatter; FSC) and granularity (side scatter; SSC) using linear scale.

6. Set up a histogram plot to detect TMRE using log scale.

7. Run an untreated sample and set the voltage and gain for FSC and SSC to ensure live cells can be detected.

8. Run a treated sample and adjust the voltage and gain for FSC and SSC to ensure all cells (live and dead) can be detected based on FSC and SSC.

    This is important to note because many flow cytometry applications routinely set gates to exclude dead cells (which are smaller and more granular than live cells) but this is likely to exclude dead cells with low $\Delta\Psi m$.

9. Set exclusion gates based on FSC and SSC to exclude cellular debris.

    Dots that come up in the bottom left corner of the FSC versus SSC plot will not stain with TMRE but should be gated out because they are unlikely to be cells due to their small size. However, it is essential to observe this population because some cells fragment into apoptotic bodies that will appear as dots in this region. Although apoptotic bodies should not be counted as cells, observing this population may yield essential information on the level of cell death in the culture.

10. Run an untreated sample and adjust the voltage and gain for the TMRE detector so that all of the cells can be detected on the right side of a histogram plot.

    Healthy cells will have high TMRE fluorescence. The gain should therefore be set to place these cells at the right hand side of the plot so that cells with low TMRE staining (low $\Delta\Psi m$) can be detected to the left of this peak. This is opposite to PI staining or annexin V staining where dead cells have higher staining than healthy cells.

11. Run the FCCP-treated sample to ensure that cells with low $\Delta\Psi m$ appear to the left of the non-FCCP-treated cells in the TMRE channel.

    FCCP is a protonophore that depolarizes mitochondria (Benz and McLaughlin 1983). FCCP therefore acts as a positive control for loss of $\Delta\Psi m$, which results in reduced TMRE staining. If the fluorescence peaks of control and FCCP treated samples are not easily distinguishable, the concentration of TMRE may not be optimal. The TMRE concentration can be titrated to obtain optimum separation of these peaks.

    FCCP is very efficient even at low concentrations. FCCP-treated samples should therefore be analyzed at the end of the experiment to avoid cross contamination of FCCP into subsequent samples.

12. Run each sample and acquire data for at least 10,000 events.

    It is not essential to collect data for 10,000 cells; however, collecting this number of events will ensure smooth peaks when presenting your data visually and will also provide statistical power when calculating the percentage cells with low TMRE fluorescence.

    It is essential to ensure correct gating of the flow cytometry plots to include live and dead cells. Gates should first be set for FSC and SSC on linear scale to ensure all cells are counted but all debris is excluded, as per the acquisition protocol. Histogram plots of cell number (y-axis) versus TMRE fluorescence (x-axis) can then be generated using log scale. These histograms can be presented visually and the x-axis should be labeled log TMRE fluorescence (Fig. 2). These plots can also be used to determine the percentage of cells with low TMRE fluorescence by setting a gate around the cells with low fluorescence. The percentage of cells within that gate will be automatically determined by the analysis software.

DISCUSSION

Cytochrome *c* release is a key event in the intrinsic apoptosis pathway that triggers activation of caspase proteases; however, measuring cytochrome *c* release can be tedious (Waterhouse and

Cite this protocol as *Cold Spring Harb Protoc*; doi:10.1101/pdb.prot087361

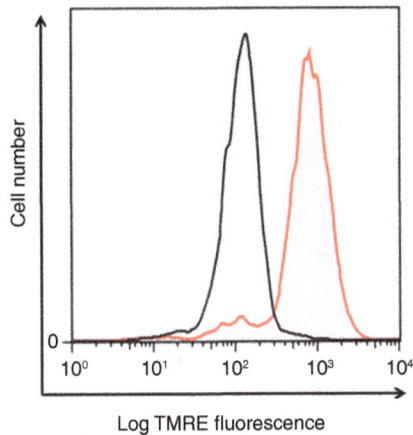

FIGURE 2. Histogram plots of TMRE-stained HeLa cells. Untreated cells (red line) and cells treated with actinomycin D (black line) stained with TMRE.

Trapani 2003). Cytochrome *c* release also causes loss of ΔΨm, which can be assayed by measuring TMRE staining using flow cytometry (Waterhouse et al. 2001b). This is a fast, simple, and effective technique that is often used as a surrogate for assessing cytochrome *c* release during apoptosis. TMRE staining is punctate in healthy mitochondria but is dim and diffuse in cells with low ΔΨm. TMRE staining can therefore also be observed using fluorescence microscopy and measured over time using time lapse fluorescence microscopy (Waterhouse et al. 2001b, 2006) Measuring ΔΨm by TMRE staining also provides information about the general health of mitochondria in living cells, which is important because some stimuli may cause loss of ΔΨm upstream of cytochrome *c* release. Additionally, blocking caspase activity downstream from cytochrome *c* release can prevent apoptosis but the cell will eventually die by a caspase-independent process. It has been proposed that this alternative process is the consequence of slow dissipation of ΔΨm after cytochrome *c* release.

Loss of ΔΨm may be a transient event. For example, ΔΨm will recover after permeability transition or if caspases are not activated following cytochrome *c* release (Waterhouse et al. 2001b). TMRE staining cannot be used to identify whether mitochondrial ΔΨm is lost and regenerated and should therefore not be performed on cells treated with caspase inhibitors to determine whether they have undergone cytochrome *c* release. TMRE can also be used to distinguish subtle changes in ΔΨm. However, subtle changes in TMRE fluorescence can be misinterpreted. This is because TMRE staining may be influenced by the number of live cells in a sample. For example, live cells in a sample that contains 50% dead cells will appear to have higher TMRE staining compared with live cells in samples that only have 5% dead cells. This is because there are more live cells to take up the dye (Vander Heiden et al. 1997; Christensen et al. 2013). However, this does not generally affect the use of TMRE to determine the overall percentage of cells with low ΔΨm.

Overall, measuring ΔΨm by TMRE staining can be useful as a surrogate marker for cytochrome *c* release and is also important for measuring the general health of mitochondria and for studying caspase-independent cell death.

RELATED TECHNIQUES

Mitochondrial depolarization (loss of ΔΨm) can be assayed by a variety of methods including staining for JC-1, MitoTracker or DioC6 (Waterhouse et al. 2001a). JC-1 is an all or nothing dye that changes fluorescence based on concentration. This dye is therefore particularly useful for distinguishing live and dead cells. MitoTracker binds covalently to mitochondria with high ΔΨm. This dye is therefore useful for staining cells that must be fixed before analysis.

ACKNOWLEDGMENTS

This work was supported by an Australian Research Council futures fellowship and project grants from the Leukaemia Foundation Queensland, the Prostrate Cancer Foundation Australia, and the Australian Department of Defence to N.J.W.

REFERENCES

Benz R, McLaughlin S. 1983. The molecular mechanism of action of the proton ionophore FCCP (carbonylcyanide p-trifluoromethoxyphenyl-hydrazone). *Biophys J* 41: 381–398.

Christensen ME, Jansen ES, Sanchez W, Waterhouse NJ. 2013. Flow cytometry based assays for the measurement of apoptosis-associated mitochondrial membrane depolarisation and cytochrome c release. *Methods* 61: 138–145.

Crowley LC, Waterhouse NJ. 2015. Triggering death of adherent cells with ultraviolet radiation. *Cold Spring Harb Protoc* doi: 10.1101/pdb.prot087148.

Crowley LC, Marfell BJ, Scott AP, Waterhouse NJ. 2015. Triggering apoptosis in hematopoietic cells with cytotoxic drugs. *Cold Spring Harb Protoc* doi: 10.1101/pdb.prot087130.

Vander Heiden MG, Chandel NS, Williamson EK, Schumacker PT, Thompson CB. 1997. Bcl-xL regulates the membrane potential and volume homeostasis of mitochondria. *Cell* 91: 627–637.

Waterhouse NJ, Trapani JA. 2003. A new quantitative assay for cytochrome c release in apoptotic cells. *Cell Death Differ* 10: 853–855.

Waterhouse NJ, Goldstein JC, Kluck RM, Newmeyer DD, Green DR. 2001a. The (Holey) study of mitochondria in apoptosis. *Methods Cell Biol* 66: 365–391.

Waterhouse NJ, Goldstein JC, von Ahsen O, Schuler M, Newmeyer DD, Green DR. 2001b. Cytochrome c maintains mitochondrial transmembrane potential and ATP generation after outer mitochondrial membrane permeabilization during the apoptotic process. *J Cell Biol* 153: 319–328.

Waterhouse NJ, Sedelies KA, Sutton VR, Pinkoski MJ, Thia KY, Johnstone R, Bird PI, Green DR, Trapani JA. 2006. Functional dissociation of DeltaPsim and cytochrome c release defines the contribution of mitochondria upstream of caspase activation during granzyme B-induced apoptosis. *Cell Death Differ* 13: 607–618.

CHAPTER 4

Biochemical Analysis of Initiator Caspase-Activating Complexes: The Apoptosome and the Death-Inducing Signaling Complex

Claudia Langlais, Michelle A. Hughes, Kelvin Cain,[1] and Marion MacFarlane[1]

MRC Toxicology Unit, Hodgkin Building, Leicester LE1 9HN, United Kingdom

Apoptosis is a highly regulated process that can be initiated by activation of death receptors or perturbation of mitochondria causing the release of apoptogenic proteins. This results in the activation of caspases, which are responsible for many of the biochemical and morphological changes associated with apoptosis. Caspases are normally inactive and require activation in a cascade emanating from an "initiator" or activating caspase, which in turn activates a downstream or "effector" caspase. Activation of initiator caspases is tightly regulated and requires the assembly of caspase-9 (via mitochondrial perturbation) or caspase-8/10 (via death receptor ligation) activating complexes, which are termed the apoptosome and the death-inducing signaling complex (DISC), respectively. These large multiprotein complexes can initially be separated according to size by gel filtration chromatography and subsequently analyzed by affinity purification or immunoprecipitation. The advantage of combining these techniques is one can first assess the assembly of individual components into a multiprotein complex, and then assess the size and composition of the native functional signaling platform within a particular cell type alongside a biochemical analysis of the enriched/purified complex. Here, we describe various methods currently used for characterization of the apoptosome and DISC.

BACKGROUND

Many key biological processes, including apoptosis, are initiated from or performed in large multi-protein complexes. Apoptosis signaling complexes that can initiate cell death include the apoptosome and the death-inducing signaling complex (DISC) (Bratton et al. 2000; Cain et al. 2002; MacFarlane 2003; Dickens et al. 2012b). Typically, apoptosis is triggered through activation of either the intrinsic (mitochondrial) or extrinsic (death receptor) pathway. Central for activation of the intrinsic pathway is the cytosolic Apaf-1/Caspase-9 apoptosome, a >700–1000-kDa complex formed following release of cytochrome *c* from mitochondria (Cain et al. 2002). In contrast, the extrinsic pathway is triggered by formation of the DISC at the plasma membrane; in this case, ligation of the death receptors CD95, TRAIL-R1, or TRAIL-R2 by their cognate ligands results in recruitment of the adaptor molecule FADD, the initiator caspase procaspase-8, and additional modulator proteins such as cFLIP (Dickens et al. 2012b).

It is now increasingly evident that the composition and stoichiometry of components within key cell death signaling platforms can determine not only the final signaling outcome but also the mode of cell death. By analyzing these complexes, we can learn how cell death is regulated as well as how key cell death signaling

[1]Correspondence: mm21@leicester.ac.uk; kc5@le.ac.uk

Cite this introduction as *Cold Spring Harb Protoc*; doi:10.1101/pdb.top070326

platforms like the apoptosome and DISC might be targeted for therapeutic benefit (MacFarlane 2009; Cain 2010). The successful application of a range of methodologies which couple characterization of complex assembly together with subsequent purification and biochemical analysis can therefore provide novel insights into how cell death signaling platforms are regulated in both normal cell physiology and disease.

Large multiprotein complexes such as the apoptosome and DISC can be separated according to their size by gel filtration chromatography and further purified by subsequent affinity purification or immunoprecipitation. By combining these methods, an indication of the size of the complex as well as the recruitment of individual components associated with the active complex can be determined, thus providing more precise and selective information on the complex itself. This combined approach has been used to purify and characterize the active apoptosome complex (Cain et al. 1999, 2000; Twiddy et al. 2006) and, more recently, we and others have used gel filtration to identify and characterize the ripoptosome (Feoktistova et al. 2011; Tenev et al. 2011). Intriguingly, this complex, depending on its protein composition, can switch between apoptotic and necrotic modes of cell death. Similarly, the complementary approach of sucrose density gradient centrifugation has been combined with immunoprecipitation/affinity purification to characterize both the active apoptosome complex, and, more recently, the native TRAIL DISC using mass spectrometry (Twiddy et al. 2004; Dickens et al. 2012a; Hughes et al. 2013). Indeed, affinity purification of the DISC using biotin-labeled ligands not only provided novel insights into the mechanisms that regulate death receptor signaling in diverse cell types and upon different treatment regimens (Harper et al. 2001, 2003a, 2003b; MacFarlane et al. 2002, 2005; Harper and MacFarlane 2008; Robinson et al. 2012), but also led us to propose a death effector domain (DED) chain DISC model and a crucial role for caspase-8 chain assembly in mediating apoptotic cell death (Dickens et al. 2012a).

ANALYZING THE APOPTOSOME AND DISC

In the accompanying protocols, we describe the various techniques and strategies we have developed over several years to successfully isolate and characterize two key initiator caspase-activating complexes, namely the apoptosome and DISC. The dATP-activation of caspases in cellular lysates has been used for many years as an in vitro model system for assembling the apoptosome complex and studying its characteristics, components, and activity (Cain et al. 1999, 2000, 2001; Beere et al. 2000; Freathy et al. 2000; Almond et al. 2001; Bratton et al. 2001; Thompson et al. 2001; Lademann et al. 2003; Twiddy et al. 2004, 2006; Hughes et al. 2013). The basis of the model, detailed in Protocol 1: In Vitro Assembly and Analysis of the Apoptosome Complex (Langlais et al. 2015), is production of a cell-free extract from a cell line of choice (e.g., the human monocytic tumor cell line THP.1) in which caspases can be processed and activated in vitro by treatment with dATP/MgCl$_2$ and cytochrome c (Liu et al. 1996; Li et al. 1997). As an alternative to apoptosome assembly based on cell lysates, a full in vitro reconstitution of the apoptosome is also feasible (Cain et al. 2001; Zou et al. 2003). In both cases, apoptosome assembly and caspase-dependent cleavage are analyzed using gel filtration in conjunction with western blotting and fluorimetric assays for effector caspase activity.

Our Protocol 2: Activation, Isolation, and Analysis of the Death-Inducing Signaling Complex (Hughes et al. 2015) is based on affinity purification using biotin-labeled antibodies and streptavidin beads for activation and capture of the native DISC complex. Triggering of receptor aggregation by the agonistic anti-CD95 antibodies (or the cognate CD95 ligand) leads to recruitment of the bipartite adaptor molecule FADD, which in turn binds the initiator caspase, procaspase-8, through its amino-terminal DED motifs (Peter and Krammer 2003; Dickens et al. 2012a; Schleich et al. 2012). On binding to FADD, procaspase-8 is activated through proximity-induced dimerization of adjacent procaspase-8 molecules, leading to proteolytic cleavage and maximal caspase-8 activation (Muzio et al. 1998; Boatright et al. 2003; Hughes et al. 2009). Activation of caspase-8 at the native DISC leads to cleavage and activation of downstream substrates, such as procaspase-3 and BID, amplification of the caspase cascade and subsequent apoptosis. Composition of the isolated DISC and caspase activation can be analyzed via western blot and fluorimetric assays for caspase-dependent cleavage.

Cite this introduction as *Cold Spring Harb Protoc*; doi:10.1101/pdb.top070326

REFERENCES

Almond JB, Snowden RT, Hunter A, Dinsdale D, Cain K, Cohen GM. 2001. Proteasome inhibitor-induced apoptosis of B-chronic lymphocytic leukaemia cells involves cytochrome c release and caspase activation, accompanied by formation of an approximately 700 kDa Apaf-1 containing apoptosome complex. *Leukemia* **15:** 1388–1397.

Beere HM, Wolf BB, Cain K, Mosser DD, Mahboubi A, Kuwana T, Tailor P, Morimoto RI, Cohen GM, Green DR. 2000. Heat-shock protein 70 inhibits apoptosis by preventing recruitment of procaspase-9 to the Apaf-1 apoptosome. *Nat Cell Biol* **2:** 469–475.

Boatright KM, Renatus M, Scott FL, Sperandio S, Shin H, Pedersen IM, Ricci JE, Edris WA, Sutherlin DP, Green DR, et al. 2003. A unified model for apical caspase activation. *Mol Cell* **11:** 529–541.

Bratton SB, MacFarlane M, Cain K, Cohen GM. 2000. Protein complexes activate distinct caspase cascades in death receptor and stress-induced apoptosis. *Exp Cell Res* **256:** 27–33.

Bratton SB, Walker G, Roberts DL, Cain K, Cohen GM. 2001. Caspase-3 cleaves Apaf-1 into an approximately 30 kDa fragment that associates with an inappropriately oligomerized and biologically inactive approximately 1.4 MDa apoptosome complex. *Cell Death Differ* **8:** 425–433.

Cain K. 2010. Chemical regulation of the apoptosome: New alternative treatments for cancer. In *Apoptosome: An up-and-coming therapeutical tool* (ed. F Cecconi, M D'Amelio), pp. 41–74. Springer, Dordrecht, The Netherlands.

Cain K, Brown DG, Langlais C, Cohen GM. 1999. Caspase activation involves the formation of the aposome, a large (approximately 700 kDa) caspase-activating complex. *J Biol Chem* **274:** 22686–22692.

Cain K, Bratton SB, Langlais C, Walker G, Brown DG, Sun XM, Cohen GM. 2000. Apaf-1 oligomerizes into biologically active approximately 700-kDa and inactive approximately 1.4-MDa apoptosome complexes. *J Biol Chem* **275:** 6067–6070.

Cain K, Langlais C, Sun XM, Brown DG, Cohen GM. 2001. Physiological concentrations of K⁺ inhibit cytochrome *c*-dependent formation of the apoptosome. *J Biol Chem* **276:** 41985–41990.

Cain K, Bratton SB, Cohen GM. 2002. The Apaf-1 apoptosome: A large caspase-activating complex. *Biochimie* **84:** 203–214.

Dickens LS, Boyd RS, Jukes-Jones R, Hughes MA, Robinson GL, Fairall L, Schwabe JW, Cain K, MacFarlane M. 2012a. A death effector domain chain DISC model reveals a crucial role for caspase-8 chain assembly in mediating apoptotic cell death. *Mol Cell* **47:** 291–305.

Dickens LS, Powley IR, Hughes MA, MacFarlane M. 2012b. The 'complexities' of life and death: Death receptor signalling platforms. *Exp Cell Res* **318:** 1269–1277.

Feoktistova M, Geserick P, Kellert B, Dimitrova DP, Langlais C, Hupe M, Cain K, MacFarlane M, Hacker G, Leverkus M. 2011. cIAPs block Ripoptosome formation, a RIP1/caspase-8 containing intracellular cell death complex differentially regulated by cFLIP isoforms. *Mol Cell* **43:** 449–463.

Freathy C, Brown DG, Roberts RA, Cain K. 2000. Transforming growth factor-β₁ induces apoptosis in rat FaO hepatoma cells via cytochrome c release and oligomerization of Apaf-1 to form a approximately 700-kd apoptosome caspase-processing complex. *Hepatology* **32:** 750–760.

Harper N, MacFarlane M. 2008. Recombinant TRAIL and TRAIL receptor analysis. *Methods Enzymol* **446:** 293–313.

Harper N, Farrow SN, Kaptein A, Cohen GM, MacFarlane M. 2001. Modulation of tumor necrosis factor apoptosis-inducing ligand-induced NF-κB activation by inhibition of apical caspases. *J Biol Chem* **276:** 34743–34752.

Harper N, Hughes M, MacFarlane M, Cohen GM. 2003a. Fas-associated death domain protein and caspase-8 are not recruited to the tumor necrosis factor receptor 1 signaling complex during tumor necrosis factor-induced apoptosis. *J Biol Chem* **278:** 25534–25541.

Harper N, Hughes MA, Farrow SN, Cohen GM, MacFarlane M. 2003b. Protein kinase C modulates tumor necrosis factor-related apoptosis-inducing ligand-induced apoptosis by targeting the apical events of death receptor signaling. *J Biol Chem* **278:** 44338–44347.

Hughes MA, Harper N, Butterworth M, Cain K, Cohen GM, MacFarlane M. 2009. Reconstitution of the death-inducing signaling complex reveals a substrate switch that determines CD95-mediated death or survival. *Mol Cell* **35:** 265–279.

Hughes MA, Langlais C, Cain K, MacFarlane M. 2013. Isolation, characterisation and reconstitution of cell death signalling complexes. *Methods* **61:** 98–104.

Hughes MA, Langlais C, Cain K, MacFarlane M. 2015. Activation, isolation, and analysis of the death-inducing signaling complex. *Cold Spring Harb Protoc* doi: 10.1101/pdb.prot087098.

Lademann U, Cain K, Gyrd-Hansen M, Brown D, Peters D, Jaattela M. 2003. Diarylurea compounds inhibit caspase activation by preventing the formation of the active 700-kilodalton apoptosome complex. *Mol Cell Biol* **23:** 7829–7837.

Langlais C, Hughes MA, Cain K, MacFarlane M. 2015. In vitro assembly and analysis of the apoptosome complex. *Cold Spring Harb Protoc* doi: 10.1101/pdb.prot087080.

Li P, Nijhawan D, Budihardjo I, Srinivasula SM, Ahmad M, Alnemri ES, Wang X. 1997. Cytochrome c and dATP-dependent formation of Apaf-1/caspase-9 complex initiates an apoptotic protease cascade. *Cell* **91:** 479–489.

Liu X, Kim CN, Yang J, Jemmerson R, Wang X. 1996. Induction of apoptotic program in cell-free extracts: Requirement for dATP and cytochrome c. *Cell* **86:** 147–157.

MacFarlane M. 2003. TRAIL-induced signalling and apoptosis. *Toxicol Lett* **139:** 89–97.

MacFarlane M. 2009. Cell death pathways—potential therapeutic targets. *Xenobiotica* **39:** 616–624.

MacFarlane M, Harper N, Snowden RT, Dyer MJ, Barnett GA, Pringle JH, Cohen GM. 2002. Mechanisms of resistance to TRAIL-induced apoptosis in primary B cell chronic lymphocytic leukaemia. *Oncogene* **21:** 6809–6818.

MacFarlane M, Kohlhaas SL, Sutcliffe MJ, Dyer MJ, Cohen GM. 2005. TRAIL receptor-selective mutants signal to apoptosis via TRAIL-R1 in primary lymphoid malignancies. *Cancer research* **65:** 11265–11270.

Muzio M, Stockwell BR, Stennicke HR, Salvesen GS, Dixit VM. 1998. An induced proximity model for caspase-8 activation. *J Biol Chem* **273:** 2926–2930.

Peter ME, Krammer PH. 2003. The CD95(APO-1/Fas) DISC and beyond. *Cell Death Differ* **10:** 26–35.

Robinson GL, Dinsdale D, MacFarlane M, Cain K. 2012. Switching from aerobic glycolysis to oxidative phosphorylation modulates the sensitivity of mantle cell lymphoma cells to TRAIL. *Oncogene* **31:** 4996–5006.

Schleich K, Warnken U, Fricker N, Ozturk S, Richter P, Kammerer K, Schnolzer M, Krammer PH, Lavrik IN. 2012. Stoichiometry of the CD95 death-inducing signaling complex: Experimental and modeling evidence for a death effector domain chain model. *Mol Cell* **47:** 306–319.

Tenev T, Bianchi K, Darding M, Broemer M, Langlais C, Wallberg F, Zachariou A, Lopez J, MacFarlane M, Cain K, et al. 2011. The Ripoptosome, a signaling platform that assembles in response to genotoxic stress and loss of IAPs. *Mol Cell* **43:** 432–448.

Thompson GJ, Langlais C, Cain K, Conley EC, Cohen GM. 2001. Elevated extracellular [K⁺] inhibits death-receptor- and chemical-mediated apoptosis prior to caspase activation and cytochrome c release. *Biochem J* **357:** 137–145.

Twiddy D, Brown DG, Adrain C, Jukes R, Martin SJ, Cohen GM, MacFarlane M, Cain K. 2004. Pro-apoptotic proteins released from the mitochondria regulate the protein composition and caspase-processing activity of the native Apaf-1/caspase-9 apoptosome complex. *J Biol Chem* **279:** 19665–19682.

Twiddy D, Cohen GM, MacFarlane M, Cain K. 2006. Caspase-7 is directly activated by the approximately 700-kDa apoptosome complex and is released as a stable XIAP-caspase-7 approximately 200-kDa complex. *J Biol Chem* **281:** 3876–3888.

Zou H, Yang R, Hao J, Wang J, Sun C, Fesik SW, Wu JC, Tomaselli KJ, Armstrong RC. 2003. Regulation of the Apaf-1/Caspase-9 apoptosome by caspase-3 and XIAP. *J Biol Chem* **278:** 8091–8098.

In Vitro Assembly and Analysis of the Apoptosome Complex

Claudia Langlais, Michelle A. Hughes, Kelvin Cain,[1] and Marion MacFarlane[1]

MRC Toxicology Unit, Hodgkin Building, Leicester LE1 9HN, United Kingdom

This protocol describes an in vitro model for studying the mechanisms of caspase activation and native apoptosome complex assembly in cell-free extracts. Active caspases in dATP-activated lysates are detected by fluorimetry using a tetrapeptide substrate (DEVD) tagged with a fluorophore (AFC), which, when released, produces a real-time readout for caspase-3 and -7 (DEVDase) activity. Gel filtration is used to isolate the apoptosome complex from the activated lysates, and assembly of Apaf-1 and caspase-9 from their monomeric forms into the multiprotein apoptosome can be confirmed via western blot. Apoptosome complex activity can be shown by incubation with exogenous procaspase-3 and -7 followed by fluorimetric bioassay (to confirm functionality of the processed effector caspases) and/or western blotting (for detection of cleaved caspase-3 and -7). A method for preparation of free procaspases for the bioassay is also described.

MATERIALS

It is essential that you consult the appropriate Material Safety Data Sheets and your institution's Environmental Health and Safety Office for proper handling of equipment and hazardous materials used in this protocol.

RECIPES: Please see the end of this protocol for recipes indicated by <R>. Additional recipes can be found online at http://cshprotocols.cshlp.org/site/recipes.

Reagents

2′-Deoxyadenosine 5′-triphosphate sodium salt (dATP) (100 mM; Sigma-Aldrich D4788)

Store aliquots at −20°C.

7-Amino-4-trifluoromethylcoumarin (AFC) for fluorimetric assay calibration

Ac-DEVD.AFC (caspase-3/-7 fluorigenic substrate) (20 mM in DMSO) (MP Biomedicals 03AFC13805)

Ac-DEVD.AFC is routinely used as a caspase-3/-7 substrate and releases the fluorophore AFC for a real-time readout of caspase activity. Alternatively, appropriate 7-amino-4-methylcoumarin (AMC) substrates are also available and can be measured using excitation/emission wavelength pairs of 380/460 nm.
Store Ac-DEVD.AFC in 20-μL aliquots at −20°C. Dilute to 2.5 mM before use.

Antibodies for western blotting

Apaf-1 (R&D Systems MAB868)

Caspase-3 and -7 (pro- and cleaved forms; many suppliers)

Caspase-9 (MBL M054–3)

Bradford assay reagents

[1]Correspondence: mm21@leicester.ac.uk; kc5@le.ac.uk

Cite this protocol as *Cold Spring Harb Protoc*; doi:10.1101/pdb.prot087080

Caspase activity assay buffer <R> (ice-cold and 37°C)

Ice-cold assay buffer is required for Steps 10 and 36. Prewarm assay buffer to 37°C for Steps 12 and 22.

Cells for preparation of lysates

Cellular lysates can be generated from primary cells or cell lines (e.g., THP.1, HeLa, or rat/human liver hepatoma). Calculate the requirement for cell lysate preparation and culture sufficient numbers according to standard protocols. As a general rule, 10^8 cells will generate ~150 µL of cellular lysate (~20 mg/mL), depending on cell size.

Column buffer for Sephacryl S-300 (salt-free) <R>

Column buffer for Superose 6 (with 50 mM NaCl) <R>

Cytochrome *c* (5 mg/mL [0.4 mM] in H_2O; Sigma-Aldrich C7752) (if needed; see Step 10)

Store aliquots at −20°C.

Gel filtration calibration kit containing standard proteins of known sizes, including blue dextran (2 MDa, to determine the void volume), thyroglobulin (669 kDa), ferritin (440 kDa), catalase (232 kDa), albumin (67 kDa), and ribonuclease A (13.7 kDa) (GE Healthcare)

$MgCl_2$ (100 mM in H_2O)

Store aliquots at −20°C.

Phosphate-buffered saline (PBS) (ice-cold)

PIPES buffer <R> (ice-cold)

Procaspases for fluorimetric analysis (bioassay) of apoptosome-associated caspase activity

Free procaspases for the bioassay can be isolated from control lysate using the method provided in Steps 28–37. These should be prepared well in advance of the bioassay and aliquots frozen at −80°C ready for use. Alternatively, recombinant procaspase-3 is commercially available (R&D Systems 731-C3).

SDS-PAGE sample loading buffer (4×) <R>

In addition, prepare 2× SDS-PAGE sample loading buffer by diluting 4× buffer with H_2O.

Equipment

96-well plates for fluorimetric assays (Scientific Laboratory Supplies 161093)

Centrifuge tubes

Centrifuges capable of running at 200g, 20,000g, and 100,000g (e.g., Beckmann Optima MAX-XP Ultracentrifuge)

Chromatographic protein separation system (e.g., AKTA [GE Healthcare] or NGC [Bio-Rad])

All gel filtration experiments should be performed at 4°C; this requires the chromatography apparatus to be located either in a cold room or a large cold cabinet/fridge.
Set an appropriate pressure limit for each column to avoid damage.

Concentrators (4 mL with 10,000 MW cutoff) (e.g., Vivaspin [Millipore] or similar) (for optional isolation of free caspases)

Gel filtration columns, as needed

HiPrep 16/60 Sephacryl S-300 HR (GE Healthcare 17-1167-01)

HiPrep 26/60 Sephacryl S-300 HR (GE Healthcare 17-1196-01)

Superose 6 10/300 GL (GE Healthcare 17-5172-01)

Incubator at 37°C

Liquid nitrogen

Microcentrifuge tubes (1.5-mL, screw-cap)

SDS-PAGE and western blotting equipment

Thermostatic plate reader capable of reading various excitation/emission wavelengths (e.g., 405/510 nm for AFC) at 37°C with kinetic capability (e.g., Wallac Victor X4)

Turn on the plate reader and set to 37°C before use. Prepare the software so measurement can start at the click of a button.

Water bath (with floating tube rack) at 37°C

METHOD

An outline of the method is provided in Figure 1.

Preparing Cellular Lysates

Cell lysates can be prepared by a variety of methods, such as freeze/thawing or using detergent-containing lysis buffers. We routinely use the freeze/thaw method for preparation of lysates for dATP-activation and analysis of the apoptosome complex (Cain et al. 1999, 2000, 2001; Thompson et al. 2001). The following basic method requires ~2.5–3 h and can be used for many different cell types.

1. Pellet the cells by centrifugation at 200–300g for 10 min at 4°C. Wash the cells twice in ice-cold PBS.

2. Discard the PBS and resuspend the cell pellet at 10^8 cells/166 µL of ice-cold PIPES buffer. Prepare aliquots of the suspension in 1.5 mL screw-cap microcentrifuge tubes.

3. Freeze/thaw the cells as follows:

    i. To prevent explosions, pierce a hole in the lid of each tube before carefully submerging in liquid nitrogen.

    ii. Freeze the cell suspensions in liquid nitrogen for 30 sec to 1 min.

    iii. Transfer the suspensions to a tube rack in a water bath at 37°C for rapid thawing. Incubate until completely thawed.

    iv. Repeat Steps 3.ii–3.iii twice for a total of three freeze/thaw cycles.

4. Centrifuge the disrupted cells at 20,000g for 30 min at 4°C. Retain the supernatant.

5. Centrifuge the supernatant for 45 min at 100,000g to pellet any remaining organelles (such as the endoplasmic reticulum).

6. Determine the protein concentration of the resulting lysate using the Bradford assay or a similar method.

    The lysate should have a minimum concentration of 10 mg/mL; otherwise, the dATP activation will not be successful.

7. Prepare 250 µL to 1 mL aliquots of the resulting lysate in 1.5 mL tubes. Store aliquots at −80°C.

    Aliquots can be used as desired for later experiments. Thawed lysates should be discarded and not refrozen.

Activating Cellular Lysates Using dATP

Depending on the method used to prepare the lysates, mitochondria will either be lysed along with the whole cell (as with the freeze/thaw method above) or remain intact. Cell lysis procedures which do not damage the mitochondria usually have low levels of cytochrome c and require both dATP and exogenous cytochrome c to activate apoptosome assembly. Cell lysates containing cytochrome c can be activated by adding dATP/ATP alone. The caspase-cleaving activity of the lysate is detected by measuring the release of AFC from the caspase-3/-7 substrate Ac-DEVD.AFC.

The following section (dATP activation and fluorimetric assay) can be completed in 1.5–2 h.

8. Calibrate the fluorimetric assay using an AFC standard curve.

    i. Prepare 4–6 AFC standards ranging from 0 to 2 µM.

    ii. Measure the fluorescence using an excitation/emission wavelength pair of 405/510 nm.

    Experimental cleavage rates can be determined by linear regression. Enzyme activities are calculated from the calibration curve and expressed as pmol/min per mg of protein or as pmol/min per fraction.

9. Immediately before the assay, thaw one vial of Ac-DEVD.AFC substrate and one aliquot of cellular lysate on ice until the samples are completely melted (4°C).

10. Dilute the lysate to 10 mg/mL using ice-cold caspase activity assay buffer. Add dATP and $MgCl_2$ each to a final concentration of 2 mm. Add additional cytochrome c to a concentration of 7 µM if necessary (Twiddy et al. 2004).

    The final volume should be ~400 µL for an experiment using the Superose 6 10/300 GL column.

Cite this protocol as *Cold Spring Harb Protoc*; doi:10.1101/pdb.prot087080

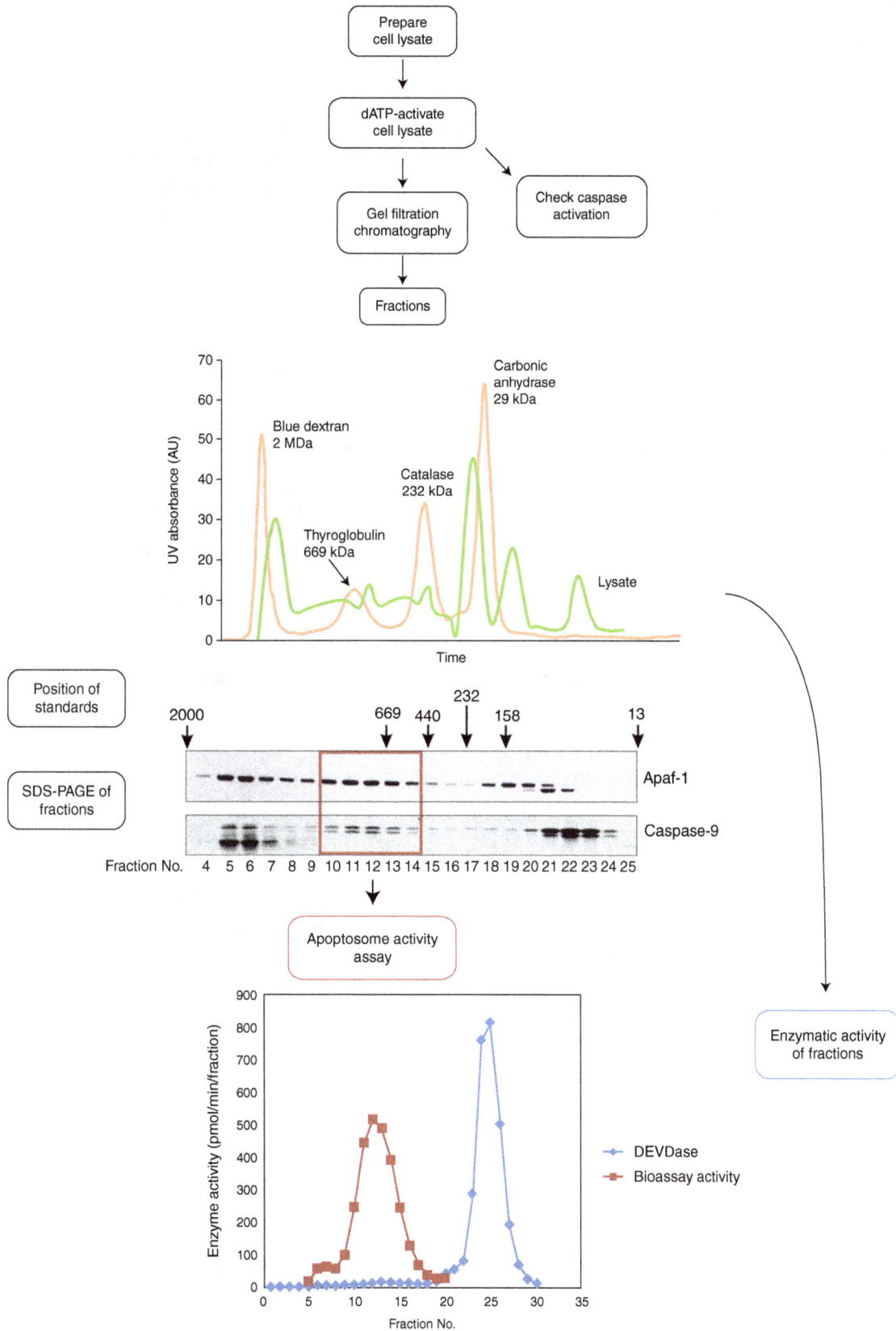

FIGURE 1. A step-by-step flowchart outlining in vitro assembly and analysis of the apoptosome complex. Cell lysates are dATP-activated and separated by gel filtration using a Superose 6 column. The gel filtration profile (*top* graph) shows the elution positions (time/volume) of standard proteins (orange line) and a cell lysate (green line) measured using UV absorbance at 280 nm. Fractions are analyzed by SDS-PAGE and western blotting with appropriate antibodies (shown here, Apaf-1 and caspase-9). Calibration standards include blue dextran (2 MDa), thyroglobulin (669 kDa), ferritin (440 kDa), catalase (232 kDa), aldolase (158 kDa), carbonic anhydrase (29 kDa), and cytochrome *c* (13 kDa). Appropriate fractions are further assayed for apoptosome complex activity by addition of procaspase-3 and -7 and measurement of resulting DEVDase activity (lower graph, red). Fractions can also be assayed directly for DEVDase activity to show the presence of free active caspase-3 and caspase-7 in lower molecular mass fractions (lower graph, blue).

11. Incubate the lysate for 30 min at 37°C.

> At this point, the lysate can briefly be stored on ice. The fluorimetric assay (Steps 12–14) followed by column loading (Steps 15–19) should be performed as soon as possible.

12. In a 96-well plate, mix 10 µL of activated lysate per well with 190 µL of prewarmed caspase activity assay buffer containing Ac-DEVD.AFC for a final concentration of 20 µM Ac-DEVD.AFC. Place the plate in the instrument as quickly as possible and immediately begin measurement.

> The reaction will start as soon as the components are mixed.

13. Record the release of AFC (caspase-3/7 activity) continuously using an excitation/emission wavelength pair of 405/510 nm for 10–20 cycles.

14. Determine the cleavage rate by linear regression and calculate enzyme activities from the calibration curve prepared in Step 8. Express caspase activity as pmol/min per mg of protein.

> See Troubleshooting.

Isolating the Apoptosome Complex by Gel Filtration Chromatography

> The following describes a standard method for analysis of apoptosome formation using a Superose 6 10/300 GL gel filtration column. This column is ideal for routine analysis of apoptosome complex formation and activity (see Discussion); however, because it is run in buffer containing 50 mM NaCl, caspases-3 and -7 will not co-elute with the apoptosome complex (n.b., caspase-9 will remain with the complex). Thus, an additional fluorimetric bioassay using exogenous free procaspases (Steps 21–25) is required to measure caspase activity (Cain et al. 2000).

> A minimum of a half-day is required to complete gel filtration chromatography and the subsequent bioassay. At least 300 µL of cell lysate (10 mg/mL) is recommended for each separation. Make sure that an appropriate pressure limit is set for each column to avoid damage to the column bed during use.

15. Program the pump system as follows: run at 0.4 mL/min, collect 0.5 mL fractions, discard the void volume (~5 mL), and collect one column volume (CV) (24 mL) in 48 fractions.

16. Equilibrate a Superose 6 column with at least two CVs of the appropriate column buffer.

17. Before sample application, calibrate the column using standard proteins of known sizes.

18. Before the column run, mix an aliquot of the dATP-activated (or control) lysate with an equal volume of 2× SDS-PAGE sample loading buffer (e.g., 50 µL of lysate and 50 µL of 2× buffer). Store this sample at −20°C until ready to proceed with SDS-PAGE (Step 26).

19. Load 250 µL (2.5 mg of protein at 10 mg/mL) of dATP-activated or control lysate onto the column and begin the run, collecting 0.5 mL fractions at 0.4 mL/min. Monitor protein elution by ultraviolet (UV) detection at 280 nm.

> Most pump systems, such as the AKTA, have a UV monitor incorporated in-line and provide an automatic printout of the chromatographic trace at 280 nm.
>
> See Troubleshooting.

20. After the run, mix 200 µL of each fraction with 68 µL of 4× SDS-PAGE sample loading buffer. Store these samples at −20°C until ready to proceed with SDS-PAGE (Step 26).

> The remaining fractions should be kept on ice before performing the bioassay (Steps 21–25). Alternatively, fractions can be stored at −20°C if bioassay is not required. Note that the bioassay will only be successful if fresh fractions are used.

Performing Fluorimetric Analysis (Bioassay) of Apoptosome-Associated Caspase Activity

> Although caspase-9 fluorigenic substrates are commercially available, their specificity is too broad to measure caspase-9 activity as the primary read-out of apoptosome activity (McStay et al. 2008). Instead, the caspase-cleaving activity of the apoptosome can be measured by addition of procaspase-3 and/or -7, which are cleaved and activated in the presence of an active apoptosome, and produce a real-time read-out using the Ac-DEVD.AFC substrate.

> The following bioassay of apoptosome-associated caspase activity can be completed in 1–2 h. The free procaspases used in the assay should be isolated from a control lysate (see Steps 28–37), or procaspase-3 can be purchased from a commercial source (see Step 21).

21. Mix 100 µL of each fraction collected in Step 19 with 25 µg of free procaspases in a 96-well plate. Incubate for 30 min at 37°C.

Cite this protocol as *Cold Spring Harb Protoc*; doi:10.1101/pdb.prot087080

If procaspases are obtained commercially rather than isolated from a control lysate, the conditions for the bioassay should be optimized for the concentration and purity of the commercial recombinant protein.

22. Add prewarmed caspase activity assay buffer containing Ac-DEVD.AFC to each well for a total volume of 200 μL and a final concentration of 20 μM Ac-DEVD.AFC per well. Place the plate in the instrument as quickly as possible and immediately begin measurement.

23. Read and analyze the plate as described in Steps 13–14.

24. Calculate and plot the activity from each fraction (pmol/fraction per min) against the fraction number to obtain a graph.

    As this assay also measures the free active caspases eluting in lower molecular mass fractions, the graph will show two peaks: one peak in fractions 10–15 (apoptosome complex) and one peak in fractions 23–28 (free active caspases). Fractions can also be assayed for DEVDase activity without prior incubation with procaspases (Fig. 1). In this case, only the free active caspases in fractions 23–28 will produce an activity peak (Cain et al. 2000).

25. (Optional) Mix an aliquot from each well of the 96-well plate with an equal volume of 2× SDS-PAGE sample loading buffer. Store this sample at −20°C until ready to proceed with SDS-PAGE (Step 27).

Analyzing Samples by SDS-PAGE and Western Blotting

26. Analyze the input lysate sample taken before gel filtration (Step 18) and the fractions (Step 20) by SDS-PAGE followed by western blotting for Apaf-1 and caspase-9. Load 20–40 μL per fraction per well for adequate antibody detection.

    Apaf-1 and caspase-9 contained in the apoptosome should be detected in fractions 10–15 (Cain et al. 2000). Importantly, caspase-9 does not cleave procaspase-3/-7 unless it is bound to the Apaf-1 apoptosome complex.

    See Troubleshooting.

27. (Optional) Confirm cleavage of the procaspases in the bioassay samples (Step 25) by SDS-PAGE and subsequent western blotting for caspase-3 and -7 (cleaved fragments).

(Optional) Preparing Procaspases for the Bioassay

We developed this technique using a large HiPrep Sephacryl S-300 26/60 column (Cain et al. 2000). If this column is unavailable, a smaller 16/60 column can be used and the sample split and run into one set of fraction tubes in two separate runs. (Adjust the flow rate and fraction size accordingly). The smaller Sephacryl S-300 16/60 column can theoretically hold a sample volume up to 5 mL, but we recommend not loading >2 mL in a single run. The elution of procaspases-3 and -7 into specific fraction numbers should be determined by western blotting (Step 34) before running a large-scale preparation.

The following procedure can be completed in 1 d and should be performed well in advance of the fluorimetric bioassay.

28. Program the pump system as follows: run at 1 mL/min, collect 5 mL fractions, discard the void volume (~16 mL), and collect at least 28 fractions.

29. Equilibrate the large Sephacryl S-300 column with at least two CVs of the appropriate salt-free column buffer.

30. Before sample application, calibrate the column using standard proteins of known sizes.

31. Thaw 3 mL of control lysate from Step 7 (~25 mg/mL) on ice until the lysate is completely melted. Pool the thawed lysates into one tube.

32. Mix an aliquot of the control lysate with an equal volume of 2× SDS-PAGE sample loading buffer (e.g., 50 μL of lysate and 50 μL of 2× buffer). Store this sample at −20°C until ready to proceed with SDS-PAGE (Step 34).

33. Load the remainder of the lysate (~75 mg of protein in total) onto the column and start the run, collecting 5 mL fractions at 1 mL/min. Monitor protein elution by UV detection at 280 nm.

34. Determine the elution fractions for free procaspases by western blotting.

    i. Mix 100 µL of each fraction with 34 µL of 4× SDS-PAGE sample loading buffer.

       This volume allows for multiple western blot analyses.

    ii. Analyze the samples by SDS-PAGE followed by western blotting for caspases-3 and -7.

       Procaspase-3 and -7 should elute in fractions 18–21, but this should be confirmed empirically.

35. Concentrate each 5 mL fraction containing the procaspases by centrifugation at 5000*g* for 5–15 min in a Vivaspin concentrator.

    The final volume for each sample should be ∼100 µL.

36. Pool the samples and determine the protein concentration by Bradford assay. Adjust the protein concentration to 12.5 mg/mL with ice-cold caspase activity assay buffer.

    An aliquot of 2 µL is equivalent to 25 µg of free procaspases for the bioassay.

37. Prepare 35 µL aliquots and freeze at −80°C.

    Each 35 µL aliquot is sufficient to assay 16 fractions from one column run.

    It is noteworthy that the prepared free procaspases will also contain procaspase-9, as well as numerous other unspecified proteins, but will not contain Apaf-1!

TROUBLESHOOTING

Problem (Step 14): dATP-activation of the cellular lysate was unsuccessful.

Solution: The lysate may be too dilute or caspases may have been activated during lysate preparation. Make sure all appropriate steps are performed at 4°C and the protein concentration of the lysate is at least 10 mg/mL. Store lysates at −80°C to prevent any unwanted caspase activation.

Problem (Step 14): The fluorimeter did not detect any enzyme activity/slope.

Solution: Make sure the fluorimeter is set up correctly and the plate with the mixed components is measured immediately. The reaction can proceed very quickly and reach a maximum plateau, which will prevent accurate measurement of the initial rate of activity. Check the concentration of all reagents and make sure they are fresh. Store dATP and MgCl$_2$ in aliquots at −20°C.

Problem (Step 19): The gel filtration chromatography system went over pressure.

Solution: This normally means that you have applied a dirty sample or the column buffer was not filtered. Replace and clean all online filters of the chromatography system and filter all buffers. If the column was damaged (i.e., the column bed was compressed because there was no pressure limit set on the system), then the column will require replacement (a very expensive mistake).

Problem (Step 26): The apoptosome complex did not elute in the same fraction numbers as described in the method.

Solution: The fraction numbers given in this method are meant as a guideline only and will vary depending on the column set up and program. As part of the initial execution of the protocol, elution of the apoptosome complex from the column should be checked by western blotting of fractions for Apaf-1 and caspase-9 and then compared with the elution position of the protein standards.

DISCUSSION

Gel filtration columns contain a packed porous matrix of chemically inert spherical particles which separate native proteins and protein complexes according to their size. Sample separation is performed isocratically, with smaller molecules diffusing into the pores of the matrix and thus being

Cite this protocol as *Cold Spring Harb Protoc*; doi:10.1101/pdb.prot087080

retained longer than larger molecules and complexes. Columns can be calibrated using commercially available proteins of known molecular mass. Western blotting of collected fractions detects proteins with different elution volumes, which can be compared to known standards. Anomalous elution usually indicates oligomerization of the proteins in question; alternatively, it indicates that the protein is part of a much larger multiprotein complex. The formation of a large complex such as the apoptosome can be shown when the monomeric components of the complex co-elute after activation and migrate as constituents of the complex of interest. Further validation of complex formation can then be achieved by immunoprecipitation from fractions containing the putative large complex (Twiddy et al. 2004; Feoktistova et al. 2011; Tenev et al. 2011).

Gel filtration methods require a chromatographic pump system and appropriate gel filtration column. The apoptosome complex can be characterized using two different gel filtration columns: the Sephacryl S-300 (Cain et al. 1999) and the Superose 6 10/300 GL (Cain et al. 2000). The difference in the resolution capacity of both columns is apparent in the high molecular mass range, as only the Superose 6 column can separate the ~700-kDa apoptosome from the larger 1.4-MDa inactive apoptosome (Cain et al. 1999, 2000).

While the Superose 6 column is ideal for routine analysis of apoptosome complex formation and activity, the running conditions for the two columns are important, as the Sephacryl S-300 column can be run without NaCl in the buffer. Under salt-free conditions, caspase-3 and -7 co-elute with the apoptosome complex and can then be detected by performing DEVDase assays and western blotting directly on the column fractions (Cain et al. 1999). In contrast, the Superose 6 column requires at least 50 mM NaCl to separate proteins correctly, as lack of salt will cause anomalous elution of protein standards and target proteins. The addition of NaCl to the column buffer results in loss of caspase-3 and -7 from the apoptosome complex, and the presence of the active apoptosome complex can then only be shown by use of a fluorimetric bioassay (Cain et al. 2000). The bioassay measures the caspase cleavage activity of the Apaf-1/caspase-9 apoptosome complex, which elutes on the column at ~700–1000 kDa. Briefly, procaspases-3 and -7 are added to the appropriate column fractions and the resulting DEVDase activity from active (cleaved) caspases-3 and -7 is measured fluorimetrically as described.

RECIPES

Caspase Activity Assay Buffer

100 mM HEPES-KOH
10% sucrose
0.1% CHAPS
10 mM dithiothreitol (DTT)

Adjust to pH 7.0. Store aliquots at −20°C.

Column Buffer for Sephacryl S-300 (Salt-Free)

5% (w/v) sucrose
0.1% (w/v) CHAPS
20 mM HEPES
5 mM dithiothreitol (DTT)

Adjust to pH 7.5. Filter at 0.45 μm.

Column Buffer for Superose 6 (with 50 mM NaCl)

5% (w/v) sucrose
0.1% (w/v) CHAPS
20 mM HEPES
5 mM dithiothreitol (DTT)

50 mM NaCl

Adjust to pH 7.5. Filter at 0.45 µm.

PIPES Buffer

50 mM PIPES/KOH
2 mM EDTA
0.1% (w/v) CHAPS
5 mM dithiothreitol (DTT)
10 µg/mL aprotinin
20 µg/mL leupeptin
10 µg/mL pepstatin A
2 mM phenylmethylsulfonyl fluoride (PMSF)

Adjust to pH 6.5.

SDS-PAGE Sample Loading Buffer (4×)

250 mM Tris–HCl (pH 6.8)
8% (w/v) sodium dodecyl sulfate (SDS)
0.2% (w/v) bromophenol blue
40% (v/v) glycerol
20% (v/v) β-mercaptoethanol

ACKNOWLEDGMENTS

We thank past and present members of the laboratory for their contributions to developing the current protocol. We acknowledge the Medical Research Council (UK) for funding our work.

REFERENCES

Cain K, Brown DG, Langlais C, Cohen GM. 1999. Caspase activation involves the formation of the aposome, a large (approximately 700 kDa) caspase-activating complex. *J Biol Chem* **274**: 22686–22692.

Cain K, Bratton SB, Langlais C, Walker G, Brown DG, Sun XM, Cohen GM. 2000. Apaf-1 oligomerizes into biologically active approximately 700-kDa and inactive approximately 1.4-MDa apoptosome complexes. *J Biol Chem* **275**: 6067–6070.

Cain K, Langlais C, Sun XM, Brown DG, Cohen GM. 2001. Physiological concentrations of K^+ inhibit cytochrome *c*-dependent formation of the apoptosome. *J Biol Chem* **276**: 41985–41990.

Feoktistova M, Geserick P, Kellert B, Dimitrova DP, Langlais C, Hupe M, Cain K, MacFarlane M, Hacker G, Leverkus M. 2011. cIAPs block Ripoptosome formation, a RIP1/caspase-8 containing intracellular cell death complex differentially regulated by cFLIP isoforms. *Mol Cell* **43**: 449–463.

McStay GP, Salvesen GS, Green DR. 2008. Overlapping cleavage motif selectivity of caspases: Implications for analysis of apoptotic pathways. *Cell Death Differ* **15**: 322–331.

Tenev T, Bianchi K, Darding M, Broemer M, Langlais C, Wallberg F, Zachariou A, Lopez J, MacFarlane M, Cain K, et al. 2011. The Ripoptosome, a signaling platform that assembles in response to genotoxic stress and loss of IAPs. *Mol Cell* **43**: 432–448.

Thompson GJ, Langlais C, Cain K, Conley EC, Cohen GM. 2001. Elevated extracellular $[K^+]$ inhibits death-receptor- and chemical-mediated apoptosis prior to caspase activation and cytochrome *c* release. *Biochem J* **357**: 137–145.

Twiddy D, Brown DG, Adrain C, Jukes R, Martin SJ, Cohen GM, MacFarlane M, Cain K. 2004. Pro-apoptotic proteins released from the mitochondria regulate the protein composition and caspase-processing activity of the native Apaf-1/caspase-9 apoptosome complex. *J Biol Chem* **279**: 19665–19682.

Cite this protocol as *Cold Spring Harb Protoc*; doi:10.1101/pdb.prot087080

Activation, Isolation, and Analysis of the Death-Inducing Signaling Complex

Michelle A. Hughes, Claudia Langlais, Kelvin Cain,[1] and Marion MacFarlane[1]

MRC Toxicology Unit, Hodgkin Building, Leicester, LE1 9HN, United Kingdom

This protocol describes activation, isolation, and analysis of the CD95 (APO-1/Fas) death-inducing signaling complex (DISC) using affinity purification. Activation is achieved using a biotin-labeled anti-CD95 antibody and the native DISC complex is captured using streptavidin beads. This approach minimizes both the number of steps involved and any potential nonspecific interactions or cross-reactivity of antibodies commonly seen in immunoprecipitations using unlabeled antibodies and protein A/G beads. Composition of the isolated complex is analyzed via western blot to identify known DISC components, and dimerization-induced autocatalytic processing of procaspase-8 at the DISC can be confirmed by detection of caspase-8 cleavage products. The potential for DISC-associated caspase-8 to activate the caspase cascade can be determined by measuring caspase-8-dependent cleavage of the fluorigenic substrate Ac-IETD.AFC, or by performing a bioassay using exogenous protein substrates.

MATERIALS

It is essential that you consult the appropriate Material Safety Data Sheets and your institution's Environmental Health and Safety Office for proper handling of equipment and hazardous materials used in this protocol.

RECIPES: Please see the end of this protocol for recipes indicated by <R>. Additional recipes can be found online at http://cshprotocols.cshlp.org/site/recipes.

Reagents

Ac-IETD.AFC (caspase-8 fluorigenic substrate) (5 mM in dimethl sulfoxide) (MP Biomedicals 03AFC14005)

Store aliquots at −20°C.

Antibodies for western blotting (see Step 27)

Caspase-8 mAb (C15) (Enzo Life Sciences ALX-804-429)

Caspase-10 mAb (MBL M059-3)

FADD mAb (BD Transduction Laboratories 610400)

FAS (C-20) rabbit polyclonal antibody (Santa Cruz Biotechnology sc-715)

FLIP mAb (NF6) (Enzo Life Sciences ALX-804-428)

Antihuman CD95 antibody, biotin-labeled (1 mg/mL) (APO-1-1BT; eBioscience BMS151BT)

We previously used APO-1-3BT (Caltag-MedSystems); however, this clone is no longer available and has been replaced by APO-1-1BT.

[1]Correspondence: mm21@leicester.ac.uk; kc5@le.ac.uk

Cite this protocol as *Cold Spring Harb Protoc*; doi:10.1101/pdb.prot087098

Bioassay substrates (as needed)

Commercially available substrates include recombinant procaspase-3 (R&D Systems 731-C3) and recombinant BID (R&D Systems 846-BD).

*Alternatively, recombinant procaspase-3 can be produced in bacteria or free procaspases can be isolated from a control lysate (see Protocol: **In Vitro Assembly and Analysis of the Apoptosome Complex** [Langlais et al. 2015]).*

Caspase activity assay buffer <R> (prewarmed to 37°C)

Cells ($\sim$1 × 10^6 cells/mL in suspension, or adherent cells at $\sim$80% confluency)

This protocol can be modified for use in a range of cell types. We successfully isolated the CD95 death-inducing signaling complex (DISC) from Jurkat T cells using the biotin-labeled antihuman CD95 antibody APO-1-3BT (Hughes et al. 2009).

The number of cells required for each DISC analysis varies depending on cell type; however, in general, 50 × 10^6 suspension cells or one T175 flask of adherent cells at $\sim$80% confluency is sufficient.

DISC lysis buffer <R> (ice-cold)

Phosphate-buffered saline (PBS) (ice-cold)

Protein A (1 mg/mL) (Sigma-Aldrich P6031)

SDS–PAGE sample loading buffer (1×) <R>

SDS–PAGE sample loading buffer (10×) <R>

Streptavidin beads (magnetic or Sepharose, as preferred)

Dynabeads M-280 streptavidin (Life Technologies 11206D)

Before use (Step 18), prewash magnetic beads (60 µL per sample) three times in 0.5 mL of DISC lysis buffer per wash, placing on a magnet for 4 min between washes. Resuspend the washed beads in DISC lysis buffer (equivalent to the starting volume) before use.

Streptavidin Sepharose High Performance (GE Healthcare Life Sciences 17-5113-01)

Before use (Step 18), prewash Sepharose beads (40–50 µL per sample) three times in 0.5 mL of DISC lysis buffer per wash, centrifuging for 1 min at 100g between washes. Resuspend the washed beads to a 50% slurry in DISC lysis buffer before use.

Equipment

96-well plates for fluorimetric assay (Scientific Laboratory Supplies 161093)

Cell scraper (for adherent cells)

Centrifuge, precooled to 4°C before use

Centrifuge tubes (15- and 50-mL)

FACS analyzer (optional; see Step 4)

Fluorimeter (e.g., Wallac Victor X4), prewarmed to 37°C before use

Magnet (e.g., DynaMag; Life Technologies) (if using magnetic beads)

Microcentrifuge, precooled to 4°C before use

Microcentrifuge tubes (2-mL)

Microscope

Rotating wheel at 4°C

SDS-PAGE and western blotting equipment (e.g., Mini-PROTEAN; Bio-Rad)

Tissue culture plates (24-well) (optional; see Step 4)

Water bath (for suspension cells) or incubator (for adherent cells) at 37°C

METHOD

An outline of the method is provided in Figure 1. It may be completed in its entirety within 3 d; this includes 2–3 h for DISC activation and generation of cell lysates, 16–17 h for isolation and subsequent washing of DISC complexes, 0.5–1 h for fluorimetric analysis of caspase-8 activity, and a further 24–36 h for DISC analysis by western blotting. A small number of treatments (i.e., up to 4) are easy to handle; however, when dealing with multiple treatments or a large number of cells, organization is key! If purified DISC complexes are to be assessed by fluorimetric assay, bioassay, and western blot (see Step 23), parallel treatments and samples are required.

Unless otherwise indicated, all steps should be performed at 4°C.

Cite this protocol as *Cold Spring Harb Protoc*; doi:10.1101/pdb.prot087098

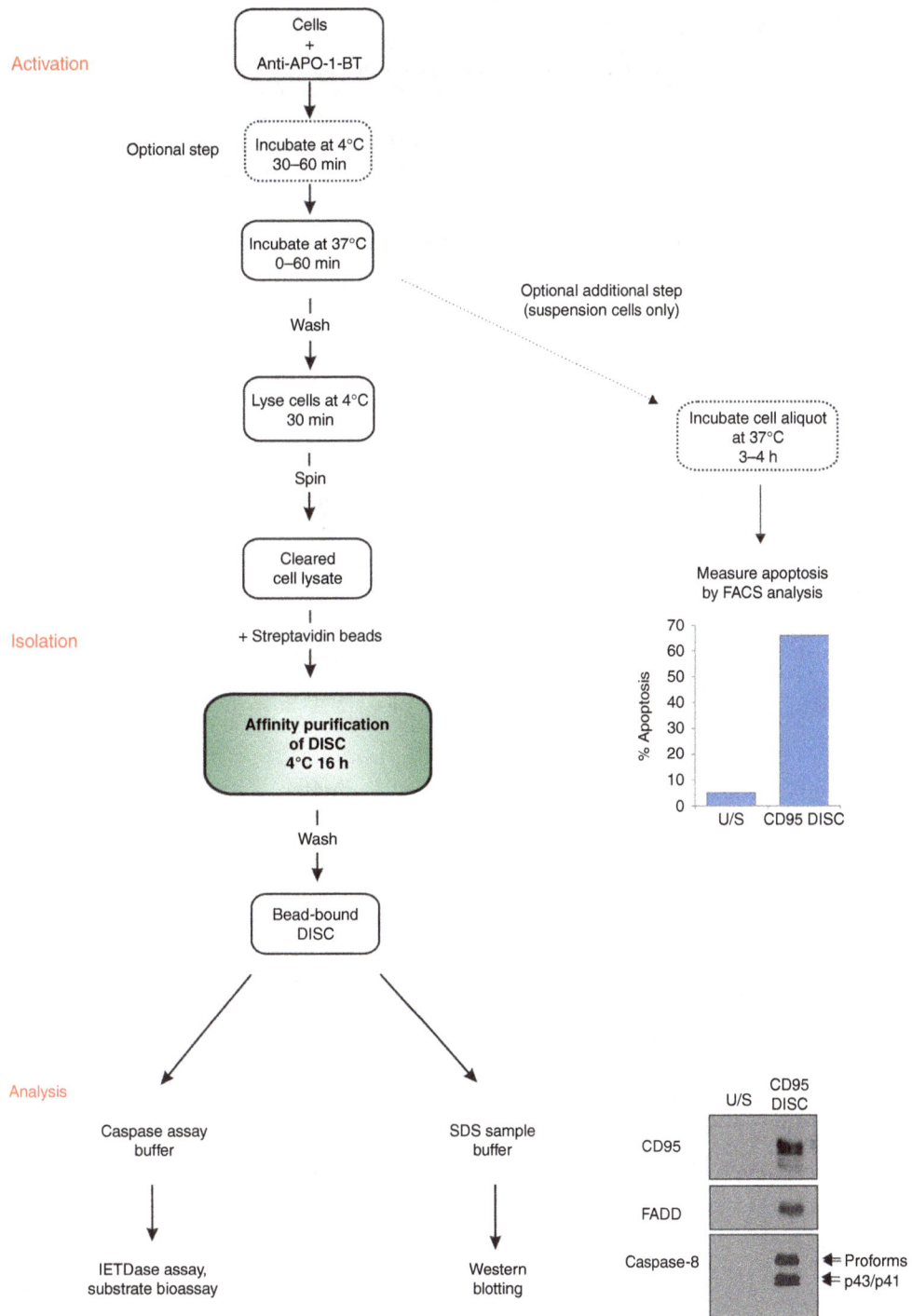

FIGURE 1. Affinity purification of the CD95 DISC. A step-by-step flowchart outlining isolation and analysis of the CD95 DISC. Cells are treated with biotin-labeled antihuman CD95 antibody at 4°C (optional) and then activated at 37°C. An aliquot of cells can be taken to assess DISC activation via FACS analysis of apoptosis (*right* graph). Following DISC formation, cleared cell lysates are incubated with streptavidin beads to capture the CD95-bound complex. Affinity-purified complexes (bead-bound DISC) can be separated by SDS–PAGE and analyzed via western blot using antibodies to known DISC components; shown here, the presence of CD95, FADD, and caspase-8 (including the caspase-8 p43/p41 cleavage products) in the stimulated sample (CD95 DISC). No bead-bound proteins are evident in the unstimulated control (U/S). In addition, caspase activity can be assessed via IETDase activity assay (described here) or bioassay using exogenous substrates (see Protocol 1: In Vitro Assembly and Analysis of the Apoptosome Complex [Langlais et al. 2015]).

Activating the DISC and Lysing Cells

In the following section, an unstimulated control should be prepared alongside experimental samples by adding APO-1-1BT and protein A after (rather than before) cell lysis. Receptors (both cell surface and intracellular) can then be precipitated as outlined in Steps 18–23. This control is important to assess any nonspecific binding to beads and identify any proteins preassociated with death receptors.

The conditions for DISC formation will differ for each cell type and hence will need to be optimized accordingly.

Treatment and Lysis of Suspension Cells

1. In 50 mL tubes, centrifuge $\sim$50 $\times$ 10^6 cells per treatment at 200g for 3 min at 4°C. Discard the supernatant and resuspend the cells in medium to a final density of 2 $\times$ 10^6 cells/mL (i.e., 25 mL per tube).

2. Add 10 µL of APO-1-1BT stock solution (final concentration 400 ng/mL) and 1.25 µL of protein A stock solution (final concentration 0.05 µL/mL) to each tube.

    The addition of protein A is required to cross-link the antibody.

3. (Optional) Prechill the samples by incubating on ice for 30–60 min.

    The prechill is an optional step to facilitate receptor loading.

4. Transfer the tubes to a water bath (submerged as much as possible) for 0–60 min at 37°C.

    The optimal time for DISC activation should be verified for each cell type/line by performing a time-course experiment. Successful DISC activation can also be confirmed by transferring a 0.5 mL aliquot of the activated cells to a 24-well plate, incubating for a further 4 h at 37°C, and assessing the levels of apoptosis by FACS analysis.

5. Centrifuge the activated cells at 200g for 3 min at 4°C and pour off the medium. Wash the cells twice with 10 mL of ice-cold PBS per wash, centrifuging as above between washes and keeping the cells on ice as much as possible.

6. Resuspend the cell pellets in 3–3.5 mL of ice-cold DISC lysis buffer. Incubate for 30 min on ice, shaking occasionally.

    It is important that cells are lysed thoroughly at this stage. Cell lysis can be monitored by light microscopy and, if required, may be optimized by varying lysis buffer incubation times.

7. Transfer the lysates to 2 mL tubes and centrifuge at 15,000g for 30 min at 4°C. Collect the lysates in 15 mL tubes.

8. For input controls for western blot analysis, remove 90 µL of each lysate and mix with 10 µL of 10$\times$ SDS–PAGE sample loading buffer. Store these samples at −80°C until ready to proceed with SDS–PAGE (Step 27), and proceed to Step 18 for immunoprecipitation of the DISC.

Treatment and Lysis of Adherent Cells

9. Remove the medium from one T175 flask (70%–80% confluent) per treatment, leaving behind 20 mL per flask.

10. Treat the cells with 400 ng/mL of APO-1-1BT and 0.05 µL/mL of protein A.

    See Step 2.

11. (Optional) Prechill the flask by laying it flat on a tray of ice for 30–60 min.

    See Step 3.

12. Transfer the flasks to an incubator at 37°C for the required length of time (0–60 min) to enable DISC formation.

    See Step 4.

13. Immediately transfer the flasks to ice, pour off the medium, and wash the cells three times with ice-cold PBS (gently swirling and pouring off the solution in between washes), keeping the cells on ice as much as possible.

Cite this protocol as *Cold Spring Harb Protoc*; doi:10.1101/pdb.prot087098

14. Remove any residual PBS by tipping up each flask and blotting on absorbent tissues. Add 3–3.5 mL of ice-cold DISC lysis buffer to each flask. Incubate for 30 min on ice, shaking occasionally.

15. Using a scraper, scrape the cells into the lysis buffer and transfer the lysates to 2 mL tubes.

16. Centrifuge the tubes at 15,000g for 30 min at 4°C and collect the lysates into 15 mL tubes.

17. For input controls for western blot analysis, remove 90 μL of each lysate and mix with 10 μL of 10× SDS–PAGE sample loading buffer. Store these samples at −20°C until ready to proceed with SDS–PAGE (Step 27), and proceed to Step 18 for immunoprecipitation of the DISC.

Immunoprecipitating the DISC Complex

18. Add 60 μL of prewashed streptavidin magnetic beads (or 40–50 μL of 50% prewashed streptavidin Sepharose beads) to each lysate sample.

19. Incubate the tubes on a rotating wheel overnight at 4°C (16–17 h).

20. On the next day, place the tubes on a magnet for 6 min (or centrifuge Sepharose beads at 100g for 1 min). Remove the supernatant and add 0.5 mL of ice-cold DISC lysis buffer to the beads.

21. Transfer the beads to 2 mL tubes and collect the beads on a magnet (or by centrifugation at 100g for 1 min). Remove the supernatant.

    Care should be taken when removing the supernatant at this step, as it is very easy to lose beads.

22. Wash the beads three times with 0.5 mL of ice-cold DISC lysis buffer per wash (separating on a magnet or centrifuging in between washes as above).

23. Resuspend the beads according to the following.

    i. To analyze caspase-8 activity by fluorimetric assay, resuspend the beads in 30–50 μL of prewarmed caspase activity assay buffer and proceed to the IETDase assay (Steps 24–26) or to a bioassay using exogenous substrates (see Protocol 1: In Vitro Assembly and Analysis of the Apoptosome Complex [Langlais et al. 2015]).

       The necessary volume of caspase assay buffer depends on the bed volume of the beads; a total volume of 50 μL is required for the IETDase assay.

       Fluorimetric analysis of caspase-8 activity must be performed immediately after DISC isolation.

    ii. To analyze samples by SDS–PAGE and western blotting, resuspend the beads in 40–50 μL of 1× SDS–PAGE sample loading buffer and proceed to Step 27.

       Beads may be stored in 1× SDS–PAGE sample loading buffer at −80°C before SDS–PAGE and western blotting.

Performing Fluorimetric Analysis (IETDase Assay) of DISC-Associated Caspase-8 Activity

For analysis of caspase-8 activity in the DISC complex, a bioassay using exogenous substrates can be performed as described in Protocol: In Vitro Assembly and Analysis of the Apoptosome Complex (Langlais et al. 2015), or caspase-8 activity can be monitored by measuring cleavage of the fluorigenic substrate Ac-IETD.AFC, as described here.

24. Transfer each sample of resuspended beads in caspase activity assay buffer (Step 23.i) to one well of a 96-well plate.

25. Immediately before use, dilute the Ac-IETD.AFC stock solution (5 mM) to 53 μM with prewarmed caspase activity assay buffer. Add 150 μL to each sample well for a final Ac-IETD.AFC concentration of 40 μM.

    Dilute only as much Ac-IETD.AFC as you will use immediately; 1 mL of a 53 μM solution is sufficient for six samples.

26. Immediately monitor the IETDase activity continuously at 37°C on a fluorimeter using excitation/emission wavelengths of 405/510 nm.

    We typically perform 50 reads with a pause of 3 sec after each set of readings.

The IETDase activity of the stimulated sample should be significantly higher than that of the unstimulated sample; typical values are 1–10 pmol/min (Hughes et al. 2013).

See Troubleshooting.

Analyzing Samples by SDS–PAGE and Western Blotting

27. Analyze the input controls (Steps 8/17) and bead-bound samples (Step 23.ii) by SDS–PAGE followed by western blotting for known DISC components (e.g., FADD and caspase-8).

    A typical DISC analyzed by SDS–PAGE and western blotting should show recruitment of FADD and pro-caspase-8 to the stimulated sample only (Fig. 1). Cleavage of caspase-8 to its p43/p41 fragments should also be evident, and if a significant amount of DISC is formed, its catalytically active p18 subunit may be observed. Similarly, the presence of other known DISC components such as FLIP$_{L/S}$ and caspase-10 can be detected.

    See Troubleshooting.

TROUBLESHOOTING

Problem (Step 26): The IETDase activity of caspase-8 at the DISC is very low (i.e., similar to the unstimulated control).

Solution: DISC activation conditions should be optimized. Alternatively, more cells may be required; in this case, the number of cells/flasks and volume of lysis buffer/beads may need to be scaled up for each treatment.

Problem (Step 27): There is little or no FADD or caspase-8 at the DISC by western blotting.

Solution: This may indicate inefficient DISC formation. It may be necessary to prechill the sample and perform a time course at 37°C to optimize DISC formation (e.g., 45 min at 4°C followed by 5, 15, 30, and 60 min at 37°C). This requires a single treatment with antihuman CD95 per time point.

RELATED TECHNIQUES

Isolation of the TRAIL DISC can also be performed using recombinant biotin-labeled TRAIL (produced as a soluble protein in *Escherichia coli* as detailed elsewhere [Harper and MacFarlane 2008]) using a method routinely performed in our laboratory (Harper et al. 2001, 2003a, 2003b; MacFarlane et al. 2002, 2005; Dickens et al. 2012; Robinson et al. 2012; Hughes et al. 2013).

RECIPES

Caspase Activity Assay Buffer

100 mM HEPES-KOH
10% sucrose
0.1% CHAPS
10 mM dithiothreitol (DTT)
Adjust to pH 7.0. Store aliquots at −20°C.

DISC Lysis Buffer

30 mM Tris/HCl (pH 7.5)
150 mM NaCl
10% glycerol
1% Triton X-100
Complete protease inhibitors (Roche)
Store buffer at 4°C.

Cite this protocol as *Cold Spring Harb Protoc*; doi:10.1101/pdb.prot087098

SDS-PAGE Sample Loading Buffer (1×)

62.5 m$_M$ Tris/HCl (pH 6.8)
2% (w/v) SDS
0.05% (w/v) bromophenol blue
10% (v/v) glycerol
5% (v/v) β-mercaptoethanol

SDS-PAGE Sample Loading Buffer (10×)

625 m$_M$ Tris–HCl (pH 6.8)
20% (w/v) sodium dodecyl sulfate (SDS)
0.5% (w/v) bromophenol blue
10% (v/v) glycerol
0.5% (v/v) β-mercaptoethanol

ACKNOWLEDGMENTS

We thank past and present members of the laboratory for their contributions to developing the current protocol. We acknowledge the Medical Research Council (UK) for funding our work.

REFERENCES

Dickens LS, Boyd RS, Jukes-Jones R, Hughes MA, Robinson GL, Fairall L, Schwabe JW, Cain K, MacFarlane M. 2012. A death effector domain chain DISC model reveals a crucial role for caspase-8 chain assembly in mediating apoptotic cell death. *Mol Cell* **47:** 291–305.

Harper N, Farrow SN, Kaptein A, Cohen GM, MacFarlane M. 2001. Modulation of tumor necrosis factor apoptosis-inducing ligand-induced NF-κB activation by inhibition of apical caspases. *J Biol Chem* **276:** 34743–34752.

Harper N, Hughes M, MacFarlane M, Cohen GM. 2003a. Fas-associated death domain protein and caspase-8 are not recruited to the tumor necrosis factor receptor 1 signaling complex during tumor necrosis factor-induced apoptosis. *J Biol Chem* **278:** 25534–25541.

Harper N, Hughes MA, Farrow SN, Cohen GM, MacFarlane M. 2003b. Protein kinase C modulates tumor necrosis factor-related apoptosis-inducing ligand-induced apoptosis by targeting the apical events of death receptor signaling. *J Biol Chem* **278:** 44338–44347.

Harper N, MacFarlane M. 2008. Recombinant TRAIL and TRAIL receptor analysis. *Methods Enzymol* **446:** 293–313.

Hughes MA, Harper N, Butterworth M, Cain K, Cohen GM, MacFarlane M. 2009. Reconstitution of the death-inducing signaling complex reveals a substrate switch that determines CD95-mediated death or survival. *Mol Cell* **35:** 265–279.

Hughes MA, Langlais C, Cain K, MacFarlane M. 2013. Isolation, characterisation and reconstitution of cell death signalling complexes. *Methods* **61:** 98–104.

Langlais C, Hughes MA, Cain K, MacFarlane M. 2015. In vitro assembly and analysis of the apoptosome complex. *Cold Spring Harb Protoc* doi:10.1101/pdb.prot087080.

MacFarlane M, Harper N, Snowden RT, Dyer MJ, Barnett GA, Pringle JH, Cohen GM. 2002. Mechanisms of resistance to TRAIL-induced apoptosis in primary B cell chronic lymphocytic leukaemia. *Oncogene* **21:** 6809–6818.

MacFarlane M, Kohlhaas SL, Sutcliffe MJ, Dyer MJ, Cohen GM. 2005. TRAIL receptor-selective mutants signal to apoptosis via TRAIL-R1 in primary lymphoid malignancies. *Cancer Res* **65:** 11265–11270.

Robinson GL, Dinsdale D, MacFarlane M, Cain K. 2012. Switching from aerobic glycolysis to oxidative phosphorylation modulates the sensitivity of mantle cell lymphoma cells to TRAIL. *Oncogene* **31:** 4996–5006.

CHAPTER 5

Biochemical Assays of the Proapoptotic Proteins Bak and Bax

Grant Dewson[1]

Cell Signalling and Cell Death Division, Walter and Eliza Hall Institute of Medical Research, Parkville, Melbourne, Victoria 3052, Australia; Department of Medical Biology, The University of Melbourne, Melbourne, Victoria 3010, Australia

The Bcl-2 family of proteins are the key mediators of apoptotic cell death in response to diverse stimuli including hypoxia, DNA damage, and growth factor withdrawal. A myriad of interactions between these proteins following a death stimulus ultimately determines the activity of Bak and Bax, the two pivotal effectors of apoptosis. Normally dormant in a healthy cell, activation of Bak and Bax is characterized by conformation change, subcellular redistribution (for Bax), and oligomerization. Assays have been developed to interrogate each of these steps on the path to mitochondrial outer membrane damage and cell death. This discussion introduces biochemical assays for investigating key points in Bak/Bax apoptotic function, with a focus on subcellular localization, conformation change, and oligomerization.

INTRODUCTION

The Bcl-2 family of proteins is the pivotal regulator of intrinsic cell death in response to multiple death stimuli. These include limiting growth factor, detachment from the extracellular matrix, and DNA damage. The family is critical for maintaining tissue homeostasis during embryonic development and immunological responses, and for restricting tumor growth (Youle and Strasser 2008). Its ultimate role is to regulate activity of Bak and Bax (Wei et al. 2001), the two key effector proteins responsible for perforating the mitochondrial outer membrane (MOM). The consequences of this attack are twofold. Paramount is prevention of the mitochondria from fulfilling their role as the powerhouses of the cell, such that the cell is ultimately destined to die. Second, breach of the outer membrane leads to the release of apoptogenic factors, including cytochrome *c* and Smac/DIABLO (Du et al. 2000; Verhagen et al. 2000). These factors facilitate the activation of caspases which allow the cell to be efficiently packaged and cleared in an immunologically "silent" fashion.

Bak and Bax have received significant research attention. It is understood that both are found in their inactive form before an apoptotic stimulus (see the review by Westphal et al. 2011). Following cellular stress, they become activated by an as yet undetermined mechanism that may involve direct interaction with a proapoptotic subclass of the Bcl-2 family, the BH3-only proteins. Following activation, Bak and Bax undergo changes in conformation which culminate in metamorphosis into their activated conformer. For Bax, this also involves subcellular redistribution from the cytosol to join Bak at the MOM (Wolter et al. 1997). The conformation changes allow Bak and Bax to self-associate to form potentially large molecular mass complexes that damage the MOM (Antonsson et al.

---

[1]Correspondence: dewson@wehi.edu.au

Cite this introduction as *Cold Spring Harb Protoc*; doi:10.1101/pdb.top070334

2001; Zhou and Chang 2008). Once Bak and/or Bax have oligomerized, the cell is on an inexorable path toward death.

ASSAYING ACTIVATION OF Bak AND Bax

The following protocols describe simple yet effective assays (requiring little prior experience) for interrogating three key steps in the activation process of Bak and Bax in apoptotic cells, isolated mitochondria or primary tissue:

1. Subcellular redistribution (Bax)

   Protocol 1: Investigating Bax Subcellular Localization and Membrane Integration (Dewson 2015a)

2. Activating conformation change

   Protocol 2: Investigating Bak/Bax Activating Conformation Change by Immunoprecipitation (Dewson 2015b)
   Protocol 3: Detection of Bak/Bax Activating Conformation Change by Intracellular Flow Cytometry (Dewson 2015c)

3. Oligomerization

   Protocol 4: Investigating the Oligomerization of Bak and Bax during Apoptosis by Cysteine Linkage (Dewson 2015d)
   Protocol 5: Blue Native PAGE and Antibody Gel Shift to Assess Bak and Bax Conformation Change and Oligomerization (Dewson 2015e)

ACKNOWLEDGMENTS

G.D. is supported by the National Health and Medical Research Council of Australia (637335), Australian Research Council (FT100100791), and the Association for International Cancer Research (10-230). The present work was made possible through Victorian State Government Operational Infrastructure Support and Australian Government NHMRC IRIISS.

REFERENCES

Antonsson B, Montessuit S, Sanchez B, Martinou JC. 2001. Bax is present as a high molecular weight oligomer/complex in the mitochondrial membrane of apoptotic cells. *J Bio Chem* **276:** 11615–11623.

Dewson G. 2015a. Investigating Bax subcellular localization and membrane integration. *Cold Spring Harb Protoc* doi: 10.1101/pdb.prot086447.

Dewson G. 2015b. Investigating Bak/Bax activating conformation change by immunoprecipitation. *Cold Spring Harb Protoc* doi: 10.1101/pdb.prot086454.

Dewson G. 2015c. Detection of Bak/Bax activating conformation change by intracellular flow cytometry. *Cold Spring Harb Protoc* doi: 10.1101/pdb.prot086462.

Dewson G. 2015d. Investigating the oligomerization of Bak and Bax during apoptosis by cysteine linkage. *Cold Spring Harb Protoc* doi: 10.1101/pdb.prot086470.

Dewson G. 2015e. Blue native PAGE and antibody gel shift to assess Bak and Bax conformation change and oligomerization. *Cold Spring Harb Protoc* doi: 10.1101/pdb.prot086488.

Du C, Fang M, Li Y, Li L, Wang X. 2000. Smac, a mitochondrial protein that promotes cytochrome *c*-dependent caspase activation by eliminating IAP inhibition. *Cell* **102:** 33–42.

Verhagen AM, Ekert PG, Pakusch M, Silke J, Connolly LM, Reid GE, Moritz RL, Simpson RJ, Vaux DL. 2000. Identification of DIABLO, a mammalian protein that promotes apoptosis by binding to and antagonizing inhibitor of apoptosis (IAP) proteins. *Cell* **102:** 43–53.

Wei MC, Zong WX, Cheng EH, Lindsten T, Panoutsakopoulou V, Ross AJ, Roth KA, MacGregor GR, Thompson CB, Korsmeyer SJ. 2001. Proapoptotic BAX and BAK: A requisite gateway to mitochondrial dysfunction and death. *Science* **292:** 727–730.

Westphal D, Dewson G, Czabotar PE, Kluck RM. 2011. Molecular biology of Bax and Bak activation and action. *Biochim Biophys Acta* **1813:** 521–531.

Wolter KG, Hsu YT, Smith CL, Nechushtan A, Xi XG, Youle RJ. 1997. Movement of Bax from the cytosol to mitochondria during apoptosis. *J Cell Biol* **139:** 1281–1292.

Youle RJ, Strasser A. 2008. The BCL-2 protein family: Opposing activities that mediate cell death. *Nat Rev Mol Cell Biol* **9:** 47–59.

Zhou L, Chang DC. 2008. Dynamics and structure of the Bax-Bak complex responsible for releasing mitochondrial proteins during apoptosis. *J Cell Sci* **121:** 2186–2196.

Cite this introduction as *Cold Spring Harb Protoc*; doi:10.1101/pdb.top070334

Investigating Bax Subcellular Localization and Membrane Integration

Grant Dewson[1]

Cell Signalling and Cell Death Division, Walter and Eliza Hall Institute of Medical Research, Parkville, Melbourne, Victoria 3052, Australia; Department of Medical Biology, The University of Melbourne, Melbourne, Victoria 3010, Australia

Bax is a pivotal effector of apoptosis responsible for permeabilization of the mitochondrial outer membrane (MOM). A key event in mitochondrial damage is the translocation of Bax from the cytosol to the MOM. A simple and effective method for assessing the cytosol vs. mitochondrial localization of Bax is digitonin fractionation, which uses a low concentration of detergent to permeabilize the plasma membrane without damaging intracellular membranes. This allows separation of the cytosol (light membranes) from the heavy membranes (with mitochondria and nuclei) by centrifugation. Localization of Bax can then be assessed by immunoblotting. To further differentiate membrane-integrated Bax from that which is peripherally associated, carbonate extraction of the membrane fraction can be performed before immunoblotting. Treatment of membranes at high pH disrupts protein–protein interactions, whereas protein–lipid interactions are largely retained, although membrane integrity is lost.

MATERIALS

It is essential that you consult the appropriate Material Safety Data Sheets and your institution's Environmental Health and Safety Office for proper handling of equipment and hazardous materials used in this protocol.

RECIPES: Please see the end of this protocol for recipes indicated by <R>. Additional recipes can be found online at http://cshprotocols.cshlp.org/site/recipes.

Reagents

Apoptotic stimulus

Carbonate extraction buffer (0.1 M Na_2CO_3, pH 11.5)

Digitonin

Immediately before use, prepare a fresh solution of 10% (w/v) digitonin in H_2O and boil to dissolve. Alternatively, resuspend in DMSO for long-term storage.

DNase I (Roche)

HCl (0.1 M)

Mouse embryonic fibroblasts (MEF) or cell line of interest, pretreated with a broad-range caspase inhibitor

[1]Correspondence: dewson@wehi.edu.au

Cite this protocol as *Cold Spring Harb Protoc*; doi:10.1101/pdb.prot086447

This protocol was validated in HeLa cells, but has been optimized for MEF. The number of cells required will depend on their expression of Bax, but a minimum of 1×10^6 cells is recommended as a starting point.
In cells that will be treated with a death stimulus, caspases should first be blocked with a broad-range caspase inhibitor, such as qVD-OPh or z-VAD-fmk (Enzyme Systems).

Nuclease buffer (10×) <R>

Permeabilization buffer <R>

Phosphate-buffered saline (PBS) (1×; Ca^{2+}- and Mg^{2+}-free) (ice-cold) <R>

Protease inhibitor cocktail tablets, without EDTA (complete; Roche) (1 tablet/50 mL)

SDS–PAGE and immunoblotting reagents, including primary antibodies for Bax, Bak, and/or cytochrome *c*

Bax translocates from the cytosol to mitochondria during apoptosis, whereas Bak is constitutively integrated into the MOM (Wolter et al. 1997; Griffiths et al. 1999); thus, immunoblotting for Bak serves as a useful positive control for membrane integration. To correlate Bax translocation with mitochondrial damage, subcellular fractions may also be immunoblotted for cytochrome c (Waterhouse et al. 2004).

SDS–PAGE sample buffer (reducing) (2×) <R>

Trypan blue (BioRad)

Equipment

Centrifuge (benchtop) at 4°C

Centrifuge tubes

Heat blocks at 37°C and 95°C

Hemocytometer

Microscope (transmitted light)

SDS–PAGE apparatus

Western transfer apparatus

METHOD

An overview of digitonin fractionation is provided in Figure 1. Depending on the number of samples, this protocol can be completed in ~45 min.

Inducing Apoptosis and Permeabilizing Cells

1. Treat cells with an apoptotic stimulus as required.

2. Harvest the cells by centrifugation at 2500*g* for 5 min. Wash once in 1 mL of ice-cold 1× PBS. Perform a cell count using a hemocytometer.

3. Centrifuge the cells at 2500*g* for 5 min at 4°C. Discard the supernatant and gently resuspend the cell pellet in permeabilization buffer supplemented with 0.025% digitonin and protease inhibitors at 1×10^7 cells per mL.

   Cells should be resuspended gently but thoroughly by pipetting up and down several times with a P1000 pipette, avoiding generation of bubbles. Disaggregation of "clumpy" cells is crucial for the efficiency of digitonin permeabilization.

4. Incubate the cells on ice for 10 min.

5. Verify permeabilization by trypan blue uptake.

   i. Remove 5 µL of permeabilized cells and add an equal volume of trypan blue.

   ii. Assess trypan blue uptake by light microscopy.

      95%–100% of cells should appear blue. If processing a large number of samples it is not necessary to confirm permeabilization of all samples, although verification of both untreated cells and cells treated with an apoptotic stimulus is recommended.

Cite this protocol as *Cold Spring Harb Protoc*; doi:10.1101/pdb.prot086447

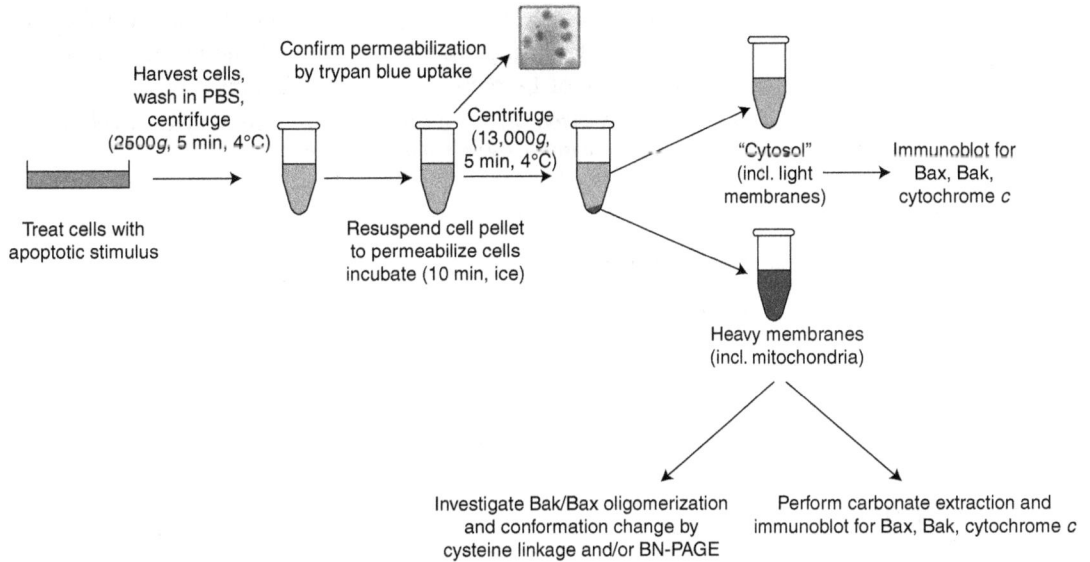

FIGURE 1. Digitonin fractionation for investigation of Bax subcellular localization and membrane integration. Recent evidence indicates there is a reversible association of Bax with mitochondria, and prosurvival proteins also show differential distribution and relocalize during apoptosis (Hsu et al. 1997; Kaufmann et al. 2003; Edlich et al. 2011). Bax subcellular localization and membrane integration in apoptotic cells can be assessed using digitonin fractionation followed by carbonate extraction of the membrane fraction and immunoblotting, as described here. The heavy membrane fraction can also be used for investigation of Bak and Bax oligomerization and/or conformation change; see Protocol 4: Investigating the Oligomerization of Bak and Bax during Apoptosis by Cysteine Linkage (Dewson 2015a) and Protocol 6: Blue Native PAGE and Antibody Gel Shift to Assess Bak and Bax Conformation Change and Oligomerization (Dewson 2015b).

6. Centrifuge the cells at 13,000g for 5 min at 4°C to separate the supernatant (cytosol) from the pellet (heavy membranes including mitochondria).

7. Transfer the supernatant (cytosolic fraction) to a new tube and add an equal volume of 2× reducing SDS–PAGE sample buffer. Proceed with the membrane fraction pellet as follows.

- If membrane integration is not to be assessed, resuspend the pellet in the same total volume of 1× reducing SDS–PAGE sample buffer as used for the cytosolic fraction and proceed directly to Step 13.

- To assess membrane integration using carbonate extraction, proceed directly to Step 8.

Performing Carbonate Extraction

Carbonate extraction of the membrane fraction of digitonin-permeabilized cells (with nuclei) is very messy due to rupture of the nuclear membrane, which causes samples to become viscous. To avoid this problem, mitochondria can be further purified from nuclei by manual cell disruption and differential centrifugation. However, the ease and final yield depends on cell type. A more straightforward method is incorporation of a DNase step before separation of the fractions, as described here.

8. Resuspend the membrane pellet in carbonate extraction buffer at 2×10^7 cells per mL. Incubate on ice for 20 min.

9. Neutralize the solution with 1/3 volume of 0.1 M HCl. Incubate for 5 min at room temperature.

10. Add 1/10 volume of 10× nuclease buffer and 1 unit of DNase I. Incubate for 10 min at 37°C.

11. Centrifuge the sample at 13,000g for 10 min.

12. Transfer the supernatant (carbonate-sensitive fraction) to a new tube. Add an equal volume of 2× reducing SDS–PAGE sample buffer. Resuspend the pellet (carbonate-resistant fraction) in the same total volume of 1× reducing SDS–PAGE sample buffer.

      If the pellet sample is too viscous, it can be resuspended in a greater volume of sample buffer. Be sure to adjust the volume of the cytosolic fraction accordingly.

Immunoblotting

13. Heat all samples for 5 min at 95°C.

14. Load the same volume of each fraction for SDS–PAGE and perform immunoblotting for Bax, Bak, and cytochrome *c* as needed.

      Volumes should be adjusted for dilution of the supernatant and membrane samples by HCl.
      See Troubleshooting.

TROUBLESHOOTING

Problem (Step 14): It is difficult to detect the protein of interest in apoptotic samples by immunoblotting.

Solution: It is important to block activated caspases during apoptosis, as they will lead to cellular destruction and loss of cells and proteins (including release of cytochrome *c* to the cytosol). qVD-OPh and z-VAD-fmk are two broad range caspase inhibitors commonly used in cell culture. Alternatively, a shorter time-point for apoptosis induction can be tested.

Problem (Step 14): Cytochrome *c* is released in nonapoptotic samples.

Solution: To maintain the integrity of the mitochondria during the fractionation process, adjust the sugar and salt concentrations of the permeabilization buffer to ensure it is norm-osmotic (300 mOsm).

RELATED TECHNIQUES

Membrane fractions can be treated with chemical cross-linkers or exogenous oxidant to monitor Bak/Bax oligomerization in the MOM during apoptosis by cysteine linkage; see Protocol 4: Investigating the Oligomerization of Bak and Bax during Apoptosis by Cysteine Linkage (Dewson 2015a).

RECIPES

Nuclease Buffer (10×)

100 mM Tris–HCl, pH 7.5
25 mM $MgCl_2$
1 mM $CaCl_2$

Permeabilization Buffer

20 mM HEPES/KOH, pH 7.5
250 mM sucrose
50 mM KCl
2.5 mM $MgCl_2$

          Cite this protocol as *Cold Spring Harb Protoc*; doi:10.1101/pdb.prot086447

Phosphate-Buffered Saline (PBS)

Reagent	Amount to add (for 1× solution)	Final concentration (1×)	Amount to add (for 10× stock)	Final concentration (10×)
NaCl	8 g	137 mM	80 g	1.37 M
KCl	0.2 g	2.7 mM	2 g	27 mM
Na_2HPO_4	1.44 g	10 mM	14.4 g	100 mM
KH_2PO_4	0.24 g	1.8 mM	2.4 g	18 mM

If necessary, PBS may be supplemented with the following:

$CaCl_2 \cdot 2H_2O$	0.133 g	1 mM	1.33 g	10 mM
$MgCl_2 \cdot 6H_2O$	0.10 g	0.5 mM	1.0 g	5 mM

PBS can be made as a 1× solution or as a 10× stock. To prepare 1 L of either 1× or 10× PBS, dissolve the reagents listed above in 800 mL of H_2O. Adjust the pH to 7.4 (or 7.2, if required) with HCl, and then add H_2O to 1 L. Dispense the solution into aliquots and sterilize them by autoclaving for 20 min at 15 psi (1.05 kg/cm^2) on liquid cycle or by filter sterilization. Store PBS at room temperature.

SDS–PAGE Sample Buffer (Reducing) (2×)

125 mM Tris–HCl, pH 6.8
20% glycerol
4% SDS
0.1% bromophenol blue
5% β-mercaptoethanol

ACKNOWLEDGMENTS

G.D. is supported by the National Health and Medical Research Council of Australia (637335), Australian Research Council (FT100100791), and the Association for International Cancer Research (10–230). The present work was made possible through Victorian State Government Operational Infrastructure Support and Australian Government NHMRC IRIISS.

REFERENCES

Dewson G. 2015a. Investigating the oligomerization of Bak and Bax during apoptosis by cysteine linkage. *Cold Spring Harb Protoc* doi: 10.1101/pdb.prot086470.

Dewson G. 2015b. Blue native PAGE and antibody gel shift to assess Bak and Bax conformation change and oligomerization. *Cold Spring Harb Protoc* doi: 10.1101/pdb.prot086488.

Edlich F, Banerjee S, Suzuki M, Cleland MM, Arnoult D, Wang C, Neutzner A, Tjandra N, Youle RJ. 2011. Bcl-x_L Retrotranslocates Bax from the mitochondria into the cytosol. *Cell* 145: 104–116.

Griffiths GJ, Dubrez L, Morgan CP, Jones NA, Whitehouse J, Corfe BM, Dive C, Hickman JA. 1999. Cell damage-induced conformational changes of the pro-apoptotic protein Bak in vivo precede the onset of apoptosis. *J Cell Biol* 144: 903–914.

Hsu Y-T, Wolter KG, Youle RJ. 1997. Cytosol-to-membrane redistribution of Bax and Bcl-X_L during apoptosis. *Proc Natl Acad Sci* 94: 3668–3672.

Kaufmann T, Schlipf S, Sanz J, Neubert K, Stein R, Borner C. 2003. Characterization of the signal that directs Bcl-x_L, but not Bcl-2, to the mitochondrial outer membrane. *J Cell Biol* 160: 53–64.

Waterhouse NJ, Steel R, Kluck R, Trapani JA. 2004. Assaying cytochrome C translocation during apoptosis. *Methods Mol Biol* 284: 307–313.

Wolter KG, Hsu YT, Smith CL, Nechushtan A, Xi XG, Youle RJ. 1997. Movement of Bax from the cytosol to mitochondria during apoptosis. *J Cell Biol* 139: 1281–1292.

Investigating Bak/Bax Activating Conformation Change by Immunoprecipitation

Grant Dewson[1]

Cell Signalling and Cell Death Division, Walter and Eliza Hall Institute of Medical Research, Parkville, Melbourne, Victoria 3052, Australia; Department of Medical Biology, The University of Melbourne, Melbourne, Victoria 3010, Australia

Activation of both Bax and Bak during apoptosis involves significant conformation change. Investigation of this phenomenon by immunoprecipitation (IP) requires a detergent such as CHAPS that does not induce significant conformation change. IP with conformation-specific Bax or Bak antibodies is observed in CHAPS only following an apoptotic stimulus, whereas the same antibodies will immunoprecipitate from both nonapoptotic and apoptotic cells in the presence of Triton X-100. Thus, the latter detergent can serve as a positive control for IP, as described here.

MATERIALS

It is essential that you consult the appropriate Material Safety Data Sheets and your institution's Environmental Health and Safety Office for proper handling of equipment and hazardous materials used in this protocol.

RECIPES: Please see the end of this protocol for recipes indicated by <R>. Additional recipes can be found online at http://cshprotocols.cshlp.org/site/recipes.

Reagents

Antibodies

Antibodies for immunoprecipitation (IP), conformation-specific (for Bak and/or Bax; see Table 1)

Antibody for immunoblotting (Jackson ImmunoResearch) (raised in a different species from IP antibodies)

Apoptotic stimulus

Cell line of interest (e.g., mouse embryonic fibroblasts [MEF] or HeLa), pretreated with a broad-range caspase inhibitor

Samples should be equilibrated using either cell number (2×10^6 is sufficient to immunoprecipitate endogenous Bak/Bax from MEF or HeLa cells) or protein concentration (0.5 mg protein per IP). The number of cells required will depend on cell type and Bak/Bax expression levels.

In cells that will be treated with a death stimulus, caspases should first be blocked with a broad-range caspase inhibitor, such as qVD-OPh or z-VAD-fmk (Enzyme Systems).

Lysis buffer for IP (ice-cold) <R>

Lysis buffer can be prepared using 1% CHAPS (preferred) or 1% digitonin; see Discussion. Lysis buffer prepared with 1% Triton X-100 is used as a positive control for IP.

---

[1]Correspondence: dewson@wehi.edu.au

Cite this protocol as *Cold Spring Harb Protoc*; doi:10.1101/pdb.prot086454

TABLE 1. Commonly used conformation-specific Bak or Bax antibodies

	Antibody (species reactivity)	Species	Source	Epitope/ immunogen	References
Bax	6A7 (h, m, r)	Mouse	Santa Cruz Biotech	aa12–24	Hsu and Youle 1998; Nechushtan et al. 1999
	NT (h, m)	Rabbit	BD Biosciences	aa1–21	Vogel et al. 2012
	N20 (h, m, r)	Rabbit	Santa Cruz Biotech	Amino terminus	Khaled et al. 1999
	Clone 3 (h)	Mouse	BD Biosciences	aa55–178	Dewson et al. 2003
Bak	Ab-1 (h)	Mouse	Millipore	Amino terminus	Griffiths et al. 1999
	NT (h, m, r)	Rabbit	Millipore	aa23–38	Oberle et al. 2010
	22–38 (h,m)	Rabbit	Sigma-Aldrich	aa23–38	

h, human; m, mouse; r, rat.

Phosphate-buffered saline (PBS) ($1\times$; Ca^{2+}- and Mg^{2+}-free) (ice-cold) <R>

Protease inhibitor cocktail tablets, without EDTA (complete; Roche) (1 tablet/50 mL)

Protein G sepharose beads (GE Lifesciences)

SDS–PAGE and immunoblotting reagents

SDS–PAGE sample buffer (reducing) ($2\times$) <R>

We recommend IPs be run on 12% Tris-glycine SDS–PAGE to aid separation of Bak and Bax from IgG light chain.

Sepharose beads (Sigma-Aldrich)

Equipment

Centrifuge (benchtop) at 4°C

Centrifuge tubes

Heat block at 95°C

Rotating wheel at 4°C

SDS–PAGE apparatus

Western transfer apparatus

METHOD

To reduce nonspecific pull-down, all steps should be performed on ice or in a cold room and all buffers should be ice-cold.

1. Treat cells with an apoptotic stimulus as required.

2. Harvest the cells by centrifugation at 2500*g* for 5 min. Wash once in 1 mL of ice-cold $1\times$ PBS.

3. Split the treated cells into two tubes for lysis. Prepare an additional two tubes containing an equivalent number of untreated cells.

4. Resuspend the treated and untreated cells in 1% CHAPS or 1% Triton X-100 lysis buffer supplemented with protease inhibitors. Incubate for 30 min on ice.

   We routinely resuspend MEF in lysis buffer at 1×10^7 cells per mL.

   Lysates may be stored at $-80°C$.

5. Centrifuge the lysates at 13,000*g* for 5 min to separate the insoluble debris. Transfer the supernatant to new tubes. Set aside a sample of lysate to be used as the whole-cell control.

6. Prepare the sepharose beads as follows.

   i. Pellet the required volume of beads (50 µL of 1:1 slurry per sample) by centrifugation at 13,000*g* for 30 sec. Discard the supernatant.

      Prepare excess beads to ensure sufficient reagent for all samples.

   ii. Resuspend the beads in 1 mL of the appropriate lysis buffer.

   iii. Pellet the beads by centrifugation at 13,000*g* for 30 sec. Discard the supernatant.

     iv. Repeat Steps 6.ii and 6.iii for a total of four washes.

     v. Resuspend the beads to produce a 1:1 slurry in the appropriate lysis buffer.

7. Add 50 µL of sepharose 1:1 slurry to each sample from Step 5, inverting the beads after each transfer. Incubate the samples for 1 h on a rotating wheel at 4°C.

    For easier pipetting, we use a P200 tip with the very end cutoff to transfer the beads.

8. Pellet the beads by centrifugation at 13,000g for 30 sec. Transfer each precleared supernatant to a new tube.

9. Add 2 µg of the appropriate conformation-specific antibody to each supernatant. Incubate the samples for 2 h on a rotating wheel at 4°C.

10. During the incubation in Step 9, prepare the protein G sepharose beads as described in Step 6.

11. Add 50 µL of protein G sepharose 1:1 slurry to each sample as described in Step 7. Incubate for 1 h on a rotating wheel at 4°C.

12. Pellet the beads by centrifugation at 13,000g for 30 sec. Discard the supernatant.

13. Wash the beads in the appropriate lysis buffer as follows.

     i. Resuspend the beads in 1 mL of lysis buffer.

     ii. Pellet the beads by centrifugation at 13,000g for 30 sec. Discard the supernatant.

     iii. Repeat Steps 13.i and 13.ii for a total of four washes.

       Ensure the beads are washed in the same lysis buffer (Triton X-100 or CHAPS) as was used for the original lysis.

14. Elute the samples from the beads as follows.

     i. Add an appropriate volume of SDS–PAGE sample buffer to each sample. Gently flick to mix.
       The appropriate elution volume will depend on the expression of Bak/Bax in the cell line and the number of cells used for IP.

     ii. Heat the samples for 5 min at 95°C.

     iii. Pellet the beads by centrifugation at 13,000g for 30 sec.

15. Retrieve the supernatant and analyze by immunoblotting.
    See Troubleshooting.

TROUBLESHOOTING

Problem (Step 15): Immunoprecipitated Bak/Bax cannot be detected in the presence of IgG light chain.

Solution: Bak (24 kDa) and Bax (21 kDa) are very similar molecular mass to IgG light chain. Thus, the antibodies used for IP and subsequent immunoblotting should be raised in different species. Alternatively, an immunoglobulin heavy chain-specific secondary antibody can be used for immunoblotting. Elution under nonreducing conditions (i.e., using sample buffer with no DTT or β-mercaptoethanol) should also be considered, as the IgG light and heavy chains will migrate higher up the gel due to retention of disulphide bonds.

Problem (Step 15): Activated Bak/Bax is pulled down in nonapoptotic samples.

Solution: It is important that all steps are performed on ice or in a cold room and buffers used are ice-cold to reduce nonspecific pull-down. Ideally, Bak/Bax IP should be limited to 1–2 h rather than overnight to limit artifactual Bak/Bax conformation change postlysis. Detectable pull-down in "healthy cells" can also be due to background cell death. Depending on the cell type under investigation, this may be difficult to solve. However, Bak/Bax activation should still be apparent if the death stimulus used is Bak/Bax-dependent.

Cite this protocol as *Cold Spring Harb Protoc*; doi:10.1101/pdb.prot086454

DISCUSSION

The activation of both Bax and Bak involves significant conformation change, including exposure of amino-terminal epitopes (Hsu and Youle 1998; Griffiths et al. 1999; Dewson et al. 2008). Seminal findings by the laboratory of Richard Youle indicated that certain detergents, such as Triton X-100 and NP-40, induce conformation change in Bax (Hsu and Youle 1997); this was later confirmed for Bak (Hsu and Youle 1998). In the current protocol, the detergent of choice is CHAPS, although we have also used digitonin effectively with both Bak and Bax (Dewson et al. 2008, 2012). Because Triton X-100 induces significant conformation change in both Bax and Bak, it is used as an IP positive control.

RECIPES

Lysis Buffer for Immunoprecipitation

20 mM Tris-Cl, pH 7.4
135 mM NaCl
1.5 mM $MgCl_2$
1 mM EGTA
10% glycerol
1% Triton X-100 or 1% CHAPS or 1% digitonin

Phosphate-Buffered Saline (PBS)

Reagent	Amount to add (for 1× solution)	Final concentration (1×)	Amount to add (for 10× stock)	Final concentration (10×)
NaCl	8 g	137 mM	80 g	1.37 M
KCl	0.2 g	2.7 mM	2 g	27 mM
Na_2HPO_4	1.44 g	10 mM	14.4 g	100 mM
KH_2PO_4	0.24 g	1.8 mM	2.4 g	18 mM
If necessary, PBS may be supplemented with the following:				
$CaCl_2 \cdot 2H_2O$	0.133 g	1 mM	1.33 g	10 mM
$MgCl_2 \cdot 6H_2O$	0.10 g	0.5 mM	1.0 g	5 mM

PBS can be made as a 1× solution or as a 10× stock. To prepare 1 L of either 1× or 10× PBS, dissolve the reagents listed above in 800 mL of H_2O. Adjust the pH to 7.4 (or 7.2, if required) with HCl, and then add H_2O to 1 L. Dispense the solution into aliquots and sterilize them by autoclaving for 20 min at 15 psi (1.05 kg/cm²) on liquid cycle or by filter sterilization. Store PBS at room temperature.

SDS–PAGE Sample Buffer (Reducing) (2×)

125 mM Tris–HCl, pH 6.8
20% glycerol
4% SDS
0.1% bromophenol blue
5% β-mercaptoethanol

ACKNOWLEDGMENTS

G.D. is supported by the National Health and Medical Research Council of Australia (637335), Australian Research Council (FT100100791), and the Association for International Cancer Research (10–230). The present work was made possible through Victorian State Government Operational Infrastructure Support and Australian Government NHMRC IRIISS.

REFERENCES

Dewson G, Kratina T, Sim HW, Puthalakath H, Adams JM, Colman PM, Kluck RM. 2008. To trigger apoptosis Bak exposes its BH3 domain and homo-dimerizes via BH3:Groove interactions. *Mol Cell* **30:** 369–380.

Dewson G, Ma S, Frederick P, Hockings C, Tan I, Kratina T, Kluck RM. 2012. Bax dimerizes via a symmetric BH3:Groove interface during apoptosis. *Cell Death Differ* **19:** 661–670.

Dewson G, Snowden RT, Almond JB, Dyer MJ, Cohen GM. 2003. Conformational change and mitochondrial translocation of Bax accompany proteasome inhibitor-induced apoptosis of chronic lymphocytic leukemic cells. *Oncogene* **22:** 2643–2654.

Griffiths GJ, Dubrez L, Morgan CP, Jones NA, Whitehouse J, Corfe BM, Dive C, Hickman JA. 1999. Cell damage-induced conformational changes of the pro-apoptotic protein Bak in vivo precede the onset of apoptosis. *J Cell Biol* **144:** 903–914.

Hsu YT, Youle RJ. 1997. Nonionic detergents induce dimerization among members of the Bcl-2 family. *J Biol Chem* **272:** 13829–13834.

Hsu Y-T, Youle RJ. 1998. Bax in murine thymus is a soluble monomeric protein that displays differential detergent-induced conformations. *J Biol Chem* **273:** 10777–10783.

Khaled AR, Kim K, Hofmeister R, Mucgge K, Durum SK. 1999. Withdrawal of IL-7 induces *bax* translocation from cytosol to mitochondria through a rise in intracellular pH. *Proc Natl Acad Sci* **96:** 14476–14481.

Nechushtan A, Smith CL, Hsu YT, Youle RJ. 1999. Conformation of the Bax C-terminus regulates subcellular location and cell death. *EMBO J* **18:** 2330–2341.

Oberle C, Huai J, Reinheckel T, Tacke M, Rassner M, Ekert PG, Buellesbach J, Borner C. 2010. Lysosomal membrane permeabilization and cathepsin release is a Bax/Bak-dependent, amplifying event of apoptosis in fibroblasts and monocytes. *Cell Death Differ* **17:** 1167–1178.

Vogel S, Raulf N, Bregenhorn S, Biniossek ML, Maurer U, Czabotar P, Borner C. 2012. Cytosolic Bax: Does it require binding proteins to keep its pro-apoptotic activity in check?. *J Biol Chem* **287:** 9112–9127.

Cite this protocol as *Cold Spring Harb Protoc*; doi:10.1101/pdb.prot086454

Protocol 3

Detection of Bak/Bax Activating Conformation Change by Intracellular Flow Cytometry

Grant Dewson[1]

Cell Signalling and Cell Death Division, Walter and Eliza Hall Institute of Medical Research, Parkville, Melbourne, Victoria 3052, Australia; Department of Medical Biology, The University of Melbourne, Melbourne, Victoria 3010, Australia

Like the commonly used immunoprecipation (IP) approach, this procedure for the detection of activated Bak or Bax by intracellular flow cytometry is based on the principle that Bak and Bax, during activation, expose occluded amino-terminal epitopes that can be recognized by conformation-specific antibodies. Flow cytometric analysis requires fewer cells and is less time-consuming than IP. Further, in contrast to IP, flow cytometry produces a quantifiable assessment of the percentage of cells containing activated Bak or Bax, which can be correlated with cell death.

MATERIALS

It is essential that you consult the appropriate Material Safety Data Sheets and your institution's Environmental Health and Safety Office for proper handling of equipment and hazardous materials used in this protocol.

RECIPES: Please see the end of this protocol for recipes indicated by <R>. Additional recipes can be found online at http://cshprotocols.cshlp.org/site/recipes.

Reagents

Antibodies

Primary antibody, conformation-specific (for Bak and/or Bax; see Table 1)

Secondary antibody, fluorochrome-conjugated (e.g., FITC or RPE conjugate; Molecular Probes)

Apoptotic stimulus

Cell line of interest (e.g., mouse embryonic fibroblasts [MEF]), pretreated with a broad-range caspase inhibitor

We routinely perform this procedure using 1×10^5 MEF.

In cells that will be treated with a death stimulus, caspases should first be blocked with a broad-range caspase inhibitor, such as qVD-OPh or zVAD-fmk (Enzyme Systems).

Intracellular FACS buffer <R>

Paraformaldehyde (PFA)

Prepare a 4% solution in 1× phosphate-buffered saline (PBS). Dissolve by heating at 60°C and add 1 M NaOH dropwise until solution clears. Dilute to 1% PFA in 1× PBS immediately before use (Step 4).

Phosphate-buffered saline (PBS) (1×; Ca^{2+}- and Mg^{2+}-free) (ice-cold) <R>

---

[1]Correspondence: dewson@wehi.edu.au

Cite this protocol as *Cold Spring Harb Protoc*; doi:10.1101/pdb.prot086462

TABLE 1. Commonly used conformation-specific Bak or Bax antibodies

	Antibody (species reactivity)	Species	Source	Epitope/ immunogen	References
Bax	6A7 (h, m, r)	Mouse	Santa Cruz Biotech	aa12–24	Hsu and Youle 1998; Nechushtan et al. 1999
	NT (h, m)	Rabbit	BD Biosciences	aa1–21	
	N20 (h, m, r)	Rabbit	Santa Cruz Biotech	Amino terminus	Vogel et al. 2012
	Clone 3 (h)	Mouse	BD Biosciences	aa55–178	Khaled et al. 1999 Dewson et al. 2003
Bak	Ab-1 (h)	Mouse	Millipore	Amino terminus	Griffiths et al. 1999
	NT (h, m, r)	Rabbit	Millipore	aa23–38	Oberle et al. 2010
	22–38 (h, m)	Rabbit	Sigma-Aldrich	aa23–38	

h, human; m, mouse; r, rat.

Equipment

Centrifuge (with fixed-angle and swing-out rotors) at 4°C
Centrifuge tubes
FACs analyzer
Flow cytometry tubes
Rotating wheel at 4°C

METHOD

1. Treat cells (1×10^5) with an apoptotic stimulus as required.

   Beginning the procedure with excess cells (1×10^5) eases sample processing and improves FACS profiles.

2. Harvest the cells by centrifugation at 2500g for 5 min. Discard the supernatant.

3. Wash the cell pellet once in ice-cold 1× PBS. Centrifuge at 2500g for 5 min and discard the supernatant.

4. Resuspend the cell pellet in 1% PFA for 15 min at room temperature to fix cells.

5. Repeat Steps 2 and 3.

   Once cells are fixed, subsequent centrifugation steps are best performed in a swing-out (rather than fixed-angle) rotor to ensure the cells pellet at the bottom of the tube.

6. Resuspend the cell pellet in 100 µL of primary antibody at 1 µg/mL in intracellular FACS buffer. Include a "no primary antibody" control for all samples. Incubate the samples on a rotating wheel for 1 h at 4°C.

7. Add 1 mL of intracellular FACS buffer to each sample.

8. Centrifuge at 2500g for 5 min and discard the supernatant.

9. Resuspend each sample in 100 µL of species-specific FITC- (or RPE)-conjugated secondary antibody at 1 µg/mL in intracellular FACS buffer. Incubate on a rotating wheel for 1 h at 4°C in the dark.

10. Repeat Steps 7 and 8.

11. Resuspend each sample in 200 µL of ice-cold 1× PBS. Transfer to flow cytometry tubes and process using the FACS analyzer.

    See Troubleshooting.

TROUBLESHOOTING

Problem (Step 11): The detectable shift in fluorescence is limited.

Cite this protocol as *Cold Spring Harb Protoc*; doi:10.1101/pdb.prot086462

Solution: The shift in fluorescence depends on Bak/Bax expression levels and the efficiency of activation. Detection can be improved using a biotinylated secondary antibody followed by signal amplification with streptavidin-FITC/RPE.

Problem (Step 11): The entire cell population shifts during apoptosis.

Solution: Given the stochastic nature of a cell death response, one should expect to detect a population of cells with activated Bak or Bax that is dependent both on the time and dose of apoptotic stimulus. If the entire population shifts due to treatment, consider the following.

- All cells have activated Bak/Bax. Based on a death assay, for example, propidium iodide uptake, perform apoptosis induction using submaximal conditions.

- There is nonspecific uptake of the secondary antibody in treated cells. Be sure to include a "no primary antibody" control for all samples and trial different conformation-specific antibodies.

Problem (Step 11): The percentage of cells with activated Bak/Bax is lower than the percentage of "dead" cells.

Solution: As Bak/Bax activation is a relatively early event during apoptosis, it should precede cell death as monitored by propidium iodide uptake or AnnexinV positivity. If this is not the case, it may indicate the death stimulus used is not Bak/Bax-dependent. Alternatively, the concentration and time of death induction can be reduced and cells pretreated with caspase inhibitors to ensure that detection of activated Bak/Bax is not obscured by late-stage caspase-mediated destruction.

RELATED TECHNIQUES

Flow cytometric analysis of Bak conformation change can be performed in conjunction with flow cytometric analysis of cytochrome *c* release (Waterhouse et al. 2004). The latter protocol involves permeabilization of the cells with digitonin before fixation and is amenable to assessing conformation change of Bak but not Bax, as the cytosolic population of Bax is lost before fixation.

RECIPES

Intracellular FACS Buffer

1× phosphate-buffered saline (Ca^{2+}- and Mg^{2+}-free) <R>
0.5% bovine serum albumin
0.1% saponin

Phosphate-Buffered Saline (PBS)

Reagent	Amount to add (for 1× solution)	Final concentration (1×)	Amount to add (for 10× stock)	Final concentration (10×)
NaCl	8 g	137 mM	80 g	1.37 M
KCl	0.2 g	2.7 mM	2 g	27 mM
Na_2HPO_4	1.44 g	10 mM	14.4 g	100 mM
KH_2PO_4	0.24 g	1.8 mM	2.4 g	18 mM
If necessary, PBS may be supplemented with the following:				
$CaCl_2 \cdot 2H_2O$	0.133 g	1 mM	1.33 g	10 mM
$MgCl_2 \cdot 6H_2O$	0.10 g	0.5 mM	1.0 g	5 mM

PBS can be made as a 1× solution or as a 10× stock. To prepare 1 L of either 1× or 10× PBS, dissolve the reagents listed above in 800 mL of H_2O. Adjust the pH to 7.4 (or 7.2, if required) with HCl, and then add H_2O to 1 L. Dispense the solution into aliquots and sterilize them by autoclaving for 20 min at 15 psi (1.05 kg/cm²) on liquid cycle or by filter sterilization. Store PBS at room temperature.

ACKNOWLEDGMENTS

G.D. is supported by the National Health and Medical Research Council of Australia (637335), Australian Research Council (FT100100791), and the Association for International Cancer Research (10-230). The present work was made possible through Victorian State Government Operational Infrastructure Support and Australian Government NHMRC IRIISS.

REFERENCES

Dewson G, Snowden RT, Almond JB, Dyer MJ, Cohen GM. 2003. Conformational change and mitochondrial translocation of Bax accompany proteasome inhibitor-induced apoptosis of chronic lymphocytic leukemic cells. *Oncogene* **22:** 2643–2654.

Griffiths GJ, Dubrez L, Morgan CP, Jones NA, Whitehouse J, Corfe BM, Dive C, Hickman JA. 1999. Cell damage-induced conformational changes of the pro-apoptotic protein Bak in vivo precede the onset of apoptosis. *J Cell Biol* **144:** 903–914.

Hsu Y-T, Youle RJ. 1998. Bax in murine thymus is a soluble monomeric protein that displays differential detergent-induced conformations. *J Biol Chem* **273:** 10777–10783.

Khaled AR, Kim K, Hofmeister R, Muegge K, Durum SK. 1999. Withdrawal of IL-7 induces *bax* translocation from cytosol to mitochondria through a rise in intracellular pH. *Proc Natl Acad Sci* **96:** 14476–14481.

Nechushtan A, Smith CL, Hsu YT, Youle RJ. 1999. Conformation of the Bax C-terminus regulates subcellular location and cell death. *EMBO J* **18:** 2330–2341.

Oberle C, Huai J, Reinheckel T, Tacke M, Rassner M, Ekert PG, Buellesbach J, Borner C. 2010. Lysosomal membrane permeabilization and cathepsin release is a Bax/Bak-dependent, amplifying event of apoptosis in fibroblasts and monocytes. *Cell Death Differ* **17:** 1167–1178.

Vogel S, Raulf N, Bregenhorn S, Biniossek ML, Maurer U, Czabotar P, Borner C. 2012. Cytosolic Bax: Does it require binding proteins to keep its pro-apoptotic activity in check? *J Biol Chem* **287:** 9112–9127.

Waterhouse NJ, Steel R, Kluck R, Trapani JA. 2004. Assaying cytochrome c translocation during apoptosis. *Methods Mol Biol* **284:** 307–313.

 Cite this protocol as *Cold Spring Harb Protoc*; doi:10.1101/pdb.prot086462

Investigating the Oligomerization of Bak and Bax during Apoptosis by Cysteine Linkage

Grant Dewson[1]

Cell Signalling and Cell Death Division, Walter and Eliza Hall Institute of Medical Research, Parkville, Melbourne, Victoria 3052, Australia; Department of Medical Biology, The University of Melbourne, Melbourne, Victoria 3010, Australia

Following conformation change, Bak and Bax self-associate to form the putative apoptotic pore in the mitochondrial outer membrane. The nature of this pore and whether it is purely proteinaceous or lipidic are still unresolved. Induction of disulfide linkage with oxidants such as copper (II)(1,10-phenanthroline)$_3$ (CuPhe) and chemical cross-linking with cell-permeable homobifunctional maleimide reagents are convenient ways to investigate Bak and Bax oligomerization in cells or isolated mitochondria. A limitation of these methods is they are based on the linkage of cysteines, and their success is reliant on the positions of the endogenous cysteines in Bak and Bax. Consequently, the protocols are more efficient and informative for human Bak than that for its murine counterpart. An additional benefit when investigating human Bak is that cysteine-based linkage assays provide information on the conformation change that precedes Bak oligomerization: Endogenous cysteines in the inactive form are in close proximity, and intramolecular linkage after treatment causes inactive Bak to migrate faster during SDS–PAGE. This intramolecular linkage is lost on activation, as the cysteines are distanced by conformation change. During apoptosis, Bak oligomerization induces the proximity of cysteines that favor intermolecular linkage. Trapped Bak oligomers can be detected with nonreducing (following oxidation with CuPhe) or reducing (following chemical cross-linking with homobifunctional maleimide reagents) SDS–PAGE and immunoblotting, as described here.

MATERIALS

It is essential that you consult the appropriate Material Safety Data Sheets and your institution's Environmental Health and Safety Office for proper handling of equipment and hazardous materials used in this protocol.

RECIPES: Please see the end of this protocol for recipes indicated by <R>. Additional recipes can be found online at http://cshprotocols.cshlp.org/site/recipes.

Reagents

Cross-linking buffer (for chemical cross-linking) <R>

Cysteine linkage reagent of interest

Chemical cross-linker (bismaleimidoethane [BMOE] or bismaleimidohexane [BMH], prepared at a concentration of 10 mM on the day of the experiment) (Pierce)

Chemical cross-linkers provide an alternative to disulfide-bonding for monitoring protein oligomerization by cysteine linkage. These reagents come in many different "flavors," with different linker arm lengths, cleavability and solubility. An excellent resource for selecting the appropriate cross-linking reagent is Piercenet

---

[1]Corresponding: dewson@wehi.edu.au

Cite this protocol as *Cold Spring Harb Protoc*; doi:10.1101/pdb.prot086470

.com. BMOE and BMH have 8 and a 12 Å linkers, respectively, and therefore will cross-link cysteines within different distance constraints. Both reagents are cell-permeable and thus can be applied directly to cells in culture. As they are uncleavable, samples are processed as normal for reducing SDS–PAGE.

Copper (II)(1,10-phenanthroline)$_3$ (CuPhe) (10 mM) <R>

Disulfide linkage using CuPhe is both rapid and efficient; because it is effectively a "zero-length" cross-linker, it can provide detailed information about the conformation or interfaces of a protein. However, due to the relative positions of the endogenous cysteines in Bak and Bax, oxidation is an informative approach only for the former, while longer-length cross-linkers are required for analysis of Bax.

Membrane fraction pellet, prepared as described in Steps 1–6 of Protocol 1: Investigating Bax Sub-cellular Localization and Membrane Integration (Dewson 2015)

As Bak and Bax oligomerize at the mitochondrial outer membrane, the cytosolic fraction (supernatant at Step 6) can be discarded. Note that permeabilization and separation of the membrane fraction from the cytosol are not absolutely necessary, as CuPhe, BMH, and BMOE are all cell-permeable; however, these steps improve detection of disulfide-linked Bak oligomers. Alternatively, whole cells may be used.

Permeabilization buffer (for disulfide linkage) <R>
Protease inhibitor cocktail tablets, without EDTA (cOmplete; Roche) (1 tablet/50 mL)
SDS–PAGE and immunoblotting reagents, including primary antibodies for Bak and/or Bax
SDS–PAGE sample buffer (nonreducing) (2×) (for disulfide linkage) <R>

Supplement nonreducing sample buffer with 100 mM EDTA before use (Step 3).

SDS–PAGE sample buffer (reducing) (2×) (for chemical cross-linking) <R>

Equipment

Heat block at 95°C
SDS–PAGE apparatus
Western transfer apparatus

METHOD

In this protocol, cells are treated to induce cysteine linkage using one of two methods: The induction of disulfide linkage by oxidation for investigation of Bak (see Steps 1–3) or the use of chemical cross-linkers for investigation of both Bak and Bax (see Steps 4–6).

Inducing Disulphide Linkage

1. Resuspend the membrane fraction pellet in permeabilization buffer (with protease inhibitors but without digitonin) at 1×10^7 cells per mL.

   The membrane fraction may be difficult to resuspend in permeabilization buffer at this stage. It is helpful to pipette up and down several times with a P200 pipette until no clumps are visible.

2. Add 10 mM CuPhe to each sample to a final concentration of 1 mM. Incubate for 30 min on ice.

3. Add an equal volume of 2× nonreducing SDS–PAGE sample buffer supplemented with 100 mM EDTA (to chelate the copper). Proceed to Step 7.

   Do not add a reducing agent (DTT or β-mercaptoethanol) at any stage, as it will inhibit/disrupt disulfide linkage.

Cross-Linking with a Chemical Cross-Linker

4. Resuspend the membrane fraction pellet in cross-linking buffer at 1×10^7 cells per mL.

5. Add 10 mM BMH or BMOE to a final concentration of 0.5 mM. Incubate for 30 min in the dark at room temperature.

   When adding BMH or BMOE in DMSO, the final DMSO concentration must be <10%.

6. Add an equal volume of 2× reducing SDS–PAGE sample buffer. Proceed to Step 7.

Cite this protocol as *Cold Spring Harb Protoc*; doi:10.1101/pdb.prot086470

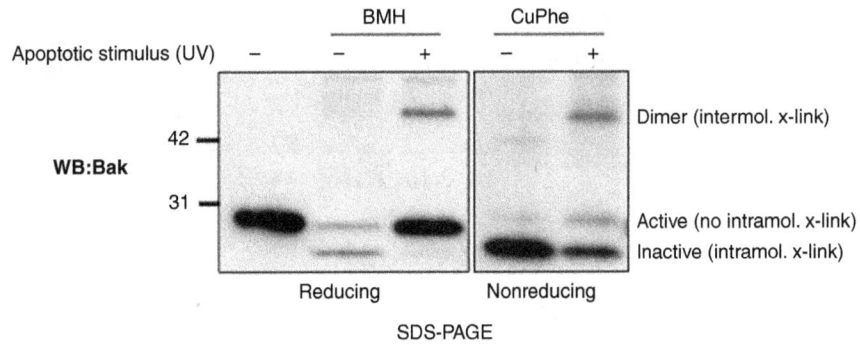

FIGURE 1. Intra- and intermolecular cysteine linkage as diagnostic of Bak activation and oligomerization status. In "healthy" cells, endogenous Cys in human Bak are proximal (C14 and C166) and can be intramolecularly cross-linked using chemical cross-linkers such as BMH or via disulphide-linkage using oxidant (CuPhe); this results in faster migration on SDS–PAGE, which is diagnostic of the inactive form (Cheng et al. 2003; Dewson et al. 2008). During apoptosis, activation of Bak displaces the two Cys so the *intra*molecular linkage is lost in favor of *inter*molecular linkage within the Bak oligomer. Apoptotic stimulus, ultraviolet (UV) irradiation; western blot (WB), Bak. x-link, cross-link. Figure modified from Dewson et al. (2008).

Immunoblotting

7. Heat the samples for 5 min at 95°C.
8. Analyze the samples by SDS–PAGE and immunoblotting (Fig. 1). If running nonreduced (CuPhe) and reduced (BMH, BMOE) samples on the same gel, leave at least one (preferably two) empty lanes between the nonreduced and reduced samples.

    Empty lanes between nonreduced and reduced samples ensure that the reducing agent does not leach across the gel during electrophoresis and disrupt disulfide-linked complexes.

RECIPES

Cross-Linking Buffer

20 mM HEPES/KOH, pH 7.5
250 mM sucrose
1 mM EDTA
50 mM KCl
2.5 mM $MgCl_2$

Copper (II)(1,10-phenanthroline)$_3$ (10 mM)

1. Dissolve 1,10-phenanthroline (Sigma-Aldrich) in 20% ethanol to prepare a stock solution of 20 mM 1,10-phenanthroline.
2. Dissolve $CuSO_4$ in H_2O to prepare a stock solution of 300 mM $CuSO_4$.
3. Using the stock solutions, prepare a solution of 10 mM phenanthroline and 30 mM $CuSO_4$ in 20% ethanol.

The concentration of the Copper (II)(1,10-phenanthroline)$_3$ solution refers to phenanthroline (10 mM). The copper concentration is 30 mM.

Permeabilization Buffer

20 mM HEPES/KOH, pH 7.5
250 mM sucrose
50 mM KCl
2.5 mM $MgCl_2$

SDS–PAGE Sample Buffer (Nonreducing) (2×)

125 mM Tris–HCl, pH 6.8
20% glycerol
4% SDS
0.1% bromophenol blue

SDS–PAGE Sample Buffer (Reducing) (2×)

125 mM Tris–HCl, pH 6.8
20% glycerol
4% SDS
0.1% bromophenol blue
5% β-mercaptoethanol

ACKNOWLEDGMENTS

G.D. is supported by the National Health and Medical Research Council of Australia (637335), Australian Research Council (FT100100791), and the Association for International Cancer Research (10–230). The present work was made possible through Victorian State Government Operational Infrastructure Support and Australian Government NHMRC IRIISS.

REFERENCES

Cheng EH, Sheiko TV, Fisher JK, Craigen WJ, Korsmeyer SJ. 2003. VDAC2 inhibits BAK activation and mitochondrial apoptosis. *Science* **301:** 513–517.

Dewson G. 2015. Investigating Bax subcellular localization and membrane integration. *Cold Spring Harb Protoc* doi: 10.1101/pdb.prot086447.

Dewson G, Kratina T, Sim HW, Puthalakath H, Adams JM, Colman PM, Kluck RM. 2008. To trigger apoptosis Bak exposes its BH3 domain and homo-dimerizes via BH3:Groove interactions. *Mol Cell* **30:** 369–380.

Cite this protocol as *Cold Spring Harb Protoc*; doi:10.1101/pdb.prot086470

Blue Native PAGE and Antibody Gel Shift to Assess Bak and Bax Conformation Change and Oligomerization

Grant Dewson[1]

Cell Signalling and Cell Death Division, Walter and Eliza Hall Institute of Medical Research, Parkville, Melbourne, Victoria 3052, Australia; Department of Medical Biology, The University of Melbourne, Melbourne, Victoria 3010, Australia

Blue native PAGE (BN-PAGE) uses Coomassie dye rather than denaturing SDS to provide a negative charge to proteins for electrophoresis. As such, it is a useful assay for investigating native supramolecular membrane complexes without the need for cross-linking. As Bak and Bax oligomers form in the mitochondrial outer membrane, and they can be efficiently monitored by BN-PAGE. Furthermore, BN-PAGE performed in conjunction with gel-shift using conformation-specific antibodies can provide additional information regarding the activation state of Bak or Bax in specific membrane complexes.

MATERIALS

It is essential that you consult the appropriate Material Safety Data Sheets and your institution's Environmental Health and Safety Office for proper handling of equipment and hazardous materials used in this protocol.

RECIPES: Please see the end of this protocol for recipes indicated by <R>. Additional recipes can be found online at http://cshprotocols.cshlp.org/site/recipes.

Reagents

Acetic acid

Anode buffer (25 mM imidazole/HCl, pH 7.0)

Antibodies

    Antibodies for gel-shift, conformation-specific (for Bax and/or Bak, e.g., anti-Bax 6A7 or anti-Bak 23–38)

    Irrelevant control antibody (of the same isotype)

    Positive control antibody (for the inactive, active, and oligomerized forms of Bak or Bax)

Blue native PAGE (BN-PAGE) loading buffer (10×) <R>

Cathode buffer I (10×) <R>

Cathode buffer II (10×) <R>

Coomassie stain <R>

Digitonin

    Immediately before use, prepare a fresh solution of 10% (w/v) digitonin in H_2O and boil to dissolve. Alternatively, resuspend in DMSO for long-term storage.

[1]Correspondence: dewson@wehi.edu.au

Cite this protocol as *Cold Spring Harb Protoc*; doi:10.1101/pdb.prot086488

Immunoblotting reagents

It is recommended that the antibody used for immunoblotting is raised in a different species from the antibodies used for gel-shift to avoid immunoreactivity with the secondary antibody.

Methanol

Mouse embryonic fibroblasts (MEF) or cell line of interest, pretreated with a broad-range caspase inhibitor

The number of cells required will depend on their expression of Bak and/or Bax, but a minimum of 1×10^6 cells is recommended as a starting point.

In cells that will be treated with a death stimulus, caspases should first be blocked with a broad-range caspase inhibitor, such as qVD-OPh or z-VAD-fmk (Enzyme Systems).

Native gradient gels (Invitrogen) and appropriate markers

Gradient gels are required for BN-PAGE.

Permeabilization buffer <R>

Phosphate-buffered saline (PBS) ($1\times$; Ca^{2+}- and Mg^{2+}-free) (ice-cold) <R>

Protease inhibitor cocktail tablets, without EDTA (cOmplete; Roche) (1 tablet/50 mL)

Solubilization buffer <R>

Tris-glycine buffer with SDS/BME (optional; see Step 11.vi) <R>

Tris-glycine/MeOH transfer buffer <R>

Supplement transfer buffer with 0.037% SDS for gel equilibration (Step 11.iv).

Equipment

Centrifuge (benchtop) at 4°C

Centrifuge tubes

Hemocytometer

Polyvinyl difluoride (PVDF) membrane

SDS–PAGE apparatus (preferably an old kit as it will be stained with Coomassie dye!)

Water bath at 65°C

Western transfer apparatus

METHOD

As needed, a reducing agent such as dithiothreitol (DTT) may be added at a concentration of 2–10 mM throughout sample preparation (Steps 3–8); see Discussion.

Inducing Apoptosis and Permeabilizing Cells

1. Treat cells with an apoptotic stimulus as required.

2. Harvest the cells by centrifugation at 2500g for 5 min. Wash once in 1 mL of ice-cold 1× PBS. Perform a cell count using a hemocytometer.

3. Centrifuge the cells at 2500g for 5 min and discard the supernatant. Gently resuspend the cell pellet in permeabilization buffer supplemented with 0.025% digitonin and protease inhibitors at 1×10^7 cells per mL.

   Although not absolutely necessary, this permeabilization step removes cytosolic proteins and therefore reduces background during immunoblotting. It also allows for the addition of conformation-specific antibodies for gel-shift assay.

4. Pellet the membranes by centrifugation at 13,000g for 5 min. Discard the supernatant (cytosolic fraction).

5. If performing gel-shift, proceed to Step 6. Otherwise, proceed directly to Step 8.

 Cite this protocol as *Cold Spring Harb Protoc*; doi:10.1101/pdb.prot086488

Antibody Gel-Shift and BN-PAGE

6. Resuspend the membranes in permeabilization buffer (with protease inhibitors but without digitonin). Add 1–2 µg of conformation-specific anti-Bak or anti-Bax antibody or a control antibody to each sample. Incubate on ice for 30 min.

   Antibody concentration and time of incubation should be determined empirically.

7. Centrifuge the samples at 13,000*g* for 5 min to pellet membranes. Discard the supernatant.

8. Resuspend the membranes in solubilization buffer with 1% digitonin by pipetting rapidly with a P200 pipette. Incubate for 30 min on ice.

   We routinely resuspend MEF at 1×10^7 cells/mL. This step is critical for the efficient solubilization of the membrane complexes and their resolution on BN-PAGE.

9. Centrifuge the samples at 13,000*g* for 5 min to pellet debris.

10. Retrieve and transfer the supernatant to a fresh tube. Add 1/0 volume of 10× BN-PAGE loading buffer to each sample.

    Do NOT heat the samples, as this will denature the proteins and their complexes and defeats the object of BN-PAGE. As complexes will potentially disassociate on storage, only prepare as much sample as will be run on the day of the experiment.

11. Proceed to BN-PAGE as described by Wittig et al. (2006) followed by immunoblotting.

    i. Assemble the native gels in the gel tank. Add anode buffer to the outer chamber of the tank so that it reaches halfway up the gel. Before loading the samples, add sufficient cathode buffer I to fill the "wells only," rather than the whole inner chamber.

       This allows you to see the samples as they are loaded and helps prevent displacement of the sample when the tank is finally filled with cathode buffer I.

    ii. Load samples alongside native markers.

       Load samples as sufficient to detect Bak/Bax by immunoblotting. This will vary depending on cell type and should be established empirically. It is important not to "overload" the gel, as this will reduce the definition of the complexes.

    iii. Run gels according to Table 1. After the Coomassie dye front has migrated a one-third of the way through the separation gel, replace cathode buffer I with cathode buffer II to allow protein visualization and aid subsequent transfer. Run the gel until the Coomassie dye front nears the bottom of the gel

       Abundant mitochondrial respiratory complexes can be directly visualized postelectrophoresis.

    iv. Before transfer, equilibrate the gel in Tris-glycine/MeOH transfer buffer supplemented with 0.037% SDS.

    v. Transfer to PVDF membrane as normal. After transfer, stain the blot with Coomassie stain. Destain in 50% methanol/25% acetic acid to more easily visualize the markers.

       Coomassie staining and destaining is only compatible with PVDF and not nitrocellulose membranes.

    vi. If necessary, incubate the blot for 30 min at 65°C in Tris-glycine buffer with SDS/BME to denature proteins and aid immunodetection.

       See Troubleshooting.

TABLE 1. Gel-running conditions for BN-PAGE

	Electrophoresis (at room temperature)	Transfer
Large gel (12 cm × 15 cm)	7 mA until Coomassie dye enters separation gel, then 14 mA for 3–5 h	400 mA for 2 h
Mini gel	8 mA for ~1.5 h	30 V for 2.5 h

TROUBLESHOOTING

Problem (Step 11): Proteins cannot be detected following immunoblotting.

Solution: The success of BN-PAGE is dependent on recognition of the native conformer of the protein of interest by the immunoblotting antibody, and so a number of antibodies should be tested. In addition, this limitation can be overcome by incubating the immunoblot with SDS and β-mercaptoethanol in Tris-glycine buffer for 30 min at 65°C to effectively denature the proteins and reveal occluded epitopes (Valentijn et al. 2008).

DISCUSSION

Bak and Bax oligomers form in the mitochondrial outer membrane (MOM) and thus can be monitored by BN-PAGE (Valentijn et al. 2008; Lazarou et al. 2010; Dewson et al. 2012). In our experience, Bak and Bax apoptotic oligomers are stable following solubilization from the MOM in 1% digitonin and subsequent BN-PAGE. It is important to note that the redox conditions within a cell vary depending on the subcellular compartment. The need for the addition of a reducing agent during the sample preparation will depend on the native environment of the complexes under investigation. For example, if the complex is exposed to the reducing environment of the cytosol, as with Bak and Bax, then addition of DTT is recommended to prevent artifactual disulphide-bonding during membrane solubilization that will alter the complexes detected on BN-PAGE. If required, DTT should be added at a concentration of 2–10 mM "throughout" sample preparation (Steps 3–8).

RECIPES

BN-PAGE Loading Buffer (10×)

5% (w/v) Coomassie blue G-250
500 mM aminocaproic acid

Cathode Buffer I (10×)

500 mM Tricine
75 mM imidazole, pH 7
0.2% (w/v) Coomassie blue G-250

Store at 4°C.

Cathode Buffer II (10×)

500 mM Tricine
75 mM imidazole, pH 7

Store at 4°C.

Coomassie Stain

50% methanol
10% acetic acid
0.25% (w/v) Coomassie blue R-250

Permeabilization Buffer

20 mM HEPES/KOH, pH 7.5
250 mM sucrose
50 mM KCl
2.5 mM $MgCl_2$

Cite this protocol as *Cold Spring Harb Protoc*; doi:10.1101/pdb.prot086488

Phosphate-Buffered Saline (PBS)

Reagent	Amount to add (for 1× solution)	Final concentration (1×)	Amount to add (for 10× stock)	Final concentration (10×)
NaCl	8 g	137 mM	80 g	1.37 M
KCl	0.2 g	2.7 mM	2 g	27 mM
Na_2HPO_4	1.44 g	10 mM	14.4 g	100 mM
KH_2PO_4	0.24 g	1.8 mM	2.4 g	18 mM

If necessary, PBS may be supplemented with the following:

$CaCl_2 \cdot 2H_2O$	0.133 g	1 mM	1.33 g	10 mM
$MgCl_2 \cdot 6H_2O$	0.10 g	0.5 mM	1.0 g	5 mM

PBS can be made as a 1× solution or as a 10× stock. To prepare 1 L of either 1× or 10× PBS, dissolve the reagents listed above in 800 mL of H_2O. Adjust the pH to 7.4 (or 7.2, if required) with HCl, and then add H_2O to 1 L. Dispense the solution into aliquots and sterilize them by autoclaving for 20 min at 15 psi (1.05 kg/cm^2) on liquid cycle or by filter sterilization. Store PBS at room temperature.

Solubilization Buffer

50 mM NaCl
5 mM aminocaproic acid
1 mM EDTA
50 mM imidazole/HCl, pH 7
1% digitonin

Tris-Glycine Buffer with SDS/BME

25 mM Tris
192 mM glycine
5% sodium dodecyl sulfate
2% β-mercaptoethanol

Tris-Glycine/MeOH Transfer Buffer

25 mM Tris
192 mM glycine
20% (v/v) methanol

ACKNOWLEDGMENTS

G.D. is supported by the National Health and Medical Research Council of Australia (637335), Australian Research Council (FT100100791), and the Association for International Cancer Research (10–230). The present work was made possible through Victorian State Government Operational Infrastructure Support and Australian Government NHMRC IRIISS.

REFERENCES

Dewson G, Ma S, Frederick P, Hockings C, Tan I, Kratina T, Kluck RM. 2012. Bax dimerizes via a symmetric BH3:groove interface during apoptosis. *Cell Death Differ* 19: 661–670.

Lazarou M, Stojanovski D, Frazier AE, Kotevski A, Dewson G, Craigen WJ, Kluck RM, Vaux DL, Ryan MT. 2010. Inhibition of Bak activation by VDAC2 is dependent on the Bak transmembrane anchor. *J Biol Chem* 285: 36876–36883.

Valentijn AJ, Upton JP, Gilmore AP. 2008. Analysis of endogenous Bax complexes during apoptosis using blue native PAGE: Implications for Bax activation and oligomerization. *Biochem J* 412: 347–357.

Wittig I, Braun HP, Schagger H. 2006. Blue native PAGE. *Nat Protoc* 1: 418–428.

CHAPTER 6

Imaging-Based Methods for Assessing Caspase Activity in Single Cells

Melissa J. Parsons,[1,2] Markus Rehm,[3,4] and Lisa Bouchier-Hayes[1,2,5]

[1]Center for Cell and Gene Therapy, Baylor College of Medicine, Houston, Texas 77030; [2]Department of Pediatrics-Hematology, Baylor College of Medicine, Houston, Texas 77030; [3]Centre for Systems Medicine, Royal College of Surgeons in Ireland, Dublin 2, Ireland; [4]Department of Physiology and Medical Physics, Royal College of Surgeons in Ireland, Dublin 2, Ireland

Caspases, a family of proteases that are essential mediators of apoptosis, are divided into two groups: initiator caspases and executioner caspases. Each initiator caspase is activated at the apex of its respective pathway, which generally leads to the cleavage and activation of executioner caspases. Executioner caspases in turn cleave numerous substrates in the cell, leading to its demise. Initiator caspases are activated when inactive monomers undergo induced proximity to form an active caspase. In contrast, executioner caspases are activated by cleavage. Based on this key difference, different imaging techniques have been developed to measure caspase activation and activity on a single-cell basis. Bimolecular fluorescence complementation (BiFC) is used to measure induced proximity of initiator caspases, whereas Förster resonance energy transfer (FRET) permits the investigation of caspase-mediated substrate cleavage in real time. Because many of the events in apoptosis, including caspase activation, are asynchronous in nature, these single-cell imaging techniques have proven to be immensely powerful in ordering and dissecting caspase pathways. When coupled with parallel detection of additional hallmark events of apoptosis, they provide detailed and quantitative kinetic and positional insights into the signal transduction pathways that regulate cell death. Here we provide a brief introduction into BiFC- and FRET-based imaging of caspase activation and activity in single cells.

INTRODUCTION

At the heart of apoptosis lies a family of enzymes called "caspases," which are cysteine-dependent aspartate-directed proteases. All caspases contain a catalytic cysteine required for their activity, and all cleave their protein substrates at specific aspartate residues (Alnemri et al. 1996). To date, 18 mammalian caspase proteins have been identified in animals and divided into initiator caspases (e.g., caspase-2, -8, and -9) and executioner caspases (caspase-3 and -7). We discuss here methods to study caspase activation and activity using single-cell imaging approaches.

Limitations of Conventional Approaches to Studying Caspase Function

Detecting the activation of specific caspases can be problematic. Because executioner caspases are activated by cleavage, monitoring the disappearance of the full-length protein and the appearance of its cleavage products by western blot is a *bona fide* measure of activation. In contrast, initiator caspases are activated not by cleavage, but rather by dimerization on recruitment to specific large molecular

[5]Correspondence: bouchier@bcm.edu

Cite this introduction as *Cold Spring Harb Protoc*; doi:10.1101/pdb.top070342

mass complexes considered to be activation platforms. Activation platforms, each specific for a particular caspase, rely on the presence of protein–protein interaction domains in the adaptor proteins and in the caspases. In the extrinsic or death-receptor pathway, for example, the ligation of death receptors leads to the formation of an activation platform for the initiator caspase-8. This activation platform is a multimeric signaling complex called the "DISC" (death-inducing signaling complex; Kischkel et al. 1995). Its death effector domains interact with the corresponding domains of caspase-8, shown in Figure 1 (left pathway). Similarly, in the extrinsic or mitochondrial pathway, the multimeric complex known as the "apoptosome" functions as an activation platform for caspase-9 (Riedl and Salvesen 2007). Caspase-9 contains a protein–protein interaction domain, called the caspase-recruitment domain (CARD). When the CARD of caspase-9 binds to the CARD of Apaf-1, the interaction facilitates induced proximity and dimerization of the caspase-9 monomers, shown in Figure 1 (right pathway). Caspase-2 appears also to be activated on an activation platform, in this case termed the "PIDDosome," which consists of PIDD and the adaptor protein RAIDD (Tinel and Tschopp 2004).

For the initiator caspases-2, -8, and -9, it has been clearly shown that cleavage of the caspase, while it does occur, is not sufficient for caspase activation (Stennicke et al. 1999; Chang et al. 2003; Baliga et al. 2004). Baliga and colleagues showed that only the dimeric species of caspase-2 is active, and that dimerization of a noncleavable mutant of caspase-2 retains only 20% of the activity of the wild-type protein (Baliga et al. 2004). Therefore, the use of western blot analysis to detect protein cleavage does not truly reflect the activation status of the initiator caspase.

Assays using fluorigenic caspase substrates and caspase inhibitors are widely used to determine caspase activities in cell populations and cell extracts. The design of these reagents is based on the predicted preferred substrate peptide motif for each caspase. These motifs are not very specific, however, as caspases substantially overlap in substrate specificities (McStay et al. 2008). Clearly, the results from such population-based analyses must be carefully controlled and interpreted with care. It has been shown that the executioner caspase, caspase-3, can cleave most fluorigenic caspase substrates with high efficiency and therefore bulk activity measurements tend to be dominated by executioner caspase contributions (McStay et al. 2008; for more discussion, see Introduction of Chapter 7: Measuring Apoptosis: Caspase Inhibitors and Activity Assays [McStay and Green 2014]).

Furthermore, the onset times of caspase activation differ considerably among cells in clonal populations, so that bulk measurements of caspase activation do not reflect intracellular signaling kinetics (Rehm et al. 2002). However, imaging-based approaches for measuring caspase activation and activity can address these shortcomings by describing apoptosis signaling with higher specificity inside single living cells in space and time.

Advantages of Imaging for Monitoring Apoptotic Events

Single-cell imaging techniques have proved immensely powerful in ordering and dissecting events during apoptosis, many of which are asynchronous in nature. One of the main advantages of single-cell imaging over more conventional methods of studying cell death pathways is its ability to precisely measure these asynchronous events directly within the complex environment of the living cell. For example, in studies of the intrinsic (mitochondrial) pathway, live cell imaging was uniquely able to show that release of cytochrome c from mitochondria proceeds in individual cells at different times in response to a cellular stress, but, in each individual cell, the release was complete and kinetically invariant (Goldstein et al. 2000). Further analysis confirmed that mitochondrial outer membrane permeabilization (MOMP) and cytochrome c release preceded executioner caspase activation in a manner that described the kinetics of these processes and their relative timing in great detail (Rehm et al. 2003). The availability of more sophisticated microscope systems that provide improved resolution, coupled with the development of enhanced fluorescent proteins and superior molecular tools, has enabled us to uncover even more and surprising details about these pathways. For example, higher resolution imaging of the release of the proteins Smac and Omi (other activators of caspases) from mitochondria that—unlike cytochrome c—are subject to proteasomal degradation on release, revealed that some mitochondria in fact remain intact. These intact mitochondria are essential for

Cite this introduction as *Cold Spring Harb Protoc*; doi:10.1101/pdb.top070342

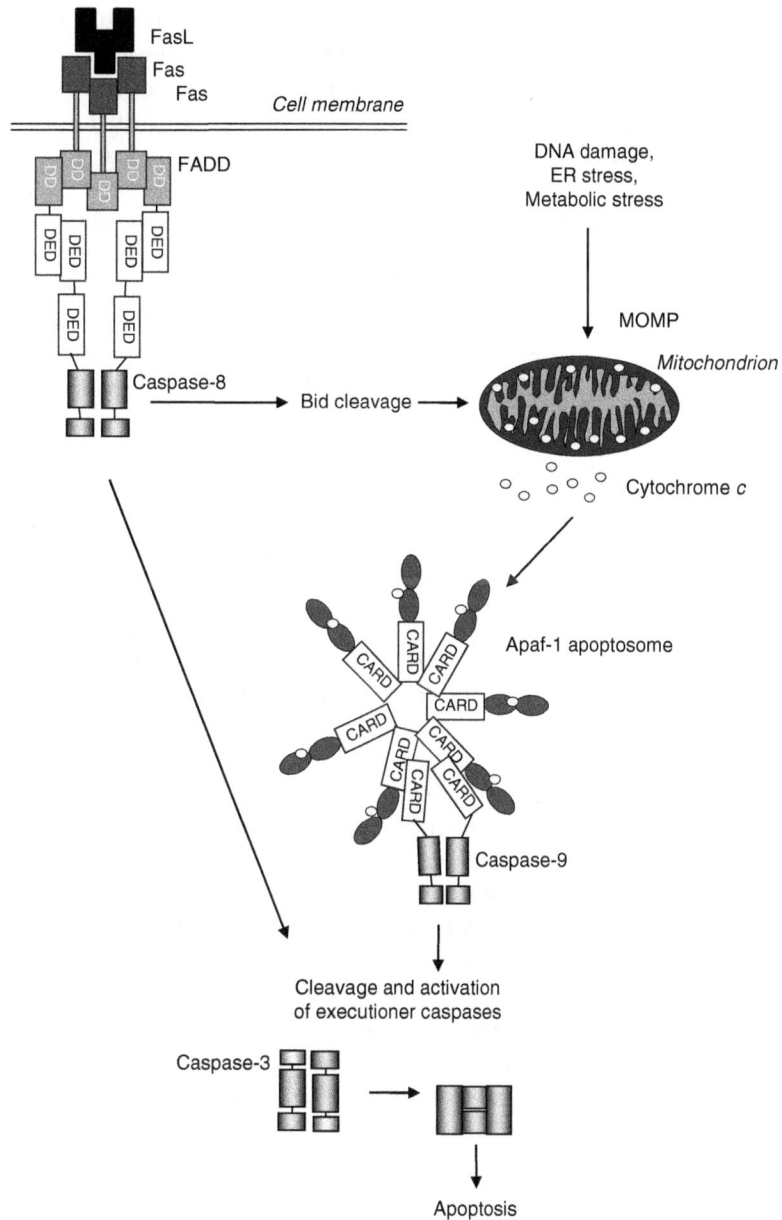

FIGURE 1. Caspase activation pathways: schematic representation of caspase activation in the extrinsic pathway and the intrinsic pathway. In the death receptor or extrinsic pathway (shown on the *left*), ligands such as FasL, TNF, and TRAIL initiate apoptosis. Ligation of death receptors on the cell surface result in assembly of the death-inducing signaling complex (DISC), the activation platform for caspase-8. DISC assembly is mediated by specific protein–protein interactions. As shown the death domain (DD) in the receptor Fas binds to the DD in the adaptor protein FADD. FADD in turn binds to caspase-8 via death effector domains (DEDs) present in both sequences. Recruitment of caspase-8 to the activation platform leads to dimerization and activation of caspase-8. In the intrinsic or mitochondrial pathway (shown on the *right*), mitochondrial outer membrane permeabilization (MOMP) initiates cytochrome *c* release from the mitochondria. Cytochrome *c* promotes the assembly of the Apaf-1 apoptosome, which is the activation platform for caspase-9. Apaf-1 binds to caspase-9 via caspase recruitment domains (CARDs) present in both sequences. This binding promotes dimerization and activation of caspase-9. Caspases-8 and -9 directly activate executioner caspases, including caspase-3. Caspase-8 also induces cleavage of the protein Bid. Cleaved Bid can then induce MOMP leading to indirect activation of downstream caspases.

recovery from death signals when caspase activation is blocked (Tait et al. 2010). Such discoveries would have been impossible without recent advances in imaging techniques.

Adapting imaging techniques for direct measurement of caspases in single cells has been a little less straightforward. The main challenge has been to distinguish activation and activity associated with one specific caspase from general caspase activity in the cell. This concern is further confounded by the differences in how initiator and executioner caspases are activated. New techniques such as caspase bimolecular fluorescence complementation (BiFC) and improvements to more established tools such as Förster resonance energy transfer (FRET)-based caspase substrates have provided superior means to study activation and activity associated with specific caspases and to measure exactly when and where an individual caspase is activated in the cell. We summarize these imaging-based approaches below and describe them in detail in the accompanying protocols (see Protocol 1: Measuring Initiator Caspase Activation by Bimolecular Fluorescence Complementation [Parsons and Bouchier-Hayes 2015] and Protocol 2: Measuring Caspase Activity by Förster Resonance Energy Transfer [Rehm et al. 2015]).

CASPASE BiFC FOR MEASURING THE INDUCED PROXIMITY OF INITIATOR CASPASES

Caspase BiFC is an imaging-based technique that measures the proximal step in initiator caspase activation: the induced proximity of caspase monomers on recruitment to activation platforms. We describe caspase BiFC in detail in an accompanying protocol (see Protocol 1: Measuring Initiator Caspase Activation by Bimolecular Fluorescence Complementation [Parsons and Bouchier-Hayes 2015]). This strategy can be used to visualize the assembly of initiator caspase activation platforms in single cells and assess the conditions under which they form, their location in cells and their kinetic relationship to other steps in apoptosis such as MOMP. This method was developed to investigate the caspase-2 activation pathway; however, the same technique can be adapted to other initiator caspases (Bouchier-Hayes et al. 2009).

In BiFC, "split" fluorescent proteins are used to measure protein–protein interactions in cells (Shyu et al. 2006). The fluorescent protein Venus, a brighter and more photostable version of yellow fluorescent protein (YFP), is often used in these studies. In this approach, the fluorescent protein is separated into two fragments; of note, each fragment on its own is not fluorescent. When each fragment is fused to one partner of an interacting protein pair, these nonfluorescent protein fragments can associate, on interaction of the partners, to reform the fluorescent molecule. Caspase BiFC adapts the technique slightly—instead of measuring direct protein–protein interactions, it detects recruitment of the caspase to its activation platform. The caspase or portion of the caspase that binds to the activation platform (the interacting prodomain that contains a CARD or DED) is fused independently to both fragments of Venus. In untreated cells, caspase monomers fused to the Venus fragments remain separate from each other. On treatment with an activating stimulus (e.g., heat shock for caspase-2), the caspase fusion proteins are recruited to the activation platform via their protein–protein interaction domains. Once aligned at the activation platform, the induced proximity of caspase monomers results in association of the Venus fragments, thereby restoring its fluorescence (Bouchier-Hayes et al. 2009; see also Fig. 2A). This fluorescence readout can be used as a measure of caspase activation, allowing us to determine when and where caspase-2 activation occurs on a cell-by-cell basis.

Split Venus is particularly suitable for real-time imaging applications of caspase BiFC because the reassociation occurs relatively quickly (it has been detected in as little as 5 min [Schmidt et al. 2003]) and occurs under physiological conditions. In contrast, a split version of mCherry (a red-shifted fluorescent protein) requires incubation at 30°C for 30 min for complementation to occur (Fan et al. 2008); mCherry is therefore not suitable for real time imaging of caspase BiFC.

Caspase BiFC experiments must be carefully controlled for specificity of the interaction between each initiator caspase with its respective activation platform, for example, by disrupting the interaction between the caspase and its adaptor protein. One main advantage of this technique is that the

 Cite this introduction as *Cold Spring Harb Protoc*; doi:10.1101/pdb.top070342

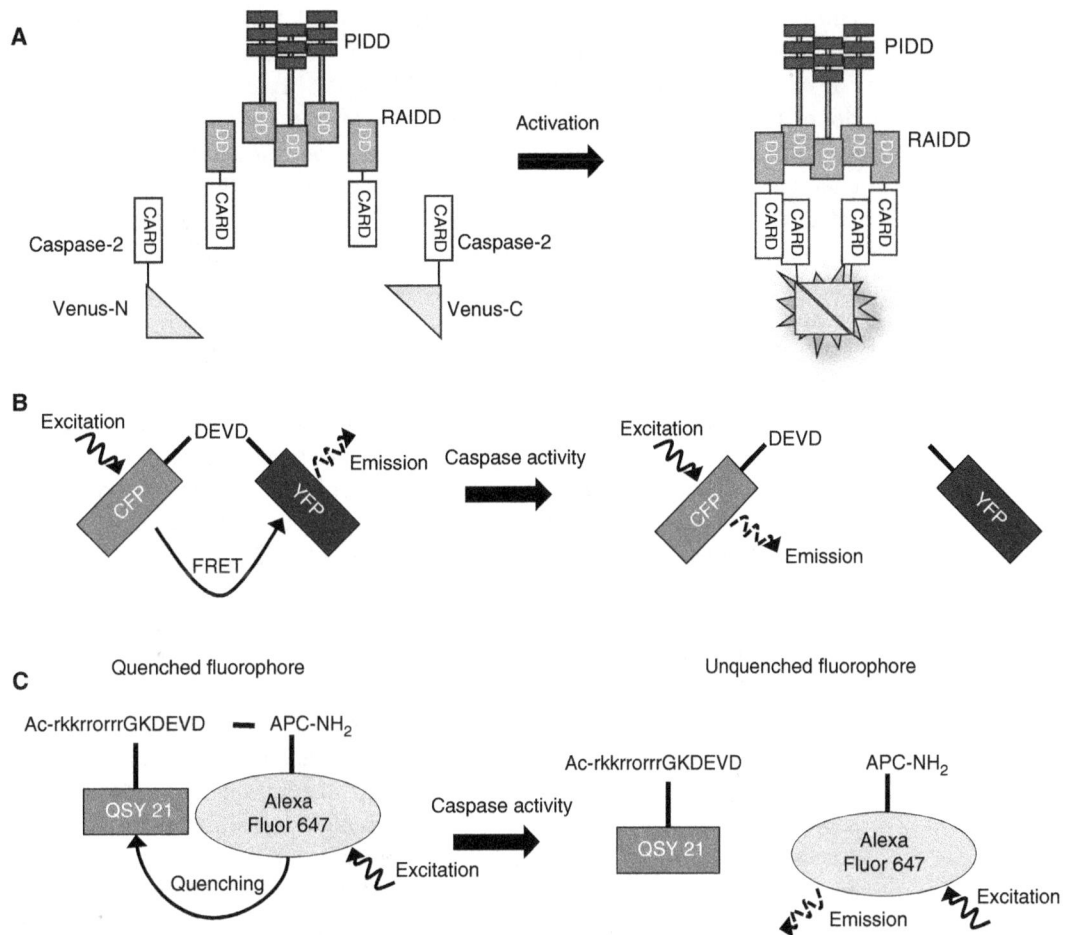

FIGURE 2. Imaging-based caspase activation assays. (*A*) Caspase BiFC. The model for measurement of caspase-2-induced proximity by the PIDDosome is shown. (*B*) FRET-based caspase substrate. (*C*) Cell-permeable caspase substrate. TcapQ647 is shown as an example.

molecular components of the activation platform do not need to be fully characterized to obtain meaningful data. Indeed, this technique can even be used to determine the relative requirements of each component of an activation platform. For example, RAIDD, the direct adaptor for caspase-2, induces robust caspase-2 BiFC. In contrast, PIDD (which recruits caspase-2 via RAIDD binding) can match the level of caspase-2 BiFC induced by RAIDD only if a small amount of RAIDD is also present, thus recapitulating the binding properties of the endogenous proteins. In the same way, the subcellular localization of initiator caspase induced proximity on recruitment to distinct activation platforms can be directly tested using this strategy.

FRET FOR MEASURING CASPASE ACTIVITY

FRET-based caspase substrates represent a second set of useful imaging tools for dissecting caspase activation pathways in real time. Unlike caspase BiFC, FRET probes can be used to measure the activity of executioner caspases as well as initiator caspase activation (Rehm et al. 2002; Albeck et al. 2008; Hellwig et al. 2008). We describe this approach in detail in an accompanying protocol (see Protocol 2: Measuring Caspase Activity by Förster Resonance Energy Transfer [Rehm et al. 2015]). Each FRET probe comprises a pair of fluorescent proteins joined by a peptide (also called the flexible linker sequence) corresponding to a caspase cleavage site (Rehm et al. 2002). The most popular FRET pair consists of the cyan fluorescent protein and yellow fluorescent protein (CFP-YFP); here, CFP is the "donor molecule" and YFP is the "acceptor molecule."

When the caspases are inactive, the CFP and YFP moieties are in close proximity and the CFP donor molecule directly transfers its excited state energy to the YFP acceptor molecule. Therefore, excitation of the donor CFP results in emission from the acceptor YFP, and the fraction of absorbed energy that is transferred to the acceptor is interpreted as the FRET efficiency. This technique is also referred to as sensitized emission-based FRET. At the same time, the transfer of energy to the acceptor reduces the CFP fluorescence quantum yield. Quantum yield, defined as the number of photons emitted per photons absorbed, is a measure of the efficiency of the fluorescence process. Caspase-mediated cleavage of the substrate peptide linking the two proteins causes their dissociation and results in a disruption of energy transfer, measured as reduced YFP emission and increased CFP emission on donor excitation (Fig. 2B). These fluorescence changes can be measured over time, with the onset of FRET disruption reflecting the time of caspase activation. Furthermore, the rate of substrate cleavage reflects the amounts of intracellular caspase activity (Rehm et al. 2002, 2006). Therefore, changes in FRET can determine both the timing and efficiency of caspase substrate cleavage, making this a very useful tool for interrogating caspase pathways in live cells.

Since the generation of the first caspase-cleavable FRET substrates, a number of further improvements have been made. CFP and YFP variants with higher quantum yields, such as Cerulean and Venus, have been developed; and the length of the flexible linker sequence between donor and acceptor has been optimized, yielding variants with 10-fold higher FRET efficiencies (Nagai and Miyawaki 2004). Traditional FRET measurements of protein–protein interactions can be prone to artifacts due to even small differences in the expression level of each fluorescent protein. In contrast, caspase FRET substrates ensure that donor and acceptor fluorophores are present at equimolar amounts, thereby significantly simplifying the analysis. Furthermore, the ratiometric analysis of fluorescence emission in sensitized emission-based FRET experiments is highly sensitive and can detect as little as 5% substrate cleavage using conventional wide-field fluorescence microscopes (Hellwig et al. 2010). The ratiometric analysis also corrects for unspecific noise that may arise from changes in cellular morphology, cell volume, and instrument focus drifts. As an alternative to ratiometric FRET analysis, FRET measurements of caspase-cleavable FRET substrates can also be conducted by fluorescence lifetime imaging microscopy (FLIM). This technique is sensitive to the lifetime of the excited state of the donor, which is shorter in the presence of the FRET acceptor. FLIM requires additional instrumentation, but, when adjusted properly, it can be more sensitive than ratiometric FRET measurements in the presence of sufficiently high fluorophore concentrations. However, FLIM measurements are less favorable for prolonged time-lapse imaging due to higher levels of phototoxicity.

The considerable overlap in the substrate specificities of each caspase means that the choice of substrate must be carefully considered when designing FRET-based caspase probes. It has been shown that changing the amino acid sequence of the linker from the caspase-3 preferred substrate, DEVD to DEVDR increased the selectivity for caspase-3 relative to caspase-8 by 20-fold (Albeck et al. 2008). Such modifications have dramatically improved the specificity of these probes. In addition, carefully designed control experiments, in which individual caspases have been depleted or post-MOMP caspase activation has been prevented, have provided the opportunity to determine the relative contribution of initiator and executioner caspases to FRET substrate cleavage in the same pathway (Rehm et al. 2002; Hellwig et al. 2008).

ADDITIONAL APPROACHES

Other approaches to measure caspase activity by live cell imaging include the use of cell-permeable fluorogenic caspase substrates and localized GFP-labeled caspase substrates that redistribute following cleavage. Several probes of these types have been developed and many are commercially or freely available. TcapQ647 is one example (Bullok and Piwnica-Worms 2005; Maxwell et al. 2009), comprising a Tat permeation peptide sequence, the caspase-3-preferred substrate DEVD, and an activat-

 Cite this introduction as *Cold Spring Harb Protoc*; doi:10.1101/pdb.top070342

able dye pair consisting of the quencher QSY and the far-red emitting fluorophore Alexa Fluor 647 (Fig. 2C). When TcapQ647 is cleaved by caspases, the fluorophore is released from the quencher, enabling detection of caspase activity. The commonly used, commercially available probe FLICA (fluorescently labeled inhibitor of caspases) is available from Immunochemistry Technologies. FLICA probes consist of a preferred caspase substrate sequence and a fluoromethyl ketone (FMK) moiety that allows the probe to bind irreversibly to the active caspase conjugated to a fluorescent tag. Therefore, the probe is retained in cells only when bound to an active caspase. One advantage of this type of probe is that it can be added directly to the cells, in contrast to BiFC and FRET approaches that require the probe to be introduced into the cell by transfection. However, often the delay in uptake of the probe into the cytosol, or the timing required for wash steps, can make precise kinetic measurements difficult.

Localized caspase substrates that redistribute on cleavage typically are constructed by fusing GFP with a caspase-cleavable linker to membrane anchors or nuclear export sequences. Caspase activity is then detected and measured by the subcellular redistribution of GFP. Such probes and their use have been described for the measurement of initiator and executioner caspases (Henderson et al. 2005; Beaudouin et al. 2013). Although their sensitivity may be lower than that of FRET-based caspase substrates, their cleavage kinetics closely correlate (Joel Beaudouin, personal communication and unpublished data). These constructs can therefore provide important insights into the subcellular activation sites of caspases.

SUMMARY

Direct imaging of caspase activation and activity in real time can be a highly informative way of investigating caspase pathways, especially when combined with complementary biochemical approaches (see Introduction of Chapter 7: Measuring Apoptosis: Caspase Inhibitors and Activity Assays [McStay and Green 2014]) and genetic models such as knockout and transgenic animals. Many challenges still remain, but the ongoing development of new probes and more user-friendly instrumentation will continue to improve these techniques, giving us superior tools to accurately dissect cellular events in caspase pathways in real time.

REFERENCES

Albeck JG, Burke JM, Aldridge BB, Zhang M, Lauffenburger DA, Sorger PK. 2008. Quantitative analysis of pathways controlling extrinsic apoptosis in single cells. *Mol Cell* 30: 11–25.

Alnemri ES, Livingston DJ, Nicholson DW, Salvesen G, Thornberry NA, Wong WW, Yuan J. 1996. Human ICE/CED-3 protease nomenclature. *Cell* 87: 171.

Baliga BC, Read SH, Kumar S. 2004. The biochemical mechanism of caspase-2 activation. *Cell Death Differ* 11: 1234–1241.

Beaudouin J, Liesche C, Aschenbrenner S, Horner M, Eils R. 2013. Caspase-8 cleaves its substrates from the plasma membrane upon CD95-induced apoptosis. *Cell Death Differ* 20: 599–610.

Bouchier-Hayes L, Oberst A, McStay GP, Connell S, Tait SW, Dillon CP, Flanagan JM, Beere HM, Green DR. 2009. Characterization of cytoplasmic caspase-2 activation by induced proximity. *Mol Cell* 35: 830–840.

Bullok K, Piwnica-Worms D. 2005. Synthesis and characterization of a small, membrane-permeant, caspase-activatable far-red fluorescent peptide for imaging apoptosis. *J Med Chem* 48: 5404–5407.

Chang DW, Xing Z, Capacio VL, Peter ME, Yang X. 2003. Interdimer processing mechanism of procaspase-8 activation. *EMBO J* 22: 4132–4142.

Fan JY, Cui ZQ, Wei HP, Zhang ZP, Zhou YF, Wang YP, Zhang XE. 2008. Split mCherry as a new red bimolecular fluorescence complementation system for visualizing protein–protein interactions in living cells. *Biochem Biophys Res Commun* 367: 47–53.

Goldstein JC, Waterhouse NJ, Juin P, Evan GI, Green DR. 2000. The coordinate release of cytochrome *c* during apoptosis is rapid, complete and kinetically invariant. *Nat Cell Biol* 2: 156–162.

Hellwig CT, Kohler BF, Lehtivarjo AK, Dussmann H, Courtney MJ, Prehn JH, Rehm M. 2008. Real time analysis of tumor necrosis factor-related apoptosis-inducing ligand/cycloheximide-induced caspase activities during apoptosis initiation. *J Biol Chem* 283: 21676–21685.

Hellwig CT, Ludwig-Galezowska AH, Concannon CG, Litchfield DW, Prehn JH, Rehm M. 2010. Activity of protein kinase CK2 uncouples Bid cleavage from caspase-8 activation. *J Cell Sci* 123: 1401–1406.

Henderson CJ, Aleo E, Fontanini A, Maestro R, Paroni G, Brancolini C. 2005. Caspase activation and apoptosis in response to proteasome inhibitors. *Cell Death Differ* 12: 1240–1254.

Kischkel FC, Hellbardt S, Behrmann I, Germer M, Pawlita M, Krammer PH, Peter ME. 1995. Cytotoxicity-dependent APO-1 (Fas/CD95)-associated proteins form a death-inducing signaling complex (DISC) with the receptor. *Embo J* 14: 5579–5588.

Maxwell D, Chang Q, Zhang X, Barnett EM, Piwnica-Worms D. 2009. An improved cell-penetrating, caspase-activatable, near-infrared fluorescent peptide for apoptosis imaging. *Bioconjug Chem* 20: 702–709.

McStay GP, Salvesen GS, Green DR. 2008. Overlapping cleavage motif selectivity of caspases: Implications for analysis of apoptotic pathways. *Cell Death Differ* 15: 322–331.

McStay GP, Green DR. 2014. Measuring apoptosis: Caspase inhibitors and activity assays. *Cold Spring Harb Protoc* doi: 10.1101/pdb.top070359.

Nagai T, Miyawaki A. 2004. A high-throughput method for development of FRET-based indicators for proteolysis. *Biochem Biophys Res Commun* **319**: 72–77.

Parsons MJ, Bouchier-Hayes L. 2015. Measuring initiator caspase activation by bimolecular fluorescence complementation. *Cold Spring Harb Protoc* doi: 10.1101/pdb.prot082552.

Rehm M, Dussmann H, Janicke RU, Tavare JM, Kogel D, Prehn JH. 2002. Single-cell fluorescence resonance energy transfer analysis demonstrates that caspase activation during apoptosis is a rapid process. Role of caspase-3. *J Biol Chem* **277**: 24506–24514.

Rehm M, Dussmann H, Prehn JH. 2003. Real-time single cell analysis of Smac/DIABLO release during apoptosis. *J Cell Biol* **162**: 1031–1043.

Rehm M, Huber HJ, Dussmann H, Prehn JH. 2006. Systems analysis of effector caspase activation and its control by X-linked inhibitor of apoptosis protein. *Embo J* **25**: 4338–4349.

Rehm M, Parsons MJ, Bouchier-Hayes L. 2015. Measuring caspase activity by Förster resonance energy transfer. *Cold Spring Harb Protoc* doi: 10.1101/pdb.prot082560.

Riedl SJ, Salvesen GS. 2007. The apoptosome: Signalling platform of cell death. *Nat Rev Mol Cell Biol* **8**: 405–413.

Schmidt C, Peng B, Li Z, Sclabas GM, Fujioka S, Niu J, Schmidt-Supprian M, Evans DB, Abbruzzese JL, Chiao PJ. 2003. Mechanisms of proinflammatory cytokine-induced biphasic NF-kappaB activation. *Mol Cell* **12**: 1287–1300.

Shyu YJ, Liu H, Deng X, Hu CD. 2006. Identification of new fluorescent protein fragments for bimolecular fluorescence complementation analysis under physiological conditions. *Biotechniques* **40**: 61–66.

Stennicke HR, Deveraux QL, Humke EW, Reed JC, Dixit VM, Salvesen GS. 1999. Caspase-9 can be activated without proteolytic processing. *J Biol Chem* **274**: 8359–8362.

Tait SW, Parsons MJ, Llambi F, Bouchier-Hayes L, Connell S, Munoz-Pinedo C, Green DR. 2010. Resistance to caspase-independent cell death requires persistence of intact mitochondria. *Dev Cell* **18**: 802–813.

Tinel A, Tschopp J. 2004. The PIDDosome, a protein complex implicated in activation of caspase-2 in response to genotoxic stress. *Science* **304**: 843–846.

Cite this introduction as *Cold Spring Harb Protoc*; doi:10.1101/pdb.top070342

Measuring Initiator Caspase Activation by Bimolecular Fluorescence Complementation

Melissa J. Parsons[1,2] and Lisa Bouchier-Hayes[1,2,3]

[1]*Center for Cell and Gene Therapy, Baylor College of Medicine, Houston, Texas 77030;* [2]*Department of Pediatrics-Hematology, Baylor College of Medicine, Houston, Texas 77030*

Initiator caspases, including caspase-2, -8, and -9, are activated by the proximity-driven dimerization that occurs after their recruitment to activation platforms. Here we describe the use of caspase bimolecular fluorescence complementation (caspase BiFC) to measure this induced proximity. BiFC assays rely on the use of a split fluorescent protein to identify protein–protein interactions in cells. When fused to interacting proteins, the fragments of the split fluorescent protein (which do not fluoresce on their own) can associate and fluoresce. In this protocol, we use the fluorescent protein Venus, a brighter and more photostable variant of yellow fluorescent protein (YFP), to detect the induced proximity of caspase-2. Plasmids encoding two fusion products (caspase-2 fused to either the amino- or carboxy-terminal halves of Venus) are transfected into cells. The cells are then treated with an activating (death) stimulus. The induced proximity (and subsequent activation) of caspase-2 in the cells is visualized as Venus fluorescence. The proportion of Venus-positive cells at a single time point can be determined using fluorescence microscopy. Alternatively, the increase in fluorescence intensity over time can be evaluated by time-lapse confocal microscopy. The caspase BiFC strategy described here should also work for other initiator caspases, such as caspase-8 or -9, as long as the correct controls are used.

MATERIALS

It is essential that you consult the appropriate Material Safety Data Sheets and your institution's Environmental Health and Safety Office for proper handling of equipment and hazardous materials used in this protocol.

RECIPES: Please see the end of this protocol for recipes indicated by <R>. Additional recipes can be found online at http://cshprotocols.cshlp.org/site/recipes.

Reagents

Activation or death stimulus (inducer of initiator caspases)

For caspase-2 activation, use vincristine or heat shock (1 h at 43°C–45°C, depending on the cell line). See Steps 13 and 19.

Cells

This protocol works well with adherent cells (e.g., the human HeLa and MCF-7 lines, as well as mouse embryonic fibroblasts [MEFs]); it has been optimized to limit toxicity that may result from the transfection reagent.

Cell growth medium (with and without serum, prewarmed to 37°C) <R>
Cell growth medium for imaging (prewarmed to 37°C) <R>

---

[3]Correspondence: bouchier@bcm.edu

Cite this protocol as *Cold Spring Harb Protoc*; doi:10.1101/pdb.prot082552

Lipofectamine 2000 (Invitrogen/Life Technologies) or other transfection reagent

The transfection reagent required will differ according to the cell type being transfected. When a new cell line is used, a number of different transfection conditions should be tested to find optimal parameters that have minimal cellular toxicity.

Mineral oil (if required; see Step 21)

Opti-MEM reduced serum medium (Gibco, 31895-088)

Plasmids

BiFC plasmids

Select the appropriate BiFC plasmid pairs to be used for transfection (see Table 1). For example, for caspase-2, use pBiFC.C2-CARD VC and pBiFC.C2-CARD VN, as well as the appropriate controls (see Discussion).

Fluorescent reporter plasmid (e.g., DsRed-Mito from Clontech)

The reporter plasmid is cotransfected with BiFC plasmids to label the transfected cells. The reporter, DsRed-Mito, targets the mitochondrial matrix, and it allows visualization of mitochondrial dynamics. Fragmented mitochondria are often an indication that the cell is dead or dying. For other goals, alternative appropriate reporter plasmids can be used. For example, if the aim of the experiment is to analyze the location of induced proximity of caspase-2, different fluorescently tagged reporter constructs that localize to additional specific organelles and structures (e.g., the nucleus, endoplasmic reticulum, or plasma membrane) can be used with or instead of DsRed-Mito.

Equipment

Glass-bottomed 3.5-cm culture dishes (or multiwell plates) with attached No. 1.5 (0.16–0.19 mm thickness) coverslips (MatTek Corporation)

For high-resolution imaging, we recommend glass-bottomed dishes with No. 1.5 coverslips that have an average thickness of 0.17 mm, which is optimal for most high-numerical aperture (NA) objectives.

Imaging system appropriate for the analysis to be undertaken

TABLE 1. Plasmids for BiFC

Plasmid name	Construct	Features
BiFC Plasmids[a]		
pBiFC-VC155	pBiFC-HA-VC155	Venus amino acids 155–238 in pCMV-HA backbone; the multiple cloning site is between HA and VC155
pBiFC-VN173	pBiFC-FLAG-VN173	Venus amino acids 1–172 in pFlag-CMV backbone; the multiple cloning site is between FLAG and VN173
Caspase-2 BiFC Plasmids[b]		
C2-FL-VC	pBiFC-HA-Caspase 2 FL (C303A)-VC155	Catalytically inactive full-length caspase-2; Venus amino acids 155–238
C2-FL-VN	pBiFC-FLAG-Caspase 2 FL (C303A)-VN173	Catalytically inactive full length caspase-2; Venus amino acids 1–172
C2-CARD-VC	pBiFC-HA-Caspase 2 CARD-VC155	Caspase-2 CARD domain (amino acids 1–122); Venus amino acids 155–238
C2-CARD-VN	pBiFC-FLAG-Caspase 2 CARD-VN173	Caspase-2 CARD domain (amino acids 1–122); Venus amino acids 1–172
C2-CARD(D83A/E87A)-VC	pBiFC-HA-Caspase 2 CARD (D83A/E87A)-VC155	RAIDD-binding mutant of the caspase-2 CARD domain (amino acids 1–122); Venus amino acids 155–238
C2-CARD(D83A/E87A)-VN	pBiFC-FLAG-Caspase 2 CARD (D83A/E87A)-VN173	RAIDD-binding mutant of the caspase-2 CARD domain (amino acids 1–122); Venus amino acids 1–172
C2-Pro-VC	pBiFC-HA-Caspase 2 prodomain-VC155	Caspase-2 amino acids 1–147; Venus amino acids 155–238
C2-Pro-VN	pBiFC-FLAG-Caspase 2 prodomain-VN173	Caspase-2 amino acids 1–147; Venus amino acids 1–172

[a]BiFC plasmids, developed by the Chang-Deng Hu laboratory (Purdue), are available from Addgene (http://www.addgene.org/pgvec1?f=c&cmd=showcol&colid=684).

[b]Caspase-2 (C2) BiFC plasmids, used to measure caspase-2 BiFC, are available by request from Lisa Bouchier-Hayes (Baylor College of Medicine; Bouchier-Hayes et al. 2009).

Cite this protocol as *Cold Spring Harb Protoc*; doi:10.1101/pdb.prot082552

Confocal imaging system suitable for time-lapse imaging

If kinetic analysis of caspase activation is studied, specialized equipment such as a laser scanning confocal or a spinning disk confocal microscope is required. This can include a motorized inverted microscope equipped with spherical aberration correction optics, a heated stage, and a CCD camera. Venus is excited using a 514-nm or a 488-nm laser line. A second laser line (e.g., 561 nm for DsRed-Mito) is required to excite the fluorescent reporter. Optional accessories include an incubator to control temperature and CO_2 concentration and an auto-focusing solution such as Definite Focus (Zeiss).

Fluorescence microscope equipped for green fluorescent protein (GFP) and red fluorescent protein (RFP) excitation/emission

RFP excitation/emission is required to visualize the fluorescent reporter. Venus fluorescence can be detected using GFP excitation/emission; therefore, the microscope does not have to specifically detect YFP excitation/ emission.

Most confocal and fluorescence microscopes are equipped with high-NA objectives (objectives that gather more light and resolve finer detail). For high spatial resolution, we recommend objectives with a magnification of 40× or higher. BiFC fluorescence can, however, also be determined confidently at a whole-cell level at 20× magnification.

Incubator at 37°C

Microcentrifuge tubes (1.5-mL)

Software for microscope control and image analysis of time-lapse data

Various options, including ZEN (Zeiss), SlideBook (3i), ImageJ (NIH), and MetaMorph (Molecular Devices), can be used. Each differs slightly, but should produce the same results. Because ImageJ is freely available and widely used, Steps 29–33 include details on how to perform the data analysis using this specific software.

METHOD

This protocol describes parameters for cells grown in a 3.5-cm dish (one well of a six-well plate); scale down appropriately if smaller surface areas are to be used. Determine the number of transfections to be performed (and the appropriate number of plates or wells to be used) in advance. This multiday procedure occurs over 4 d (see Fig. 1). The transfection procedure on the second day takes ~30 min to set up. Imaging is performed on the third and/or fourth days, and requires up to 1 h to set up, depending on equipment and expertise. The time-lapse experiment can be automated to run over 16 h (usually overnight). The setup and basic fluorescence analysis (e.g., scoring cells as positive for BiFC) requires basic tissue culture and microscopy expertise. More advanced expertise in confocal microscopy is required for in-depth time-lapse kinetic and positional analysis.

Transfection of Cells with BiFC Plasmid Pairs

1. Plate 1×10^5 cells in cell growth medium on each 3.5-cm glass-bottomed dish. Incubate the cells overnight at 37°C.

   See Troubleshooting.

2. On the next day, prepare the transfection reagent. For six 3.5-cm dishes, deliver 12 µL of Lipofectamine 2000 into 750 µL of Opti-MEM in a sterile 1.5-mL microcentrifuge tube. Incubate the mixture for 5 min at room temperature.

   It is important to add the Lipofectamine 2000 to the Opti-MEM (in that order) so that the Lipofectamine 2000 does not adhere to the surface of the plastic tube.

3. Prepare the BiFC plasmids to be transfected. Dilute each BiFC plasmid stock to an appropriate concentration, such that the volume to be used in each well will be 1–2 µL. Set up a separate tube for each treatment condition, and transfer the appropriate amount of BiFC plasmid pair into each tube.

   The amount of each plasmid should be optimized with respect to both the BiFC pair selected (see Table 1) and the cell type. For caspase-2, typically 20–40 ng of each plasmid in the C2-CARD pair is used for each 3.5-cm dish of HeLa cells. The same range of amounts is typically used for the C2-Pro pair as well. However, 100–200 ng of each plasmid in the C2-FL pair is used because its expression is somewhat lower. MEFs require higher amounts (250–500 ng of each plasmid in a 3.5-cm dish), likely because they show a lower transfection efficiency.

4. Dilute the stock of fluorescent reporter plasmid to the appropriate concentration, and add 1–2 µL to each tube of BiFC plasmid (from Step 3) for cotransfection.

   The recommended amount of DsRed-Mito is 10 ng per 3.5-cm dish.

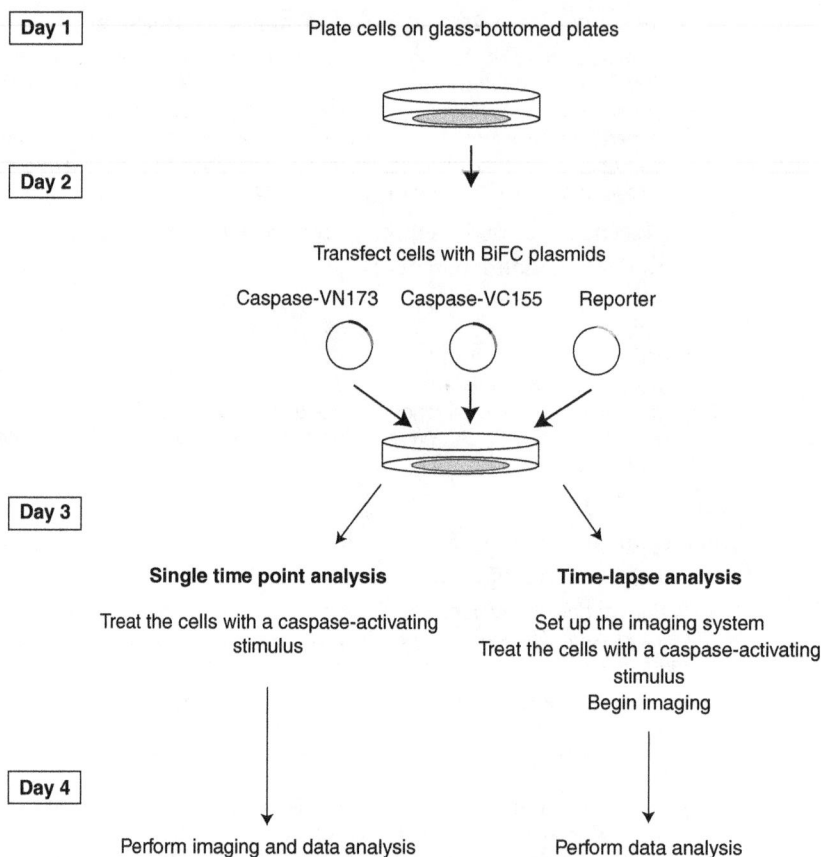

FIGURE 1. Flowchart outlining the steps and timing of a caspase BiFC experiment.

5. Dilute each plasmid mix with Opti-MEM to a total volume of 100 μL.

6. Transfer, dropwise, 100 μL of the Lipofectamine 2000 solution from Step 2 into each tube of plasmid solution.

7. Incubate the plasmid mixtures for 20 min at room temperature to allow the complexes to form.

8. Exchange the medium on each dish of cells (from Step 1) for 800 μL of serum-free cell growth medium that has been prewarmed to 37°C.

   Opti-MEM can be substituted for serum-free cell growth medium in this step, especially when using delicate cell lines that are prone to Lipofectamine 2000 toxicity.

9. Add the plasmid complexes (from Step 7) dropwise to each plate of cells.

10. Incubate the transfection complexes for 3–5 h at 37°C.

11. Remove the serum-free medium from the cells, and replace it with complete growth medium that has been prewarmed to 37°C.

12. Incubate the cells for 24 h at 37°C to allow expression of the caspase BiFC components.

    To carry out single time point acquisition of data, continue to Step 13. To set up analysis by time-lapse microscopy, proceed to Step 17.

Analysis of Caspase Activation at a Single Time Point

Treatment of Cells with Stimulus

13. Treat the transfected cells with the activation or death stimulus of choice for the appropriate period of time (usually 24 h).

Cite this protocol as *Cold Spring Harb Protoc*; doi:10.1101/pdb.prot082552

TABLE 2. Common inducers of initiator caspases

Predicted caspase to be activated	Treatment	Concentration
Caspase-2	Heat shock	43°C–45°C for 1 h[1]
	Vincristine	1 μM
Caspase-8	TNF/cycloheximide	10 ng/mL/10 μg/mL
	Anti-Fas antibody/cycloheximide	500 μg/mL/10 μg/mL
Caspase-9	Actinomycin D	500 nM–1 μM
	Staurosporine	1–2 μM

[1]The required temperature should be empirically determined for each cell line.

Caspase-2 can be activated by using 1 μM vincristine in cell growth medium (containing supplements for imaging) or by treating the cells with heat shock (1 h at 45°C for HeLa cells). Because each cell line and stimulus can behave quite differently, we recommend that the conditions for each treatment be determined experimentally. For common inducers of other initiator caspases, see Table 2.

Imaging and Data Analysis

14. At the appropriate time point after treatment with the stimulus, quantify the number of cells that are BiFC-positive. Use an epifluorescence microscope to count at least 100 cells in each of three different areas of each plate. First, count only the red cells, and then record the number of red cells that are also green (Venus-positive).

    The red cells are those that have likely taken up the BiFC plasmids. The green fluorescence is an indication that induced proximity of caspase-2 has occurred. Often cells that are not red will be green (see Fig. 2A), but to maintain an objective count, do not include them.

    See Troubleshooting.

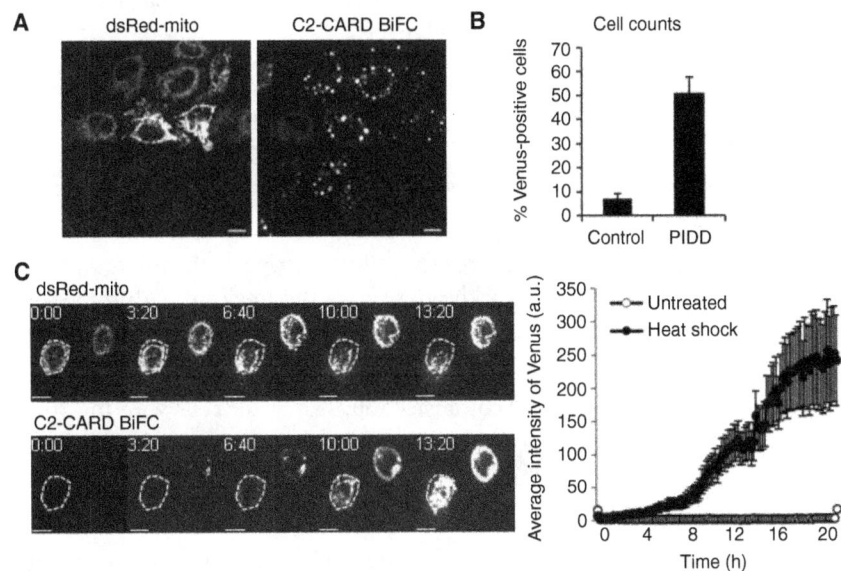

FIGURE 2. Sample data from single-time-point (A and B) and time-lapse (C) caspase BiFC experiments. (A) The pattern of fluorescence for DsRed-Mito, used as for a reporter for transfection, is shown on the *left*. The cells on the *right* were transfected with the C2-CARD BiFC pair and with an expression plasmid for p53-induced protein with death domain (PIDD), which is known to activate caspase-2. The image shows caspase-2 BiFC induced by PIDD expression. (B) Graph showing the percentage of cells from the experiment in A that underwent induced proximity of caspase-2. The total number of transfected cells (DsRed-positive cells) was counted, and the number of those cells that were also Venus-positive was expressed as a percentage of the total number of transfected cells. (C) Frames from a movie of caspase-2 BiFC in response to heat shock are shown on the *left*. The dashed circles represent the region of interest (ROI) that was drawn around the cell. The intensity of the pixels within the ROI was measured for each frame, and the data were displayed in the graph on the *right*.

15. Calculate the percentage of cells showing induced proximity of the caspase (i.e., the percentage of transfected cells that are Venus-positive).

    For an example, see Figure 2A,B.

16. Repeat the experiment (Steps 1–15) a sufficient number of times (typically, at least three times) to confirm your observations.

Analysis of Caspase Activation Using Time-Lapse Microscopy

Treatment of Cells with Stimulus and Imaging

Control experiments should be performed on untreated cells, either simultaneously or using the same conditions, to ensure that the imaging conditions are not inducing phototoxicity and that mitosis can ensue.

17. Remove the medium from the cells and replace it with cell growth medium (containing supplements for imaging) that has been prewarmed to 37°C.

    The cell growth medium for imaging contains HEPES (a buffer) and 2-mercaptoethanol (which prevents unwanted accumulation of reactive oxygen species [ROS] that can be toxic to the cells). If the microscope has an attached incubator, cells can be kept in a CO_2-enriched atmosphere for the duration of the experiment, and HEPES can be omitted from the medium.

18. Turn on the microscope and set the temperature to 37°C at least 1 h before imaging the cells. Use a temperature controller that sits on the microscope stage or preferably a microscope with an incubator enclosure to maintain the cells at a constant temperature for the duration of the time-lapse experiment.

    Setting the temperature at this point is required to give the microscope enough time to reach thermal equilibrium.

19. Treat the cells with the activation or death stimulus of choice at least 30 min before imaging.

    If caspase activation is expected to occur very rapidly on stimulus addition, then wait to add the stimulus just before image capture (at Step 27).

    Caspase-2 can be activated by using 1 μM vincristine in cell growth medium (containing supplements for imaging) or by treating the cells with heat shock (1 h at 45°C for HeLa cells). Because each cell line and stimulus can behave quite differently, we recommend that the conditions for each treatment be determined experimentally. For common inducers of other initiator caspases, see Table 2.

20. Carefully place the dish (or slide chamber) onto the microscope stage using the correct adaptor.

21. Turn on the CO_2 source to the microscope. If the microscope does not have an incubator with a humidified CO_2 workhead, overlay the medium with mineral oil to prevent evaporation of the medium from the dish.

22. Locate the transfected cells by looking for cells that express the reporter. If the reporter DsRed-Mito was used for cotransfection, search for cells with the 561-nm laser on.

    Transfected cells can also be located by eye in widefield mode. If the transfection efficiency was low, this approach may be easier for identifying transfected cells.

23. Focus the cells using a 40× (or 60×/63×) oil objective.

24. Using a positive control, empirically determine the least amount of laser light and shortest exposure times required to detect the Venus signal.

    Venus is excited using the commonly available 488- or 514-nm Argon laser line or by the 488–489-nm line of a diode pumped solid state (DPSS) laser. At the start of the time-lapse, no Venus fluorescence should be observed. Therefore, it is important to determine in a separate experiment (or by using a positive control within the experiment) the laser light and exposure time required to image the cells when they have undergone caspase BiFC. Choose settings for which the Venus signal is clearly resolved and not saturated. Some trial and error is required to determine the optimal settings, but once identified, the same settings can usually be applied to a series of experiments, provided no changes are made to the lasers (if a laser is replaced or realigned, the settings must be optimized again). Imaging cells with low levels of laser light and short exposure times will minimize phototoxicity.

 Cite this protocol as *Cold Spring Harb Protoc*; doi:10.1101/pdb.prot082552

25. Empirically determine the least amount of laser light and shortest exposure times required to detect the signal of the reporter plasmid.

> *The reporter DsRed can be excited with a 561-nm DPSS laser.*

26. If using a motorized XY stage and a microscope with multifield capabilities, choose multiple field positions.

> *This feature allows different fields of view and therefore multiple populations of cells (e.g., treated vs. untreated cells) to be assessed side by side.*

27. If the cells were not treated with the activation or death stimulus in Step 19, treat them now. Check each field position again to confirm that the cells are still in the correct focal plane and position.

> *If the stimulus is delivered at this step, great care must be taken not to disrupt the position of the plate when adding the drug. If the focal plane and/or position have changed, adjust the position of the objective (in XY and Z) accordingly before starting image capture.*

28. Set the time interval between image capture, and begin imaging.

> *The time interval depends on how fast caspase activation is expected to occur in response to the stimulus and the duration of the time lapse. Because there is increased potential for phototoxicity with an increased number of images, larger intervals are recommended for longer experiments. For caspase-2 BiFC in response to heat shock, imaging over a 16- to 20-h period with 10-min intervals between frames yields good results (see Fig. 2C for an example). Standard proapoptotic stimuli such as staurosporine or actinomycin D are also typically imaged over a 16-h period with 5- to 10-min intervals between frames. If caspase activation occurs more rapidly, the time between frames can be reduced. Control experiments in untreated cells should be performed to determine the maximum number of images that can be taken over a given time period without compromising viability of the cells.*
>
> *See Troubleshooting.*

Data Analysis

Time-lapse data of the induction of the induced proximity of initiator caspases, as measured by BiFC, can be displayed graphically. Cells are measured for changes in the average intensity of Venus fluorescence within the cell, where increases in Venus intensity are representative of induced proximity. These types of data can be analyzed as described in Steps 29–33. The example in Figure 2C shows the increase in the average intensity of Venus over time, which represents increased caspase-2-induced proximity in response to heat shock.

29. Using the appropriate tool, draw a region of interest (ROI) around each of the cells that is to be analyzed. Select a small region in an area of the image where there are no cells for the background measurement.

> *In ImageJ, go to Analyze → Tools → ROI manager. Draw ROIs using the polygon tool and add to the ROI manager by pressing "t" or clicking "Add." Once all ROIs have been generated, save them as a file (ROI manager → More → Save).*

30. Use the software to measure the average intensity of pixels in each ROI in each frame of the movie in the Venus emission channel. Export the results to an Excel spreadsheet.

> *This measure represents the brightness of Venus or the extent of caspase BiFC.*
>
> *In the ROI manager of ImageJ, go to Measure → Results → Set Measurements. Select Mean gray value. To record the average intensities of multiple ROIs for the entire stack select the "multi measure" option (ROI manager → More → Multimeasure).*

31. For each cell, subtract the background fluorescence value at each time point.

32. Average the values at each time point, and plot the results.

> *This analysis generates a read-out for a population of cells. Single-cell traces can also be generated that can be especially useful for analyzing asynchronous caspase activation events.*

33. Add error bars to each data point by calculating the standard error of the mean (SEM) of the cells for each time point.

> *SEM measures the precision of the sample mean and is used as a statistic here to show the sampling distribution.*

TROUBLESHOOTING

Problem (Step 1): Cells do not adhere well to the glass.

Solution: Precoat the glass with collagen, fibronectin, or poly-L-lysine to improve the attachment of the cells. Cells do not adhere to glass surfaces as well as they do to plastic surfaces. However, standard plastic tissue culture dishes and glass slides are too thick for use in imaging with many objectives because of their short working distance. Also, plastic is not sufficiently transparent for fluorescence and frequently is auto-fluorescent. Some newer epifluorescence microscopes are equipped with long-distance objectives that will pass through standard plastic tissue culture plates. Therefore, the characteristics of the objectives available should be assessed when planning the experiment.

Problem (Step 14): High-background Venus fluorescence is observed in transfected cells that have not been exposed to a death stimulus.

Solution: Titrate the amount of the caspase BiFC plasmid pair to a level where, in the absence of treatment, the Venus fluorescence is negligible.

Problem (Step 14): Floating cells interfere with image analysis.

Solution: Add a caspase inhibitor such as qVD-OPH (20 μM) to inhibit apoptosis and to prevent cells that have undergone apoptosis from detaching. Add qVD-OPH directly to the cell growth medium (containing supplements for imaging) at the same time as the death or activation stimulus. However, if downstream events such as Annexin V binding or mitochondrial outer membrane permeabilization (MOMP) are simultaneously being measured, the use of qVD-OPH is not recommended.

Problem (Step 14): When DsRed-Mito is used as the reporter, a dull mitochondrial pattern visible in many cells when excited by GFP. This pattern is usually only seen by eye under the GFP filter when using widefield microscopy.

Solution: This problem is caused by immature DsRed-Mito. When DsRed is synthesized, it develops a dim green fluorescence by forming the same chromophore that is present in GFP. A second oxidation reaction then generates the red chromophore. Do not count these cells as positive. It is best to remove these cells from the count as being ambiguous. This problem is particularly evident when DsRed-Mito expression is very strong. If the problem persists, reduce the amount of DsRed-Mito plasmid that is transfected. The immature DsRed-mito is not usually detected by confocal analysis using a YFP laser because there is less crossover between the excitation/emission spectra of YFP and DsRed.

Problem (Step 28): The movie loses focus.

Solution: Even small changes in temperature can lead to significant focal drift. The cells should be allowed to equilibrate to the desired temperature for at least 30 min before focusing on the cells. Some microscopes are equipped with autofocus solutions (e.g., Definite Focus [Zeiss]) that counteract drifting in the z-direction. These features significantly overcome focal drift problems.

DISCUSSION

The advantages of BiFC for the study of initiator caspase activation are numerous. By far the most significant advantage is the ability to measure caspase activation in a single cell, which allows for precise ordering of molecular events on a cell-by-cell basis. Because the split Venus fragments will not fluoresce until caspase monomers linked to each half of Venus are brought together in close proximity, BiFC also allows for precise temporal and spatial visualization of caspase activation. Additionally, unlike cell-permeable caspase substrates, which have overlapping specificities (McStay et al. 2008; see also Introduction of Chapter 7: Measuring Apoptosis: Caspase Inhibitors and Activity Assays [McStay and Green

Cite this protocol as *Cold Spring Harb Protoc;* doi:10.1101/pdb.prot082552

2014]), the nature of BiFC affords the researcher unmatched specificity, as the initiator caspase in question is being used to drive assembly of the intact Venus. This technique has been used successfully to show that the induced proximity of caspase-2 in response to heat shock, cytoskeletal disruption, and alpha toxin occurs in the cytoplasm (Bouchier-Hayes et al. 2009; Imre et al. 2012). It has not been established whether the same strategy would work for other initiator caspases (e.g., caspase-8 or -9), but in principle, this should be feasible as long as the correct controls are used.

It is critical to optimize each caspase BiFC pair independently and to select the correct negative controls for each caspase. High-level coexpression of the VN173 or the VC155 fragments as well as of many BiFC fusion proteins, produces detectable fluorescence (Shyu et al. 2006). Therefore, the expression of the VN173 or VC155 fragments alone (or the coexpression of both fragments) cannot be used as negative controls. It is critically essential to include controls in which the interaction interface of the caspase is mutated to disrupt binding to its adaptor protein. Constructs can be created with BiFC plasmids using standard molecular cloning techniques (e.g., see Table 1). For example, two residues in the caspase-2 CARD, E87 and D83, have been reported to mediate the binding of caspase-2 to its adaptor protein RAIDD (Duan and Dixit 1997). Accordingly, mutation of both of these residues to alanine results in a substantial impairment of caspase-2 BiFC (Bouchier-Hayes et al. 2009). Similarly, R13A or R56A mutations in the CARD of caspase-9 have been reported to disrupt caspase-9 binding to Apaf-1 (Qin et al. 1999). In the caspase-8 DED, mutation of both F122 and L123 to glycine abolishes its binding to FADD (Yang et al. 2005). A suitable secondary control is to express the caspase BiFC pair in cells deficient for the adaptor protein that recruits the caspase to its activation platform (e.g., RAIDD$^{-/-}$ MEF for caspase-2, FADD$^{-/-}$ MEF for caspase-8, or Apaf-1$^{-/-}$ MEF for caspase-9).

The caspase BiFC strategy can be used to follow the assembly of activation platforms for initiator caspases over time but can also be used to dissect the molecular requirements for distinct caspase activation platforms. For example, although PIDD and RAIDD appear to form the core components of the caspase-2 activation platform, studies using genetic models suggest that PIDD and RAIDD are not always essential for caspase-2 activation, and alternate caspase-2 activation platforms may exist (Berube et al. 2005; Manzl et al. 2009). Caspase-2 BiFC can determine in a context-specific manner if and how caspase-2 is activated in the absence of PIDD or RAIDD. This strategy also provides a unique way to monitor where in the cell distinct activation platforms localize in response to different stimuli, observations that can provide valuable clues to caspase function. In addition, the caspase BiFC system does not contribute any artificial additional activity. Because they typically comprise either the pro-domain of the caspase or a catalytically inactive version, the constructs used are inactive and therefore will not induce apoptosis or other downstream events on their own. Surprisingly, at least in the case of caspase-2, the prodomain or CARD BiFC constructs do not act as dominant negative inhibitors of the pathway, because caspase-2 BiFC did not prevent or delay downstream events such as MOMP (Bouchier-Hayes et al. 2009). This situation allows simultaneous measurement of the downstream consequences of initiator caspase activation.

Imaging of other cellular events, such as MOMP, DNA fragmentation, or phosphatidyl serine externalization, can be easily incorporated into caspase BiFC experiments. However, it is important to note that, at this point, the technique has been optimized to use Venus only as the fluorescent molecule. Therefore, when analyzing additional events, care should be taken not to overlap with Venus fluorescence excitation and emission spectra. The ability to simultaneously analyze more than one initiator caspase in the same cell would be a huge advance for the existing technique. Unfortunately, complementation using other split fluorescent proteins, such as Cerulean or mLumin, have not yet been optimized for initiator caspases (Shyu et al. 2006; Chu et al. 2009) but the availability of these tools represents an exciting future direction for this imaging strategy.

RELATED TECHNIQUES

See Protocol 2: Measuring Caspase Activity by Förster Resonance Energy Transfer (Rehm et al. 2015) for a FRET-based method to measure initiator (and executioner) caspase activity. For general discussions of caspase assays, see Introduction: Imaging-Based Methods for Assessing Caspase Activity in Single Cells (Parsons et al. 2015) and Introduction of Chapter 7: Measuring Apoptosis: Caspase Inhibitors and Activity Assays (McStay and Green 2014).

RECIPES

Cell Growth Medium

For HeLa cells

Fetal bovine serum (FBS)	10% (v/v)
Penicillin/streptomycin (10,000 units/mL of penicillin; 10,000 µg/mL of streptomycin, Gibco, 15140-122)	1% (v/v)
L-Glutamine (200 mM)	1% (v/v)

For MEF cells, combine the above with the following

Nonessential amino acids (NEAA) (Gibco, 11140-076)	1% (v/v)
Sodium pyruvate (100 mM)	1% (v/v)

Prepare in Dulbecco's modified Eagle's medium (DMEM) or Roswell Park Memorial Institute (RPMI) medium. Store at 4°C for up to 1 mo.

Cell Growth Medium for Imaging

Prepare cell growth medium with the following supplements.

HEPES (pH 7.2–7.5)	20 mM
2-Mercaptoethanol	55 µM

Phenol red–free medium can be used to reduce artifacts caused by autofluorescence, but it is not essential. Add supplements fresh, before using medium.

ACKNOWLEDGMENTS

The work described here was supported by an award from the Texas Children's Hospital Research Pilot to Lisa Bouchier-Hayes.

REFERENCES

Berube C, Boucher LM, Ma W, Wakeham A, Salmena L, Hakem R, Yeh WC, Mak TW, Benchimol S. 2005. Apoptosis caused by p53-induced protein with death domain (PIDD) depends on the death adapter protein RAIDD. *Proc Natl Acad Sci* 102: 14314–14320.

Bouchier-Hayes L, Oberst A, McStay GP, Connell S, Tait SW, Dillon CP, Flanagan JM, Beere HM, Green DR. 2009. Characterization of cytoplasmic caspase-2 activation by induced proximity. *Mol Cell* 35: 830–840.

Chu J, Zhang Z, Zheng Y, Yang J, Qin L, Lu J, Huang ZL, Zeng S, Luo Q. 2009. A novel far-red bimolecular fluorescence complementation system that allows for efficient visualization of protein interactions under physiological conditions. *Biosens Bioelectron* 25: 234–239.

Duan H, Dixit VM. 1997. RAIDD is a new 'death' adaptor molecule. *Nature* 385: 86–89.

Imre G, Heering J, Takeda AN, Husmann M, Thiede B, zu Heringdorf DM, Green DR, van der Goot FG, Sinha B, Dotsch V, et al. 2012. Caspase-2 is an initiator caspase responsible for pore-forming toxin-mediated apoptosis. *EMBO J* 31: 2615–2628.

Manzl C, Krumschnabel G, Bock F, Sohm B, Labi V, Baumgartner F, Logette E, Tschopp J, Villunger A. 2009. Caspase-2 activation in the absence of PIDDosome formation. *J Cell Biol* 185: 291–303.

McStay GP, Salvesen GS, Green DR. 2008. Overlapping cleavage motif selectivity of caspases: Implications for analysis of apoptotic pathways. *Cell Death Differ* 15: 322–331.

McStay GP, Green DR. 2014. Measuring apoptosis: Caspase inhibitors and activity assays. *Cold Spring Harb Protoc* doi: 10.1101/pdb.top070359.

Parsons MJ, Rehm M, Bouchier-Hayes L. 2015. Imaging-based methods for assessing caspase activity in single cells. *Cold Spring Harb Protoc* doi: 10.1101/pdb.top070342.

Rehm M, Parsons MJ, Bouchier-Hayes L. 2015. Measuring caspase activity by Förster resonance energy transfer. *Cold Spring Harb Protoc* doi: 10.1101/pdb.prot082560.

Qin H, Srinivasula SM, Wu G, Fernandes-Alnemri T, Alnemri ES, Shi Y. 1999. Structural basis of procaspase-9 recruitment by the apoptotic protease-activating factor 1. *Nature* 399: 549–557.

Shyu YJ, Liu H, Deng X, Hu CD. 2006. Identification of new fluorescent protein fragments for bimolecular fluorescence complementation analysis under physiological conditions. *Biotechniques* 40: 61–66.

Yang JK, Wang L, Zheng L, Wan F, Ahmed M, Lenardo MJ, Wu H. 2005. Crystal structure of MC159 reveals molecular mechanism of DISC assembly and FLIP inhibition. *Mol Cell* 20: 939–949.

Cite this protocol as *Cold Spring Harb Protoc*; doi:10.1101/pdb.prot082552

Measuring Caspase Activity by Förster Resonance Energy Transfer

Markus Rehm,[3,4,5] Melissa J. Parsons,[1,2] and Lisa Bouchier-Hayes[1,2,5]

[1]Center for Cell and Gene Therapy, Baylor College of Medicine, Houston, Texas 77030; [2]Department of Pediatrics-Hematology, Baylor College of Medicine, Houston, Texas 77030; [3]Centre for Systems Medicine, Royal College of Surgeons in Ireland, Dublin 2, Ireland; [4]Department of Physiology and Medical Physics, Royal College of Surgeons in Ireland, Dublin 2, Ireland

Förster resonance energy transfer (FRET) occurs across very short distances (in the nanometer range) between donor and acceptor fluorophores that overlap in their emission and absorption spectra. FRET-compatible green fluorescent protein (GFP) variants that are fused to short peptide linkers containing caspase cleavage sites can be used to measure caspase activity. In the intact probes, the donor and acceptor fluorophores are in close proximity, and FRET is highly efficient. On caspase activation, proteolysis of the linker occurs, and the donor is separated from the acceptor. This results in a disruption of resonance energy transfer and an increase in donor fluorescence quantum yield; this event is typically referred to as sensitized emission or donor unquenching. A number of highly sensitive FRET probes based on the cyan fluorescent protein–yellow fluorescent protein (CFP-YFP) pair, or improved variants thereof, have been developed to detect intracellular caspase activities. In this protocol we describe how to use FRET-based caspase substrates and time-lapse imaging to measure caspase activity in cells undergoing apoptosis.

MATERIALS

It is essential that you consult the appropriate Material Safety Data Sheets and your institution's Environmental Health and Safety Office for proper handling of equipment and hazardous materials used in this protocol.

RECIPES: Please see the end of this protocol for recipes indicated by <R>. Additional recipes can be found online at http://cshprotocols.cshlp.org/site/recipes.

Reagents

Apoptosis-inducing stimulus appropriate for the caspase (see Table 1)

Cells (adherent cell lines) transfected with the appropriate reporters and controls (see Discussion, Table 1, and Protocol 1: Measuring Initiator Caspase Activation by Bimolecular Fluorescence Complementation [Parsons and Bouchier-Hayes 2015])

Cells expressing a FRET-based caspase reporter (e.g., CFP-YFP probes)

Cells expressing a noncleavable FRET reporter (e.g., CFP-YFP pairs with uncleavable linkers)

Cells expressing only the donor or only the acceptor fluorophore (e.g., CFP and YFP)

[5]Correspondence: mrehm@rcsi.ie; bouchier@bcm.edu

Cite this protocol as Cold Spring Harb Protoc; doi:10.1101/pdb.prot082560

TABLE 1. Examples of FRET-based caspase sensors and their experimentally tested behavior

Construct	Apoptosis-inducing stimulus	Cell system	FRET disruption	Main contributing caspases	References
CFP-DEVG-YFP	Any	COS-7, HeLa	No	Non-cleavable	Rehm et al. 2002; Tyas et al. 2000
CFP-DEVD-YFP	Staurosporine (0.1–3 μM)	HeLa, MCF-7, HCT-116, DLD-1, COS-7	Yes	Caspase-3, -7	O'Connor et al. 2008; Rehm et al. 2002, 2006; Takemoto et al. 2003; Tyas et al. 2000
CFP-DEVD-YFP	Etoposide (10 μM)	HeLa	Yes	Caspase-3, -7	Rehm et al. 2002
CFP-DEVD-YFP	TRAIL (10–100 ng/mL)/ cycloheximide (1–2.5 μg/mL)	HeLa	Yes	Caspase-8, -10, -3, -7	Albeck et al. 2008
CFP-DEVDR-YFP	TRAIL (10–100 ng/mL)/ cycloheximide (1–2.5 μg/mL)	HeLa	Yes	Caspase-3, -7	Albeck et al. 2008
CFP-IETD-YFP	TRAIL (10–1000 ng/mL)/ cycloheximide (1 μg/mL)	HeLa, MCF-7	Yes	Caspase-8, -10, -3, -7	Hellwig et al. 2008, 2010
CFP-IETD-YFP	TNFα (100 ng/mL)/cycloheximide (1 μg/mL)	HeLa	Yes	Caspase-8, -10, -3, -7	Hellwig et al. 2010
CFP-IETD-IETD-YFP	TRAIL (10–100 ng/mL)/ cycloheximide (1–2.5 μg/mL)	HeLa	Yes	Caspase-8, -10, -3, -7	Albeck et al. 2008

A number of laboratories have independently generated and distributed highly efficient FRET-based caspase sensors that are optimized for initiator and executioner caspase activity measurements (see Nagai and Miyawaki 2004 and references listed in Table 1).

Cell growth medium (prewarmed to 37°C) <R>

Cell growth medium for imaging (prewarmed to 37°C) (optional; see Step 4) <R>

Mineral oil (if required; see Step 6)

Equipment

Epifluorescence or confocal imaging system suitable for time-lapse imaging and equipped for CFP and YFP excitation/emission as well as for acquisition of YFP emission on CFP excitation (FRET channel) with optimized filters

A conventional epifluorescence or confocal microscope equipped with a heated stage and objective heater is required. Optional accessories include heated incubation chamber, autofocus solution, or a motorized stage for multiposition measurements. The setups should be optimized for detection sensitivity; hard-coated filters (such as Semrock, Chroma); and sensitive detectors such as electron multiplying charge coupled device (EMCCD) cameras are recommended. Some vendors offer detection systems and light path components for the parallel acquisition of CFP and FRET channel emission signals. These systems can significantly reduce phototoxicity.

Glass-bottomed culture dishes with four chambers (e.g., Nunc Lab-Tek Chamber Slide System or equivalent)

Incubator at 37°C

Software for image acquisition and analysis

Options include the commercially available software programs MetaMorph (Molecular Devices) and SlideBook (3i), which have optional plug-ins that assist in FRET analysis (FRET module in MetaMorph; FRD module in SlideBook). Alternatively, the freely available ImageJ software (NIH) requires the installation of the "Image Calculator Plus" and the "BG subtraction from ROI" plug-ins (http://rsbweb.nih.gov/ij/). Equivalent plug-ins are also available from Albeck et al. (2008). The RiFRET plug-in for ImageJ is particularly suited for time-lapse experiments and is freely available from Roszik et al. (2009; http://www.biophys.dote.hu/rifret/). Because ImageJ is freely available and widely used, we have included additional information in Steps 22–35 on how to perform the processing steps in this software environment.

METHOD

This multiday experiment occurs over 3 d (see Fig. 1). The imaging requires up to an hour to set up, depending on the equipment and expertise. The time-lapse experiment can be automated to run for 8–48 h (depending on the time required for apoptosis induction by different drugs).

Cite this protocol as *Cold Spring Harb Protoc*; doi:10.1101/pdb.prot082560

Phase 1—Instrument adjustments and control measurements
Multiple iterations may be required to identify optimal conditions; total duration 3–14 d

Day 1   Plate cells expressing FRET probe (e.g., cleavable CFP-YFP probe) and control constructs (e.g., only CFP and only YFP) on glass-bottomed plates (Steps 1–2)

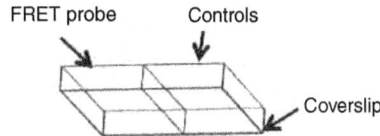

Days 2–3   Prepare for time-lapse imaging (Steps 3–7)
Perform instrument adjustments and control measurements (Steps 8–16 and Step 23)

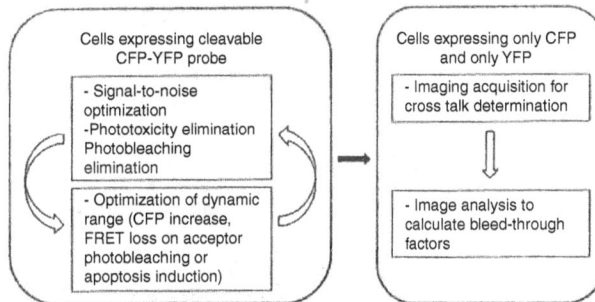

Phase 2—Routine FRET measurements

Day 1   Plate cells expressing FRET probes (e.g., cleavable and uncleavable CFP-YFP probes) (Steps 1–2)

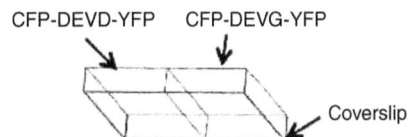

Day 2   Prepare for time-lapse imaging (Steps 3–7)
Begin routine FRET measurements (Steps 17–21)

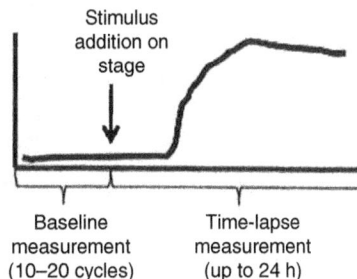

Day 3   Perform image processing and data analysis (Steps 22–35)

FIGURE 1. Flowchart outlining the steps for measuring caspase activity by FRET.

Preparation of Cells for Time-Lapse Imaging

1. Plate 2×10^4 cells expressing a FRET-based caspase reporter or an appropriate control construct in cell growth medium in each well of a four-chamber glass-bottomed plate.

   Scale up or down appropriately according to the surface area of the dish or well.
   See Discussion.

2. Incubate the cells for 24 h at 37°C to allow the cells to adhere to the glass surface.

3. Turn on the microscope and set the temperature to 37°C at least 1 h before imaging the cells. Use a temperature controller that sits on the microscope stage or preferably a microscope with an

incubator enclosure to maintain the cells at a constant temperature for the duration of the time-lapse experiment.

Setting the temperature at this point is required to give the microscope enough time to reach thermal equilibrium.

4. Remove the medium from the cells and replace it with cell growth medium (containing supplements for imaging) that has been prewarmed to 37°C.

The supplements in the cell growth medium for imaging are HEPES (a buffer) and 2-mercaptoethanol (which prevents unwanted accumulation of reactive oxygen species [ROS] that can be toxic to the cells). The inclusion of these reagents in the medium may not be necessary. Appropriately adjusted instrumentation and acquisition parameters may reduce phototoxicity and/or photobleaching to negligible amounts (see Steps 12–14), but 2-mercaptoethanol can be added as a ROS scavenger if needed. If the microscope has an attached incubator, cells can be kept in unsupplemented medium in a CO_2-enriched atmosphere for the duration of the experiment.

5. Place the dish (or chamber slide) on the microscope stage using the correct adaptor at least 30 min before imaging the cells.

6. Turn on the CO_2 source to the microscope. If the microscope does not have an incubator with a humidified CO_2 workhead, overlay the medium with mineral oil to prevent evaporation of the medium from the dish.

7. Locate and focus on the cells of interest.

For CFP-YFP probes, it is easiest to identify suitable regions using the YFP fluorescence channel. If required, multiple positions can be selected when working with motorized stages. This feature allows multiple treatments and multiple fields of view to be assessed as part of a single experiment. For single-cell quantitative kinetic analysis, choose objectives of 20× or higher. For a more simplistic single-cell-based scoring of effector caspase activation, even 4× objectives may be suitable (Rehm et al. 2009).

To continue with instrument adjustment and measurement of controls, continue with Step 8. To carry out routine FRET measurements, proceed to Step 17.

Instrument Adjustment and Control Measurements

This section describes important steps for instrument adjustment and for running important controls before carrying out routine FRET measurements. These steps include the optimization of automated image acquisition, determination of channel cross talk, and the reduction or elimination of phototoxicity and photobleaching.

8. Adjust the instrument for automated acquisition of three channels (donor excitation/emission, donor excitation/acceptor emission [FRET], and acceptor excitation/emission).

It is advisable to also capture a transmission light image (differential interference contrast [DIC] or phase contrast) to document cellular morphology.

9. Determine suitable acquisition settings using cells expressing the cleavable caspase FRET probe. As a general guideline, keep excitation intensities low and compensate with prolonged exposure times if needed.

Strong signal changes are expected in the donor and FRET channels (signal increases and decreases, respectively).

10. Adjust the donor channel to provide low signals in untreated cells at baseline, and adjust the FRET channel to provide high intensities. Adjust the acceptor channel (YFP) to provide signals that allow an easy segmentation of cells from the background noise (cells should be clearly visible but signals should not be saturated).

When capturing images with digital cameras, select a bit depth of 12 bit rather than 8 bit to allow for sufficient dynamic range.

11. Test the channels to avoid signal saturation in the donor channel or complete loss of signal in the FRET channel on FRET disruption.

i. To trigger FRET disruption, use standard agents that reliably induce apoptosis in your cell system (see Table 1).

Cite this protocol as Cold Spring Harb Protoc; doi:10.1101/pdb.prot082560

Alternatively, acceptor photobleaching can be performed when using laser-scanning microscopes. In this case, bleach YFP with the 514-nm laser line at 100% intensity, and observe the changes in signal in the donor and FRET channels.

ii. If signals in the donor channel become saturated or if signals in the FRET channel are lost on FRET disruption, readjust the acquisition settings (excitation intensity and exposure time).

Avoid both scenarios because the full-probe cleavage event cannot be followed otherwise.

12. Using the settings determined in Steps 8–11, run a time-lapse experiment with untreated cells expressing the caspase FRET probe.

The temporal resolution (time interval between images) requirment depends on the events that need to be observed. Effector caspase activation and substrate cleavage can occur in minutes, whereas initiator caspase activities may last over hours. Capturing images every 1–2 min is typically sufficient to capture rapid downstream events, whereas longer intervals can be used for prolonged upstream events.

13. Analyze the data for phototoxicity.

Spontaneous cell death either should not occur or should not exceed the amount of residual cell death observed outside the field of view (this can be easily checked at the end of the experiment). For long-term experiments (>18 h), cell division in the field of view should be detectable in proliferating cells. FRET substrate cleavage should not be detectable; see Troubleshooting.

14. Analyze the data for photobleaching.

Photobleaching can be determined by comparing cellular fluorescence intensities early and late during the experiment in each channel. Intensity changes should be insignificant.

15. Repeat Steps 12–14, adjusting the imaging settings until phototoxicity and photobleaching is minimized.

These settings can be maintained for all subsequent experiments. If required, FRET measurements can be corrected for photobleaching using more advanced protocols described in the literature (Zal and Gascoigne 2004).

16. Image the donor-only and the acceptor-only cells. Use the settings determined in Step 15 to capture images in the donor, FRET, and acceptor channels.

These images will be used to calculate bleed-through (see Step 23). See Troubleshooting.

Routine FRET Measurements for the Detection of Caspase Activity

The optimized imaging settings determined in Steps 8–16 can be used to carry out routine FRET measurements as described here.

17. Select one or multiple fields of view (see Step 7).

18. Initiate automated time-lapse imaging using the settings determined in Steps 8–16.

19. Image the cells for 10–20 loops.

This step will establish the baseline before treatment.

20. Treat the cells with the selected apoptosis inducing stimulus, and document the time/loop at which the stimulus was added. Analyze the acquired images according to Steps 22–35.

Treating the cells on the stage rather than working with pretreated cells is particularly important when investigating upstream initiator caspase activation. For example, initiator caspases-8 and -10 can be activated within minutes after addition of death ligand (Albeck et al. 2008; Hellwig et al. 2008).

See Troubleshooting.

21. Repeat the experiment a sufficient number of times (typically three) to validate your findings. In addition, conduct a control experiment with cells expressing a noncleavable FRET probe to exclude the possibility of FRET disruption occurring independently of caspase activity.

Alternatively, control experiments can be conducted in the presence of sufficiently high concentrations of a caspase inhibitor such as zVAD-fmk.

Image Processing and Data Analysis

A range of different FRET indices have been defined (Zal and Gascoigne 2004). These are all ratiometric analyses that correct for differences in probe expression between cells and for nonspecific signal changes that affect all measurement channels (e.g., changes in cellular morphology, volume, minor focus drifts, or fluctuations). The most popular indices are the donor/acceptor (CFP/YFP) ratio, the FRET/acceptor ratio (FRET/YFP), and the FRET/donor or donor/FRET (CFP/FRET) ratio. Donor/acceptor and FRET/acceptor ratios can be used together, with the donor/acceptor ratio expected to increase and the FRET/acceptor ratio expected to decrease on FRET substrate cleavage (Hellwig et al. 2008). Likewise, donor/FRET (CFP/FRET) ratios are frequently used and provide a large dynamic range because both channels are sensitive to substrate cleavage. For simplicity, this section assumes the use of a CFP-YFP pair and the donor (CFP)/FRET ratio, but other pairs and indices can be substituted. This part of the protocol is optimized for conventional wide-field epifluorescence microscopy using CCD camera-based image acquisition, but can also be applied to the analysis of images acquired by laser scanning confocal microscopy. Images generated by ratiometry can be used to observe the substrate cleavage events (Fig. 2A), and extracted single-cell-based traces of FRET substrate cleavage provide detailed quantitative insight into caspase activation times and activities in relation to other signaling events (Fig. 2B).

22. Subtract background noise from CFP, FRET, and YFP channels.

    i. Load the images or the image stack for each of the fluorescence channels.

       See Troubleshooting.

    ii. Draw and save a region of interest (ROI) that will serve as a background region. Make sure that the background region is large enough to capture the average background intensity appropriately. Also ensure that cells do not enter this background region in any of the images (scroll through the image stack to ensure this is the case).

       Cell movement and detached or floating (dead) cells may distort the data.

    iii. Subtract the average intensity of the background from each respective image in the CFP, YFP, and FRET channels to obtain background-corrected stacks, and save these.

    iv. Repeat the background subtraction for the images captured from cells expressing only CFP or YFP.

       In ImageJ, run the "BG subtraction from ROI" plug-in. Alternatively, apply the "rolling ball" background subtraction (ImageJ Menu → Process → Subtract background). Refer to the online user guide to adjust the rolling ball radius correctly for your type of images. As a general guideline, a sufficiently large radius is required; therefore, select a radius that exceeds the size of the largest cells.

23. Calculate the bleed-through from donor and acceptor into the FRET channel by analyzing images from cells expressing only CFP or YFP.

    i. After background subtraction, compare the intensities between the channels. Draw and save ROIs that capture CFP-expressing cells, and calculate the ratio of the average intensities measured in the FRET and CFP channels to obtain the donor bleed-through factor:

    $$\text{FRET/CFP} = \text{Donor bleed-through.}$$

       In ImageJ, open the ROI manager (ImageJ menu → Analyze → Tools → ROI manager) to assist in storing ROIs and measuring mean intensities.

    ii. Repeat Step 23.i for the FRET and YFP channel signals for a cell expressing only YFP to obtain the acceptor bleed-through.

       These ratios describe the fraction of donor and acceptor fluorescence that cross talk into the FRET channel. For an optimally configured imaging system, the acceptor bleed-through should be negligible for CFP-YFP caspase substrates. This procedure should be performed only once before setting up a series of experiments, as long as no changes are made to the image acquisition settings.

24. Correct the FRET channel for donor and acceptor cross talk.

    i. Multiply the CFP channel images by the bleed-through factor (see Step 23.i), and subtract the resulting images from the FRET channel according to the following formula:

    $$[\text{FRET} - (\text{Donor bleed-through}) \times \text{CFP}] = \text{FRET}_{\text{Donor corrected.}}$$

A

B

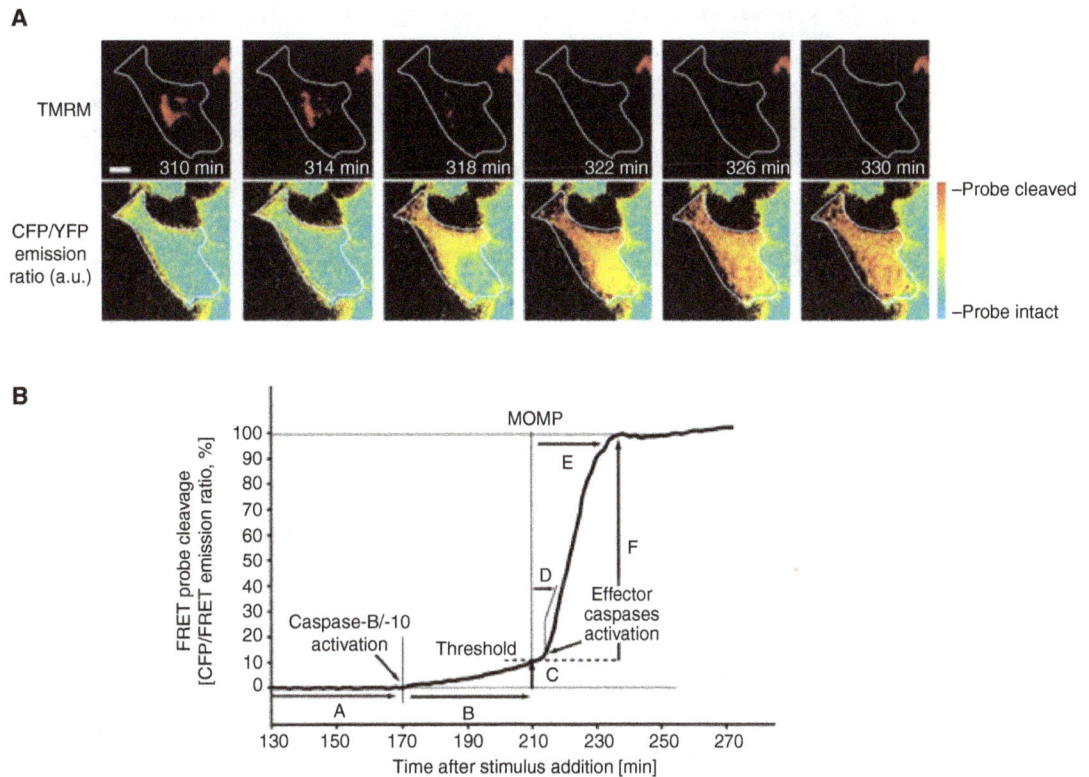

FIGURE 2. Measurement of caspase activation and activity by FRET probe cleavage. (*A*) Representative ratiometry images of FRET disruption in a HeLa cell expressing a CFP-DEVD-YFP FRET probe. FRET disruption following treatment with 1 μM staurosporine is shown as the CFP/YFP emission ratio using a pseudocolor scale. Note that probe cleavage in the cytosol precedes cleavage in the nucleus. Loss of TMRM fluorescence indicates mitochondrial depolarization. The time stamp indicates minutes after drug addition. The scale bar represents 5 μm. Panel modified and reproduced from Rehm et al. (2006) with kind permission from Nature Publishing Group. (*B*) The graph shows a representative result of an epifluorescence time-lapse experiment tracking cleavage of a CFP-DEVD-YFP FRET probe. The response of a single HeLa cell treated with 100 ng TRAIL/1 μg/mL cycloheximide is displayed. Images were acquired at 2-min intervals. Multiple parameters can be quantified from this trace as indicated by bold letters. (*A*) The time from drug addition to the activation of initiator caspases-8/-10, (*B*) the duration of caspases-8/-10 activity until induction of MOMP, (*C*) the amount of cleaved substrate observed at the time of MOMP (MOMP threshold; detectable, e.g., by parallel measurement of MOMP markers), (*D*) the delay between MOMP and effector caspase-3/-7 activation, (*E,F*) the duration and amount of substrate cleavage during apoptosis execution. Additional parameters that are not shown here can be determined as well, including the time of cell shrinkage and blebbing (as determined directly from images) as well as the time of plasma membrane permeabilization (as determined by fluorescence loss caused by FRET probe leakage into the surrounding medium).

> *For image multiplication in ImageJ, enter the donor bleed-through value in the "Multiply" graphical user interface (GUI) (ImageJ menu → Process → Math → Multiply). For image subtraction, define the operation in the "Image Calculator" GUI (ImageJ menu → Process → Image calculator).*

    ii. If necessary, correct the resulting FRET$_{Donor\ corrected}$ images for acceptor bleed-through.

    iii. Save the corrected FRET images.

25. Segment cells from the background in the background corrected-acceptor images by loading the background-corrected YFP images and setting an intensity threshold that optimally covers the cells and excludes background areas.

> *The brightness of YFP and the less phototoxic direct excitation of the acceptor typically provide the images with the best signal-to-noise ratios for cell segmentation.*
>
> *In ImageJ, select "dark background" in the Threshold interface and adjust the sliders appropriately (ImageJ Menu → Image → Adjust → Threshold). Ensure that the threshold is optimal for the entire stack. If needed, cutoff values can be defined manually by pressing the "Set" button.*

26. **Convert the thresholded acceptor images to a stack of binary masks.**

The generation of a binary image is a standard option in all recommended image processing software packages. For ImageJ, the binary conversion can be found in the Process menu entry (ImageJ Menu → Process → Binary → Make Binary). Select black background for this operation. Note that in ImageJ and MetaMorph, the default binary image will be an 8-bit image with the background set to an intensity of zero and the cell regions set to 255. Divide the binary image stack by 255 to obtain a new binary image stack in which cell region intensities have the value 1. In ImageJ, go to ImageJ Menu → Process → Math → Divide. The binary menu in ImageJ also provides options to dilate or erode the mask, if needed.

27. **Multiply the binary masks with the CFP$_{Background\ corrected}$ and FRET$_{Donor\ corrected}$ image stacks.**

This operation sets noncellular regions to zero and significantly improves the quality of the subsequently generated ratiometric images and plotted traces of single-cell FRET probe cleavage.

In ImageJ, this multiplication is performed using the Image Calculator GUI (ImageJ Menu → Process → Image Calculator).

28. **Divide the new CFP image stack by the new FRET image stack to obtain a ratiometric CFP/FRET image stack. Include a scaling factor if necessary.**

$$CFP/FRET \times scaling\ factor$$

It is usually necessary to include a scaling factor during the division to stretch the results across the bit depth of the resulting image. This is because the initial CFP channel intensity is usually lower than the initial FRET channel intensity, resulting in a value <1 after division. This would be interpreted as a pixel intensity of zero. Depending on the relative brightness of the CFP and FRET images, scaling factors between 100 and 10,000 typically provide good results. If scaling is not performed, the resulting pixel intensities may be too low to visualize and analyze FRET disruption properly.

For Image J, this operation requires the "Calculator Plus" plug-in. Open the plug-in, select the "Divide" operation and assign CFP and FRET stacks as nominator and denominator channels, respectively. Adjust the scaling factor "k1" as needed (keep k2 = 0). Save the resulting stack.

29. **Use the resulting ratiometric image stack to display FRET substrate cleavage as changes in signal intensity.**

Brightness/contrast adjustment of the ratiometric stack will allow FRET substrate cleavage to be displayed as an increase in cellular intensity. A pseudocolor lookup table (LUT) (e.g., "Royal") can be assigned to assist in visualizing intensity changes (see Fig. 2A). To carry out this task in ImageJ, go to ImageJ menu → Lookup Tables.

Images from selected time points can be converted to 24-bit RGB and saved for presentation purposes. In ImageJ, go to ImageJ Menu → Image → Type → RGB color.

30. **To plot single-cell kinetics of caspase FRET substrate cleavage, draw ROIs around the cells that are to be analyzed.**

The quality of the resulting traces strongly depends on the quality of the ROIs, cell density, and cell movement. Dynamic regions and object-tracking options that are provided in some software packages often fail in these types of analyses. Static regions need to be drawn to optimally capture the cell for the duration of the event (plus additional baselines before and after). Rapid cleavage events of effector caspase substrates (that occur within minutes) during apoptosis execution, therefore, are easy to analyze. In contrast, analysis of the slow and submaximal substrate cleavage by initiator caspases requires regions that optimally capture individual cells for durations of several hours. Regions may be larger than the cells to allow for cell movement.

In ImageJ, multiple ROIs can effectively be managed using the ROI manager (ImageJ Menu → Analyze → Tools → ROI manager). Once all ROIs have been generated, save these as a file (ROI manager → More → Save).

31. **Set a threshold in the ratiometric stack to segment the cells from the background.**

The threshold sliders need to be adjusted to exclude the zero-value background area.

In ImageJ, go to ImageJ Menu → Adjust → Threshold.

32. **For all ROIs, record the average signal intensities for the entire stack.**

For ImageJ, the ROI manager provides this as the "multi measure" option (ROI manager → More → Multi measure). Save and export the results.

Cite this protocol as *Cold Spring Harb Protoc*; doi:10.1101/pdb.prot082560

33. Plot the data against time to visualize substrate cleavage kinetics. Display additionally recorded processes or events as needed.

Excel and similar programs are the most widely used options. Several image analysis software packages also allow direct plotting and exporting of the data. (Semi-)automated options are preferred when analyzing large numbers of cells for convenience. Plotted traces can be normalized if needed.

34. Collect quantitative information from individual traces as needed. The parameters that can be measured depend on the experimental design and the measurement of parallel events.

Figure 2B provides an example of single-cell measurement of FRET probe cleavage by initiator and executioner caspases. Multiple parameters such as onset of probe cleavage, duration, and amount of probe cleavage can be quantified for an individual cell. Data collected from multiple cells then allows evaluation of cell-to-cell variability for these parameters.

35. Display representative traces as needed, and combine these with an overall quantitative analysis of the results.

The quantified processes in different experimental conditions can be statistically compared. Detailed examples for initiator and executioner caspase activity analysis can be found in the literature (e.g., Albeck et al. 2008; Hellwig et al. 2008; Laussmann et al. 2012; Rehm et al. 2002, 2006; Spencer et al. 2009).

TROUBLESHOOTING

Problem (Step 13): FRET changes are detected in control measurements with the uncleavable FRET substrate.

Solution: FRET disruption could be caused by acceptor photobleaching. Readjust the imaging settings to reduce photobleaching (shorten excitation time and reduce excitation intensities). Boost signal-to-noise by camera pixel binning if required. If basal rates of FRET disruption, which may be caused by probe degradation, are still detected, correct traces of treated cells for cleavage rates are measured in control measurements. Furthermore, necrotic cell death may result in unspecific probe cleavage and FRET disruption on cell rupture.

Problem (Step 16): Bleed-through of donor emission is observed in FRET channel.

Solution: Bleed-through usually occurs when using CFP-YFP FRET pairs because CFP has a long emission shoulder and results in detectable signal contribution to the FRET channel. Furthermore, the FRET channel can be contaminated by acceptor emission from direct excitation of the bright YFP/Venus fluorophore. This can be corrected for by calculating the bleed-through factor (see Step 23).

Problem (Step 20): FRET disruption is not detected.

Solution: Use other methods (e.g., Annexin V labeling followed by flow cytometry) to ensure that cells are undergoing caspase-dependent apoptosis in the time frame of the time-lapse experiment.

Problem (Step 20): It is unclear whether substrate cleavage has gone to completion.

Solution: At the end of the experiment, photobleach the acceptor. Acceptor photobleaching detects donor unquenching and can be used to determine whether all of the substrate has been cleaved. Donor unquenching occurs only if a portion of the FRET substrate is still intact. This step is best performed using a confocal microscope. Use a 514-nm Argon ion laser line for YFP bleaching to avoid donor bleaching.

Problem (Step 20): The unexpected death of imaged cells is observed.

Solution: At the end of the experiment, compare cell death in the field of view with cell death outside of the fields of view for each treatment condition. If death is only observed in the imaged field, it is likely that the stimulus used induces unwanted phototoxicity. Certain stimuli such as doxorubicin act as photosensitizers and may induce excessive cell death in the field of view. Control for this by further adjusting the imaging parameters as described in Steps 8–11.

Problem (Step 22.i): A high level of pixel-to-pixel variability (noise) is observed in the images.

Solution: Before background subtraction, convolve all images with an image kernel (typically a 5×5 Gaussian). Alternatively, try a "Gaussian blur," which is often available as a preset filter. This reduces the spatial resolution of the measurement; however, the more homogeneous signals of both cells and background allow a significantly improved object segmentation and ratiometry.

DISCUSSION

Experimental Considerations

Numbers of Cells

When considering the scale required for each experiment, the number of cells required depends on the process under investigation. Because FRET substrate cleavage during apoptosis execution typically is kinetically invariant, analyzing data from 10 to 20 responding cells per condition will provide a very precise description of the event. If the time of onset is of interest, higher cell numbers are required because of the temporal asynchrony in caspase activation. This is particularly important also when working with low, submaximal concentrations of apoptosis inducers; in these cases, the experiment may require the analysis of hundreds of cells (Rehm et al. 2009; Spencer et al. 2009). It is therefore worthwhile to record estimates on the number of cells observed and responding (dying) during each experiment. Because image analysis and data plotting are only conducted subsequently, these records will allow you to keep track of whether a sufficient number of events is being analyzed for a statistically solid analysis of cell-to-cell variability in responses.

Controls

The design of each FRET-based experiment should incorporate a number of important controls. Cross talk and cross-excitation of two fluorophores are intrinsic problems of FRET measurements; these events occur when the spectra of fluorophores used in the experiment are overlapping, thus making it difficult to distinguish the activity of each on its own. Therefore, the use of control cells expressing only the donor (e.g., CFP) and only the acceptor (e.g., YFP) is essential to determine the contribution of donor and acceptor cross talk into the FRET channel. Inclusion of a control experiment with cells expressing a noncleavable version of the FRET probe is essential to exclude caspase-independent FRET disruption. As an alternative, control experiments can be conducted in the presence of sufficiently high concentrations of a caspase inhibitor such as zVAD-fmk. All of the probes described in this protocol are also suitable for flow cytometric measurements. However, these population-based FRET-based measurements typically must be interpreted as activity measurements of multiple caspases, each capable of cleaving the substrate caused by the overlapping specificities of their substrate motifs (see Introduction of Chapter 7: Measuring Apoptosis: Caspase Inhibitors and Activity Assays [McStay and Green 2014]).

Applications

Many advances in cell death research have been made using FRET-based caspase probes (Spencer and Sorger 2011). Before the availability of these tools, population-based analysis—such as immunoblotting of effector caspase processing—indicated that caspase activation proceeds over hours. In contrast, single-cell data showed that effector caspases are activated rapidly and cleave the entire pool of substrate within minutes (Rehm et al. 2002). Estimates of when caspase activation occurred after the initial step of mitochondrial outer membrane permeabilization (i.e., post-MOMP) were based on the timing of measurable caspase-dependent events (e.g., phosphatidyl serine externalization as measured by Annexin V binding), which can occur as much as 2 h

Cite this protocol as *Cold Spring Harb Protoc*; doi:10.1101/pdb.prot082560

after MOMP has reached completion (Goldstein et al. 2000). More direct measurements of FRET changes on cleavage of the caspase-3 preferred substrate DEVD showed that caspase activation occurs within minutes of the release of second mitochondria-derived activator of caspases (Smac) or of cytochrome-c and loss of mitochondrial membrane potential (Dussmann et al. 2003; Rehm et al. 2003).

FRET substrates such as IETD-based CFP-YFP probes (probes containing the cleavage sequence motif IETD) have also been used to study both initiator and effector caspase activity. The use of these probes has provided quantitative and kinetic insight into apoptosis initiation and its transition into a rapid cell death execution phase (Albeck et al. 2008; Hellwig et al. 2008). When considering these types of experiments, however, it is important to take into account the potentially poor specificity of the FRET probes. For example, caspases-3/-7 (the most efficient DEVDases) efficiently cleave after the caspase-8/-10-preferred motif IETD. Furthermore, the relative abundance of different caspases within the cells significantly contributes toward observable activities. In HeLa cells, the intracellular IETDase activity of effector caspases greatly exceeds the activity of caspases-8/-10 (Hellwig et al. 2008). There-fore, results from these experiments should be considered with these limitations in mind, and addi-tional control experiments may be required to determine which caspases contribute to probe cleavage (Rehm et al. 2002; Albeck et al. 2008; Hellwig et al. 2008).

One major advantage of time-lapse analysis of cleavage of FRET-based caspase substrates is that it can be used to separate activities of different caspases over time. Measuring IETDase activity together with a marker for MOMP allows the separation of the initiation and execution phases in time. For example, the upstream IETDase activity of caspases-8/-10 and the subsequently added IETDase activity of executioner caspases can be determined in the same cell demonstrating the importance of a temporal analysis of caspase activities (Fig. 2B). One limitation of the protocol for FRET-based analysis of caspase activity described here is that, because of the rapid intracellular diffusion of free FRET probes, a spatial analysis to determine where caspases are activated first is only feasible when comparing substrate cleavage between subcellular compartments that are separated by diffusion barriers. For example, during staurosporine-induced apoptosis, DEVD FRET substrate cleavage in the cytosol of HeLa cells precedes cleavage in the nucleus (Fig. 2A).

It should also be noted that measured substrate cleavage may not simply be compared with or correlated to classical in vitro measurements of caspase activities. Whereas in vitro experiments typically measure Michaelis–Menten kinetics to determine the maximal enzyme activity at substrate saturation, the in cellulo scenario differs in that enzyme activities are not constant over time and that substrate amounts are limited. Intracellular FRET measurements therefore provide physiolog-ical substrate cleavage kinetics that are more accurately described by mass action kinetics (Rehm et al. 2006).

The determination of intracellular caspase activities by FRET substrate cleavage also played a fundamental role in a number of systems biological studies that provided unprecedented insight into apoptosis signal transduction and the control of cell death decisions. For example, DEVD-FRET substrate cleavage data confirmed mathematical predictions that the relative abundance of XIAP and other downstream apoptosis proteins can set a death decision threshold that effectively suppresses apoptosis execution after MOMP (Rehm et al. 2006). Similarly, systems approaches have been successfully used to develop a mathematical model that predicts the kinetic relationship between caspase-8 activation (through cleavage of an IETD-based FRET probe), MOMP (release of a mito-chondrial intermembrane space protein), and caspase-3 activation (through cleavage of a DEVD-based FRET probe) under various conditions and revealed that the inhibitor of apoptosis, XIAP, sequesters executioner caspases in an inactive state before MOMP in the extrinsic pathway (Albeck et al. 2008; Aldridge et al. 2011). This advance contributed to growing evidence that indicates that levels of XIAP largely determine how dependent a cell is on the mitochondrial pathway for inducing apoptosis in response to death receptor ligation (Jost et al. 2009). With time-lapse fluorescence microscopy becoming increasingly available in many cell biological laboratories and research centers, FRET-based imaging of caspase activities is likely to become a more widely used method in the coming years.

Chapter 6

RELATED TECHNIQUES

Another imaging-based method for analyzing initiator caspase activation is described in Protocol 1: Measuring Initiator Caspase Activation by Bimolecular Fluorescence Complementation (Parsons and Bouchier-Hayes 2015). For general discussions of caspase assays, see Introduction: Imaging-Based Methods for Assessing Caspase Activity in Single Cells (Parsons et al. 2015) and Introduction of Chapter 7: Measuring Apoptosis: Caspase Inhibitors and Activity Assays (McStay and Green 2014).

RECIPES

Cell Growth Medium

For HeLa cells

Fetal bovine serum (FBS)	10% (v/v)
Penicillin/streptomycin (10,000 units/mL of penicillin; 10,000 µg/mL of streptomycin, Gibco, 15140-122)	1% (v/v)
L-Glutamine (200 mM)	1% (v/v)

For MEF cells, combine the above with the following

Nonessential amino acids (NEAA) (Gibco, 11140-076)	1% (v/v)
Sodium pyruvate (100 mM)	1% (v/v)

Prepare in Dulbecco's modified Eagle's medium (DMEM) or Roswell Park Memorial Institute (RPMI) medium. Store at 4°C for up to 1 mo.

Cell Growth Medium for Imaging

Prepare cell growth medium with the following supplements.

HEPES (pH 7.2–7.5)	20 mM
2-Mercaptoethanol	55 µM

Phenol red–free medium can be used to reduce artifacts caused by autofluorescence, but it is not essential. Add supplements fresh, before using medium.

ACKNOWLEDGMENTS

Work described here was supported by an award from the Texas Children's Hospital Research Pilot to L.B.-H. M.R. received support from the Science Foundation Ireland, the Irish Health Research Board, the National Biophotonics and Imaging Platform Ireland (HEA PRTLI Cycle 4), and the European Union (FP7 APO-DECIDE, FP7 SYS-MEL).

REFERENCES

Albeck JG, Burke JM, Aldridge BB, Zhang M, Lauffenburger DA, Sorger PK. 2008. Quantitative analysis of pathways controlling extrinsic apoptosis in single cells. *Mol Cell* **30:** 11–25.

Aldridge BB, Gaudet S, Lauffenburger DA, Sorger PK. 2011. Lyapunov exponents and phase diagrams reveal multi-factorial control over TRAIL-induced apoptosis. *Mol Syst Biol* **7:** 553.

Dussmann H, Rehm M, Kogel D, Prehn JH. 2003. Outer mitochondrial membrane permeabilization during apoptosis triggers caspase-independent mitochondrial and caspase-dependent plasma membrane potential depolarization: A single-cell analysis. *J Cell Sci* **116:** 525–536.

Goldstein JC, Waterhouse NJ, Juin P, Evan GI, Green DR. 2000. The coordinate release of cytochrome c during apoptosis is rapid, complete and kinetically invariant. *Nat Cell Biol* **2:** 156–162.

Hellwig CT, Kohler BF, Lehtivarjo AK, Dussmann H, Courtney MJ, Prehn JH, Rehm M. 2008. Real time analysis of tumor necrosis factor-related

apoptosis-inducing ligand/cycloheximide-induced caspase activities during apoptosis initiation. *J Biol Chem* **283:** 21676–21685.

Hellwig CT, Ludwig-Galezowska AH, Concannon CG, Litchfield DW, Prehn JH, Rehm M. 2010. Activity of protein kinase CK2 uncouples Bid cleavage from caspase-8 activation. *J Cell Sci* **123:** 1401–1406.

Jost PJ, Grabow S, Gray D, McKenzie MD, Nachbur U, Huang DC, Bouillet P, Thomas HE, Borner C, Silke J, et al. 2009. XIAP discriminates between type I and type II FAS-induced apoptosis. *Nature* **460:** 1035–1039.

Laussmann MA, Passante E, Hellwig CT, Tomiczek B, Flanagan L, Prehn JH, Huber HJ, Rehm M. 2012. Proteasome inhibition can impair caspase-8 activation upon submaximal stimulation of apoptotic tumor necrosis factor-related apoptosis inducing ligand (TRAIL) signaling. *J Biol Chem* **287:** 14402–14411.

Cite this protocol as *Cold Spring Harb Protoc*; doi:10.1101/pdb.prot082560

McStay GP, Green DR. 2014. Measuring apoptosis: Caspase inhibitors and activity assays. *Cold Spring Harb Protoc* doi: 10.1101/pdb.top070359.

Nagai T, Miyawaki A. 2004. A high-throughput method for development of FRET-based indicators for proteolysis. *Biochem Biophys Res Commun* **319:** 72–77.

O'Connor CL, Anguissola S, Huber HJ, Dussmann H, Prehn JH, Rehm M. 2008. Intracellular signaling dynamics during apoptosis execution in the presence or absence of X-linked-inhibitor-of-apoptosis-protein. *Biochim Biophys Acta* **1783:** 1903–1913.

Parsons MJ, Bouchier-Hayes L. 2015. Measuring initiator caspase activation by bimolecular fluorescence complementation. *Cold Spring Harb Protoc* doi: 10.1101/pdb.prot082552.

Parsons MJ, Rehm M, Bouchier-Hayes L. 2015. Imaging-based methods for assessing caspase activity in single cells. *Cold Spring Harb Protoc* doi: 10.1101/pdb.top070342.

Rehm M, Dussmann H, Janicke RU, Tavare JM, Kogel D, Prehn JH. 2002. Single-cell fluorescence resonance energy transfer analysis demonstrates that caspase activation during apoptosis is a rapid process. Role of caspase-3. *J Biol Chem* **277:** 24506–24514.

Rehm M, Dussmann H, Prehn JH. 2003. Real-time single cell analysis of Smac/DIABLO release during apoptosis. *J Cell Biol* **162:** 1031–1043.

Rehm M, Huber HJ, Dussmann H, Prehn JH. 2006. Systems analysis of effector caspase activation and its control by X-linked inhibitor of apoptosis protein. *EMBO J* **25:** 4338–4349.

Rehm M, Huber HJ, Hellwig CT, Anguissola S, Dussmann H, Prehn JH. 2009. Dynamics of outer mitochondrial membrane permeabilization during apoptosis. *Cell Death Differ* **16:** 613–623.

Roszik J, Lisboa D, Szollosi J, Vereb G. 2009. Evaluation of intensity-based ratiometric FRET in image cytometry–approaches and a software solution. *Cytometry Part A* **75:** 761–767.

Spencer SL, Gaudet S, Albeck JG, Burke JM, Sorger PK. 2009. Non-genetic origins of cell-to-cell variability in TRAIL-induced apoptosis. *Nature* **459:** 428–432.

Spencer SL, Sorger PK. 2011. Measuring and modeling apoptosis in single cells. *Cell* **144:** 926–939.

Takemoto K, Nagai T, Miyawaki A, Miura M. 2003. Spatio-temporal activation of caspase revealed by indicator that is insensitive to environmental effects. *J Cell Biol* **160:** 235–243.

Tyas L, Brophy VA, Pope A, Rivett AJ, Tavare JM. 2000. Rapid caspase-3 activation during apoptosis revealed using fluorescence-resonance energy transfer. *EMBO Rep* **1:** 266–270.

Zal T, Gascoigne NR. 2004. Photobleaching-corrected FRET efficiency imaging of live cells. *Biophys J* **86:** 3923–3939.

Measuring Apoptosis: Caspase Inhibitors and Activity Assays

Gavin P. McStay[1,3] and Douglas R. Green[2]

[1]*Department of Life Sciences, New York Institute of Technology, Old Westbury, New York 11568;* [2]*Department of Immunology, St. Jude Children's Research Hospital, Memphis, Tennessee 38105*

Caspases are proteases that initiate and execute apoptotic cell death. These caspase-dependent events are caused by cleavage of specific substrates that propagate the proapoptotic signal. A number of techniques have been developed to follow caspase activity in vitro and from apoptotic cellular extracts. Many of these techniques use molecules that are based on optimal peptide motifs for each caspase and on our understanding of caspase cleavage events that occur during apoptosis. Although these approaches are useful, there are several drawbacks associated with them. The optimal peptide motifs are not unique recognition sites for each caspase, so techniques that use them may yield information about more than one caspase. Furthermore, caspase cleavage does not take into account the different caspase activation mechanisms. Recently, probes having greater specificity for individual caspases have been developed and are being used successfully. This introduction provides background on the various caspases and introduces a set of complementary techniques to examine the activity, substrate specificity, and activation status of caspases from in vitro or cell culture experiments.

INTRODUCTION TO CASPASES

Every cell and every organism that undergoes apoptotic cell death uses caspases. This family of proteases plays critical roles in inflammation and cell death, and is essential for the initiation and execution of apoptosis. Cellular signals, such as death ligands or mitochondrial outer membrane permeabilization (MOMP), activate the first of the two classes of caspases: Initiator caspases. Activated initiator caspases either cleave and activate executioner caspases, or cleave other proteins that transmit the apoptotic signal to the executioner caspases. Executioner caspases are responsible for cleaving a myriad of substrates, some of which mediate the apoptotic phenotype in dying cells.

Caspases are cysteine-dependent aspartic-acid-directed proteases. Caspase substrates are cleaved at the carboxy-terminal peptide bond of the aspartic acid residue in the peptide motif, causing a loss or gain of function of the target protein, generally favoring the proapoptotic signaling pathway. The first mammalian caspase to be identified was caspase-1. It was detected by its ability to process the inflammatory cytokine interleukin 1β (Cerretti et al. 1992; Thornberry et al. 1992). The cleavage site in interleukin 1β after aspartic acid residue 116 was the first hint regarding the specificity of caspase cleavage motifs. Since this initial discovery, the caspase family has grown substantially (Kumar 2007).

[3]Correspondence: gmcstay@nyit.edu

Cite this introduction as *Cold Spring Harb Protoc;* doi:10.1101/pdb.top070359

ACTIVATION OF CASPASES

Caspases become activated in one of the two ways, depending on whether they are initiator or executioner caspases. Initiator caspases (e.g., caspase-2, -8, -9, and -10) are activated by dimerization on large multiprotein complexes. Caspase-8 and -10 activation occurs via the death-inducing signaling complex (DISC) that is formed in response to death ligands (Medema et al. 1997; Sprick et al. 2002). Caspase-9 activation occurs on the apoptosome, which is formed when cytochrome *c* and adenosine triphosphate bind to apoptotic protease activating factor-1 (APAF-1) causing oligomerization (Li et al. 1997). Caspase-2 is activated by the PIDDosome, a poorly understood complex containing the adaptor proteins p53 inducible protein with a death domain (PIDD) and RIP-associated ICH/CED-3 homologous protein with a death domain (RAIDD) (Tinel and Tschopp 2004). In each case, the induced dimerization event results in the formation of an active site on one or both of the caspase monomers. Once dimerized, the caspase is able to cleave target substrates (Boatright et al. 2003).

Executioner caspases (e.g., caspase-3, -6, -7) exist as inactive dimers that are activated upon proteolytic cleavage of both monomers at loops containing specific cleavage sites (Berger et al. 2006a; Denault et al. 2006; Edgington et al. 2012). Cleavage of these loops results in structural changes that serve to form the active sites, allowing the executioner caspases to cleave downstream substrates (Chai et al. 2001; Riedl et al. 2001). Activated executioner caspases can be detected using antibodies that recognize caspase proforms or cleavage fragments (see Protocol 3: Detection of Caspase Activity Using Antibody-Based Techniques [McStay and Green 2014a]).

During apoptosis, all caspases undergo cleavage at similar sites located between the large and small catalytic domains. In addition, in the case of initiator caspases, there is often additional cleavage between the pro-domain and the catalytic domains. These cleavage events are mediated by autocleavage and other caspases. The function of these cleavage events differs slightly between initiator and executioner caspases. Cleavage of initiator caspases-2 and -8 stabilizes the enzyme (Baliga et al. 2004; Oberst et al. 2010). In the case of caspase-9, cleavage dissociates the enzyme from the activation platform, whereby the caspase is replaced by another pro-caspase-9 molecule in a "tick-over" mechanism (Malladi et al. 2009). With executioner caspases, cleavage is necessary and sufficient to induce their activation (Riedl and Shi 2004). This has been shown by comparing inactive and active caspase-7 structures, which revealed that cleavage of its loops freed up the caspase active site (Chai et al. 2001; Riedl et al. 2001).

CASPASE SUBSTRATES

Using combinatorial peptide libraries and purified recombinant caspases, characterization of caspase substrates has revealed the presence of a four-amino-acid consensus motif upstream of the cleavage site that is recognized for substrate proteolysis. The motif is designated as P1, P2, P3, and P4, where P1 is the amino acid closest to the cleavage site. Positions P2 and P4 contain amino acids with similar properties and positions P1 and P3 are held by specific amino acids. Based on structural studies of caspase active sites, caspases-3, -6, and -7 only accommodate small residues such as alanine or valine at P2. Initiator caspases are able to tolerate bulkier residues at this position. The P3 site is occupied by a conserved arginine residue that favors glutamate binding in the substrate pocket. Caspases-2, -3, and -7 accommodate an aspartate residue at P4 with high specificity, while in the other apoptotic caspases the P4 site tolerates a variety of amino acids. The P4 site of caspase-1 can accommodate large hydrophobic amino acids, such as tryptophan (Chereau et al. 2003). The vast majority of caspase substrates contain an aspartate residue at the P1 position. Cleavage occurs by hydrolysis of the peptide bond following this residue. Table 1 contains a summary of the motif preferences of each caspase based on structural features. These are merely preferences and do not imply that other sites cannot be cleaved by these enzymes. Indeed, misconceptions surrounding these preferences have led to misinterpretations in the literature, as discussed in more detail below.

Cite this introduction as *Cold Spring Harb Protoc*; doi:10.1101/pdb.top070359

TABLE 1. Caspase cleavage motifs

Caspase	Prediction from structure[a]	Determination of cleavage motif		
		Substrate cleavage site in cell culture[b]	Peptide cleavage by isolated caspase[c]	Peptide cleavage more effective by listed caspase[b]
1	WEHD	N/A	WEHD	N/A
2	DEXD	XXXD	VDVAD	3
3	DE(A/V)D	DEVD	DEVD	3
6	(L/I/V)E(A/V)D	(E/I/V)EVD	VEID	3
7	DE(A/V)D	D(E/Q)VD	DEVD	3
8	(L/I/V)EXD	LEVD	IETD	3, 6
9	(L/I/V)EXD	XEVD	LEHD	3, 6, 8
10	(L/I/V)EXD	XEVD	IETD	3, 6

X represents any amino acid.
[a]Based on Chereau et al. 2003.
[b]Based on McStay et al. 2008.
[c]Based on Thornberry et al. 1997.

Approximately 1000 caspase substrates have been identified (Mahrus et al. 2008; Crawford et al. 2012; Shimbo et al. 2012) with functions that encompass many cellular processes. Poly-ADP-ribose polymerase was the first apoptotic caspase substrate to be identified (Tewari et al. 1995). Recently, caspase substrates have been identified by large-scale proteomic or in silico approaches (Table 2). Caspase cleavage of a substrate can yield one of the several outcomes: The substrate loses function, gains a function, changes its location, or is subjected to altered regulation. These changes in the substrate are subsequently responsible for the apoptotic phenotype—for example, phosphatidylserine exposure on the external face of the plasma membrane, loss of mitochondrial membrane potential, and chromatin condensation. Proteomic approaches have been used to determine large numbers of caspase substrates during apoptosis (Van Damme et al. 2005; Mahrus et al. 2008). The substrates identified by these methods have been classified by annotation into specific pathways or functions, giving us some idea of processes altered by active caspase cleavage. These putative substrates must be validated using some kind of experimental screen, such as is described in Protocol 5: Verification of a Putative Caspase Substrate [McStay and Green 2014b]). On a smaller scale, putative caspase cleavage motifs in proteins can be identified by in silico prediction models (Scott et al. 2008).

The consensus substrate motifs have been used to generate short peptide-based inhibitors and substrates for each known caspase (Talanian et al. 1997; Thornberry et al. 1997; Garcia-Calvo et al. 1998). These reagents, which are commercially available, have been used extensively to test the involvement of caspases in apoptosis. Despite their utility, these reagents lack the selectivity they

TABLE 2. Caspase substrate databases and substrate prediction servers

Name	URL	Laboratory	Substrate number	Species covered	Reference
CASBAH	http://bioinf.gen.tcd.ie/casbah/	Seamus Martin	777	H, M, R, A, D, C, V	Luthi and Martin 2007
CASVM	http://www.casbase.org/casvm/index.html	Shoba Ranganathan	272	H, V	Wee et al. 2007
CasCleave	http://sunflower.kuicr.kyoto-u.ac.jp/~sjn/Cascleave/	Jiangning Song	370		Song et al. 2010
CasPredictor[a]	http://icb.usp.br/~farmaco/ Jose/ CaSpredictorfiles	Jose Ernesto Belizario	160		Garay-Malpartida et al. 2005
PMAP-CutDB[b]	http://www.proteolysis.org/proteases/	Jeffrey Smith		H, M, R, A, D, C	Igarashi et al. 2009
MEROPS[b]	http://merops.sanger.ac.uk/index.shtml	Neil Rawlings & Alan Barret		H, M, R, A, D, C	Rawlings et al. 2012

[a]Database closed, contact J.E. Belizario for further information.
[b]A large protease and substrate database with large amounts of information on caspases and substrates from many species.
H, human; M, mouse; R, rat; A, Arabidopsis; D, Drosophila; C, C. elegans; V, virus

are claimed to have for individual caspases in whole cell and in vivo assays where multiple caspases are present (Garcia-Calvo et al. 1998; Ekert et al. 1999), as well as in some in vitro assays (Berger et al. 2006b; McStay et al. 2008; Pereira and Song 2008; Benkova et al. 2009). This lack of specificity may be attributed to the expression, activation, substrate specificity, and/or activity of caspases that prevent a substrate-based tool from selecting for a single caspase when multiple caspases are present. Let us look at these reasons more closely.

Caspase protein expression levels shows cell-type variability, as was showed in a study that measured caspase protein levels in a selection of cancer cell lines (Svingen et al. 2004). Because there are no comprehensive reports on protein expression of caspases in tissues or cell lines, one should empirically determine caspase protein expression relative to a standard amount of protein or a housekeeping protein in the cell line or tissue of interest. Problems can arise when using a peptide-based substrate to study a caspase with specific-substrate cleavage ability, if that caspase is present at lower levels than another caspase that is both more highly expressed has promiscuous substrate cleavage ability. The result may be that the more abundant caspase is responsible for a significant amount of substrate cleavage although the substrate was designed to probe the activity of a different caspase (Fig. 1A).

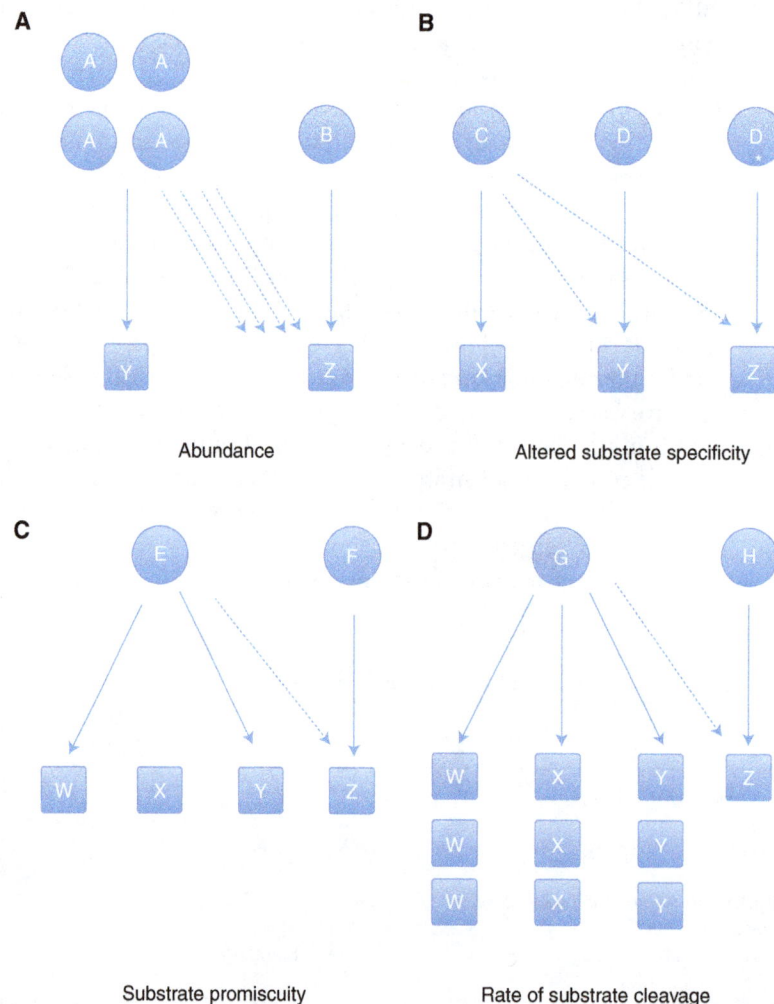

FIGURE 1. Properties of caspases that confound interpretation of assays using substrate motif-based activity probes and inhibitors. (A) Caspase A has a greater expression level than caspase B. (B) Caspase D has altered substrate specificity when in an intermediate activation step (D*) compared to the fully mature enzyme. (C) Caspase E is able to cleave many different substrates while caspase F cleaves only one particular substrate. (D) Caspase G is able to cleave many substrates in the same amount of time caspase H cleaves one substrate.

Cite this introduction as *Cold Spring Harb Protoc*; doi:10.1101/pdb.top070359

The substrate specificity of some caspases can change depending on the activation state. Since there are two caspase activation mechanisms, inference of activity by use of substrate-based tools also needs to consider the mechanism and state of activation. As was discussed previously, initiator caspases go through a multistep process in which one or both active sites can be in the active conformation. In the case of caspase-8, these different forms display different substrate preferences (Hughes et al. 2009). With executioner caspases, cleavage of a specific loop on both monomers is required to induce a conformational change in the enzyme that makes the active site available to the whole repertoire of substrates at the same time (Fig. 1B).

Initiator caspases are activated in large complexes in response to a proapoptotic stimulus. These caspases do not have to cleave their limited suite of substrates at a rapid rate. Executioner caspases, on the other hand, cleave on the order of a thousand substrates and must do so quickly to ensure the cell dismantles itself in an orderly and timely fashion and is engulfed by phagocytic cells to avoid activation of inflammation. Therefore, the degree of substrate specificity—limited for initiator caspases and broad for executioner caspases—can affect experiments using peptide-based substrate probes. The promiscuity of an executioner caspase might swamp the signal of an initiator caspase (Fig. 1C).

As noted, caspases cleave substrates at different rates. For example, caspase-3 is much more active than any other caspase at cleaving its favored substrate (Stennicke et al. 2000). This variation in cleavage rate can also confound interpretation of assays using putative caspase-specific substrates when multiple caspases are present (Fig. 1D).

INHIBITORS OF CASPASES

Caspase inhibitors are available as either pan-caspase or caspase-specific inhibitors. Depending on the assay used to measure apoptosis, these inhibitors will not always prevent cell death. This may be due to toxicity associated with caspase inhibitors, cell death that is caspase independent, or ineffective employment of the caspase inhibitor. The most common caspase inhibitors are short peptides with a moiety that covalently binds to the catalytic cysteine. These moieties include fluoromethyl ketone (FMK), chloromethyl ketone and 2,6-difluorophenoxymethyl ketone (OPH). FMK moieties are metabolized to fluoroacetate, an inhibitor of aconitase, which leads to mitochondrial toxicity in vivo and possibly necrotic features in vitro (Van Noorden 2001; Chauvier et al. 2007). The OPH moiety is less toxic and tolerated by many different cell types and is less toxic in vivo (Chauvier et al. 2007).

Caspase inhibitor studies have revealed two types of death pathways. First, initiation of the intrinsic pathway of apoptosis in the presence of caspase inhibitors does not result in apoptotic features, but cells lack clonogenic survival. This phenomenon occurs because mitochondrial outer membrane permeabilization (MOMP) is a point from which cells often cannot recover, cytochrome c is released into the cytosol, and mitochondrial-dependent energy production is severely impaired. Under most circumstances, cells are unable to circumvent this problem and will eventually die due to depletion of ATP and decreased mitochondrial functions necessary for survival, even if caspases downstream from MOMP are inhibited. Second, cells exposed to tumor necrosis factor-α in the presence of caspase inhibitors do not undergo apoptosis; instead they engage a pathway called necroptosis. Inhibition of the catalytic activity of caspase-8 prevents cleavage of receptor interacting protein-1 (RIPK1) kinase, which engages RIPK3 kinase to initiate downstream signaling that results in caspase-independent cell death (Oberst et al. 2011). Finally, to inhibit apoptosis, caspase inhibitors need to gain access to the cytosol and be present at an appropriate concentration. By monitoring characteristic executioner caspase-dependent events, such as caspase-3 activity assays or cleavage, it can be seen whether an appropriate concentration of inhibitor was used. To claim that a cell death pathway acts independently of caspases, these caspase-dependent events must be completely inhibited as the cell proceeds to die. Therefore, even if cells are treated with caspase inhibitors, it does not mean that they will necessarily survive over the long term.

NONCASPASE PROTEASES INVOLVED IN CELL DEATH

Proteases other than caspases have been implicated in some cell death paradigms that can impinge on the canonical apoptotic pathway. These include the serine protease family, of which one directly engages apoptosis. Granzyme B is a serine protease secreted from cytotoxic cells to kill infected or malignant cells. Granzyme B enters the target cell where it cleaves and activates caspase-3 directly, leading to apoptosis (Darmon et al. 1995). Also, human Granzyme B cleavage of BID causes MOMP and apoptosis (Cullen et al. 2007). Cathepsins are another family of proteases postulated to be involved in apoptosis. These are cysteine proteases that reside mainly in the lysosome and are mostly responsible for proteolysis in lysosomal degradation pathways. In some forms of cell death, cathepsins are released from lysosomes into the cytosol, where they can cleave a variety of substrates (Repnik et al. 2012). There are many cathepsin-specific inhibitors, but these proteases are also sensitive to caspase inhibitors like zVAD-FMK (Rozman-Pungercar et al. 2003). Calpains are calcium-activated cysteine proteases that have been implicated in some apoptotic pathways (Storr et al. 2011). Some inhibitors of cathepsins also inhibit calpains, including zVAD-FMK (Rozman-Pungercar et al. 2003). Unlike calpains and cathepsins, Granzyme B is not inhibited by any known caspase inhibitors.

ASSAYS TO STUDY CASPASES

Caspases degrade specific substrates after apoptosis or inflammation is initiated. Therefore, tracking caspase activity during these cellular processes should reveal the roles caspases are playing in specific signal transduction pathways. The conventional way to follow caspase activity is by monitoring cleavage of either model peptide substrates or physiological substrates of the enzyme. These assays can be performed on apoptotic cell extracts, in vitro-activated caspases (see Protocol 1: Preparation of Cytosolic Extract and Activation of Caspases by Cytochrome *c* [McStay and Green 2014c]), or on purified caspases. Using the first two systems, the contributions of multiple caspases found within a cell at a given time can be analyzed. Working with a purified caspase permits more detailed analysis of the function of that caspase. The caspases involved in specific apoptosis scenarios can be identified using substrate-based inhibitor studies (see Protocol 2: Assaying Caspase Activity In Vitro and Protocol 4: Identification of Active Caspases Using Affinity-Based Probes [McStay and Green 2014d,e]). Although the available caspase inhibitors show limited selectivity for an individual caspase, by performing complementary assays the identification of an activated caspase in the apoptotic cascade can be confirmed. These assays also aid in determining the mechanisms of caspase activation that differ among the classes of caspases and among individual caspases from the same class. The combination of these two properties of caspases is important for further understanding the contributions caspases make in many different apoptosis scenarios.

When caspase cleavage motif-based tools are used, complementary assays must be performed to validate the involvement of a suspected caspase. Caspase-3 cleaves multiple caspase cleavage motifs, and analysis of the contribution of this caspase should be determined by using the preferred caspase cleavage motif DEVD. Experiments using inhibitors must be supported with genetic approaches that lower the amount of caspase, either by using interfering RNA or cells derived from knockout animals that lack the specific caspase. Such complementary approaches have been used to identify the emerging role of caspase-8 as an activator of interleukin 1β following a variety of extracellular inflammatory stimuli (Maelfait et al. 2008; Bossaller et al. 2012; Gringhuis et al. 2012; Vince et al. 2012). These substrate motif-based reagents still represent the only source of commercially available substrates or inhibitors. A new series of activity-based probes based on natural and nonnatural amino acids have been generated, which show enhanced specificity for individual caspases in some cases (Berger et al. 2006a; Edgington et al. 2009, 2012). The activity-based probe developed for caspase-8 cross-reacts with caspase-3. The probe developed for caspase-9 cross-reacts with caspase-3 and -8. Further re-

Cite this introduction as *Cold Spring Harb Protoc*; doi:10.1101/pdb.top070359

finement of these probes and ultimately the design of true caspase-specific probes will require an approach that integrates information about structure and mechanisms of activation.

USE OF COMMERCIAL KITS TO MONITOR CASPASES

Many life science companies sell kits that provide all the necessary reagents to monitor caspases in models of apoptosis. Generally, these kits supply buffers for cell lysis and caspase assays, along with a fluorogenic substrate and inhibitors to the caspase in question. Some kits are marketed for individual caspases and others are for multiple caspases. The kits putatively for individual caspases generally contain reagents for that particular caspase, and thus have more limited utility. The kits that have broad specificity contain substrate-based reagents of all caspases and therefore allow for direct comparison among substrates, inhibitors, and caspases at the same time. It is also possible to purchase recombinant caspases either alone or as part of some assay kits. These may be worthwhile if only a limited numbers of assays are planned. For extensive studies with recombinant caspases, it is more economical to prepare recombinant caspases in-house. This rule may also hold true for caspase inhibitors and fluorogenic substrates, which may be synthesized at a core facility when a large number of experiments are planned.

CONCLUSION

The biochemical steps of caspase activation and cleavage of caspase substrates have been extensively studied since the first apoptotic caspase and substrate were identified. These studies have generated a variety of reagents and assays to interrogate the roles of caspases and their substrates in many paradigms of apoptosis. These reagents include small molecule substrates and inhibitors that can be used to determine activation of caspases and cleavage of substrates after proapoptotic stimuli. In isolation, these assays provide clues to the involvement of a particular caspase in a particular apoptosis scenario. However, these assays by themselves do not provide all the necessary information about the activation of a caspase or whether a single caspase was responsible for the observed apoptotic process. By combining these reagents and assays, while complementing them with more recently developed experimental tools, a more complete understanding of which caspases are involved and how the caspases are activated in apoptotic pathways can be achieved.

REFERENCES

Baliga BC, Read SH, Kumar S. 2004. The biochemical mechanism of caspase-2 activation. *Cell Death Differ* 11: 1234–1241.

Benkova B, Lozanov V, Ivanov IP, Mitev V. 2009. Evaluation of recombinant caspase specificity by competitive substrates. *Anal Biochem* 394: 68–74.

Berger AB, Witte MD, Denault JB, Sadaghiani AM, Sexton KM, Salvesen GS, Bogyo M. 2006a. Identification of early intermediates of caspase activation using selective inhibitors and activity-based probes. *Mol Cell* 23: 509–521.

Berger AB, Sexton KB, Bogyo M. 2006b. Commonly used caspase inhibitors designed based on substrate specificity profiles lack selectivity. *Cell Res* 16: 961–963.

Boatright KM, Renatus M, Scott FL, Sperandio S, Shin H, Pedersen IM, Ricci JE, Edris WA, Sutherlin DP, Green DR, et al. 2003. A unified model for apical caspase activation. *Mol Cell* 11: 529–541.

Bossaller L, Chiang PI, Schmidt-Lauber C, Ganesan S, Kaiser WJ, Rathinam VA, Mocarski ES, Subramanian D, Green DR, Silverman N, et al. 2012. Cutting Edge: FAS (CD95) mediates noncanonical IL-1β and IL-18 maturation via caspase-8 in an RIP3-independent manner. *J Immunol* 189: 5508–5512.

Cerretti DP, Kozlosky CJ, Mosley B, Nelson N, Van Ness K, Greenstreet TA, March CJ, Kronheim SR, Druck T, Cannizzaro LA, et al. 1992. Mo-

lecular cloning of the interleukin-1 beta converting enzyme. *Science* 256: 97–100.

Chai J, Wu Q, Shiozaki E, Srinivasula SM, Alnemri ES, Shi Y. 2001. Crystal structure of a procaspase-7 zymogen: Mechanisms of activation and substrate binding. *Cell* 107: 399–407.

Chauvier D, Ankri S, Charriaut-Marlangue C, Casimir R, Jacotot E. 2007. Broad-spectrum caspase inhibitors: From myth to reality? *Cell Death Differ* 14: 387–391.

Chereau D, Kodandapani L, Tomaselli KJ, Spada AP, Wu JC. 2003. Structural and functional analysis of caspase active sites. *Biochemistry* 42: 4151–4160.

Crawford ED, Seaman JE, Agard N, Hsu GW, Julien O, Mahrus S, Nguyen H, Shimbo K, Yoshihara HA, Zhuang M, et al. 2012. The DegraBase: A database of proteolysis in healthy and apoptotic human cells. *Mol Cell Proteomics* 12: 813–824.

Cullen SP, Adrain C, Luthi AU, Duriez PJ, Martin SJ. 2007. Human and murine granzyme B exhibit divergent substrate preferences. *J Cell Biol* 176: 435–444.

Darmon AJ, Nicholson DW, Bleackley RC. 1995. Activation of the apoptotic protease CPP32 by cytotoxic T-cell-derived granzyme B. *Nature* 377: 446–448.

Denault JB, Bekes M, Scott FL, Sexton KM, Bogyo M, Salvesen GS. 2006. Engineered hybrid dimers: Tracking the activation pathway of caspase-7. *Mol Cell* 23: 523–533.

Edgington LE, Berger AB, Blum G, Albrow VE, Paulick MG, Lineberry N, Bogyo M. 2009. Noninvasive optical imaging of apoptosis by caspase-targeted activity-based probes. *Nat Med* 15: 967–973.

Edgington LE, van Raam BJ, Verdoes M, Wierschem C, Salvesen GS, Bogyo M. 2012. An optimized activity-based probe for the study of caspase-6 activation. *Chem Biol* 19: 340–352.

Ekert PG, Silke J, Vaux DL. 1999. Caspase inhibitors. *Cell Death Differ* 6: 1081–1086.

Garay-Malpartida HM, Occhiucci JM, Alves J, Belizário JE. 2005. CaSPredictor: A new computer-based tool for caspase substrate prediction. *Bioinformatics* 21Suppl 1: i169–i176.

Garcia-Calvo M, Peterson EP, Leiting B, Ruel R, Nicholson DW, Thornberry NA. 1998. Inhibition of human caspases by peptide-based and macromolecular inhibitors. *J Biol Chem* 273: 32608–32613.

Gringhuis SI, Kaptein TM, Wevers BA, Theelen B, van der Vlist M, Boekhout T, Geijtenbeek TB. 2012. Dectin-1 is an extracellular pathogen sensor for the induction and processing of IL-1β via a noncanonical caspase-8 inflammasome. *Nat Immunol* 13: 246–254.

Hughes MA, Harper N, Butterworth M, Cain K, Cohen GM, MacFarlane M. 2009. Reconstitution of the death-inducing signaling complex reveals a substrate switch that determines CD95-mediated death or survival. *Mol Cell* 35: 265–279.

Igarashi Y, Heureux E, Doctor KS, Talwar P, Gramatikova S, Gramatikoff K, Zhang Y, Blinov M, Ibragimova SS, Boyd S, et al. 2009. PMAP: Databases for analyzing proteolytic events and pathways. *Nucleic Acids Res* 37 (Database issue): D611–D618.

Kumar S. 2007. Caspase function in programmed cell death. *Cell Death Differ* 14: 32–43.

Li P, Nijhawan D, Budihardjo I, Srinivasula SM, Ahmad M, Alnemri ES, Wang X. 1997. Cytochrome *c* and dATP-dependent formation of Apaf-1/caspase-9 complex initiates an apoptotic protease cascade. *Cell* 91: 479–489.

Luthi AU, Martin SJ. 2007. The CASBAH: A searchable database of caspase substrates. *Cell Death Differ* 14: 641–650.

Maelfait J, Vercammen E, Janssens S, Schotte P, Haegman M, Magez S, Beyaert R. 2008. Stimulation of Toll-like receptor 3 and 4 induces interleukin-1β maturation by caspase-8. *J Exp Med* 205: 1967–1973.

Mahrus S, Trinidad JC, Barkan DT, Sali A, Burlingame AL, Wells JA. 2008. Global sequencing of proteolytic cleavage sites in apoptosis by specific labeling of protein N termini. *Cell* 134: 866–876.

Malladi S, Challa-Malladi M, Fearnhead HO, Bratton SB. 2009. The Apaf-1* procaspase-9 apoptosome complex functions as a proteolytic-based molecular timer. *EMBO J* 28: 1916–1925.

McStay GP, Green DR. 2014a. Detection of caspase activity using antibody-based techniques. *Cold Spring Harb Protoc* doi: 10.1101/pdb.prot080291.

McStay GP, Green DR. 2014b. Verification of a putative caspase substrate. *Cold Spring Harb Protoc* doi: 10.1101/pdb.prot080317.

McStay GP, Green DR. 2014c. Preparation of cytosolic extract and activation of caspases by cytochrome *c*. *Cold Spring Harb Protoc* doi: 10.1101/pdb.prot080275.

McStay GP, Green DR. 2014d. Assaying caspase activity in vitro. *Cold Spring Harb Protoc* doi: 10.1101/pdb.prot080283.

McStay GP, Green DR. 2014e. Identification of active caspases using affinity-based probes. *Cold Spring Harb Protoc* doi: 10.1101/pdb.prot080309.

McStay GP, Salvesen GS, Green DR. 2008. Overlapping cleavage motif selectivity of caspases: Implications for analysis of apoptotic pathways. *Cell Death Differ* 15: 322–331.

Medema JP, Scaffidi C, Kischkel FC, Shevchenko A, Mann M, Krammer PH, Peter ME. 1997. FLICE is activated by association with the CD95 death-inducing signaling complex (DISC). *EMBO J* 16: 2794–2804.

Oberst A, Pop C, Tremblay AG, Blais V, Denault JB, Salvesen GS, Green DR. 2010. Inducible dimerization and inducible cleavage reveal a requirement for both processes in caspase-8 activation. *J Biol Chem* 285: 16632–16642.

Oberst A, Dillon CP, Weinlich R, McCormick LL, Fitzgerald P, Pop C, Hakem R, Salvesen GS, Green DR. 2011. Catalytic activity of the caspase-8-FLIP_L complex inhibits RIPK3-dependent necrosis. *Nature* 471: 363–367.

Pereira NA, Song Z. 2008. Some commonly used caspase substrates and inhibitors lack the specificity required to monitor individual caspase activity. *Biochem Biophys Res Commun* 377: 873–877.

Rawlings ND. Barrett AJ, Bateman A. 2012. MEROPS: The database of proteolytic enzymes, their substrates and inhibitors. *Nucleic Acids Res* 40(Database issue): D343–D350.

Repnik U, Stoka V, Turk V, Turk B. 2012. Lysosomes and lysosomal cathepsins in cell death. *Biochim Biophys Acta* 1824: 22–33.

Riedl SJ, Shi Y. 2004. Molecular mechanisms of caspase regulation during apoptosis. *Nat Rev Mol Cell Biol* 5: 897–907.

Riedl SJ, Fuentes-Prior P, Renatus M, Kairies N, Krapp S, Huber R, Salvesen GS, Bode W. 2001. Structural basis for the activation of human procaspase-7. *Proc Natl Acad Sci* 98: 14790–14795.

Rozman-Pungercar J, Kopitar-Jerala N, Bogyo M, Turk D, Vasiljeva O, Stefe I, Vandenabeele P, Bromme D, Puizdar V, Fonovic M, et al. 2003. Inhibition of papain-like cysteine proteases and legumain by caspase-specific inhibitors: When reaction mechanism is more important than specificity. *Cell Death Differ* 10: 881–888.

Scott FL, Fuchs GJ, Boyd SE, Denault JB, Hawkins CJ, Dequiedt F, Salvesen GS. 2008. Caspase-8 cleaves histone deacetylase 7 and abolishes its transcription repressor function. *J Biol Chem* 283: 19499–19510.

Shimbo K, Hsu GW, Nguyen H, Mahrus S, Trinidad JC, Burlingame AL, Wells JA. 2012. Quantitative profiling of caspase-cleaved substrates reveals different drug-induced and cell-type patterns in apoptosis. *Proc Natl Acad Sci* 109: 12432–12437.

Song J, Tan H, Shen H, Mahmood K, Boyd SE, Webb GI, Akutsu T, Whisstock JC. 2010. Cascleave: Towards more accurate prediction of caspase substrate cleavage sites. *Bioinformatics* 26: 752–760.

Sprick MR, Rieser E, Stahl H, Grosse-Wilde A, Weigand MA, Walczak H. 2002. Caspase-10 is recruited to and activated at the native TRAIL and CD95 death-inducing signalling complexes in a FADD-dependent manner but can not functionally substitute caspase-8. *EMBO J* 21: 4520–4530.

Stennicke HR, Renatus M, Meldal M, Salvesen GS. 2000. Internally quenched fluorescent peptide substrates disclose the subsite preferences of human caspases 1, 3, 6, 7 and 8. *Biochem J* 350 Pt 2: 563–568.

Storr SJ, Carragher NO, Frame MC, Parr T, Martin SG. 2011. The calpain system and cancer. *Nat Rev Cancer* 11: 364–374.

Svingen PA, Loegering D, Rodriquez J, Meng XW, Mesner PW Jr, Holbeck S, Monks A, Krajewski S, Scudiero DA, Sausville EA, et al. 2004. Components of the cell death machine and drug sensitivity of the National Cancer Institute Cell Line Panel. *Clin Cancer Res* 10: 6807–6820.

Talanian RV, Quinlan C, Trautz S, Hackett MC, Mankovich JA, Banach D, Ghayur T, Brady KD, Wong WW. 1997. Substrate specificities of caspase family proteases. *J Biol Chem* 272: 9677–9682.

Tewari M, Quan LT, O'Rourke K, Desnoyers S, Zeng Z, Beidler DR, Poirier GG, Salvesen GS, Dixit VM. 1995. Yama/CPP32β, a mammalian homolog of CED-3, is a CrmA-inhibitable protease that cleaves the death substrate poly(ADP-ribose) polymerase. *Cell* 81: 801–809.

Thornberry NA, Bull HG, Calaycay JR, Chapman KT, Howard AD, Kostura MJ, Miller DK, Molineaux SM, Weidner JR, Aunins J, et al. 1992. A novel heterodimeric cysteine protease is required for interleukin-1β processing in monocytes. *Nature* 356: 768–774.

Thornberry NA, Rano TA, Peterson EP, Rasper DM, Timkey T, Garcia-Calvo M, Houtzager VM, Nordstrom PA, Roy S, Vaillancourt JP, et al. 1997. A combinatorial approach defines specificities of members of the caspase family and granzyme B. Functional relationships established for key mediators of apoptosis. *J Biol Chem* 272: 17907–17911.

Tinel A, Tschopp J. 2004. The PIDDosome, a protein complex implicated in activation of caspase-2 in response to genotoxic stress. *Science* 304: 843–846.

Van Damme P, Martens L, Van Damme J, Hugelier K, Staes A, Vandekerckhove J, Gevaert K. 2005. Caspase-specific and nonspecific in vivo protein processing during Fas-induced apoptosis. *Nat Methods* 2: 771–777.

Van Noorden CJ. 2001. The history of Z-VAD-FMK, a tool for understanding the significance of caspase inhibition. *Acta Histochem* 103: 241–251.

Vince JE, Wong WW, Gentle I, Lawlor KE, Allam R, O'Reilly L, Mason K, Gross O, Ma S, Guarda G, et al. 2012. Inhibitor of apoptosis proteins limit RIP3 kinase-dependent interleukin-1 activation. *Immunity* 36: 215–227.

Wee LJ, Tan TW, Ranganathan S. 2007. CASVM: Web-server for SVM based prediction of caspase substrates cleavage sites. *Bioinformatics* 23: 3241–3243.

Cite this introduction as *Cold Spring Harb Protoc*; doi:10.1101/pdb.top070359

Preparation of Cytosolic Extracts and Activation of Caspases by Cytochrome c

Gavin P. McStay[1,3] and Douglas R. Green[2]

[1]Department of Life Sciences, New York Institute of Technology, Old Westbury, New York 11568; [2]Department of Immunology, St. Jude Children's Research Hospital, Memphis, Tennessee 38105

It can be useful to explore the caspase activation process in an in vitro setting. In this protocol, cytosolic extracts prepared from cell culture are incubated with cytochrome c and adenosine triphosphate (dATP), leading to the oligomerization of apoptotic protease activating factor-1 (APAF-1) and the formation of the apoptosome. The apoptosome serves as an activation platform for caspase-9, which binds to the apoptosome through heterodimeric caspase recruitment domain (CARD) interactions and then dimerizes. This leads to cleavage of the executioner, caspase-3. These extracts contain highly active caspases that can be analyzed using a variety of biochemical assays.

MATERIALS

It is essential that you consult the appropriate Material Safety Data Sheets and your institution's Environmental Health and Safety Office for proper handling of equipment and hazardous materials used in this protocol.

RECIPES: Please see the end of this protocol for recipes indicated by <R>. Additional recipes can be found online at http://cshprotocols.cshlp.org/site/recipes.

Reagents

Adherent cell line of interest, grown in ten 15-cm dishes

Using fewer dishes of cells generally results in a lower yield of cytosolic extract caused by inefficiencies during cell lysis. Alternatively, suspension cell lines such as Jurkat can be used. See Step 1.

Cytochrome c from horse heart (Sigma-Aldrich; 1 mM in H_2O)
dATP (10 mM in H_2O)
Homogenization buffer for caspase activation <R>
PBS (A), prechilled on ice <R>
Trypan blue (0.4% in PBS)

Equipment

Cell scraper
Centrifugal filters for microcentrifuge (0.22-μm)
Centrifuge (benchtop) at 4°C
Conical tubes (50-mL)
Graduated pipette (5-mL) for Steps 5–6

---

[3]Correspondence: gmcstay@nyit.edu

Cite this protocol as *Cold Spring Harb Protoc*; doi:10.1101/pdb.prot080275

Incubator at 37°C
Microcentrifuge at 4°C
Micropipette with 1-mL tip for Step 8
Spectrophotometer or equivalent (see Step 15)
Syringe (3 mL) and needle (22 gauge)
Tubes (1.5-mL) for microcentrifuge and ultracentrifuge
Ultracentrifuge (benchtop) at 4°C

METHOD

A flowchart of this protocol is shown in Figure 1.

Preparation of Cytosolic Extract

1. Grow ten 15-cm diameter plates of adherent cells until they are confluent.

   Alternatively, grow at least 300 mL of suspension cells (e.g., Jurkat). Centrifuge the cells to pellet them. Wash the cell pellet with 50 mL of ice-cold phosphate-buffered saline (PBS), centrifuge again, and skip to Step 5.

2. Rinse each dish with 5 mL of ice-cold PBS. Remove the PBS.

3. Scrape adherent cells off the dishes and transfer to a 50-mL conical tube. Fill the tube with ice-cold PBS to 50 mL.

4. Centrifuge the cells at 250g for 5 min at 4°C.

5. Discard the supernatant. Add 2 mL of ice-cold PBS in a 5-mL graduated pipette to the cell pellet. Resuspend the pellet in the PBS.

6. Determine the pellet volume by drawing the cell suspension into the graduated pipette.

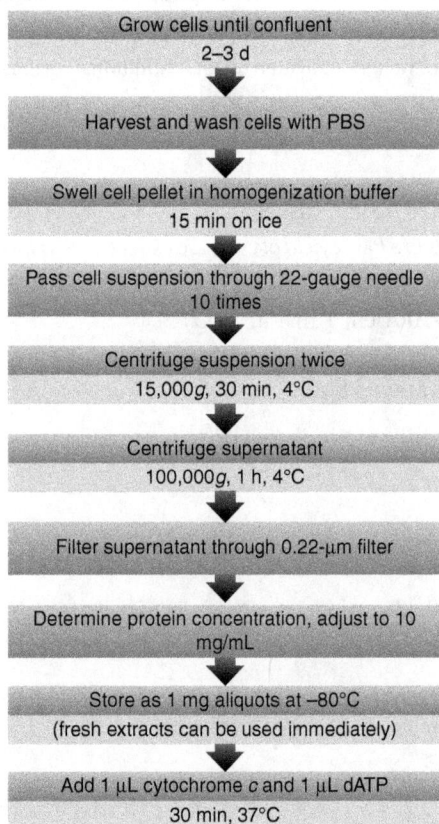

FIGURE 1. Preparation of cytosolic extracts and activation of caspases by cytochrome *c*. This procedure requires 2–3 d for cells to grow, and 3–4 h for extract preparation.

Cite this protocol as *Cold Spring Harb Protoc*; doi:10.1101/pdb.prot080275

The volume of cells equals the volume of PBS-plus-cells minus the volume of PBS. A cell pellet volume >0.5 mL is anticipated. Volumes smaller than this tend to result in lower lysis efficiency and a poorer yield.

7. Return the cell suspension to the 50-mL conical tube. Wash the cells once more by filling the conical tube to 50 mL with ice-cold PBS and centrifuging as in Step 4.

8. Remove as much of the supernatant as possible. Add a volume of homogenization buffer equal to the cell pellet volume and resuspend by passing the suspension through a 1-mL pipette tip until the suspension is smooth. Incubate the cell suspension for 15 min on ice.

9. Using a 3-mL syringe, pass the cell suspension 10 times through a 22-gauge needle.

10. Check the extent of lysis by staining 10 µL of the suspension with 10 µL of Trypan blue solution.

    Greater than 80% of the nuclei should stain blue. If this is not the case, continue homogenizing. See Troubleshooting.

11. Transfer the homogenized cell suspension to 1.5-mL microcentrifuge tubes. Centrifuge at 15,000g for 30 min at 4°C.

12. Transfer the supernatant to a fresh 1.5-mL microcentrifuge tube and centrifuge again at 15,000g for 30 min at 4°C.

13. Transfer the supernatant to a 1.5-mL ultracentrifuge tube. Centrifuge at 100,000g for 1 h at 4°C.

14. Pass the supernatant through a 0.22-µm centrifugal filter.

15. Determine the protein concentration using absorbance at 280 nm or another suitable method.

16. Adjust the sample concentration to 10 mg/mL with homogenization buffer.

    See Troubleshooting.

17. Aliquot 100 µL (1 mg) of cytosolic extract into microcentrifuge tubes. Freeze and store them at −80°C.

In Vitro Activation of Apoptosome and Downstream Caspases

18. Rapidly thaw an aliquot of frozen cytosolic extract by rolling the tube between your hands.

    Once thawed the cytosolic extract cannot be refrozen because apoptosome formation becomes less efficient with freeze–thaw cycles.

19. As a starting point, add 1 µL of cytochrome c and 1 µL of dATP to 10 µL of the thawed extract. Prepare other experimental samples and controls as appropriate (see Discussion).

20. Incubate the reactions and controls at 37°C for 30 min (different lengths of time can be used for kinetic analysis).

TROUBLESHOOTING

Problem (Step 10): Cell lysis does not occur.

Solution: For efficient cell lysis the cells must swell in the homogenization buffer. Be sure to remove all of the excess PBS (Step 8) remaining after the second wash and make sure determination of the pellet volume at Step 6 is accurate.

Problem (Step 16): Protein concentration is <10 mg/mL.

Solution: For downstream applications the extract should be at a concentration of 10 mg/mL, because cytochrome c-dependent caspase activity requires this concentration of protein to ensure efficient apoptosome assembly. The extract can be concentrated using a low-molecular-weight cut-off centrifugal concentrator. If the protein concentration is >10 mg/mL, dilute with homogenization buffer.

DISCUSSION

This method for preparing cytosolic extract is ideal for in vitro activation of the apoptosome by cytochrome *c*. The activated extract can be used for a variety of downstream applications, such as measuring caspase activity (see Protocol 2: Assaying Caspase Activity In Vitro [McStay and Green 2014a]), western blotting (see Protocol 3: Detection of Caspase Activity Using Antibody-Based Techniques [McStay and Green 2014b]), immunoprecipitation (McStay et al. 2008) or gel filtration and western blotting to analyze apoptosome formation (Cain et al. 2001).

Every set of experiments should include an unactivated sample as a control for basal caspase activity. Other experimental samples that should be considered include:

- Titrations of cytochrome *c* and/or dATP to determine dose dependency of caspase activity.

- Activated cytosolic extract with caspase inhibitor (i.e., cytochrome *c*, dATP, and caspase inhibitor) to determine the specificity of caspase activity (see Protocol 2: Assaying Caspase Activity In Vitro [McStay and Green 2014a]).

- Other conditions, such as the addition of small molecules or purified proteins.

RELATED TECHNIQUES

Similar techniques have been developed for in vitro activation of other initiator caspases. Caspase-8 activation by the death-inducing signaling complex (DISC) can be modeled using the intracellular domain of CD95 with an amino-terminal glutathione *S*-transferase (GST) tag (Hughes et al. 2009). Procedures for activating caspase-2 are less well developed. Caspase-2 can be activated and cleaved by incubation of a cytosolic extract at 37°C. This is associated with both adaptor proteins associated with caspase-2 activation, RAIDD and PIDD (Tinel and Tschopp 2004). Cells treated with pro-apoptotic stimuli provide a source of activated caspases. This is a heterogeneous source of all of the caspases expressed in the cell type of interest, which may exist as inactive or active caspases, depending on the apoptotic stimulus. Immunoprecipitation of caspases from these apoptotic lysates can be performed and analysis of the purified caspase is possible with some downstream applications.

Caspases can also be expressed recombinantly and purified to assess their individual activity. The initiator and executioner caspases are expressed differently due to the properties of the two types of enzyme. Initiator caspases need to be expressed as the carboxy-terminal portion of the protein, free of the amino-terminal pro-domain as this renders the protein insoluble. Executioner caspases can be expressed as full-length polypeptides as they are highly soluble when overexpressed in bacteria. At the high concentrations achieved in bacteria the caspases undergo autocleavage and thus autoactivation. Therefore, after cell lysis and purification the enzymes are catalytically active and ready for downstream applications (Stennicke and Salvesen 1999; Roschitzki-Voser et al. 2012). Purified recombinant caspases are also available commercially.

RECIPES

Homogenization Buffer for Caspase Activation

10 mM HEPES (pH 7.0)
5 mM MgCl$_2$
0.67 mM DTT
1 protease inhibitor cocktail tablet (Roche) for the appropriate volume of homogenization buffer

The homogenization buffer without DTT or protease inhibitors can be stored at 4°C for up to 6 mo. Once DTT and protease inhibitors are added, the homogenization buffer should be stored at −20°C in 10-mL aliquots; it will last for up to 1 yr under these conditions.

Cite this protocol as *Cold Spring Harb Protoc*; doi:10.1101/pdb.prot080275

PBS (A)

Reagent	Amount to add for 1 L	Final concentration
NaCl	8.0 g	137.0 mM
KCl	0.2 g	2.7 mM
KH_2PO_4	0.2 g	1.5 mM
NaH_2PO_4	1.14 g	8.0 mM

Adjust pH to 7.2 with HCl ($\sim$4–6 droplets of 6 M HCl).

ACKNOWLEDGMENTS

Supported by grants from the National Institutes of Health.

REFERENCES

Cain K, Langlais C, Sun XM, Brown DG, Cohen GM. 2001. Physiological concentrations of K^+ inhibit cytochrome *c*-dependent formation of the apoptosome. *J Biol Chem* **276:** 41985–41990.

Hughes MA, Harper N, Butterworth M, Cain K, Cohen GM, MacFarlane M. 2009. Reconstitution of the death-inducing signaling complex reveals a substrate switch that determines CD95-mediated death or survival. *Mol Cell* **35:** 265–279.

McStay GP, Green DR. 2014a. Assaying caspase activity in vitro. *Cold Spring Harb Protoc* doi: 10.1101/pdb.prot080283.

McStay GP, Green DR. 2014b. Detection of caspase activity using antibody-based techniques. *Cold Spring Harb Protoc* doi: 10.1101/pdb.prot080291.

McStay GP, Salvesen GS, Green DR. 2008. Overlapping cleavage motif selectivity of caspases: Implications for analysis of apoptotic pathways. *Cell Death Differ* **15:** 322–331.

Roschitzki-Voser H, Schroeder T, Lenherr ED, Frolich F, Schweizer A, Donepudi M, Ganesan R, Mittl PR, Baici A, Grutter MG. 2012. Human caspases in vitro: Expression, purification and kinetic characterization. *Protein Expr Purif* **84:** 236–246.

Stennicke HR, Salvesen GS. 1999. Caspases: Preparation and characterization. *Methods* **17:** 313–319.

Tinel A, Tschopp J. 2004. The PIDDosome, a protein complex implicated in activation of caspase-2 in response to genotoxic stress. *Science* **304:** 843–846.

Protocol 2

Assaying Caspase Activity In Vitro

Gavin P. McStay[1,3] and Douglas R. Green[2]

[1]*Department of Life Sciences, New York Institute of Technology, Old Westbury, New York 11568;* [2]*Department of Immunology, St. Jude Children's Research Hospital, Memphis, Tennessee 38105*

Monitoring the activity of a caspase, either as an isolated protein or in a complex mixture (e.g., a cytosolic extract), can be achieved by measuring substrate cleavage. Chromogenic or fluorogenic substrates are available for many caspases. These substrates usually consist of the four-amino-acid motif that is optimal for each caspase and a moiety that, when cleaved, generates either a chromophore or a fluorophore that can be detected using spectrophotometric or fluorimetric means. In this protocol, we describe how to use these substrates to monitor caspase activity in samples containing active caspases (e.g., apoptotic cells). Caspase inhibitors, which contain a moiety that covalently attaches to the active site of the caspase, can be used in these assays. These assays will ascertain whether caspases are involved in a specific process (e.g., whether caspases are activated after an apoptotic stimulus) and are particularly informative if a purified caspase is used. However, the substrates and inhibitors are not specific for a particular caspase in an environment containing multiple caspases. So, if cytosolic or apoptotic cell extracts are used in these assays, additional experiments must be performed to identify exactly which caspases are involved.

MATERIALS

It is essential that you consult the appropriate Material Safety Data Sheets and your institution's Environmental Health and Safety Office for proper handling of equipment and hazardous materials used in this protocol.

RECIPE: Please see the end of this protocol for recipes indicated by <R>. Additional recipes can be found online at http://cshprotocols.cshlp.org/site/recipes.

Reagents

7-Amino-4-trifluoromethyl coumarin (AFC)-conjugated substrate (EMD Millipore; 20 mM in dimethylsulfoxide [DMSO])

Caspase assay buffer <R>

Caspase inhibitor in DMSO (e.g., zVAD-FMK [EMD Millipore] or Q-VD-OPH [MP Biomedicals]) (optional; see Step 2)

Source of active caspase (activated extract, apoptotic extract, or purified caspase)

> *To prepare an activated extract from any cell type, see Protocol: **Preparation of Cytosolic Extracts and Activation of Caspases by Cytochrome c** [McStay and Green 2014]). The extract should be incubated with cytochrome c and dATP for 30 min at 37°C as described in Steps 18–20 of that protocol before proceeding with Step 1 of the protocol below.*

Equipment

96-Well plate (flat bottom, black)

Fluorescent plate reader (see Step 4)

[3]Correspondence: gmcstay@nyit.edu

Cite this protocol as *Cold Spring Harb Protoc*; doi:10.1101/pdb.prot080283

Software for rate analysis (see Steps 5 and 6)

Software may be included with fluorescent plate reader. Microsoft Excel can also be used.

METHOD

A flowchart of this protocol is shown in Figure 1.

1. Add 10 μL of caspase-containing sample to each well of a 96-well plate.

   Ten microliters of sample should equal ~100 ng of active caspase, 100 μg of cytochrome c-activated cytosolic extract, or 100 μg of apoptotic lysate. To analyze purified caspase, it may be necessary to use a range of concentrations to account for different levels of caspase activity or variations among preparations. A very active caspase, like caspase-3, may require < 100 ng, whereas a caspase with poor activity, like caspase-9, may require more than this amount.

2. (Optional) Add caspase inhibitor(s) to the sample, and incubate for 10 min at room temperature. Use a range of inhibitor concentrations from 10 to 100 μM.

3. Add caspase assay buffer containing 100 μM of the AFC-conjugated substrate to each sample well.

4. Place the plate in a fluorescent plate reader and incubate at 37°C. Set the plate reader excitation at 400 nm and emission at 505 nm. Set the plate reader to mix and take a reading every minute for 30 min.

5. Identify the linear portion of the fluorophore-generating reaction and determine its slope, which is the initial rate of the reaction (Fig. 2).

   The initial rate of the reaction is a measure of the activity of the caspase. There may be an initial lag phase— which represents an equilibration phase in the assay—that should not be included in the analysis of initial velocity. See Troubleshooting.

6. Calculate the specific activity of the caspase by dividing the initial rate of the reaction (from Step 5) by the amount of enzyme.

   This calculation can be used for a purified caspase. However, the amount of a specific caspase present in a cytosolic extract cannot be accurately determined, and several caspases in an extract may contribute to the activity observed. Thus, for cytosolic extracts, this equation can be used to express activity against a particular substrate. For example, using the approximate values from Figure 2, the activity is 10,000 RFLU/100 μg protein = 0.167 RFLU/sec/μg.

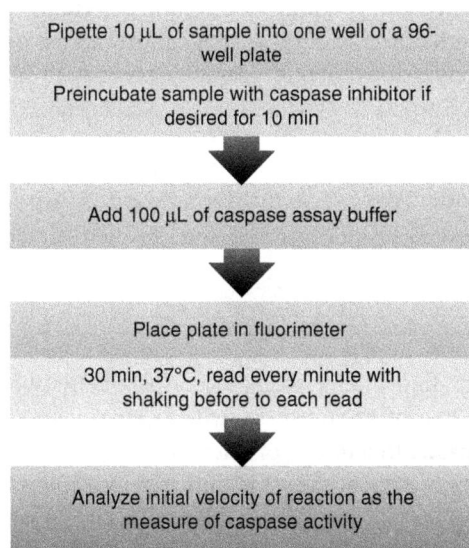

FIGURE 1. Assaying caspase activity in vitro. This procedure takes ~1 h to complete.

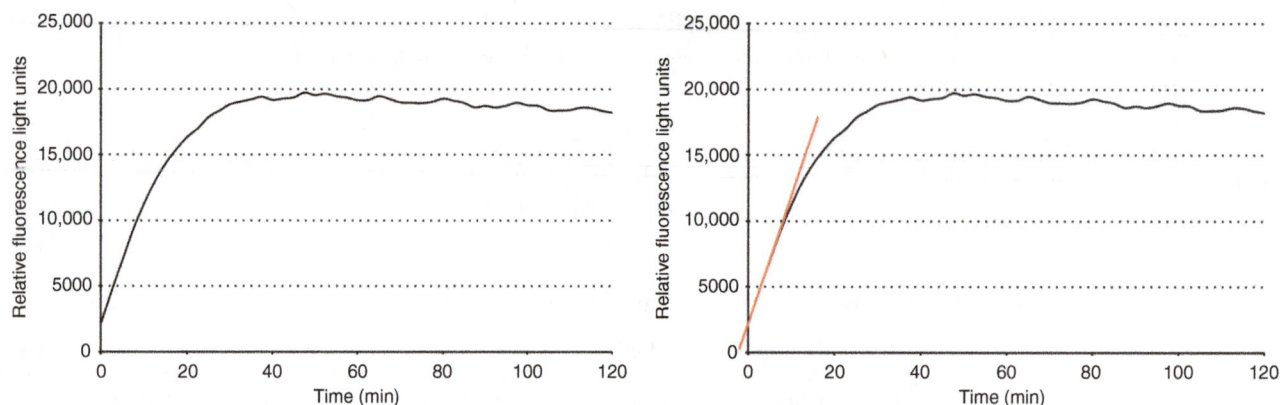

FIGURE 2. Jurkat cell cytosolic extract (100 µg) was incubated with 100 µM cytochrome *c* and 1 mM dATP for 30 min at 37°C. Caspase activity of the activated extract was monitored by following release of AFC from DEVD-AFC. Measurements were taken every 2.5 min with shaking. The graph shows the relative fluorescence light units (RFLU) as a function of time. The right-hand graph indicates the linear portion of the reaction that should be used to determine the rate of the reaction (in red). The gradient of this line gives the value of the rate in change in RFLU/min.

TROUBLESHOOTING

Problem (Step 5): The initial rate of reaction is too high to determine.
Solution: Dilute samples by factors of 10 and measure the initial rates of these reactions. When two dilutions are found that are 10-fold different in initial rate, analyze the other samples with these same dilutions.

Problem (Step 5): No caspase activity is seen.
Solutions: Consider the following.

- Ensure that the scale measuring substrate cleavage is at the appropriate setting. Some programs are set to automatically change the scale during the run, while other programs need to be adjusted manually. To do so, take the equivalent concentration of free fluorophore/chromophore and measure this on the plate reader. Adjust the gain for the reading so that this signal is visible on the chart, but is not above the limits of the newly set gain.

- Exogenous addition of small molecules or purified proteins to the cytosolic extracts can interfere with apoptosome assembly and downstream caspase activation. This problem generally arises from the addition of additional salt into activation reactions which can interfere with apoptosome assembly (Cain et al. 2001). To avoid interference from these exogenously added molecules dilute them with or dialyze them against homogenization buffer for caspase assays (see Protocol 1: Preparation of Cytosolic Extract and Activation of Caspases by Cytochrome *c* [McStay and Green 2014]).

Problem (Step 5): There is no inhibition of caspase activity with caspase inhibitor.
Solution: Titrate the active caspase and the inhibitor so that the amount of caspase is within the range of inhibitor used.

DISCUSSION

Analyzing substrate-cleaving activity by fluorimetry yields the caspase activity of a purified caspase, activated extract, or apoptotic extract. Except when using purified caspase, however, these assays are not entirely specific for individual caspases. In a more complex mixture, like a cytosolic extract or an apoptotic cell extract, many caspases are present at the same time. For example, cleavage of an LEHD-based substrate (used as a caspase-9-specific substrate) in an apoptotic cell extract does not necessarily mean caspase-9 has been activated. This substrate is more efficiently cleaved by caspase-3 and caspase-

6 (McStay et al. 2008). To determine which caspase is responsible for the activity, further examination of the caspases present in the extracts needs to be performed, such as by immunoprecipitation followed by measurements of the activity of the isolated caspases.

As with other motif-based tools of caspases, interpretations from results using caspase-specific inhibitors must be made while considering the specific context of the assay and the assay performed. Caspases in isolation will show varying degrees of inhibition by both pan-caspase and caspase-cleavage motif-based inhibitors. However, once these inhibitors are used in assays using cellular extracts or whole cells, problems with interpretation will arise. In situations where multiple caspases are present, the actual caspase inhibited by the inhibitor cannot be inferred directly from the assay. This is the reason for the development of more specific caspase inhibitors (Berger et al. 2006; Edgington et al. 2009, 2012) and also the use of techniques to isolate caspases to determine activity (Tu et al. 2006; McStay et al. 2008).

RELATED TECHNIQUES

Caspase inhibitors can be used in cell culture, and even in whole animals. In these cases, the inhibitor must be modified to increase cell permeability, usually by methylating hydrophilic residues, such as aspartic acid. The inhibitor must be incubated for a longer time to ensure uptake by the cell and hydrolysis of the methyl group to produce the active inhibitor. Once the cells have been preincubated with the inhibitor, apoptotic stimuli can be applied. After a certain amount of treatment time, cells or tissues are harvested and analyzed for apoptotic features, such as annexin V staining and caspase immunoblotting, among many other parameters that analyze apoptotic events at the cellular or biochemical level. It is important to ensure that the concentration of caspase inhibitor employed in each assay is sufficient to inhibit executioner caspase-dependent events.

RECIPE

Caspase Assay Buffer

20 mM PIPES (pH 7.4)
100 mM NaCl
1 mM EDTA
0.1% CHAPS
10% sucrose
10 mM dithiothreitol (add fresh)

Store at 4°C.

ACKNOWLEDGMENTS

Supported by grants from the National Institutes of Health.

REFERENCES

Berger AB, Witte MD, Denault JB, Sadaghiani AM, Sexton KM, Salvesen GS, Bogyo M. 2006. Identification of early intermediates of caspase activation using selective inhibitors and activity-based probes. *Mol Cell* 23: 509–521.

Cain K, Langlais C, Sun XM, Brown DG, Cohen GM. 2001. Physiological concentrations of K⁺ inhibit cytochrome *c*-dependent formation of the apoptosome. *J Biol Chem* 276: 41985–41990.

Edgington LE, Berger AB, Blum G, Albrow VE, Paulick MG, Lineberry N, Bogyo M. 2009. Noninvasive optical imaging of apoptosis by caspase-targeted activity-based probes. *Nat Med* 15: 967–973.

Edgington LE, van Raam BJ, Verdoes M, Wierschem C, Salvesen GS, Bogyo M. 2012. An optimized activity-based probe for the study of caspase-6 activation. *Chem Biol* 19: 340–352.

McStay GP, Green DR. 2014. Preparation of cytosolic extract and activation of caspases by cytochrome *c*. *Cold Spring Harb Protoc* doi: 10.1101/pdb.prot080275.

McStay GP, Salvesen GS, Green DR. 2008. Overlapping cleavage motif selectivity of caspases: Implications for analysis of apoptotic pathways. *Cell Death Differ* 15: 322–331.

Tu S, McStay GP, Boucher LM, Mak T, Beere HM, Green DR. 2006. In situ trapping of activated initiator caspases reveals a role for caspase-2 in heat shock-induced apoptosis. *Nat Cell Biol* 8: 72–77.

Detection of Caspase Activity Using Antibody-Based Techniques

Gavin P. McStay[1,3] and Douglas R. Green[2]

[1]*Department of Life Sciences, New York Institute of Technology, Old Westbury, New York 11568;* [2]*Department of Immunology, St. Jude Children's Research Hospital, Memphis, Tennessee 38105*

A number of antibodies have been generated that recognize caspases from mammalian model organisms. These include antibodies that recognize specific caspase pro-forms and others that bind caspase cleavage fragments. These antibodies are excellent reagents for identifying which executioner caspases have been activated following application or induction of a specific apoptotic stimulus. This approach is more difficult to use with initiator caspases, however, because cleavage does not necessarily correlate with caspase activation. In this protocol, cultured cells are treated with a proapoptotic stimulus, and then protein lysates are prepared from the treated cells. The proteins are then separated by gel electrophoresis and transferred to a suitable membrane. The fragment-specific antibodies that recognize executioner caspases are used in a western analysis to determine the extent of activation and to aid in identifying which caspases have been activated.

MATERIALS

It is essential that you consult the appropriate Material Safety Data Sheets and your institution's Environmental Health and Safety Office for proper handling of equipment and hazardous materials used in this protocol.

RECIPES: Please see the end of this protocol for recipes indicated by <R>. Additional recipes can be found online at http://cshprotocols.cshlp.org/site/recipes.

Reagents

Antibody to caspase, recognizing either full-length protein or a cleavage fragment
Antibody recognizing primary antibody, conjugated to horseradish peroxidase
Bradford protein assay materials (see Step 6)
Cell line to be treated with an apoptotic stimulus
Dry ice–methanol bath (optional; see Step 4)
Enhanced chemiluminescent reagent (Thermo Scientific)
Homogenization buffer for caspase activation (optional; see Step 4) <R>
Nonfat dry milk (5%) in TBS-T
NP-40 lysis buffer <R>

   Add protease inhibitors (e.g., complete protease inhibitor cocktail tablets [Roche]) to the NP-40 lysis buffer according to the manufacturer's instructions.

Pan-caspase inhibitor (e.g., methylated zVAD-FMK)
PBS (A) <R>

[3]Correspondence: gmcstay@nyit.edu

Cite this protocol as *Cold Spring Harb Protoc*; doi:10.1101/pdb.prot080291

Proapoptotic stimulus

The apoptotic stimulus can be a reagent like etoposide, a ligand like anti-Fas antibody, or a physical perturbation like UV radiation.

SDS–PAGE running buffer <R>
SDS–PAGE sample buffer (4×) <R>
TBS-T buffer (e.g., Abcam or Cell Signaling Technology)
Trypsin–EDTA (e.g., trypsin–EDTA in PBS [Life Technologies])
Western transfer buffer <R>

Equipment

Autoradiography film, film cassette, and film processor
Centrifuge (benchtop) for 15-mL conical tubes
Heating block set to 95°C
Microcentrifuge at 4°C
Nitrocellulose membrane
SDS–PAGE equipment
Western transfer apparatus

METHOD

A flowchart of this protocol is shown in Figure 1.

1. Treat cells with a proapoptotic stimulus.

   Inclusion of negative controls is important at this stage. Include two types of negative controls: (1) cells not treated with the proapoptotic stimulus and (2) cells pretreated with a pan-caspase inhibitor that is able to enter cells (e.g., methylated zVAD-FMK). In the latter control, incubate cells and the pan-caspase inhibitor for 1 h before treatment with proapoptotic stimulus. The concentration of the inhibitor depends on which inhibitor is included (e.g., use 100 µM methylated zVAD-FMK or 20 µM QVD-OPH).

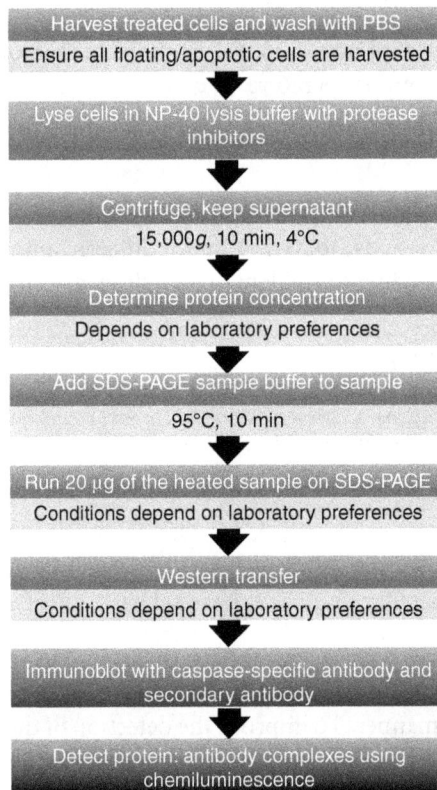

Harvest treated cells and wash with PBS
Ensure all floating/apoptotic cells are harvested

Lyse cells in NP-40 lysis buffer with protease inhibitors

Centrifuge, keep supernatant
15,000g, 10 min, 4°C

Determine protein concentration
Depends on laboratory preferences

Add SDS-PAGE sample buffer to sample
95°C, 10 min

Run 20 µg of the heated sample on SDS-PAGE
Conditions depend on laboratory preferences

Western transfer
Conditions depend on laboratory preferences

Immunoblot with caspase-specific antibody and secondary antibody

Detect protein: antibody complexes using chemiluminescence

FIGURE 1. Detection of caspase activity using antibody-based techniques. This protocol requires 1–1.5 d to complete.

2. Harvest treated cells by collecting the medium (containing detached apoptotic cells) and releasing adherent cells from the dish with trypsin–EDTA solution. Combine all of the cells in a centrifuge tube. Collect the cells by centrifugation at 400g for 5 min at room temperature.

3. Wash the cells in PBS and collect them by centrifugation as in Step 2.

4. Discard the supernatant. Add 100 µL of NP-40 lysis buffer with protease inhibitors per 5×10^5 cells. Lyse the cells by pipetting the cell pellet up and down until no clumps remain. Incubate the cells on ice for 10 min.

   Alternatively, cytosolic extracts can be prepared by resuspending cell pellets in homogenization buffer for caspase activation, followed by three freeze–thaw cycles in a dry ice-methanol bath.

5. Centrifuge the lysate at 15,000g for 10 min at 4°C.

6. Determine the protein concentration of the supernatant using the Bradford assay or another suitable method.

7. Add 4× SDS–PAGE sample buffer to an appropriate volume of cell lysate containing 20 µg of protein. Heat at 95°C for 10 min.

8. Run the heated lysate on a 12% SDS–PAGE gel at 200 V until the gel front dye reaches the bottom of the gel.

9. Transfer the separated proteins onto nitrocellulose using standard western transfer techniques and western transfer buffer.

10. Block the membrane with 5% nonfat dry milk in TBS-T for 30 min at room temperature. Rock the membrane gently during blocking.

11. Wash the membrane in TBS-T. Incubate the membrane in primary antibody (dilution depends on the antibody being used) in 5% nonfat dry milk in TBS-T solution for 2 h, rocking at room temperature (or overnight at 4°C).

12. Remove the antibody solution. Wash the membrane three times for 5 min each by rocking in TBS-T at room temperature.

13. Incubate the membrane in secondary antibody (1:10,000 dilution) in 5% nonfat dry milk in TBS-T solution for 2 h with rocking at room temperature.

14. Remove the antibody solution. Wash the membrane three times for 5 min each by rocking in TBS-T at room temperature.

15. Expose the membrane to enhanced chemiluminescent reagent for 1 min.

16. Cover the membrane with plastic and expose to X-ray film for different amounts of time (seconds to minutes) to detect bands corresponding to full length and cleaved caspases.

   Other methods for chemiluminescence capture are suitable, such as a gel documentation system that has chemiluminescent detection capability.

   The exposed western blot should show smaller protein bands corresponding to the cleaved form of the caspases (~15 kDa–20 kDa for executioner caspases and ~40 kDa for initiator caspases). If these are true caspase cleavage products, then the bands should be absent from both the untreated cells and the cells pretreated with pan-caspase inhibitor before treatment with proapoptotic stimulus (Fig. 2). See Troubleshooting.

TROUBLESHOOTING

Problem (Step 16): No cleaved caspases are seen in apoptotic extracts.
Solution: Some caspases (such as caspase-7 and caspase-9) once cleaved become susceptible to degradation in an ubiquitin-dependent manner. To improve the detection of these cleavage products, take samples at early time points (several hours instead of overnight treatment). It is also possible to use a proteasome inhibitor, such as MG132, to prevent degradation. There are also some

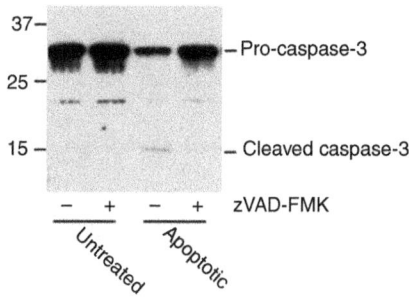

FIGURE 2. Caspase-3 western blot of Jurkat cells preincubated with 100 μM zVAD-FMK for 10 min before the induction of apoptosis overnight. The untreated samples show caspase-3 as full-length procaspase at 30 kDa. In the apoptotic sample there is less full-length caspase-3 and a 15 kDa band appears, representing cleaved caspase-3 fragments. In cells preincubated with zVAD-FMK before treatment with the proapoptotic stimulus, only full-length caspase-3 is seen, indicating that cleavage was caspase dependent.

circumstances in which caspases remain uncleaved when apoptosis is triggered. This is especially true for initiator caspases (see Discussion).

DISCUSSION

Detection of caspase cleavage products by western blotting is a straightforward method for implicating a specific caspase in a particular death pathway. However, this approach is crude, because it does not take into account the feedback that occurs in apoptotic pathways or the mechanisms of activation of initiator caspases. Monitoring cleavage events of executioner caspases provides a direct read-out of activation. The disappearance of the full-length protein, coupled with the appearance of the cleaved fragments containing the large and small catalytic subunits, indicates that one or more executioner caspases have been activated (Fig. 3). The same methodology does not yield comparable results with initiator caspases. Although cleavage of an initiator caspase is sometimes *necessary* for its activity, it is not *sufficient* to activate it. Thus, simply monitoring the cleavage of an initiator caspase does not tell us whether the caspase dimerized—and was therefore truly activated—before cleavage. In general, however, it is true that if an initiator caspase is not cleaved, then it is unactivated (Fig. 3). Cleaved initiator caspases typically function as substrates of executioner caspases as part of the controlled demolition of a cell (Inoue et al. 2009).

RELATED TECHNIQUES

Antibodies that recognize specific cleaved fragments of caspases are available, and these are commonly used in ELISAs or immunohistochemical staining of tissue to detect active caspases.

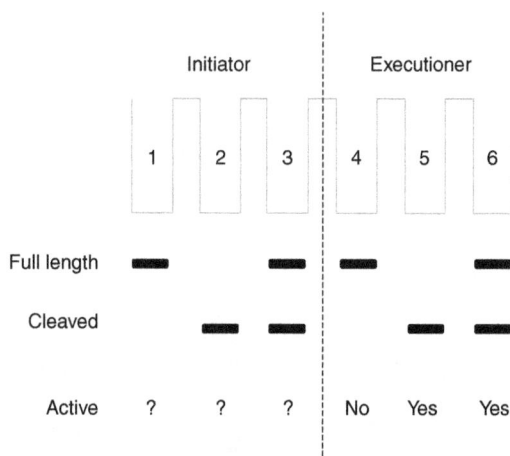

FIGURE 3. Schematic representation of a caspase immunoblot depicting problems of interpreting activity based on caspase cleavage patterns. Initiator caspases in lanes *1, 2,* and *3,* whether uncleaved, cleaved, or a mixture, do not reveal whether the caspase was activated. Executioner caspases in lanes *4, 5,* and *6* show that uncleaved executioner caspases are not active and cleaved executioner caspases are active.

RECIPES

Homogenization Buffer for Caspase Activation

10 mM HEPES (pH 7.0)
5 mM MgCl$_2$
0.67 mM DTT
1 protease inhibitor cocktail tablet (Roche) for the appropriate volume of homogenization buffer

The homogenization buffer without DTT or protease inhibitors can be stored at –4°C for up to 6 mo. Once DTT and protease inhibitors are added, the homogenization buffer should be stored at –20°C in 10-mL aliquots; it will last for up to 1 yr under these conditions.

NP-40 Lysis Buffer

NaCl (150 mM)
NP-40 (1.0%)
Tris-Cl (50 mM, pH 8.0)

For 1 L of NP-40 lysis buffer, combine 30 mL of 5 M NaCl, 100 mL of 10% NP-40, 50 mL of 1 M Tris (pH 8.0), and 820 mL of H$_2$O. Store at 4°C.

Triton X-100 can be used with similar results. Useful variations include lowering the detergent concentration, raising the salt concentration, or switching to other detergents such as saponin, digitonin, or CHAPS.

PBS (A)

Reagent	Amount to add for 1 L	Final concentration
NaCl	8.0 g	137.0 mM
KCl	0.2 g	2.7 mM
KH$_2$PO$_4$	0.2 g	1.5 mM
NaH$_2$PO$_4$	1.14 g	8.0 mM

Adjust pH to 7.2 with HCl (~4–6 droplets of 6 M HCl).

SDS-PAGE Running Buffer

25 mM Tris-Cl
192 mM glycine
0.1% SDS

SDS-PAGE Sample Buffer (4×)

40% glycerol
200 mM Tris-Cl (pH 6.8)
8% SDS
4% β-mercaptoethanol
0.02% bromophenol blue

Western Transfer Buffer

25 mM Tris-Cl
192 mM glycine
20% (v/v) methanol

Cite this protocol as *Cold Spring Harb Protoc*; doi:10.1101/pdb.prot080291

ACKNOWLEDGMENTS

Supported by grants from the National Institutes of Health.

REFERENCES

Inoue S, Browne G, Melino G, Cohen GM. 2009. Ordering of caspases in cells undergoing apoptosis by the intrinsic pathway. *Cell Death Differ* **16:** 1053–1061.

Identification of Active Caspases Using Affinity-Based Probes

Gavin P. McStay[1,3] and Douglas R. Green[2]

[1]*Department of Life Sciences, New York Institute of Technology, Old Westbury, New York 11568;* [2]*Department of Immunology, St. Jude Children's Research Hospital, Memphis, Tennessee 38105*

Small-molecule inhibitors of caspases can be modified with moieties such as biotin or fluorescent molecules. After the inhibitor molecule has bound to an active caspase, the caspase itself becomes labeled and can be isolated using affinity purification. This protocol describes the use of the biotinylated pan-caspase inhibitor VAD-FMK and streptavidin beads to isolate active caspases. These caspases are then separated by gel electrophoresis and identified with caspase-specific antibodies using western blotting techniques. Other caspase inhibitors bound with biotin or other labels can be substituted in this assay; labeled inhibitors are available commercially as either pan-caspase or caspase-specific probes.

MATERIALS

It is essential that you consult the appropriate Material Safety Data Sheets and your institution's Environmental Health and Safety Office for proper handling of equipment and hazardous materials used in this protocol.

RECIPES: Please see the end of this protocol for recipes indicated by <R>. Additional recipes can be found online at http://cshprotocols.cshlp.org/site/recipes.

Reagents

Biotin-VAD-FMK (e.g., Abcam or Santa Cruz Biotechnology)
Cell line to be exposed to caspase inhibitor and apoptosis stimulus
CHAPS lysis buffer <R>
PBS (A), room temperature and ice-cold <R>
Pro-apoptotic stimulus

> *The apoptotic stimulus can be a reagent like etoposide, a ligand like anti-Fas antibody, or a physical perturbation like ultraviolet radiation.*

Streptavidin peroxidase (GE Healthcare)
Streptavidin resin

> *Wash the streptavidin beads in CHAPS lysis buffer before use in Step 8.*

SDS–PAGE sample buffer (1×) <R>
Streptavidin peroxidase (e.g., Abcam or Life Technologies)
Western transfer buffer <R>

Equipment

Centrifuge (benchtop) at 4°C
Heating block set to 95°C

---

[3]Correspondence: gmcstay@nyit.edu

Cite this protocol as *Cold Spring Harb Protoc*; doi:10.1101/pdb.prot080309

Nitrocellulose
SDS–PAGE equipment
Tube rotator
Water bath, boiling
Western transfer apparatus

METHOD

A flowchart of this protocol is shown in Figure 1.

1. Incubate the cells with 50 μM biotin-VAD-FMK for 2 h.

   The optimal cell number should be determined for each cell type. 50×10^6 Jurkat cells in 2 mL of cell culture medium work well.

2. Treat the cells with a proapoptotic stimulus for the desired amount of time.

   Include a sample that is not treated with the proapoptotic stimulus as a negative control.

3. Wash the cells with 1 mL of ice-cold PBS, and centrifuge at 250g for 5 min at 4°C.

4. Remove the supernatant, resuspend the pellet in 1 mL of ice-cold PBS, and repeat the centrifugation.

5. Lyse the cells in 500 μL CHAPS lysis buffer by pipetting the cell pellet up and down until no clumps remain.

6. Centrifuge the lysate at 15,000g for 10 min at 4°C.

7. Transfer the supernatant to a fresh tube and boil it for 5 min.

8. Incubate the boiled lysate with 40 μL of prewashed streptavidin beads for 3 h at 4°C on a tube rotator.

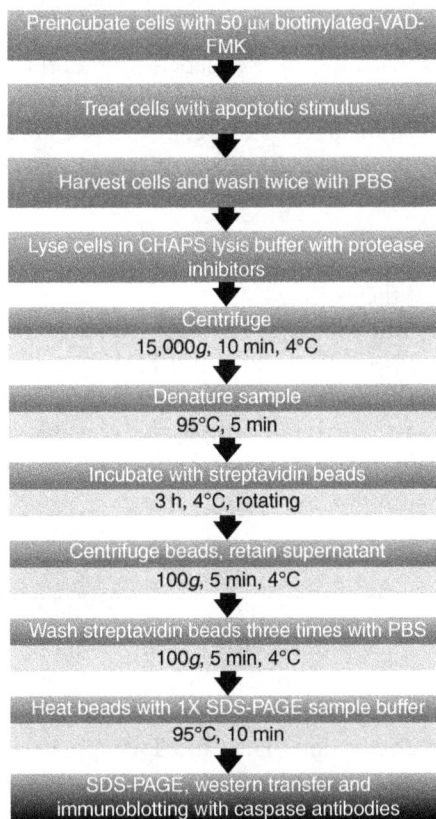

FIGURE 1. Identification of caspase activity by affinity-based probes. This protocol takes 2–3 d to complete.

9. Centrifuge the mixture at 100*g* for 5 min at 4°C.

10. Transfer the supernatant to a fresh tube and keep as the flowthrough sample.

11. Wash the pelleted beads three times each with 1 mL of room temperature PBS by centrifuging as in Step 9, removing the supernatant, and retaining the pellet.

12. Resuspend the streptavidin beads in 25 µL of 1× SDS–PAGE sample buffer and heat at 95°C for 10 min.

13. Immediately load the sample onto a 12% SDS–PAGE gel and run the gel at 200 V until the dye front reaches the bottom of the gel.

14. Transfer the separated proteins onto nitrocellulose using standard western transfer techniques and western transfer buffer.

15. Immunoblot with an antibody recognizing the caspase of interest (see Protocol 3: Detection of Caspase Activity Using Antibody-Based Techniques [McStay and Green 2014]).

    The exposed western blot should contain one or more bands corresponding to the caspase of interest that are present only in the samples treated with proapoptotic stimulus. For executioner caspases, these bands will represent the cleaved forms of the caspase, while for the initiator caspases these bands will represent a mixture of full-length and cleaved polypeptides. These bands will be absent from the untreated control.

    See Troubleshooting.

16. As a loading control for the western blot, reprobe the western blot with streptavidin peroxidase at a 1:10,000 dilution.

    Streptavidin peroxidase will recognize some large molecular mass bands corresponding to biotinylated mitochondrial proteins that should be present in all of the samples.

TROUBLESHOOTING

Problem (Step 15): Caspase fails to bind to the streptavidin beads.

Solution: Successful labeling of caspases with biotin-VAD-FMK depends on the amount and type of caspases that are activated. Executioner caspases are highly soluble, promiscuous, and have two accessible active sites, which ensures that they are well labeled by biotin-VAD-FMK. Initiator caspases, on the other hand, are activated in large complexes, as dimers, with specificity for only a few substrates, and potentially in low abundance. These characteristics make it harder to label initiator caspases. To increase the specificity of binding of an initiator caspase, use a large number of cells that undergo quick and nearly synchronous apoptosis. This will help to boost the amount of activated caspase.

Problem (Step 15): Bound caspase is not of the expected size.

Solution: This problem arises most often when looking at initiator caspases. Under some circumstances, cleaved forms of initiator caspases bind biotin-VAD-FMK, probably because the active site forms faster than biotin-VAD-FMK can bind to the active site. Another possibility is that in a newly activated initiator caspase the active site is not completely accessible to biotin-VAD-FMK. Both of these possibilities would explain why a full-length initiator caspase is not associated with biotin-VAD-FMK after an apoptotic stimulus.

DISCUSSION

Implicating a specific caspase in a particular apoptosis scenario is important both for understanding the mechanisms of apoptosis and for elucidating apoptotic pathways in cells and in vivo. The affinity-based probe biotin-VAD-FMK has been used successfully in vitro to characterize caspase-7 activation (Denault and Salvesen 2003) and caspase-2 activation (Tinel and Tschopp 2004), in cells to identify caspases activated by multiple apoptotic stimuli (Tu et al. 2006) and caspase-2 activation in neurons (Tizon et al. 2010; Ribe et al. 2012), and in vivo to detect caspase-9 activation in a mouse stroke model

(Akpan et al. 2011). A second generation of activity-based probes has been developed. These probes have greater specificity for individual caspases, and they are available coupled to biotin for affinity labeling (Berger et al. 2006) or to fluorescent dyes to detect activity in vitro and in vivo (Edgington et al. 2009, 2012).

RECIPES

CHAPS Lysis Buffer

150 mM KCl
50 mM HEPES (pH 7.4)
0.1% CHAPS
1 protease inhibitor cocktail tablet (Roche) per 50 mL

Store the buffer without protease inhibitors at 4°C for up to 6 mo. Buffer with protease inhibitor should be divided into 5-mL aliquots and stored at −20°C for up to 1 yr.

PBS (A)

Reagent	Amount to add for 1 L	Final concentration
NaCl	8.0 g	137.0 mM
KCl	0.2 g	2.7 mM
KH_2PO_4	0.2 g	1.5 mM
NaH_2PO_4	1.14 g	8.0 mM

Adjust pH to 7.2 with HCl (∼4–6 droplets of 6 M HCl).

SDS-PAGE Sample Buffer (1×)

10% glycerol
50 mM Tris-Cl (pH 6.8)
2% SDS
1% β-mercaptoethanol
0.005% bromophenol blue

Western Transfer Buffer

25 mM Tris-Cl
192 mM glycine
20% (v/v) methanol

ACKNOWLEDGMENTS

Supported by grants from the National Institutes of Health.

REFERENCES

Akpan N, Serrano-Saiz E, Zacharia BE, Otten ML, Ducruet AF, Snipas SJ, Liu W, Velloza J, Cohen G, Sosunov SA, et al. 2011. Intranasal delivery of caspase-9 inhibitor reduces caspase-6-dependent axon/neuron loss and improves neurological function after stroke. *J Neurosci* **31:** 8894–8904.

Berger AB, Witte MD, Denault JB, Sadaghiani AM, Sexton KM, Salvesen GS, Bogyo M. 2006. Identification of early intermediates of caspase activation using selective inhibitors and activity-based probes. *Mol Cell* **23:** 509–521.

Denault JB, Salvesen GS. 2003. Human caspase-7 activity and regulation by its N-terminal peptide. *J Biol Chem* **278:** 34042–34050.

Edgington LE, Berger AB, Blum G, Albrow VE, Paulick MG, Lineberry N, Bogyo M. 2009. Noninvasive optical imaging of apoptosis by caspase-targeted activity-based probes. *Nat Med* **15:** 967–973.

Edgington LE, van Raam BJ, Verdoes M, Wierschem C, Salvesen GS, Bogyo M. 2012. An optimized activity-based probe for the study of caspase-6 activation. *Chem Biol* **19:** 340–352.

McStay GP, Green DR. 2014. Detection of caspase activity using antibody-based techniques. *Cold Spring Harb Protoc* doi: 10.1101/pdb.prot080291.

Ribe EM, Jean YY, Goldstein RL, Manzl C, Stefanis L, Villunger A, Troy CM. 2012. Neuronal caspase 2 activity and function requires RAIDD, but not PIDD. *Biochem J* 444: 591–599.

Tinel A, Tschopp J. 2004. The PIDDosome, a protein complex implicated in activation of caspase-2 in response to genotoxic stress. *Science* 304: 843–846.

Tizon B, Ribe EM, Mi W, Troy CM, Levy E. 2010. Cystatin C protects neuronal cells from amyloid-β-induced toxicity. *J Alzheimers Dis* 19: 885–894.

Tu S, McStay GP, Boucher LM, Mak T, Beere HM, Green DR. 2006. In situ trapping of activated initiator caspases reveals a role for caspase-2 in heat shock-induced apoptosis. *Nat Cell Biol* 8: 72–77.

Cite this protocol as *Cold Spring Harb Protoc*; doi:10.1101/pdb.prot080309

Verification of a Putative Caspase Substrate

Gavin P. McStay[1,3] and Douglas R. Green[2]

[1]*Department of Life Sciences, New York Institute of Technology, Old Westbury, New York 11568;* [2]*Department of Immunology, St. Jude Children's Research Hospital, Memphis, Tennessee 38105*

Proteomic approaches have been adopted to survey the degradome of caspases during apoptosis. These approaches provide a comprehensive list of substrates and give clues to which pathways are altered during apoptosis by activated caspases. However, substrates identified by large-scale proteomic screening need to be validated as bona fide caspase targets. This ensures that conclusions derived from the screen are based on real substrates and not on artifacts of the proteomic screen. The validation method described in this protocol uses radiolabeled versions of the putative substrates synthesized using in vitro transcription/translation methods. These are incubated with purified caspases to determine whether they are genuine caspase substrates.

MATERIALS

It is essential that you consult the appropriate Material Safety Data Sheets and your institution's Environmental Health and Safety Office for proper handling of equipment and hazardous materials used in this protocol.

RECIPE: Please see the end of this protocol for recipes indicated by <R>. Additional recipes can be found online at http://cshprotocols.cshlp.org/site/recipes.

Reagents

Caspase of interest, purified
cDNA of putative caspase substrate encoded on a plasmid with a compatible promoter (SP6, T7)
[^{35}S]-Methionine and [^{35}S]-cysteine
Rabbit reticulocyte lysate in vitro transcription/translation kit (SP6 or T7; Promega)
SDS–PAGE sample buffer (4×) <R>

Equipment

Autoradiography film, film cassette, and film processor
Gel drier
Gel transfer apparatus (optional; see Step 6)
Heating block set to 95°C
Incubator set to 37°C
Nitrocellulose or PVDF membrane (optional; see Step 6)
SDS–PAGE equipment

---

[3]Correspondence: gmcstay@nyit.edu

Cite this protocol as *Cold Spring Harb Protoc*; doi:10.1101/pdb.prot080317

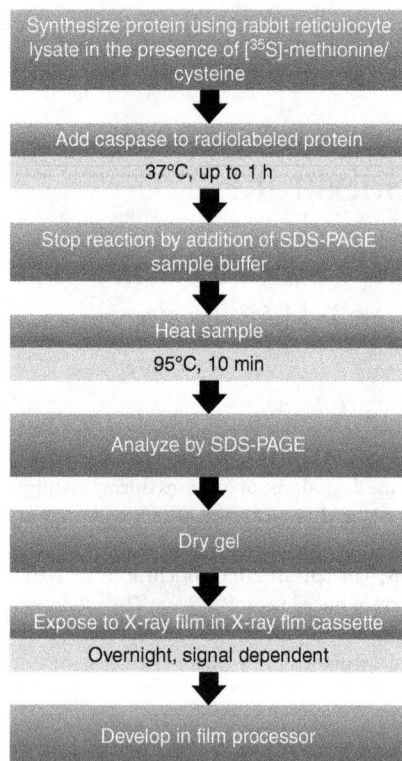

FIGURE 1. Verification of a putative caspase substrate. This procedure requires 1–2 d to complete.

METHOD

A flowchart of this protocol is shown in Figure 1.

1. Synthesize putative caspase substrate in the presence of radiolabeled [^{35}S]-methionine and [^{35}S]-cysteine using a commercial rabbit reticulocyte lysate in vitro transcription/translation kit. Follow the manufacturer's instructions.

2. Verify that a polypeptide of the correct size was generated by running the reaction on a SDS–PAGE gel, drying the gel, and exposing it to X-ray film.

3. Incubate the radiolabeled caspase substrate with varying amounts of purified caspase for 30 min at 37°C.

   Include a control sample containing no caspase and samples of caspase preincubated with an inhibitor. As a starting point, use 1 μL substrate and 1 μL caspase in a total volume of 10 μL using caspase assay buffer (see Protocol: Assaying Caspase Activity In Vitro [McStay and Green 2014a]).

4. Stop the reaction by adding 4× SDS–PAGE sample buffer. Heat the samples for 10 min at 95°C.

5. Analyze the samples by SDS–PAGE, using a gel whose percentage acrylamide is appropriate for the size of the synthesized protein. Run the gel at 200 V until the dye front reaches the bottom of the gel.

6. Dry the gel or transfer to a nitrocellulose or PVDF membrane.

7. Expose the gel (or membrane) to X-ray film overnight.

8. Develop the X-ray film.

   On the exposed film, one or more cleavage products should be visible only in the samples treated with caspase. This band(s) should not be present in the untreated sample or the sample treated with inhibited caspase. See Troubleshooting.

TROUBLESHOOTING

Problem (Step 8): No cleavage of putative substrate is evident.

Cite this protocol as *Cold Spring Harb Protoc*; doi:10.1101/pdb.prot080317

Solution: If the putative substrate is not cleaved by the active caspase, it may be that it is not a genuine caspase substrate. To control for the reaction conditions, use canonical caspase-3 substrates such as poly(ADP-ribose) polymerase (PARP) or inhibitor of caspase-activated DNase (ICAD). If these substrates are cleaved under the same conditions, then the putative substrate is not a true substrate. If the control substrates are not cleaved, then the reaction conditions need to be optimized and the samples tested again. Test the caspase for activity by using a fluorogenic substrate (see Protocol: **Assaying Caspase Activity In Vitro** [McStay and Green 2014a]).

DISCUSSION

Validation of a putative caspase substrate is important so that the manifestations of apoptosis can be associated with specific caspase cleavage events. Mutating the conserved aspartate at the cleavage site to an alanine residue renders a substrate resistant to caspase cleavage. This mutant can provide insights into the apoptotic process with which the caspase-cleaved substrate is associated. For example, once the substrate ICAD is cleaved by caspase-3, ICAD no longer inhibits the action of CAD (caspase-activated DNase), the endonuclease responsible for cleaving nuclear DNA into the discrete 180-bp fragments typically seen in apoptotic DNA preparations. If the aspartate residues of the caspase cleavage sites of ICAD are mutated to alanine, ICAD can no longer be cleaved by caspase-3. In this situation ICAD is still able to inhibit CAD and DNA laddering does not occur (Sakahira et al. 1998). Validation of a true caspase cleavage event must also be confirmed by the use of caspase inhibitors in these assays (see Protocol 3: Detection of Caspase Activity Using Antibody-Based Techniques [McStay and Green 2014b]). This will show that the caspase is responsible for the cleavage event and not a nonspecific protease.

RELATED TECHNIQUES

Caspase substrate cleavage can be followed by western blotting (similar to Protocol 4: Identification of Active Caspases Using Affinity-Based Probes [McStay and Green 2014c]) by taking samples from apoptotic extracts and comparing them to untreated extracts. If the noncleavable substrate can be introduced into cells, it should be possible to determine the role of the caspase cleavage event. This approach was taken for a caspase-3 substrate in complex I of the mitochondrial electron transport chain. Introduction of the noncleavable mutant delayed mitochondrial membrane potential loss and annexin V exposure (Ricci et al. 2004). Noncleavable mutants can be generated by site-directed mutagenesis.

RECIPE

SDS-PAGE Sample Buffer (4×)

40% glycerol
200 mM Tris-Cl (pH 6.8)
8% SDS
4% β-mercaptoethanol
0.02% bromophenol blue

ACKNOWLEDGMENTS

Supported by grants from the National Institutes of Health.

REFERENCES

McStay GP, Green DR. 2014a. Assaying caspase activity in vitro. *Cold Spring Harb Protoc* doi: 10.1101/pdb.prot080283.

McStay GP, Green DR. 2014b. Detection of caspase activity using antibody-based techniques. *Cold Spring Harb Protoc* doi: 10.1101/pdb.prot080291.

McStay GP, Green DR. 2014c. Identification of active caspases using affinity-based probes. *Cold Spring Harb Protoc* doi: 10.1101/pdb.prot080309.

Ricci JE, Munoz-Pinedo C, Fitzgerald P, Bailly-Maitre B, Perkins GA, Yadava N, Scheffler IE, Ellisman MH, Green DR. 2004. Disruption of mitochondrial function during apoptosis is mediated by caspase cleavage of the p75 subunit of complex I of the electron transport chain. *Cell* 117: 773–786.

Sakahira H, Enari M, Nagata S. 1998. Cleavage of CAD inhibitor in CAD activation and DNA degradation during apoptosis. *Nature* 391: 96–99.

Cite this protocol as *Cold Spring Harb Protoc*; doi:10.1101/pdb.prot080317

CHAPTER 8

Methods for Probing Lysosomal Membrane Permeabilization

Marja Jäättelä and Jesper Nylandsted[1]

Unit for Cell Death and Metabolism, Center for Autophagy, Recycling and Disease, Danish Cancer Society Research Center, Strandboulevarden 49, DK-2100 Copenhagen, Denmark

Cell death triggered by lysosomal membrane permeabilization (LMP) is gaining increased interest as target for cancer therapy, but the death pathway also plays an important role in normal physiology (e. g., during involution of the mammary gland). LMP-induced cell death is triggered by release of hydrolases including cysteine cathepsin proteases from the lysosomal lumen into the cytosol. Limited release of proteases to the cytoplasm induces apoptosis or apoptosis-like cell death, whereas massive LMP results in rapid cellular necrosis. Here we introduce three complementary methods for quantifying and visualizing LMP: (i) monitoring LMP by immunocytochemistry, (ii) visualizing LMP by fluorescent dextran release, and (iii) quantification of LMP by activity measurements of lysosomal enzymes in digitonin-extracted cytosol.

BACKGROUND

Lysosomes are spherical acidic organelles found dispersed in the cytoplasm in all mammalian cells (except blood erythrocytes). They were first discovered 60 years ago by Belgian scientist Christian De Duve who also introduced lysosomes as potential "suicide bags" in cell death (de Duve et al. 1955; de Duve 1983). Their main function—with their cocktail of hydrolic enzymes—involve digestion and recycling of macromolecules, damaged and worn-out organelles, and degradation of extracellular material provided by endocytosis, phagocytosis, and autophagy (Saftig and Klumperman 2009; Pryor and Luzio 2009; Kolter and Sandhoff 2010). Lysosomes contain over 50 luminal hydrolases, which are optimally active at pH 4.5–5 and include proteases, lipases, phosphatases glycosidases, phosphatases, and nucleases. Many of these hydrolases can also function outside lysosomes at neutral pH and play important roles in tissue remodeling, membrane repair, and programmed cell death (PCD) (Gerasimenko et al. 2001; Kroemer and Jäättelä 2005; Vasiljeva and Turk 2008).

de Duve's discovery ignited research into pharmaceutical strategies to either destabilize or stabilize lysosomal membranes that would prove advantageous for the treatment of cancer and degenerative disorders, respectively (Firestone et al. 1979; de Duve 1983). Although cell necrosis and tissue autolysis induced by lysosomal membrane permeabilization (LMP) was well established, the importance of regulated LMP in PCD was first recognized and accepted in the 1990s. This delayed acceptance stemmed from structural studies by electron microscopy of lysosomes in PCD that appeared intact even though hydrolases leaked to the cytosol (Brunk et al. 2001). Furthermore, examination of caspase-mediated PCD using caspase inhibitors such as zVAD-fmk (methyl ketone peptide inhibitor)

[1]Correspondence: jnl@cancer.dk

Cite this introduction as *Cold Spring Harb Protoc*; doi:10.1101/pdb.top070367

proved also to inhibit several lysosomal cathepsins that function as effectors of lysosomal cell death (Schotte et al. 1999; Foghsgaard et al. 2001). However, subsequent studies have shown that LMP induction can trigger classical apoptosis or apoptosis-like PCD in addition to uncontrolled necrosis (Brunk et al. 1997; Kagedal et al. 2001; Boya et al. 2003; Cirman et al. 2004). The mode of cell death probably depends on the amount of lysosomal permeabilization and the mode of cell death might also explain the different death phenotypes following LMP. Limited release of lysosomal proteases to the cytoplasm induces apoptosis or apoptosis-like cell death, whereas massive LMP results in rapid cellular necrosis (Brunk et al. 1997; Kagedal et al. 2001). Cysteine cathepsins B and L and aspartate cathepsin D are among the most well-defined cytosolic protease effectors in LMP-induced PCD (Stoka et al. 2007; Kirkegaard and Jäättelä 2009). Release of these proteases to the cytosol can activate the intrinsic apoptosis pathway by cleavage-mediated activation of proapoptotic Bcl-2 family proteins or caspase-independent apoptosis-like cell death in apoptosis-resistant cells (Kroemer and Jäättelä 2005). The impact of other lysosomal hydrolases (e.g., nucleases and lipases) and consequences of LMP-induced damage to lysosomes on cell viability remain to be examined.

The translocation of proteases from lysosomes to the cytosol is the hallmark of LMP, although the exact trigger of LMP remains elusive. Reactive oxygen species (ROS) released from, for example, damaged mitochondria can directly induce injury to the lysosomal membrane leading to a less-controlled leakage of lysosomal proteases from the affected lysosome. Thus, susceptibility of distinct lysosomes to ROS-induced LMP is dependent on the spatial distribution of the lysosomes in relation to the source of ROS, such as damaged mitochondria (Boya and Kroemer 2008). In addition, sensitization of cells to LMP can occur due to enhanced levels of lysosomal iron, which in response to oxidative stress catalyzes the production of highly reactive prooxidants via Fenton reactions (Link et al. 1993; Kurz et al. 2011). How transient pores or channels are formed in the lysosomal membrane during LMP and if minor holes in the membrane can self-heal or are repaired to restore function, has yet to be established.

LMP and lysosomal proteases participate in the execution of PCD induced by various apoptotic stimuli including cytotoxic drugs not primarily designed to target lysosomes, death receptor activation, and the tumor suppressor protein p53 (Kroemer and Jäättelä 2005). Notably, most cancer cells have defects in their classical apoptosis machinery but are still able to undergo PCD by lysosomal cell death. Cancer cells are especially dependent on effective lysosomal function to cope with accelerated proliferation and metabolism, and dramatic changes in lysosomal composition, volume, hydrolase levels, and cellular distribution occur during transformation and cancer progression (Moin et al. 1998; Palermo and Joyce 2008; Kallunki et al. 2013). The realization that changes associated with cancer transformation, including invasive growth, simultaneously make lysosomes more vulnerable to LMP and has sparked a new interest in lysosomes as targets for cancer therapy (Fehrenbacher et al. 2004). To this end, cancer cells counteract LMP either by translocation of the stress protein, heat shock protein 70 (Hsp70) to lysosomes, where it stabilizes lysosomal membranes by enhancing the activity of acid sphingomyelinase (Nylandsted et al. 2004; Kirkegaard et al. 2010), or/and by up-regulation of cytosolic protease inhibitors (Silverman et al. 1998; Suminami et al. 1998). In addition to the role of lysosomal cell death in cancer, LMP was recently shown to be important for normal mammary gland involution, giving LMP a broader physiological function (Kreuzaler et al. 2011).

TECHNIQUES FOR ASSESSING LMP

Historically, LMP was assessed using the impermeable substrate β-glycerophosphate, which does not readily penetrate the lysomal membrane unless permeability is altered. The degree of LMP was visualized by Gomori acid phosphatase staining and this approach led Bitensky and coworkers to develop a "lysosomal fragility test" to estimate LMP and early cell injury (Bitensky 1963). Since then, additional methods have been introduced, including immunocytochemistry, fluorescent dextran release, and enzymatic quantification of lysosomal proteases released to the cytosol.

With the improvements of antibodies and bright fluorochromes for immunocytochemistry, LMP can be monitored simply by staining for lysosomal cathepsins as described in Protocol 3: A Method to Monitor Lysosomal Membrane Permeabilization by Immunocytochemistry (Groth-Pedersen et al. 2015). In healthy cells, lysosomal cathepsins are localized in the lumen, which appear as punctate staining pattern, whereas an LMP-inducing insult results in the release of cathepsins into the cytosol and a diffuse staining pattern throughout the cytoplasm (Nylandsted et al. 2004). Co-staining of other relevant death components (e.g., activated Bax, activated caspases, or cytochrome c release) can be used to address the order of events in the studied death pathway.

LMP can also be observed in real time by time-lapse imaging of lysosomes loaded with fluorescent dextran as described in Protocol 2: Visualizing Lysosomal Membrane Permeabilization by Fluorescent Dextran Release (Ellegaard et al. 2015). This approach allows the researcher to visualize LMP directly by monitoring dextran release to the cytosol and enables one to determine membrane pore sizes by size-exclusion using different sizes of dextrans.

In order to quantify LMP more precisely, we developed the method presented in Protocol 1: Quantification of Lysosomal Membrane Permeabilization by Cytosolic Cathepsin and β-N-Acetyl-Glucosaminidase Activity Measurements (Jäättelä and Nylandsted 2015). The method is based on digitonin extraction of cytosol followed by measurement of lysosomal hydrolase activities in the extracted cytosol and total cellular lysate. We hope these protocols, alongside the continuing development of better methods to detect and quantify LMP (e.g., assays based on the translocation of sugar-binding galectin proteins to damaged lysosomes; see Aits et al. 2015), will attract more researchers to this exciting field with many unanswered questions.

REFERENCES

Aits S, Jäättelä M, Nylandsted J. 2015. Methods for the quantification of lysosomal membrane permeabilization: A hallmark of lysosomal cell death. In *Lysosomes and Lysosomal Diseases* (ed Platt F, Platt N), pp. 261–286. Academic Press, Waltham.

Bitensky L. 1963. Modifications to the Gomori acid phosphatase technique for controlled temperature frozen sections. *Q J Micr Sci* **104:** 193–196.

Boya P, Kroemer G. 2008. Lysosomal membrane permeabilization in cell death. *Oncogene* **27:** 6434–6451.

Boya P, Andreau K, Poncet D, Zamzami N, Perfettini JL, Metivier D, Ojcius DM, Jäättelä M, Kroemer G. 2003. Lysosomal membrane permeabilization induces cell death in a mitochondrion-dependent fashion. *J Exp Med* **197:** 1323–1334.

Brunk UT, Dalen H, Roberg K, Hellquist HB. 1997. Photo-oxidative disruption of lysosomal membranes causes apoptosis of cultured human fibroblasts. *Free Radic Biol Med* **23:** 616–626.

Brunk UT, Neuzil J, Eaton JW. 2001. Lysosomal involvement in apoptosis. *Redox Rep* **6:** 91–97.

Cirman T, Oresic K, Mazovec GD, Turk V, Reed JC, Myers RM, Salvesen GS, Turk B. 2004. Selective disruption of lysosomes in HeLa cells triggers apoptosis mediated by cleavage of Bid by multiple papain-like lysosomal cathepsins. *J Biol Chem* **279:** 3578–3587.

de Duve C. 1983. Lysosomes revisited. *Eur J Biochem* **137:** 391–397.

de Duve C, Pressman BC, Gianetto R, Wattiaux R, Appelmans F. 1955. Tissue fractionation studies. 6. Intracellular distribution patterns of enzymes in rat-liver tissue. *Biochem J* **60:** 604–617.

Ellegaard A-M, Jäättelä M, Nylandsted J. 2015. Visualizing lysosomal membrane permeabilization by fluorescent dextran release. *Cold Spring Harb Protoc* doi: 10.1101/pdb.prot086173.

Fehrenbacher N, Gyrd-Hansen M, Poulsen B, Felbor U, Kallunki T, Boes M, Weber E, Leist M, Jäättelä M. 2004. Sensitization to the lysosomal cell death pathway upon immortalization and transformation. *Cancer Res* **64:** 5301–5310.

Firestone RA, Pisano JM, Bonney RJ. 1979. Lysosomotropic agents. 1. Synthesis and cytotoxic action of lysosomotropic detergents. *J Med Chem* **22:** 1130–1133.

Foghsgaard L, Wissing D, Mauch D, Lademann U, Bastholm L, Boes M, Elling F, Leist M, Jäättelä M. 2001. Cathepsin B acts as a dominant

execution protease in tumor cell apoptosis induced by tumor necrosis factor. *J Cell Biol* **153:** 999–1010.

Gerasimenko JV, Gerasimenko OV, Petersen OH. 2001. Membrane repair: Ca(2+)-elicited lysosomal exocytosis. *Curr Biol* **11:** R971–R974.

Groth-Pedersen L, Jäättelä M, Nylandsted J. 2015. A method to monitor lysosomal membrane permeabilization by immunocytochemistry. *Cold Spring Harb Protoc* doi: 10.1101/pdb.prot086181.

Jäättelä M, Nylandsted J. 2015. Quantification of lysosomal membrane permeabilization by cytosolic cathepsin and β-N-acetyl-glucosaminidase activity measurements. *Cold Spring Harb Protoc* doi: 10.1101/pdb.prot086165.

Kagedal K, Zhao M, Svensson I, Brunk UT. 2001. Sphingosine-induced apoptosis is dependent on lysosomal proteases. *Biochem J* **359:** 335–343.

Kallunki T, Olsen OD, Jäättelä M. 2013. Cancer-associated lysosomal changes: Friends or foes. *Oncogene* **32:** 1995–2004.

Kirkegaard T, Jäättelä M. 2009. Lysosomal involvement in cell death and cancer. *Biochim Biophys Acta* **1793:** 746–754.

Kirkegaard T, Roth AG, Petersen NH, Mahalka AK, Olsen OD, Moilanen I, Zylicz A, Knudsen J, Sandhoff K, Arenz C, et al. 2010. Hsp70 stabilizes lysosomes and reverts Niemann-Pick disease-associated lysosomal pathology. *Nature* **463:** 549–553.

Kolter T, Sandhoff K. 2010. Lysosomal degradation of membrane lipids. *FEBS Lett* **584:** 1700–1712.

Kreuzaler PA, Staniszewska AD, Li W, Omidvar N, Kedjouar B, Turkson J, Poli V, Flavell RA, Clarkson RW, Watson CJ. 2011. Stat3 controls lysosomal-mediated cell death in vivo. *Nat Cell Biol* **13:** 303–309.

Kroemer G, Jäättelä M. 2005. Lysosomes and autophagy in cell death control. *Nat Rev Cancer* **5:** 886–897.

Kurz T, Eaton JW, Brunk UT. 2011. The role of lysosomes in iron metabolism and recycling. *Int J Biochem Cell Biol* **43:** 1686–1697.

Link G, Pinson A, Hershko C. 1993. Iron loading of cultured cardiac myocytes modifies sarcolemmal structure and increases lysosomal fragility. *J Lab Clin Med* **121:** 127–134.

Moin K, Cao L, Day NA, Koblinski JE, Sloane BF. 1998. Tumor cell membrane cathepsin B. *Biol Chem* **379:** 1093–1099.

Nylandsted J, Gyrd-Hansen M, Danielewicz A, Fehrenbacher N, Lademann U, Hoyer-Hansen M, Weber E, Multhoff G, Rohde M, Jäättelä M. 2004.

Heat shock protein 70 promotes cell survival by inhibiting lysosomal membrane permeabilization. *J Exp Med* **200:** 425–435.

Palermo C, Joyce JA. 2008. Cysteine cathepsin proteases as pharmacological targets in cancer. *Trends Pharmacol Sci* **29:** 22–28.

Pryor PR, Luzio JP. 2009. Delivery of endocytosed membrane proteins to the lysosome. *Biochim Biophys Acta* **1793:** 615–624.

Saftig P, Klumperman J. 2009. Lysosome biogenesis and lysosomal membrane proteins: Trafficking meets function. *Nat Rev Mol Cell Biol* **10:** 623–635.

Schotte P, Declercq W, Van Huffel S, Vandenabeele P, Beyaert R. 1999. Non-specific effects of methyl ketone peptide inhibitors of caspases. *FEBS Lett* **442:** 117–121.

Silverman GA, Bartuski AJ, Cataltepe S, Gornstein ER, Kamachi Y, Schick C, Uemura Y. 1998. SCCA1 and SCCA2 are proteinase inhibitors that map to the serpin cluster at 18q21.3. *Tumour Biol* **19:** 480–487.

Stoka V, Turk V, Turk B. 2007. Lysosomal cysteine cathepsins: Signaling pathways in apoptosis. *Biol Chem* **388:** 555–560.

Suminami Y, Nawata S, Kato H. 1998. Biological role of SCC antigen. *Tumour Biol* **19:** 488–493.

Vasiljeva O, Turk B. 2008. Dual contrasting roles of cysteine cathepsins in cancer progression: Apoptosis versus tumour invasion. *Biochimie* **90:** 380–386.

Cite this introduction as *Cold Spring Harb Protoc*; doi:10.1101/pdb.top070367

Quantification of Lysosomal Membrane Permeabilization by Cytosolic Cathepsin and β-*N*-Acetyl-Glucosaminidase Activity Measurements

Marja Jäättelä and Jesper Nylandsted[1]

Unit for Cell Death and Metabolism, Center for Autophagy, Recycling and Disease, Danish Cancer Society Research Center, DK-2100, Copenhagen, Denmark

Programmed cell death involving lysosomal membrane permeabilization (LMP) is an alternative cell death pathway induced under various cellular conditions and by numerous cytotoxic stimuli. The method presented here to quantify LMP takes advantage of the detergent digitonin, which creates pores in cellular membranes by replacing cholesterol. The difference in cholesterol content between the plasma membrane (high) and lysosomal membrane (low) allows titration of digitonin to a concentration that permeabilizes the plasma membrane but leaves lysosomal membranes intact. The extent of LMP is determined by measuring the cytosolic activity of lysosomal hydrolases (e.g., cysteine cathepsins) and/or β-*N*-acetyl-glucosaminidase in the digitonin-extracted cytoplasm and comparing it to the total cellular enzyme activity. Digitonin extraction of the cytosol can be combined with precipitation of protein and/or western blot analysis for detection of lysosomal proteins (e.g., cathepsins).

MATERIALS

It is essential that you consult the appropriate Material Safety Data Sheets and your institution's Environmental Health and Safety Office for proper handling of equipment and hazardous materials used in this protocol.

RECIPES: Please see the end of this protocol for recipes indicated by <R>. Additional recipes can be found online at http://cshprotocols.cshlp.org/site/recipes.

Reagents

β-*N*-acetyl-glucosaminidase reaction buffer (NAG RB) <R>

Caspase reaction buffer (caspase RB), freshly prepared (optional; see Step 14) <R>

Caspase-3 activity can be measured simultaneously (on the same plate) with cathepsin activity, provided that the substrates are coupled to the same fluorophore (e.g., Ac-DEVD–7-amino-trifluoromethylcoumarin [AFC]), but this measurement is not required for analysis of lysosomal membrane permeabilization (LMP).

Cathepsin reaction buffer (cathepsin RB), freshly prepared <R>

Cell line of interest and appropriate growth medium

The current protocol has been tested on various cell lines treated with a range of different cytotoxic stimuli and works well for most standard cancer cell lines (including MCF-7, HeLa, and U2OS) using the following conditions: 15 min extraction, lifting frequency 50–60/min, digitonin concentration 15–25 µg/mL (Foghsgaard et al. 2001; Nylandsted et al. 2004; Groth-Pedersen et al. 2007). However, the optimal extraction time and lifting frequency

---

[1]Correspondence: jnl@cancer.dk

Cite this protocol as *Cold Spring Harb Protoc*; doi:10.1101/pdb.prot086165

may differ for other cell lines and should be determined empirically. For example, immortalized mouse embryonic fibroblasts (MEFs) are processed as follows: 10 min extraction, lifting frequency 110/min, digitonin concentration 17 µg/mL (Gyrd-Hansen et al. 2006).

Cytotoxic agent (compound or siRNA) for induction of LMP

Digitonin extraction (DE) buffer, freshly prepared <R>

For digitonin optimization (Step 2) and subsequent measurement of LMP (Step 10), prepare 200 µL of DE buffer per well (plus an additional 20 µL to adjust for pipetting loss). (This volume can be scaled up accordingly for larger well formats [e.g., six-well plates].)

Digitonin stocks (5 and 50 mg/mL in H_2O)

Digitonin usually precipitates and must be redissolved by heating and occasional vortexing. Immediately before use (Step 2), heat the digitonin stocks to 75°C for 5–10 min to dissolve any precipitates.

HCl (1 M)

LDH cytotoxicity detection kit (Roche or similar)

Total protein assay kit (if needed; see Step 16)

Equipment

Absorbance plate reader (e.g., VersaMax; Molecular Devices)

Incubator for cell culture

Microwell plates, black (96-well)

Rocking table with adjustable lifting frequency

Spectrofluorometer (e.g., Spectramax Gemini; Molecular Devices)

Tissue culture plates (24- and 96-well)

Vacuum suction pump

METHOD

Determining Optimal Digitonin Concentration for Extraction of Lysosome-Free Cytoplasm

The following procedure is necessary to determine the optimal digitonin concentration for permeabilization of the plasma membrane with minimal impact on the lysosomal membrane. LDH activities in the extracts are used as internal standards to which the lysosomal hydrolase activities are normalized. The procedure takes ~1 h and must be performed for each cell line separately. It should be performed regularly, as reagents (e.g., digitonin stocks) and cellular conditions may change over time.

Note that the ability of digitonin to permeabilize cellular membranes depends not only on digitonin concentration but also on the total amount of digitonin per cell; thus, the volume of DE buffer per cell should remain constant.

1. Seed 5×10^4 cells/well in a 24-well plate and allow to adhere overnight in an incubator.

    At least 12 wells should be used for the optimization.

    LMP can also be measured in nonadherent cells; in this case, it is advantageous to scale down the cell number and seed cells in 96-well rather than 24-well plates to prevent cells from floating.

2. On the next day (immediately before use), prepare the following 12 dilutions of digitonin in DE buffer: 0, 5, 10, 15, 20, 25, 30, 35, 40, 45, 50, and 200 µg/mL digitonin.

    A digitonin solution of 200 µg/mL is used for total/complete permeabilization of cells.

3. Place the plate on ice and begin the extraction as follows:

    i. Working with sets of six wells at a time, remove the medium from each well using a vacuum suction pump. (If using nonadherent cells, carefully aspirate the medium from the wells and leave ~20 µL of medium on the cells.)

       Wells should be processed six at a time to prevent the cells from drying out when medium is removed.

    ii. Add 200 µL of one digitonin dilution to each well. Start a timer when digitonin is added to each set of six wells.

       The digitonin solution should be added to the side of the well to avoid flushing off the attached cells.

       Exact timing is critical to ensure that each well is extracted equally.

4. Incubate the cells on ice for 15 min on a rocking table (lifting frequency ~50–60/min).

5. Transfer 180 µL of each extract into one well of a labelled 96-well plate on ice.

6. Perform the cathepsin assay as follows:

    i. Mix 50 µL of extract with 50 µL of cathepsin RB in one well of a black 96-well plate.

    ii. Preincubate the plate for 5 min at 30°C in the fluorometer.

    iii. Measure the kinetics of cathepsin activity (i.e., the V_{max} of the liberation of AFC) for 20 min at 30°C (excitation, 400 nm; emission, 489 nm; 45 sec interval).

7. During the cathepsin assay, perform the LDH assay as follows:

    i. Transfer 30 µL of extract per well to a 96-well plate and equilibrate to room temperature for 5–10 min.

    ii. Add 30 µL of mixed LDH reagent to each well and allow the reaction to run for 2–10 min. Stop each reaction with 20 µL of 1 M HCl.

       Make sure all samples have equal reaction time before ending the reaction, typically when the samples with highest LDH content are medium to intense red.

    iii. Assay LDH activity by measuring the optical density (OD) 490 nm (OD_{490}) in an absorbance plate reader.

8. To determine the optimal digitonin concentration for cytosolic extraction, graph and compare the cathepsin release and LDH raw values (Fig. 1).

   The concentration that produces the best possible permeabilizion of the plasma membrane (LDH release) with minimal cathepsin release from the lysosomes is optimal.

Measuring LMP

In the following procedure, the level of LMP in response to a given treatment is measured by digitonin extraction of the cytoplasmic fraction followed by hydrolase and/or NAG activity measurements. Measurement of LMP by digitonin extraction takes 2–3 h for two 24-well plates.

9. For each cellular condition (e.g., treatment with siRNA or cytotoxic compound), seed cells in triplicate wells of a 24-well plate. Seed parallel triplicate wells for total cellular cathepsin measurements (Fig. 2A).

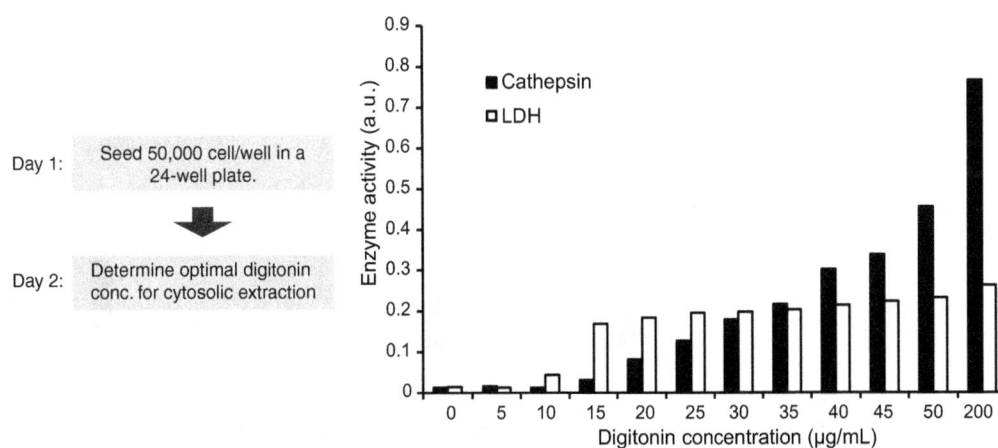

FIGURE 1. Example of optimization experiment used to determine the digitonin concentration to be used for cytoplasmic extraction of MCF-7. Cells (5×10^4 per well) were extracted for 15 min on a rocking table (lifting frequency 50–60/min) with the indicated digitonin concentrations. The level of plasma membrane permeabilization (LDH activity) and lysosome permeabilization (cathepsin activity) were measured. Cells were completely permeabilized at 200 µg/mL digitonin (total). A digitonin concentration that only permeabilizes the plasma membrane and leaves lysosomes intact is ideal. Here, 15–18 µg/mL is optimal for MCF-7 cells.

Cells should be seeded at a density that will on the day of analysis result in the density used in the digitonin optimization experiment (e.g., ~5 × 10⁴ cells).

10. On the day of analysis (immediately before use), prepare digitonin solutions in DE buffer for cytoplasmic extraction (using the optimal concentration obtained in Step 8) and total cell extraction.

    In our laboratory, we use digitonin solutions of 17 µg/mL and 200 µg/mL, respectively, for cytoplasmic and total cell extractions in MEF. See Figure 1, for example, using MCF-7 cells.

11. Place the plate on ice and begin the digitonin extraction as described in Step 3. Mark the time of digitonin addition on the plate lid for each set of six wells.

    Swift pipetting is necessary to avoid too much variation among wells.

12. Incubate the plate on ice for 15 min on a rocking table (lifting frequency ~50–60/min).

    The optimal time and lifting frequency may vary between cell lines.

13. Transfer 180 µL of each extract into one well of a labelled 96-well plate on ice.

    The cytoplasmic/total extracts can now be used to measure lysosomal cysteine cathepsin, NAG, LDH, and caspase 3-like activities.

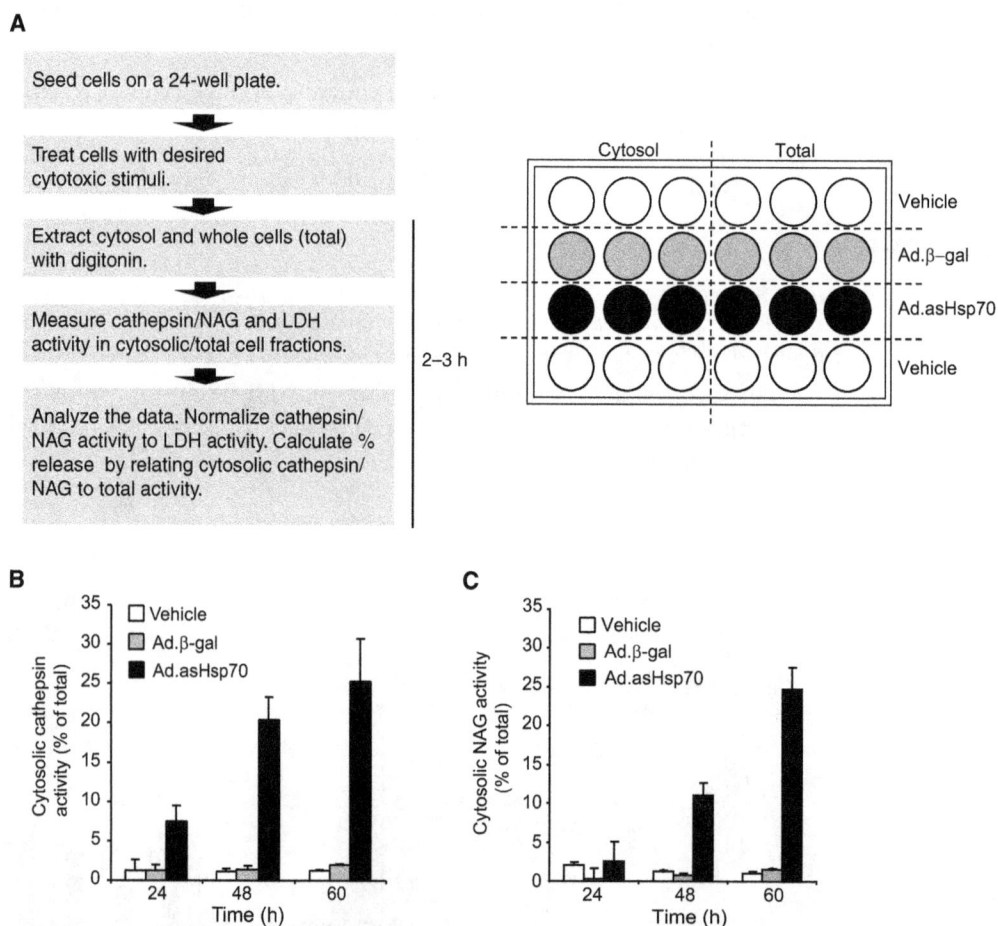

FIGURE 2. (A) Flowchart of the digitonin extraction procedure for measuring LMP. Sample LMP extraction setup in a typical 24-well format with vehicle control, β-galactosidase adenovirus control (Ad. β-gal) or adenovirus containing antisense Hsp70 cDNA (Ad.asHsp70) as the LMP inducer. Cells are seeded in triplicate for both cytosolic and total cellular extraction, and digitonin is added to sets of six wells and timed to ensure all wells are extracted equally. (B) A representative example of lysosomal cathepsin B/L release and (C) NAG release obtained from MCF-7 cells transduced for 24–60 h with Ad.asHsp70, which depletes Hsp70 and triggers LMP. Values are normalized to LDH and presented as % cytosolic release of total cellular cathepsin or NAG activity. Numbers represent mean of triplicate measurements and error bars indicate SD values (© Nylandsted et al. 2004. Originally published in *J Exp Med* **200:** 425–435.)

14. Perform cathepsin and caspase-3 assays as follows:

    i. Mix 50 μL of extract with 50 μL of cathepsin RB per well in a black 96-well plate.

    ii. (Optional) Mix 50 μL of extract with 50 μL of caspase RB per well in the same black 96-well plate.

    iii. Preincubate the plate for 5 min at 30°C in the fluorometer.

    iv. Measure as described in Step 6.iii.

       Cathepsin (and caspase) activity in cytosolic and total extracts is calculated as the average rate (slope of the curve; fluorescence units/sec).

15. Measure lysosomal NAG release as follows:

    i. Mix 30 μL of extract with 100 μL of NAG RB per well in a black 96-well plate.

    ii. Preincubate the plate for 3–5 min at 30°C in the fluorometer.

    iii. Measure the liberation of methylumbelliferone for 20 min at 30°C (excitation, 356 nm; emission, 444 nm; 45 sec interval).

       NAG activity in extracts is calculated as the average rate (slope of the curve; fluorescence units/sec).

16. Perform the LDH assay for each well as described in Step 7.

    If the treatment of the cells alters the cellular LDH activity, hydrolase activities can be normalized to total protein content using a commercial kit designed for this purpose.

Data Analysis

17. Normalize each lysosomal hydrolase/NAG value to the corresponding LDH value from the same well (Fig. 2A).

    See Troubleshooting.

18. Calculate the mean values of the triplicate measurements for cytoplasmic and total protease levels (Fig. 2A).

19. Calculate the percentage of released enzyme activity by relating the LDH-corrected cytoplasmic activity of the lysosomal hydrolase/NAG measured for each well with the mean value from triplicate samples for corresponding total cellular activity of the hydrolase/NAG (Fig. 2B,C).

    See Troubleshooting.

TROUBLESHOOTING

Problem (Step 17): Digitonin overextraction results in high cathepsin/NAG background levels (>10%).

Solution: This is a common difficulty associated with LMP measurements and can be adjusted by further fine-tuning the digitonin extraction (e.g., by using a narrower range of digitonin concentrations during optimization to determine an optimal lower concentration). In addition, cell density is critical to achieving the best digitonin:cell ratio for cytoplasmic extraction; this can be optimized by keeping the digitonin concentration constant and varying the cell density. Finally, optimization of extraction time may be helpful with some cell types.

Problem (Step 17): Because of the shorter half-life of cathepsins in the cytosol as compared with the lysosome, the obtained values are, in fact, lower than the actual release of the hydrolases.

Solution: This may especially become a problem in long-term experiments and should be considered if the total cathepsin/LDH ratios decline extensively. The cytosolic half-life of NAG is longer than that of cathepsins and thus NAG measurements can give more accurate values in long-term assays. Additionally, determining the kinetics of the LMP by multiple measurements at different time points after the stimulus may help with interpretation of results.

Problem (Step 19): Cytotoxic stimuli with a direct impact on lipids in the plasma membrane interfere with the digitonin extraction procedure, producing misleading results.

Solution: LMP should be estimated using fluorescent dextran-loaded lysosomes (see Protocol 2: Visualizing Lysosomal Membrane Permeabilization by Fluorescent Dextran Release [Ellegaard et al. 2015]) or cathepsin immunostaining (see Protocol 3: A Method to Monitor Lysosomal Membrane Permeabilization by Immunocytochemistry [Groth-Pedersen et al. 2015]).

RELATED TECHNIQUES

When using treatments that influence cellular cholesterol content or have detergent-like properties (e.g., cationic amphiphilic drugs), the above method must be used with care, as such treatments may interfere with the digitonin extraction procedure. In these cases, alternative approaches should be used to estimate LMP; see Protocol 2: Visualizing Lysosomal Membrane Permeabilization by Fluorescent Dextran Release (Ellegaard et al. 2015) and Protocol 3: A Method to Monitor Lysosomal Membrane Permeabilization by Immunocytochemistry (Groth-Pedersen et al. 2015). Furthermore, in challenging conditions where the level of LMP is low because of rupture of only a few lysosomes per cell, it is not feasible to measure LMP by this method. It may instead prove advantageous to estimate LMP by transmission electron microscopy-based methods (e.g., by using BSA gold-labeled lysosomes to monitor LMP at a single-cell level [Gyrd-Hansen et al. 2006]).

RECIPES

β-*N*-Acetyl-Glucosaminidase Reaction Buffer

Reagent	Final concentration
Sodium citrate buffer (pH 4.5)	0.2 M
4-Methylumbelliferyl-2-acetamido-2-deoxy-β-D-glucopyranoside substrate (Sigma-Aldrich)	300 µg/mL

Store in aliquots at −20°C.

Caspase Reaction Buffer

Reagent	Final concentration
HEPES	100 mM
Glycerol	20%
EDTA	0.5 mM
CHAPS	0.1%
Pefabloc	0.5 mM
Dithiothreitol (DTT)	8 mM
Ac-DEVD–7-amino-trifluoromethylcoumarin (AFC) substrate (Enzo Life Sciences)	50 µM

Prepare a stock solution that contains the first four ingredients, adjust pH to 7.5, and store it at 4°C. Immediately before use, add the last three ingredients.

Cathepsin Reaction Buffer

Reagent	Final concentration
Sodium acetate	50 mM
EDTA	4 mM
Pefabloc	0.5 mM
Dithiothreitol (DTT)	8 mM
Z-FR-AFC substrate (MP Bio)	50 µM

Prepare a stock solution that contains the first two ingredients, adjust pH to 6.0, and store it at 4°C. Immediately before use, add the last three ingredients.

Cite this protocol as *Cold Spring Harb Protoc*; doi:10.1101/pdb.prot086165

Digitonin Extraction Buffer

Reagent	Final concentration
Sucrose	250 mM
HEPES	20 mM
KCl	10 mM
MgCl$_2$	1.5 mM
EDTA	1 mM
EGTA	1 mM
Pefablock	0.5 mM

Prepare a stock solution that contains the first six ingredients, adjust pH to 7.5, and store it at 4°C. Immediately before use, add the Pefablock.

REFERENCES

Ellegaard A-M, Jäättelä M, Nylandsted J. 2015. Visualizing lysosomal membrane permeabilization by fluorescent dextran release. *Cold Spring Harb Protoc* doi: 10.1101/pdb.prot086173.

Foghsgaard L, Wissing D, Mauch D, Lademann U, Bastholm L, Boes M, Elling F, Leist M, Jaattela M. 2001. Cathepsin B acts as a dominant execution protease in tumor cell apoptosis induced by tumor necrosis factor. *J Cell Biol* 153: 999–1010.

Groth-Pedersen L, Ostenfeld MS, Hoyer-Hansen M, Nylandsted J, Jaattela M. 2007. Vincristine induces dramatic lysosomal changes and sensitizes cancer cells to lysosome-destabilizing siramesine. *Cancer Res* 67: 2217–2225.

Groth-Pedersen L, Jäättelä M, Nylandsted J. 2015. A method to monitor lysosomal membrane permeabilization by immunocytochemistry. *Cold Spring Harb Protoc* doi: 10.1101/pdb.prot086181.

Gyrd-Hansen M, Farkas T, Fehrenbacher N, Bastholm L, Hoyer-Hansen M, Elling F, Wallach D, Flavell R, Kroemer G, Nylandsted J, et al. 2006. Apoptosome-independent activation of the lysosomal cell death pathway by caspase-9. *Mol Cell Biol* 26: 7880–7891.

Nylandsted J, Gyrd-Hansen M, Danielewicz A, Fehrenbacher N, Lademann U, Hoyer-Hansen M, Weber E, Multhoff G, Rohde M, Jaattela M. 2004. Heat shock protein 70 promotes cell survival by inhibiting lysosomal membrane permeabilization. *J Exp Med* 200: 425–435.

Visualizing Lysosomal Membrane Permeabilization by Fluorescent Dextran Release

Anne-Marie Ellegaard, Marja Jäättelä, and Jesper Nylandsted[1]

Unit for Cell Death and Metabolism, Center for Autophagy, Recycling and Disease, Danish Cancer Society Research Center, DK-2100, Copenhagen, Denmark

Lysosomal membrane permeabilization (LMP) is an effective programmed cell death pathway triggered in response to a variety of cytotoxic stimuli and cellular conditions. In the method presented here, LMP is monitored by first taking advantage of the steady endocytic capacity of cells to load fluorescent dextran into lysosomes, and then simply observing the translocation of lysosomally localized dextran into the cytosol after an LMP-inducing insult. Fluorescent dextran in healthy cells appears in punctate structures representing intact lysosomes, whereas after LMP, a diffuse staining pattern throughout the cytoplasm is observed. Using this method, LMP can be followed in real time using time-lapse imaging. The size of pores formed in the membrane during LMP by size exclusion can also be determined using dextrans of different sizes and colors.

MATERIALS

It is essential that you consult the appropriate Material Safety Data Sheets and your institution's Environmental Health and Safety Office for proper handling of equipment and hazardous materials used in this protocol.

Reagents

Cell line of interest (adherent) and appropriate growth medium

Concanamycin A (if needed; see Step 4)

Cytotoxic agent (compound or siRNA) for induction of lysosomal membrane permeabilization (LMP)

A potent LMP inducer is required for real-time imaging; see Discussion.

Dextran, fluorescently labeled (e.g., conjugated to Alexa Fluor 488/594 [10- or 70-kDa, anionic, fixable] or fluorescein isothiocyanate (FITC) [Ex 488; 70-kDa])

Various sizes of fluorescently labeled dextran are commercially available (Molecular Probes) and can be used to estimate pore sizes created during LMP by dextran size exclusion (e.g., by comparing the release of 10-, 70-, and 500-kDa dextran in different colors) (Bidere et al. 2003; Fig. 1B). FITC conjugated to dextran is used to monitor early changes in lysosomal pH during the LMP process: Neutralization of lysosomal pH upon LMP can result in a severalfold increase in fluorescence intensity measurable by flow cytometry or from microscopy pictures using dedicated software (e.g., ImageJ or MetaMorph) (see www.invitrogen.com/site/us/en/home/References/Molecular-Probes-The-Handbook/pH-Indicators/Probes-Useful-at-Near-Neutral-pH.html).

Prepare a stock solution (5 μg/μL) of the desired fluorescent dextran in serum-free medium and store at 4°C.

Dulbecco's phosphate-buffered saline (DPBS) (without calcium and magnesium; Gibco 14190-136)

---

[1]Correspondence: jnl@cancer.dk

Cite this protocol as *Cold Spring Harb Protoc*; doi:10.1101/pdb.prot086173

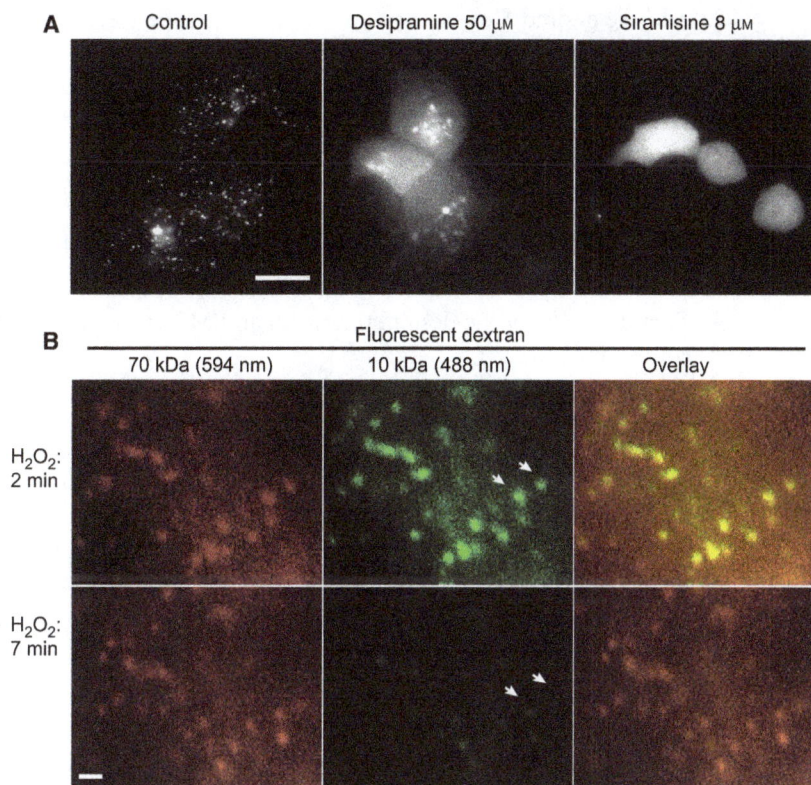

FIGURE 1. LMP visualized by fluorescent dextran release. (*A*) Representative pictures of MCF-7 breast carcinoma cells loaded with a single 10-kDa Alexa Fluor 488-conjugated dextran and treated for 20 h with indicated cationic amphiphilic drugs to induce LMP. Note the diffuse staining pattern in the cytosol after LMP induction. (Reprinted from Petersen et al. 2013; © Elsevier, Inc.) (*B*) Pictures obtained from a time-lapse movie of HeLa cells loaded with 10-kDa Alexa Fluor 488 and 70-kDa Alexa Fluor 594-conjugated dextran and incubated with hydrogen peroxide to induce LMP. Only 10-kDa dextran is released to the cytosol and disappears from lysosomes after 7 min of exposure to hydrogen peroxide, indicating that the pores formed in the lysosomal membrane are restricting release of 70-kDa dextran. Scale bar, 1 µm.

L-Leucyl-L-leucine-methyl ester
Paraformaldehyde (4% in DPBS) (optional; see Step 6)

Equipment

Glass chamber slides or cover slips for live cell imaging (e.g., Lab-Tek chambered coverglass; Nalge Nunc)
Imaging software (e.g., ImageJ or MetaMorph)
Incubator for cell culture
Inverted fluorescence microscope or/and confocal microscope (optimally equipped with time-lapse imaging setup)

METHOD

The following protocol applies to dextran loading and subsequent imaging of dextran release during LMP in adherent cells. Loading of cells with fluorescent dextran, including a 2 h chase period, can be performed within 2–8 h, although a longer incubation time (16 h) is recommended to increase the number of stained lysosomes.

1. Seed cells in chamber slides or on glass coverslips and allow to adhere.

2. Add the desired fluorescent dextran to the medium (e.g., 50–200 µg/mL of Alexa Fluor 488- or 594-dextran [10-kDa] or 75–100 µg/mL of FITC-dextran). Incubate the cells for 2–16 h.

> The confluency of cells should be 10%–60%. Optimal dextran concentration and incubation time varies between cell lines and should be determined empirically. We have found that treatment with 50–100 µg/mL of Alexa Fluor 488/594-dextran (10-kDa) for 3–6 h works well for HeLa and MCF-7 cells. Alternatively, dextran can be loaded faster by incubating cells directly in the stock solution (5 µg/µL) for 1–2 h. To conserve valuable dextran, the stock solution can be collected and re-used.

3. After dextran loading, wash the cells twice in DPBS and chase for 2 h in fresh medium.

> Dextran uptake into lysosomes can now be inspected live using an inverted fluorescence microscope.

4. Treat the cells for the desired time with an LMP-inducing cytotoxic stimulus. Include a positive control for LMP.

> A 2–6 h treatment with L-leucyl-L-leucine-methyl ester (1–2.5-mM) is a suitable control for most cell lines.
>
> The fluorescence intensity of FITC-dextran is reduced in normal lysosomes at acidic pH 4–5. When using FITC-dextran to monitor LMP-associated pH changes, cells should be incubated for 2 h with concanamycin A (10-nM) as a positive control; this inhibits the activity of the lysosomal V-H$^+$-ATPase pump and increases lysosomal pH and FITC fluorescence up to eightfold without causing LMP.

5. After the proper length of incubation, inspect the cells under a fluorescence microscope.

> The degree of LMP (dextran release to the cytosol) can be imaged and quantified manually or automatically using dedicated imaging software (e.g., ImageJ) (Fig. 1).
>
> See Troubleshooting.

6. (Optional) Fix cells loaded with anionic fixable Alexa Fluor 488/594 dextran in 4% paraformaldehyde for 30 min for later imaging/estimation of cellular LMP.

TROUBLESHOOTING

Problem (Step 5): It is difficult to recognize/monitor LMP by microscopy using the dextran release procedure.

Solution: The degree of LMP triggered by a particular cytotoxic insult may be relatively weak. Increasing the concentration of dextran and/or applying sensitive imaging software dedicated to measuring differences in cellular fluorescence (e.g., MetaMorph software) may prove advantageous.

DISCUSSION

Data Analysis

In this protocol, fluorescence microscopy is used to detect a single dextran conjugate, which changes its staining pattern from a punctate to a cytosolic pattern upon LMP (Fig. 1A) (Petersen et al. 2013). For each experiment, the proportion of cells with LMP (appearance of cytosolic dextran) can be obtained by counting 10 randomly chosen areas with a minimum of 50 cells for each condition and calculating the percentage of cells with LMP out of the total for each condition. Using dextran conjugates of two different sizes (10- and 70-kDa) and colors, timing of dextran release and pore size can be determined by time-lapse imaging. For example, exposure of HeLa cells to hydrogen peroxide creates pores in the lysosomal membrane which allow 10-kDa dextran to be released while retaining 70-kDa dextran in the lumen (Fig. 1B). Lysosomal pH changes can be measured either by flow cytometry (for adherent and nonadherent cells) or intensity measurements of fluorescence images obtained using FITC-dextran. Automated image analysis (e.g., using ImageJ) can be used to measure dextran intensity in lysosomes throughout the process.

Real-Time Imaging

In principle, it should be possible to follow LMP via lysosomal dextran release in real time by time-lapse video microscopy. However, lysosomes are highly dynamic organelles that regularly move in and out of the focus plane. Because most LMP-inducing insults do not work instantly, it can be challenging to capture LMP events in real time. Real-time imaging is possible using potent LMP inducers such as O-methyl-serine dodecylamide hydrochloride, L-leucyl-L-leucine-methyl ester (which accumulates in lysosomes and has detergent-like properties [Li et al. 2000]), or hydrogen peroxide (Fig. 1B).

REFERENCES

Bidere N, Lorenzo HK, Carmona S, Laforge M, Harper F, Dumont C, Senik A. 2003. Cathepsin D triggers Bax activation, resulting in selective apoptosis-inducing factor (AIF) relocation in T lymphocytes entering the early commitment phase to apoptosis. *J Biol Chem* **278:** 31401–31411.

Li W, Yuan X, Nordgren G, Dalen H, Dubowchik GM, Firestone RA, Brunk UT. 2000. Induction of cell death by the lysosomotropic detergent MSDH. *FEBS Lett* **470:** 35–39.

Petersen NH, Olsen OD, Groth-Pedersen L, Ellegaard AM, Bilgin M, Redmer S, Ostenfeld MS, Ulanet D, Dovmark TH, Lønborg A, et al. 2013. Transformation-associated changes in sphingolipid metabolism sensitize cells to lysosomal cell death induced by inhibitors of acid sphingomyelinase. *Cancer Cell* **24:** 379–393.

A Method to Monitor Lysosomal Membrane Permeabilization by Immunocytochemistry

Line Groth-Pedersen,[1,2] Marja Jäättelä,[2] and Jesper Nylandsted[2,3]

[1]Department of Pediatrics and Adolescent Medicine, Rigshospitalet University Hospital, DK-2100 Copenhagen, Denmark; [2]Unit for Cell Death and Metabolism, Center for Autophagy, Recycling and Disease, Danish Cancer Society Research Center, DK-2100 Copenhagen, Denmark

Programmed cell death involving lysosomal membrane permeabilization (LMP) is a common phenomenon—more the rule than the exception under various cytotoxic stimuli and stressful cellular conditions. The protocol presented here is based on immunocytochemical staining of cathepsin B or L to visualize translocation from the lysosomal lumen to the cytosol. In healthy cells, cathepsins appear in localized punctate structures representing intact lysosomes, whereas LMP results in a diffuse staining pattern throughout the cytoplasm. LMP can be triggered upstream, downstream, or independently of the classical apoptotic death pathway involving mitochondrial outer membrane permeabilization (MOMP). Co-staining with antibodies recognizing the active form of Bax allows investigation of the order of events between LMP and MOMP in death signaling.

MATERIALS

It is essential that you consult the appropriate Material Safety Data Sheets and your institution's Environmental Health and Safety Office for proper handling of equipment and hazardous materials used in this protocol.

RECIPES: Please see the end of this protocol for recipes indicated by <R>. Additional recipes can be found online at http://cshprotocols.cshlp.org/site/recipes.

Reagents

Antifade mounting medium

Cell line of interest and appropriate growth medium

Cytotoxic agent (compound or siRNA) for induction of LMP

Dulbecco's phosphate-buffered saline (DPBS) (without calcium and magnesium; Gibco 14190-136)

Immunofluorescence buffer 1 (IF-Buffer-1) <R>

Immunofluorescence buffer 2 (IF-Buffer-2) <R>

Immunofluorescence buffer 3 (IF-Buffer-3) <R>

Fetal calf serum (FCS)

L-Leucyl-L-leucine-methyl ester

Paraformaldehyde (4% in DPBS) and/or methanol (ice-cold)

Primary antibodies for staining

Mouse anti-human cathepsin B (Calbiochem)

[3]Correspondence: jnl@cancer.dk

Cite this protocol as Cold Spring Harb Protoc; doi:10.1101/pdb.prot086181

Mouse anti-human cathepsin L (BD Transduction Laboratories)
Rabbit anti-human Bax, active conformation (Cell Signaling Technology)
Rabbit anti-Lamp-1 (Abcam)

Secondary antibodies

Alexa Fluor 488-conjugated anti-mouse
Alexa Fluor 594-conjugated anti-rabbit

Triton X-100 (0.2% in DPBS) (if needed; see Step 4)

Equipment

Cover slides
Cytospin benchtop centrifuge (if needed; see Step 3)
Fluorescence or confocal microscope
Glass coverslips

For easy handling, cells can be seeded on coverslips that fit into the wells of a 24-well plate.

Humidified chamber (if needed; see Step 6)
Incubator for cell culture
Parafilm (optional; see Step 5)

METHOD

The following immunostaining procedure can be performed within 4–5 h. Cathepsin B or L staining can be combined with staining of a lysosomal membrane protein such as Lamp-1, as described here, or expression of a lysosomal membrane protein coupled to GFP (e.g., CD63-eGFP), as shown in Figure 1A. Co-staining of cathepsin B or L with an antibody recognizing the active form of Bax can be used to investigate cross-talk between LMP and MOMP (Fig. 1B).

1. Seed cells on coverslips at 25%–50% confluency (according to the length of the planned incubation) and allow to adhere.

   We typically use 2.5×10^4 cells and a 48-h incubation period.

2. Treat the cells with the LMP-inducing cytotoxic agent for the desired length of time. Include a positive control for LMP.

   A 2- to 6-h treatment with L-leucyl-L-leucine-methyl ester (1–2.5 mM) is a suitable control for most cell lines.

3. Rinse the treated cells twice in DPBS. Fix the cells in 4% paraformaldehyde for 20 min at room temperature, or in ice-cold methanol for 3 min.

   When using nonadherent cells or treatments that cause early detachment (e.g., cytoskeleton-disrupting drugs like vincristine), cells can be centrifuged onto glass slides before fixation using a Cytospin centrifuge (600g for 5 min).

4. Rinse the fixed cells once in DPBS. Permeabilize paraformaldehyde-fixed cells in 0.2% Triton X-100 for 2 min.

   Cells can now be stored in DPBS at 4°C for later staining.

5. Rinse the cells twice in DPBS. Block the cells by incubating the coverslips for 20 min in IF-Buffer-1 containing 5% FCS.

   To conserve antibody, coverslips can now be gently transferred to a flat area (e.g., glass plate wrapped in parafilm) for subsequent staining.

6. Wash the cells once in IF-Buffer-1 for 5 min at room temperature. Overlay the coverslips with ~100 µL of IF-Buffer-1 containing antibodies to cathepsin L (or B) and Lamp-1(or Bax) (1:200). Incubate for 1 h at room temperature.

   For longer incubations, coverslips should be placed in a humidified chamber to prevent the antibody solution from drying.

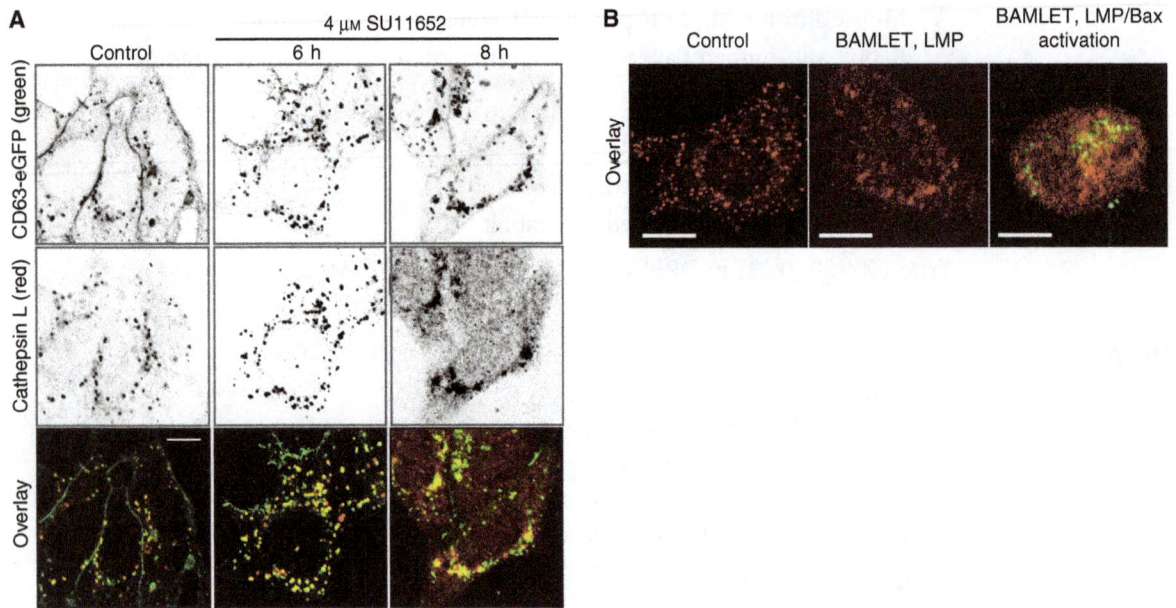

FIGURE 1. (*A*) Confocal images of HeLa cells expressing CD63 (lysosomal-associated membrane protein 3) coupled to eGFP and stained for cathepsin L. After 8 h of treatment with 4 μM SU11652, which accumulates in lysosomes and induces LMP, cathepsin L is released from lysosomes (marked by CD63-eGFP) into the cytosol and appears as a diffuse staining pattern in the cytoplasm (*right* panel) with decreased colocalization with CD63. (Reprinted from Ellegaard et al. 2013; © American Association for Cancer Research [AACR].) (*B*) Example of LMP with and without Bax activation in MCF-7 cells induced by a complex of bovine lactalbumin and oleic acid (BAMLET). Representative confocal pictures of MCF-7 cells treated with 100 μg/mL of BAMLET for 3 h (1 h without serum followed by 2 h with serum) and stained for cathepsin L (red) and the active conformation of Bax (green). The *middle* panel shows a cell with cathepsin L release indicative of early LMP before Bax activation and the *right* panel shows cathepsin L release and Bax activation. Scale bars, 20 μm. (Reprinted from Rammer et al. 2010; © American Association for Cancer Research [AACR].)

7. Wash the cells three times for 5 min per wash at room temperature in IF-Buffer-2.

8. Overlay the cells with ~100 μL of IF-Buffer-2 containing Alexa Fluor 488/594-conjugated secondary antibodies (1:1000). Incubate for 1 h at room temperature in the dark.

9. Wash the cells three times for 5 min per wash at room temperature in IF-Buffer-3.

10. Rinse the coverslips in H$_2$O and remove any excess liquid with a paper towel. Mount the coverslips on cover slides (cell side down) with a drop of antifade mounting medium.

11. Allow the mounting medium to solidify (usually 6–12 h) and then examine the slides by fluorescence microscopy.

    Slides may be stored at 4°C.

12. Analyze the cells for vesicular versus diffuse staining of lysosomal cathepsin B or L and/or absence versus presence of active Bax. For each condition, count 50 to 100 randomly chosen cells and calculate the number of cells out of the total containing released cathepsin B or L. Likewise, calculate the percentage of cells with cathepsin B or L and active translocated Bax out of the total (Groth-Pedersen et al. 2007).

    See Troubleshooting.

TROUBLESHOOTING

Problem (Step 12): Cathepsin B or L staining appears nonspecific with high background.
Solution: Use freshly prepared reagents, especially 4% paraformaldehyde (or thaw a fresh stock solution from −20°C). Optimize the fixation procedure further by reducing the fixation time,

changing the permeabilization agent to saposin, or combining 4% paraformaldehyde fixation with subsequent methanol permeabilization/fixation.

DISCUSSION

The protocol described here is useful for assessing LMP combined with other cell death markers under various cellular conditions. LMP can, depending on the cytotoxic stimuli and cell type, be activated in an MOMP- and apoptosome-independent manner still involving caspase-9 (Gyrd-Hansen et al. 2006). It can also be induced before MOMP, as in the case of HeLa cells treated with vincristine, which depolymerizes microtubules (Groth-Pedersen et al. 2007). It can even be triggered independently of the intrinsic apoptosis pathways, as in LMP-induced death via Hsp70 depletion (Nylandsted et al. 2000, 2004). This method can be modified to use antibodies to other abundant luminal lysosomal proteins (e.g., NAG) or relevant death components (e.g., caspases or cytochrome c), and co-staining can be used to address several different proteins involved in the death pathway of interest. In these cases, the method should be applied using narrow time intervals (15- to 30-min time points around the time of LMP induction) to address the order of events for a given death inducer and establish the governing cell death mechanism.

RECIPES

Immunofluorescence Buffer 1

Prepare a solution of 1% BSA and 0.3% Triton X-100 in Dulbecco's phosphate-buffered saline (without calcium and magnesium; Gibco 14190-136). Store at −20°C.

Immunofluorescence Buffer 2

Prepare a solution of 0.25% BSA and 0.1% Triton X-100 in Dulbecco's phosphate-buffered saline (without calcium and magnesium; Gibco 14190-136). Store at 4°C.

Immunofluorescence Buffer 3

Prepare a solution of 0.05% Tween 20 in Dulbecco's phosphate-buffered saline (without calcium and magnesium; Gibco 14190-136). Store at 4°C.

REFERENCES

Ellegaard AM, Groth-Pedersen L, Oorschot V, Klumperman J, Kirkegaard T, Nylandsted J, Jäättelä M. 2013. Sunitinib and SU11652 inhibit acid-sphingomyelinase, destabilize lysosomes, and inhibit multidrug resistance. *Mol Cancer Ther* 12: 2018–2030.

Groth-Pedersen L, Ostenfeld MS, Hoyer-Hansen M, Nylandsted J, Jäättelä M. 2007. Vincristine induces dramatic lysosomal changes and sensitizes cancer cells to lysosome-destabilizing siramesine. *Cancer Res* 67: 2217–2225.

Gyrd-Hansen M, Farkas T, Fehrenbacher N, Bastholm L, Hoyer-Hansen M, Elling F, Wallach D, Flavell R, Kroemer G, Nylandsted J, et al. 2006. Apoptosome-independent activation of the lysosomal cell death pathway by caspase-9. *Mol Cell Biol* 26: 7880–7891.

Nylandsted J, Rohde M, Brand K, Bastholm L, Elling F, Jäättelä M. 2000. Selective depletion of heat shock protein 70 (Hsp70) activates a tumor-specific death program that is independent of caspases and bypasses Bcl-2. *Proc Natl Acad Sci* 97: 7871–7876.

Nylandsted J, Gyrd-Hansen M, Danielewicz A, Fehrenbacher N, Lademann U, Hoyer-Hansen M, Weber E, Multhoff G, Rohde M, Jäättelä M. 2004. Heat shock protein 70 promotes cell survival by inhibiting lysosomal membrane permeabilization. *J Exp Med* 200: 425–435.

Rammer P, Groth-Pedersen L, Kirkegaard T, Daugaard M, Rytter A, Szyniarowski P, Hoyer-Hansen M, Povlsen LK, Nylandsted J, Larsen JE, et al. 2010. BAMLET activates a lysosomal cell death program in cancer cells. *Mol Cancer Ther* 9: 24–32.

Strategies for Assaying Lysosomal Membrane Permeabilization

Urška Repnik,[1,2,4] Maruša Hafner Česen,[1] and Boris Turk[1,3,4]

[1]Department of Biochemistry and Molecular and Structural Biology, J. Stefan Institute, SI-1000 Ljubljana, Slovenia; [2]Department of Biosciences, University of Oslo, NO-0371 Oslo, Norway; [3]Center of Excellence CIPKEBIP, SI-1000 Ljubljana, Slovenia

Late endosomal organelles have an acidic pH and contain hydrolytic enzymes to degrade cargo delivered either from the extracellular environment by endocytosis or from within the cell itself by autophagy. In the event of lysosomal membrane permeabilization (LMP), the contents of late endosomes and lysosomes can be released into the cytosol and then initiate apoptosis. Compounds that can trigger LMP are therefore candidates for the induction of apoptosis, in particular in anticancer therapy. Alternatively, drug-delivery systems, such as nanoparticles, can have side effects that can include LMP, which has toxic consequences for the cells. To determine when, to what extent, and with what consequences LMP occurs is therefore of paramount importance for the evaluation of new potentially LMP-inducing compounds. In this introduction, we provide an overview of some basic assays for assessing LMP, such as staining with lysosomotropic dyes and measurement of cysteine cathepsin activity, and discuss additional strategies for the detection of the release of endogenous lysosomal molecules or preloaded exogenous tracers into the cytosol.

INTRODUCTION

The endosomal system is a dynamic set of organelles that has many important functions, and its perturbations have diverse effects on intracellular membrane trafficking and cell metabolism. Its unique characteristics include connection with the extracellular environment and possession of low pH at its distal end (Huotari and Helenius 2011). Late endosomes and lysosomes (hereafter collectively referred to as lysosomes) contain acidic hydrolases (Pillay et al. 2002; Schroder et al. 2010). The risk for the survival of cells in situations in which the contents of lysosomes are released into the cytosol was recognized soon after their discovery (de Duve 2005). Later, it was shown that dipeptidyl methyl esters, such as Leu-Leu-O-methylester (LLOMe), and dipeptidyl-β-naphthylamides, such as Gly-Phe-β-naphthylamide, can trigger lysosomal membrane permeabilization (LMP) (Goldman and Kaplan 1973; Reeves 1979). LMP and leaky lysosomes have been associated with neurological disorders and aging (Terman and Brunk 2004). LMP is unlikely to occur in healthy cells and tissues; moreover, there are enzyme inhibitors in the cytosol that can safeguard the cell from a limited LMP event (Repnik et al. 2012). In contrast, LMP often accompanies the late stages of apoptosis, in particular after generation of reactive oxygen species (ROS) in dysfunctional mitochondria (Špes et al. 2012; Huai et al. 2013).

Over the years, a number of compounds have been identified that can induce LMP when they accumulate in the endocytic pathway (Repnik et al. 2013, 2014). This has raised interest in the study of LMP from the perspective of drug delivery (Lloyd 2000), and even more so with the aim to induce LMP to trigger apoptosis in anticancer therapy (Erdal et al. 2005; Kirkegaard and Jäättelä 2009). Lately,

---

[4]Correspondence: urska.repnik@ibv.uio.no; boris.turk@ijs.si

Cite this introduction as *Cold Spring Harb Protoc*; doi:10.1101/pdb.top077479

LMP has been associated with the toxicity of nanomaterials (Stern et al. 2012). Although the list of LMP-inducing compounds has grown (Boya 2012), it should be pointed out that the evidence for LMP is not always convincing. There are two major concerns. First, whether a compound can indeed induce LMP, and second the position of the observed LMP in the time-line of intracellular events. Both are of paramount importance for evaluating the potential of pharmacological compounds. Proving LMP is not an easy task. Various approaches have been applied, but the results sometimes seem overinterpreted, and additional controls or alternative assays would be advisable to resolve concerns.

Strictly speaking, the term LMP covers a wide range of the extent of membrane-permeabilization scenarios, although it often refers only to a narrower definition of membrane permeabilization that results in the release of lysosomal enzymes into the cytosol. An important consequence of such release is the initiation of the apoptotic signaling cascade (Stoka et al. 2001; Cirman et al. 2004; Droga-Mazovec et al. 2008) or of the distinct process of necrosis (Jacobson et al. 2013; Lima et al. 2013).

In total, there are more than 50 different acid hydrolases and several activator proteins in the lumen of lysosomes (Schroder et al. 2010). Most of them are glycosylated, which substantially increases their molecular mass. Mature forms of cathepsins are 25–30 kDa (Turk et al. 2012), which corresponds to a hydrodynamic radius of ~2.5 nm (Erickson 2009). Many other hydrolases are even larger—acid ceramidase and acid phosphatase are ~50 kDa, and alpha-N-acetylglucosaminidase is ~80 kDa as a monomer, but it also exists as a homodimer. The release of such large molecules would require commensurate gateways in the lysosomal membrane. In contrast, the loss of a proton gradient over the lysosomal membrane can occur even with only minor destabilization. LMP may be only transient and is not necessarily irreversible (Steinberg et al. 2010); however, little is known about the mechanisms of lysosomal membrane repair following LMP. The nature of membrane destabilization during LMP is still poorly understood and could depend on organized pore-like structures or structurally less-discrete transient defects, which nonetheless increase membrane permeability (Repnik et al. 2014). This introduction describes and discusses several approaches to detect LMP and aims not only to present the existing methods but also to stimulate the search for improvements and the application of novel techniques.

THE CHOICE OF ASSAYS FOR LYSOSOMAL MEMBRANE PERMEABILIZATION

There is not a single, universally applicable, method to determine LMP, so we propose a workflow rather than a list of independent methods (Fig. 1).

Retention of Lysosomotropic Dyes

Lysosomotropic dyes, known also as acidotropic dyes, such as acridine orange or LysoTracker reagents (Invitrogen), accumulate in acidic compartments. This kind of staining is a good early screen, especially in combination with flow-cytometric analysis, because it can be quantified and allows rapid analysis of several thousand cells, which makes results reliable [see Protocol 1: The Use of Lysosomotropic Dyes to Exclude Lysosomal Membrane Permeabilization (Repnik et al. 2015a)]. However, with this method, one cannot discriminate whether the loss of the ability to retain acidic dyes is a consequence of LMP or, alternatively, the loss of the proton gradient due to the inhibition of the vacuolar ATPase (v-ATPase). Taken together, loss of staining with lysosomotropic dyes alone does not prove LMP; nevertheless, the ability of organelles to accumulate such dyes to the same extent as control cells argues for the integrity of the lysosomal membrane and the functional v-ATPase.

Measuring Cathepsin Activity

For the next step, we suggest the measurement of the cysteine cathepsin activity with a small fluorogenic substrate such as Z-Phe-Arg-AMC (~600 Da) [see Protocol 2: Measuring Cysteine Cathepsin Activity to Detect Lysosomal Membrane Permeabilization (Repnik et al. 2015b)]. Upon cleavage, the release of fluorescent 7-amino-4-methylcoumarin (AMC) can be quantified using a spectrofluorom-

Cite this introduction as *Cold Spring Harb Protoc*; doi:10.1101/pdb.top077479

Protocol 1

Loss of pH gradient over lysosomal membranes (staining with lysosomotropic dyes)

YES   NO ⇒ no LMP

Protocol 2

Activity of lysosomal enzymes after digitonin lysis of the plasma membrane

In the presence of lysed cells	NO	YES	YES
In the cell-free extract	NO	NO	YES
	⇓	⇓	⇓
	MINOR destabilization, increased permeability for ions	MEDIUM destabilization, increased permeability for small molecules	MAJOR destabilization, increased permeability for lysosomal luminal enzymes

(e.g., at least 600 Da in the case of Z-FR-AMC)

Protocol 3

Release of endogenous lysosomal proteins into the cytosol	Release of preloaded exogenous lysosomal tracers into the cytosol
Western blotting of cell-free cytosolic extracts	Sulforhodamine B (~600 Da)
Whole-mount immunofluorescence labeling	Dextran (2/10/40/70 kDa)
On-section immunofluorescence or immunogold labeling	BSA-gold (5 nm)

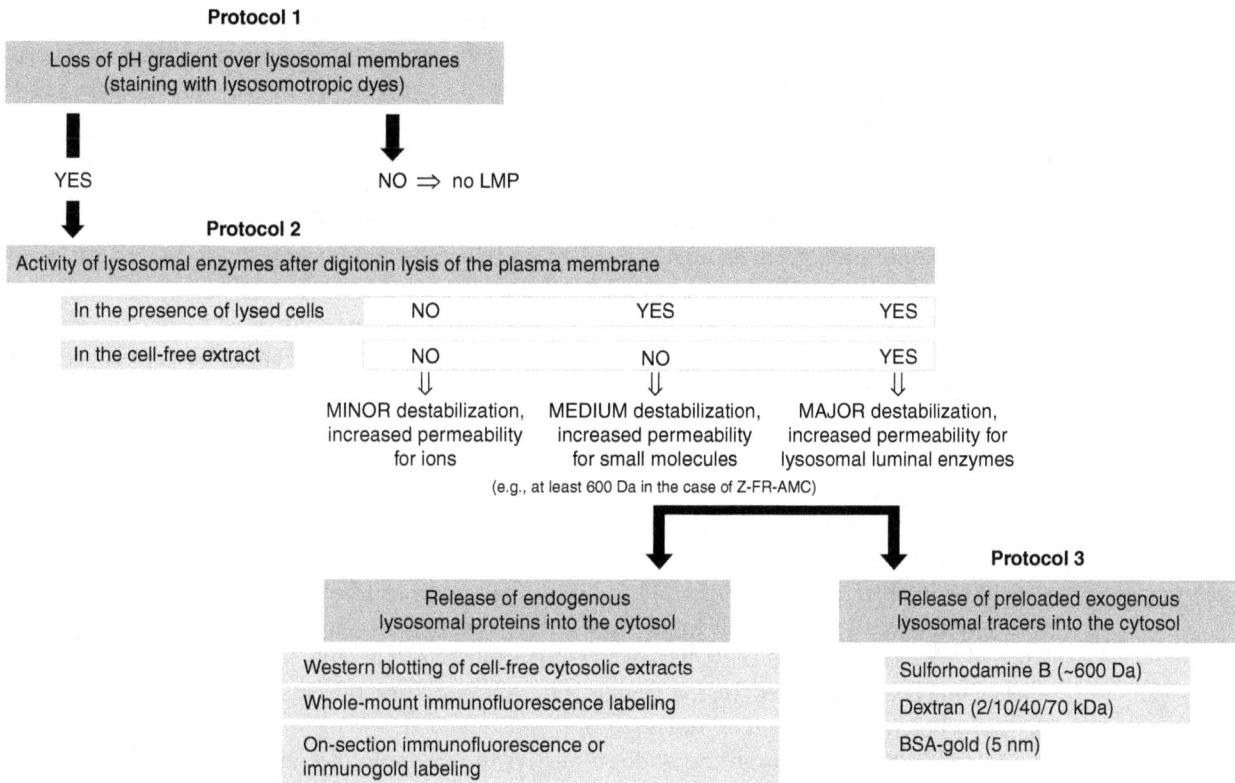

FIGURE 1. Workflow for studying LMP. In the first part of the workflow (Protocol 1), cells are analyzed for their ability to retain lysosomotropic dyes such as acridine orange or LysoTracker reagents are detected accumulating in acidic compartments—the associated staining is a good initial screen for the integrity of lysosomal membranes and for v-ATPase activity, especially in combination with flow-cytometric analysis (for further details, see Protocol 1: The Use of Lysosomotropic Dyes to Exclude Lysosomal Membrane Permeabilization [Repnik et al. 2015a]). The next featured step (Protocol 2) measures cysteine cathepsin activity by means of a small fluorogenic substrate such as Z-Phe-Arg-AMC (Z-FR-AMC; ~600 Da) (see Protocol 2: Measuring Cysteine Cathepsin Activity to Detect Lysosomal Membrane Permeabilization [Repnik et al. 2015b]). With just these two protocols, LMP can be reliably determined, in addition to the extent of membrane destabilization. Finally, western blot, immunofluorescence- and immunogold-based methods have their pros and cons, but, as an alternative to localizing endogenous lysosomal molecules in the cytosol, a recommended technique (Protocol 3) is the analysis of the release of exogenous tracers preloaded into lysosomes (see Protocol 3: Studying Lysosomal Membrane Permeabilization by Analyzing the Release of Preloaded BSA–Gold Particles into the Cytosol [Repnik et al. 2015c]).

eter. If LMP destabilizes the lysosomal membrane to such an extent that lysosomal enzymes are translocated into the cytosol, then their activity could be detected in the cell-free extract. Alternatively, if a small fluorogenic substrate is added directly to lysed cells, it might reach the lumen of lysosomes through destabilized lysosomal membranes and be cleaved there, indicating moderate LMP.

With these two protocols alone, one can reliably determine LMP as well as the extent of membrane destabilization. Following this approach, it has recently been shown that siramesine is not an LMP-inducing agent (Hafner Česen et al. 2013), in contrast to earlier reports (Ostenfeld et al. 2005, 2008). Additionally, both protocols are appropriate for high-throughput screening of different compounds or different experimental conditions. The sample preparation is straightforward, the reagents are robust and reliable, unlike antibodies (Griffiths and Lucocq 2014), and the fluorescence signal can be quantified.

Immunolabeling and Release of Tracers

In our experience, translocation of lysosomal lumenal proteins into the cytosol is not easy to prove with immunolabeling. The low sensitivity of the methods seems to be an important limitation. Only a fraction of molecules is labeled with antibodies, and the signal is only weakly amplified compared with the accumulation of the fluorescent reporter after hydrolysis of the fluorogenic substrate. There are two basic strategies to analyze subcellular localization of a particular protein by immunolabeling. One can prepare cytosolic cell-free extracts (Ivanova et al. 2008) and analyze the presence of selected molecules in it by means of western blotting. Alternatively, the localization of the molecules can be directly analyzed by whole-mount immunofluorescence labeling.

A crucial factor in any kind of immunolabeling is the actual specificity of an antibody, and it is this area that has not been given enough consideration (Griffiths and Lucocq 2014). In western blotting, some control over the specificity of immunolabeling is provided by the estimation of the molecular mass of the labeled proteins, whereas, in whole-mount immunofluorescence, demonstrating the specificity of immunolabeling is a particularly challenging task. In addition, some sample preparation steps, such as fixation and postfixation detergent-mediated permeabilization, can significantly affect the results, including the labeling pattern (Griffiths et al. 1993).

Moreover, fluorescence imaging has advantages as well as pitfalls (North 2006). A fluorescence signal is strong when it is concentrated, so immunolabeling for cathepsins in lysosomes gives a strong signal. However, after their release into the cytosol, they become diluted and the signal should therefore fade. A strong cytosolic signal indicating the translocation of cathepsins into the cytosol should be interpreted with caution. In contrast, immunogold labeling combined with electron microscopy enables imaging with higher resolution, the signal is not affected by bleaching and the "digital" character of gold particles enables quantification (Griffiths and Lucocq 2014). However, the labeling is still dependent on the quality of the antibody, and the labeling intensity is usually lower than in whole-mount immunofluorescence.

As an alternative to localizing endogenous lysosomal molecules in the cytosol, analysis of the release of exogenous tracers preloaded into lysosomes is discussed in the final protocol (see Protocol 3: Studying Lysosomal Membrane Permeabilization by Analyzing the Release of Preloaded BSA–Gold Particles into the Cytosol [Repnik et al. 2015c]) (Fig. 1).

CONCLUDING REMARKS

Because of the inherent dynamics of biological systems, the hallmarks of LMP are usually transient, and therefore consideration must be given to the timing of LMP. Compounds whose primary effect is to induce LMP are likely to act early after the treatment, and LMP can occur quite synchronously or at least in a narrow time-frame in the majority of cells. Several compounds have been reported to induce LMP 48 h after the treatment, in which case LMP is likely mediated not directly by the compound but indirectly, by earlier cellular responses. Therefore, it can occur asynchronously in different cells and over a considerable timespan. Whatever the method, the experiments should aim to establish when LMP occurs in relation to other cellular events and determine whether it is a primary effect of the compound or occurs secondarily in response to injuries sustained by other organelles, such as generation of ROS in mitochondria (Huai et al. 2013). Therefore, in parallel to evaluating LMP, assays should be planned to probe the effects of the treatment on the function and integrity of other organelles (e.g., mitochondria, endoplasmic reticulum, Golgi apparatus, and nucleus).

REFERENCES

Boya P. 2012. Lysosomal function and dysfunction: Mechanism and disease. *Antioxid Redox Sig* 17: 766–774.

Cirman T, Oresic K, Mazovec GD, Turk V, Reed JC, Myers RM, Salvesen GS, Turk B. 2004. Selective disruption of lysosomes in HeLa cells triggers apoptosis mediated by cleavage of Bid by multiple papain-like lysosomal cathepsins. *J Biol Chem* 279: 3578–3587.

de Duve C. 2005. The lysosome turns fifty. *Nat Cell Biol* 7: 847–849.

Droga-Mazovec G, Bojič L, Petelin A, Ivanova S, Romih R, Repnik U, Salvesen GS, Stoka V, Turk V, Turk B. 2008. Cysteine cathepsins trigger caspase-dependent cell death through cleavage of bid and antiapoptotic Bcl-2 homologues. *J Biol Chem* 283: 19140–19150.

Erdal H, Berndtsson M, Castro J, Brunk U, Shoshan MC, Linder S. 2005. Induction of lysosomal membrane permeabilization by compounds that activate p53-independent apoptosis. *Proc Natl Acad Sci* 102: 192–197.

Cite this introduction as *Cold Spring Harb Protoc*; doi:10.1101/pdb.top077479

Erickson HP. 2009. Size and shape of protein molecules at the nanometer level determined by sedimentation, gel filtration, and electron microscopy. *Biol Proced Online* **11:** 32–51.

Goldman R, Kaplan A. 1973. Rupture of rat liver lysosomes mediated by L-amino acid esters. *Biochim Biophysica Acta* **318:** 205–216.

Griffiths G, Lucocq JM. 2014. Antibodies for immunolabeling by light and electron microscopy: Not for the faint hearted. *Histochem Cell Biol* **142:** 347–360.

Griffiths G, Parton RG, Lucocq J, van Deurs B, Brown D, Slot JW, Geuze HJ. 1993. The immunofluorescent era of membrane traffic. *Trends Cell Biol* **3:** 214–219.

Hafner Česen M, Repnik U, Turk V, Turk B. 2013. Siramesine triggers cell death through destabilisation of mitochondria, but not lysosomes. *Cell Death Dis* **4:** e818.

Huai J, Vogtle FN, Jockel L, Li Y, Kiefer T, Ricci JE, Borner C. 2013. TNFα-induced lysosomal membrane permeability is downstream of MOMP and triggered by caspase-mediated NDUFS1 cleavage and ROS formation. *J Cell Sci* **126:** 4015–4025.

Huotari J, Helenius A. 2011. Endosome maturation. *EMBO J* **30:** 3481–3500.

Ivanova S, Repnik U, Bojič L, Petelin A, Turk V, Turk B. 2008. Lysosomes in apoptosis. *Methods Enzymol* **442:** 183–199.

Jacobson LS, Lima H Jr, Goldberg MF, Gocheva V, Tsiperson V, Sutterwala FS, Joyce JA, Gapp BV, Blomen VA, Chandran K, et al. 2013. Cathepsin-mediated necrosis controls the adaptive immune response by Th2 (T helper type 2)-associated adjuvants. *J Biol Chem* **288:** 7481–7491.

Kirkegaard T, Jäättelä M. 2009. Lysosomal involvement in cell death and cancer. *Biochim Biophys Acta* **1793:** 746–754.

Lima H Jr, Jacobson LS, Goldberg MF, Chandran K, Diaz-Griffero F, Lisanti MP, Brojatsch J. 2013. Role of lysosome rupture in controlling Nlrp3 signaling and necrotic cell death. *Cell Cycle* **12:** 1868–1878.

Lloyd JB. 2000. Lysosome membrane permeability: Implications for drug delivery. *Adv Drug Deliv Rev* **41:** 189–200.

North AJ. 2006. Seeing is believing? A beginners' guide to practical pitfalls in image acquisition. *J Cell Biol* **172:** 9–18.

Ostenfeld MS, Fehrenbacher N, Hoyer-Hansen M, Thomsen C, Farkas T, Jäättelä M. 2005. Effective tumor cell death by σ-2 receptor ligand siramesine involves lysosomal leakage and oxidative stress. *Cancer Res* **65:** 8975–8983.

Ostenfeld MS, Hoyer-Hansen M, Bastholm L, Fehrenbacher N, Olsen OD, Groth-Pedersen L, Puustinen P, Kirkegaard-Sorensen T, Nylandsted J, Farkas T, et al. 2008. Anti-cancer agent siramesine is a lysosomotropic detergent that induces cytoprotective autophagosome accumulation. *Autophagy* **4:** 487–499.

Pillay CS, Elliott E, Dennison C. 2002. Endolysosomal proteolysis and its regulation. *Biochem J* **363:** 417–429.

Reeves JP. 1979. Accumulation of amino acids by lysosomes incubated with amino acid methyl esters. *J Biol Chem* **254:** 8914–8921.

Repnik U, Stoka V, Turk V, Turk B. 2012. Lysosomes and lysosomal cathepsins in cell death. *Biochim Biophys Acta* **1824:** 22–33.

Repnik U, Hafner Česen M, Turk B. 2013. The endolysosomal system in cell death and survival. *Cold Spring Harb Perspect Biol* **5:** a008755.

Repnik U, Hafner Česen M, Turk B. 2014. Lysosomal membrane permeabilization in cell death: Concepts and challenges. *Mitochondrion* **19:** 49–57.

Repnik U, Hafner Česen M, Turk B. 2015a. The use of lysosomotropic dyes to exclude lysosomal membrane permeabilization. *Cold Spring Harb Protoc* doi: 10.1101/pdb.prot087106.

Repnik U, Hafner Česen M, Turk B. 2015b. Measuring cysteine cathepsin activity to detect lysosomal membrane permeabilization. *Cold Spring Harb Protoc* doi: 10.1101/pdb. prot087114.

Repnik U, Hafner Česen M, Turk B. 2015c. Studying lysosomal membrane permeabilization by analyzing the release of preloaded BSA–gold particles into the cytosol. *Cold Spring Harb Protoc* doi: 10.1101/pdb. prot087122.

Schroder BA, Wrocklage C, Hasilik A, Saftig P. 2010. The proteome of lysosomes. *Proteomics* **10:** 4053–4076.

Špes A, Sobotič B, Turk V, Turk B. 2012. Cysteine cathepsins are not critical for TRAIL- and CD95-induced apoptosis in several human cancer cell lines. *Biol Chem* **393:** 1417–1431.

Steinberg BE, Huynh KK, Brodovitch A, Jabs S, Stauber T, Jentsch TJ, Grinstein S. 2010. A cation counterflux supports lysosomal acidification. *J Cell Biol* **189:** 1171–1186.

Stern ST, Adiseshaiah PP, Crist RM. 2012. Autophagy and lysosomal dysfunction as emerging mechanisms of nanomaterial toxicity. *Part Fibre Toxicol* **9:** 20.

Stoka V, Turk B, Schendel SL, Kim TH, Cirman T, Snipas SJ, Ellerby LM, Bredesen D, Freeze H, Abrahamson M, et al. 2001. Lysosomal protease pathways to apoptosis. Cleavage of bid, not pro-caspases, is the most likely route. *J Biol Chem* **276:** 3149–3157.

Terman A, Brunk UT. 2004. Aging as a catabolic malfunction. *Int J Biochem Cell Biol* **36:** 2365–2375.

Turk V, Stoka V, Vasiljeva O, Renko M, Sun T, Turk B, Turk D. 2012. Cysteine cathepsins: From structure, function and regulation to new frontiers. *Biochim Biophys Acta* **1824:** 68–88.

The Use of Lysosomotropic Dyes to Exclude Lysosomal Membrane Permeabilization

Urška Repnik,[1,2,4] Maruša Hafner Česen,[1] and Boris Turk[1,3,4]

[1]Department of Biochemistry and Molecular and Structural Biology, J. Stefan Institute, SI-1000 Ljubljana, Slovenia; [2]Department of Biosciences, University of Oslo, NO-0371 Oslo, Norway; [3]Center of Excellence CIPKEBIP, SI-1000 Ljubljana, Slovenia

Progressive lowering of pH is characteristic for the endocytic pathway and enables efficient degradation of molecules by hydrolytic enzymes at its distal end. The existence of the proton gradient over the endosomal/lysosomal membranes depends on the action of the vacuolar ATPase (v-ATPase). During lysosomal membrane permeabilization (LMP), protons leak through the destabilized membrane, resulting in loss of the pH gradient. Here, we present a protocol showing how this effect can be detected by staining cells with lysosomotropic dyes, which accumulate in acidic organelles after protonation. During LMP, cells lose the ability to retain these dyes and therefore appear pale. Among the most commonly used lysosomotropic dyes are LysoTracker reagents and acridine orange. Cells can be analyzed with a fluorescence microscope; however, flow-cytometric analysis enables fast, objective, and reliable evaluation of differences between samples. Advantages of the technique include the fact that sample preparation is relatively simple and can be scaled-up to test several different compounds or conditions. However, as we will discuss, cells treated with v-ATPase inhibitors also lose the pH gradient across lysosomal membranes and cannot be stained with lysosomotropic dyes, although this is not accompanied by LMP. Therefore, merely observing loss of staining is not in itself a proof of LMP.

MATERIALS

It is essential that you consult the appropriate Material Safety Data Sheets and your institution's Environmental Health and Safety Office for proper handling of equipment and hazardous material used in this protocol.

RECIPES: Please see the end of this protocol for recipes indicated by <R>. Additional recipes can be found online at http://cshprotocols.cshlp.org/site/recipes.

Reagents

Acridine orange (1 mg/mL stock in distilled water)
Bafilomycin A1 (optional; see Step 2)
Cell-detachment reagent (e.g., TrypLE Select, Invitrogen/Life Technologies)
Cell source

   Stocks (e.g., HeLa cells) to be used both as an experimental sample and as a reference sample (untreated cells).

Imaging medium <R>
Leu-Leu-O-methylester (LLOMe) (optional; see Step 2)
LysoTracker reagents (1 mM stocks in DMSO)
   LysoTracker Green DND-26 (Molecular Probes/Life Technologies, L7526)

[4]Correspondence: urska.repnik@ibv.uio.no; boris.turk@ijs.si

Cite this protocol as *Cold Spring Harb Protoc*; doi:10.1101/pdb.prot087106

LysoTracker Red DND-99 (Molecular Probes/Life Technologies, L7528)
> *See Discussion.*

Phosphate-buffered saline (PBS[D] <R>)

Equipment

Centrifuge
CO_2 incubator
Flow cytometer (with a blue 488 nm laser, such as FACSCalibur, BD Biosciences)
Fluorescence-activated cell sorter (FACS) tubes (3.5 mL, for flow-cytometric analysis)
> *Note that the volumes given in the text below are relevant for the FACSCalibur flow cytometer (BD Biosciences).*

Fluorescence microscope (wide-field or confocal)
Glass-bottom cell-culture dishes (for light-microscopic analysis)
Pipette
Plates (24-well)

METHOD

Preparation of Cells

> *The following protocol is for adherent cells; note that, as the cells can detach upon treatment, the supernatant containing detached cells should also be collected during sample preparation. However, the protocol works equally well for cells in suspension—this is simpler as no cell detachment is needed.*

1. Seed cells in a 24-well plate and allow to attach overnight at their normal growth temperature in a CO_2 incubator.

   > *The number of cells seeded and the volume of the cell culture supernatant should correspond to the growth area of the dish relative to other types of dishes used. With practice, two 24-plates can be prepared in parallel.*

2. Perform the treatment necessary for the experimental aims. Also run a positive control for reduced staining: For example, apply bafilomycin A1 (30 nM) for a minimum of 30 min to inhibit the vacuolar ATPase (v-ATPase) and/or the lysosomal membrane permeabilization (LMP)-inducing compound LLOMe (1–5 mM) for 15 min.

3. Treat the cells in the 24-well plate with a lysosomotropic dye by adding it to the cell culture medium and incubating for the appropriate period of time at 37°C (e.g., for mammalian cells, use 40 nM LysoTracker Green for 5–10 min or 1 µg/mL acridine orange for 15 min).

4. After incubation with the dye, analyze cells with a fluorescence microscope or with a flow cytometer (see the next steps). For microscopic analysis, wash cells 3× with warm PBS and add imaging medium. For flow cytometry analysis, begin to prepare a cell suspension. To this end, transfer the cell culture supernatant into a 3.5-mL FACS tube, wash the well with 0.5 mL of warm (at room temperature) PBS, and add the wash to the supernatant in the FACS tube.

5. Add a small volume (sufficient just to cover the cells) of a cell-detachment reagent (e.g., TrypLE Select) to the wells with attached cells and incubate at 37°C untill the cells detach.

6. Transfer the detached cells to the FACS tube containing the culture supernatant from Step 4.

7. Centrifuge the cells (e.g., at 300g for 5 min at room temperature). With a pipette, carefully remove and discard the supernatant.

8. Resuspend the pellet in 300–500 µL of warm (room temperature) PBS. Analyze immediately with a flow cytometer.

FACS Analysis

9. Select settings for forward and side scatter (FSC, SSC) so that the cell population in the FSC/SSC plot is central and can be delineated from the debris and dead cells. Set a gate around the population of viable cells.

10. Open a histogram plot to measure fluorescence. For LysoTracker Green, select the green channel (530/30 bandpass filter), and, for acridine orange, select the far-red channel (>670 nm).

11. Set the voltage on the detectors so that the main peak of stained untreated control cells lies between values of 10^2 and 10^3, which enables the detection of increased and decreased fluorescence in treated samples.

    If cells are stained only with the lysosomotropic dye, no compensation to correct for the overlap between emission spectra of different fluorochromes is required.

12. Collect between 5000 and 10,000 events (cells), and analyze the data (see Fig. 1).

    Set a gate around live cells in a FSC/SSC plot (Fig. 1A) to exclude debris and dead cells, which have a reduced FSC signal. Fluorescence data can be presented as frequency histograms of fluorescence intensity (Fig. 1, B and D). For semiquantitative analysis the geometric mean rather than simple mean values can be read, especially when the histograms are asymmetric. The geometric mean data can be statistically analyzed and presented in a bar plot (Fig. 1, C and E). To allow a direct comparison of geometric mean values, samples must be stained in parallel and analyzed with the same settings. Results can also be presented relative to untreated control cells (as a fraction of the fluorescence intensity of control cells). When cells in the population behave in more than one way, the percentage of pale cells can be determined and the geometric mean value of this subpopulation can be read.

DISCUSSION

Choice of Dye

Various LysoTracker reagents with different excitation and emission spectra are available. LysoTracker Green is suitable for flow-cytometric analysis because it absorbs at 488 nm, and most of the flow cytometry machines are equipped with an appropriate blue laser. For fluorescence microscopy, in our experience, LysoTracker Red is more convenient because it is more stable, whereas LysoTracker Green bleaches quickly and LysoTracker Blue is rather faint. Alternatively, acridine orange dye can be used, which is particularly suitable for the analysis with a flow cytometer because the blue laser and the far-red detector are compatible with the analysis of its metachromatic emission. This is characteristic of protonated acridine orange molecules in acidic organelles, which form aggregates whose emission maximum changes to far-red, compared to orthochromatic green emission maximum of the monomeric dye (Darzynkiewicz and Kapuscinski 1990). The terms orthochromatic and metachromatic imply that spectral properties of the dye on binding to the target do not change and do change, respectively.

In our experience, prolonged incubation with LysoTracker dyes can result in nonselective staining of cellular structures other than lysosomes. We incubate for 15 min with LysoTracker Red and acridine orange, and only 5–10 min with LysoTracker Green. Even if stained cells are analyzed with flow cytometry, stained samples should be inspected in the fluorescence microscope during the protocol optimization.

Lysosomes of dead cells are not stained, thus, especially in samples taken late after the experimental treatment, cell viability should also be analyzed to discriminate live cells, with pale lysosomes, from dead cells. Dead cells can be distinguished from viable cells on the basis of altered light-scatter properties; they generally show a decrease in forward-scatter and an increase in side-scatter signals (Loken and Herzenberg 1975). If this is the case, they should be excluded from the analysis gate in a FSC/SSC plot. An increased fraction of dead cells in the sample can indicate damage to organelles other than lysosomes, and this should be investigated to understand the role of LMP in cell death signaling.

LMP can be transient, in which case the lysosomal membrane regains its integrity and rapidly restores the acidic pH (Steinberg et al. 2010). Transient LMP can therefore be showed during a limited period of time. Finally, note that if experimentally tested compounds are fluorescent and their

Cite this protocol as *Cold Spring Harb Protoc*; doi:10.1101/pdb.prot087106

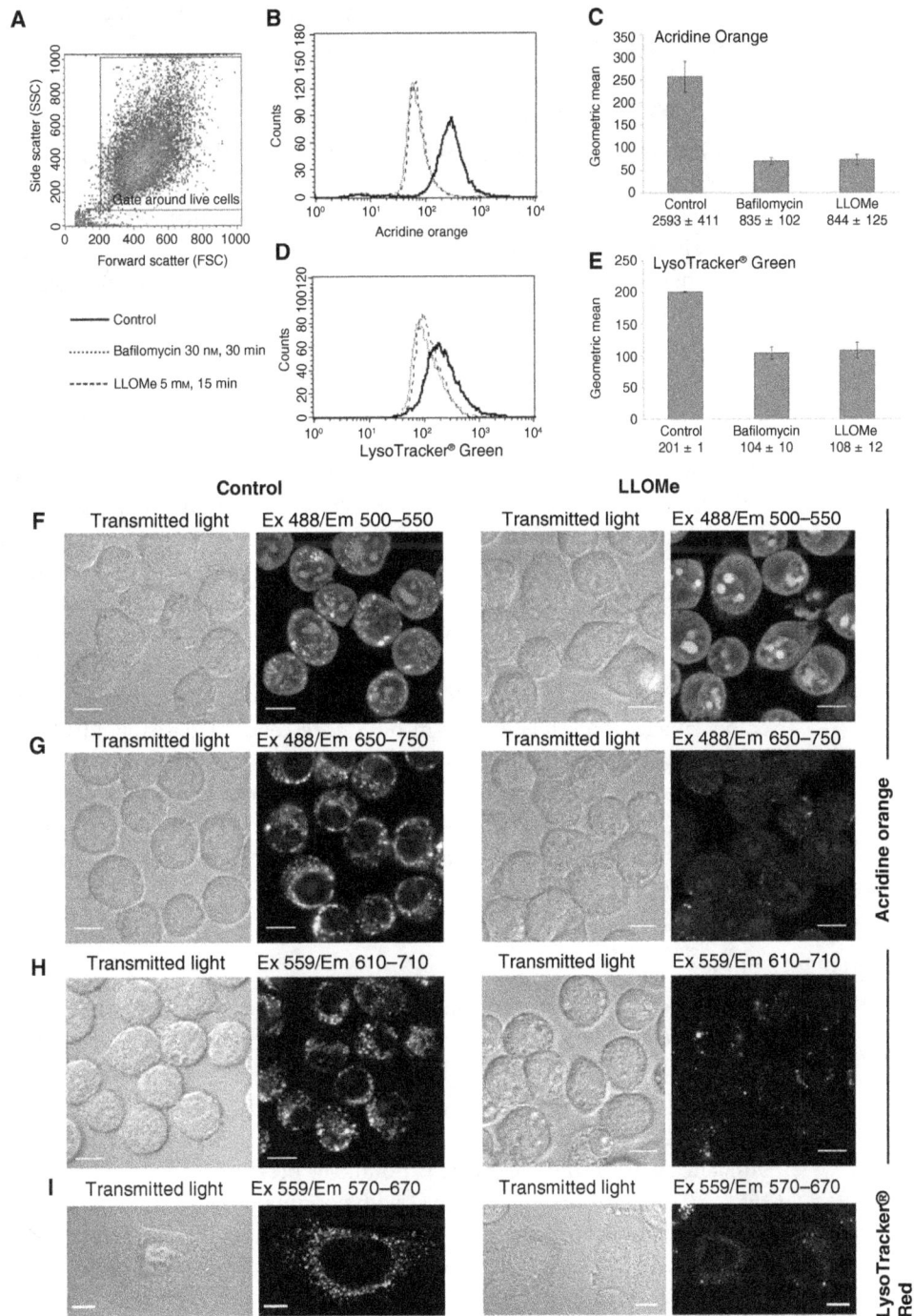

FIGURE 1. Lysosomal membrane permeabilization (LMP) causes the loss of the pH gradient over lysosomal membranes, and as a result cells show reduced staining with lysosomotropic dyes compared with untreated control cells. (*A–E*) HeLa cells, untreated or cells treated with 30 nM bafilomycin A1 (v-ATPase inhibitor) for 30 min, or with 5 mM Leu-Leu-*O*-methylester (LLOMe; a known LMP inducer) for 15 min, were stained with acridine orange or LysoTracker Green and analyzed with a flow cytometer. Before the fluorescence analysis, the gate was set around live cells in a density plot of forward and side scatter (FSC/SSC) to exclude debris. Results of the acridine orange or Lysotracker Green staining are presented as frequency histograms of fluorescence intensity (*B,C*) and as mean ±s.D. values for the geometric mean of fluorescence intensity (*D,E*). The experiment was performed in duplicate. (*F–I*) RAW264.7 cells, stained with acridine orange (*F–H*), and HeLa cells, stained with Lysotracker Red (*I*), were analyzed with a confocal laser scanning microscope. Cells stained with acridine orange were analyzed with three different combinations of excitation and emission light. Paired untreated control and LLOMe-treated samples were analyzed at the same imaging settings. Scale bar, 10 μm.

emission spectra overlap with those of lysosomotropic dyes, the cells treated with them cannot be analyzed with this protocol.

Experimental Tactics and Caveats

Cells with decreased fluorescence in comparison with untreated controls, also termed "pale cells," represent cells with lost or reduced ability to retain acidic dyes. This can be the consequence of either LMP or the inhibition of the v-ATPase—and use of this method alone does not allow the two causes to be discriminated. In contrast, an increased fluorescent signal indicates an increase in the volume of the endosomal compartments that occurs owing to increased number or size of the acidic organelles. For example, chloroquine and NH_4Cl increase the volume of the endosomes owing to the so-called proton-sponge effect.

In our experience, acridine orange gives a stronger signal and a stronger lysosomal signal:background ratio compared with LysoTracker Green (Fig. 1, B and D), which makes it a more sensitive dye. Acridine orange is a metachromatic dye and, as such, emits two colors. When excited with blue light, monomeric acridine orange emits orthochromatic green. When acridine orange accumulates in lysosomes at a high concentration, it precipitates into oligomeric structures that show a shift in the emission to metachromatic red at 640 nm (Darzynkiewicz and Kapuscinski 1990).

To detect LMP, it is possible to stain cells with acridine orange before the treatment and monitor the loss of red fluorescence and the concomitant increase in green fluorescence when the dye is released into the cytosol (Huai et al. 2013). However, there are several limitations to this approach. First, in the case of incubations lasting several hours, acridine orange that is loaded in lysosomes will interfere with their function. Second, the increase in green fluorescence is considerably smaller than the decrease in red fluorescence (Huai et al. 2013), which lowers the sensitivity of the assay. Third, this method is not compatible with lysosomotropic compounds, whose accumulation in the lysosomal lumen depends on protonation and thereby low pH. Acridine orange, like LysoTracker, is a lysosomotropic compound and its accumulation in lysosomes raises pH, which is likely to interfere with the accumulation of a lysosomotropic LMP-inducing compound, such as LLOMe.

An Experimental Example

To show the versatility of acridine orange for acidic organelle staining, we have labeled untreated control RAW264.7 cells or cells that were treated with the LMP inducer LLOMe (5 mM) with acridine orange for 15 min and analyzed them with a confocal laser scanning microscope. When excited with a 488 nm laser, acidic organelles in control cells emit green light. Nuclei also emit green, and there is considerable green background in the cytosol. In cells with destabilized lysosomal membranes, the punctate signal in the cytoplasm is no longer present, whereas nuclei and the cytoplasm are stained (Fig. 1F). If lysosomes were stained with acridine orange before LMP was induced, the dye released into the cytosol should increase the green cytosolic signal. However, this increase is relatively small and difficult to show given the strong background signal. The metachromatic emission is shown in Fig. 1G. When excited with blue light, the far-red signal represents acidic organelles. This detection corresponds to the detection analyzed with a flow cytometer (Fig. 1, B and C). Acidic organelles stained with acridine orange can also be observed with green excitation—for example, with a 559 nm laser. The emitted light is weaker compared with the excitation at 488 nm, but the signal is also predominantly localized to the acidic organelles (Fig. 1H). After LMP, the punctate staining is lost. Similarly, untreated control HeLa cells stained with Lysotracker Red show a punctate pattern representing acidic organelles, whereas cells that have experienced LMP appear pale (Fig. 1I).

Concluding Remarks

In summary, although staining with lysosomotropic dyes is a rather simple method, it is nevertheless extremely powerful. Sample preparation is straightforward and flow-cytometric analysis is fast, objective, statistically strong, and allows high throughput. This all makes this protocol a good starting point to screen different compounds or different experimental conditions and perhaps the simplest approach to determine the exact time-point when LMP occurs.

RECIPES

Imaging Medium

Phenol-red-free cell growth medium (e.g., RPMI or DMEM)
Fetal bovine serum (10%)
Glutamine (2 mM)
HEPES (25 mM, pH 7.4)

PBS(D)

Reagent	Amount to add for 1 L	Final concentration
NaCl	8.00 g	137.0 mM
KCl	0.20 g	2.7 mM
KH$_2$PO$_4$	0.20 g	1.5 mM
Na$_2$HPO$_4$	1.14 g	8.0 mM

To prepare a 1× solution, combine all components in ~800 mL of H$_2$O and stir to dissolve. Adjust the pH to 7.4 and bring to a final volume of 1 L. Sterilize by autoclaving. (A 10× stock solution of PBS(D) can be prepared and diluted to 1× as needed.)

ACKNOWLEDGMENTS

We thank the NorMIC Imaging Platform at the Department of Biosciences, University of Oslo, and especially Cathrine Heyward for discussion and assistance with the imaging.

REFERENCES

Darzynkiewicz Z, Kapuscinski J. 1990. Acridine orange: A versatile probe of nucleic acids and other cell constituents. In *Flow cytometry and sorting* (ed. Melamed MR, Mendelsohn M, Lindmo T), pp. 291–314. Wiley-Liss, New York.

Huai J, Vogtle FN, Jockel L, Li Y, Kiefer T, Ricci JE, Borner C. 2013. TNFα-induced lysosomal membrane permeability is downstream of MOMP and triggered by caspase-mediated NDUFS1 cleavage and ROS formation. *J Cell Sci* 126: 4015–4025.

Loken MR, Herzenberg LA. 1975. Analysis of cell populations with a fluorescence-activated cell sorter. *Ann New York Acad Sci* 254: 163–171.

Steinberg BE, Huynh KK, Brodovitch A, Jabs S, Stauber T, Jentsch TJ, Grinstein S. 2010. A cation counterflux supports lysosomal acidification. *J Cell Biol* 189: 1171–1186.

Protocol 2

Measuring Cysteine Cathepsin Activity to Detect Lysosomal Membrane Permeabilization

Urška Repnik,[1,2,4] Maruša Hafner Česen,[1] and Boris Turk[1,3,4]

[1]Department of Biochemistry and Molecular and Structural Biology, J. Stefan Institute, SI-1000 Ljubljana, Slovenia; [2]Department of Biosciences, University of Oslo, NO-0371 Oslo, Norway; [3]Center of Excellence CIPKEBIP, SI-1000 Ljubljana, Slovenia

During lysosomal membrane permeabilization (LMP), lysosomal lumenal contents can be released into the cytosol. Small molecules are more likely to be released, and cysteine cathepsins, with mature forms possessing a mass of 25–30 kDa, are among the smallest lumenal lysosomal enzymes. In addition, specific substrates for cysteine cathepsins are available to investigators, and therefore the measurement of the cathepsin activity as a hallmark of LMP works well. Here, we present a protocol for measuring the activity of these enzymes after selective plasma membrane permeabilization with a low concentration of digitonin and after total cell membrane lysis with a high concentration of digitonin. A fluorogenic substrate can be added either directly to the well with lysed cells to show LMP or to the cell-free extract to show that the lysosomal membrane has been sufficiently destabilized to allow the translocation of lysosomal enzymes. Although the content of lysosomal cysteine cathepsins differs between cell lines, this method has general applicability, is sensitive, and has high throughput. The presented protocol shows how to measure cysteine cathepsin activity in the presence of lysed cells and also in cell-free extracts. Depending on the aim of the study, one or both types of measurements can be performed.

MATERIALS

It is essential that you consult the appropriate Material Safety Data Sheets and your institution's Environmental Health and Safety Office for proper handling of equipment and hazardous material used in this protocol.

RECIPES: Please see the end of this protocol for recipes indicated by <R>. Additional recipes can be found online at http://cshprotocols.cshlp.org/site/recipes.

Reagents

β-Nicotinamide adenine dinucleotide, reduced (NADH) (5 mM, freshly prepared in 100 mM phosphate buffer, pH 7.4)

Cell source

   Stocks (e.g., U87-MG cells) to be used both as an experimental sample and as a reference sample (untreated cells).

Digitonin

   Prepare a stock solution of 10 mg/mL digitonin in ethanol just before the experiment and heat it carefully to dissolve.

[4]Correspondence: urska.repnik@ibv.uio.no; boris.turk@ijs.si

Cite this protocol as Cold Spring Harb Protoc; doi:10.1101/pdb.prot087114

Dithiothreitol (DTT)

Leu-Leu-*O*-methylester (LLOMe) (for positive controls)

Phosphate buffer (100 mM, pH 6.0) <R>

Phosphate buffer (100 mM, pH 7.4) <R>

Phosphate-buffered saline (PBS[D] <R>)

Sodium pyruvate (10 mM solution, freshly prepared in 100 mM phosphate buffer, pH 7.4)

Z-FR-AMC (e.g., Z-Phe-Arg-AMC · HCl, Bachem, I-1160; 50 mM stock solution in DMSO)

Equipment

96-well black microtiter plate with transparent base (e.g., Nunc165305) for fluorescence measurements (e.g., cysteine cathepsin activity)

96-well transparent microtiter plate for absorbance measurements (e.g., LDH activity)

CO_2 incubator

Ice

Multichannel automatic pipettes

Orbital shaker

Plate reader (for absorbance and fluorescence measurements)

Suction pump

METHOD

Preparation of Cells

1. Seed cells onto a 96-well microtiter plate and allow them to attach overnight at their normal growth temperature in a CO_2 incubator. Plan three wells (triplicate) for each condition.

   For the purpose of experimental design, investigators should use untreated cells as a reference sample, and, as a positive control for lysosomal membrane permeabilization (LMP), apply an LMP-inducing compound, LLOMe (1–5 mM) e.g., for 15 min.

2. After the desired experimental treatment, remove culture medium by turning the plate upside down above a sink. Rinse once with PBS by adding PBS to the wells, turning the plate upside down above the sink again to remove the bulk of the PBS, and then removing the remaining liquid with a suction pump.

   Continue to Step 3 to measure cysteine cathepsin activity to determine LMP, or proceed to Step 7 to measure lactate dehydrogenase (LDH) activity to determine the low digitonin concentration that selectively permeabilizes the plasma membrane.

Measurement of Cysteine Cathepsin Activity

3. Add 75 µL of digitonin at the appropriate concentration (15–50 µg/mL digitonin for the selective permeabilization of the plasma membrane or 200 µg/mL digitonin for the total permeabilization of the cell membrane) in 100 mM phosphate buffer (pH 6.0) to each appropriate well of the plate. Permeabilize the membranes by incubating the plate for 10 min on ice with shaking on an orbital shaker at 70 rpm.

4. Add 75 µL of 5 mM DTT and 30 µM cathepsin-specific fluorogenic substrate Z-FR-AMC in 100 mM phosphate buffer (pH 6.0) to appropriate wells of the plate, kept on ice, as follows:

   • To measure enzyme activity in the presence of lysed cells (Fig. 1Bi), add buffer with the substrate directly to the well.

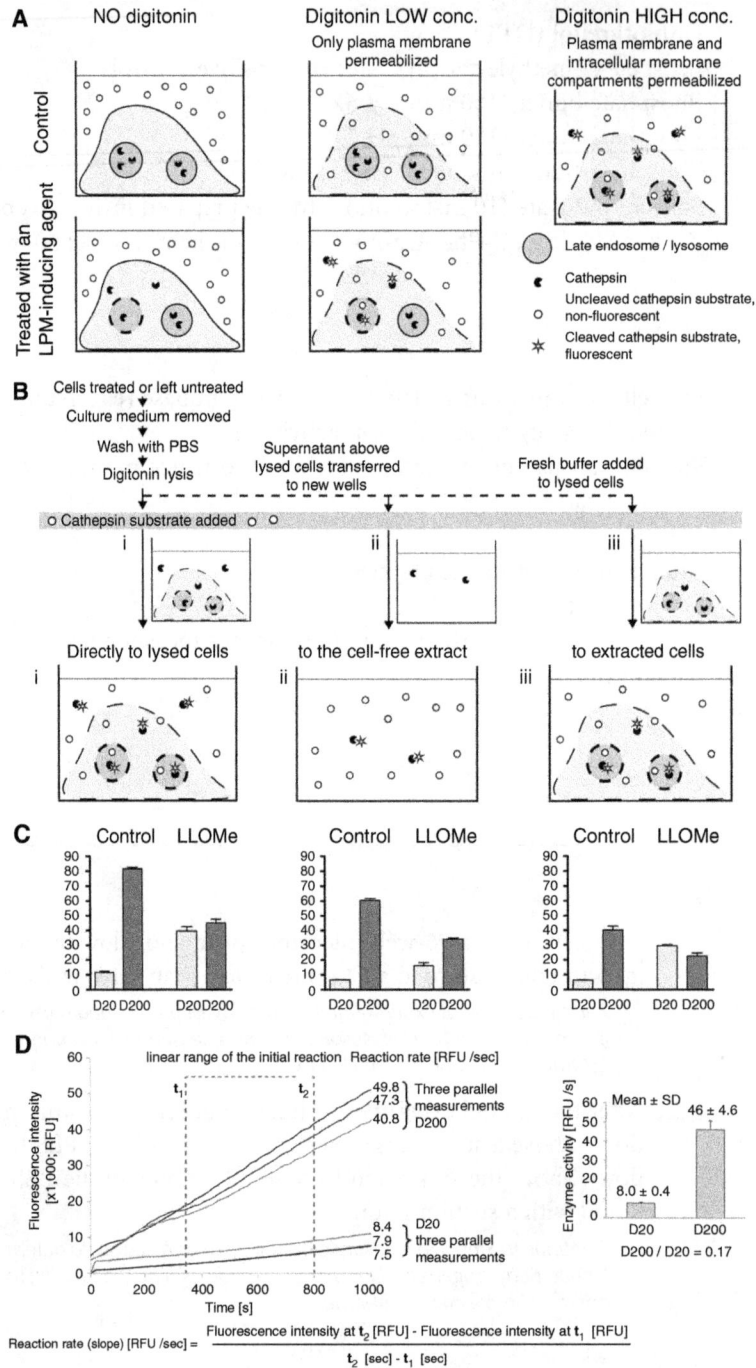

FIGURE 1. Cysteine cathepsin activity in cells with lysosomal membrane permeabilization (LMP) can be detected after selective plasma membrane lysis with a low concentration of digitonin. (*A*) Schematic representation of untreated control cells and cells treated with an LMP-inducing agent exposed to no, a low or high concentration of digitonin. (*B*) Schematic representation of the procedure to measure cysteine cathepsin activity (i) directly in lysed cells, (ii) in the cell-free extract, or (iii) in the extracted cells. (*C*) Human glioblastoma U87-MG cells growing on a 96-well plate were left untreated (control) or were treated with the LMP-inducing compound 5 mM Leu-Leu-*O*-methylester (LLOMe) for 15 min. Then they were lysed either with 20 (D20) or 200 (D200) µg/mL digitonin. Cysteine cathepsin activity was measured with the cathepsin-specific fluorogenic substrate Z-FR-AMC, as explained in *B*. The experiment was performed in triplicate, and initial rates of reactions were calculated as explained in *D*. Bars represent the mean + s.D. enzyme activity. (*D*) Schematic representation of data analysis for cysteine cathepsin activity in cells lysed with a low (D20) or high (D200) concentration of digitonin. The reaction rate was calculated, from the linear range of the primary data, as the slope of the line (see the equation). Results of parallel measurements are presented as the mean ± s.D., or as a fraction of total activity (D20/D200), calculated from the mean values. RFU, relative fluorescence units; t, time (sec).

- To measure enzyme activity in the cell-free extract (Fig. 1Bii), after the incubation with digitonin, transfer the lysis buffer above the cells (cell-free extract) to a new well on the same plate (best into the well directly below or above) and then add the substrate.

- To measure residual activity of the enzymes in the extracted cells (Fig. 1Biii), add 75 µL of fresh phosphate buffer (pH 6.0) to the emptied wells and then add the substrate.

5. Put the plate in a plate reader. For Z-FR-AMC, select 370 nm for excitation and 460 nm for the emission wavelength. Select shaking function before the measurement starts (dual orbital for 10 s). Measure for 20–30 min at 37°C with the shortest possible interval between the cycles.

    Measurement settings (e.g., gain) differ among cell lines and should be determined in a preliminary experiment.

6. Analyze the data as follows:

    - Draw plots of fluorescence intensity over time and determine the time interval of the linear range of the reaction to calculate the initial rate of the reaction (slope function) (Fig. 1D).

    - Present the results for the enzyme activity as the mean ± s.d. of the initial reaction rate (relative fluorescence units per second, RFU/sec) in parallel for the cytosolic and total enzyme activity or as a ratio between the mean values of cytosolic and total initial reaction rates (the fraction of the total activity) (Fig. 1D).

        When total activity differs among samples, we suggest presenting values for the reaction rate rather than values for fractions of the total activity.

Measurement of LDH Activity

7. Treat the cells on the plate wells with a range of digitonin concentrations. For example, use 75 µL/well of 0, 10, 20, 30, 40, 50, 200 µg/mL digitonin in 100 mM phosphate buffer (pH 7.4).

8. Permeabilize the membranes by incubating the plate for 10 min on ice with shaking on an orbital shaker at 70 rpm.

9. Add 75 µL of 100 mM phosphate buffer (pH 7.4) containing 2 mM sodium pyruvate and 1 mM NADH to each well.

10. Measure the absorbance in a plate reader at 340 nm for 20 min at 37°C. Calculate the rate (slope) of the reaction (i.e., the decrease in the absorbance), which indicates oxidation of NADH.

    For the calculation of the reaction rate, see Step 6.

DISCUSSION

Optimizing the Protocol and the Use of Controls

Enzyme activity in this protocol is measured after selective plasma membrane permeabilization with a low concentration of digitonin and after total cell membrane lysis with a high concentration of digitonin (Fig. 1A). The low concentration of digitonin that selectively lyses the plasma membrane and leaves lysosomal membranes intact should be determined experimentally for each cell type. To achieve this, parallel measurements of the activity of LDH (a cytosolic enzyme) and cathepsins (lysosomal enzymes) need to be performed in untreated control cells exposed to increasing digitonin concentrations. The catalytic action of LDH can be measured spectrophotometrically as reduced absorbance at 340 nm arising from the oxidation of NADH to NAD^+ (Chacon et al. 1997; Caruso et al. 2006; Ivanova et al. 2008). The appropriate concentration is the one at which LDH reaches the plateau, whereas cathepsin activity is still low.

In cells treated with an LMP-inducing agent, cathepsin and LDH activities should also be analyzed in samples not exposed to digitonin. Increased cathepsin or LDH activities in these samples would indicate that the plasma membrane has lost its integrity and that the LMP-inducing agent does not act selectively on lysosomal membranes.

Parallel measurements can be performed using the entire microtiter plate, but the preparation (pipetting) must be fast to avoid significant differences in the progress of reaction between the first and the last wells on the plate, so the use of a multichannel pipette is strongly recommended. Untreated control cell samples can be analyzed at both ends of the plate as an internal control.

To compare the values of cathepsin activity between different samples, the number of cells should be as similar as possible. Note that the compounds applied might affect the cell cycle and thereby cell proliferation, which is relevant in particular for treatments of long duration. The number of cells can be determined with LDH measurements, as suggested previously (Ivanova et al. 2008), or with alternative assays. A separate plate must be prepared for these assays. If, or when, cells start detaching, this method is no longer optimal and alternative strategies to prepare the cytosolic extracts should be considered.

We use acidic pH for the lysis because some cathepsins are sensitive to neutral pH. In our experience, using an acidic buffer increases the sensitivity of the assay but does not affect the overall result. The main advantage of this method is that the processing time between cell lysis and the measurement is short and thus cathepsin activity is largely preserved.

Why Measure Cathepsin Activity?

We recommend the measurement of cysteine cathepsin activity because these enzymes are soluble proteins and are among the smallest lysosomal enzymes (Turk et al. 2012), thereby making them highly likely to be translocated through the destabilized membrane. Although fluorogenic substrates exist that are preferentially cleaved by one and not certain other cysteine cathepsins (Ivanova et al. 2008), these substrates should be used with caution if the activity of a particular cysteine cathepsin is to be tested in complex cell extracts. The reason behind this is that the selectivity of these substrates depends on the concentration of individual cysteine cathepsins, which can be controlled in in vitro studies using recombinant enzymes, but not in complex samples, such as cell extracts.

Note that this protocol can be modified for the measurement of other lysosomal enzymes, such as N-acetyl-β-D-glucosaminidase (NAG), which is a considerably larger enzyme compared with cathepsins. A fresh 15 mg/mL solution of 4-methylumbelliferyl-N-acetyl-β-D-glucosaminide, a NAG substrate, is prepared in H_2O. This is diluted to 0.3 mg/mL in 100 mM phosphate buffer (pH 6.0) and added to digitonin-lysed cells. After a 30 min incubation at 37°C, 100 mM glycine buffer (pH 10) is added to raise the pH because the fluorescent signal is much stronger at basic pH. The plate is analyzed in a single measurement, with the excitation at 360 nm and the emission at 440 nm.

Concluding Remarks

The results presented in Fig. 1C indicate that only a fraction of cysteine cathepsin molecules is translocated into the cytosol and the supernatant above the lysed cells. A considerable fraction of the enzyme molecules thus remains inside the lysed cells, but is nonetheless accessible to small substrate molecules—for example, Z-FR-AMC has a mass of ∼600 Da, and can transit through destabilized lysosomal membranes. Therefore, measuring cysteine cathepsin activity in the presence of lysed cells is a sensitive test of LMP. However, to show that destabilization of the lysosomal membrane is sufficient to allow the translocation of the enzymes into the cytosol, the enzyme activity must be measured in the cell-free extract. Together these and other data indicate that, although the protocol is straightforward, the information obtained is complex and should be interpreted with caution.

Cite this protocol as *Cold Spring Harb Protoc*; doi:10.1101/pdb.prot087114

RECIPES

PBS(D)

Reagent	Amount to add for 1 L	Final concentration
NaCl	8.00 g	137.0 mM
KCl	0.20 g	2.7 mM
KH_2PO_4	0.20 g	1.5 mM
Na_2HPO_4	1.14 g	8.0 mM

To prepare a 1× solution, combine all components in ~800 mL of H_2O and stir to dissolve. Adjust the pH to 7.4 and bring to a final volume of 1 L. Sterilize by autoclaving. (A 10× stock solution of PBS(D) can be prepared and diluted to 1× as needed.)

Phosphate Buffer (100 mM, pH 6.0)

Reagent	Volume
Monobasic sodium phosphate (200 mM)	43.8 mL
Dibasic sodium phosphate (200 mM)	6.2 mL
Distilled water	50 mL

Mix thoroughly to give 100 mL of 100 mM stock solution. Check pH and store for up to 3 mo at 4°C.

Phosphate Buffer (100 mM, pH 7.4)

Reagent	Volume
Monobasic sodium phosphate (200 mM)	9.5 mL
Dibasic sodium phosphate (200 mM)	40.5 mL
Distilled water	50 mL

Mix thoroughly to give 100 mL of 100 mM stock solution. Check pH and store for up to 3 mo at 4°C.

REFERENCES

Caruso JA, Mathieu PA, Joiakim A, Zhang H, Reiners JJ Jr. 2006. Aryl hydrocarbon receptor modulation of tumor necrosis factor-alpha-induced apoptosis and lysosomal disruption in a hepatoma model that is caspase-8-independent. *J Biol Chem* **281:** 10954–10967.

Chacon E, Acosta D, Lemasters JJ. 1997. Primary cultures of cardiac myocytes as in vitro model for pharmacological and toxicological assess-ments. In *In vitro methods in pharmaceutical research* (ed. Castel JV, Gomez-Lechon MJ) Academic, London.

Ivanova S, Repnik U, Bojič L, Petelin A, Turk V, Turk B. 2008. Lysosomes in apoptosis. *Methods Enzymol* **442:** 183–199.

Turk V, Stoka V, Vasiljeva O, Renko M, Sun T, Turk B, Turk D. 2012. Cysteine cathepsins: From structure, function and regulation to new frontiers. *Biochim Biophys Acta* **1824:** 68–88.

Studying Lysosomal Membrane Permeabilization by Analyzing the Release of Preloaded BSA–Gold Particles into the Cytosol

Urška Repnik,[1,2,4] Maruša Hafner Česen,[1] and Boris Turk[1,3,4]

[1]Department of Biochemistry and Molecular and Structural Biology, J. Stefan Institute, SI-1000 Ljubljana, Slovenia; [2]Department of Biosciences, University of Oslo, NO-0371 Oslo, Norway; [3]Center of Excellence CIPKEBIP, SI-1000 Ljubljana, Slovenia

In addition to techniques involving assaying the release of endogenous lysosomal molecules into the cytosol, the endocytic system can be preloaded with exogenous fluorescent or electron-dense tracers. These tracers will translocate into the cytosol upon lysosomal membrane permeabilization and have the advantage of being detectable directly without additional labeling. Another benefit is that the tracers can be made more abundant than most endogenous lysosomal molecules, which facilitates their detection. Tracers that can be analyzed with fluorescence microscopy include low-molecular-mass molecules such as sulforhodamine B and also fluorescent polymers of dextran that are available in a wide range of molecular masses. This protocol shows how, for electron-microscopic analysis, cells can be fed with colloidal gold or ferrofluid particles complexed to bovine serum albumin. Although electron microscopy entails a high-resolution analysis, which can be advantageous, we caution how it is important to note that particulate tracers are larger than many endogenous lysosomal molecules and might be released only upon extensive membrane permeabilization.

MATERIALS

It is essential that you consult the appropriate Material Safety Data Sheets and your institution's Environmental Health and Safety Office for proper handling of equipment and hazardous material used in this protocol.

RECIPES: Please see the end of this protocol for recipes indicated by <R>. Additional recipes can be found online at http://cshprotocols.cshlp.org/site/recipes.

Reagents

Bovine serum albumin (BSA) (1% in PBS)
BSA–gold 5-nm particles <R> (see also Step 1)
Cell culture medium
Cell source (e.g., macrophages)
Epoxy resin
Glutaraldehyde
HEPES (200 mM) or phosphate buffer (100 mM, pH 7.4) <R>

[4]Correspondence: urska.repnik@ibv.uio.no; boris.turk@ijs.si

Cite this protocol as Cold Spring Harb Protoc; doi:10.1101/pdb.prot087122

Equipment

Freeze substitution device or dry ice (McDonald 2014) (optional; see Step 6)

Fume hood

High-pressure freezing machine for cryofixation (Kaech and Ziegler 2014) (optional; see Step 5)

Polymerization oven

Suction pump

Ultramicrotome

Transmission electron microscope

> The equipment listed above, typically found in an electron microscopy laboratory, is the minimum needed to analyze samples (Bozzola 2014).

METHOD

> Note that analyzing the release of BSA–gold 5-nm particles from lysosomes requires ultrastructural analysis using a transmission electron microscope. Owing to space limitations in this protocol, sample preparation cannot be described in detail for the whole procedure. Readers should assimilate advice given in the references below for general procedures concerning sample preparation, which is usually best undertaken in laboratories equipped for electron microscopy. The points below that are featured are specifically related to the analysis of lysosomal membrane permeabilization (LMP) and include information about BSA–gold particles and the interpretation of micrographs.

Tracer Loading

1. Use a commercial source of BSA–gold or self-prepare colloidal gold and complex it with BSA by following published procedures (Slot and Geuze 1985).

2. Dilute BSA–gold with complete culture medium to obtain the desired OD_{600} (see below) and filter-sterilize it. Use a small volume of diluted BSA–gold that is just sufficient to cover the cell monolayer.
   - Incubate cells possessing high pinocytotic activity (e.g., macrophages) with BSA–gold at OD_{600} 5 for 1 h.
   - Incubate cells with lower pinocytotic activity with BSA–gold at OD_{600} 5 for 4 h or at OD_{600} 1 overnight.

3. After the incubation, remove the remaining BSA–gold with a suction pump and wash the cells three times with 1% BSA in PBS.

4. Add culture medium and incubate cells for another 5–10 min so that any gold particles attached to the plasma membrane are internalized.

   > This strategy will load BSA–gold into the whole of the endocytic pathway. To load the late-endosomal compartments only, incubate cells in the absence of BSA–gold for a minimum of 1 h to allow gold particles to move downstream.

Fixation and Microscopy

5. Fix the cells using one of the following methods.
   - Use chemical fixation with 1% glutaraldehyde in 200 mM HEPES or 100 mM phosphate buffer (pH 7.4). Incubate the cells with the fixative for at least 1 h at room temperature.

     > This is considered strong fixation, compatible with ultrastructural analysis; however, artifacts are common.

   - Use a high-pressure freezing machine.

     > This technique preserves the native ultrastructure, but it is technically more demanding (Szczesny et al. 1996).

6. Prepare cells for analysis by transmission electron microscopy.
   - For chemically fixed samples, use conventional protocols for contrasting and dehydration at room temperature (Bozzola 2014).
   - For samples fixed with high-pressure freezing, perform freeze-substitution (McDonald 2014).

7. Embed samples in epoxy resin so that ultrathin sections (thickness <100 nm) can be prepared with an ultramicrotome and analyzed with a transmission electron microscope.

DISCUSSION

Exogenous Tracers

The main advantage of looking for the release of preloaded tracers is that they can be detected directly, independently of immunolabeling, where the major concerns are the actual specificity of an antibody and the background signal. Another advantage over the endogenous lysosomal molecules is that the concentration of preloaded exogenous tracers in lysosomes can be considerably higher and thereby they can give a stronger signal, even upon translocation into the cytosol. The markers can be low-molecular-mass molecules such as sulforhodamine B (Steinberg et al. 2010) or fluorescent polymers of dextran, which are available in a wide range of molecular masses (Bidere et al. 2003). Note that dextran molecules can have larger hydrodynamic radii than proteins of the corresponding molecular mass (Armstrong et al. 2004; Erickson 2009). For electron microscopy, tracers should be electron dense, such as colloidal gold particles or ferrofluids, and sufficiently large to be identified against the complex reference space. In general, ideal markers should not interfere with the lysosomal function and affect the lumenal pH. Therefore, membrane-permeable lysosomotropic dyes, such as acridine orange, are not suitable tracers. The alternative is that tracers are internalized by pinocytosis (Griffiths et al. 1989). Fluorescent tracers have the advantage of allowing live imaging, whereas only fixed samples can be analyzed with electron microscopy. If gold particles are coated with BSA and rhodamine (Anes et al. 2006) then samples can be analyzed with correlative light and electron microscopy (CLEM).

The sensitivity of the protocol outlined above depends on the size and concentration of preloaded tracers. To prevent aggregation, the colloidal gold particles are coated with BSA molecules, which increases their overall size by ∼7 nm, such that particles of 5 nm become 12 nm (Fig. 1I) and particles of 15 nm have an approximate size of 22 nm (Fig. 1J). The longer the period the gold particles reside in lysosomes, the more likely it is that they will aggregate because BSA gets degraded. Because of the increased size of the aggregates, aggregates are less likely to be released into the cytosol following LMP. Note too that macrophages are relatively easy to load with fluid-phase tracers, whereas most other cell types require a longer incubation time and/or a higher concentration of tracers.

With conventional resin-embedding procedures, which include chemical fixation and dehydration with an alcohol series at room temperature, artifacts in membrane preservation, such as breaks, are common. Membrane artifacts are unlikely to affect the release of BSA–gold particles from lysosomes, but they can be misinterpreted as evidence of membrane damage associated with LMP. Therefore, it is important to emphasize that the effect of LMP on the integrity of lysosomal membranes cannot be judged purely by the state of ultrastructural preservation of these membranes.

Data Analysis

Even with optimal sample preparation not all membranes are distinctly visible because of the oblique section plane and/or insufficient contrast. Therefore, additional aspects of the reference space need to be considered to decide whether gold particles are contained within a membrane compartment or whether they are localized in the cytosol, for example, the presence of ribosomes gives a distinct texture to the cytosol. Not all gold particles may be released from lysosomes so that a gradient in their concentration may be observed. Depending on the abundance and distribution of gold particles, the volume fraction of leaky lysosomes or the relative density of gold particles in lysosomes and in the cytosol can be determined by stereological analysis (Griffiths 1993).

Cite this protocol as *Cold Spring Harb Protoc*; doi:10.1101/pdb.prot087122

Transmitted light   10 kDa Dextran-Alexa 546   Transmitted light   10 kDa Dextran-Alexa 546

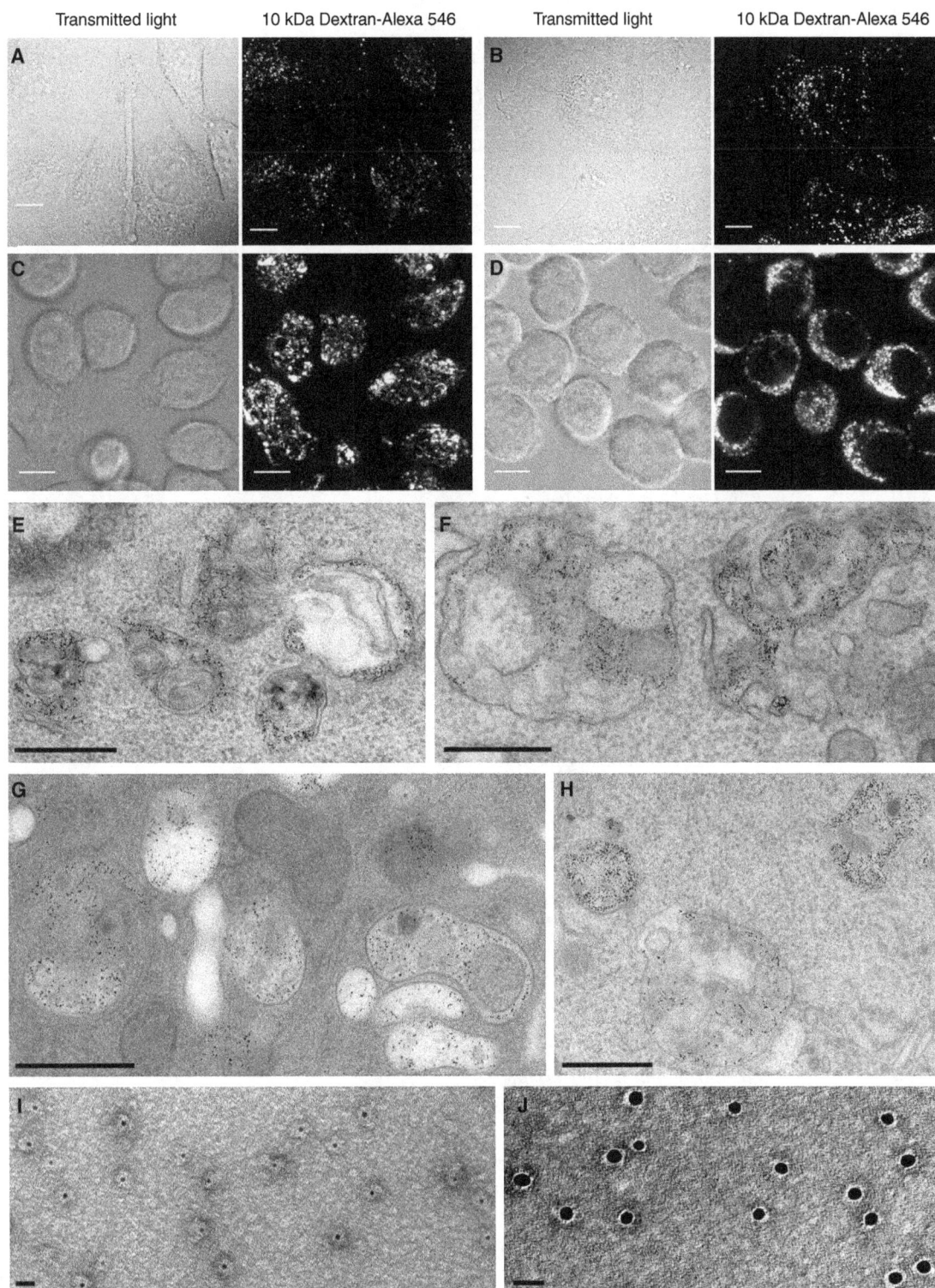

FIGURE 1. Preloaded exogenous lysosomal tracers for the evaluation of LMP. (A–D) HeLa cells (A,B) were preloaded with 500 µg/mL 10-kDa dextran–Alexa Fluor 546 for 4 h, and RAW264.7 macrophages (C,D) with 200 µg/mL for 2 h. Untreated control cells (A,C) as well as cells treated with LMP-inducing compound 5 mM Leu-Leu-O-methylester (LLOMe) for 30 min (B,D) display punctate staining. (E–H) RAW264.7 macrophages were preloaded with BSA–gold 5-nm particles, OD_{600} 5, for 1 h. Cell were fixed with 1% glutaraldehyde (E,F) or high-pressure freezing (G,H). In untreated control cells (E,G) and in cells treated with 5 mM LLOMe for 30 min (F,H), gold particles are enclosed in late endosomes/lysosomes. (I,J) BSA–gold 5-nm (I) and BSA–gold 15-nm (J) particles were negatively stained with 4% uranyl acetate. The pale coats that surround the black gold particles indicate BSA molecules, which increase the overall size of the particles. Scale bars: 10 µm (A–D); 500 nm (E–H); 20 nm (I,J).

Concluding Remarks

We have been unable to show a release of 10-kDa dextran (Fig. 1A–D) or 5-nm gold particles (Fig. 1E–H) from lysosomes when inducing LMP with Leu-Leu-O-methylester (LLOMe; a known LMP inducer)—indeed, shortly after the treatment, lysosomes appear similar to those in untreated control cells. Nevertheless, immunofluorescence and immunogold-labeling experiments have been published showing that the contents of lysosomes can spread throughout the cytosol upon induction of LMP (Johansson et al. 2010). The method described in this protocol could be used to strengthen this concept. Although the release of 5-nm gold particles from lysosomes would require extensive LMP, if showed, the results should be reliable. This is so because there are no obvious crucial steps in the sample processing that could interfere with the gold particle re-distribution compared with techniques dependent on nonspecific immunolabeling or silver enhancement.

RECIPES

BSA-Gold 5-nm Particles

Reagent	Volume
Solution A	
Trisodium citrate (1%)	8 mL
Tannic acid (1%)	4 mL
Potassium carbonate (25 mM)	4 mL
Distilled water	24 mL
Solution B	
Gold(III) chloride trihydrate (5%)	400 µL
Distilled water	160 mL

Heat both solutions to 60°C, and then plunge solution A into the vigorously stirred solution B. Continue to mix and heat the colloidal gold solution until it boils, and then cool it on ice. Complex the gold with BSA by rapidly adding 3 mL of 1% BSA to the colloidal gold solution while stirring at room temperature. Before ultracentrifugation, add 2 mL of 10% BSA to stabilize. Ultracentrifuge at 100,000g for 1 h at 4°C, and collect the loose pellet. Add 200 mM phosphate buffer (pH 7.4) to a final concentration of 10 mM. Dilute an aliquot of the BSA-gold 50–100×, and measure the optical density at 600 nm (OD$_{600}$). Store at 4°C until use.

Phosphate Buffer (100 mM, pH 7.4)

Reagent	Volume
Monobasic sodium phosphate (200 mM)	9.5 mL
Dibasic sodium phosphate (200 mM)	40.5 mL
Distilled water	50 mL

Mix thoroughly to give 100 mL of 100 mM stock solution. Check pH and store for up to 3 mo at 4°C.

ACKNOWLEDGMENTS

We thank the Electron Microscopy laboratory and the NorMIC Imaging Platform at the Department of Biosciences, University of Oslo, and Andreas Brech, Institute for Cancer Research, University of Oslo.

Cite this protocol as *Cold Spring Harb Protoc*; doi:10.1101/pdb.prot087122

REFERENCES

Anes E, Peyron P, Staali L, Jordao L, Gutierrez MG, Kress H, Hagedorn M, Maridonneau-Parini I, Skinner MA, Wildeman AG, et al. 2006. Dynamic life and death interactions between *Mycobacterium smegmatis* and J774 macrophages. *Cell Microbiol* **8:** 939–960.

Armstrong JK, Wenby RB, Meiselman HJ, Fisher TC. 2004. The hydrodynamic radii of macromolecules and their effect on red blood cell aggregation. *Biophys J* **87:** 4259–4270.

Bidere N, Lorenzo HK, Carmona S, Laforge M, Harper F, Dumont C, Senik A. 2003. Cathepsin D triggers Bax activation, resulting in selective apoptosis-inducing factor (AIF) relocation in T lymphocytes entering the early commitment phase to apoptosis. *J Biol Chem* **278:** 31401–31411.

Bozzola JJ. 2014. Conventional specimen preparation techniques for transmission electron microscopy of cultured cells. In *Electron microscopy: Methods and protocols* (ed. Kuo J), pp. 1–19. Humana Press, New York.

Erickson HP. 2009. Size and shape of protein molecules at the nanometer level determined by sedimentation, gel filtration, and electron microscopy. *Biol Proced Online* **11:** 32–51.

Griffiths G. 1993. Quantitative aspects of immunocytochemistry. In *Fine structure immunocytochemistry*, pp. 371–445. Springer, Berlin.

Griffiths G, Back R, Marsh M. 1989. A quantitative analysis of the endocytic pathway in baby hamster kidney cells. *J Cell Biol* **109:** 2703–2720.

Johansson AC, Appelqvist H, Nilsson C, Kagedal K, Roberg K, Ollinger K. 2010. Regulation of apoptosis-associated lysosomal membrane permeabilization. *Apoptosis* **15:** 527–540.

Kaech A, Ziegler U. 2014. High-pressure freezing: Current state and future prospects. In *Electron microscopy: Methods and protocols* (ed. Kuo J), pp. 151–171. Humana Press, New York.

McDonald K. 2014. High-pressure freezing for preservation of high resolution fine structure and antigenicity for immunolabeling. In *Electron microscopy methods and protocols* (ed. Hajibagheri N), pp. 77–97. Humana Press, New York.

Slot JW, Geuze HJ. 1985. A new method of preparing gold probes for multiple-labeling cytochemistry. *Eur J Cell Biol* **38:** 87–93.

Steinberg BE, Huynh KK, Brodovitch A, Jabs S, Stauber T, Jentsch TJ, Grinstein S. 2010. A cation counterflux supports lysosomal acidification. *J Cell Biol* **189:** 1171–1186.

Szczesny PJ, Walther P, Muller M. 1996. Light damage in rod outer segments: The effects of fixation. *Curr Eye Res* **15:** 807–814.

CHAPTER 10

Techniques to Distinguish Apoptosis from Necroptosis

Maria Feoktistova,[1,4] Fredrik Wallberg,[2,4] Tencho Tenev,[2,4] Peter Geserick,[1] Martin Leverkus,[1,3,4,5] and Pascal Meier[2,4,5]

[1]Section of Molecular Dermatology, Department of Dermatology, Venereology, and Allergology, Medical Faculty Mannheim, University Heidelberg, 68167 Mannheim, Germany; [2]The Breakthrough Toby Robins Breast Cancer Research Centre, Institute of Cancer Research, Chester Beatty Laboratories, London SW3 6JB, United Kingdom; [3]Department of Dermatology & Allergology, University Hospital of RWTH Aachen University, 52074 Aachen, Germany

The processes by which cells die are as tightly regulated as those that govern cell growth and proliferation. Recent studies of the molecular pathways that regulate and execute cell death have uncovered a plethora of signaling cascades that lead to distinct modes of cell death, including "apoptosis," "necrosis," "autophagic cell death," and "mitotic catastrophe." Cells can readily switch from one form of death to another; therefore, it is vital to have the ability to monitor the form of death that cells are undergoing. A number of techniques are available that allow the detection of cell death and when combined with either knockdown approaches or inhibitors of specific signaling pathways, such as caspase or RIP kinase pathways, they allow the rapid dissection of divergent cell death pathways. However, techniques that reveal the end point of cell death cannot reconstruct the sequence of events that have led to death; therefore, they need to be complemented with methods that can distinguish all forms of cell death. Apoptotic cells frequently undergo secondary necrosis under in vitro culture conditions; therefore, novel methods relying on high-throughput time-lapse fluorescence video microscopy are necessary to provide temporal resolution to cell death events. Further, visualizing the assembly of multiprotein signaling hubs that can execute apoptosis or necroptosis helps to explore the underlying processes. Here we introduce a suite of techniques that reliably distinguish necrosis from apoptosis and secondary necrosis, and that enable investigation of signaling platforms capable of instructing apoptosis or necroptosis.

INTRODUCTION

Cells in the process of dying show different morphological features depending on the cell death process occurring. These features provide clues to the underlying molecular pathways that regulate and execute the different forms of cell death (Galluzzi et al. 2012). Although biochemical assays for monitoring cell death phenomena have become routine, a systematic methodology has not yet been adopted to distinguish apoptosis from necrosis (or necroptosis). Apoptosis is morphologically characterized by membrane blebbing, cell shrinkage and fragmentation, nuclear condensation, formation of apoptotic bodies, and activation of caspases (Galluzzi et al. 2012). The appearance of the major biochemical and morphological hallmarks of apoptosis are mediated by caspases (Luthi and Martin 2007). Activation of caspases can be triggered in response to immune-mediated signals via the "extrinsic" pathway, which is controlled by the death receptor family and their ligands. Alternatively,

---

[4]These authors contributed equally to this work.

[5]Correspondence: pascal.meier@icr.ac.uk; mleverkus@ukaachen.de

Cite this introduction as Cold Spring Harb Protoc; doi:10.1101/pdb.top070375

developmental cues and cellular stresses promote apoptosis via activation of the "intrinsic" pathway (Meier and Vousden 2007). Cells that die by apoptosis, at least in vivo, generally do not release their intracellular contents. However, apoptotic cells do not always die silently because they can trigger the secretion of chemotactic factors and other immunologically active proteins that can influence the immune response toward dying cells (Cullen et al. 2013). In contrast to apoptosis, necrosis is morphologically characterized by cell rounding, cytoplasmic swelling, presence of dilated organelles, and absence of caspase activation and chromatin condensation. Cells that die by necrosis spill their intracellular contents, and hence trigger an inflammatory response. Necroptosis refers to a regulated form of necrosis. It is biochemically defined as a form of cell death that is dependent on the serine–threonine kinase receptor-interacting protein 1 (RIPK1), RIPK3, and the mixed lineage kinase domain-like protein (MLKL) (Sun et al. 2012). This pathway plays important roles in a variety of physiological and pathological conditions, including development, response to tissue damage, and antiviral immunity (Vandenabeele et al. 2010).

STRATEGIES TO DISTINGUISH BETWEEN APOPTOTIC AND NECROPTOTIC CELL DEATH

Although the morphological features of apoptosis and necroptosis are clearly distinct, under tissue culture conditions apoptotic cells frequently undergo secondary necrosis (Berghe et al. 2010), which complicates the analysis of specific cell death modalities. Therefore, different strategies have to be used to analyze and distinguish apoptotic from necroptotic cell death. Chemical inhibitors of caspases, RIPK1, and MLKL are useful tools to determine whether cells die by apoptosis or necroptosis. Experimentally, apoptosis triggered by the intrinsic cell death pathway can be delayed (yet rarely completely blocked) by treatment with pan-caspase chemical inhibitors, such as N-benzyloxycarbonyl-Val-Ala-Asp-fluoromethylketone (zVAD-fmk) (Galluzzi et al. 2012) or quinolyl-valyl-O-methylaspartyl-[2,6-difluorophenoxy]-methyl-ketone (Q-VD-OPh) (Caserta et al. 2003). Although such pharmacological compounds are useful tools to inhibit caspases, it should be noted that the point of no return in mitochondria-mediated apoptosis is not caspase activation but irreversible dissipation of the mitochondrial membrane potential. Loss of mitochondrial membrane permeability (MOMP) leads to a release of toxic proteins from the intermembrane space into the cytosol that ultimately activates a positive feedback circuit that amplifies the apoptotic signal (Ekert et al. 2004). Activation of caspases, following MOMP, accelerates the execution phase of cell death. While chemical inhibition of caspases rarely confers long-term cytoprotective effects, or truly prevents cell death induced by death stimuli that funnel through mitochondria, zVAD-fmk and Q-VD-OPh can block most hallmarks of apoptosis. The cells still die, but via a different cell death modality, affecting morphology and immunogenicity of the dying cell. In contrast to cell death triggered via the mitochondrial pathway, zVAD-fmk and Q-VD-OPh can block most, if not all, apoptosis-inducing extrinsic signals. Thus, in combination with other techniques zVAD-fmk and Q-VD-OPh are useful tools to experimentally assess the modality of cell death, and to establish the relative contribution of caspases to cell death. The latter can be estimated by the extent of short-term (24–48-h) cytoprotection. While zVAD-fmk can delay apoptosis, it either has no effect or can even exacerbate necroptosis and can, therefore, be used as additional evidence for this type of cell death modality. For instance, treatment with zVAD-fmk stimulates necroptosis in L929 cells (Vandenabeele et al. 2010). This is because zVAD-fmk inhibits caspase-8/FLIP$_L$ heterodimers, which cleave and inactivate RIPK1 and RIPK3, thereby suppressing the formation of necroptotic complexes (Geserick et al. 2009; Kaiser et al. 2011; Oberst et al. 2011; Welz et al. 2011) (Fig. 1). Moreover, in L929 cells, treatment with zVAD-fmk drives autocrine production of tumor necrosis factor (TNF) (Hitomi et al. 2008), which promotes the assembly of necroptosis-inducing complexes.

Pharmacological inhibitors, such as necrostatin-1 (a RIPK1 inhibitor [Degterev et al. 2008]) and necrosulfonamide [an MLKL inhibitor (Sun et al. 2012)] are frequently used to suppress necroptosis. However, necrostatin-1 not only blocks necroptosis, but can also modulate RIPK1-mediated apoptosis (Tenev et al. 2011). Necrostatin-1 is, therefore, not a particularly specific tool to distinguish necroptosis from apoptosis, and conclusions cannot be drawn by using it as a stand-alone approach.

Cite this introduction as *Cold Spring Harb Protoc*; doi:10.1101/pdb.top070375

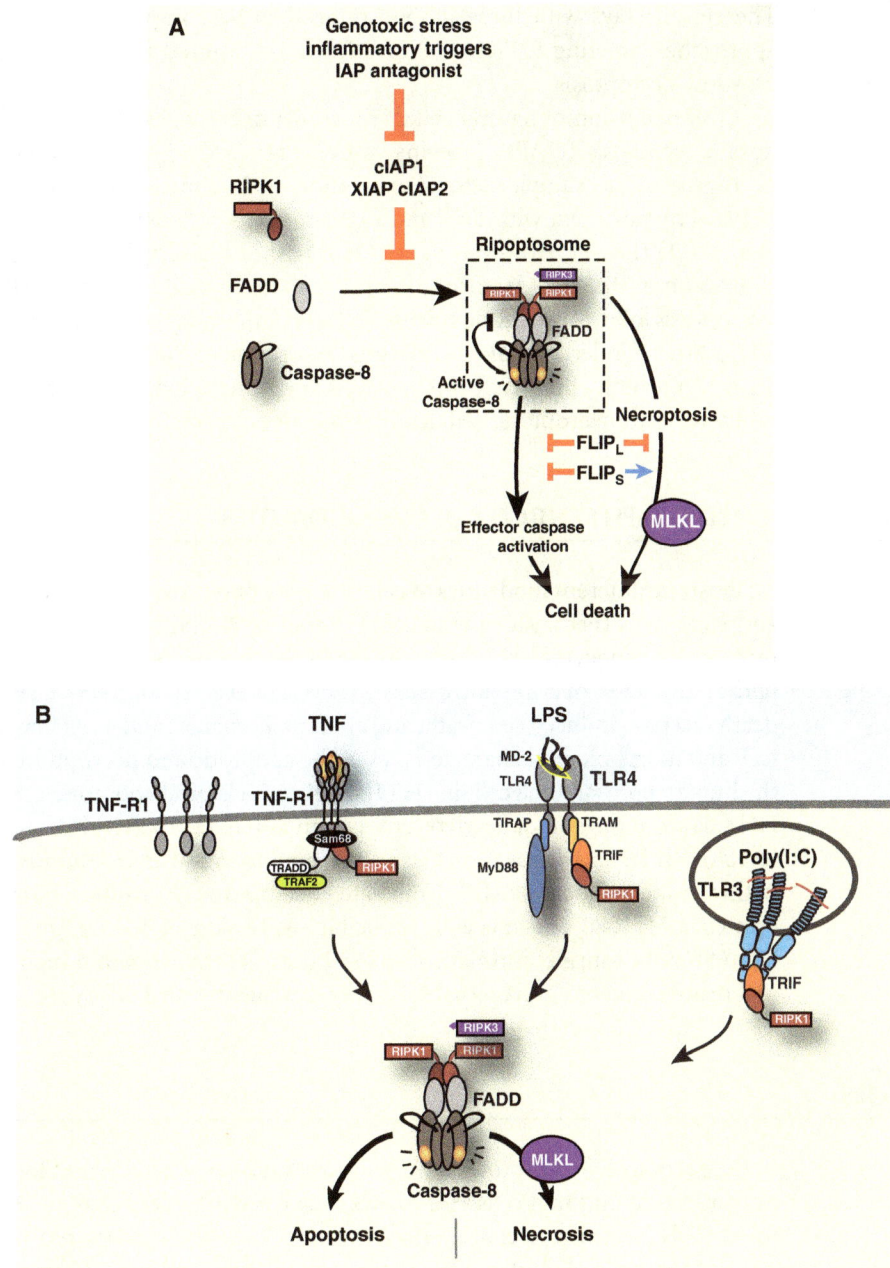

FIGURE 1. Cell death following formation of the RIPK1-dependent platform, also termed ripoptosome, complex-IIB, or necrosome. (A) cIAP1, cIAP2, and XIAP target RIPK1 and components of the ripoptosome (caspase-8 and cFLIP$_L$) for Ub-mediated inactivation. Following genotoxic stress, cytokine signaling-induced depletion of cIAPs, or SMAC-mimetic (SM) treatment, cIAP1, cIAP2, and XIAP levels rapidly decline and/or are inactivated. This allows formation and accumulation of the ripoptosome. In the presence of high levels of RIPK3 this can lead to necroptosis. cFLIP also regulates ripoptosome-mediated cell death. cFLIP$_L$ prevents apoptosis and necroptosis, while FLIP$_S$ inhibits apoptosis but promotes necroptosis. (B) Under steady-state conditions, the majority of RIPK1 appears to be "inaccessible," preventing it from binding to partner proteins. Cytokine receptor stimulation can convert a small fraction of RIPK1 into an "accessible," binding competent configuration. In the presence of cIAPs and XIAP, binding competent RIPK1 is targeted for Ub-mediated inactivation, most likely via proteasomal degradation. Under conditions where IAP levels are low, however, unmodified and binding-competent RIPK1 accumulates and can form the ripoptosome. In the presence of high levels of cFLIP$_L$, the ripoptosome is dissolved via caspase-8-cFLIP$_L$-mediated cleavage of RIPK1. When cFLIP$_L$ levels are low, the ripoptosome can promote caspase-dependent or caspase-independent cell death.

Therefore, assays with these inhibitors need to be complemented with knockdown or knockout approaches targeting RIPK1, RIPK3, MLKL, or caspases to substantiate their contribution to apoptosis or necroptosis.

Overexpression of natural cell death regulators (e.g., cFLIP, XIAP, CrmA) and/or cellular inhibitors of apoptosis (cIAPs) proteins can also be used to analyze whether cells die by apoptosis or necroptosis. For example, ectopic expression of the short or long isoform of cFLIP blocks apoptosis induced by treatment with TNF and IAP-antagonists (Feoktistova et al. 2011; Oberst et al. 2011; Pop et al. 2011) (Fig. 1). However, in cells that harbor high levels of RIPK3 and that die by necroptosis, expression of the short isoform of cFLIP (cFLIP$_S$) readily enhances RIPK1/RIPK3/MLKL-mediated necroptosis following treatment with TNF and IAP-antagonists (Geserick et al. 2009; Feoktistova et al. 2011, 2012). Under the same conditions, expression of the long isoform (cFLIP$_L$) blocks necroptosis. Thus, while both cFLIP isoforms suppress apoptotic cell death, expression of cFLIP$_L$, but not cFLIP$_S$ interferes with necroptotic cell death in the absence of cIAPs.

CELL LINES FOR STUDYING DIFFERENT CELL DEATH MODALITIES

To illustrate different modalities of cell death, we have used different cell lines: HT1080, L929, HaCaT, and HeLa cells (Feoktistova et al. 2011; Tenev et al. 2011). HaCaT cells (Boukamp et al. 1988) are primary keratinocyte-derived transformed cells that express low levels of cFLIP and XIAP, but that harbor high levels of cIAPs and caspases (Leverkus et al. 2000, 2003). HaCaT cells are highly sensitive to death receptor-induced cell death, and are thus a suitable model cell line to study molecular, biochemical, and morphological characteristics of the apoptotic and necroptotic signaling machinery. Cells of the human fibrosarcoma cell line HT1080 display classical apoptotic cell death features in response to DNA damaging agents or death receptor stimulation. HT1080 and HeLa cells do not express detectable levels of RIPK3 and hence die exclusively by apoptosis and are, therefore, useful to investigate cell death pathways independent of RIPK3-mediated necroptosis. To study necroptosis in HT1080 cells we have created a HT1080 cell line with inducible expression of RIPK3 (HT1080$^{ind-RIPK3}$). In contrast to HT1080 cells, murine fibroblast-like L929 cells naturally contain high levels of endogenous RIPK3, and readily undergo necroptosis following treatment with TNF (Vercammen et al. 1998).

PROTOCOLS

In the accompanying protocols, we present a number of techniques for the analysis of apoptosis and necroptosis in cultured cells. See Protocol 1: Crystal Violet Assay for Determining Viability of Cultured Cells (Feoktistova et al. 2015a), Protocol 2: Analysis of Apoptosis and Necroptosis by Fluorescence-Activated Cell Sorting (Wallberg et al. 2015a), Protocol 3: Time-Lapse Imaging of Cell Death (Wallberg et al. 2015b), and Protocol 4: Ripoptosome Analysis by Caspase-8 Coimmunoprecipitation (Feoktistova et al. 2015b). These techniques enable the visualization and quantification of the effects on cell death of the tools and manipulations described above. Time-lapse video microscopy can be used to provide further insight into the morphological characteristics of dying cells. Biochemical studies that address the kinetics of the assembly of intracellular cell death platforms provide further insight into the molecular mechanisms of cell death. Additional and complementary techniques are discussed elsewhere (Krysko et al. 2008).

REFERENCES

Berghe TV, Vanlangenakker N, Parthoens E, Deckers W, Devos M, Festjens N, Guerin CJ, Brunk UT, Declercq W, Vandenabeele P. 2010. Necroptosis, necrosis and secondary necrosis converge on similar cellular disintegration features. *Cell Death Differ* 17: 922–930.

Boukamp P, Petrussevska RT, Breitkreutz D, Hornung J, Markham A, Fusenig NE. 1988. Normal keratinization in a spontaneously immortalized aneuploid human keratinocyte cell line. *J Cell Biol* 106: 761–771.

Caserta TM, Smith AN, Gultice AD, Reedy MA, Brown TL. 2003. Q-VD-OPh, a broad spectrum caspase inhibitor with potent antiapoptotic properties. *Apoptosis* **8:** 345–352.

Cullen SP, Henry CM, Kearney CJ, Logue SE, Feoktistova M, Tynan GA, Lavelle EC, Leverkus M, Martin SJ. 2013. Fas/CD95-induced chemokines can serve as "Find-Me" signals for apoptotic cells. *Mol Cell* **49**(6): 1034–1038.

Degterev A, Hitomi J, Germscheid M, Che'en IL, Korkina O, Teng X, Abbott D, Cuny GD, Yuan C, Wagner G. 2008. Identification of RIP1 kinase as a specific cellular target of necrostatins. *Nat Chem Biol* **4:** 313–321.

Ekert PG, Read SH, Silke J, Marsden VS, Kaufmann H, Hawkins CJ, Gerl R, Kumar S, Vaux DL. 2004. Apaf-1 and caspase-9 accelerate apoptosis, but do not determine whether factor-deprived or drug-treated cells die. *J Cell Biol* **165:** 835–842.

Feoktistova M, Geserick P, Leverkus M. 2015a. Crystal violet assay for determining viability of cultured cells. *Cold Spring Harb Protoc* doi: 10.1101/pdb.prot087379.

Feoktistova M, Geserick P, Leverkus M. 2015b. Ripoptosome analysis by caspase-8 coimmunoprecipitation. *Cold Spring Harb Protoc* doi: 10.1101/pdb.prot087403.

Feoktistova M, Geserick P, Kellert B, Dimitrova DP, Langlais C, Hupe M, Cain K, MacFarlane M, Hacker G, Leverkus M. 2011. cIAPs block Ripoptosome formation, a RIP1/caspase-8 containing intracellular cell death complex differentially regulated by cFLIP isoforms. *Mol Cell* **43:** 449–463.

Feoktistova M, Geserick P, Panayotova-Dimitrova D, Leverkus M. 2012. Pick your poison: The Ripoptosome, a cell death platform regulating apoptosis and necroptosis. *Cell Cycle* **11:** 460–467.

Galluzzi L, Vitale I, Abrams JM, Alnemri ES, Baehrecke EH, Blagosklonny MV, Dawson TM, Dawson VL, El-Deiry WS, Fulda S, et al. 2012. Molecular definitions of cell death subroutines: Recommendations of the Nomenclature Committee on Cell Death 2012. *Cell Death Differ* **19:** 107–120.

Geserick P, Hupe M, Moulin M, Wong WW, Feoktistova M, Kellert B, Gollnick H, Silke J, Leverkus M. 2009. Cellular IAPs inhibit a cryptic CD95-induced cell death by limiting RIP1 kinase recruitment. *J Cell Biol* **187:** 1037–1054.

Hitomi J, Christofferson DE, Ng A, Yao J, Degterev A, Xavier RJ, Yuan J. 2008. Identification of a molecular signaling network that regulates a cellular necrotic cell death pathway. *Cell* **135:** 1311–1323.

Kaiser WJ, Upton JW, Long AB, Livingston-Rosanoff D, Daley-Bauer LP, Hakem R, Caspary T, Mocarski ES. 2011. RIP3 mediates the embryonic lethality of caspase-8-deficient mice. *Nature* **471:** 368–372.

Krysko DV, Vanden Berghe T, D'Herde K, Vandenabeele P. 2008. Apoptosis and necrosis: Detection, discrimination and phagocytosis. *Methods* **44:** 205–221.

Leverkus M, Neumann M, Mengling T, Rauch CT, Brocker EB, Krammer PH, Walczak H. 2000. Regulation of tumor necrosis factor-related apoptosis-inducing ligand sensitivity in primary and transformed human keratinocytes. *Cancer Res* **60:** 553–559.

Leverkus M, Sprick MR, Wachter T, Mengling T, Baumann B, Serfling E, Brocker EB, Goebeler M, Neumann M, Walczak H. 2003. Proteasome inhibition results in TRAIL sensitization of primary keratinocytes by removing the resistance-mediating block of effector caspase maturation. *Mol Cell Biol* **23:** 777–790.

Luthi AU, Martin SJ. 2007. The CASBAH: A searchable database of caspase substrates. *Cell Death Differ* **14:** 641–650.

Meier P, Vousden KH. 2007. Lucifer's labyrinth–Ten years of path finding in cell death. *Mol Cell* **28:** 746–754.

Oberst A, Dillon CP, Weinlich R, McCormick LL, Fitzgerald P, Pop C, Hakem R, Salvesen GS, Green DR. 2011. Catalytic activity of the caspase-8-FLIP(L) complex inhibits RIPK3-dependent necrosis. *Nature* **471:** 363–367.

Pop C, Oberst A, Drag M, Van Raam BJ, Riedl SJ, Green DR, Salvesen GS. 2011. FLIP(L) induces caspase 8 activity in the absence of interdomain caspase 8 cleavage and alters substrate specificity. *Biochem J* **433:** 447–457.

Sun L, Wang H, Wang Z, He S, Chen S, Liao D, Wang L, Yan J, Liu W, Lei X, et al. 2012. Mixed lineage kinase domain-like protein mediates necrosis signaling downstream of RIP3 kinase. *Cell* **148:** 213–227.

Tenev T, Bianchi K, Darding M, Broemer M, Langlais C, Wallberg F, Zachariou A, Lopez J, MacFarlane M, Cain K, et al. 2011. The Ripoptosome, a signaling platform that assembles in response to genotoxic stress and loss of IAPs. *Mol Cell* **43:** 432–448.

Vandenabeele P, Galluzzi L, Vanden Berghe T, Kroemer G. 2010. Molecular mechanisms of necroptosis: an ordered cellular explosion. *Nat Rev Mol Cell Biol* **11:** 700–714.

Vercammen D, Beyaert R, Denecker G, Goossens V, Van Loo G, Declercq W, Grooten J, Fiers W, Vandenabeele P. 1998. Inhibition of caspases increases the sensitivity of L929 cells to necrosis mediated by tumor necrosis factor. *J Exp Med* **187:** 1477–1485.

Wallberg F, Tenev T, Meier P. 2015a. Analysis of apoptosis and necroptosis by fluorescence-activated cell sorting. *Cold Spring Harb Protoc* doi: 10.1101/pdb.prot087387.

Wallberg F, Tenev T, Meier P. 2015b. Time-Lapse Imaging of Cell Death. *Cold Spring Harb Protoc* doi: 10.1101/pdb.prot087395.

Welz PS, Wullaert A, Vlantis K, Kondylis V, Fernandez-Majada V, Ermolaeva M, Kirsch P, Sterner-Kock A, van Loo G, Pasparakis M. 2011. FADD prevents RIP3-mediated epithelial cell necrosis and chronic intestinal inflammation. *Nature* **477:** 330–334.

Crystal Violet Assay for Determining Viability of Cultured Cells

Maria Feoktistova,[1] Peter Geserick,[1] and Martin Leverkus[1,2,3]

[1]*Section of Molecular Dermatology, Department of Dermatology, Venereology and Allergology, Medical Faculty Mannheim, University of Heidelberg, Heidelberg 68167, Germany;* [2]*Department of Dermatology & Allergology, University Hospital of RWTH Aachen University, 52074 Aachen, Germany*

Adherent cells detach from cell culture plates during cell death. This characteristic can be used for the indirect quantification of cell death and to determine differences in proliferation upon stimulation with death-inducing agents. One simple method to detect maintained adherence of cells is the staining of attached cells with crystal violet dye, which binds to proteins and DNA. Cells that undergo cell death lose their adherence and are subsequently lost from the population of cells, reducing the amount of crystal violet staining in a culture. This protocol describes a quick and reliable screening method that is suitable for the examination of the impact of chemotherapeutics or other compounds on cell survival and growth inhibition. However, characterization of the cause of reduced crystal violet staining requires additional methods detailed elsewhere.

MATERIALS

It is essential that you consult the appropriate Material Safety Data Sheets and your institution's Environmental Health and Safety Office for proper handling of equipment and hazardous materials used in this protocol.

RECIPES: Please see the end of this protocol for recipes indicated by <R>. Additional recipes can be found online at http://cshprotocols.cshlp.org/site/recipes.

Reagents

Adherent cell line of interest

Cell culture medium

Use the appropriate cell culture medium according to the cell type/line being cultured.

Crystal violet staining solution (0.5%) <R>

Drug(s) appropriate for experimental goals (e.g., CD95L, poly[I:C])

Methanol

Equipment

96-well plate reader that is able to read absorbance at 570 nm (e.g., VICTOR3 Multilabel Plate Counter from PerkinElmer)

96-well tissue culture plates

Bench rocker (2D or 3D)

Filter paper

Incubator (37°C, 5% CO_2, humidified)

Vacuum aspirator

---

[3]Correspondence: mleverkus@ukaachen.de

Cite this protocol as *Cold Spring Harb Protoc*; doi:10.1101/pdb.prot087379

METHOD

See Figure 1 for an overview of the crystal violet assay. Here, we describe the method using a 96-well format. However, the same method can be used with other plate sizes by simply scaling cell numbers and volumes accordingly. Perform Steps 1–3 under a laminar flow hood.

1. Seed cells in a 96-well plate. To three wells, add medium without cells. Ensure that the volume of the culture medium is ≥100 µL/well to avoid evaporation effects. Incubate the cells for 18–24 h at 37°C to enable adhesion of cells to wells.

   Usually 1–2 × 10⁴ cells/well should be seeded to achieve confluence of ~40%–50%. Initial cell confluence depends on the proliferation and cell size of the cell line used; therefore adjust accordingly. The wells without cells will serve as controls for nonspecific binding of the crystal violet dye.

2. Aspirate the medium from the wells, and add ≥100 µL/well of fresh medium supplemented with drugs appropriate for the experimental goals (e.g., 2–20 µg/mL of poly[I:C] for HaCaT cells). Do not let the cells dry out when changing medium; cells that have dried out will not undergo cell death induction. Treat cells in triplicate wells for each condition. Do not treat cells in three of the wells (these will serve as controls). Incubate cells for the desired time in the desired conditions (e.g., in case of HaCaT and poly(I:C) treatment it is for 18–24 h at 37°C).

   Different concentrations of other stimulants (e.g., death ligands such as TRAIL [1–1000 ng/mL] or TNF [1–1000 ng/mL]) should be analyzed in preliminary experiments to determine dose-dependent cell death induction before co- and prestimulation with respective inhibitors (e.g., caspase inhibitor: zVAD-fmk [10 µM]; IAP antagonist: GT12911e [100 nM]).

3. Aspirate the medium, and wash the cells twice in a gentle stream of tap water. Tilt the plate to prevent the stream of water from hitting the cell monolayer directly. After the wells have filled with water, immediately aspirate the water from the wells. After washing, invert the plate on filter paper and tap the plate gently to remove any remaining liquid.

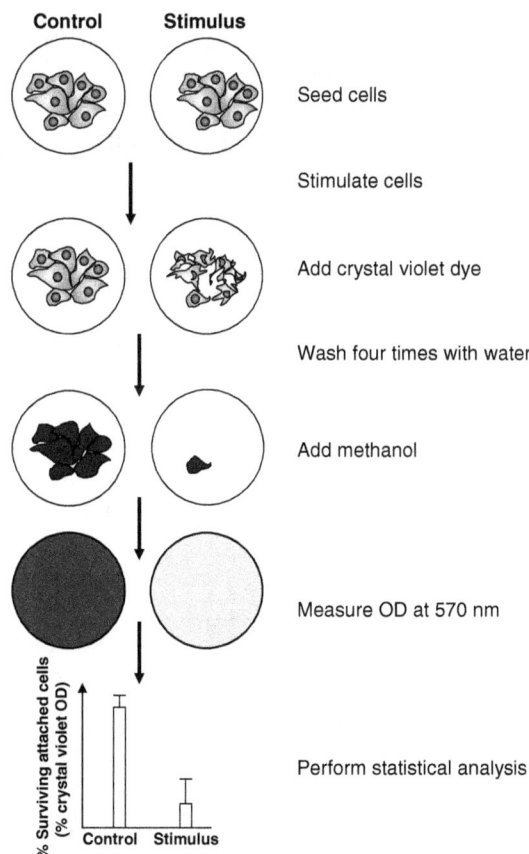

FIGURE 1. Schematic workflow for crystal violet assay.

Some cells (e.g., keratinocytes) adhere strongly to the surface of cell culture plates (Boukamp et al. 1988), whereas others (e.g., HeLa or 293T cells) adhere weakly and are, therefore, easily flushed off the plate surface during washing. Although the washing step increases the sensitivity of the assay, it can be skipped for cells that do not adhere well to the plate.

4. Add 50 µL of 0.5% crystal violet staining solution to each well, and incubate for 20 min at room temperature on a bench rocker with a frequency of 20 oscillations per minute.

5. Wash the plate four times in a stream of tap water as described in Step 3. After washing, invert the plate on filter paper and tap the plate gently to remove any remaining liquid. Air-dry the plate without its lid for at least 2 h at room temperature.

   It is recommended to dry the plate at room temperature for 16–24 h.

6. Add 200 µL of methanol to each well, and incubate the plate with its lid on for 20 min at room temperature on a bench rocker with a frequency of 20 oscillations per minute.

7. Measure the optical density of each well at 570 nm (OD_{570}) with a plate reader.

8. Subtract the average OD_{570} of the wells without cells from the OD_{570} of each well on the plate.

   The OD_{570} of the wells without cells, but processed identically to wells with cells, defines the background of the staining method.

9. Set the average OD_{570} of nonstimulated cells to 100%. Then determine the percentage of stimulated cells that are viable (attached) by comparing the average OD_{570} values of stimulated cells with the OD_{570} values of the nonstimulated cells.

   See Troubleshooting.

10. Calculate the mean and the standard error of the mean for at least three independent experiments.

TROUBLESHOOTING

Problem (Step 9): Wells treated with the same conditions stain unequally with crystal violet.
Solution: Before plating adherent cells, make sure that the cells are in a homogenous, single-cell suspension. If they are not, clumps of cells can be seeded in wells, resulting in unequal cell densities. After crystal violet staining, this will give increase to OD_{570} measurements with large standard errors that inaccurately reflect loss of viability.

DISCUSSION

Crystal violet staining is a quick and versatile assay for screening cell viability under diverse stimulation conditions (Geserick et al. 2009). However, it is potentially compromised by proliferative responses that occur at the same time as cell death responses. Therefore, chemical inhibitors of caspases and/or of necroptosis may be incorporated into the assay (Degterev et al. 2008; Sun et al. 2012). Alternatively, molecular studies (e.g., overexpression or knockdown) can be performed to more specifically address the nature of cell death (Feoktistova et al. 2011).

RECIPE

Crystal Violet Staining Solution (0.5%)

0.5 g crystal violet powder (Sigma-Aldrich)
80 mL distilled H_2O
20 mL methanol

Dissolve crystal violet powder in H_2O and then add methanol. No sterilization procedures are required. Store solution in the dark at room temperature. Use within 2 mo.

Cite this protocol as *Cold Spring Harb Protoc*; doi:10.1101/pdb.prot087379

ACKNOWLEDGMENTS

We thank all current and previous members of the Section of Molecular Dermatology, Medical Faculty of Mannheim, University of Heidelberg for their discussions and suggestions. The work in the laboratory of M.L. was supported by grants from the Deutsche Forschungsgemeinshaft (Le953/ 5-1, 6-1, 8-1), the Wilhelm-Sander-Stiftung (2008.072.1), and the Mildred-Scheel-Stiftung (Projekt 109891).

REFERENCES

Boukamp P, Petrussevska RT, Breitkreutz D, Hornung J, Markham A, Fusenig NE. 1988. Normal keratinization in a spontaneously immortalized aneuploid human keratinocyte cell line. *J Cell Biol* **106:** 761–771.

Degterev A, Hitomi J, Germscheid M, Ch'en IL, Korkina O, Teng X, Abbott D, Cuny GD, Yuan C, Wagner G, et al. 2008. Identification of RIP1 kinase as a specific cellular target of necrostatins. *Nat Chem Biol* **4:** 313–321.

Feoktistova M, Geserick P, Kellert B, Dimitrova DP, Langlais C, Hupe M, Cain K, MacFarlane M, Hacker G, Leverkus M. 2011. cIAPs block Ripoptosome formation, a RIP1/caspase-8 containing intracellular cell death complex differentially regulated by cFLIP isoforms. *Mol Cell* **43:** 449–463.

Geserick P, Hupe M, Moulin M, Wong WW, Feoktistova M, Kellert B, Gollnick H, Silke J, Leverkus M. 2009. Cellular IAPs inhibit a cryptic CD95-induced cell death by limiting RIP1 kinase recruitment. *J Cell Biol* **187:** 1037–1054.

Sun L, Wang H, Wang Z, He S, Chen S, Liao D, Wang L, Yan J, Liu W, Lei X, et al. 2012. Mixed lineage kinase domain-like protein mediates necrosis signaling downstream of RIP3 kinase. *Cell* **148:** 213–227.

Analysis of Apoptosis and Necroptosis by Fluorescence-Activated Cell Sorting

Fredrik Wallberg, Tencho Tenev, and Pascal Meier[1]

The Breakthrough Toby Robins Breast Cancer Research Centre, Institute of Cancer Research, Chester Beatty Laboratories, London SW3 6JB, United Kindom

Fluorescence-activated cell sorting (FACS) is a laser-based, biophysical technology that allows simultaneous multiparametric analysis. For the analysis of dying cells, fluorescently labeled Annexin V (Annexin V^{FITC}) and propidium iodide (PI) are the most commonly used reagents. Instead of PI, 4′,6-diamidino-2-phenylindole (DAPI) can also be used. DAPI is a fluorescent stain that binds strongly to A-T-rich regions in DNA. DAPI and PI only inefficiently pass through an intact cell membrane and, therefore, preferentially stain dead cells. DAPI can be combined with Annexin V^{FITC} and the potentiometric fluorescent dye, tetramethylrhodamine methyl ester (TMRM), which measures mitochondrial permeability transition and mitochondrial membrane depolarization. TMRM is a cell-permeable fluorescent dye that is sequestered to active mitochondria, and hence labels live cells. On apoptosis or necroptosis the TMRM signal is lost. The advantage of using Annexin V^{FITC}/DAPI/TMRM is that the entire cell population is labeled, and it is easy to distinguish living (TMRM + /Annexin V^{FITC}-/DAPI-) from dying or dead cells (apoptosis: TMRM-/Annexin V^{FITC} + /DAPI-; necrosis: TMRM-/Annexin V^{FITC} + /DAPI+). This is important because cell debris (fluorescent negative particles) must be avoided to establish the correct parameters for the FACS analysis, otherwise incorrect statistical values will be obtained. To obtain information on the cell concentration or absolute cell counts in a sample, it is recommended to add an internal microsphere counting standard to the flow cytometric sample. This protocol describes the FACS analysis of cell death in HT1080 and L929 cells, but it can be readily adapted to other cell types of interest.

MATERIALS

It is essential that you consult the appropriate Material Safety Data Sheets and your institution's Environmental Health and Safety Office for proper handling of equipment and hazardous materials used in this protocol.

RECIPES: Please see the end of this protocol for recipes indicated by <R>. Additional recipes can be found online at http://cshprotocols.cshlp.org/site/recipes.

Reagents

4′,6-Diamidino-2-phenylindole (DAPI)
Annexin V^{FITC}
$CaCl_2$ (1 M)
CountBright Absolute Counting Beads (Life Technologies)
Doxycycline

---

[1]Correspondence: pmeier@icr.ac.uk

Cite this protocol as *Cold Spring Harb Protoc*; doi:10.1101/pdb.prot087387

Dulbecco's modified Eagle's medium (DMEM) with phenol red (Life Technologies, 41966-029)

Fetal bovine serum (FBS)

HT1080^{indRIPK3} cells

Parental HT1080 cells (ATCC) were transduced with lentiviral particles carrying RIPK3 cDNA expressed under the control of a doxycycline regulatable promoter. For RIPK3 induction, cells should be cultured in media containing doxycycline (DOX) (100 ng/mL) for at least 1 wk prior to the experiment.

IAP-antagonist (SMAC mimetic, SM; 1 mM in DMSO)

L929 cells

Necrostatin-1 (Sigma-Aldrich; 40 mM in DMSO)

Recombinant human TNF (Alexis; 100 µg/mL in DMEM)

Tetramethylrhodamine methyl ester (TMRM) (Life technologies)

Trypsin

zVAD-fmk (20 mM in DMSO)

Equipment

Cell culture plates (six-well)

BD LSR II FACS system (BD Bioscience) or equivalent

The system needs to be equipped with three lasers: 404, 488, and 561 nm. FITC is excited by the 488 nm laser and emission light is measured with filters 505 LP 525/50 BP. DAPI is excited by the 404 nm laser and emission light is collected with filters 450/50 BP (Kapuscinski 1995). TMRM is excited by a 561 nm laser and emission light is collected with filters 595/40 BP. The CountBright absolute counting beads can be excited by any laser and detected by any filter.

Incubator (37°C, 10% CO_2, humidified)

Polystyrene round-bottom FACS tubes (5 mL)

Tissue culture equipment, including a laminar flow hood

METHOD

Perform Steps 1–3 under a laminar flow hood using aseptic techniques.

1. Seed 3×10^5 cells/well (HT1080^{indRIPK3} or L929) in 2 mL of DMEM supplemented with 10% FBS into six-well plates. Culture cells at 37°C under a 10% CO_2 atmosphere.

2. Once the cells have reached 50%–60% confluency, treat them as follows. Ensure sufficient wells are left untreated for use as controls, as described in Step 6.

   i. Treat wells of HT1080^{indRIPK3} cells for 12 h with the following combinations of reagents. Use three wells per condition.

      - 20 µM zVAD-fmk
      - 10 ng/mL TNF and 500 nM SM
      - 10 ng/mL TNF and 500 nM SM and 20 µM zVAD-fmk
      - 10 ng/mL TNF and 500 nM SM and 20 µM zVAD-fmk and 40 µM necrostatin-1

   ii. Treat wells of L929 cells for 2 h with the following combinations of reagents. Use three wells per condition.

      - 10 ng/mL TNF + 500 nM SM + 20 µM zVAD-fmk
      - 10 ng/mL TNF + 500 nM SM + 20 µM zVAD-fmk + 40 µM necrostatin-1

        Cells can be supplemented with other drugs or treatment combinations appropriate for the experimental goals. For treatments up to 12 h there is no need to change the medium.

3. Aspirate the supernatants into prelabeled FACS tubes.

   This allows the collection of detached dead cells.

4. Trypsinize cells and carefully resuspend them with their own medium (collected in Step 3). Return each cell suspension to its respective FACS tube.

5. Add 2.5 mM CaCl$_2$, 5 μL/mL Annexin V^{FITC}, 1.43 μM DAPI, 20 nM TMRM, and CountBright beads (25000/mL), and proceed to Step 7.

   CountBright absolute counting beads are a calibrated suspension of microspheres that are brightly fluorescent across a wide range of excitation and emission wavelengths, and contain a known concentration of microspheres. For absolute counts, a specific volume of microsphere suspension is added to a specific volume of sample, so that the ratio of sample volume to microsphere volume is known. The volume of sample analyzed can be calculated from the number of microspheres counted, and can be used to correct for cell loss during analysis and to determine cell concentration.

6. Prepare controls as follows.

   i. Leave one well of HT1080^{indRIPK3} cells and one well of L929 cells without any dye.

   ii. Treat HT1080^{indRIPK3} and L929 cells with each of the following single dyes. Use one well of cells per dye.

   - 5 μL/mL Annexin V^{FITC}
   - 1.43 μM DAPI
   - 20 nM TMRM
   - CountBright beads (25,000/mL)

7. Incubate the cells from Steps 5 and 6 for at least 20 min at 37°C in the incubator to allow the dyes to adhere.

   After this, cells can be kept on ice.

8. Analyze samples using an LSRII FACS system, or equivalent.

   i. Analyze the unstained control cells first and adjust the voltage and threshold.

   ii. Analyze the single-stained controls, and set up a compensation matrix.

   iii. Use the nonstained and single-stained control samples for gating.

   The exact parameters and conditions for sorting will vary among FACS systems and facilities and should be discussed with experts in your facility.

9. Use three technical replicates per experimental condition and three biologically independent experiments for statistical analysis using Student's t-test.

   For sample results, see Figure 1.

DISCUSSION

FACS provides a fast, objective, and quantitative method of recording of the number of dying cells in a population and is, therefore, routinely used to study cell death (Christensen et al. 2013). Annexin V is used as a probe to detect cells in which phosphatidylserine (PS) is exposed at the outer leaflet of the plasma membrane (Tait et al. 1989; Andree et al. 1990; Vermes et al. 1995). PS externalization occurs in early apoptotic cells, whereas living cells remain Annexin V negative. Although Annexin V can be used to detect apoptosis, it should be noted that necroptotic cells also become Annexin V-positive as Annexin V can bind to internal PS following cell rupture. However, when combined with PI the double-labeling procedure allows a further distinction of necrotic (Annexin V+/PI+) from early apoptotic (Annexin V+/PI−) cells (Vermes et al. 1995). Although cells that are Annexin V+/PI− can be considered to die by apoptosis at early time points, it should be noted that at late time points apoptotic cells can become Annexin V+/PI+. Therefore, it is important to combine this assay with a time-course study, and/or the use of specific inhibitors. Although the FACS method is a powerful approach to study cell death, it is not without its caveats. A particular problem is the fact that dying cells can disintegrate and the resulting cell debris cannot be captured by FACS analysis. To circumvent

Cite this protocol as *Cold Spring Harb Protoc*; doi:10.1101/pdb.prot087387

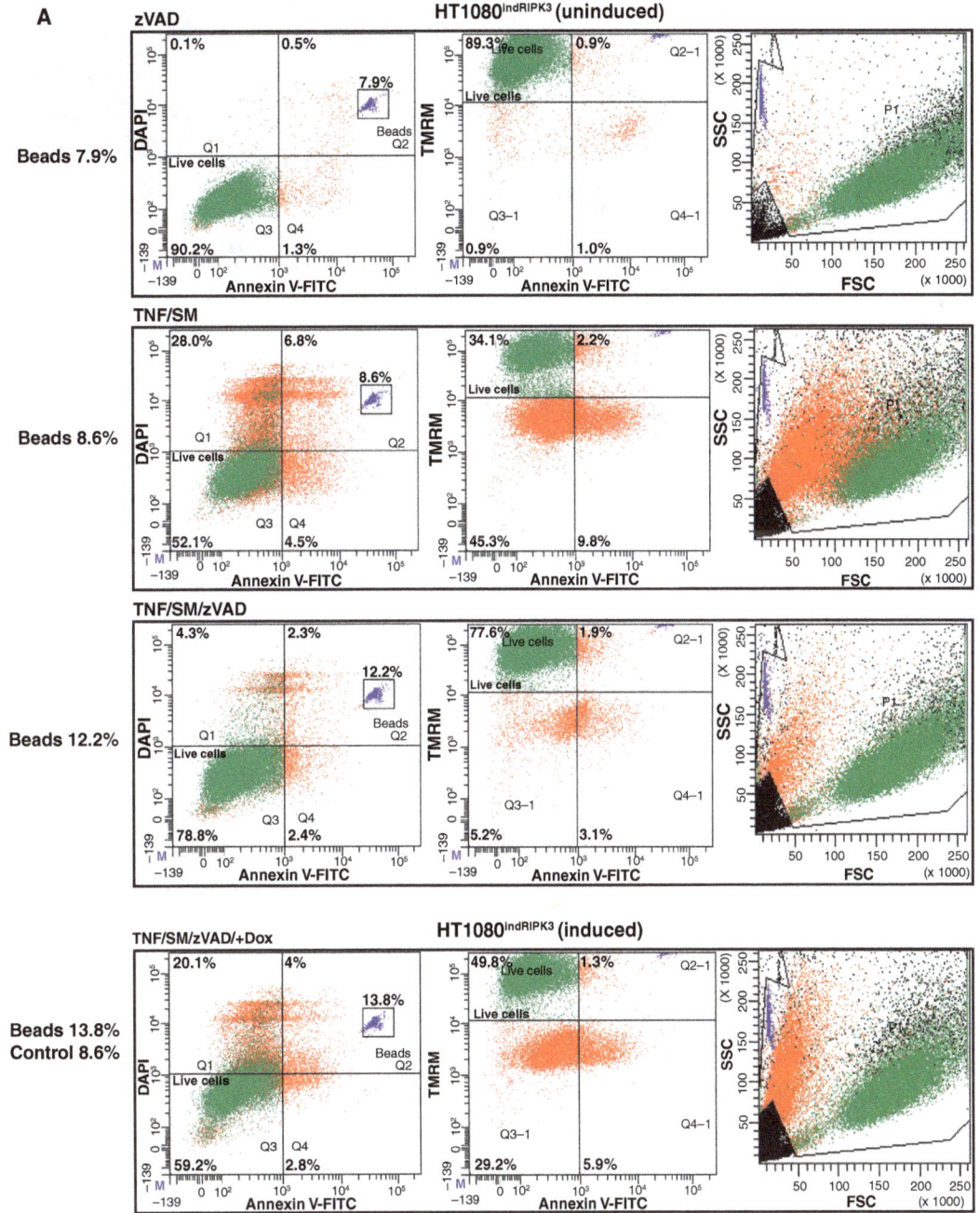

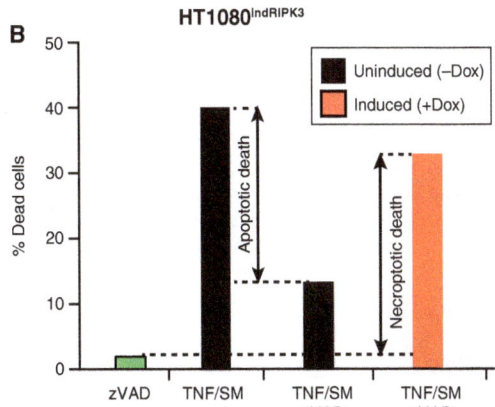

FIGURE 1. (*See following page for legend.*)

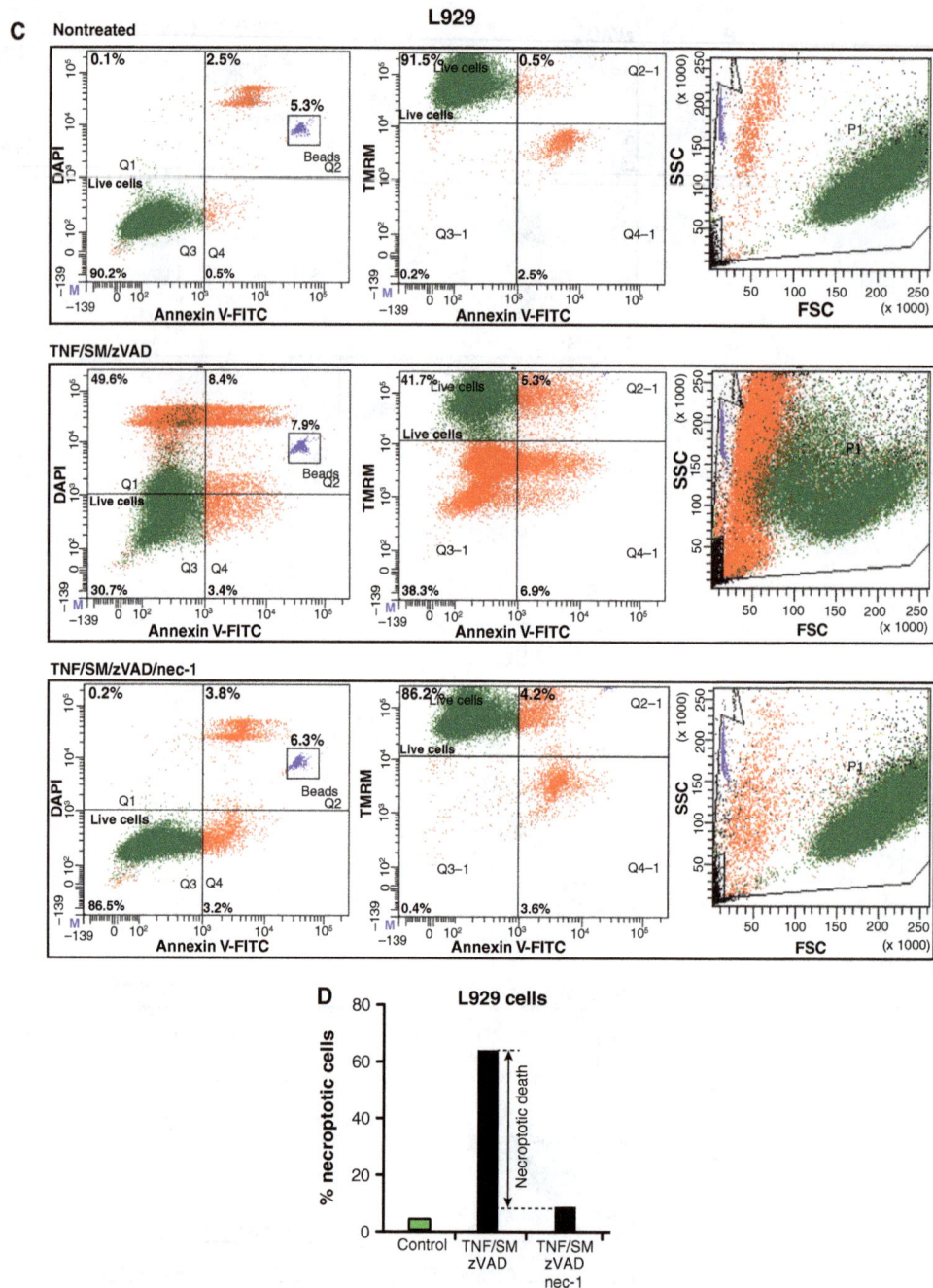

FIGURE 1. FACS analysis of cells undergoing necroptosis. (*A,B*) HT1080[indRIPK3] cells were grown in the absence or presence of 100 ng/mL Dox to induce RIPK3 expression. Cells were seeded into six-well plates and cultured for 24 h before being treated with 20 μM zVAD-fmk, 10 ng/mL TNF + 500 nM SM, or 10 ng/mL TNF + 500 nM SM + 20 μM zVAD-fmk. Prior to flow cytometry, cells were trypsinized and incubated with 5 μL/mL Annexin V[FITC], 1.43 μM DAPI, 20 nM TMRM and CountBright™ absolute counting beads (25,000/mL). (*B*) Graphic representation of cell death from (*A*) expressed as % of Annexin V[FITC] and DAPI-positive (dead) cells, and corrected for the loss of cells using the microsphere counting beads. (*C,D*) L929 cells were seeded into 6-well plates for 24 h before being treated for 2 h with control, 10 ng/mL TNF + 500 nM SM + 20 μM zVAD-fmk, or 10 ng/mL TNF + 500 nM SM + 20 μM zVAD-fmk + 40 μM necrostatin-1. Prior to flow cytometry, cells were trypsinized and incubated with 5 μL/mL Annexin V[FITC], 1.43 μM DAPI, 20 nM TMRM and beads (25,000/mL). (*D*) Graphic representation of cell death expressed as a % of Annexin V[FITC] and DAPI-positive (dead) cells, and corrected for the loss of cells using the microsphere counting beads.

Cite this protocol as *Cold Spring Harb Protoc*; doi:10.1101/pdb.prot087387

this problem, CountBright absolute counting beads can be used to determine how many cells were lost due to cell disintegration. When using microspheres, it is important to take into consideration that the untreated controls will continue to proliferate, which might affect the cell count. Hence, the respective treatment times should not be too long. Despite its advantages and ease of use, FACS is best complemented with time-lapse video microscopy as the combination of both methodologies more accurately captures the modality of cell death. This is especially true for some cell lines that stain poorly for Annexin V when undergoing *bona fide* apoptosis (Fadeel et al. 1999; Lee et al. 2013). For example, PS exposure reportedly does not occur in autophagy-deficient cells succumbing to apoptosis (Qu et al. 2007). Nevertheless, such cells clearly lose cell mass through membrane blebbing and nuclear shrinkage, and at later stages they undergo secondary necrosis and become PI positive.

RELATED TECHNIQUES

In addition to the method described here, there are many other FACS-based assays that can be used, such as evaluation of the percentage of sub-G0 cells after permeabilization followed by PI or 7-aminoactinomycin D (7-AAD) staining (Tenev et al. 2001; Zembruski et al. 2012).

ACKNOWLEDGMENTS

We thank Katiuscia Bianchi and Hugh Paterson for technical support and insightful discussions and suggestions. We acknowledge National Health Service funding to the National Institute for Health Research Biomedical Research Centre.

REFERENCES

Andree HA, Reutelingsperger CP, Hauptmann R, Hemker HC, Hermens WT, Willems GM. 1990. Binding of vascular anticoagulant α (VACα) to planar phospholipid bilayers. *J Biol Chem* **265:** 4923–4928.

Christensen ME, Jansen ES, Sanchez W, Waterhouse NJ. 2013. Flow cytometry based assays for the measurement of apoptosis-associated mitochondrial membrane depolarisation and cytochrome *c* release. *Methods* **61(2):** 138–145.

Fadeel B, Gleiss B, Hogstrand K, Chandra J, Wiedmer T, Sims PJ, Henter JI, Orrenius S, Samali A. 1999. Phosphatidylserine exposure during apoptosis is a cell-type-specific event and does not correlate with plasma membrane phospholipid scramblase expression. *Biochem Biophys Res Commun* **266:** 504–511.

Kapuscinski J. 1995. DAPI: A DNA-specific fluorescent probe. *Biotech Histochem* **70:** 220–233.

Lee SH, Meng XW, Flatten KS, Loegering DA, Kaufmann SH. 2013. Phosphatidylserine exposure during apoptosis reflects bidirectional trafficking between plasma membrane and cytoplasm. *Cell Death Differ* **20:** 64–76.

Qu X, Zou Z, Sun Q, Luby-Phelps K, Cheng P, Hogan RN, Gilpin C, Levine B. 2007. Autophagy gene-dependent clearance of apoptotic cells during embryonic development. *Cell* **128:** 931–946.

Tait JF, Gibson D, Fujikawa K. 1989. Phospholipid binding properties of human placental anticoagulant protein-I, a member of the lipocortin family. *J Biol Chem* **264:** 7944–7949.

Tenev T, Marani M, McNeish I, Lemoine NR. 2001. Pro-caspase-3 overexpression sensitises ovarian cancer cells to proteasome inhibitors. *Cell Death Differ* **8:** 256–264.

Vermes I, Haanen C, Steffens-Nakken H, Reutelingsperger C. 1995. A novel assay for apoptosis. Flow cytometric detection of phosphatidylserine expression on early apoptotic cells using fluorescein labelled Annexin V. *J Immunol Methods* **184:** 39–51.

Zembruski NC, Stache V, Haefeli WE, Weiss J. 2012. 7-Aminoactinomycin D for apoptosis staining in flow cytometry. *Anal Biochem* **429:** 79–81.

Time-Lapse Imaging of Cell Death

Fredrik Wallberg, Tencho Tenev, and Pascal Meier[1]

The Breakthrough Toby Robins Breast Cancer Research Centre, Institute of Cancer Research, Chester Beatty Laboratories, London SW3 6JB, United Kingdom

The best approach to distinguish between necrosis and apoptosis is time-lapse video microscopy. This technique enables a biological process to be photographed at regular intervals over a period, which may last from a few hours to several days, and can be applied to cells in culture or in vivo. We have established two time-lapse microscopy methods based on different ways of calculating cell death: semiautomated and automated. In the semiautomated approach, cell death can be visualized by staining with combinations of Alexa Fluor 647-conjugated Annexin V and Sytox Green (SG), or Annexin V[FITC] and Propidium iodide (PI). The automated method is similar except that all cells are labeled with dyes. This allows faster quantification of data. To this end Cell Tracker Green is used to label all cells at time zero in combination with PI and Alexa Fluor 647-conjugated Annexin V. Necrotic cell death is accompanied by either simultaneous labeling with Annexin V and PI or SG (double-positive), or direct PI or SG staining. Additionally, necrotic cells display characteristic morphology, such as cytoplasmic swelling. In contrast to necrosis where membrane permeabilization is an early event, cells that die by apoptosis lose their membrane permeability relatively late. Therefore, the time between Annexin V staining and PI or SG uptake (double-positive) can be used to distinguish necrosis from apoptosis. This protocol describes the analysis of cell death by time-lapse imaging of HT1080 and L929 cells stained with these dyes, but it can be readily adapted to other cell types of interest.

MATERIALS

It is essential that you consult the appropriate Material Safety Data Sheets and your institution's Environmental Health and Safety Office for proper handling of equipment and hazardous materials used in this protocol.

RECIPES: Please see the end of this protocol for recipes indicated by <R>. Additional recipes can be found online at http://cshprotocols.cshlp.org/site/recipes.

Reagents

Alexa Fluor 647-conjugated Annexin V (Life Technologies)

Cell Tracker Green (CTG) (Life Technologies, 12106-029)

Doxycycline (Dox)

Dulbecco's modified eagle medium (DMEM) without phenol red (Life Technologies 21063-029)

> *Add sodium pyruvate (Sigma-Aldrich S8636) to a final concentration of 1mM. Media without phenol red should be used for fluorescence imaging. Phenol red, which serves as a pH indicator dye, tends to produce a high background signal in fluorescent images, especially in the red emission channel.*

Fetal bovine serum (FBS)

---

[1]Correspondence: pascal.meier@icr.ac.uk

Cite this protocol as *Cold Spring Harb Protoc*; doi:10.1101/pdb.prot087395

HT1080^{indRIPK3} cells

Parental HT1080 cells (ATCC) were transduced with lentiviral particles carrying RIPK3 cDNA expressed under the control of a Dox regulatable promoter. For RIPK3 induction, cells should be cultured in media containing doxycycline (DOX) (100 ng/mL) for at least 1 wk before the experiment.

L929 cells

Necrostatin-1 (Sigma-Aldrich; 40 mM in dimethyl sulfoxide [DMSO])

Phosphate-buffered saline (PBS:172 mM NaCl, 1.8 mM KH$_2$PO$_4$, 10 mM Na$_2$HPO$_4$, 10 mM Na$_2$HPO$_4$, 3.3 mM KCl, pH 7.2)

Propidium iodide (PI)

Recombinant human TNF (Alexis; 100 µg/mL in DMEM)

SMAC mimetic (SM) (1 mM in DMSO)

Sytox Green (SG) (Life Technologies)

zVAD-fmk (20 mM in DMSO)

Equipment

Cell culture plates (12- and 24-well)

Glass bottomed dishes give sharper images in time-lapse microscopy, but it should be noted that some cell types may have difficulties attaching and spreading in glass dishes.

Cell Profiler image analysis software (http://www.cellprofiler.org/)

Cell profiler is an open source software that is designed to handle large sets of fluorescent 2D images.

Cylinders of premixed 10% CO$_2$

Incubator (37°C, 10% CO$_2$, humidified)

Microscope

Environmentally controlled Olympus IX70 inverted wide-field microscope equipped with LUCPLFLN (×20) or UPLFLN (×10) objectives, with motorized stage, shutters, and filter wheels, all driven via a Pro-Scan controller (Prior Scientific). Wide-field and fluorescent images are captured on a Hamamatsu Orca R2 camera using a 84000v2 DAPI/FITC/TRITC/Cy5 Quad filter set (Chroma). The system is controlled and coordinated by HClimageLive, version 4.2.6.14 (Hamamatsu) "Simple-PCI" software version 6.6 (Compix). Environmental control is maintained using a heated incubator jacket around the upper half of the microscope (Solent Scientific), permitting experiments to be performed at a constant temperature of 37°C and to be perfused with premixed 10% CO$_2$ in air. Of note, although a wide variety of inverted microscopes could equally be used to perform this technique, they should be equipped with comparable fluorescent optics and environmental control.

If available, use an adjustable fluorescent lamp. This allows the intensity of excitation light to be minimized to keep photo-toxicity low.

Tissue culture equipment, including a laminar flow hood

METHOD

Preparing the Cells

Perform Steps 1 and 2 under a laminar flow hood using aseptic techniques.

1. Seed cells (HT1080^{indRIPK3} or L929) in DMEM supplemented with 10% FBS into 12-well (1 × 10^5 cells and 2 mL of medium per well) or 24-well (0.5 × 10^5 cells and 1 mL of medium per well) plates. Culture cells at 37°C under a 10% CO$_2$ atmosphere. Once the cells have reached 50%–60% confluency, proceed to Step 2 for the automated method or Step 3 for the semiautomated method.

   It is advisable to use generous volumes of medium in time-lapse microscopy to minimize the effect of evaporation. It is inevitable that a certain amount of evaporation occurs during the course of the experiment. Typically, a 12-well plate should have 2 mL medium per well, and a 24-well plate should have 1 mL medium per well.

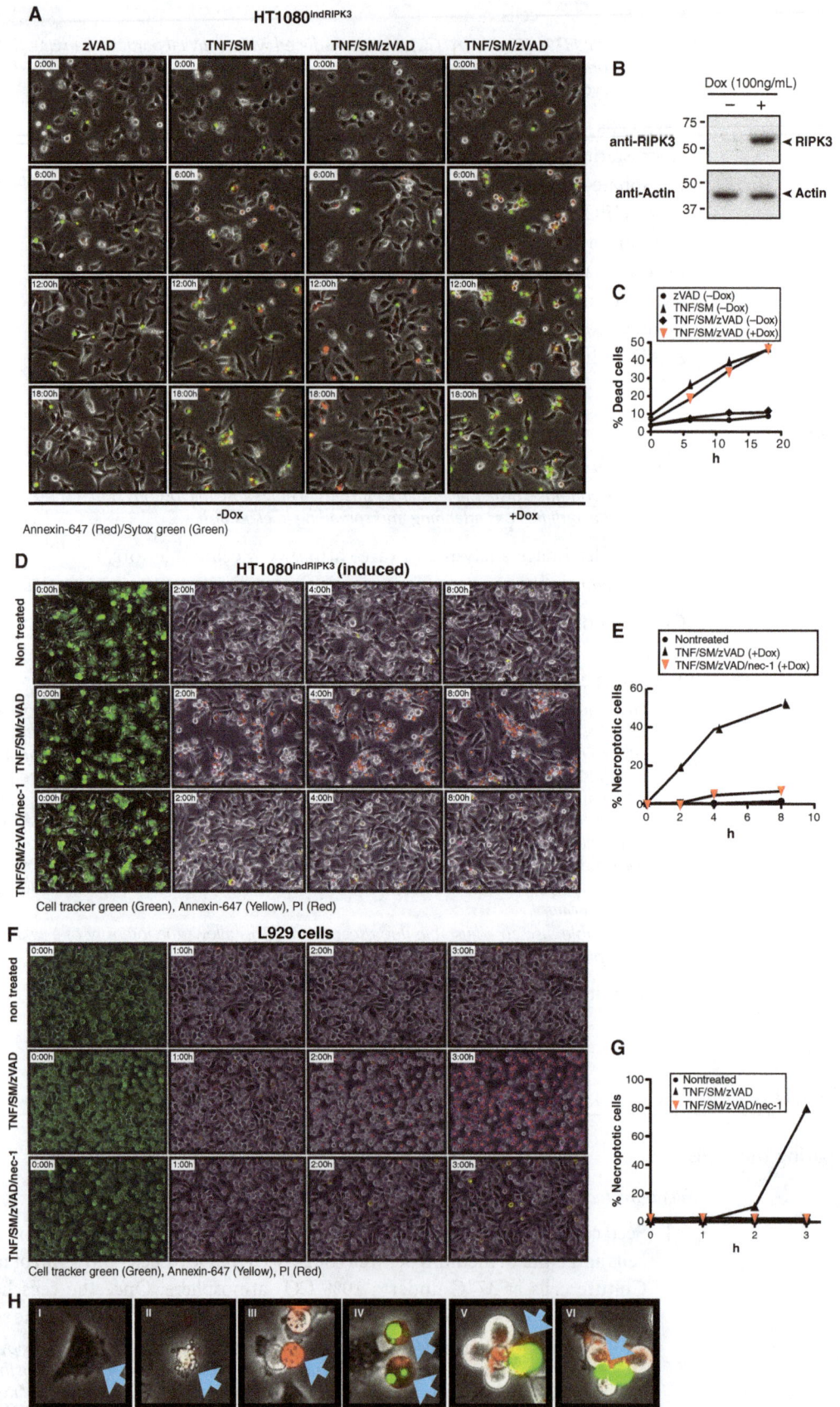

FIGURE 1. (*See following page for legend.*)

2. If using the automated method, proceed as follows:

   i. Add 0.5 µM CTG and for 30 min incubate at 37°C under a 10% CO_2 atmosphere.

      If dyes or other drugs used in these experiments are dissolved in DMSO or other organic solvents, great care should be taken when adding to cells because exposure of cells to solvents can kill cells locally. It is always important to add these solutions to the side of the well, followed by swirling. This also ensures an even distribution of the dye/compound.

   ii. Remove medium and rinse cells at least twice with PBS.

   iii. Add fresh, prewarmed phenol red-free DMEM supplemented with 10% FBS (2 mL per well of a 12-well and 1 mL per well of a 24-well plate).

      Cells can round up when medium is changed. Therefore, it is advisable to replace the medium well in advance of the experiment.

      This step is not required for the semiautomated method of quantification.

3. Treat HT1080^{indRIP3} cells with the following combinations of reagents. Use three wells per experimental condition.

   • Control (no reagents)

   • 20 µM zVAD-fmk

   • 10 ng/mL TNF + 500 nM SM

   • 10 ng/mL TNF + 500 nM SM + 20 µM zVAD-fmk

      The effects of these reagents are presented in Figure 1A–C. Cells can be supplemented with other drugs or treatment combinations appropriate for the experimental goals.

4. Treat Dox-treated HT1080^{indRIP3} cells and L929 cells with the following combinations of reagents:

   • Control (no reagents)

   • 10 ng/mL TNF + 500 nM SM + 20 µM zVAD-fmk

FIGURE 1. Time-lapse video microscopy of necroptotic and apoptotic cell death. (*A*) HT1080^{indRIPK3} cells were grown in the absence or presence of 100 ng/mL Dox to induce RIPK3 expression. Cells were seeded into 12-well plates and cultured for 24 h before being treated with 20 µM zVAD-fmk, 10 ng/mL TNF + 500 nM SM, or 10 ng/mL TNF + 500 nM SM + 20 µM zVAD-fmk. To monitor the type of cell death, cells were treated with Alexa Fluor 647-conjugated Annexin V (5 µL/mL) and 0.5 µM Sytox green (SG). Images of cells were taken at the indicated time points. (*B*) Western blot analysis of HT1080^{indRIPK3} cell lysates demonstrating inducible expression of RIPK3 following treatment with 100 ng/mL Dox. (*C*) Graphic representation of different types of cell death overtime using semiautomated quantification. One image every hour was exported as a single image and the number of Annexin V/SG-positive cells was counted using Cell profiler. The total number of cells at time zero (bright field) was calculated using ImageJ. Percentage of necroptotic cells was calculated by dividing the number of Annexin V/SG-positive cells by the total number of cells, and multiplying by 100. More than one field was used to calculate % of cell death. (*D*) HT1080^{indRIPK3} cells were grown in the presence of 100 ng/mL Dox to induce RIPK3 expression. Cells were seeded into 12-well plates and cultured for 24 h. Before time-lapse microscopy, cells were treated with 0.5 µM Cell Tracker Green (CTG) for 30 min. To remove all nonbound dye from the wells, medium was changed and cells were washed twice with PBS before adding phenol red-free medium and being treated with 10 ng/mL TNF + 500 nM SM + 20 µM zVAD-fmk or 10 ng/mL TNF + 500 nM SM + 20 µM zVAD-fmk + 40 µM necrostatin-1. To monitor the type of cell death, Alexa Fluor 647-conjugated Annexin V (5 µL/mL) and 1.5 µM PI were added to the medium. Images of cells were taken at the indicated time points. (*E*) Graphic representation of the automated quantification. CTG positive cells (total number of cells) and Alexa Fluor 647-conjugated Annexin V/PI positive cells (dead cells) were counted in Cell profiler as separate objects. The total number of Alexa Fluor 647-conjugated Annexin V/PI-positive cells was divided by the total number of cells (CTG-positive cells), and multiplied by 100 to obtain a percentage of dead cells in an image. (*F,G*) L929 cells were seeded in a 12-well plate and left for 24 h. Cells were treated as in *D* and the extent of necroptosis was calculated using the automated quantification method as in (*E*). Note, necroptotic cell death was determined by the presence of RIPK3 expression and the fact that z-VAD-fmk was not able to block cell death. Necroptosis was accompanied by simultaneous uptake of Alexa Fluor 647-conjugated Annexin V/SG or direct SG staining. Additionally, cells were devoid of blebbing and lacked nuclear condensation. (*H*) Morphology and Alexa Fluor 647-Annexin V/SG uptake of HT1080-RIPK3 cells visualized by time-lapse microscopy. (I) Healthy cell; (II) unstained dead blebbing cell; (III) dead cell, positive for Alexa Fluor 647-Annexin V only; (IV) dead cells, positive for Alexa Fluor 647-Annexin V and SG; (V) dead cell, positive for Alexa Fluor 647-Annexin V and SG; (VI) dead cell, positive for SG only.

- 10 ng/mL TNF + 500 nM SM + 20 μM zVAD-fmk + 40 μM necrostatin-1

   The effects of these reagents are presented in Figure 1D–G.

5. Add 2.5 mM $CaCl_2$, 5 μL/mL Alexa Fluor 647-conjugated Annexin V, and 0.5 μM SG (semi-automated counting method) or 1.5 μM PI (automated counting method). Incubate for at least 20 min at 37°C to allow the dyes to adhere. After this, image the cells as described below.

   When using Annexin V it is important to include 2.5 mM $CaCl_2$ in the culture medium because Annexin V only binds to PS in the presence of $CaCl_2$. We have not encountered any problems when culturing cells in medium containing 2.5 mM $CaCl_2$.

   See Troubleshooting.

Performing Time-Lapse Microscopy

6. Disable any Internet or network connections before starting a time-lapse experiment.

   Active network connections can cause the computer to restart automatically (e.g., while installing updates) and thereby terminate the experiment.

7. Turn on CO_2. Set the flow rate to a level so that it is possible to see individual bubbles going through the water bottle used to humidify the gas.

   We use cylinders of premixed 10% CO_2 in air for gassing the culture vessels on the microscope stage. This avoids the requirement for expensive gas-mixing apparatus and assures that cells cannot be exposed to too much CO_2. It is important to humidify the CO_2/air mixture before it reaches the culture vessel to minimize problems due to evaporation of medium. Humidify the CO_2/air mixture by bubbling it through a water bottle that is set up inside the microscope incubator jacket. This ensures that the humidified gas mixture reaches the cells at the correct temperature, thus avoiding cooling of the specimen and producing unwanted condensation.

8. Ensure that the temperature of the environmentally controlled microscope is at or slightly below 37°C.

9. Place the plate on the microscope stage. Let the plate sit for at least 1 h to stabilize the temperature and pH before data recording.

   This period of equilibration is required to minimize problems due to focus drift.

   See Troubleshooting.

10. Choose the objective and set up the condenser, phase rings, and the correction collar of the objective.

    Adjustable correction collars, often fitted to long-working distance objectives, allow the user to compensate for focusing through different thickness (e.g., plastic or glass) in the base of the culture vessel. As a general rule, the correction collar should be set to 1.1 mm for focusing through standard tissue culture plastic dishes, and 0.17 mm for glass-bottomed plates. However, these are only approximations, and the user may wish to establish the optimal settings for any particular culture vessel.

11. Start the software—HClimageLive version 4.2.6.14 (Hamamatsu) Simple PC1. Select the acquisition parameters. Click "Capture" and adjust the settings on the live image. Ensure that the software or hardware autofocus system is performing adequately (if in use).

    Always choose imaging positions that are close to the center of the well. Bright-field image illumination may be uneven in areas close to the edge of wells.

    To compensate for chromatic aberration in the imaging system, it may be necessary to introduce a focus offset into one or more channels to ensure that all fluorescent and transmitted-light images are parfocal. Sharp focus is first established on the bright-field channel. A focus offset can then be introduced by means of the acquisition software into each fluorescence channel in turn until complete parfocality is achieved for all channels. Set up each of the multiple stage positions together with the number and interval of the time points to be recorded.

    See Troubleshooting.

12. Start the time-lapse experiment.

    To avoid photo-toxicity keep the exposure time to a minimum. If the signal is weak, it is less damaging to the cells to use high camera gain, rather than high fluorescence excitation. If possible, avoid using maximum gain, because this usually produces poor image quality.

Cite this protocol as *Cold Spring Harb Protoc*; doi:10.1101/pdb.prot087395

It is important to optimize the frequency of image capture to avoid any dye-related photo-toxicity. We routinely use 10 min intervals. However, for some conditions this can be too frequent. When using dyes that stain the entire cell, such as CTG, lower imaging frequencies should be chosen to avoid photo-toxicity.

If a compound or dye needs to be added during the course of a time-lapse experiment, the dye/compound should be added to the well in a large volume to prevent toxic effects of any solvent, e.g., DMSO, used to dissolve the compound. Do not swirl the plate to mix. It is important to ensure that final volumes of all wells are equal.

Analyzing the Data

Data Analysis for Automated Quantification

13. Batch export single parameter images from the time-lapse movies to the *tiff* format.

14. Load images into Cell Profiler using the module LoadImages. (This should be the first module in the Cell Profiler pipeline.)

    Adjust the Cell Profiler pipeline in test mode using a few images. This is a much faster way to anticipate what changes need to be made to the pipeline compared with working on all images at once.

    It is easier to quantify the results as a series of single images rather than in movie format. This enables single images to be analyzed/processed using separate software.

15. Transform the color images to grayscale with the module ColorToGray. Use one module for cell tracker images and one for PI images.

16. Use the module IdentifyPrimaryObjects to identify and count the cells in every cell tracker image. Add one more IdentifyPrimaryObjects module to the pipeline and identify and count all dead cells in every PI image.

17. Use the module CalculateMath to divide the number of dead cells by the number of cells and multiply by 100.

18. Export data to Excel with the ExportToSpreadsheet module.

Data Analysis for Semi-Automated Quantification

19. Count the total number of cells at time zero in the bright field image using the ImageJ program.

20. Transform the color images to grayscale with the module ColorToGray.

21. Use the module IdentifyPrimaryObjects to identify and count all dead cells in every PI image.

22. In Excel, divide the number of dead cells by the total number of cells obtained in Step 19 and multiply by 100.

TROUBLESHOOTING

Problem (Step 5): The dyes are cytotoxic.

Solution: Use as many controls as possible and assess whether dyes might be cytotoxic. Always include unstained and single-stained controls. We have not seen any toxicity effects when using PI. However, we did notice that treatment with CTG showed some photo-toxicity in a concentration dependent manner. Therefore, it is important to carefully titrate the concentration of dyes, particularly when using CTG.

Problem (Step 9): The growth rate of cells is affected by the microscope environmental chamber.

Solution: It is important to ensure that the growth rate of cells is not affected by the environmental chamber of the microscope; therefore, it is recommended to compare the growth curve of cells growing in the environmentally controlled setting of the time-lapse microscope with one of cells growing in an incubator.

Problem (Step 9): The microscope stage fluctuates in temperature.
Solution: It is vital to ensure that the temperature of the environmentally controlled microscope does not fluctuate. As such, good room ventilation is essential to ensure stable room temperature. This might be problematic in small rooms due to light sources emitting a lot of heat. A change in room temperature often causes a drift in focus, and over a longer time-lapse experiment this can cause significant problems.

Problem (Step 9): The microscope vibrates.
Solution: Minimize vibrations by using antivibration tables.

Problem (Step 11): The fluorescent signal from the specimen is low.
Solution: Software "binning" can be used to enhance signal detection. Binning combines the signal from blocks of neighboring pixel elements in the camera sensor array to produce higher sensitivity. While software binning enhances signal detection, it will reduce image resolution. "Binning 2×2" thus increases the camera sensitivity fourfold, but also decreases the image resolution by a factor of 4.

Problem (Step 11): Difficulty in choosing settings that is appropriate for all wells.
Solution: Depending on the capture software used, it may not be possible to choose individual settings for individual images. If settings between wells need to be changed, one either needs to make a compromise, or take two images with different settings per well. For example, green fluorescent protein (GFP) can have very different intensities in different wells. To compensate for this, it is possible to take two GFP images with different exposure settings at every position. During analysis of the data the user can decide which exposure gives the best image on a well-by-well basis. The user should keep in mind; however, that this procedure is likely to increase the level of photo-toxicity.

DISCUSSION

Distinguishing necroptosis from apoptosis can be challenging because apoptotic cells can undergo secondary necrosis, in which case Annexin V-binding can occur in cells that have lost their membrane integrity due to membrane rupture following secondary necrosis. Under such conditions, the Annexin V probe detects internal PS. The combination of Annexin V with noncell permeable DNA stains, such as PI or SG allows the accurate distinction between necroptosis from apoptosis. PI and SG readily enter dead cells with compromised membrane integrity and subsequently show fluorescence enhancement of hundreds of fold upon binding to DNA. Annexin V^{FITC} and PI or Alexa Fluor 647-conjugated Annexin V and SG have different excitation spectrums, which allow them to be used in combination. Cells that die by necrosis immediately become Annexin V/PI or Annexin/SG double-positive, while cells that die by apoptosis first become Annexin V positive, and only after a considerable lag period turn Annexin V/PI or Annexin/SG double-positive or only PI or SG positive (which indicates secondary necrosis). Of note, not all cell types stain well for Annexin V when undergoing *bona fide* apoptosis (Fadeel et al. 1999; Lee et al. 2013).

The use of pharmacological inhibitors, such as zVAD-fmk or necrostatin-1, can provide clues as to the cell death modality, although it must be noted that necrostatin-1-mediated inhibition of RIPK1 can also affect apoptosis (Tenev et al. 2011). Therefore, the combination of morphological observation combined with the time between Annexin V staining and PI or SG uptake (double-positive staining) enables unambiguous distinction between necrosis and apoptosis. At present, time-lapse video microscopy represents one of the best approaches to study different cell death modalities.

We have observed that fluorescently labeled proteins, such as GFP, frequently display a heterogeneous intensity profile. This may exceed the dynamic range of the camera, and may complicate the optimization of the exposure time, gain, etc. For example, if low settings are chosen to capture the brightest cells, signals from dimly labeled cells may be lost, while if settings are chosen to capture dimly labeled cells, it might not be possible to visualize internal details of cells that are very bright. This problem could be avoided by FACS sorting cells before imaging. For multicolor fluorescence imaging it is important to choose dye combinations with minimum spectral overlap. Single stained controls can be examined to determine if there are spectral overlaps between dyes.

ACKNOWLEDGMENTS

We thank Katiuscia Bianchi and Hugh Paterson for technical support and insightful discussions and suggestions. We acknowledge National Health Service funding to the National Institute for Health Research Biomedical Research Centre.

REFERENCES

Fadeel B, Gleiss B, Hogstrand K, Chandra J, Wiedmer T, Sims PJ, Henter JI, Orrenius S, Samali A. 1999. Phosphatidylserine exposure during apoptosis is a cell-type-specific event and does not correlate with plasma membrane phospholipid scramblase expression. *Biochem Biophys Res Commun* 266: 504–511.

Lee SH, Meng XW, Flatten KS, Loegering DA, Kaufmann SH. 2013. Phosphatidylserine exposure during apoptosis reflects bidirectional trafficking between plasma membrane and cytoplasm. *Cell Death Differ* 20: 64–76.

Tenev T, Bianchi K, Darding M, Broemer M, Langlais C, Wallberg F, Zachariou A, Lopez J, MacFarlane M, Cain K, et al. 2011. The Ripoptosome, a signaling platform that assembles in response to genotoxic stress and loss of IAPs. *Mol Cell* 43: 432–448.

Ripoptosome Analysis by Caspase-8 Coimmunoprecipitation

Maria Feoktistova,[1] Peter Geserick,[1] and Martin Leverkus[1,2,3]

[1]Section of Molecular Dermatology, Department of Dermatology, Venereology, and Allergology, Medical Faculty Mannheim, University Heidelberg, Heidelberg 68167, Germany; [2]Department of Dermatology & Allergology, University Hospital of RWTH Aachen University, 52074 Aachen, Germany

The biochemical signaling of cell death pathways is executed at a number of different intracellular and/or membrane-bound high-molecular mass complexes. It is crucial to be able to detect the formation, differences in assembly, and differential composition of such complexes to understand their contribution to the execution phase of apoptotic or necroptotic cell death. We describe here the use of caspase-8 coimmunoprecipitation in the spontaneously transformed keratinocyte cell line, HaCaT, to study the formation and composition of the Ripoptosome, a complex that is based on the serine–threonine kinase receptor-interacting protein 1 (RIPK1). However, the method can be adapted for use with other antibodies and cell lines. This protocol determines whether cells form the Ripoptosome complex, which is important for both apoptosis and necroptosis execution. Caspase-8 is an indispensible Ripoptosome component; therefore, caspase-8 antibodies are used to pull down the respective complex. However, the method cannot discriminate whether this complex triggers apoptosis (through the RIPK1 → FADD → caspase-8 activation pathway), necroptosis (through the RIPK1 → RIPK3 → MLKL activation pathway) or nondeath signaling. The actual signaling output (death or nondeath signaling) depends on the stoichiometry of the respective molecules as well as on the activity of FLIP, caspase-8, or other factors.

MATERIALS

It is essential that you consult the appropriate Material Safety Data Sheets and your institution's Environmental Health and Safety Office for proper handling of equipment and hazardous materials used in this protocol.

RECIPES: Please see the end of this protocol for recipes indicated by <R>. Additional recipes can be found online at http://cshprotocols.cshlp.org/site/recipes.

Reagents

Anti-caspase-8 antibody (Santa Cruz clone C-20)

Cell culture medium

Use the appropriate cell culture medium according to the cell type/line being cultured. For HaCaT cells, we use DMEM (Gibco/Life Technologies) containing 10% fetal calf serum (FCS).

Cells of interest

In this protocol, we use the spontaneously transformed keratinocyte cell line, HaCaT (Boukamp et al. 1988).

DISC lysis buffer <R>

IAP antagonist (e.g., GT 12911e from Tetralogics)

---

[3]Correspondence: mleverkus@ukaachen.de

Cite this protocol as *Cold Spring Harb Protoc*; doi:10.1101/pdb.prot087403

Laemmli loading buffer (5×) <R>
PBS for caspase-8 IP (10×) <R>

Additional reagents (e.g., trypsin) may be necessary for harvesting cells. See Step 2.

Protein G beads (Roche)
Quick Start Bradford protein assay (Bio-Rad) or equivalent kit for measuring protein concentration
Western blot analysis reagents

Equipment

Cell scraper
Centrifuge 5430R (Eppendorf)
Dry heating block
Ice bucket
Incubator (37°C, 5% CO_2, humidified)
Microcentrifuge tubes (1.5- and 2.0-mL)
Rocker platform (optional; see Step 2)
Syringe needles (0.4 × 19 mm)
T-175 cell culture flask
Test tube rotator
Tissue culture equipment, including a laminar flow hood
Vacuum pump
Western blot analysis apparatus

METHOD

Please refer to Figure 1, which depicts the work flow of the protocol.

Preparation of Protein Samples

1. To induce Ripoptosome formation, treat 1×10^7 cells (one T175 flask of HaCaT or HeLa cells at ∼70%–80% confluency) with 100 nM of IAP antagonist for 4 h at 37°C in the cell culture medium recommended for the cell type of choice.

   Each experiment must have a negative control of unstimulated cells in which caspase-8-IP does not co-purify other proteins. This sample defines the background signal. Treatment with zVAD-fmk greatly stabilizes the Ripoptosome complex (Geserick et al. 2009); therefore, it is recommended to stimulate cells in the presence of 10 μM of zVAD-fmk, unless it is not necessary to assess the cleavage of caspase substrates in the complex (in this case, the negative control sample should be treated with zVAD-fmk alone).

   Other conditions can also be used for Ripoptosome formation, e.g., inducible RIPK1 overexpression (Feok-tistova et al. 2012). We recommend conducting a time-course study and dose–response curve for your cells and stimuli of interest. First, take the highest concentration of the stimulus used in prior studies and perform a time-course study; we recommend at 2, 4, and 6 h. After defining the time point where the Ripoptosome is robustly formed, titrate down the stimulus, to identify the best concentration.

2. Collect cells via one of the following methods:

   - Trypsinize cells and collect them via centrifugation at 400*g* for 5 min at room temperature. Resuspend cell pellet in 1 mL of PBS, and transfer to a 2 mL microcentrifuge tube. Centrifuge at 400*g* for 5 min at room temperature, and discard supernatant. Add 1 mL of DISC lysis buffer.

   - Wash cells three to four times with ice-cold PBS (use ∼20 mL/flask). For each wash, tilt the plate or flask gently back and forth several times by hand and remove the PBS. After the last wash, leave the flasks tilted on the table for 1–2 min to let the residual PBS collect, and remove it with a pipette. Add 1 mL of DISC lysis buffer to the flask. Spread the DISC lysis buffer

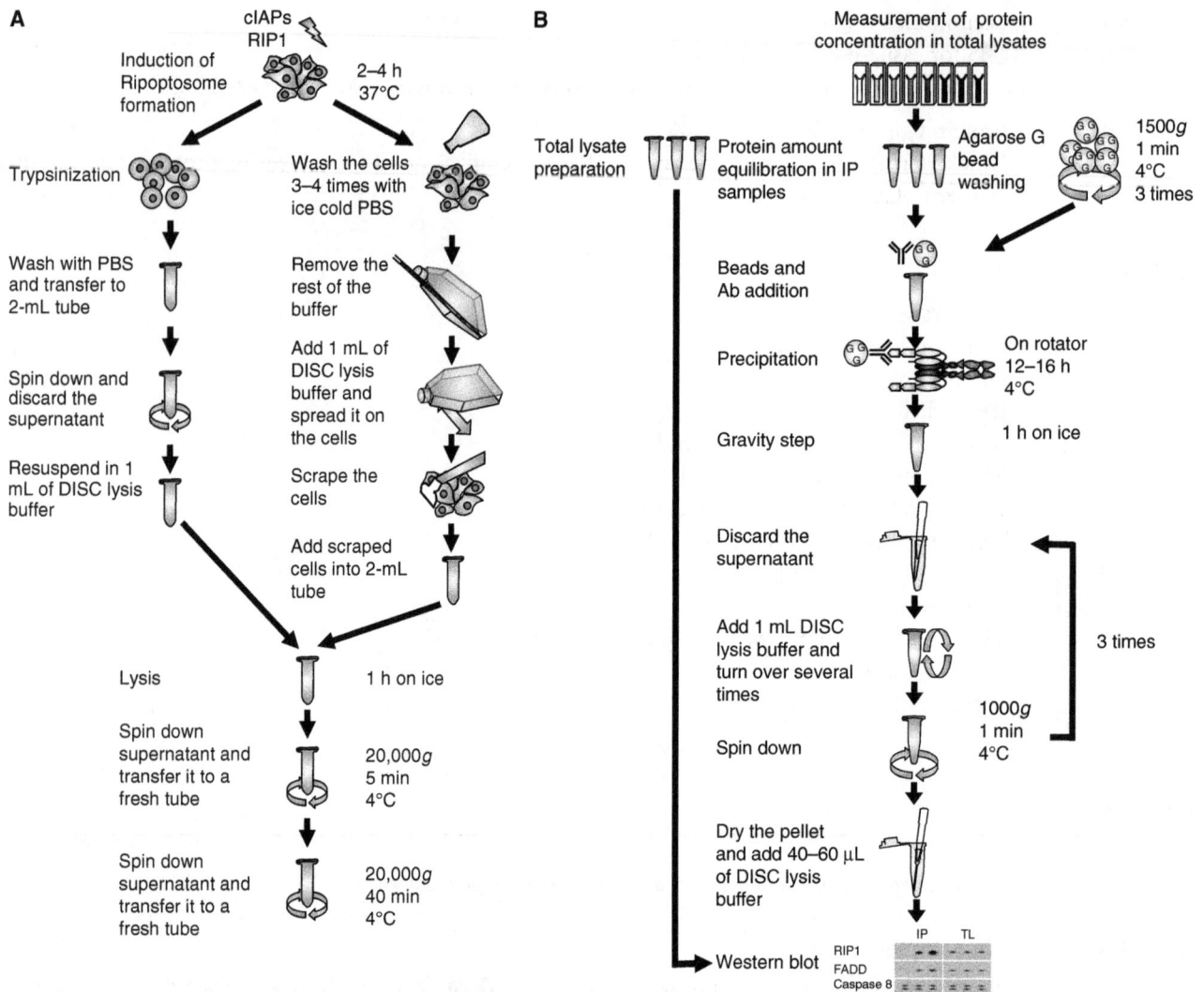

FIGURE 1. Ripoptosome analysis. (*A*) Schematic work flow for lysate preparation. (*B*) Schematic work flow for caspase-8 coimmunoprecipitation.

over the cells by gentle shaking. Harvest cells with a cell scraper and transfer to a 2 mL micro-centrifuge tube.

We prefer to use trypsinization rather than scraping because this gives higher protein yields.

3. Incubate cells in DISC lysis buffer on ice for 1 h.

4. Centrifuge the cell debris at 20,000*g* for 5 min at 4°C. Transfer the supernatant to a fresh 2 mL microcentrifuge tube. Discard the pellet.

5. Centrifuge the supernatant at 20,000*g* for 40 min at 4°C. Keep the resulting supernatant for use in Step 6. Discard the pellet, which contains cell debris.

6. Measure the protein concentration of the lysates using a Bradford protein assay.

The protein concentration is usually very high, which may make it necessary to dilute a small aliquot of the lysate for accurate measurement.

7. Prepare total lysate samples for western blot analysis and store them at −20°C until use in Step 20.

The exact amount of protein loaded per lane will depend on the actual size of your lanes; we usually load 5 µg per lane.

Store the remaining total lysate samples at −20°C for use in Step 8.

Immunoprecipitation of Caspase-8

8. Prepare samples for immunoprecipitation (IP) in 1.5 mL microcentrifuge tubes. Use 4 mg of protein in a volume of 1 mL. Dilute the samples in DISC lysis buffer.

    If your samples have a protein concentration of <4 mg/mL, then use the maximum amount of the sample with the lowest protein concentration and use the same amount of protein for all your samples. Use DISC lysis buffer to adjust the volume of your samples. Do not to exceed 1.2 mL in volume.

    Store the remaining total lysate samples at −20°C.

9. Add 1 μg of anticaspase-8 antibody per 4 mg of protein lysate to each tube.

10. Prepare the Protein G agarose beads.

    i. Collect beads (40 μL of beads per sample, or a minimum of 10 μL beads per 1 mg protein) by centrifugation at 1500g for 1 min at 4°C. Discard the supernatant.

    ii. Add DISC lysis buffer. Gently invert the tube several times to resuspend the beads in the buffer. Collect beads by centrifugation at 1500g for 1 min at 4°C, and discard the supernatant.

        The amount of DISC lysis buffer is dependent on the initial bead volume. For each 40 μL of initial bead volume, add 100 μL of DISC lysis buffer.

    iii. Repeat Step 10.ii two more times.

    iv. Resuspend the beads in DISC lysis buffer by several gentle inversions of the tube.

        These washing steps are required because the beads are stored in ethanol and therefore require washing before being added to samples. The beads become very sticky during the washing steps; therefore, do not touch them with a pipette tip and do not resuspend by pipetting.

11. Add 100 μL of bead suspension to each IP sample.

    Remember that the beads are sticky and are quickly pelleted by gravity; therefore, resuspend the beads immediately before aliquoting by gently inverting the tube several times. Do not mix the beads with the IP sample by pipetting otherwise beads will accumulate in the pipette tip which reduces the total amount of beads in the sample.

12. Incubate overnight at 4°C on a tube rotator at ∼10 rpm.

    If IP incubation is longer than 16 h the background tends to increase.

13. Place the IP samples on ice and leave for 1 h for the beads to sediment by gravity.

14. Discard the supernatant by pipetting using a 1 mL tip, then a 200 μL tip.

    This allows removal of most of the fluid without disturbing the sedimented beads.

15. Wash the beads.

    i. Add 1 mL of DISC lysis buffer. Mix by gently inverting the tube several times.
        Do not mix by pipetting.

    ii. Centrifuge at 400g for 1 min at 4°C, and discard the supernatant.

    iii. Repeat Steps 15.i–15.ii two more times.

16. Dry the pellet.

    i. Remove any residual supernatant from Step 15 with a syringe needle (0.4 mm) connected to a vacuum pump.

    ii. Centrifuge at 400g for 1 min at 4°C.

    iii. Repeat Step 16.i once.

17. Add 60 μL of DISC lysis buffer and 15 μL of 5× Laemmli loading buffer to each pellet.

18. Dissociate the proteins from the beads by heating the samples for 10 min at 95°C.

19. Centrifuge for 1 min at 5000g.

    Store the samples at −20°C until use in Step 20. Note that although the immunoprecipitated proteins will dissociate from the beads when heated in Step 18, the beads and proteins are stored together in one tube.

When the sample is loaded onto a gel in Step 20, the beads will not be transferred to the gel because the pipette tips are too small to take up the beads.

Western Blotting

20. Analyze IP and total lysate samples by western blot analysis.

We recommend loading the IP and total lysate samples on the same gel. Always keep an empty lane between IP and total lysate samples because the total lysate signal is usually much stronger than those from IP samples (except caspase-8), which might interfere with the detection of IP sample signals. Use the caspase-8 signal as a control for equal loading. The signal for caspase-8 should be much stronger in the IP than in the total lysate.

See Troubleshooting.

TROUBLESHOOTING

Problem (Step 20): There is no signal on the western blot.
Solution: Proceed as follows.

1. Check the signal in total lysate samples. If there is no signal, there is a problem with the western blotting procedure.

2. Check the caspase-8 signal in the IP.

   i. If the signal is absent or unequal among samples, there is a problem with the IP. Check that the amount of input protein is the same in all samples.

   ii. If the signal is equal among samples but lower than those of the total cell lysates, check that the correct amount of antibody was added to samples. Also check that the amount of the protein in the samples was not too low.

   iii. If the caspase-8 signal is satisfactory, but associated proteins cannot be detected, there may be a complex formation or stability problem; check Ripoptosome stimulation conditions. Remember also that the IP signal for some proteins can be very weak; therefore, use highly sensitive chemoluminescence reagents for detection (e.g., ECL+) and, if necessary, make very long exposures.

DISCUSSION

The number of characterized cell death pathways has increased enormously over the past decade. The biochemical signaling of these pathways is executed at a number of different intracellular and/or membrane-bound high-molecular mass complexes, which include the death-inducing signaling complex (DISC) (Lavrik et al. 2005), TNF complex II (Micheau and Tschopp 2003), the RIG-I complex (Rajput et al. 2011), the PIDDosome (Tinel and Tschopp 2004), the necrosome (Vandenabeele et al. 2010), and the Ripoptosome (Feoktistova et al. 2011; Tenev et al. 2011).

The method described here allows the detection of the Ripoptosome but also other caspase-8-RIPK1 complexes, which are formed following certain death stimuli (e.g., complex II after TNF stimulation). It also allows the degree of complex formation following various treatments to be compared, and enables complex compositions to be studied. In addition, immunoprecipitations using caspase-8 as a bait allow the differentiation of the Ripoptosome from other RIPKI- containing complexes such as the necrosome. Although described for caspase-8 IP, the method can be adapted to target any other bait for which good antibodies exist. Each new antibody, however, requires thorough testing, best performed in parallel with the caspase-8 antibody. It should be noted that antibodies have the potential to interfere with complex stability and may lead to disassembly of the complex, which preclude analysis of the native complex.

Although the procedure is described for adherent cells, it is also applicable to nonadherent cells. One of the method's limitations is that it cannot distinguish among different caspase-8 containing complexes, which include the death receptor signaling complex (DISC) and TNF complex II, as well as the Ripoptosome, which in principle can form in parallel. In our experiments we excluded the possibility that the RIPK1-caspase-8 complex (here Ripoptosome) is the TNF complex II because we detect complex formation (Ripoptosome) in the absence of TNF signaling (Feoktistova et al. 2011). However, if more than one RIPK1-caspase-8 containing complex is forming under a given condition, we would be unable to differentiate between them. Therefore, we recommend using at least two different baits to define the formation of a particular cell death platform, or to check for molecules exclusively involved in specific signaling pathways. A prominent example is the absence of caspase-8 and FADD in the TNF complex I, whereas RIPK1 exclusively associates with caspase-8 in TNF complex II. If the experimental question is aimed at discriminating different complexes formed in parallel, then a DISC IP (detecting receptor-associated proteins) should be performed first. The cell lysate can then be investigated for additional intracellular complexes that do not associate with receptor proteins, such as the Ripoptosome and the TNF complex II.

RECIPES

DISC Lysis Buffer

30 mM Tris–HCl (pH 7.0)
120 mM NaCl
10% Glycerol
1% Triton X-100
Complete Protease Inhibitor Cocktail Tablets (Roche)

Prepare 50-mL aliquots of the buffer and store them at −20°C. The buffer is stable for at least 1 yr under these conditions.

Laemmli Loading Buffer (5×, pH 6.8)

2.5 mL 2 M Tris-HCL (pH 6.8)
2 g sodium dodecyl sulfate
100 mg bromophenol blue
10 mL glycerol
1.542 g DTT

Add up to 20 mL of dH_2O. Generate 100-µL aliquots of the buffer, and store them at −20°C for up to 3 mo.

PBS for Caspase-8 IP (10×, pH 7.4)

64.8 mM Na_2HPO_4
1.37 M NaCl
26.9 mM KCl
14.7 mM $KHPO_4$

Adjust the pH to 7.4 with HCl. Sterilize by autoclaving. Store the solution at room temperature, and dilute to 1× when required.

ACKNOWLEDGMENTS

We thank all current and previous members of the Section of Molecular Dermatology, Medical Faculty of Mannheim, University of Heidelberg and within the RWTH Aachen for their discussions and suggestions. The work in the laboratory of M.L. was supported by grants from the Deutsche Forschungsgemeinschaft (Le953/6-1, 8-1, RTG 2099/1 (Project 9 and 10), and the Mildred-Scheel-Stiftung (Projekt 109891).

REFERENCES

Boukamp P, Petrussevska RT, Breitkreutz D, Hornung J, Markham A, Fusenig NE. 1988. Normal keratinization in a spontaneously immortalized aneuploid human keratinocyte cell line. *J Cell Biol* **106:** 761–771.

Feoktistova M, Geserick P, Kellert B, Dimitrova DP, Langlais C, Hupe M, Cain K, MacFarlane M, Hacker G, Leverkus M. 2011. cIAPs block Ripoptosome formation, a RIP1/caspase-8 containing intracellular cell death complex differentially regulated by cFLIP isoforms. *Mol Cell* **43:** 449–463.

Feoktistova M, Geserick P, Panayotova-Dimitrova D, Leverkus M. 2012. Pick your poison: The Ripoptosome, a cell death platform regulating apoptosis and necroptosis. *Cell Cycle* **11:** 460–467.

Geserick P, Hupe M, Moulin M, Wong WW, Feoktistova M, Kellert B, Gollnick H, Silke J, Leverkus M. 2009. Cellular IAPs inhibit a cryptic CD95-induced cell death by limiting RIP1 kinase recruitment. *J Cell Biol* **187:** 1037–1054.

Lavrik I, Golks A, Krammer PH. 2005. Death receptor signaling. *J Cell Sci* **118:** 265–267.

Micheau O, Tschopp J. 2003. Induction of TNF receptor I-mediated apoptosis via two sequential signaling complexes. *Cell* **114:** 181–190.

Rajput A, Kovalenko A, Bogdanov K, Yang SH, Kang TB, Kim JC, Du J, Wallach D. 2011. RIG-I RNA helicase activation of IRF3 transcription factor is negatively regulated by caspase-8-mediated cleavage of the RIP1 protein. *Immunity* **34:** 340–351.

Tenev T, Bianchi K, Darding M, Broemer M, Langlais C, Wallberg F, Zachariou A, Lopez J, MacFarlane M, Cain K, et al. 2011. The Ripoptosome, a signaling platform that assembles in response to genotoxic stress and loss of IAPs. *Mol Cell* **43:** 432–448.

Tinel A, Tschopp J. 2004. The PIDDosome, a protein complex implicated in activation of caspase-2 in response to genotoxic stress. *Science* **304:** 843.

Vandenabeele P, Galluzzi L, Vanden Berghe T, Kroemer G. 2010. Molecular mechanisms of necroptosis: An ordered cellular explosion. *Nat Rev Mol Cell Biol* **11:** 700–714.

Cite this protocol as *Cold Spring Harb Protoc*; doi:10.1101/pdb.prot087403

Techniques for the Detection of Autophagy in Primary Mammalian Cells

Daniel Puleston,[1] Kanchan Phadwal,[2] Alexander Scarth Watson,[2] Elizabeth J. Soilleux,[3] Suganthi Chittaranjan,[4] Svetlana Bortnik,[4] Sharon M. Gorski,[5] Nicholas Ktistakis,[6] and Anna Katharina Simon[1,7,8]

[1]MRC Human Immunology Unit, Weatherall Institute of Molecular Medicine, John Radcliffe Hospital, Oxford OX3 9DS, United Kingdom; [2]BRC Translational Immunology Lab, Nuffield Department of Medicine, Oxford University, Oxford OX3 9DS, United Kingdom; [3]Nuffield Department of Clinical Laboratory Sciences, Oxford OX3 9DS, United Kingdom; [4]The Genome Sciences Centre, BC Cancer Agency, Vancouver V5Z 1L3, Canada; [5]Department of Molecular Biology and Biochemistry, Simon Fraser University, Burnaby V5A 1S6, Canada; [6]Babraham Institute, Cambridge CB22 3AT, United Kingdom; [7]BRC Translational Immunology Lab, Oxford University, Oxford OX3 9DS, United Kingdom

Autophagy is a lysosomal catabolic pathway responsible for the degradation of cytoplasmic constituents. Autophagy is primarily a survival pathway for recycling cellular material in times of nutrient starvation, and in response to hypoxia, endoplasmic reticulum stress, and other stresses, regulated through the mammalian target of rapamycin pathway. The proteasomal pathway is responsible for degradation of proteins, whereas autophagy can degrade cytoplasmic material in bulk, including whole organelles such as mitochondria (mitophagy), bacteria (xenophagy), or lipids (lipophagy). Although signs of autophagy can be present during cell death, it remains controversial whether autophagy can execute cell death in vivo. Here, we will introduce protocols for detecting autophagy in mammalian primary cells by using western blots, immunofluorescence, immunohistochemistry, flow cytometry, and imaging flow cytometry.

INTRODUCTION

There are three main types of autophagy: chaperone-mediated autophagy, microautophagy, and macroautophagy (Mizushima and Komatsu 2011). The accompanying protocols detect processes related to macroautophagy, referred to as autophagy hereafter.

Over 32 related proteins have been identified that mediate the completion of the double-membraned autophagosome before fusion to the lysosome and degradation of its contents (see Fig. 1 for an overview). The formation of the autophagosome is initiated by a protein complex, which includes Ulk1 (Atg1) and Atg13, that in mammals can only be activated in the absence of signaling from the nutrient-sensing kinase mammalian target of rapamycin (mTOR). Also key to autophagosome development in humans are the class III phosphoinositide 3-kinases Vps34 and Vps35, which along with Beclin-1 (human homolog of Atg6) mediate the "nucleation" step. Elongation of the isolation membrane is mediated by two ubiquitin-like conjugation systems that are key to the autophagosome expansion: one system where Atg7 (an E1-like protein) and Atg10 (E2-like) act to conjugate Atg5–Atg12, and another where the Atg5–Atg12 conjugate (an E3-like protein) acts in concert with

[8]Correspondence: katja.simon@ndm.ox.ac.uk

Cite this introduction as Cold Spring Harb Protoc; doi:10.1101/pdb.top070391

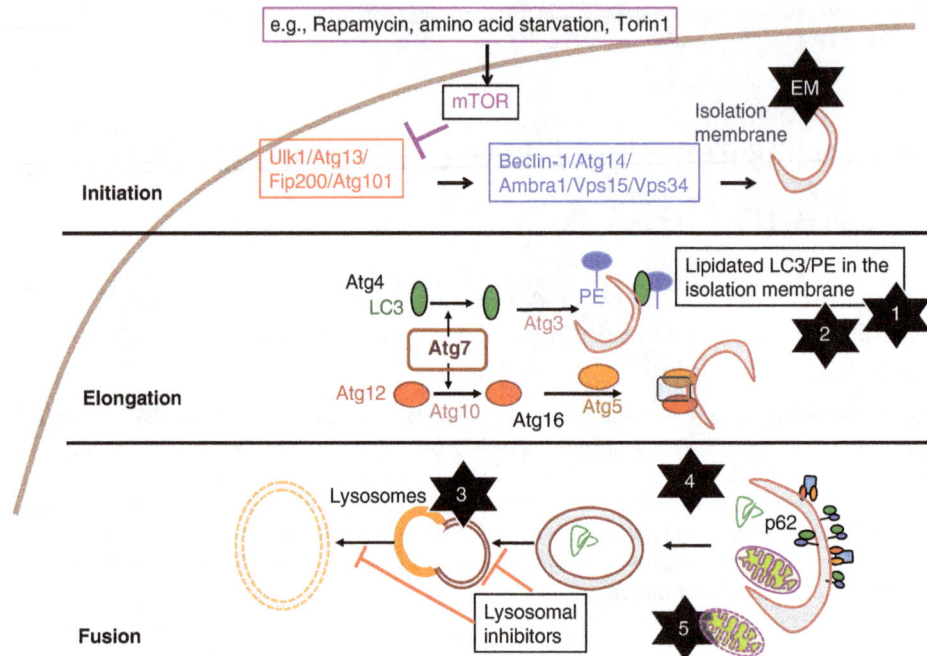

FIGURE 1. Overview of the autophagy pathway, its key molecular players and protocols. The numbered star symbols relate to the accompanying protocols; "EM" refers to electron microscopy (not described), which can detect double-membrane vesicles. Protocols 1 and 2 detect lipidated LC3; Protocol 3 quantifies the fusion of lysosomes with autophagosomes; Protocol 4 detects the presence of nondegraded adaptor protein p62 on tissue sections, an indication of reduced levels of autophagy; and Protocol 5 measures mitochondrial damage, a consequence of reduced autophagy. Abbreviation: PE, phosphatidylethanolamine. (1) Protocol 1: Monitoring Autophagic Flux by Using Lysosomal Inhibitors and Western Blotting of Endogenous MAP1LC3B (Chittaranjan et al. 2015); (2) Protocol 2: Monitoring the Localization of MAP1LC3B by Indirect Immunofluorescence (Ktistakis 2015); (3) Protocol 3: Analyzing the Colocalization of MAP1LC3 and Lysosomal Markers in Primary Cells (Phadwal 2015); (4) Protocol 4: Detection of p62 on Paraffin Sections by Immunohistochemistry (Watson and Soilleux 2015); (5) Protocol 5: Detection of Mitochondrial Mass, Damage, and Reactive Oxygen Species by Flow Cytometry (Puleston 2015).

Atg7 and Atg3 (E2-like) to conjugate Atg8 to phosphatidylethanolamine (so-called "lipidation" of LC3) for insertion into the membranes of the growing autophagosome (Mizushima and Komatsu 2011).

DETECTION OF AUTOPHAGY DEPENDENT ON LC3

In mammals, eight members of the Atg8 family have been identified. The best-known member of the family is LC3 (*MAP1LC3A* or *MAP1LC3B*), which is a key molecule in the majority of "classical" autophagy-detection assays and can be detected in its endogenous form in primary cells. The lipidation of LC3 causes LC3 (LC3-II) to run faster than its nonlipidated form (LC3-I) in a western blot (see Protocol 1: Monitoring Autophagic Flux by Using Lysosomal Inhibitors and Western Blotting of Endogenous MAP1LC3B [Chittaranjan et al. 2015]). LC3-I is uniformly distributed in the cell when autophagy levels are low, whereas, upon induction of autophagy, lipidation of LC3 causes its relocalization to the autophagosome, which can be visualized and quantified by counting LC3 spots through using immunofluorescence microscopy (see Protocol 2: Monitoring the Localization of MAP1LC3B by Indirect Immunofluorescence [Ktistakis 2015]). The third "classical" technique is the visualization of double-membrane vesicles by standard electron microscopy. Although this technique is not elaborated here, the reader is referred to excellent reviews elsewhere on this method (Eskelinen 2008).

However, there are caveats to all three techniques when applied to primary cells: detecting autophagy in one cell type among a mixture of primary cells (different cell types have different levels of autophagy) require cell sorting. Similar to other cellular stresses, sorting with magnetic beads or by flow cytometry can induce autophagy, thereby making it difficult to show a further induction. Protocol 3: Analyzing the Colocalization of MAP1LC3 and Lysosomal Markers in Primary Cells (Phadwal 2015) avoids sorting by simultaneously identifying cell types and measuring autophagy using the ImageStream, which is an imaging cytometer. With the ImageStream, it is possible to quantify either endogenous LC3 puncta or LC3–GFP puncta while detecting surface markers. LC3 puncta can also be colocalized with a lysosomal dye or antibody to detect autolysosomes. Note that commercially available kits have recently been developed for detecting autophagy that can be used with conventional flow cytometers (see Protocol 3: Analyzing the Colocalization of MAP1LC3 and Lysosomal Markers in Primary Cells [Phadwal 2015] for details).

INHIBITORS AND INDUCERS

The three protocols mentioned above require the addition of lysosomal inhibitors to stop the autophagic flux and thereby allow the accumulation of autophagosomes or autolysosomes to be detected. We present examples of bafilomycin A and chloroquine, both inhibiting the fusion of the autophagosome with the lysosome. We also use the inhibitors E64d and pepstatin, which inhibit lysosomal proteases that are responsible for the degradation of proteins that occurs within lysosomes. For the induction of autophagy, we have used 1 or 2 h of starvation (amino acid deprivation) or drugs such as rapamycin and Torin 1 (ATP competitive inhibitor of mTOR). The initiation of autophagy can be inhibited by wortmannin, an inhibitor of phosphoinositide 3-kinases. As a cautionary note, it is important to bear in mind that many of these compounds do not exclusively target the autophagic pathway.

OTHER METHODS FOR DETECTING THE CONSEQUENCES OF THE MODULATION OF AUTOPHAGY FOR TISSUES AND FOR PRIMARY CELLS BY FLOW CYTOMETRY

As described elsewhere (see Protocol 1: Monitoring Autophagic Flux Using Ref(2)P, the *Drosophila* p62 Ortholog [DeVorkin and Gorski 2014]), p62 (Sequestosome-1) is an adaptor protein that targets the ubiquitylated cargo to the LC3-decorated autophagosome by means of its LC3-binding domain. When autophagic flux is arrested, p62 accumulates. This can be used to assess the levels of autophagy in tissue sections, both paraffin embedded and frozen (see Protocol 4: Detection of p62 on Paraffin Sections by Immunohistochemistry [Watson and Soilleux 2015]). Note that an increase p62 mRNA can cause the accumulation of p62 protein and should be measured, if possible, concomitantly.

Decreases in the levels of autophagy are often accompanied by reduced mitophagy, leading to the accumulation of damaged mitochondria and reactive oxygen species (ROS). Traditionally, these have been detected by immunofluorescence by using fluorescent probes. We present a protocol that simultaneously quantifies mitochondrial damage or ROS, and surface markers, by flow cytometry using the same probes (see Protocol 5: Detection of Mitochondrial Mass, Damage, and Reactive Oxygen Species by Flow Cytometry [Puleston 2015]).

DIFFERENCES BETWEEN PRIMARY AND TRANSFORMED CELLS

Primary cells typically have high basal levels of autophagy and show a much reduced level of increase in autophagy (e.g., in response to starvation) compared with that of cell lines. Each cell type will require a different stimulus for the induction of autophagy. Moreover, the levels of autophagy differ between cell types, even related ones such as B and T lymphocytes. Additionally, as expected, a large variation in autophagy between different healthy donors can be observed. As autophagy can be induced by many stimuli, including starvation, exercise, and circadian rhythm, even the timing of tissue donation needs to be considered. For knockdown experiments, although the key autophagy machinery and its molecules most probably remain the same, adaptor molecules can differ between cell types. Transformed cell lines often carry monoallelic deletions that encompass autophagy genes, so knockdown of those particular genes can be very efficient. This obviously does not apply to primary cell lines.

Finally, note that the autophagy research community has contributed to a comprehensive publication of guidelines for the detection of autophagy (Klionsky et al. 2008), with an update planned for 2015, as well as a useful glossary (Klionsky et al. 2011).

ACKNOWLEDGMENTS

We thank the National Institute for Health Research/Biomedical Research Centre Oxford for funding A.K.S. and K.P., the Wellcome Trust for funding D.P., and the Lady Tata Memorial Trust and Natural Sciences and Engineering Research Council of Canada for supporting A.S.W.

REFERENCES

Chittaranjan S, Bortnik S, Gorski SM. 2015. Monitoring autophagic flux by using lysosomal inhibitors and western blotting of endogenous MAP1LC3B. *Cold Spring Harb Protoc* doi: 10.1101/pdb.prot086256.

DeVorkin L, Gorski SM. 2014. Monitoring autophagic flux using Ref(2)P, the *Drosophila* p62 ortholog. *Cold Spring Harb Protoc* doi: 10.1101/pdb.prot080333.

Eskelinen EL. 2008. To be or not to be? Examples of incorrect identification of autophagic compartments in conventional transmission electron microscopy of mammalian cells. *Autophagy* 4: 257–260.

Klionsky DJ, Abeliovich H, Agostinis P, Agrawal DK, Aliev G, Askew DS, Baba M, Baehrecke EH, Bahr BA, Ballabio A, et al. 2008. Guidelines for the use and interpretation of assays for monitoring autophagy in higher eukaryotes. *Autophagy* 4: 151–175.

Klionsky DJ, Baehrecke EH, Brumell JH, Chu CT, Codogno P, Cuervo AM, Debnath J, Deretic V, Elazar Z, Eskelinen EL, et al. 2011. A compre-

hensive glossary of autophagy-related molecules and processes (2nd edition). *Autophagy* 7: 1273–1294.

Ktistakis NT. 2015. Monitoring the localization of MAP1LC3B by indirect immunofluorescence. *Cold Spring Harb Protoc* doi: 10.1101/pdb.prot086264.

Mizushima N, Komatsu M. 2011. Autophagy: Renovation of cells and tissues. *Cell* 147: 728–741.

Phadwal K. 2015. Analyzing the colocalization of MAP1LC3 and lysosomal markers in primary cells. *Cold Spring Harb Protoc* doi: 10.1101/pdb.prot086272.

Puleston D. 2015. Detection of mitochondrial mass, damage, and reactive oxygen species by flow cytometry. *Cold Spring Harb Protoc* doi:10.1101/pdb.prot086298.

Watson AS, Soilleux EJ. 2015. Detection of p62 on paraffin sections by immunohistochemistry. *Cold Spring Harb Protoc* doi: 10.1101/pdb.prot086280.

Cite this introduction as *Cold Spring Harb Protoc*; doi:10.1101/pdb.top070391

Monitoring Autophagic Flux by Using Lysosomal Inhibitors and Western Blotting of Endogenous MAP1LC3B

Suganthi Chittaranjan,[1] Svetlana Bortnik,[1,2] and Sharon M. Gorski[1,2,3,4]

[1]The Genome Sciences Centre, BC Cancer Agency, Vancouver V5Z 1L3, Canada; [2]Interdisciplinary Oncology Program, The University of British Columbia, Vancouver V5Z 1L3, Canada; [3]Department of Molecular Biology and Biochemistry, Simon Fraser University, Burnaby V5A 1S6, Canada

Assays that monitor autophagic flux, or degradative completion of autophagy, are crucial for the assessment of the dynamic autophagy process in a variety of systems. Such assays help to distinguish between an increase in autophagosomes resulting from induced autophagic activity versus an increase in autophagosomes due to reduced lysosomal turnover. The majority of flux assays use autophagy protein MAP1LC3B (microtubule-associated proteins 1A/1B light chain 3B, here referred to as LC3B) as a marker for autophagy, and most are based on the use of reporters. Here, we describe a method, suitable for monitoring flux in primary cells and/or when reporters are not available or desirable, that uses lysosomal inhibitors and the analysis of endogenous LC3B-II (the lipidated form of LC3B that is associated with autophagosomes) by western blotting. A common application of this method, detailed here, is to test whether a treatment of interest (e.g., chemotherapy drug) induces autophagic flux in the cells of interest. If it is found that there is no difference in LC3B-II levels between treatment with lysosomal inhibitor alone versus drug plus lysosomal inhibitor, then this suggests that the drug is not inducing autophagic flux. Elevated levels of LC3B-II in treatments with drug plus lysosomal inhibitor, compared with drug treatment alone and inhibitor treatment alone, indicate that the drug is probably leading to an increase in autophagic flux.

MATERIALS

It is essential that you consult the appropriate Material Safety Data Sheets and your institution's Environmental Health and Safety Office for proper handling of equipment and hazardous materials used in this protocol.

RECIPE: Please see the end of this protocol for recipes indicated by <R>. Additional recipes can be found online at http://cshprotocols.cshlp.org/site/recipes.

Reagents

Drug or treatment of interest (e.g., epirubicin hydrochloride)
Dry ice
Growth medium (e.g., DMEM complete for MDA-MB-231 cells <R>)
Lysosomal inhibitors: bafilomycin A_1 (Sigma-Aldrich) or chloroquine diphosphate (Sigma-Aldrich)
NuPAGE LDS Sample Buffer (4×) (Invitrogen/Life Technologies NP007)
Odyssey blocking buffer (if using an Odyssey infrared [IR] system)

[4]Correspondence: sgorski@bcgsc.ca

Cite this protocol as Cold Spring Harb Protoc; doi:10.1101/pdb.prot086256

PBS (1×, pH 7.4)

Prepare 1 L in advance by diluting 10× PBS from Sigma-Aldrich (P3813) and storing for up to 6 mo at 4°C. Prewarm to 37°C before use on live cells.

PBST (1× phosphate-buffered saline [PBS] containing 0.1% Tween 20)
Primary antibody to detect β-actin (Abcam ab6276)
Primary antibody to detect LC3B-II (Anti-LC3B antibody; Abcam ab48394)
Primary cells or cell line of interest (e.g., the MDA-MB-231 breast cancer cell line)
Protein quantitation assay kit (e.g., Pierce BCA Protein Assay Kit)
Radioimmunoprecipitation assay (RIPA) Lysis Buffer System (Santa Cruz SC-24948)
Scanning system or horseradish peroxidase (HRP) for chemiluminescence
SDS-PAGE gel (e.g., NuPAGE 10% Tris–Bis gel from Invitrogen/Life Technologies)
Secondary antibodies conjugated to a reporter appropriate for the detection system (see Step 20)
Skim milk blocking buffer (2%)

Prepare fresh by dissolving 2 g skim milk powder in 100 mL of 1× PBS.

Trypsin–EDTA (Invitrogen/Life Technologies 25300-062) (ethylenediaminetetraacetic acid)

Equipment

BD Falcon six-well plates

If using T25 flasks as an alternative, increase medium and cells accordingly.

Biosafety cabinet
Detection system for scanning the western blot (see Step 23)
Heating block, set at 70°C
Incubator (37°C with controlled humidity and 5% CO_2)
Microcentrifuge, benchtop
Microcentrifuge tubes (1.5 mL)
Mini Trans-Blot Electrophoretic Transfer Cell (Bio-Rad)
Powerpack (100–200 V)

Required to run gels and ensure protein transfer.

Polyvinylidene difluoride (PVDF) membrane (e.g., Millipore or Bio-Rad)
Rocker
Software for data analysis (e.g., Image Quant)
SureLock XCell (for NuPAGE gels)

METHOD

For a summary of the whole procedure (Steps 1–25) and the associated timelines, see Figure 1. The procedure described in Steps 1–10 requires the use of standard cell culture techniques and western blot analysis. Careful handling of cells and diligent maintenance of consistent culture conditions are required to ensure reproducibility.

Establishing the Saturating Concentration of Lysosomal Inhibitor for the Cells of Interest

1. Plate sufficient cells in 1–3 mL of medium per well in a BD Falcon six-well plate to yield 90%–100% confluency by 48 h.

   The cell number for initial plating can vary depending on the doubling time of the cell type; for MDA-MB-231 cells, we plate 2.5×10^5 cells per well to yield 50%–60% confluency by 24 h and ~90% confluency by 48 h.

2. At 48 h after plating, add varying concentrations of a lysosomal inhibitor such as bafilomycin A_1 (typically 5–75 nM; see Fig. 2A,B) or chloroquine diphosphate (typically 5–50 μM) and incubate for 5 h.

   Note that the required concentrations of bafilomycin A_1 or chloroquine can vary according to the cell line and incubation time; typically, we use a time-period of 2–6 h. Do not treat cells with the inhibitors for more

 Cite this protocol as *Cold Spring Harb Protoc*; doi:10.1101/pdb.prot086256

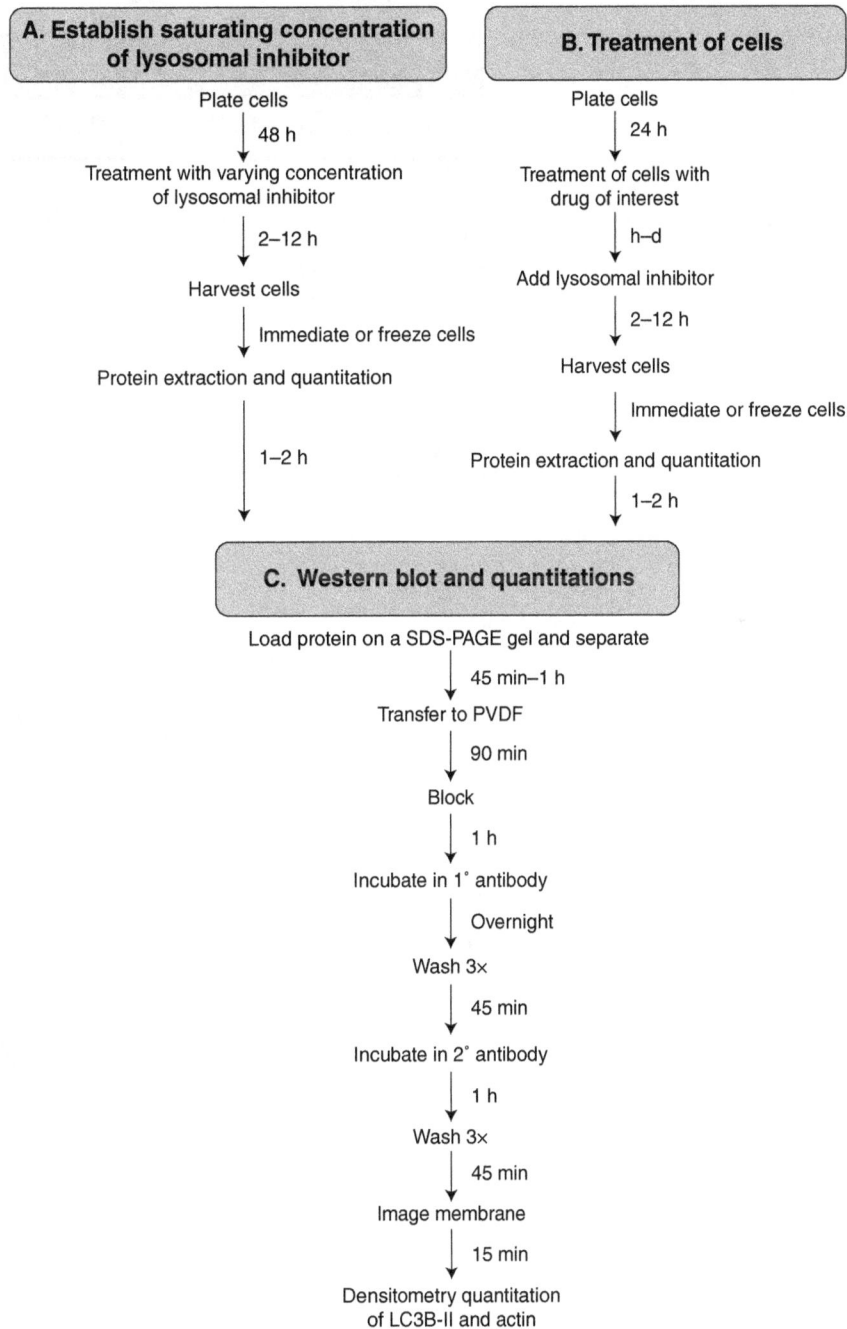

FIGURE 1. Measuring autophagic flux using lysosomal inhibitors and endogenous LC3B western blotting. First, a saturating concentration of lysosomal inhibitor must be determined for the cells of interest (*A*). Second, cells are treated with the test drug alone, lysosomal inhibitor alone, and a combination of drug plus lysosomal inhibitor (*B*). Third, the levels of endogenous LC3B-II are determined by western blot analysis (*C*).

> than ~12 h owing to the development of nonspecific effects, and, for a 12 h incubation period, try using lower concentrations of inhibitors.

3. To harvest cells, remove the medium, rinse the cells three times with ~1 mL of 1× PBS, trypsinize with 200 μL of trypsin–EDTA for 2–3 min, add 1 mL of medium, and collect the cells in a 1.5-mL microcentrifuge tube.

4. Centrifuge the cells at 1000*g* for 5 min at room temperature, discard the medium, and wash the cells twice with 1 mL of 1× PBS, centrifuging at 800–1000*g* for 3–5 min at room temperature between washes.

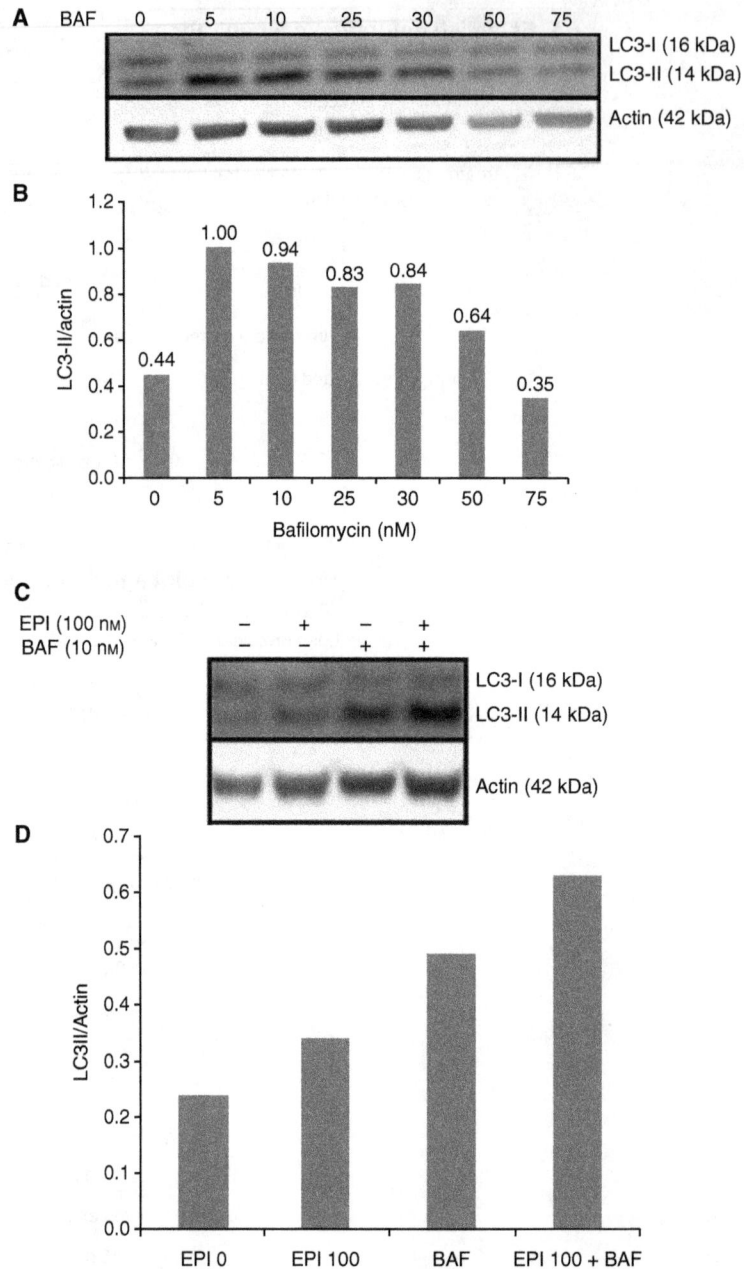

FIGURE 2. Investigation of autophagic flux in cells treated with the anthracycline chemotherapy drug epirubicin (EPI). (*A,B*) Establishing a saturating concentration for the lysosomal inhibitor bafilomycin A₁ (BAF) in human breast ade-nocarcinoma MDA-MB-231 cells. (*A*) Western blot analysis of the levels of LC3B-II in MDA-MB-231 cells. The cells were plated out, then 48 h later treated with increasing concentrations of BAF and finally incubated for 5 h before harvesting. The treatments and western blot analysis were performed as described in this protocol. Actin was used as a loading control. (*B*) Representative quantitation of the relative levels of LC3B-II and actin proteins determined by means of Image Quant software. The signals from LC3B-II and actin were comparable at 5–10 nM BAF but decreased at concentrations of ≥25 nM. In this example, the saturating levels of BAF were determined to be 5–10 nM. (*C,D*) EPI treatment increases autophagic flux in MDA-MB-231 cells. (*C*) Western blot analysis of LC3B and actin. MDA-MB-231 cells were treated with EPI for 24 h, and the saturating concentration of the lysosomal inhibitor BAF (determined in *B* above) was added for the final 5 h before harvesting. Control cells received no treatment. (*D*) Representative bar chart display of results from Image Quant analysis of the ratio of LC3B-II to actin. The increase in LC3B-II observed in the cells treated with EPI plus BAF indicates that the EPI treatment increases autophagic flux at this time-point.

Cite this protocol as *Cold Spring Harb Protoc*; doi:10.1101/pdb.prot086256

5. Discard the PBS from the last wash, and immediately flash-freeze the pellets in dry ice.

    Proceed immediately to Step 6 or store the cell pellet at −80°C.

6. Perform western blot analysis to detect the levels of LC3B-II (see Steps 11–23) to deduce the saturating concentrations of lysosomal inhibitor (Fig. 2A,B).

    Note that the saturating concentrations of lysosomal inhibitors can vary dramatically among different cell lines and must be determined empirically for each line investigated. It is important to ensure that the saturating concentration and incubation time selected do not affect cell viability.

    See Troubleshooting.

Treatment of Cells in the Presence and Absence of Lysosomal Inhibitor

7. Plate the cells of interest as described in Step 1. For each time-point that will be analyzed, ensure that control wells are set up for "vehicle alone" (i.e., no drug or treatment) and "vehicle plus lysosome inhibitor."

8. At 24 h after plating, add the drug or treatment of interest to the appropriate wells and incubate the cells for an additional 24 h at 37°C in an incubator (or alternative period of time, as required by the particular experiment).

9. Several hours (e.g., 5 h in our example) before harvesting the cells, add the chosen lysosomal inhibitor at its saturating concentration (determined in Step 6; e.g., 5–10 nM bafilomycin A_1 for MDA-MB-231 [see Fig. 2A,B]) to the appropriate wells.

10. Harvest the cells and flash-freeze the pellet as described in Steps 3–5.

Protein Extraction, Western Blotting, and Quantitation

11. Prepare 1× RIPA lysis buffer system with protease inhibitors according to the manufacturer's protocol.

12. Add 30–100 μL of 1× RIPA lysis buffer system according to the pellet size (use ~3–5 volumes of buffer per volume of pellet).

13. Incubate the cells on an end-to-end rocker for 1 h at 4°C.

14. Centrifuge the lysate at 15,000g for 15 min at 4°C. Transfer the supernatant that contains the protein to a new prelabeled tube.

15. Quantitate the level of protein with a protein quantitation assay kit (e.g., BCA Protein Assay).

16. Mix preferred amount of protein with the LDS Sample Buffer (1× final) and heat to 70°C for 5 min in the heating block. Load and run protein samples on an SDS-PAGE gel (e.g., NuPAGE 10% Bis–Tris with SureLock XCell apparatus). Use 40–50 μg of protein for IR detection or 10–20 μg of protein for chemiluminescence detection. Transfer to a PVDF membrane (e.g., Millipore membrane, which gives a low background for IR) using a Mini Trans-Blot apparatus and powerpack.

17. Block nonspecific binding by incubating the membrane in 2% (w/v) skim milk blocking buffer in 1× PBS for 1 h at room temperature.

18. Incubate with primary antibody in 2% (w/v) skim milk blocking buffer in 1× PBS overnight at 4°C with gentle rocking.

    Antibodies against both LC3B and actin can be used at the same time. The antibody against LC3B can be incubated for 4 h at room temperature, but the signals might be considerably lower.

19. Wash the membrane three times in 1× PBST for at least 15 min each time at room temperature.

20. Dilute the appropriate secondary antibodies in the appropriate blocking buffer, and then incubate the membrane for 1 h at room temperature.

    - For IR, dilute IR-conjugated secondary antibodies in Odyssey blocking buffer. Use different IR conjugates for detection of LC3B and actin (e.g., use rabbit IR700 for LC3B and mouse IR800 for actin).

Typically, we use a 1:10,000 dilution for IR700 and 1:5000 for IR800.

From this step onwards, the membrane should be protected from light if the IR method is used.

- For chemiluminescence, dilute secondary antibodies conjugated to HRP 1:5000 to 1:10,000 in 2% (w/v) skim milk blocking buffer in 1× PBS.

21. Wash the membrane three times in PBST at room temperature for at least 15 min each time.

22. Rinse the membrane twice with 1× PBS at room temperature to remove any traces of Tween 20 from the PBST washes.

23. Scan the prepared membrane.

    - For IR, use the Odyssey system. (LI-COR)

        See Troubleshooting.

    - For chemiluminescence, add developer to the membrane per the manufacturer's instructions before scanning with the appropriate imaging apparatus (e.g., Bio-Rad or Fuji scanner with chemiluminescence capability). If an imaging apparatus is not available, use X-ray film, but this is not ideal for quantitation owing to its poor dynamic range.

        A representative image of a western blot and plotted data for MDA-MB-231 cells treated with epirubicin for 24 h, with 10 nM bafilomycin A1 added for the final 5 h, is shown in Figure 2C,D.

Data Analysis

24. Use Image Quant to determine the intensity levels of LC3B-II and actin (alternative comparable imaging software can be used).

25. Normalize the levels of LC3B-II in each sample to those of actin (e.g., Fig. 2B,D).

    If there is no change in the level of LC3B-II between treatment with lysosomal inhibitor alone versus drug (or treatment of interest) plus lysosomal inhibitor, then this suggests that the drug is not inducing autophagic flux. In contrast, a finding of elevated levels of LC3B-II in a treatment comprising drug plus lysosomal inhibitor, compared with drug treatment alone and inhibitor treatment alone, will indicate that the drug is probably resulting in elevated autophagic flux.

TROUBLESHOOTING

Problem (Step 6): A saturating concentration of inhibitor cannot be acquired (or the cells die at the saturating concentration).

Solution: There are two approaches that can be considered:

- Try reducing the duration of the incubation with inhibitor; and/or

- use an alternative inhibitor (e.g., chloroquine vs. bafilomycin A_1).

Problem (Step 23): A high level of background staining is present, especially for the IR700.

Solution: Consider:

- Reducing the concentration of secondary antibody or using the concentrations recommended by the manufacturer; and/or

- washing membranes for a longer period of time or using an alternative membrane. Note that, in our hands, Millipore or Bio-Rad membranes yield good results, with a low background.

DISCUSSION

Advantages of the Method

There are three main advantages to the method described here. First, it enables the determination of whether autophagosome synthesis and subsequent lysosomal degradation are probably occurring in a

Cite this protocol as *Cold Spring Harb Protoc*; doi:10.1101/pdb.prot086256

given system. Although other approaches can measure the levels of LC3B-II, LC3B puncta (see Protocol 2: Monitoring the Localization of LC3B by Indirect Immunofluorescence [Ktistakis 2015]), or GFP–LC3 puncta (Kabeya et al. 2000), it is important to keep in mind that those readouts correlate with the numbers of autophagosomes in cells at a particular point in time. The sole reliance on LC3B-II levels or GFP–LC3 puncta can lead to erroneous interpretations of autophagy activity levels (Tanida et al. 2005). For example, low amounts of LC3-II can be indicative of either low levels of autophagosome formation or, alternatively, high autophagic flux (i.e., degradative completion of autophagy, where autophagosomes are turned over rapidly). Similarly, an increase in GFP–LC3 can be indicative of either an increase in autophagosome formation or, instead, can result from blockade of fusion between autophagosomes and lysosomes. The latter leads to an accumulation of autophagosomes in the cell, detectable as an increase in GFP–LC3 puncta or the levels of LC3-II. For these reasons, it is important to include assays, like the protocol described here, that assess autophagic flux by including an evaluation of effects in both the absence and presence of a lysosomal inhibitor. A second advantage of this method is that it relies on endogenous LC3B and thus avoids the need to transfect cells with a reporter-based system. Finally, the method is relatively simple to perform, requiring no specialized equipment or reagents.

Limitations and Caveats

There are also drawbacks to this western blot method using lysosomal inhibitors and measuring endogenous LC3B-II. For example, in cases of high basal autophagy, the difference between the treated and untreated samples can be subtle and difficult to detect. In those cases, additional methods of autophagic flux evaluation that monitor degradation of autophagy substrates (e.g., p62; see Protocol 4: Detection of p62 on Paraffin Sections by Immunohistochemistry [Watson and Soilleux 2015] and Protocol 1: Monitoring Autophagic Flux Using Ref(2)P, the *Drosophila* p62 Ortholog [DeVorkin and Gorski 2014]) can be helpful.

Second, it is also important to note that the saturating concentration of the lysosomal inhibitor established at a particular time point and on a certain cell number (and cell type) cannot be generalized and used for multiple time points. Ideally, Steps 1–6 (establishing the saturating concentration for the lysosomal inhibitor) should be performed for each condition (time point and cell number) tested in the experiment.

A third limitation of the method described above is that it is primarily applicable to in vitro studies. The inability of most commonly used lysosomal inhibitors to reach their saturating concentrations in vivo without causing unacceptable toxicity (McAfee et al. 2012) is one of the reasons for not using this assay in vivo.

Excellent and detailed discussions of additional considerations related to this method are available elsewhere (Rubinsztein et al. 2009; Klionsky et al. 2012). In all cases, we highly recommend combining this protocol with other methods to monitor autophagy to interpret the effects of a particular drug or treatment on the levels of autophagy (see Klionsky et al. 2012). In particular, as LC3B can be associated with other membranes, we recommend conducting a complementary method that includes visualization of autophagic structures, such as electron microscopy (see Klionsky et al. 2012) or monitoring LC3 tandem constructs (e.g., RFP–GFP–LC3) (Kimura et al. 2007).

RELATED TECHNIQUES

Complementary techniques for monitoring autophagic flux include three reporter-based assays: (1) the free-GFP assay (Shintani and Klionsky 2004; Hosokawa et al. 2006), (2) the lysosomal inhibitor plus GFP–LC3 assay (Gutierrez et al. 2007), and (3) the tandem fluorescent-tagged LC3 assay (Kimura et al. 2007). Another complementary technique involves analysis of the levels of the autophagic adaptor protein p62 by western blot analysis or immunohistochemistry (see Protocol 4: Detection of p62 on Paraffin Sections by Immunohistochemistry [Watson and Soilleux 2015]), but see also Protocol 1: Monitoring Autophagic Flux Using Ref(2)P, the *Drosophila* p62 Ortholog (DeVorkin and

Gorski 2014) and Klionsky and colleagues' guidelines for the use and interpretation of assays for monitoring autophagy (Klionsky et al. 2012) for further discussion related to that technique.

RECIPE

DMEM Complete for MDA-MB-231 Cells

Reagents	Final concentrations
DMEM (Invitrogen/Life Technologies 11995-073)	1×
Heat-inactivated fetal bovine serum (FBS)	10%
Insulin (Sigma-Aldrich)	10 µg/L
L-Glutamine (Invitrogen/Life Technologies)	2 mM
Sodium pyruvate (Invitrogen/Life Technologies)	1 mM
Nonessential amino acids (100×; Invitrogen/Life Technologies 11140050)	1×

Combine reagents under sterile conditions and store for up to 6 mo at 4°C. Prewarm to 37°C before use.

ACKNOWLEDGMENTS

We acknowledge members of the Gorski laboratory for helpful discussions and thank Canadian Institutes of Health Research (CIHR) GPG102167 for financial support. S.M.G. is supported in part by a CIHR New Investigator Award, and S.B. is supported by a CIHR Frederick Banting and Charles Best Canada Graduate Scholarship Doctoral Award.

REFERENCES

DeVorkin L, Gorski SM. 2014. Monitoring autophagic flux using Ref(2)P, the *Drosophila* p62 ortholog. *Cold Spring Harb Protoc* doi: 10.1101/pdb.prot080333.

Gutierrez MG, Saka HA, Chinen I, Zoppino FC, Yoshimori T, Bocco JL, Colombo MI. 2007. Protective role of autophagy against *Vibrio cholera* cytolysin, a pore-forming toxin from *V. cholerae*. *Proc Natl Acad Sci* 104: 1829–1834.

Hosokawa N, Hara Y, Mizushima N. 2006. Generation of cell lines with tetracycline-regulated autophagy and a role for autophagy in controlling cell size. *FEBS Lett* 580: 2623–2629.

Kabeya Y, Mizushima N, Ueno T, Yamamoto A, Kirisako T, Noda T, Kominami E, Ohsumi Y, Yoshimori T. 2000. LC3, a mammalian homologue of yeast Apg8p, is localized in autophagosome membranes after processing. *EMBO J* 19: 5720–5728.

Kimura S, Noda T, Yoshimori T. 2007. Dissection of the autophagosome maturation process by a novel reporter protein, tandem fluorescent-tagged LC3. *Autophagy* 3: 452–460.

Klionsky DJ, Abdalla FC, Abeliovich H, Abraham RT, Acevedo-Arozena A, Adeli K, Agholme L, Agnello M, Agostinis P, Aguirre-Ghiso JA, et al. 2012. Guidelines for the use and interpretation of assays for monitoring autophagy. *Autophagy* 8: 445–544.

Ktistakis NT. 2015. Monitoring the localization of LC3B by indirect immunofluorescence. *Cold Spring Harb Protoc* doi: 10.1101/pdb.prot086264.

McAfee Q, Zhang Z, Samanta A, Levi SM, Ma XH, Piao S, Lynch JP, Uehara T, Sepulveda AR, Davis LE, et al. 2012. Autophagy inhibitor Lys05 has single-agent antitumor activity and reproduces the phenotype of a genetic autophagy deficiency. *Proc Natl Acad Sci* 109: 8253–8258.

Rubinsztein DC, Cuervo AM, Ravikumar B, Sarkar S, Korolchuk V, Kaushik S, Klionsky DJ. 2009. In search of an autophagomometer. *Autophagy* 5: 585–589.

Shintani T, Klionsky DJ. 2004. Cargo proteins facilitate the formation of transport vesicles in the cytoplasm to vacuole targeting pathway. *J Biol Chem* 279: 29889–29894.

Tanida I, Minematsu-Ikeguchi N, Ueno T, Kominami E. 2005. Lysosomal turnover, but not a cellular level, of endogenous LC3 is a marker for autophagy. *Autophagy* 1: 84–91.

Watson AS, Soilleux EJ. 2015. Detection of p62 on paraffin sections by immunohistochemistry. *Cold Spring Harb Protoc* doi: 10.1101/pdb.prot086280.

Monitoring the Localization of MAP1LC3B by Indirect Immunofluorescence

Nicholas T. Ktistakis[1]

Babraham Institute, Cambridge CB22 3AT, United Kingdom

The autophagy protein MAP1LC3B (microtubule-associated proteins 1A/1B light chain 3B, hereafter referred to as LC3B), which is one of several mammalian homologs of yeast Atg8, is one of the most popular markers for autophagosome formation because its distribution changes from cytosolic/diffuse to punctate upon the induction of autophagy. In many settings, plasmids encoding fluorescently tagged LC3B are introduced into cells, and the subsequent autophagy response is monitored. However, for a variety of reasons, it would be desirable also to have a protocol to monitor the localization of endogenous LC3B under various conditions. This protocol provides such a methodology for the staining of endogenous LC3B by indirect immunofluorescence, such that autophagy responses can be monitored in mammalian cells.

MATERIALS

It is essential that you consult the appropriate Material Safety Data Sheets and your institution's Environmental Health and Safety Office for proper handling of equipment and hazardous materials used in this protocol.

Reagents

Bovine serum albumin (BSA; 2%, prepared in phosphate-buffered saline [PBS])

Distilled water, deionized

Mammalian cells and appropriate cell-culture media for normal growth and for experimental treatment (see Steps 1–2)

Methanol

    Store overnight at −20°C.

Mounting medium (e.g., Aqua-Poly/Mount medium, Polysciences)

PBS (pH 7.4)

Primary antibody against LC3B (antiLC3B antibody produced in rabbit, Sigma-Aldrich L7543)

Secondary antibody (e.g., widely available goat antirabbit antibody conjugated to fluorescein isothiocyanate)

[1]Correspondence: nicholas.ktistakis@babraham.ac.uk

Cite this protocol as *Cold Spring Harb Protoc*; doi:10.1101/pdb.prot086264

Equipment

Aspirator
Beakers (250 mL)
Cell incubator
Confocal (or wide-field) microscope

The two most widely used in our laboratory are a Zeiss Manual Upright Axioimager A2 running a Zeb Blue imaging system and an Olympus FV1000 point scanning confocal.

Forceps, fine
Microscope coverslips
Microscope slides
Tissue-culture plates, empty

Two types are required: one the same as those used to grow the cells for methanol fixation, and another the same as those used to grow the cells for subsequent staining steps.

METHOD

For a summary of the whole procedure (Steps 1–18) and the associated timelines, see Figure 1.

Cell Culture and Treatments

1. Grow the mammalian cells of interest under standard conditions. Then, on the day before the procedure, place microscope coverslips inside the wells of standard tissue-culture plates, and plate the required numbers of cells on them.

   For example, we plate 300,000 HEK-293 cells per 35-mm dish coverslip.

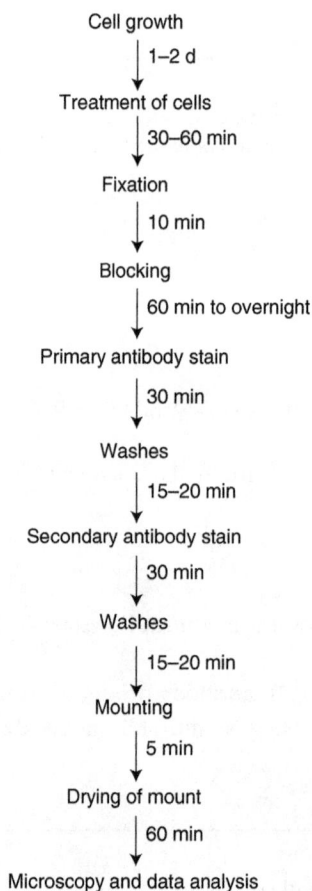

Cell growth
↓ 1–2 d
Treatment of cells
↓ 30–60 min
Fixation
↓ 10 min
Blocking
↓ 60 min to overnight
Primary antibody stain
↓ 30 min
Washes
↓ 15–20 min
Secondary antibody stain
↓ 30 min
Washes
↓ 15–20 min
Mounting
↓ 5 min
Drying of mount
↓ 60 min
Microscopy and data analysis

FIGURE 1. Timeline for performing the protocol.

Cite this protocol as *Cold Spring Harb Protoc*; doi:10.1101/pdb.prot086264

2. The following day, use an aspirator to remove the growth medium. Switch the cells on cover-slips to the appropriate cell-culture medium for application of the experimental treatment (Fig. 2).

- To investigate the effect of amino acid starvation, apply a medium containing 140 mM NaCl, 1 mM CaCl$_2$, 1 mM MgCl$_2$, 5 mM glucose, 1% BSA and 20 mM HEPES (pH 7.4) to the cells for 60 min.

  Other appropriate media for amino acid starvation include Hank's balanced salt solution and Earle's balanced salt solution.

- To investigate the effects of the inhibition of the nutrient-sensing kinase mammalian target of rapamycin (mTOR), add the inhibitor directly to complete medium (e.g., use 250 nM Torin 1 in complete medium, and incubate for 60 min).

- To reveal the extent of basal autophagy by blocking autophagosome–lysosome fusion, use 100 nM bafilomycin A1 in complete medium, and incubate for 60 min.

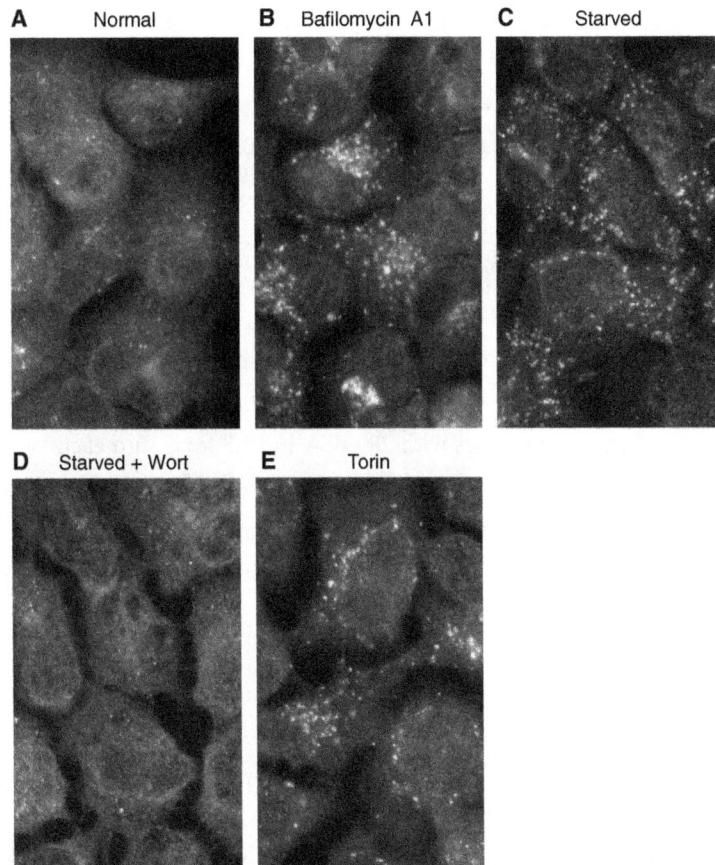

FIGURE 2. Immunofluorescence of the autophagy marker protein LC3B in human embryonic kidney HEK-293 cells. Cells were (A) left untreated (control), (B) treated for 60 min with 100 nM bafilomycin A1 in complete medium (which inhibits autophagosome–lysosome fusion, causing LC3B to build up), (C) starved for 60 min, thus inducing autophagy, (D) starved for 60 min in the presence of 100 nM wortmannin (inhibiting phosphoinositide 3-kinases and hence the nucleation step of autophagosome formation), or (E) treated in complete medium containing the mTOR kinase inhibitor Torin 1 (a kind gift of Nathaniel Gray, Harvard University) for 60 min. At the end of the treatment, the cells were stained for LC3B, as described in the text. Note the low punctate basal signal in the control cells, which increases with starvation and is blocked by wortmannin.

Indirect Immunofluorescence of LC3B

Fixation

3. Fill one empty tissue-culture plate with ~5 mL of 2% BSA in PBS.

4. Fill four 250 mL beakers with ~200 mL each of PBS at room temperature.

5. Fill another empty tissue-culture plate with ~5 mL of methanol at −20°C and place on ice.

6. Take the plate containing cells from the cell incubator and place it on the bench.

7. With a fine forceps, carefully and quickly lift each coverslip out of the dish, successively dip it in the first two beakers containing PBS, and then place in methanol for 10 min.

8. After 10 min, lift coverslip from methanol and successively dip it in the last two beakers containing PBS.

    Make sure to use different beakers of PBS than the ones used in Step 7.

9. Place the coverslips in a plate containing 2% BSA in PBS.

10. Allow the cells on coverslips to be blocked in the 2% BSA–PBS solution for at least 1 h, or even overnight.

    For 1-h treatments, we use room temperature, whereas, for overnight, we incubate at 4°C.
    This step wets the glass and reduces the background signal.

Antibody Staining

Perform Steps 11–18 at room temperature.

11. Wash the coverslips twice with 2% BSA in PBS for 5 min each.

12. Dilute the primary antibody against LC3B 1:200 in 2% BSA in PBS and add it to the coverslips for 30 min.

    To minimize the volume of reagent used, consider draining the well of all liquid, then center the coverslip away from the edges and add a drop of solution (12, 14, or 30 µL for 11- 14- or 22-mm round coverslips, respectively).

13. Wash three times with 2% BSA in PBS for 5 min each.

14. Dilute the secondary antibody 1:100 in 2% BSA in PBS and add to the coverslip for 30 min.

15. Wash three times with 2% BSA in PBS for 5 min each.

16. Wash quickly in distilled deionized water and mount the coverslips on glass microscope slides.

17. Allow mounting medium to set for at least 60 min.

18. View by fluorescence microscopy with a confocal or wide-field microscope.

    See Troubleshooting.

TROUBLESHOOTING

Problem (Step 18): There is too strong a punctate signal in control cells.
Solution: Some cells (e.g., murine embryonic fibroblasts [MEFs]) have a naturally high basal level of autophagy, but, for most others, better growing conditions should reduce basal autophagy. For example, the cells could be split before reaching 100% confluency, or alternative sources of serum that improve basal growth could be tried. For some isolates of MEFs, it has been noted that basal autophagy is reduced by growth with 10% CO_2 instead of 5% CO_2—but this is not a global property of MEFs.

Problem (Step 18): There is a high level of background staining.
Solution: There are several possible ways to proceed.

Cite this protocol as *Cold Spring Harb Protoc*; doi:10.1101/pdb.prot086264

- Allow the blocking step (Step 10) to exceed 1 h.
- Try increasing the BSA concentration to 5%.
- Dilute the antibody further (it is an affinity-purified antibody and it does vary from batch to batch).

RELATED TECHNIQUES

For a related western-based approach for assaying endogenous LC3B, see Protocol 1: Monitoring Autophagic Flux by Using Lysosomal Inhibitors and Western Blotting of Endogenous LC3B (Chittaranjan et al. 2015).

ACKNOWLEDGMENTS

N.T.K. and colleagues are supported by the Biotechnology and Biological Sciences Research Council.

REFERENCES

Chittaranjan S, Bortnik S, Gorski SM. 2015. Monitoring autophagic flux by using lysosomal inhibitors and western blotting of endogenous LC3B. *Cold Spring Harb Protoc* doi: 10.1101/pdb.prot086256.

Analyzing the Colocalization of MAP1LC3 and Lysosomal Markers in Primary Cells

Kanchan Phadwal[1]

BRC Translational Immunology Laboratory, Nuffield Department of Medicine, Oxford University, Oxford OX3 9DS, United Kingdom

This technique evaluates the colocalization of the autophagy protein MAP1LC3 (microtubule-associated proteins 1A/1B light chain 3B, here referred to as LC3) with lysosomes (autolysosomes) in primary cells in a high-throughput manner. It uses an imaging fluorescence-activated cell sorting cytometer called the ImageStream to concomitantly detect surface molecules, making possible the identification of cells in mixed cell populations (e.g., in blood or bone marrow). It can be applied to clinical samples and to rare cell populations because only a few cells are needed for detection.

MATERIALS

It is essential that you consult the appropriate Material Safety Data Sheets and your institution's Environmental Health and Safety Office for proper handling of equipment and hazardous material used in this protocol.

RECIPE: Please see the end of this protocol for recipes indicated by <R>. Additional recipes can be found online at http://cshprotocols.cshlp.org/site/recipes.

Reagents

Antibodies against cell-surface markers (e.g., CD3, CD19, CD8, CD14, and CD45)

Bafilomycin A1 (optional; see Step 1)

Cell source and suitable growth medium

> *Use any primary cells in cell suspension (e.g., peripheral blood mononuclear cells [PBMCs] or cell lines in suspension). Note that adherent cells need to be detached before staining. The protocol below is for PBMCs that have been isolated from whole blood using a density gradient. The PBMCs can either be used directly after isolation or from frozen (−80°C).*

Chloroquine

E64d (lysosomal protease inhibitor)

Fixative (4% paraformaldehyde or Intracellular Fixation and Permeabilization Buffer [eBioscience 88-8823-88])

Fluorescence-activated cell sorting (FACS) buffer for ImageStream <R>

Hank's balanced salt solution (HBSS; starvation medium; Life Technologies 24020)

Live/dead marker (e.g., Fixable Violet Dead Cell Stain kit, Molecular Probes L34955)

Lyso-ID Red detection kit for lysosomes (Enzo Life Sciences ENZ-51005-500)

Pepstatin A (PepA; lysosomal protease inhibitor)

---

[1]Correspondence: kphadwal@staffmail.ed.ac.uk

Cite this protocol as *Cold Spring Harb Protoc*; doi:10.1101/pdb.prot086272

Primary monoclonal or polyclonal antibodies against LC3

This can be an unconjugated antibody against LC3 or a directly conjugated monoclonal antibody against LC3 appropriate for flow cytometry. Ensure that they are made in a species different from that used for the antibodies against surface markers.

Rapamycin (optional; see Step 1)

Secondary antibodies

These should be generated against the species of primary antibody and be conjugated to fluorescein isothiocyanate (FITC) or phycoerythrin.

Equipment

Aluminum foil

Centrifuge for 15 mL conical tubes

Conical tubes (15 mL)

IDEAS software (Amnis-Merck Millipore)

ImageStream 100 (IS100) or ImageStream X (ISX) (Amnis-Merck Millipore)

This protocol was developed on the IS100.

Incubator (set at 37°C)

Microcentrifuge tubes (0.5 mL)

Plate shaker

Round-bottom plates (96-well)

Six-well plates

Tissue paper

METHOD

Experimental Method

Maintain all buffers, dyes, and antibodies at 4°C during the staining process.

1. Induce autophagy in PBMCs by aliquoting them into the wells of a six-well plate (5×10^6 cells per well). Perform all treatments and controls in volumes of 5 mL for 2 h (or appropriate time) at 37°C.

   Include a control (full medium alone), full medium plus 10 µg/mL E64d or 10 µg/mL PepA, starved (HBSS medium), and HBSS plus E64d/PepA or HBSS plus 20 µM chloroquine or 10 nM bafilomycin. As an alternative to starvation, 1 µM rapamycin can also be included with and without E64d or PepA.

2. Harvest the cells by centrifuging them at 1500 rpm in 15 mL conical tubes for 8 min at 4°C.

3. Resuspend the cells in 100 µL of cold FACS buffer for ImageStream, and distribute them in a round-bottom plate (96-well).

   For single-color controls, dedicate one well to each fluorochrome, to be stained with one fluorochrome-conjugated antibody only, and leave one well unstained to allow setting of the laser voltages. These controls need to be run to create a compensation file to compensate for cross talk occurring between the different fluorochromes used in the assay.

4. Centrifuge the plate for 8 min at 4°C at 1500 rpm, flick buffer into sink, and wipe dish on tissue paper.

5. Add the first layer of stain—live/dead marker (1:2000 dilution in FACS buffer for ImageStream)—and incubate on a plate shaker for 7 min at room temperature.

6. Fill wells with cold FACS buffer for ImageStream. Centrifuge the plate at 300*g* for 8 min, flick buffer into sink, and wipe dish on tissue paper.

   Do not worry about cell loss from a short flick because cells stick tightly to the bottom of the well.

7. Conduct the second layer of staining—directly conjugated antibody against the surface marker (e.g., a 1:100 dilution of CD8-PE)—and incubate on a plate shaker for 15–20 min at room temperature. Then wash the cells as in Step 6.

8. Perform the third layer of staining—the Lyso ID Red stain (1:600 dilution in the manufacturer's assay buffer)—and incubate for 10 min at 37°C. Then wash the cells as in Step 6.

9. Add 100 μL of cell fixative (e.g., from Intracellular Fixation and Permeabilization Buffer kit) to each well, and incubate for 20 min at room temperature.

10. Wash twice with ice-cold 1× permeabilization buffer (e.g., from Intracellular Fixation and Permeabilization Buffer kit). Centrifuge the plate at 300g for 8 min, flick buffer into sink, and wipe dish on tissue paper.

11. Perform primary and secondary antibody treatments.

    i. Add 100 μL of primary antibody against LC3 (diluted 1:300 in 1× permeabilization buffer) to each well, and incubate on a plate shaker for 20 min at room temperature. Wash with ice-cold FACS buffer for ImageStream as in Step 6.

    ii. If anti-LC3 is not directly conjugated, add 100 μL of secondary antibody (diluted 1:400 or 1:500 in 1× permeabilization buffer) to each well, and incubate on a plate shaker for 20 min at room temperature. Wash with ice-cold FACS buffer for ImageStream as in Step 6.

12. Add 60 μL of ice-cold FACS buffer for ImageStream to each well, and then use a pipette to transfer the cells from the wells of a 96-well plate to 0.5 mL tubes. Keep the tubes on ice, covered with aluminium foil to stop the photobleaching of fluorochromes, until the samples are ready to run on the ImageStream.

    The IS100 can only take 0.5-mL tubes.
    At this point, the cells can be kept at 4°C, covered with aluminium foil, for a maximum of 6 d.

13. Follow the manufacturer's instructions for analyzing the cells on the ImageStream.

    See Troubleshooting.

Data Analysis

Analyze the acquired data using the IDEAS software. For a flow chart outlining the data analysis steps, see Figure 1. For sample results, see Figure 2.

14. Gate on "in focus," "single," and "live" cells, using either the building blocks available in the IDEAS software or the readymade colocalization wizard.

    High Gradient RMS (root mean square of the rate of change of the image intensity profile) means better-focused cells. Clusters of cells, doublets, and large cells (e.g., monocytes) should be gated out.

    Gated cells will appear in the image gallery with the fluorochrome used to stain them.

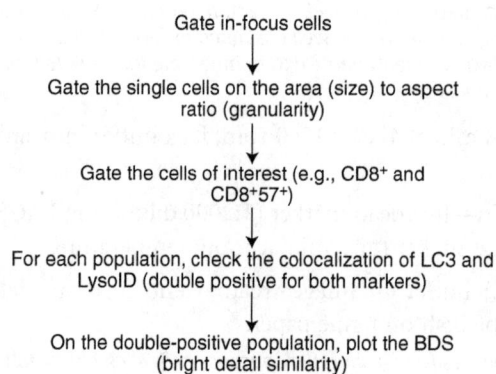

Gate in-focus cells
↓
Gate the single cells on the area (size) to aspect ratio (granularity)
↓
Gate the cells of interest (e.g., CD8⁺ and CD8⁺57⁺)
↓
For each population, check the colocalization of LC3 and LysoID (double positive for both markers)
↓
On the double-positive population, plot the BDS (bright detail similarity)

FIGURE 1. ImageStream data analysis outline.

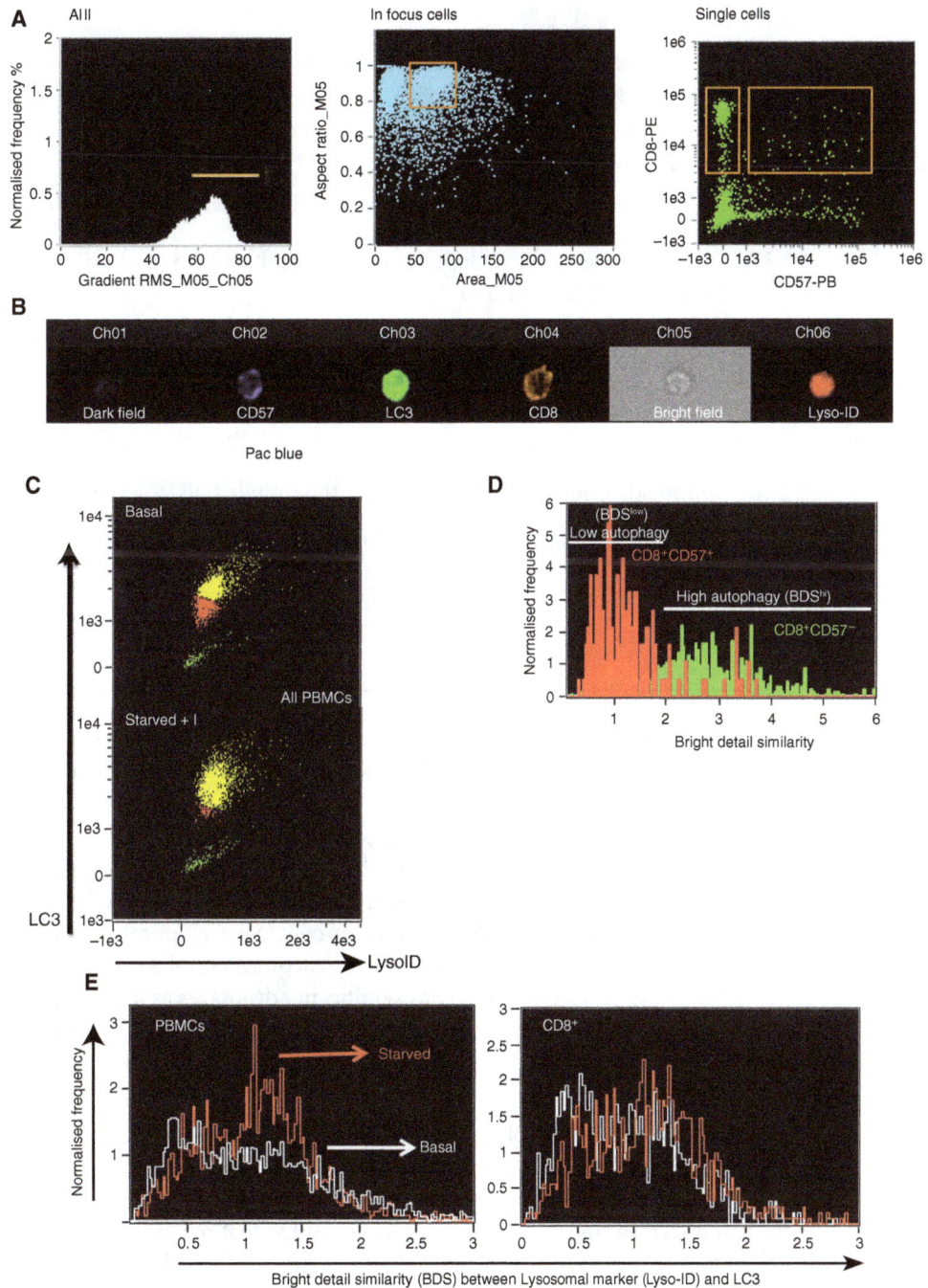

FIGURE 2. (A) Acquired T cells (*left* panel) gated on "cells in focus" (*middle* panel) and "single cells"/"cell type" (CD57/CD8-expressing) (*right* panel). (B) Image of a single human T cell (Ch01, Ch05) stained with antibody against CD57 conjugated to Pacific Blue (Ch02), with FITC-labeled antibody to autophagy protein LC3 (Ch03) CD8 conjugated to PE (Ch04) and with the proprietary dye Lyso-ID Red (Ch06) (Enzo Life Sciences). (C) PBMCs double-positive (bright, yellow) for LC3 and Lyso-ID from a healthy human donor under basal conditions (*upper*) or after 2 h of starvation in the presence of lysosomal protease inhibitors (I; i.e., E64d and pepstatin A) (*lower*). Note the increased frequency of autophagy in starved cells in which LC3 accumulates in the presence of the inhibitors. (D) Overlay of BDS for T cells (CD8$^+$CD57$^-$ / CD8$^+$CD57$^+$). CD8$^+$CD57$^+$ (area under red curve) shows low BDS and reduced autophagy when compared with CD8$^+$CD57$^-$ (area under green curve). (E) BDS overlay between basal (white line) and starved plus lysosomal protease inhibitors (red line) in all PBMCs (*left*) or CD8$^+$ T cells (*right*). The mean BDS signal both in all PBMCs and CD8$^+$ T cells shifts under the treatment with starvation and inhibitors (area under red histogram).

15. Gate on "cell type."

> *The cell type is identified by the antibodies against surface markers (e.g., use CD8 as a T-cell marker and CD57 as an aging T-cell marker).*

16. Gate on the two markers to be colocalized.

> *For example, identify cells staining double positive for LC3 and Lyso-ID. There is a colocalization wizard available in IDEAS, which is very user-friendly. If three markers need to be colocalized, it is necessary to use manual colocalization.*

17. On this double-positive population, plot results for bright detail similarity (BDS), either as mean BDS or gated BDS >1.5 or 2.

> *BDS is a feature available on IDEAS. It derives from the nonmean normalized Pearson's correlation coefficient (r) calculated for pairs of values taken from different data sources (pixel intensities in different channels) and is an estimate of the degree of colocalization. The difference in the mean BDS would tell the difference between the treatments.*
>
> *See Troubleshooting.*

18. (Optional) Count LC3 punctae using the inbuilt IDEAS wizard for spot counting.

> *These data can be compared with the LC3/LysoID BDS data for different treatments.*

TROUBLESHOOTING

Problem (Step 13): No LC3 staining or excess lysosomal staining is observed.
Solution: All the dyes and antibodies should be titrated for the particular cell type used.

Problem (Step 13): There are very few cells in the sample.
Solution: If working on a rare cell population, it is essential to increase the initial cell number and of course acquire more cells. At least 10,000–20,000 cells should be run per sample and 2000–5000 should be acquired for the single-color controls. Cell numbers are a serious issue while running cells on an ImageStream. The IS100 can acquire samples only at a speed of 100 cells/sec, but the later version (ISX or better) has significant advantages as it can acquire at a rate of 1000 cells/sec. There must be two to three million cells in the final suspension to acquire at least 10,000–20,000 cells in 15–20 min (this protocol was developed on the IS100).

Problem (Step 17): No colocalization is obtained in the presence of inhibitors.
Solution: A saturating concentration of inhibitor may not have been reached, or the cells may have died at the saturating concentration. The time and concentration required for inhibitors and inducers should be worked out for each individual cell type. Shorter incubation periods work well (longer incubations with inhibitors can be toxic to cells and might kill them). Only lysosomal protease inhibitors (E64d and PepA) will show an increase in colocalization because they inhibit degradation in, but not fusion with, lysosomes; in contrast, chloroquine and bafilomycin will show reduced colocalization in this assay as they change the lysosomal pH and therefore inhibit the fusion with lysosomes. Note that this is different from the situation for data based on western blotting or immunofluorescence, where all of these inhibitors should in theory increase the LC3II signal or LC3 puncta (see Protocol 1: Monitoring Autophagic Flux by Using Lysosomal Inhibitors and Western Blotting of Endogenous MAP1LC3B [Chittaranjan et al. 2015]) and Protocol 2: Monitoring the Localization of MAP1LC3B by Indirect Immunofluorescence [Ktistakis 2015]).

Problem (Step 17): There is no induction of autophagy in the experimental cell lines or primary cells.
Solution: Note the following points.

Cite this protocol as *Cold Spring Harb Protoc;* doi:10.1101/pdb.prot086272

- The induction of autophagy works consistently on PBMCs from young donors (age < 30 yr), but, if the donor is older or the cell lines have been overconfluent, it is less likely that induction of autophagy will occur.

- Some cell lines can require longer periods of starvation.

- For induction of autophagy, also consider using Torin 1 (ATP competitive inhibitor of mammalian target of rapamycin; see Protocol 2: Monitoring the Localization of MAP1LC3B by Indirect Immunofluorescence [Ktistakis 2015]).

Problem (Step 17): Spillage of Lyso-ID is evident.

Solution: Spillage of Lyso-ID can occur after fixation when samples are kept too long before acquisition. To avoid the issue, perform data acquisition immediately or within 12 h of staining. Alternatively, substitute Lyso-ID for an antibody directed against lysosomes; however, note that these results, for unknown reasons, might not be as reproducible.

DISCUSSION

Although the staining procedure and sample-acquisition methodologies described above are straightforward for someone with experience of flow cytometry, the data-analysis steps ideally require some training sessions. The technique has good potential for use in a variety of investigations, as evidenced by several published studies (Kwok et al. 2011; Lopez-Herrera et al. 2012; Phadwal et al. 2012).

The methodology outlined above measures the number of autolysosomes—that is, the turnover of intra-autolysosomal lipid-modified LC3-II—by quantifying the cells that are double-positive for LC3 and lysosomal markers (only the LC3-II will colocalize with the lysosomal marker) and showing colocalization (BDS) between these markers. The proximity of these two vesicles is also a good sign of their imminent fusion—that is, for autophagic flux. Inhibitors control for lysosomal defects and can be used to determine how much autophagosomal delivery to lysosomes takes place. If the lysosomal protease inhibitors E64d or PepA increase the BDS signal, this suggests that autophagic flux is induced. If the inhibitors result in no further increase, this could indicate that the lysosomal function is already suboptimal—that is, further inhibition does not augment this defect. The addition of chloroquine or bafilomycin should abolish all colocalization, which is a good indication that staining and data analysis are correct. Performing these two controls is really important.

This assay can give insight into the autophagic flux of a population (as it measures the delivery of cargo to lysosomes) if combined with evidence of subsequent degradation and recycling of cargo. The assay is high-throughput and automated, thus avoiding the subjective bias associated with some existing assays, and it correlates well with LC3 immunoblot and confocal results. The "turnover" in the ImageStream assay compared with LC3 turnover (measured by levels of LC3-I vs. LC3-II) in western blots has been found to be similar in cell lines, and the slight differences between these two assays might be due to the different read-outs—one is measuring LC3 turnover, whereas the other measures the number of cells with high autolysosome throughput.

Successful quantification of colocalization of LC3 labeled with green fluorescent protein with lysosomal markers in primary cells (from transgenic mice) has been shown with this technique and found to be similar to that of endogenous LC3 colocalization (K Phadwal, unpubl.).

The ImageStream technology can also measure colocalization of LC3 with the cargo to be degraded, such as mitochondria (mitophagy). Colocalization of three molecules is possible, but this requires advanced data analysis.

Recently, kits have become commercially available that can be used on conventional flow cytometers (see Introduction: Techniques for the Detection of Autophagy in Primary Mammalian Cells [Puleston et al. 2015]), which also work on primary cells. Two kits have been tested on primary cells: first, the LC3 kit from Merck Millipore that detects membrane-bound LC3-II after washing out the nonmembrane-bound LC3-I by using mild detergent and, second, Cyto-ID from Enzo Life Sciences that selectively labels preautophagosomes, autophagosomes, and autolysosomes while only minimally staining lysosomes (K Phadwal, personal observation). The precise labeling mechanism of Cyto-ID has not been revealed by the manufacturer (as also for the Lyso-ID Red detection kit for lysosomes). Finally, note that both the LC3 and Cyto-ID kits work well on human and murine macrophages and activated T cells.

RELATED TECHNIQUES

Related methodologies include intracellular LC3 staining by FACS (Eng et al. 2010); confocal staining for LC3 (see Protocol 2: Monitoring the Localization of MAPLC3B by Indirect Immunofluorescence [Ktistakis 2015]) and for LAMP1/LAMP2 (Eskelinen et al. 2002); and fluorescent microscopy for lysosomes or LC3 (Wang et al. 2013).

RECIPE

FACS Buffer for ImageStream

Phosphate-buffered saline (PBS)	1×
Fetal calf serum (FCS)	1%
Sodium azide (NaN₃)	0.02%

Store and use the buffer at 4°C. Discard after 3 mo.

ACKNOWLEDGMENTS

This work was funded by the Biomedical Research Centre Oxford (National Institute for Health Research), the Wellcome Trust, Natural Sciences and Engineering Research Council of Canada and the Biotechnology and Biological Sciences Research Council.

REFERENCES

Chittaranjan S, Bortnik S, Gorski SM. 2015. Monitoring autophagic flux by using lysosomal inhibitors and western blotting of endogenous MAP1LC3B. Cold Spring Harb Protoc doi: 10.1101/pdb.prot086256.

Eng KE, Panas MD, Karlsson Hedestam GB, McInerney GM. 2010. A novel quantitative flow cytometry-based assay for autophagy. Autophagy 6: 634–641.

Eskelinen EL, Illert AL, Tanaka Y, Schwarzmann G, Blanz J, Von Figura K, Saftig P. 2002. Role of LAMP2 in lysosome biogenesis and autophagy. Mol Biol Cell 13: 3355–3368.

Ktistakis NT. 2015. Monitoring the localization of MAP1LC3B by indirect immunofluorescence. Cold Spring Harb Protoc doi: 10.1101/pdb.prot086264.

Kwok AS, Phadwal K, Turner BJ, Oliver PL, Raw A, Simon AK, Talbot K, Agashe VR. 2011. HspB8 mutation causing hereditary distal motor neuropathy impairs lysosomal delivery of autophagosomes. J Neurochem 187: 5268–5276.

Lopez-Herrera G, Tampella G, Pan-Hammarström Q, Herholz P, Trujillo-Vargas CM, Phadwal K, Simon AK, Moutschen M, Etzioni A, Mory A, et al. 2012. Deleterious LRBA mutations in a novel syndrome of immune deficiency and autoimmunity. Am J Hum Genet 90: 986–1001.

Phadwal K, Alegre-Abarrategui J, Watson AS, Pike L, Anbalagan S, Hammond EM, Wade-Martins R, McMichael A, Klenerman P, Simon AK. 2012. A novel method for autophagy detection in primary cells: Impaired levels of macroautophagy in immunosenescent T cells. Autophagy 8: 677–689.

Puleston D, Phadwal K, Watson AS, Soilleux EJ, Chittaranjan S, Bortnik S, Gorski SM, Ktistakis N, Simon AK. 2015. Techniques for the detection of autophagy in primary mammalian cells. Cold Spring Harb Protoc doi: 10.1101/pdb.top070391.

Wang L, Chen M, Yang J, Zhang Z. 2013. LC3 fluorescent puncta in autophagosomes or in protein aggregates can be distinguished by FRAP analysis in living cells. Autophagy 9: 756–769.

Cite this protocol as Cold Spring Harb Protoc; doi:10.1101/pdb.prot086272

Detection of p62 on Paraffin Sections by Immunohistochemistry

Alexander S. Watson[1,3] and Elizabeth J. Soilleux[2,3]

[1]BRC Translational Immunology Laboratory, Nuffield Department of Medicine, Oxford OX3 9DS, United Kingdom; [2]Nuffield Department of Clinical Laboratory Sciences, Oxford OX3 9DS, United Kingdom

The study of autophagy in human disease is a rapidly expanding field. Diagnostic paraffin sections of a variety of patient tissues, including bone marrow, are available to researchers—yet are unsuitable for traditional autophagy quantification methods such as western blot or electron microscopy. This protocol outlines the immunohistochemical detection of the protein p62 (sequestosome-1, encoded by the gene *SQSTM1*)—an indicator of autophagic degradative activity—in slide-mounted paraffin sections such as bone marrow samples cut by a trephine. The p62 protein is an autophagic cargo adaptor, capable of binding to ubiquitylated proteins as well as autophagosome membrane proteins (LC3B and GABA(A) receptor-associated protein [GABARAP] family members) and hypothesized thus to target protein aggregates for lysosomal degradation. p62 itself is degraded by autophagy, remaining at low levels when autophagy is induced, and has been shown to accumulate when autophagy is deficient. Qualitative assessment and comparison of p62 staining between healthy and disease sections or disease subtypes will help target further investigation into the potential roles for autophagy in a variety of disorders.

MATERIALS

It is essential that you consult the appropriate Material Safety Data Sheets and your institution's Environmental Health and Safety Office for proper handling of equipment and hazardous material used in this protocol.

Reagents

Citroclear (TCS Biosciences HC5005)

Distilled water (deionized)

EnVision+ Dual Link System-HRP (DAB+) (Dako K4065)

Ethanol (100% and 50% stock solutions)

Hematoxylin solution (Sigma-Aldrich GHS316)

Horse serum

Immu-Mount (Thermo Scientific SD9990412)

Peptide corresponding to amino acids 387–436 of human p62 (H-CPPEADPRLIESLSQMLSMGFS-DEGGWLTRLLQTKNYDIGAALDTIQYSKH-OH) (Enzo Life Sciences PRN3424)

Phosphate-buffered saline (PBS; 1×, pH 7.4)

Polyclonal rabbit anti-human p62 antibody (Enzo Life Sciences BML-PW9860)

[3]Correspondence: alec.watson@alumni.ubc.ca; elizabeth.soilleux@ndcls.ox.ac.uk

Cite this protocol as *Cold Spring Harb Protoc*; doi:10.1101/pdb.prot086280

Slides with paraffin-embedded tissue sections

This protocol assumes that the tissues have already been fixed, paraffin-embedded, sectioned, and mounted on slides.

Target Retrieval Solution (10×; Dako S1699)

Equipment

Beaker (glass)
Light microscope (with camera)
Microscope coverslips (60 mm)
Microscope slides
Microwave (possessing variable power output)
PAP Pen (Liquid Blocker Super Pap Pen; Daido Sangyo)
Pencil
Slide staining box

This should be a flat box with parallel ridges, allowing slides to be placed flat for antibody staining while being held above the base of the box, where liquid will collect from the slide washes; provision of a lid for the box is ideal to prevent excess evaporation (see also washing rack below).

Squirt bottles

Ideally need two, one for PBS and one for distilled water washes.

Washing rack

This should hold multiple slides, together with a container for slide immersion (it can vary, but the rack should hold slides vertically close together, for efficiency, and the container should allow for complete immersion of the slides in wash solutions).

METHOD

Before starting, consider what the primary comparison will be for each experiment (e.g., comparing p62 levels between healthy and particular disease patients) as it is necessary that the samples to be compared should be stained together. For a summary of the whole procedure and the associated timelines, see Figure 1.

Primary Antibody and Peptide Preparation

1. Make up two p62 antibody mixes of volume suitable for 100 μL per slide (i.e., two separate tubes, each with 100 μL per intended sample, plus a dead volume of 200 μL). Start with a dilution of 1:800 in PBS containing 2% horse serum for bone marrow trephine samples.

   Staining more than 10 samples at once is not recommended owing to the accompanying increased handling time for incubations.

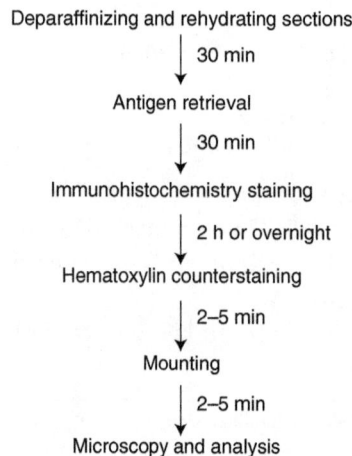

Deparaffinizing and rehydrating sections
↓ 30 min
Antigen retrieval
↓ 30 min
Immunohistochemistry staining
↓ 2 h or overnight
Hematoxylin counterstaining
↓ 2–5 min
Mounting
↓ 2–5 min
Microscopy and analysis

FIGURE 1. Timeline for performing the protocol.

Cite this protocol as *Cold Spring Harb Protoc*; doi:10.1101/pdb.prot086280

2. Keep one antibody mix as described, and to the other add a 400× molar excess of competing (cognate) peptide (here, peptide corresponding to amino acids 387–436 of human p62) to act as a negative control. Incubate both mixes for 1 h on ice before adding to microscope slides (at Step 12).

> *If desired, an additional control per sample can be stained with rabbit serum at the same protein concentration as that of the antibody.*

Deparaffinizing and Rehydrating Sections

3. With a pencil, label paraffin section slides Peptide (+) or Peptide (−).

4. Place slides horizontally in glass/metal slide staining holder/rack, then place in sequence (5 min each) in the following solutions:

   i. Citroclear I
   ii. Citroclear II (second aliquot of citroclear solution to avoid paraffin build-up)
   iii. 100% ethanol I
   iv. 100% ethanol II
   v. 50% ethanol
   vi. Distilled water

Antigen Retrieval

5. Microwave, at highest power (∼800 W) until boiling (∼3 min), a glass beaker or container (must be able to fit slide rack) containing ∼300 mL of 1× Target Retrieval Solution (diluted from 10× in distilled water).

> *Perform this step while the slides are in the final washing solution (Step 4.vi), leaving the slides in the distilled water until ready.*

6. Transfer the slide rack to the hot 1× Target Retrieval Solution and microwave for 15 min on medium or low power (∼150 W).

7. Transfer the slides to a container of PBS at room temperature to cool for 10 min.

> *The retrieval buffer can be saved for 1–2 more uses in refrigerator.*

Immunohistochemistry Staining (Primary and Secondary)

8. Draw circles around the tissue sections on the slides with a PAP pen (as small as possible; do quickly so that the samples do not dry) and place in a slide staining box.

9. Apply two to three drops of Dual Endogenous Enzyme Block (from EnVision+ Dual Link System-HRP [DAB+]). Incubate the slide for 5 min.

> *During breaks in this procedure, make up the chromogen solution to be used in Step 16 by adding 1 mL of DAB+ Substrate Buffer by eye to a tube, and then adding one drop of DAB+ Chromogen before mixing well.*

10. Remove the blocking solution with distilled water projected from a squirt bottle.

> *Do not aim at the tissue—instead, squirt above the PAP circle and let the liquid wash over the sample, while ensuring never to let the tissue dry out.*

11. Move the slides into a washing rack, wash twice for 3 min each in distilled water, and tap to dry.

12. Place the slides in a staining box, and add 100 µL of antibody or the control antibody–peptide mix per slide. Place a lid over the staining box and incubate for 1 h at room temperature (or leave in a cold room overnight).

> *It is advisable to put some liquid in the bottom of the slide box and cover to prevent evaporation.*

13. Remove incubation mixes with PBS, directing squirt bottle as in Step 10, and wash the slides with PBS in washing rack as in Step 11.

14. Place the slides in a staining box, and add two to three drops of labeled polymer-HRP per slide. Incubate for 30 min while covered as in Step 12.

15. Remove the labeled polymer-HRP with a squirt bottle as in Step 10, and wash slides in PBS as in Step 13.

16. Place the slides in a staining box, add ∼100 µL of the mixed chromogen solution per slide, and incubate for 10 min.

17. Remove the chromogen solution with a squirt bottle of distilled water as in Step 10.

    At this point, there is no need to wash using a holder (as in Step 11).

Hematoxylin Counterstaining

18. Add ∼100 µL of hematoxylin solution for couterstaining, leave for 30 sec, and then wash with water squirts.

19. Move the slides into a vertical washing rack, wash five times by adding fresh deionized water, swirling, and then dumping out the water each time.

    This will "blue" the samples.

Mounting and Imaging

20. Tap the slides to dry. Lay out microscope coverslips on the benchtop, and apply Immu-Mount to each coverslip (approximately one drop is sufficient). Then carefully drop each slide on top of a coverslip.

    Avoid getting bubbles between the tissue and the coverslip.

21. Leave the slides to dry in a cold-room overnight before imaging on a light microscope.

    It is advisable to store in cool and dark conditions, but image within 1–2 mo as sample drying can distort staining over longer periods.

    See Troubleshooting.

TROUBLESHOOTING

Problem (Step 21): The staining for p62 is either too strong or too weak to differentiate donors.
Solution: Titrate the p62 antibody, using a recommend dilution range from 1:200 to 1:1600 (concomitantly, adjust the control peptide concentration, as appropriate), until differences are visible between individuals.

Problem (Step 21): There is a high degree of variation between primary tissue-section donors.
Solution: For primary bone marrow sections, we have found that the decalcification method affects p62 staining—it is important to ensure that the procedures used for decalcification are consistent to enable a proper comparison. Bone marrow trephine specimens (for which the protocol is designed) are decalcified before processing to paraffin and therefore can vary in the way they have been handled before clinical diagnosis and thus before their use in research. It is infinitely preferable to obtain trephines that have been decalcified in 10% EDTA, pH 7.4.

Problem (Step 21): There is a high level of background staining in both the antibody and peptide–antibody control samples.
Solution: We have found some tissues to have a high level of background when analyzed with the Dako kit. Comparing the peptide–antibody control with the antibody-alone sample should allow identification of specific staining at a suitable level of antibody dilution.

DISCUSSION

Assessing the levels of p62 protein as a marker for autophagic degradation (Bjørkøy et al. 2005) has been explored in several assay formats, including western blot, immunofluorescence, and electron

Control (with peptide)   Experimental (no peptide)

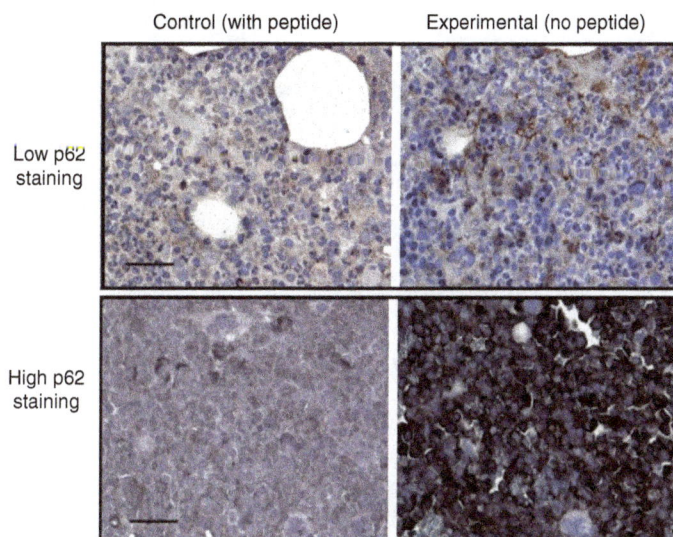

FIGURE 2. p62 staining on paraffin-embedded bone marrow trephine samples and qualitative comparison. Each donor (*upper* and *lower*) was stained either with antibody against p62 preincubated with blocking p62 peptide (controls, *left*) or with anti-p62 antibody alone (experimental samples, *right*). The control column shows very little positive brown staining (blue represents hematoxylin-stained nuclei), as expected, confirming the low level of background nonspecific staining levels. In the experimental column, the *lower* donor expresses higher levels of p62 protein than the *upper* donor, indicating defective autophagy. Scale bars, 50 µm.

microscopy (Bjørkøy et al. 2009). This protocol provides a complementary method for assessing p62 protein in patient bone marrow trephine specimens, important for pathological study, and could be adapted for similar antibody–cognate–peptide combinations. Other related techniques include many standard immunostaining protocols, in particular the assessment of LC3 protein by immunohisto-chemistry (Rosenfeldt et al. 2012), which could be used alongside p62 staining on tissue sections. Finally, this protocol could be transferred to, and optimized for, an automated immunostaining machine.

In our experience, when conducting data analysis, qualitative comparison of p62 staining on different sections is often sufficient to discern repeatable differences. Taking into account the level of nonspecific staining in the peptide-incubated controls, we have found that the level of p62 protein is visibly different between donors (e.g., Fig. 2). Further analysis and quantification of staining with image-analysis software might be possible; however, we have not explored this.

ACKNOWLEDGMENTS

We thank Enzo Life Sciences (Farmingdale, NY) for the kind provision of the p62 peptide, and also the Lady Tata Memorial Trust and Natural Sciences and Engineering Research Council of Canada for supporting A.S.W.

REFERENCES

Bjørkøy G, Lamark T, Brech A, Outzen H, Perander M, Overvatn A, Stenmark H, Johansen T. 2005. p62/SQSTM1 forms protein aggregates degraded by autophagy and has a protective effect on huntingtin-induced cell death. *J Cell Biol* **171**: 603–614.

Bjørkøy G, Lamark T, Pankiv S, Øvervatn A, Brech A, Johansen T. 2009. Monitoring autophagic degradation of p62/SQSTM1. *Methods Enzymol* **452**: 181–197.

Rosenfeldt MT, Nixon C, Liu E, Mah LY, Ryan KM. 2012. Analysis of macroautophagy by immunohistochemistry. *Autophagy* **8**: 963–969.

Detection of Mitochondrial Mass, Damage, and Reactive Oxygen Species by Flow Cytometry

Daniel Puleston[1]

Human Immunology Unit, Weatherall Institute of Molecular Medicine, John Radcliffe Hospital, Oxford OX3 9DS, United Kingdom

The reagents and procedures highlighted here will give the investigators an indication of the health status and volume of mitochondria in primary cells and cell lines through the use of a number of cell-permeable dyes. Mitochondrial volume can be monitored by using the probe Mito-Tracker Green FM. This reagent labels mitochondria in a manner that is independent of the membrane potential, therefore providing a readout relating purely to the mitochondrial mass of the cell. In contrast, MitoTracker Red CMXRos, tetramethylrhodamine methyl ester, and 10-N-nonyl acridine orange label mitochondria in a manner dependent on the membrane potential, thus giving an indication of mitochondrial stress. Using MitoSOX Red, it is also possible to analyze the production of the mitochondrial superoxide anion. Like the MitoTracker probes, MitoSOX Red is taken up passively by cells. In the mitochondria, the probe is oxidized by superoxide, resulting in the emission of red fluorescence.

MATERIALS

It is essential that you consult the appropriate Material Safety Data Sheets and your institution's Environmental Health and Safety Office for proper handling of equipment and hazardous material used in this protocol.

RECIPE: Please see the end of this protocol for recipes indicated by <R>. Additional recipes can be found online at http://cshprotocols.cshlp.org/site/recipes.

Reagents

Antibodies against cell-surface markers (optional; see Step 6)

Cell-viability stain

For identification of live cells, we recommend Live/Dead Fixable Violet Dead Cell Stain Kit (Life Technologies L34955). The use of cell-viability dyes is essential when analyzing mitochondrial volume and health in primary cells. Mitochondria undergo numerous changes during apoptosis that affect their membrane potential (Mito-Tracker Red and 10-N-Nonyl acridine orange [NAO] bind in a manner that is dependent on the mitochondrial membrane potential) and ability to conduct respiration (thereby affecting production of superoxide anions). Thus, inclusion of dead cells in the final analysis could therefore lead to misinterpretation of the data set.

Fetal calf serum (FCS; 5% in phosphate-buffered saline [PBS])

Fluorescence-activated cell sorting (FACS) buffer <R>

Mitochondrial dye of interest (see Table 1)

---

[1]Correspondence: daniel.puleston@ndm.ox.ac.uk

Cite this protocol as *Cold Spring Harb Protoc*; doi:10.1101/pdb.prot086298

TABLE 1. Summary of the parameters of useful mitochondrial dyes

Mitochondrial dye	Purpose	Working concentration	Diluent	Incubation time (at 37°C)	Fixable	Excitation/ emission (nm)
MitoTracker Green FM	Mitochondrial volume indicator	150 nM	PBS + 5% FCS	25 min	No	490/516
10-N-Nonyl acridine orange (NAO)	Mitochondrial stress indicator	100 nM	PBS	15 min	No	495/517
MitoTracker Red CMXRos	Mitochondrial stress indicator	200 nM	PBS + 5% FCS	25 min	Yes	579/599
MitoSOX Red	Mitochondrial superoxide indicator	5 µM	PBS	15 min	No	510/580
Tetramethylrhodamine, methyl ester (TMRM)	Mitochondrial stress indicator	100 nM	PBS	15 min	No	548/573

MitoSOX Red (Life Technologies M36008)

Prepare a 5-mM stock solution by resuspending the contents of the manufacturer's vial (50 µg) in 13 µL of high-quality dimethyl sulfoxide (DMSO). Before use, dilute as in Table 1 and keep on ice, protected from light.

MitoTracker Green FM (Life Technologies M-7514)

Prepare a 1-mM stock solution by resuspending the contents of the manufacturer's vial (50 µg) in 74.4 µL of high-quality DMSO. Before use, dilute as in Table 1 and keep on ice, protected from light.

MitoTracker Red CMXRos (Life Technologies M-7512)

Prepare a 1-mM stock solution by resuspending the contents of the manufacturer's vial (50 µg) in 94.1 µL of high-quality DMSO. Before use, dilute as in Table 1 and keep on ice, protected from light.

NAO (Life Technologies A-1372)

Tetramethylrhodamine methyl ester (TMRM; Life Technologies T-668)

Paraformaldehyde (2%–4% in PBS, freshly made) (optional; see Steps 9–11)

PBS

Equipment

Centrifuge
Flow cytometer (equipped with a 488 nm laser)
Incubator (for cell culture, at 37°C)
Mesh or filter (50 or 70 µm)
Pipettes
Round-bottom 96-well plate

METHOD

For a summary of the whole procedure and the associated timelines, see Figure 1.

Cell Preparation

1. Prepare a single-cell suspension.

   • If working with primary organs, use mechanical disruption of the tissue through a 50- or 70-µm mesh or filter into PBS containing 5% FCS.

   • If using cultured cells, remove them from the incubator and gently pipette them in and out. Filter them, if necessary, to form a single-cell suspension.

   Note that adherent cells will need to be trypsinized or unstuck first.

Chapter 11

Harvest cells and create
single cell suspension

Centrifuge

Stain pelleted cells with
viability dye

Incubate for 15 min
then wash ×2

Stain cells with antibodies
to relevant cell surface maker

Incubate for 20 min
then wash

Stain cells with mitochondrial
probe

Incubate
then wash ×2

Analyse cells on flow
cytometer

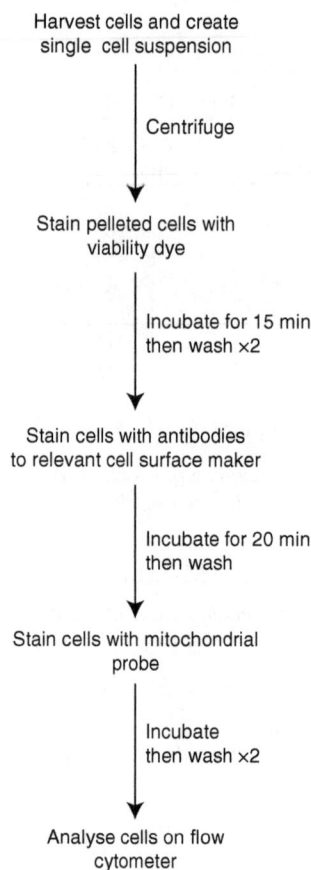

FIGURE 1. Timeline for performing the protocol.

2. Transfer $\leq 1 \times 10^6$ cells to each well of a round-bottom 96-well plate.

3. Pellet the cells by centrifuging at 300g for 5 min at 4°C and then discard the supernatant.

4. Stain cells in the 96-well plate by adding 100 µL of cell-viability Live/Dead Fixable Violet Dead Cell Stain at a dilution of 1:1000 in PBS, ensuring that the cell pellet is completely resuspended, and incubating the plates for 10 min at room temperature while protecting them from light.

5. Wash the cells twice by topping the wells with ~200 µL of FACS buffer (maintained at 4°C) and centrifuging (300g for 5 min at 4°C). Discard the supernatants.

6. (Optional) If working with a heterogeneous cell solution, it may be necessary to label the cells to identify individual subsets. To achieve this, add antibodies, diluted in the appropriate buffer, against the cell-surface markers of choice. Incubate the cells for 20–30 min at 4°C while protecting from light, and then wash the cells once as in Step 5.

Mitochondrial Staining

7. Label cells in the 96-well plate with 100 µL of the chosen mitochondrial stain (e.g., MitoSOX Red, MitoTracker Green FM, MitoTracker Red CMXRos, or NAO), dissolved in DMSO and diluted in PBS or PBS–FCS (see Table 1). Incubate the plate at 37°C for the appropriate incubation time (see Table 1).

8. Wash the cells twice as in Step 5.

   If using MitoTracker Red, continue to Step 9. If using another stain, proceed to Step 12.

Fixation

Steps 9–11 pertain to cells stained with MitoTracker Red only.

Cite this protocol as *Cold Spring Harb Protoc*; doi:10.1101/pdb.prot086298

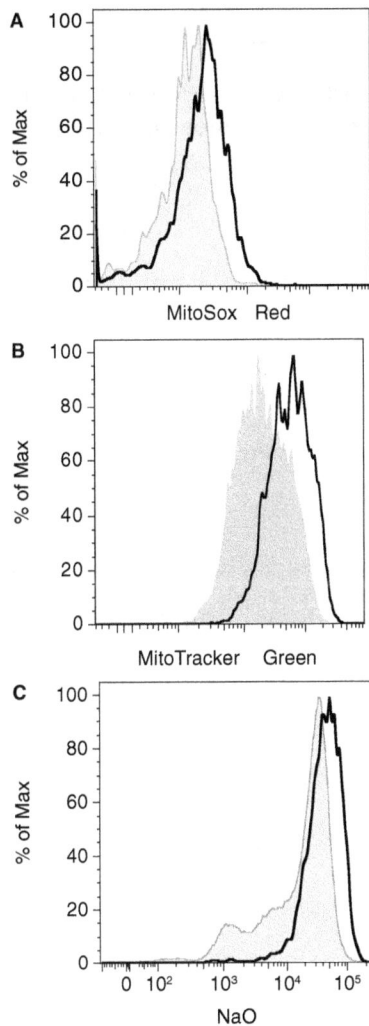

FIGURE 2. Use of fluorescent probes as indicators of mitochondrial health in murine bone-marrow-derived cells. (*A*) MitoSOX Red fluorescence and (*B*) MitoTracker Green fluorescence in T-cell-receptor-positive cells expressing the co-receptor CD8 (TCR⁺ CD8⁺ cells) and (*C*) NaO Green fluorescence in hematopoietic progenitors, all from 8-wk-old wild-type C57BL/6 mice (shaded histograms; controls) and C57BL/6 mice harboring a conditional deletion of the autophagy gene *Atg7* in the respective cells (nonshaded histograms). These data indicate that the mitochondria from the conditional autophagy mutant have a defect in mitochondrial function and also show increased steady-state levels of reactive oxygen species.

9. Add 200 μL of freshly made paraformaldehyde (2%–4% in PBS).

10. Incubate for 15 min at room temperature.

11. Wash the cells twice as in Step 5.

Analysis

12. Resuspend the cells in FACS buffer and analyze on a flow cytometer.

    For sample results, see Figure 2.

    See Troubleshooting.

TROUBLESHOOTING

Problem (Step 12): Compensation issues arise during flow-cytometric analysis.

Solution: MitoTracker Green and NAO, having fluorescein-like fluorescence, will typically "bleed" into the phycoerythrin (PE) channels. Similarly, MitoTracker Red and MitoSOX Red will appear in PE–Texas-Red in most modern flow cytometers and bleed strongly into the cyanine (Cy) signal of PE–Cy5 and PE–Cy7. Therefore, the use of these fluorophores, where possible, should be avoided when using the mitochondrial dyes featured in this protocol.

RELATED TECHNIQUES

For a general introduction to the techniques available for the detection of autophagy and autophagic flux, see Introduction: Techniques for the Detection of Autophagy in Primary Mammalian Cells (Puleston et al. 2015).

RECIPE

Fluorescence-Activated Cell Sorting Buffer

2% fetal calf serum (FCS)

0.02% sodium azide (NaN_3)

Prepare in phosphate-buffered saline (PBS). Filter-sterilize and store for up to 1 mo at 5°C.

ACKNOWLEDGMENTS

This work was funded by the Wellcome Trust and the Allan and Nesta Ferguson Charitable Trust.

REFERENCES

Puleston D, Phadwal K, Watson AS, Soilleux EJ, Chittaranjan S, Bortnik S, Gorski SM, Ktistakis N, Simon AK. 2015. Techniques for the detection of autophagy in primary mammalian cells. *Cold Spring Harb Protoc* doi: 10.1101/pdb.top070391.

Measuring Apoptosis in Mammals In Vivo

Andrea Newbold,[1] Ben P. Martin,[1] Carleen Cullinane,[2,3] and Michael Bots[4,5]

[1]Gene Regulation Laboratory, Cancer Therapeutics Program, Peter MacCallum Cancer Centre, East Melbourne 3002, Victoria, Australia; [2]Translational Research Laboratory, Cancer Therapeutics Program, Peter MacCallum Cancer Centre, East Melbourne 3002, Victoria, Australia; [3]Sir Peter MacCallum Department of Oncology, The University of Melbourne, Parkville 3010, Victoria, Australia; [4]Laboratory for Experimental Oncology and Radiobiology (LEXOR), Center for Experimental Molecular Medicine, Academic Medical Center, 1105 AZ, Amsterdam, The Netherlands

Apoptosis is a mode of cell death that is essential in multicellular organisms for the removal of superfluous, damaged, or potentially dangerous cells during development, infection, or normal tissue homeostasis. To prevent inflammation, cells undergoing apoptosis produce "find-me" signals that trigger the recruitment of phagocytes, which clear the apoptotic cells on recognition of "eat-me" signals. Despite the loss of billions of cells per day by apoptosis in the human body, the number of apoptotic cells found in healthy tissue is surprisingly low and reflects the efficiency of this process. However, in certain conditions (e.g., in cancer cells responding to chemotherapy), the number of apoptotic cells is too high to be efficiently cleared by phagocytes, and apoptotic cells can be observed. In these situations, the detection of apoptosis may be helpful in monitoring disease progression as well as in predicting the responses of tumors to anticancer therapies. Here we introduce various methods for monitoring apoptotic cells in vivo using a murine model of B-cell lymphoma and a solid tumor xenograft.

INTRODUCTION

Apoptosis is a mode of cell death that is essential in multicellular organisms for maintaining normal tissue homeostasis by eliminating superfluous, damaged, infected, and potentially dangerous cells (Taylor et al. 2008). Crucial for the induction of apoptosis is the activation of members of a family of cysteine proteases called caspases. Upon activation, these caspases cleave hundreds of substrates important for normal cellular function (Lüthi and Martin 2007). These activities and losses of function ultimately lead to morphological and biochemical changes characteristic of apoptosis: Nuclear condensation, shedding of apoptotic bodies, mitochondrial membrane permeabilization, DNA fragmentation, and the exposure of the phospholipid phosphatidylserine (Taylor et al. 2008; Kroemer et al. 2009).

Compared with other modes of cell death, apoptotic cells are quickly recognized and removed by phagocytes, thus preventing the release of proinflammatory signals into the surroundings (Taylor et al. 2008; Ravichandran 2011). Removal of the apoptotic cells is mainly mediated by "professional" phagocytes, such as macrophages and dendritic cells, which are recruited to apoptotic cells via the release of "find-me" signals (Ravichandran 2011). These signals serve as chemoattractants for phagocytes and include, among others, lysophosphatidylcholine and the chemokine (C-X3-C motif) ligand 1 (CX3CL1). In addition to signals involved in recruiting phagocytes, apoptotic cells also express "eat-

[5]Correspondence: m.bots@amc.uva.nl

Cite this introduction as Cold Spring Harb Protoc; doi:10.1101/pdb.top070417

me" signals. These signals, such as calriticulin and phosphatidylserine, promote recognition of apoptotic cells by phagocytes, which eventually lead to their engulfment and degradation.

Under normal circumstances, the process of engulfment of apoptotic cells is very efficient; although billions of cells die by apoptosis every day in the human body, the number of apoptotic cells detectable in vivo is limited (Elliott and Ravichandran 2010). It is only in certain conditions that the number of apoptotic cells is too high to be cleared efficiently by phagocytes and apoptotic cells can be visualized (Elliott and Ravichandran 2010). And it is those conditions—for example, certain neurodegenerative diseases—for which the detection of apoptosis in vivo may be very helpful in monitoring disease progression. Further, in therapeutic strategies aimed to eradicate potentially dangerous cells via apoptosis, such as in the treatment of cancer, detection of apoptosis may be beneficial. Thus, methods for detection of apoptosis not only allow for better monitoring of the response of cancer patients to chemotherapy, but can also facilitate the development of novel anticancer drugs in animal models.

DETECTING APOPTOSIS IN VIVO

Dozens of techniques are currently available for the identification of apoptotic cells in vitro (Galluzzi et al. 2009; Kepp et al. 2011), whereas the number of techniques used to discriminate between healthy and apoptotic cells in vivo is quite limited. One useful tool for assessing cell viability in vivo is positron emission tomography (PET), a noninvasive imaging modality that involves labeling biological compounds with positron-emitting radioisotopes. Depending on the particular PET tracer used, various biological processes can be visualized. In two of the accompanying protocols, we describe the use of PET imaging for assessing apoptosis in vivo (see Protocol 1: Fluorodeoxyglucose-Based Positron Emission Tomography Imaging to Monitor Drug Responses in Hematological Tumors [Newbold et al. 2014a] and Protocol 2: Fluorodeoxyglucose-Based Positron Emission Tomography Imaging to Monitor Drug Responses in Solid Tumors [Newbold et al. 2014b].) Because cancer cells are assumed to be highly dependent on glucose uptake, the glucose analogue fluorodeoxyglucose is a widely used PET tracer in oncology (Kelloff et al. 2005; Vander Heiden et al. 2009); it has been clinically used in the staging of cancer patients (Cullinane et al. 2005; Kelloff et al. 2005). The level of glucose uptake is generally considered to correlate with viable tumor mass. A potential complication of this technique in assessing apoptosis in vivo is the fact that a decreased uptake of glucose (associated with decreased tumor mass) may result from the induction of apoptosis, but also (or alternatively) from other processes such as cell cycle arrest. The development of more specific markers for detecting apoptotic cells, such as radiolabeled caspase inhibitors or annexin V, may solve this problem in the near future (Blankenberg et al. 1998; Nguyen et al. 2009).

Other methods commonly used to assess characteristic features of apoptosis in vivo include immunohistochemistry for evaluating DNA fragmentation or caspase activation, and flow cytometry for detecting DNA fragmentation and the expression of phosphatidylserine (Galluzzi et al. 2009; Kepp et al. 2011). We provide step-by-step protocols for detecting DNA fragmentation in apoptotic cells using immunohistochemical techniques and flow cytometry (see Protocol 3: Detection of Apoptotic Cells Using Immunohistochemistry [Newbold et al. 2014c] and Protocol 4: Detection of Apoptotic Cells Using Propidium Iodide Staining [Newbold et al. 2014d], respectively). Independent of the method chosen to probe for apoptosis, it is important to note that both the induction of apoptosis and the clearance of apoptotic cells are dynamic processes. Therefore, tracking apoptotic cells in vivo using either single end point determinations or assays that focus on a single characteristic apoptotic feature can be not only challenging, but also may not accurately reflect the true nature or sequence of events. Determining apoptosis in vivo using multiple time-points as well as different types of assays is thus strongly recommended.

In the associated protocols, we describe assays we commonly use to detect apoptotic tumor cells in vivo. Because we are primarily interested in the responses of hematological tumors to anticancer

Cite this introduction as *Cold Spring Harb Protoc*; doi:10.1101/pdb.top070417

drugs, we routinely work with mouse models of B-cell lymphoma or acute myeloid leukemia. We believe, however, that these methods are relevant in other (solid) tumor models as well.

ACKNOWLEDGMENTS

Work mentioned here was supported by a grant from the Dutch Cancer Society to M.B. (AMC2009-4457) and a grant from Pfizer Inc. to C.C.

REFERENCES

Blankenberg FG, Katsikis PD, Tait JF, Davis RE, Naumovski L, Ohtsuki K, Kopiwoda S, Abrams MJ, Darkes M, Robbins RC, et al. 1998. In vivo detection and imaging of phosphatidylserine expression during programmed cell death. *Proc Natl Acad Sci* **95**: 6349–6354.

Cullinane C, Dorow DS, Kansara M, Conus N, Binns D, Hicks RJ, Ashman LK, McArthur GA, Thomas DM. 2005. An in vivo tumor model exploiting metabolic response as a biomarker for targeted drug development. *Cancer Res* **65**: 9633–9636.

Elliott MR, Ravichandran KS. 2010. Clearance of apoptotic cells: Implications in health and disease. *J Cell Biol* **189**: 1059–1070.

Galluzzi L, Aaronson SA, Abrams J, Alnemri ES, Andrews DW, Baehrecke EH, Bazan NG, Blagosklonny MV, Blomgren K, Borner C, et al. 2009. Guidelines for the use and interpretation of assays for monitoring cell death in higher eukaryotes. *Cell Death Differ* **16**: 1093–1107.

Kelloff GJ, Hoffman JM, Johnson B, Scher HI, Siegel BA, Cheng EY, Cheson BD, O'Shaughnessy J, Guyton KZ, Mankoff DA, et al. 2005. Progress and promise of FDG-PET imaging for cancer patient management and oncologic drug development. *Clin Cancer Res* **11**: 2785–2808.

Kepp O, Galluzzi L, Lipinski M, Yuan J, Kroemer G. 2011. Cell death assays for drug discovery. *Nat Rev Drug Discov* **10**: 221–237.

Kroemer G, Galluzzi L, Vandenabeele P, Abrams J, Alnemri ES, Baehrecke EH, Blagosklonny MV, El-Deiry WS, Golstein P, Green DR, et al. 2009. Classification of cell death: Recommendations of the Nomenclature Committee on Cell Death 2009. *Cell Death Differ* **16**: 3–11.

Lüthi AU, Martin SJ. 2007. The CASBAH: A searchable database of caspase substrates. *Cell Death Differ* **14**: 641–650.

Newbold A, Martin BP, Cullinane C, Bots M. 2014a. Fluorodeoxyglucose-based positron emission tomography imaging to monitor drug responses in hematological tumors. *Cold Spring Harb Protoc* doi: 10.1101/pdb.prot082511.

Newbold A, Martin BP, Cullinane C, Bots M. 2014b. Fluorodeoxyglucose-based positron emission tomography imaging to monitor drug responses in solid tumors. *Cold Spring Harb Protoc* doi: 10.1101/pdb.prot082529.

Newbold A, Martin BP, Cullinane C, Bots M. 2014c. Detection of apoptotic cells using immunohistochemistry. *Cold Spring Harb Protoc* doi: 10.1101/pdb.prot082537.

Newbold A, Martin BP, Cullinane C, Bots M. 2014d. Detection of apoptotic cells using propidium iodide staining. *Cold Spring Harb Protoc* doi: 10.1101/pdb.prot082545.

Nguyen Q-D, Smith G, Glaser M, Perumal M, Arstad E, Aboagye EO. 2009. Positron emission tomography imaging of drug-induced tumor apoptosis with a caspase-3/7 specific [18F]-labeled isatin sulfonamide. *Proc Natl Acad Sci* **106**: 16375–16380.

Ravichandran KS. 2011. Beginnings of a good apoptotic meal: The find-me and eat-me signaling pathways. *Immunity* **35**: 445–455.

Taylor RC, Cullen SP, Martin SJ. 2008. Apoptosis: Controlled demolition at the cellular level. *Nat Rev Mol Cell Biol* **9**: 231–241.

Vander Heiden MG, Cantley LC, Thompson CB. 2009. Understanding the Warburg effect: The metabolic requirements of cell proliferation. *Science* **324**: 1029–1033.

Fluorodeoxyglucose-Based Positron Emission Tomography Imaging to Monitor Drug Responses in Hematological Tumors

Andrea Newbold,[1] Ben P. Martin,[1] Carleen Cullinane,[2,3] and Michael Bots[4,5]

[1]Gene Regulation Laboratory, Cancer Therapeutics Program, Peter MacCallum Cancer Centre, East Melbourne 3002, Victoria, Australia; [2]Translational Research Laboratory, Cancer Therapeutics Program, Peter MacCallum Cancer Centre, East Melbourne 3002, Victoria, Australia; [3]Sir Peter MacCallum Department of Oncology, The University of Melbourne, Parkville 3010, Victoria, Australia; [4]Laboratory for Experimental Oncology and Radiobiology (LEXOR), Center for Experimental Molecular Medicine, Academic Medical Center, 1105 AZ, Amsterdam, The Netherlands

Positron emission tomography (PET) can be used to monitor the uptake of the labeled glucose analog fluorodeoxyglucose (^{18}F-FDG), a process that is generally believed to reflect viable tumor cell mass. The use of ^{18}F-FDG PET can be helpful in documenting over time the reduction in tumor mass volume in response to anticancer drug therapy in vivo. In this protocol, we describe how to monitor the response of murine B-cell lymphomas to an inducer of apoptosis, the anticancer drug vorinostat (a histone deacetylase inhibitor). B-cell lymphoma cells are injected into recipient mice and, on tumor formation, the mice are treated with vorinostat. The tracer ^{18}F-FDG is then injected into the mice at several time points, and its uptake is monitored using PET. Because the uptake of ^{18}F-FDG is not a direct measure of apoptosis, an additional direct method proving that apoptotic cells are present should also be performed.

MATERIALS

It is essential that you consult the appropriate Material Safety Data Sheets and your institution's Environmental Health and Safety Office for proper handling of equipment and hazardous materials used in this protocol.

RECIPES: Please see the end of this protocol for recipes indicated by <R>. Additional recipes can be found online at http://cshprotocols.cshlp.org/site/recipes.

Reagents

Apoptosis inducer (e.g., vorinostat)

Prepare vorinostat by dissolving at 10 mM in dimethylsulfoxide. Store at −20°C, and avoid multiple freeze–thaw cycles.

C57BL/6 mice (8 wk–10 wk of age; see Step 7)
Complete Anne Kelso medium <R>
Eµ-Myc B-cell lymphoma cells (cryopreserved)
Fluorodeoxyglucose (^{18}F-FDG)
Isoflurane
Phosphate-buffered saline (PBS) <R>

---

[5]Correspondence: m.bots@amc.uva.nl

Cite this protocol as *Cold Spring Harb Protoc*; doi:10.1101/pdb.prot082511

Equipment

Automated hematology analyzer (Advia 120 or equivalent)
Centrifuge (benchtop)
Conical tubes (50 mL and 15 mL)
Hemocytometer
PET scanner (for small animals)
Syringes (1 mL) with 25-gauge needles
Veterinary anesthetic machine

METHOD

Figure 1 summarizes the sequence of steps in this method. The procedure itself is straightforward, but when performing a large (≥12-animal) PET experiment, it is advisable to plan the experiment carefully to adequately manage the number of manipulations on the day of imaging. In addition, because a large amount of radioactivity is required for such an experiment, the development of workflows and shielding is essential to minimize radiation exposure.

Transplantation of B-Cell Lymphoma Cells and Tumor Development

1. Thaw cryopreserved Eµ-Myc B-cell lymphoma cells in complete Anne Kelso medium, wash the cells twice with PBS, and perform a cell count.

2. Resuspend 2×10^6 Eµ-Myc B-cell lymphoma cells in 1 mL of PBS and transfer 200 µL of the cell suspension (4×10^5 cells) into syngeneic recipient mice by tail vein injection.

3. During tumor development over the next 7–14 d, monitor the disease burden by palpating the inguinal or brachial lymph nodes and by measuring the concentration of white blood cells (WBC) in ∼25 µL of blood from the tail vein using an Advia 120 automated hematology analyzer or equivalent.

4. Once lymph nodes are clearly palpable (usually corresponding to a peripheral WBC concentration of $\sim 50 \times 10^6$ cells/mL or greater), randomize the mice into treatment groups.

5. Administer the apoptosis inducer to the mice (e.g., inject vorinostat intraperitoneally at 200 mg/kg).

Delivery of Tracer and Measurement of Uptake

In Steps 6–12, ^{18}F-FDG uptake is monitored at several time points (e.g., at 4 h, 8 h, 12 h, and 24 h) after administration of vorinostat.

6. At 4.5 h before imaging the mice, begin a fasting regime.

7. After 3 h of fasting, anesthetize the mice. Use 2.5% isoflurane in air containing 50% oxygen, and deliver at a flow rate of 200 mL/min.

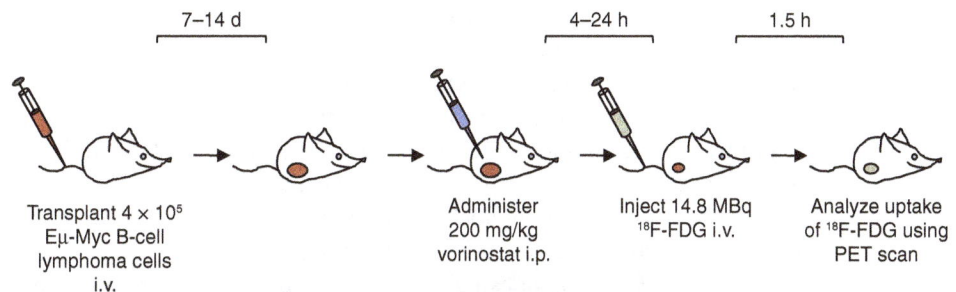

FIGURE 1. Flowchart outlining the steps of ^{18}F-FDG-PET monitoring of the apoptotic response in hematological tumors. i.v., Intravenous; i.p., intraperitoneal.

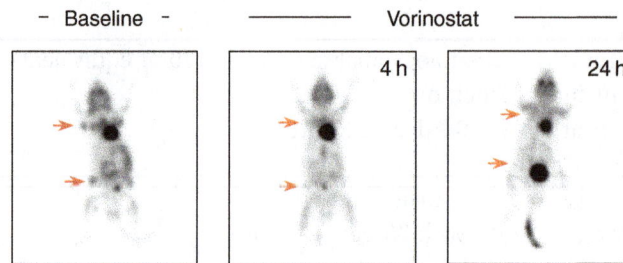

FIGURE 2. ^{18}F-FDG PET images of mice bearing Eμ-Myc lymphomas scanned at the baseline and at 4 h and 24 h after a dose of 200 mg/kg vorinostat. The arrows indicate the tumor burden in both the brachial and the inguinal lymph nodes.

> *C57BL/6 mice are very sensitive to anesthetics (they may require ≤2% isoflurane); if the tumor burden is too high, animals may die under anesthesia. Other mouse strains such as BALB/c nude and SCID mice are more robust under anesthesia.*

8. Inject each mouse with 14.8 MBq of ^{18}F-FDG via the tail vein.

> *The amount of tracer administered to the animals is critically important, and should be performed by an experienced researcher.*

9. Maintain the mice under anesthesia for an additional 20 min before allowing the animals to recover.

10. At 1.5 h after administration of the tracer, anesthetize the mice (as in Step 7).

11. Scan the mice for the uptake of ^{18}F-FDG on a small animal PET scanner for 5 min.

> *The accumulation of tracer in the tails of the animals indicates that some of the tracer has not been delivered to the animal's body.*

12. Continue to measure ^{18}F-FDG uptake by repeating Steps 7–11 at additional time points (e.g., at 8 h, 12 h, and 24 h after injection of the apoptosis inducer).

> *For examples of ^{18}F-FDG PET images of mice after treatment with vorinostat, see Figure 2.*

Data Analysis

13. After completing the acquisitions of images, reconstruct the PET images using the software on the PET scanner.

14. On each transaxial image slice, use the region-of-interest (ROI) software to draw an ROI around the tumor and a representative background region.

15. Calculate the tumor-to-background ratio by dividing the maximum pixel intensity within a tumor ROI by the average pixel intensity within the background ROI (e.g., see Cullinane et al. 2011).

RELATED TECHNIQUES

Although widely used as a surrogate indication for viable tumor mass, ^{18}F-FDG measures glucose metabolism. Thus, the decreased uptake of ^{18}F-FDG is not always correlated with cell death, but could alternatively result from cell cycle arrest or a change in the expression of glucose transporters. To prove that the response to anticancer drug therapy is attributable to apoptosis, the results obtained with this protocol should be confirmed with additional assays (e.g., TUNEL or annexin V staining). The use of tracers specifically designed to detect apoptotic cells, if they are available, is informative, as is conducting an additional apoptosis-specific readout to confirm the induction of apoptosis once mice show decreased uptake of ^{18}F-FDG (Blankenberg et al. 1998; Nguyen et al. 2009).

^{18}F-FDG PET imaging can also be used to monitor the response of solid tumors to apoptosis-inducing drugs (see Protocol 2: Fluorodeoxyglucose-Based Positron Emission Tomography Imaging to Monitor Drug Responses in Solid Tumors [Newbold et al. 2014]).

RECIPES

Complete Anne Kelso Medium

Dulbecco's modified Eagle's medium (DMEM), high-glucose
10% fetal calf serum (FCS)
100 units/mL penicillin G
100 µg/mL streptomycin
0.1 mM L-asparagine
50 µM 2-mercaptoethanol

Store at 4°C for up to several months.

Phosphate-Buffered Saline (PBS)

Reagent	Amount to add (for 1× solution)	Final concentration (1×)	Amount to add (for 10× stock)	Final concentration (10×)
NaCl	8 g	137 mM	80 g	1.37 M
KCl	0.2 g	2.7 mM	2 g	27 mM
Na$_2$HPO$_4$	1.44 g	10 mM	14.4 g	100 mM
KH$_2$PO$_4$	0.24 g	1.8 mM	2.4 g	18 mM

If necessary, PBS may be supplemented with the following:

CaCl$_2$•2H$_2$O	0.133 g	1 mM	1.33 g	10 mM
MgCl$_2$•6H$_2$O	0.10 g	0.5 mM	1.0 g	5 mM

PBS can be made as a 1× solution or as a 10× stock. To prepare 1 L of either 1× or 10× PBS, dissolve the reagents listed above in 800 mL of H$_2$O. Adjust the pH to 7.4 (or 7.2, if required) with HCl, and then add H$_2$O to 1 L. Dispense the solution into aliquots and sterilize them by autoclaving for 20 min at 15 psi (1.05 kg/cm^2) on liquid cycle or by filter sterilization. Store PBS at room temperature.

ACKNOWLEDGMENTS

Work described here was supported by a grant from the Dutch Cancer Society to M.B. (AMC2009-4457) and a grant from Pfizer Inc. to C.C.

REFERENCES

Blankenberg FG, Katsikis PD, Tait JF, Davis RE, Naumovski L, Ohtsuki K, Kopiwoda S, Abrams MJ, Darkes M, Robbins RC, et al. 1998. In vivo detection and imaging of phosphatidylserine expression during programmed cell death. *Proc Natl Acad Sci* 95: 6349–6354.

Cullinane C, Dorow D, Jackson S, Solomon B, Bogatyreva E, Binns D, Young R, Arango ME, Christensen JG, McArthur GA, et al. 2011. Differential (18)F-FDG and 3′-deoxy-3′-(18)F-fluorothymidine PET responses to pharmacologic inhibition of the c-MET receptor in preclinical tumor models. *J Nucl Med* 52: 1261–1267.

Newbold A, Martin BP, Cullinane C, Bots M. 2014. Fluorodeoxyglucose-based positron emission tomography imaging to monitor drug responses in solid tumors. *Cold Spring Harb Protoc* doi: 10.1101/pdb.prot082529.

Nguyen Q-D, Smith G, Glaser M, Perumal M, Arstad E, Aboagye EO. 2009. Positron emission tomography imaging of drug-induced tumor apoptosis with a caspase-3/7 specific [18F]-labeled isatin sulfonamide. *Proc Natl Acad Sci* 106: 16375–16380.

Fluorodeoxyglucose-Based Positron Emission Tomography Imaging to Monitor Drug Responses in Solid Tumors

Andrea Newbold,[1] Ben P. Martin,[1] Carleen Cullinane,[2,3] and Michael Bots[4,5]

[1]Gene Regulation Laboratory, Cancer Therapeutics Program, Peter MacCallum Cancer Centre, East Melbourne 3002, Victoria, Australia; [2]Translational Research Laboratory, Cancer Therapeutics Program, Peter MacCallum Cancer Centre, East Melbourne 3002, Victoria, Australia; [3]Sir Peter MacCallum Department of Oncology, The University of Melbourne, Parkville 3010, Victoria, Australia; [4]Laboratory for Experimental Oncology and Radiobiology (LEXOR), Center for Experimental Molecular Medicine, Academic Medical Center, 1105 AZ, Amsterdam, The Netherlands

Positron emission tomography (PET) is used to monitor the uptake of the labeled glucose analogue fluorodeoxyglucose (^{18}F-FDG) by solid tumor cells, a process generally believed to reflect viable tumor cell mass. The use of ^{18}F-FDG exploits the high demand for glucose in tumor cells, and serves to document over time the response of a solid tumor to an inducer of apoptosis. The apoptosis inducer crizotinib is a small-molecule inhibitor of c-Met, a receptor tyrosine kinase that is often dysregulated in human tumors. In this protocol, we describe how to monitor the response of a solid tumor to crizotinib. Human gastric tumor cells (GTL-16 cells) are injected into recipient mice and, on tumor formation, the mice are treated with crizotinib. The tracer ^{18}F-FDG is then injected into the mice at several time points, and its uptake is monitored using PET. Because ^{18}F-FDG uptake varies widely among different tumor models, preliminary experiments should be performed with each new model to determine its basal level of ^{18}F-FDG uptake. Verifying that the basal level of uptake is sufficiently above background levels will assure accurate quantitation. Because ^{18}F-FDG uptake is not a direct measure of apoptosis, it is advisable to carry out an additional direct method to show the presence of apoptotic cells.

MATERIALS

It is essential that you consult the appropriate Material Safety Data Sheets and your institution's Environmental Health and Safety Office for proper handling of equipment and hazardous materials used in this protocol.

Reagents

Apoptosis inducer (e.g., the c-Met inhibitor crizotinib)

Crizotinib is prepared at Pfizer Laboratories at a concentration of 5 mg/mL; dose at 50 mg/kg. As a control for crizotinib treatment, use the vehicle (solvent or solution) in which crizotinib is dissolved.

Fluorodeoxyglucose (^{18}F-FDG)

Formalin

Human gastric tumor cell line GTL-16

Isoflurane

[5]Correspondence: m.bots@amc.uva.nl

Cite this protocol as *Cold Spring Harb Protoc*; doi:10.1101/pdb.prot082529

Liquid nitrogen
Matrigel (50%)
RPMI 1640 culture medium containing 10% fetal bovine serum (FBS)
SCID mice (6–8 wk of age)

Equipment

Centrifuge (benchtop)
Conical tubes (50- and 15-mL)
Electronic calipers
Electronic shaver
Hemocytometer
Insulin syringes with a permanently attached needle (29G)
Light microscope
PET scanner (for small animals)
Syringes (1-mL) with 20-gauge plastic gavage tubes
Veterinary anesthetic machine

METHOD

Figure 1 summarizes the sequence of steps in this method. Because a large amount of radioactivity is required for a large (≥12-animal) PET experiment, the development of workflows and shielding is essential to minimize radiation exposure. Expected timeframes are indicated for a 24-mouse experiment (i.e., for three groups of eight mice). The growth of subcutaneous tumors in vivo can vary widely. Therefore, we recommend injecting tumor cells into 20% more mice than will be required for the ^{18}F-FDG PET imaging analysis to ensure that a sufficient number of mice with similarly sized tumors will be available for randomization.

Transplantation of Tumor Cells and Tumor Development

1. One day before implantation, shave the fur from the right side of each mouse.
   Shaving the mice will take ~45 min.

2. Culture the tumor cell line GTL-16 in RPMI 1640 culture medium containing 10% FBS until the cells are ~90% confluent. Harvest the cells and prepare a cell suspension of 20×10^6 cells/mL in 50% Matrigel.

3. Using an insulin syringe, transfer 200 µL of the cell suspension by subcutaneous injection into the right flank of each mouse.

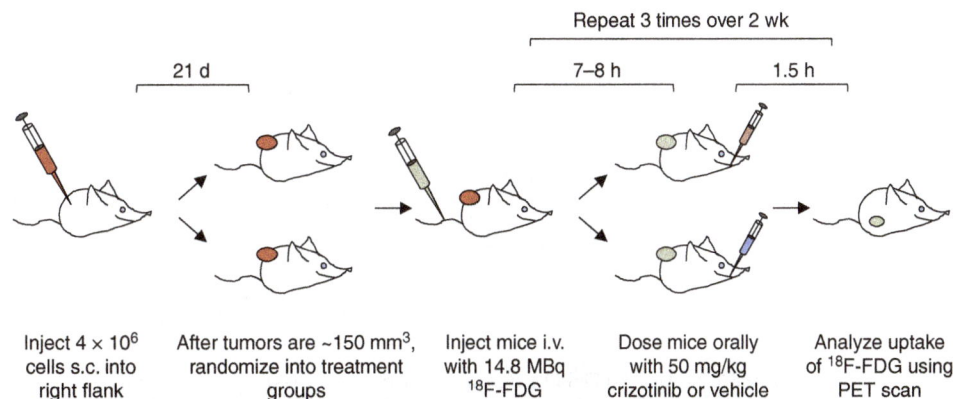

FIGURE 1. Flowchart outlining the steps of ^{18}F-FDG-PET monitoring of the apoptotic response in solid tumors. s.c., Subcutaneous; i.v., intravenous.

The FDG signal arising from the bladder can be very high if animals do not void before scanning. Therefore, to ensure adequate separation of signal from the bladder and from the tumor, implant the tumors into the mid- to forward-flank area. Preparing the cells and injecting them into the mice will take ~2.5 h.

4. Monitor tumor development over the next 2–3 wk by palpating the right flank area. When the tumors become palpable, use electronic calipers to measure the tumors and calculate the tumor volume. Do this every 2–3 d until the average tumor volume reaches ~150 mm^3.

   Monitoring the mice and measuring tumor growth will require ~1 h per wk over 3 wk.

Delivery of Tracer and Measurement of Uptake

Randomizing mice into treatment groups, fasting, PET imaging, and drug dosing will require ~7–8 h.

5. Randomize the mice into treatment groups of at least eight mice per group and move them into clean housing boxes.

6. Begin a fasting regime and maintain it for a minimum of 3 h.

7. After at least 3 h of fasting, anesthetize the mice. Use 2.5% isoflurane in air containing 50% oxygen, and deliver it at a flow rate of 200 mL/min.

8. To introduce the tracer, inject each mouse with 14.8 MBq of ^{18}F-FDG via the tail vein.

   The amount of tracer administered to the animals is critically important, and should be performed by an experienced researcher.

9. Maintain the mice under anesthesia for an additional 20 min before allowing the animals to recover.

10. At 90 min after injection of the tracer, anesthetize the mice as in Step 7.

11. Position a mouse on the bed of the PET scanner and scan it for 10 min.

12. Remove the scanned mouse from anesthesia and return it to its housing cage to recover.

13. Repeat Steps 11–12 for each mouse, until all mice have been scanned.

14. Dose each group of mice by oral gavage with crizotinib at 50 mg/kg or with the vehicle at 0.1 mL/10 g of body weight.

15. Repeat the dosing (Step 14) every day for 2 wk.

16. At three separate time points during the 2-wk period, repeat Steps 6–13.

    For examples of ^{18}F-FDG PET images of mice after treatment with the vehicle and with crizotinib, see Figure 2.

17. Following the final scan, kill the mice and harvest the tumors. Snap-freeze one half of each tumor sample in liquid nitrogen and fix the other half in formalin for biomarker analysis.

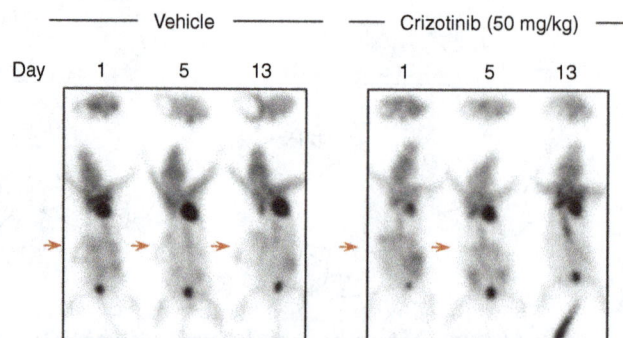

FIGURE 2. Serial ^{18}F-FDG PET transaxial (*upper*) and maximum-intensity-projection (*lower*) images of representative mice from groups treated either with vehicle (*left* panel) or with 50 mg/kg crizotinib (*right* panel). The arrows indicate the positions of GTL-16 gastric tumor cells. (This research was originally published in *JNM* [Cullinane et al. 2011] © by the Society of Nuclear Medicine and Molecular Imaging, Inc.)

Data Analysis

18. After completing the acquisitions of images, reconstruct the PET images using the software on the PET scanner.

19. On each transaxial image slice, use the region-of-interest (ROI) software to draw a ROI around the tumor and around a representative background region.

20. Calculate the tumor-to-background ratio by dividing the maximum pixel intensity within a tumor ROI by the average pixel intensity within the background ROI.

    For an example, see Cullinane et al. (2011).

RELATED TECHNIQUES

Although widely used as readout for viable tumor mass, ^{18}F-FDG PET measures glucose metabolism. Decreased uptake of ^{18}F-FDG may not be caused by apoptosis; it could instead be the result of cell cycle arrest or a change in the expression of glucose transporters. Because ^{18}F-FDG uptake is not a specific marker of apoptosis, apoptosis-specific biomarker assays (e.g., TUNEL or annexin V staining) should be performed in parallel to confirm the presence of tumor cell apoptosis (Zou et al. 2007; Cullinane et al. 2011).

^{18}F-FDG PET imaging can also be used to monitor the response of other tumor types to apoptosis-inducing drugs (see Protocol 1: Fluorodeoxyglucose-Based Positron Emission Tomography Imaging to Monitor Drug Responses in Hematological Tumors [Newbold et al. 2014]).

REFERENCES

Cullinane C, Dorow DS, Jackson S, Solomon B, Bogatyreva E, Binns D, Young R, Arango ME, Christensen JG, McArthur GA, et al. 2011. Differential ^{18}F-FDG and 3′-deoxy-3′-^{18}F-fluorothymidine PET responses to pharmacologic inhibition of the c-MET receptor in preclinical tumor models. *J Nucl Med* **52:** 1261–1267.

Newbold A, Martin BP, Cullinane C, Bots M. 2014. Fluorodeoxyglucose-based positron emission tomography imaging to monitor drug responses in hematological tumors. *Cold Spring Harb Protoc* doi: 10.1101/pdb.prot082511.

Zou HY, Li Q, Lee JH, Arango ME, McDonnell SR, Yamazaki S, Koudriakova TB, Alton G, Cui JJ, Kung P-P, et al. 2007. An orally available small-molecule inhibitor of c-Met, PF-2341066, exhibits cytoreductive antitumor efficacy through antiproliferative and antiangiogenic mechanisms. *Cancer Res* **67:** 4408–4417.

Detection of Apoptotic Cells Using Immunohistochemistry

Andrea Newbold,[1] Ben P. Martin,[1] Carleen Cullinane,[2,3] and Michael Bots[4,5]

[1]Gene Regulation Laboratory, Cancer Therapeutics Program, Peter MacCallum Cancer Centre, East Melbourne 3002, Victoria, Australia; [2]Translational Research Laboratory, Cancer Therapeutics Program, Peter MacCallum Cancer Centre, East Melbourne 3002, Victoria, Australia; [3]Sir Peter MacCallum Department of Oncology, The University of Melbourne, Parkville 3010, Victoria, Australia; [4]Laboratory for Experimental Oncology and Radiobiology (LEXOR), Center for Experimental Molecular Medicine, Academic Medical Center, 1105 AZ, Amsterdam, The Netherlands

Immunohistochemistry is commonly used to show the presence of apoptotic cells in situ. In this protocol, B-cell lymphoma cells are injected into recipient mice and, on tumor formation, the mice are treated with the apoptosis inducer vorinostat (a histone deacetylase inhibitor). Tumor samples are fixed and sectioned, and fragmented DNA (a feature of apoptotic cells) is end-labeled by terminal deoxynucleotidyl transferase dUTP nick-end labeling (TUNEL). Immunohistochemical methods are then used to detect the labeled DNA and identify B-cell lymphoma cells in the last stage of apoptosis. Because the assay can lead to false-positive results, it is advisable to carry out an additional assay (e.g., immunohistochemistry for active caspase-3) to confirm the presence of apoptotic cells.

MATERIALS

It is essential that you consult the appropriate Material Safety Data Sheets and your institution's Environmental Health and Safety Office for proper handling of equipment and hazardous materials used in this protocol.

RECIPES: Please see the end of this protocol for recipes indicated by <R>. Additional recipes can be found online at http://cshprotocols.cshlp.org/site/recipes.

Reagents

ApopTag Peroxidase In Situ Apoptosis Detection Kit (Millipore) or equivalent

The kit includes equilibration buffer, reaction buffer, TdT enzyme, stop/wash buffer, and horseradish peroxidase (HRP)-conjugated antidigoxygenin antibody. Prepare a working solution of TdT in reaction buffer according to the manufacturer's directions.

Apoptosis inducer (e.g., vorinostat)

Prepare vorinostat by dissolving at 10 mM in DMSO. Store at −20°C, and avoid multiple freeze–thaw cycles.

C57BL/6 mice (8–10 wk of age)
Complete Anne Kelso medium <R>
Diaminobenzidine (DAB)

---

[5]Correspondence: m.bots@amc.uva.nl

Cite this protocol as Cold Spring Harb Protoc; doi:10.1101/pdb.prot082537

Dimethylsulfoxide (DMSO)

Ethanol (70%, 95%, and 100%)

Eµ-Myc B-cell lymphoma cells (cryopreserved)

Formalin, neutral-buffered (10%)

Hematoxylin

Hydrogen peroxide in PBS (3% solution)

Mounting medium

Paraffin

Phosphate-buffered saline (PBS) <R>

Proteinase K

Water (distilled)

Xylene

Equipment

Automated hematology analyzer (Advia 120 or equivalent)

Centrifuge (benchtop)

Conical tubes (50- and 15-mL)

Coplin jars

Coverslips

Dissection tools (small surgical scissors and curved microforceps)

Glass slides (Superfrost Plus)

Hemocytometer

Humidified chamber (at room temperature [for Step 19] and at 37°C [for Step 16])

Light microscope

Microtome

Pen (hydrophobic) for marking slides

Syringes (1-mL) with 25-gauge needles

METHOD

Figure 1 summarizes the sequence of steps in this method. Perform all steps at room temperature unless otherwise stated.

Transplantation of B-Cell Lymphoma Cells and Induction of Apoptosis

Preparing the lymphoma cells and carrying out the injections will take ~60 min.

1. Thaw cryopreserved Eµ-Myc B-cell lymphoma cells in complete Anne Kelso medium, wash the cells twice with PBS, and perform a cell count.

2. Resuspend 2×10^6 Eµ-Myc B-cell lymphoma cells in 1 mL of PBS and transfer 200 µL of the cell suspension (4×10^5 cells) into syngeneic recipient mice by tail vein injection.

3. During tumor development over the next 7–14 d, monitor the disease burden by palpating the inguinal or brachial lymph nodes and by measuring the concentration of white blood cells (WBCs) in ~25 µL of blood from the tail vein using an Advia 120 automated hematology analyzer or equivalent.

4. Once lymph nodes are clearly palpable (corresponding to a peripheral WBC concentration of $\sim 50 \times 10^6$ cells/mL or greater), weigh the mice and administer the apoptosis inducer (e.g., inject vorinostat intraperitoneally at 200 mg/kg). As a control, inject one mouse with an equal volume of vehicle only (e.g., for the vorinostat control, use DMSO).

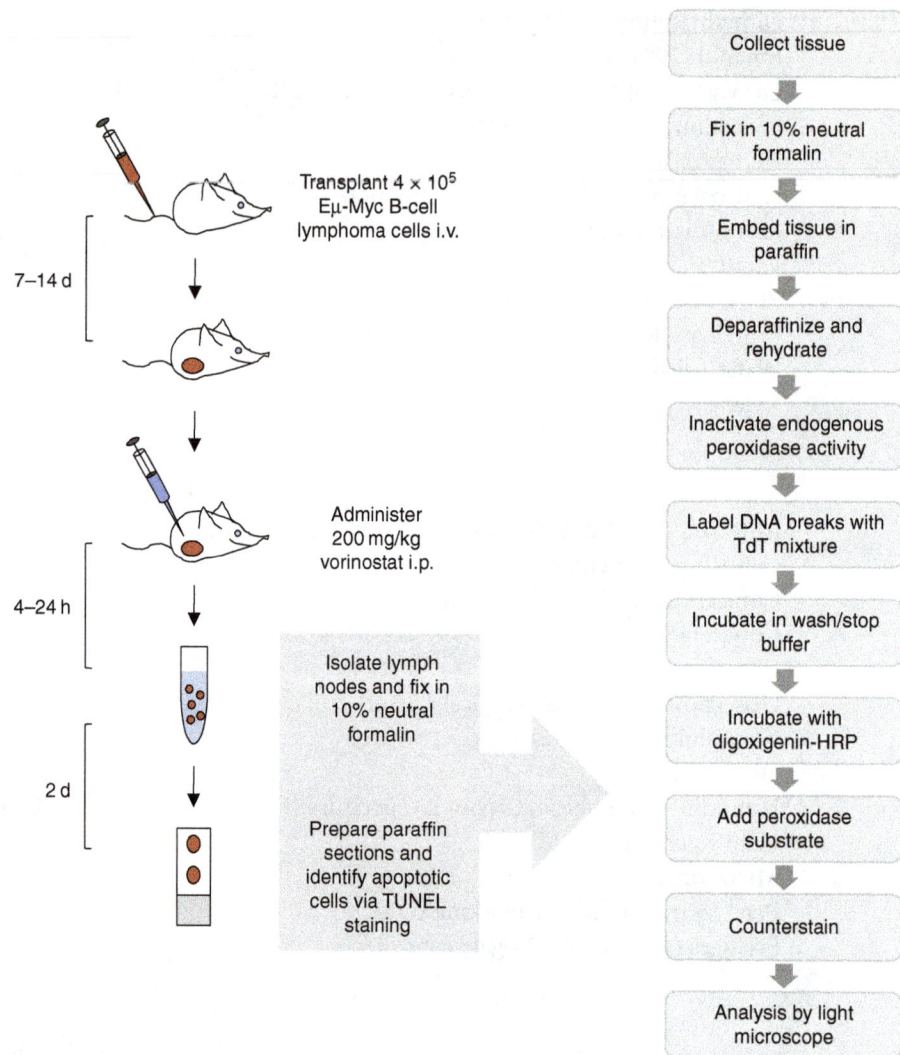

FIGURE 1. Flowchart outlining the steps for detecting fragmented DNA in apoptotic cells using TUNEL. i.v., Intravenous; i.p., intraperitoneal.

Collection and Fixation of the Tumor Tissue

5. At each of several time points after administration of vorinostat (e.g., at 4, 8, 12, and 24 h), kill a set of the mice by cervical dislocation.

6. Using small scissors and curved forceps, collect lymph nodes (e.g., inguinal, brachial, auxiliary, and mesenteric) and/or spleen samples.

    Collecting the tissues will take ~10 min per animal.

7. Transfer each tissue sample into a separate 15-mL Falcon tube filled with 10% neutral-buffered formalin, and fix the tissues for at least 12 h.

    The fixation of tissue can require from 12 to 24 h.

8. Embed the formalin-fixed tissues in paraffin, prepare 4-μm sections, and transfer them onto Superfrost Plus glass slides.

9. Deparaffinize and rehydrate the formalin-fixed, paraffin-embedded tissue sections in a Coplin jar:

    i. Wash the slides in three changes of xylene for 5 min each time.

Cite this protocol as *Cold Spring Harb Protoc*; doi:10.1101/pdb.prot082537

ii. Wash the slides twice in 100% ethanol for 5 min each time.

iii. Wash the slides once in 95% ethanol and once in 70% ethanol for 3 min each time.

iv. Rinse the slides in one change of PBS for 5 min.

10. Mark the tissue sections on the slides by circling each section using a hydrophobic pen.

11. Prepare a freshly diluted solution of proteinase K (20 μg/mL). Expose the DNA by pretreating each tissue section with ~50 μL of the diluted enzyme for 15 min.

   The specific volume of diluted enzyme is not important. In this step and below, use a sufficient volume of the appropriate solution (~50–60 μL) to cover the section.

12. Rinse the slides twice in distilled water in a Coplin jar for 2 min each time.

13. Inactivate endogenous peroxidase by incubating the slides in 3% hydrogen peroxide in PBS in a Coplin jar for 5 min.

14. Rinse the slides twice with PBS in a Coplin jar for 5 min each time.

Labeling and Detection of Cells with Fragmented DNA

15. Gently tap the excess liquid off of each slide, immediately apply ~60 μL of equilibration buffer to each section, and incubate at room temperature for at least for 30 sec.

16. Gently tap the excess liquid off of each slide and immediately apply ~50 μL of the working solution of TdT to each section. Incubate the slides in a humidified chamber at 37°C for 1 h.

   Nucleotides in the reaction buffer (both digoxigenin-conjugated nucleotides and unlabeled nucleotides) are added to the DNA by TdT.

   We recommend including positive and negative controls. As a positive control for DNA fragmentation, DNase I-treated sections can be used; as a negative control, sections can be treated with reaction buffer only (by omitting TdT from the working solution) in this step.

17. Transfer the slides into a Coplin jar containing stop/wash buffer, agitate for 15 sec, and incubate at room temperature for another 10 min.

18. Rinse the slides three times in PBS for 2 min each time.

19. Gently tap the excess liquid off each slide and apply ~50 μL HRP-conjugated anti-digoxigenin antibody to each section. Incubate the slides in a humidified chamber at room temperature for 30 min.

20. In a Coplin jar, rinse the slides four times in PBS at room temperature for 2 min each time.

21. Apply ~60 μL of DAB (peroxidase substrate) to cover each of the sections and incubate for 1–3 min at room temperature.

22. Monitor color development by viewing the slides under a light microscope.

   To obtain optimal specific staining, develop the slides for the minimum time necessary.

23. Once the sections are optimally stained, rinse the slides three times in distilled water in a Coplin jar for 1 min each time.

24. Counterstain the sections in hematoxylin for 4–10 sec at room temperature and rinse the slides in water until the water is clear.

25. Dehydrate the sections in a Coplin jar:

   i. Wash the slides once in 95% ethanol for 5 min.

   ii. Wash the slides twice in 100% ethanol for 3 min each time.

   iii. Clear the slides twice in xylene for 5 min each time.

26. Mount the sections on the slides under a coverslip with mounting medium.

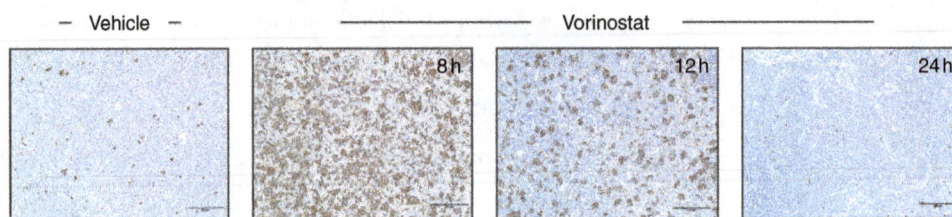

FIGURE 2. Histological analysis of apoptosis via TUNEL staining in lymph nodes derived from Eµ-Myc lymphoma-bearing mice treated with 200 mg/kg vorinostat for 8, 12, or 24 h.

27. View the sections under a light microscope to identify areas of staining.

The labeled ends of DNA fragments are typically located in apoptotic nuclei. Therefore, it is advisable to focus on the nuclear areas of the cells when analyzing the staining patterns.

For an example of TUNEL staining in lymph nodes, see Figure 2.

See Troubleshooting.

TROUBLESHOOTING

Problem (Step 27): The staining appears to be nonspecific and/or false-positive results are suspected.

Solution: The activity of endogenous nucleases may lead to false-positive results. These adverse effects can be diminished by fixing the tissue immediately after collecting it (Step 7) and/or by treating the sections with proteinase K (Step 11) for the minimum time necessary.

Nonspecific staining can be caused by endogenous peroxidase activity or by using a mixture of TdT that is too concentrated. When preparing the working solution of TdT in the reaction buffer, try lowering the concentration of TdT.

RELATED TECHNIQUES

In certain cases, the TUNEL assay may lead to false-positive results (Galluzzi et al. 2009). It is therefore advisable to carry out at least one additional immunohistochemistry assay (e.g., to detect active caspase-3) or a flow cytometry cell death assay. Flow cytometry assays that detect the binding of annexin V in combination with propidium iodide are described in Galluzzi et al. (2009) and Kepp et al. (2011); see also Protocol 4: Detection of Apoptotic Cells Using Propidium Iodide Staining (Newbold et al. 2014).

RECIPES

Complete Anne Kelso Medium

Prepare in high glucose Dulbecco's modified Eagle's medium (DMEM)
10% fetal calf serum (FCS)
100 units/mL penicillin G
100 µg/mL streptomycin
0.1 mM L-asparagine
50 µM 2-mercaptoethanol

Store at 4°C for up to several months.

Cite this protocol as *Cold Spring Harb Protoc*; doi:10.1101/pdb.prot082537

Phosphate-Buffered Saline (PBS)

Reagent	Amount to add (for 1× solution)	Final concentration (1×)	Amount to add (for 10× stock)	Final concentration (10×)
NaCl	8 g	137 mM	80 g	1.37 M
KCl	0.2 g	2.7 mM	2 g	27 mM
Na_2HPO_4	1.44 g	10 mM	14.4 g	100 mM
KH_2PO_4	0.24 g	1.8 mM	2.4 g	18 mM
If necessary, PBS may be supplemented with the following:				
$CaCl_2 \cdot 2H_2O$	0.133 g	1 mM	1.33 g	10 mM
$MgCl_2 \cdot 6H_2O$	0.10 g	0.5 mM	1.0 g	5 mM

PBS can be made as a 1× solution or as a 10× stock. To prepare 1 L of either 1× or 10× PBS, dissolve the reagents listed above in 800 mL of H_2O. Adjust the pH to 7.4 (or 7.2, if required) with HCl, and then add H_2O to 1 L. Dispense the solution into aliquots and sterilize them by autoclaving for 20 min at 15 psi (1.05 kg/cm^2) on liquid cycle or by filter sterilization. Store PBS at room temperature.

ACKNOWLEDGMENTS

Work described here was supported by a grant from the Dutch Cancer Society to M.B. (AMC2009-4457) and a grant from Pfizer Inc. to C.C.

REFERENCES

Galluzzi L, Aaronson SA, Abrams J, Alnemri ES, Andrews DW, Baehrecke EH, Bazan NG, Blagosklonny MV, Blomgren K, Borner C, et al. 2009. Guidelines for the use and interpretation of assays for monitoring cell death in higher eukaryotes. *Cell Death Differ* **16:** 1093–1107.

Kepp O, Galluzzi L, Lipinski M, Yuan J, Kroemer G. 2011. Cell death assays for drug discovery. *Nat Rev Drug Discov* **10:** 221–237.

Newbold A, Martin BP, Cullinane C, Bots M. 2014. Detection of apoptotic cells using propidium iodide staining. *Cold Spring Harb Protoc* doi: 10.1101/pdb.prot082545.

Detection of Apoptotic Cells Using Propidium Iodide Staining

Andrea Newbold,[1] Ben P. Martin,[1] Carleen Cullinane,[2,3] and Michael Bots[4,5]

[1]Gene Regulation Laboratory, Cancer Therapeutics Program, Peter MacCallum Cancer Centre, East Melbourne 3002, Victoria, Australia; [2]Translational Research Laboratory, Cancer Therapeutics Program, Peter MacCallum Cancer Centre, East Melbourne 3002, Victoria, Australia; [3]Sir Peter MacCallum Department of Oncology, The University of Melbourne, Parkville 3010, Victoria, Australia; [4]Laboratory for Experimental Oncology and Radiobiology (LEXOR), Center for Experimental Molecular Medicine, Academic Medical Center, 1105 AZ, Amsterdam, The Netherlands

Flow cytometry assays are often used to detect apoptotic cells in in vitro cultures. Depending on the experimental model, these assays can also be useful in evaluating apoptosis in vivo. In this protocol, we describe a propidium iodide (PI) flow cytometry assay to evaluate B-cell lymphomas that have undergone apoptosis in vivo. B-cell lymphoma cells are injected into recipient mice and, on tumor formation, the mice are treated with the apoptosis inducer vorinostat (a histone deacetylase inhibitor). Tumor samples collected from the lymph nodes and/or the spleen are used to prepare a single-cell suspension that is exposed to a hypotonic solution containing the fluorochrome PI. The DNA content of the cells, now labeled with PI, is analyzed by flow cytometry. Nuclear DNA content is lost during apoptosis, resulting in a hypodiploid (or sub-G_1) DNA profile during flow cytometry. In contrast, healthy cells display a sharp diploid DNA profile.

MATERIALS

It is essential that you consult the appropriate Material Safety Data Sheets and your institution's Environmental Health and Safety Office for proper handling of equipment and hazardous materials used in this protocol.

RECIPES: Please see the end of this protocol for recipes indicated by <R>. Additional recipes can be found online at http://cshprotocols.cshlp.org/site/recipes.

Reagents

ACK lysis buffer <R>

Apoptosis inducer (e.g., vorinostat)

Prepare vorinostat by dissolving at 10 mm in dimethylsulfoxide (DMSO). Store at −20°C, and avoid multiple freeze–thaw cycles.

C57BL/6 mice (8–10 wk of age)

Complete Anne Kelso medium <R>

DMSO

Eμ-Myc B-cell lymphoma cells

Nicoletti buffer <R>

Phosphate-buffered saline (PBS) containing 10% fetal calf serum (FCS)

PBS <R>

---

[5]Correspondence: m.bots@amc.uva.nl

Cite this protocol as *Cold Spring Harb Protoc*; doi:10.1101/pdb.prot082545

Equipment

Automated hematology analyzer (Advia 120 or equivalent)
Cell strainer (nylon, 70 µm)
Centrifuge (benchtop, Eppendorf 5810)
Conical tubes (50 and 15 mL)
Culture dishes (24-well)
Dissection tools (small scissors and curved forceps)
Flow cytometer
Hemacytometer
Light microscope
Microcentrifuge (Eppendorf 5415D)
Microcentrifuge tubes (1.5 mL)
Syringes (1 mL) with 25-gauge needles

METHOD

Figure 1 summarizes the sequence of steps in the method. Timeframes indicated are for a six-mouse experiment.

Transplantation of B-Cell Lymphoma Cells and Induction of Apoptosis

1. Thaw cryopreserved Eµ-Myc B-cell lymphoma cells in complete Anne Kelso medium, wash the cells twice with PBS, and perform a cell count.

2. Resuspend 2×10^6 Eµ-Myc B-cell lymphoma cells in 1 mL of PBS and transfer 200 µL of the cell suspension (4×10^5 cells) into syngeneic recipient mice by tail vein injection.

   Preparing and injecting the lymphoma cells will take ~60 min.

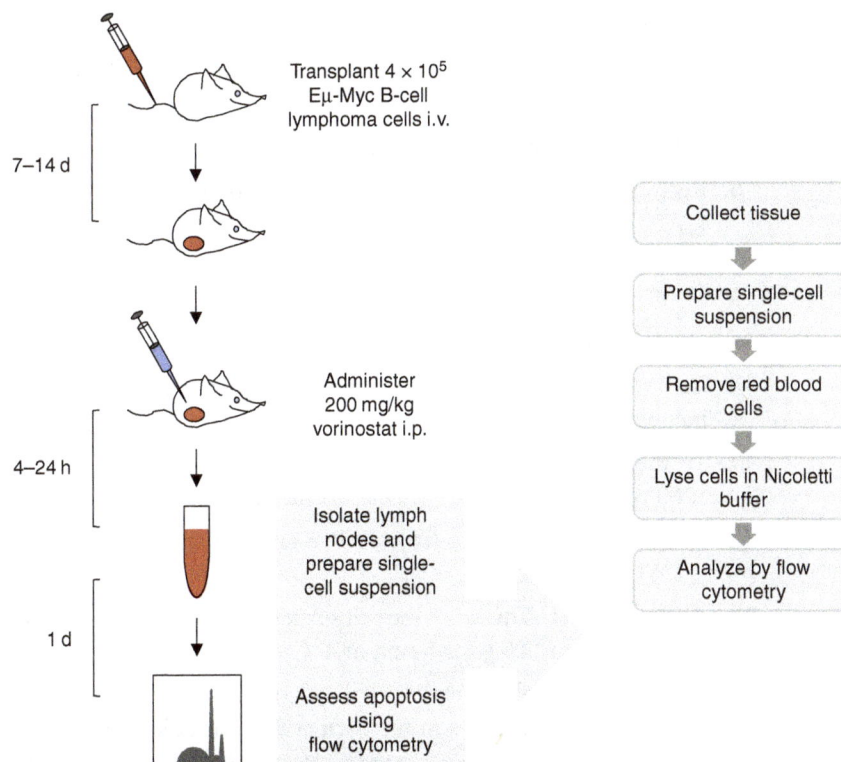

FIGURE 1. Flowchart outlining the steps for detecting fragmented DNA in apoptotic cells using propidium iodide (PI) staining. i.v., Intravenous; i.p., intraperitoneal.

3. During tumor development over the next 7–14 d, monitor the disease burden by palpating the inguinal or brachial lymph nodes and by measuring the concentration of white blood cells (WBC) in ~25 μL of blood from the tail vein using an Advia 120 automated hematology analyzer or equivalent.

4. Once lymph nodes are clearly palpable (corresponding to a peripheral WBC concentration of ~50 × 10⁶ cells/mL or greater), weigh the mice and administer the apoptosis inducer to the mice (e.g., inject vorinostat intraperitoneally at 200 mg/kg).

As a control, inject one mouse with vehicle only (use DMSO for a vorinostat control). Injecting the apoptosis inducers will take ~2 min per animal.

Collection of Tumor Tissue and Preparation of Single-Cell Suspension

Steps 5–9 will require ~15 min per animal.

5. At each of several time-points after administration of vorinostat (e.g., at 4, 8, 12, and 24 h), kill the appropriate number of mice by cervical dislocation.

6. Using small scissors and curved forceps, collect lymph nodes (e.g., inguinal, brachial, auxiliary, and mesenteric) and/or spleen into the wells of a 24-well culture dish, each containing 1 mL of PBS.

7. Prepare a single-cell suspension of lymph node tissue.

   i. Use a 1-mL plunger to dissociate the lymph nodes in the wells of the culture dish.

   ii. Pass the cell suspension through a nylon cell strainer into a 15-mL tube containing 5 mL of PBS.

   iii. Wash the strainer twice with 4 mL of PBS.

8. Prepare a single-cell suspension of spleen tissue.

   i. Transfer the spleen onto a nylon cell strainer.

   ii. Use a 1-mL plunger to gently push the spleen through the strainer into a 50-mL tube containing 10 mL of PBS.

   iii. Wash strainer twice with 10 mL of PBS.

9. Recover the cells by centrifuging at 1200 rpm or 290 g for 4 min at 4°C. Discard the supernatant.

Lysis of Cells and PI Staining

Steps 10–14 should take <30 min.

10. Remove the red blood cells by gently mixing the cells in 5 mL of ice-cold ACK lysis buffer and incubating on ice for 4 min.

11. To the cell solution, add 25 mL of ice-cold PBS containing 10% FCS and centrifuge the cells at 1200 rpm for 4 min at 4°C. Discard the supernatant.

12. Wash the cells once with ice-cold PBS and resuspend them in ice-cold PBS at a concentration of ~2 × 10⁶ cells/mL.

13. Aliquot 0.5-mL samples of the cell suspension into microcentrifuge tubes and centrifuge the cells at 3000 rpm or 800 g for 5 min at 4°C.

14. Resuspend the cell pellets in 1 mL of cold Nicoletti buffer and mix by gently pipetting up and down. Keep the samples in the dark at 4°C for at least 1 h (and no longer than 24 h).

In this step, the cells are lysed and the DNA is labeled with the fluorochrome PI, which permits the analysis of cellular DNA content by flow cytometry (Riccardi and Nicoletti 2006). Lysing cells at high cell density (>10⁶ cells/mL) may reduce the sharpness of the DNA profiles obtained during flow cytometry and can hamper the analysis.

Cite this protocol as *Cold Spring Harb Protoc*; doi:10.1101/pdb.prot082545

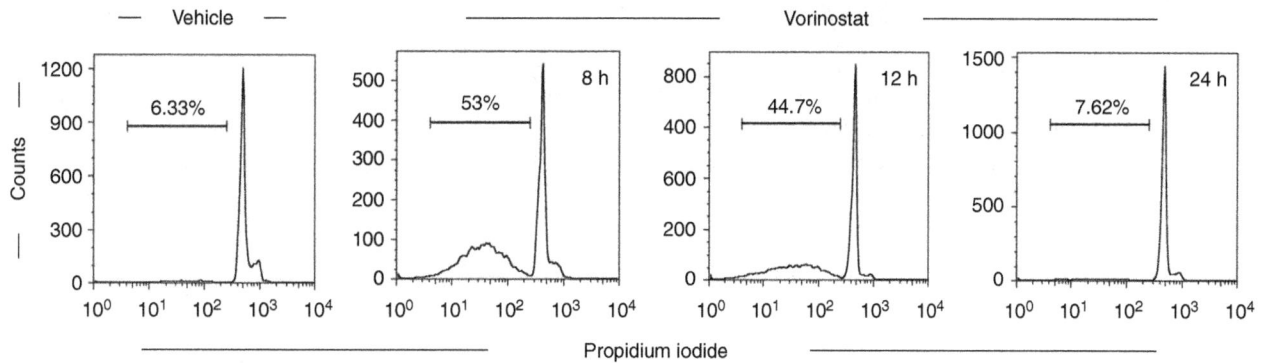

FIGURE 2. Flow cytometry analysis of apoptosis in lymph nodes from Eµ-Myc lymphoma-bearing mice treated with 200 mg/kg vorinostat for 8, 12, or 24 h. The percentage of apoptotic nuclei after elimination of residual debris is indicated for each case.

Flow Cytometry and Data Analysis

15. Read the PI fluorescence (FL3) of each sample in a flow cytometer. Adjust the FL3 voltage in such a way that the G_1 peak falls at $\sim 10^3$ units (log scale). Acquire the samples at low speed, and collect at least 20,000–30,000 events. To selectively observe the cells of interest, eliminate, or "gate-out" the residual debris (lower left corner) and analyze the hypodiploid nuclei.

    Because apoptotic cells lose nuclear DNA content, they display a hypodiploid (or sub-G_1) DNA profile in the flow cytometry profile and are therefore easily distinguished from healthy cells that display a sharp diploid DNA profile. For the results of a typical flow cytometry analysis, see Figure 2.

DISCUSSION

This protocol, as well as other flow cytometry assays that rely on the binding of annexin V in combination with PI (Galluzzi et al. 2009; Kepp et al. 2011), is especially useful for detecting apoptosis in tissues that are easily processed into single-cell suspensions, such as lymph nodes, spleen, and bone marrow. The detection of apoptotic cells in solid organs by using these approaches is far more difficult. In these cases, a great deal of effort is required to prepare high-quality single-cell suspensions, and during this time, apoptosis may proceed. It should be recognized, therefore, that the relative extent of apoptosis detected following the preparation of cell suspensions may be higher than that which actually occurred in vivo. Furthermore, the presence of necrotic regions in solid tumors may result in significant debris and hamper the analysis. Therefore, appropriate controls should be included in the experiment to ensure the detection of drug-induced apoptosis.

RECIPES

ACK Lysis Buffer

Reagent	Quantity (for 1000 mL)	Final concentration
NH_4Cl	8.26 g	150 mM
$KHCO_3$	1 g	10 mM
Na_2EDTA	37.2 mg	0.1 mM

Dissolve all reagents in 850 mL of H_2O. Adjust the pH to 7.2–7.4, and add H_2O to 1000 mL. Store for up to 6 mo at room temperature.

Complete Anne Kelso Medium

Dulbecco's modified Eagle's medium (DMEM), high-glucose
10% fetal calf serum (FCS)
100 units/mL penicillin G
100 µg/mL streptomycin
0.1 mM L-asparagine
50 µM 2-mercaptoethanol

Store at 4°C for up to several months.

Nicoletti Buffer

Reagent	Quantity (for 100 mL)	Final concentration
$Na_3C_6H_5O_7$	0.1 g	0.1%
Propidium iodide	5 mg	50 µg/mL
Triton X-100 (10%)	1 mL	0.1%

Dissolve the first two reagents in 85 mL of H_2O and then add the 10% Triton X-100. Bring the total volume to 100 mL with H_2O. Store in the dark at 4°C; the solution will keep for several months under these conditions.

Phosphate-Buffered Saline (PBS)

Reagent	Amount to add (for 1× solution)	Final concentration (1×)	Amount to add (for 10× stock)	Final concentration (10×)
NaCl	8 g	137 mM	80 g	1.37 M
KCl	0.2 g	2.7 mM	2 g	27 mM
Na_2HPO_4	1.44 g	10 mM	14.4 g	100 mM
KH_2PO_4	0.24 g	1.8 mM	2.4 g	18 mM
If necessary, PBS may be supplemented with the following:				
$CaCl_2 \cdot 2H_2O$	0.133 g	1 mM	1.33 g	10 mM
$MgCl_2 \cdot 6H_2O$	0.10 g	0.5 mM	1.0 g	5 mM

PBS can be made as a 1× solution or as a 10× stock. To prepare 1 L of either 1× or 10× PBS, dissolve the reagents listed above in 800 mL of H_2O. Adjust the pH to 7.4 (or 7.2, if required) with HCl, and then add H_2O to 1 L. Dispense the solution into aliquots and sterilize them by autoclaving for 20 min at 15 psi (1.05 kg/cm^2) on liquid cycle or by filter sterilization. Store PBS at room temperature.

ACKNOWLEDGMENTS

Work described here was supported by a grant from the Dutch Cancer Society to M.B. (AMC2009-4457) and a grant from Pfizer Inc. to C.C.

REFERENCES

Galluzzi L, Aaronson SA, Abrams J, Alnemri ES, Andrews DW, Baehrecke EH, Bazan NG, Blagosklonny MV, Blomgren K, Borner C, et al. 2009. Guidelines for the use and interpretation of assays for monitoring cell death in higher eukaryotes. *Cell Death Differ* 16: 1093–1107.

Kepp O, Galluzzi L, Lipinski M, Yuan J, Kroemer G. 2011. Cell death assays for drug discovery. *Nat Rev Drug Discov* 10: 221–237.
Riccardi C, Nicoletti I. 2006. Analysis of apoptosis by propidium iodide staining and flow cytometry. *Nat Protoc* 1: 1458–1461.

CHAPTER 13

Studying Apoptosis in *Drosophila*

Donna Denton[1] and Sharad Kumar[1]

Centre for Cancer Biology, University of South Australia, Adelaide, South Australia 5001, Australia

The apoptotic machinery is highly conserved throughout evolution, and central to the regulation of apoptosis is the caspase family of cysteine proteases. Insights into the regulation and function of apoptosis in mammals have come from studies using model organisms. *Drosophila* provides an exceptional model system for identifying the function of conserved mechanisms regulating apoptosis, especially during development. The characteristic patterns of apoptosis during *Drosophila* development have been well described, as has the apoptotic response following DNA damage. The focus of this discussion is to introduce methodologies for monitoring apoptosis during *Drosophila* development and also in *Drosophila* cell lines.

INTRODUCTION

Historically, the terms "programmed cell death" (PCD) and "apoptosis" have been used interchangeably, but it is now clear that multiple additional modes of cell death occur during animal development and in the adult. These cell deaths can be classed as caspase-dependent and caspase-independent and include apoptosis, programmed necrosis, and autophagic cell death (reviewed in Galluzzi et al. 2012). Some of the cell-death modalities play specific roles, whereas the caspase-dependent apoptotic pathways mediate the majority of PCD in mammals, including control of tissue shape and size as well as the elimination of obsolete and damaged cells (reviewed in Galluzzi et al. 2012; Denton et al. 2012).

Here, we discuss protocols that specifically determine (caspase-dependent) apoptosis in the fruit fly *Drosophila*. In the accompanying protocols (see below), we will describe the in vivo detection of apoptotic cells by means of terminal deoxynucleotidyl transfer-mediated dUTP nick-end labeling (TUNEL), acridine orange staining, and immunostaining with an antibody against active caspases. Furthermore, we will discuss how cell lines can provide a useful complement to the in vivo studies—by way of example, a method will be presented for examining the transcriptional response to a death stimulus by application of quantitative PCR. Finally, we will illustrate how, following the induction of apoptosis in cell lines or whole organisms, the activation of caspases can be measured by using synthetic peptide substrates.

CELL DEATH IN *DROSOPHILA*

Drosophila has emerged as a powerful model for identifying conserved regulatory mechanisms controlling cell death and has aided in our understanding of the roles of apoptosis during development. The reproducible patterns of PCD during *Drosophila* development have been well described, including

[1]Correspondence: sharad.kumar@unisa.edu.au; Donna.Denton@unisa.edu.au

Cite this introduction as *Cold Spring Harb Protoc*; doi:10.1101/pdb.top070433

the formation of the embryonic nervous system, morphogenesis of the eye, and removal of obsolete larval tissues during metamorphosis and of surplus cells during oogenesis (Wolff and Ready 1991; Abrams et al. 1993; Jiang et al. 2000; Pritchett et al. 2009). In addition, in response to genotoxic stress, apoptosis in the fly is used to remove damaged cells.

During embryogenesis, apoptosis can be first detected in a small number of cells, and this becomes more widespread at later stages in tissues such as the central nervous system and during morphogenesis of segments (Abrams et al. 1993; Lohmann et al. 2002). Apoptosis is also necessary in later developmental stages during the morphogenesis that gives rise to adult structures, including the compound eye. The adult compound eye comprises ∼750 individual units called ommatidia, and differentiation initiates from the eye imaginal disc during larval stages. Following the specification of the majority of the ommatidial cell types, the remaining cells are removed by apoptosis (Wolff and Ready 1991; Brachmann and Cagan 2003).

An increase in PCD occurs also during *Drosophila* metamorphosis to delete cells no longer required. The obsolete larval tissues, including salivary glands and midgut, are removed in response to pulses of the steroid hormone ecdysone during metamorphosis (reviewed in Baehrecke 2000). Ecdysone acts by binding to a heterodimeric receptor comprising the ecdysone receptor and the protein ultraspiracle (EcR–Usp) to regulate the expression of cell-death genes. Complete removal of the salivary glands requires both apoptosis and autophagy as inhibiting both pathways results in a greater delay in tissue destruction than inhibiting either pathway alone (Lee and Baehrecke 2001; Berry and Baehrecke 2007). Additional modes of cell-death regulation are required for egg formation during various stages of oogenesis (reviewed in Pritchett et al. 2009). Nutrient limitation induces nurse cell death during mid-oogenesis and requires both apoptotic and autophagic pathways. During the late stages of oogenesis, maturation of the egg chamber requires developmental cell death of the nurse cells, a process in which autophagy functions upstream of apoptosis.

As well as the roles during development, apoptosis is required for the removal of potentially harmful cells such as those that have acquired DNA damage. During specific stages of embryogenesis and in the proliferating larval imaginal disc, cells that have sustained damaged DNA undergo apoptosis in a process that eliminates the damaged cells. In response to irradiation in the larval wing imaginal disc, cell-cycle arrest occurs within 30 min, persisting for ∼6 h, during which DNA repair is complete within the first 3 h. Apoptosis can be detected at 4 h postirradiation, coinciding with upregulation of proapoptotic genes and continues for a further 20 h (Jaklevic and Su 2004).

In addition to apoptosis, there are other forms of cell death that occur during *Drosophila* development—indeed, the removal of the larval midgut can still occur in the absence of the apoptotic machinery, through a process that requires autophagy (Denton et al. 2009). The detection of autophagy in *Drosophila* is described elsewhere.

THE APOPTOSIS MACHINERY OF *DROSOPHILA*

A crucial step in the initiation of apoptosis is activation of caspases, a family of cysteine proteases that are synthesized as inactive zymogens. Following death signals, caspase activation involves oligomerization and/or proteolytic cleavage into two subunits that constitute the active enzyme (Kumar 2007). Based on their structure and function, caspases can be classified as "initiator" caspases, containing a long Amino-terminal prodomain, and "effector" (executioner) caspases, which possess a short Amino-terminal prodomain (Kumar 2007; Shalini et al. 2014).

Until the appropriate cell-death signal, caspases are maintained in an inactive form by the *Drosophila* inhibitor of apoptosis protein (DIAP1), an important negative regulator that prevents cell death by inhibiting caspase processing and activity (Fig. 1) (Hay et al. 1995; Lee et al. 2011). The cell-death response of several signaling pathways is mediated through the transcriptional activation of the inhibitor of apoptosis protein (IAP) antagonists Reaper, Hid, and Grim (RHG) that act to block DIAP1 function, promoting its ubiquitylation and degradation, thereby alleviating caspase inhibition (reviewed by Steller 2008). The *reaper*, *hid*, and *grim* genes are essential for initiation of cell death as

Cite this introduction as *Cold Spring Harb Protoc*; doi:10.1101/pdb.top070433

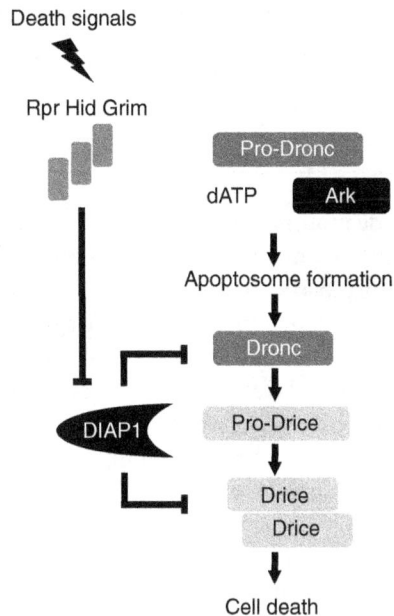

FIGURE 1. The *Drosophila* apoptotic machinery. In the absence of apoptotic signals in *Drosophila*, the *Drosophila* inhibitor of apoptosis protein 1 (DIAP1) binds to and inhibits both the initiator caspase Dronc and effector caspases (Drice and Dcp1). During apoptosis, the IAP antagonists Reaper, Hid, and Grim (RHG) bind to DIAP1, promoting its ubiquitylation and degradation, thus alleviating the inhibition of the caspases. Removal of DIAP1 enables activation of Dronc by association with Ark to form an active apoptosome, leading to activation of the effector caspase Drice and execution of the apoptotic process.

deletion of all three genes prevents apoptosis during embryogenesis (White et al. 1994). Other IAP antagonists include Sickle and Jafrac2, which act to promote cell death (Christich et al. 2002; Srinivasula et al. 2002; Tenev et al. 2002).

There are seven caspase family members in *Drosophila*—initiator caspases Dronc (in UniProt, named as caspase Nc), Dredd (caspase-8) and Strica (Dream), and effector caspases Drice (*Drosophila* ICE), Dcp-1 (caspase-1), Decay and Damm (Song et al. 1997; Fraser and Evan 1997; Chen et al. 1998; Dorstyn et al. 1999a, 1999b; Harvey et al. 2001; Doumanis et al. 2001; reviewed in Kumar and Doumanis 2000). The majority of developmental and stress-induced apoptosis in *Drosophila* requires the initiator caspase Dronc (a CARD-domain containing ortholog of mammalian caspase-9) and the effector caspase Drice (orthologous to mammalian caspase-3), and loss-of-function mutations of the genes encoding these caspases show severe developmental defects, often resulting in lethality (Chew et al. 2004; Daish et al. 2004; Xu et al. 2005, 2006; Kondo et al. 2006; Muro et al. 2006). Activation of Dronc requires dimerization and recruitment by the single CED4/Apaf-1 ortholog, Ark forming the apoptosome, important for most developmental apoptosis (Fig. 1) (Rodriguez et al. 1999; Mills et al. 2006; Srivastava et al. 2007; Dorstyn and Kumar 2008).

Despite the crucial role in mammals of mitochondrial outer membrane permeabilization and cytochrome *c* release, which is required for the formation of a large multimeric complex known as the apoptosome, mitochondria have only a limited role in *Drosophila* apoptosis. Structural studies of the *Drosophila* apoptosome indicate that it forms in the absence of cytochrome *c* (Yu et al. 2006; Yuan et al. 2011). Furthermore, cell death occurs normally in *Drosophila* cells depleted of cytochrome *c* (Dorstyn et al. 2002, 2004), and loss-of functions of the *Drosophila* Bcl-2 members Debcl (Colussi et al. 2000; Galindo et al. 2009) and Buffy (Quinn et al. 2003), either individually or together, have no effects on caspase activation or apoptosis (Sevrioukov et al. 2007). Caspase activation in *Drosophila* is primarily regulated by DIAP1, which keeps caspases in check until RHG proteins bind and degrade DIAP1, allowing activation of Dronc through interaction with the Ark apoptosome.

The remaining *Drosophila* caspases play redundant roles or have noncell-death functions. In the germline, the initiator caspase Strica and the effector caspase Dcp-1 function redundantly with other caspases during midoogenesis PCD (Laundrie et al. 2003; Kondo et al. 2006; Xu et al. 2006; Baum et al. 2007). A noncell-death function for the initiator caspase Dredd is required for cleavage of the NF-κB transcription factor Relish during the innate immune response (Leulier et al. 2000). Furthermore, in a model for retinal degeneration, Dredd and activation of Relish are required for neuronal cell death

(Chinchore et al. 2012). Finally, a role for the effector caspases Decay and Damm in PCD has not been established (Kondo et al. 2006).

PROTOCOLS FOR DETECTING APOPTOSIS IN *DROSOPHILA*

The characteristics of apoptotic cell death include morphological changes, such as membrane blebbing, cytoplasmic and nuclear condensation, and DNA fragmentation, as well as molecular changes, including activation of caspases. There are several protocols to detect apoptotic cells during *Drosophila* development. These include using the TUNEL technique to detect DNA fragmentation (see Protocol 1: Terminal Deoxynucleotidyl Transferase (TdT)-Mediated dUTP Nick-End Labeling (TUNEL) for Detection of Apoptotic Cells in *Drosophila* [Denton and Kumar 2015a]), staining with acridine orange, a vital dye that is taken up by dying cells (see Protocol 2: Using the Vital Dye Acridine Orange to Detect Dying Cells in *Drosophila* [Denton and Kumar 2015b]), immunostaining to detect active caspases (see Protocol 3: Immunostaining Using an Antibody against Active Caspase-3 to Detect Apoptotic Cells in *Drosophila* [Denton and Kumar 2015c]), as well as biochemical determination of caspase activity in cell or tissue lysates (see Protocol 5: Using Synthetic Peptide Substrates to Measure *Drosophila* Caspase Activity [Denton and Kumar 2015d]). *Drosophila* cell lines can also be used to complement in vivo studies (see Protocol 4: Analyzing the Response of RNAi-Treated *Drosophila* Cells to Death Stimuli by Quantitative Real-Time Polymerase Chain Reaction [Denton and Kumar 2015e]).

ACKNOWLEDGMENTS

We thank the members of our laboratory for their comments on the accompanying protocols. The *Drosophila* cell death research in our laboratory is supported by the National Health and Medical Research Project Grant (1041807) and a Senior Principal Research Fellowship (1002863) to S.K.

REFERENCES

Abrams JM, White K, Fessler LI, Steller H. 1993. Programmed cell death during *Drosophila* embryogenesis. *Development* 117: 29–43.

Baehrecke EH. 2000. Steroid regulation of programmed cell death during *Drosophila* development. *Cell Death Differ* 7: 1057–1062.

Baum JS, Arama E, Steller H, McCall K. 2007. The *Drosophila* caspases Strica and Dronc function redundantly in programmed cell death during oogenesis. *Cell Death Differ* 14: 1508–1517.

Berry DL, Baehrecke EH. 2007. Growth arrest and autophagy are required for salivary gland cell degradation in *Drosophila*. *Cell* 131: 1137–1148.

Brachmann CB, Cagan RL. 2003. Patterning the fly eye: The role of apoptosis. *Trends Genet* 19: 91–96.

Chen P, Rodriguez A, Erskine R, Thach T, Abrams JM. 1998. Dredd, a novel effector of the apoptosis activators *reaper, grim* and *hid* in *Drosophila*. *Dev Biol* 201: 202–216.

Chew SK, Akdemir F, Chen P, Lu WJ, Mills K, Daish T, Kumar S, Rodriguez A, Abrams JM. 2004. The apical caspase *dronc* governs programmed and unprogrammed cell death in *Drosophila*. *Dev Cell* 7: 897–907.

Chinchore Y, Gerber GF, Dolph PJ. 2012. Alternative pathway of cell death in *Drosophila* mediated by NF-κB transcription factor Relish. *Proc Natl Acad Sci* 109: E605–E612.

Christich A, Kauppila S, Chen P, Sogame N, Ho SI, Abrams JM. 2002. The damage-responsive *Drosophila* gene *sickle* encodes a novel IAP binding protein similar to but distinct from *reaper, grim*, and *hid*. *Curr Biol* 12: 137–140.

Colussi PA, Quinn LM, Huang DC, Coombe M, Read SH, Richardson H, Kumar S. 2000. Debcl, a proapoptotic Bcl-2 homologue, is a component of the *Drosophila melanogaster* cell death machinery. *J Cell Biol* 148: 703–714.

Daish TJ, Mills K, Kumar S. 2004. *Drosophila* caspase DRONC is required for specific developmental cell death pathways and stress-induced apoptosis. *Dev Cell* 7: 909–915.

Denton D, Shravage B, Simin R, Mills K, Berry DL, Baehrecke EH, Kumar S. 2009. Autophagy, not apoptosis, is essential for midgut cell death in *Drosophila*. *Curr Biol* 19: 1741–1746.

Denton D, Nicolson S, Kumar S. 2012. Cell death by autophagy: Facts and apparent artefacts. *Cell Death Differ* 19: 87–95.

Denton D, Kumar S. 2015a. Terminal deoxynucleotidyl transferase (TdT)-mediated dUTP nick-end labeling (TUNEL) for detection of apoptotic cells in *Drosophila*. *Cold Spring Harb Protoc* doi: 10.1101/pdb.prot086199.

Denton D, Kumar S. 2015b. Using the vital dye acridine orange to detect dying cells in *Drosophila*. *Cold Spring Harb Protoc* doi: 10.1101/pdb.prot086207.

Denton D, Kumar S. 2015c. Immunostaining using an antibody against active caspase-3 to detect apoptotic cells in *Drosophila*. *Cold Spring Harb Protoc* doi: 10.1101/pdb.prot086215.

Denton D, Kumar S. 2015d. Using synthetic peptide substrates to measure *Drosophila* caspase activity. *Cold Spring Harb Protoc* doi: 10.1101/pdb.prot086231.

Denton D, Kumar S. 2015e. Analyzing the response of RNAi-treated *Drosophila* cells to death stimuli by quantitative real-time polymerase chain reaction. *Cold Spring Harb Protoc* doi: 10.1101/pdb.prot086223.

Dorstyn L, Kumar S. 2008. A biochemical analysis of the activation of the *Drosophila* caspase DRONC. *Cell Death Differ* 15: 461–470.

Dorstyn L, Colussi PA, Quinn LM, Richardson H, Kumar S. 1999a. DRONC, a novel ecdysone-inducible *Drosophila* caspase. *Proc Natl Acad Sci* 96: 4307–4312.

Cite this introduction as *Cold Spring Harb Protoc*; doi:10.1101/pdb.top070433

Dorstyn L, Read SH, Quinn LM, Richardson H, Kumar S. 1999b. DECAY, a novel *Drosophila* caspase related to mammalian caspase-3 and caspase-7. *J Biol Chem* **274:** 30778–30783.

Dorstyn L, Read S, Cakouros D, Huh JR, Hay BA, Kumar S. 2002. The role of cytochrome *c* in caspase activation in *Drosophila melanogaster* cells. *J Cell Biol* **156:** 1089–1098.

Dorstyn L, Mills K, Lazebnik Y, Kumar S. 2004. The two cytochrome *c* species, DC3 and DC4, are not required for caspase activation and apoptosis in *Drosophila* cells. *J Cell Biol* **167:** 405–410.

Doumanis J, Quinn L, Richardson H, Kumar S. 2001. STRICA, a novel *Drosophila* caspase with an unusual serine/threonine-rich prodomain, interacts with DIAP1 and DIAP2. *Cell Death Differ* **8:** 387–394.

Fraser AG, Evan GI. 1997. Identification of a *Drosophila melanogaster* ICE/CED-3-related protease, drICE. *EMBO J* **16:** 2805–2813.

Galindo KA, Lu WJ, Park JH, Abrams JM. 2009. The Bax/Bak ortholog in *Drosophila*, Debcl, exerts limited control over programmed cell death. *Development* **136:** 275–283.

Galluzzi L, Vitale I, Abrams JM, Alnemri ES, Baehrecke EH, Blagosklonny MV, Dawson TM, Dawson VL, El-Deiry WS, Fulda S, et al. 2012. Molecular definitions of cell death subroutines: Recommendations of the Nomenclature Committee on Cell Death 2012. *Cell Death Differ* **19:** 107–120.

Harvey NL, Daish T, Mills K, Dorstyn L, Quinn LM, Read SH, Richardson H, Kumar S. 2001. Characterization of the *Drosophila* caspase, DAMM. *J Biol Chem* **276:** 25342–25350.

Hay BA, Wassarman DA, Rubin GM. 1995. *Drosophila* homologs of baculovirus inhibitor of apoptosis proteins function to block cell death. *Cell* **83:** 1253–1262.

Jaklevic BR, Su TT. 2004. Relative contribution of DNA repair, cell cycle checkpoints, and cell death to survival after DNA damage in *Drosophila* larvae. *Curr Biol* **14:** 23–32.

Jiang C, Lamblin AF, Steller H, Thummel CS. 2000. A steroid-triggered transcriptional hierarchy controls salivary gland cell death during *Drosophila* metamorphosis. *Mol Cell* **5:** 445–455.

Kondo S, Senoo-Matsuda N, Hiromi Y, Miura M. 2006. DRONC coordinates cell death and compensatory proliferation. *Mol Cell Biol* **26:** 7258–7268.

Kumar S. 2007. Caspase function in programmed cell death. *Cell Death Differ* **14:** 32–43.

Kumar S, Doumanis J. 2000. The fly caspases. *Cell Death Differ* **7:** 1039–1044.

Laundrie B, Peterson JS, Baum JS, Chang JC, Fileppo D, Thompson SR, McCall K. 2003. Germline cell death is inhibited by P-element insertions disrupting the dcp-1/pita nested gene pair in *Drosophila*. *Genetics* **165:** 1881–1888.

Lee CY, Baehrecke EH. 2001. Steroid regulation of autophagic programmed cell death during development. *Development* **128:** 1443–1455.

Lee TV, Fan Y, Wang S, Srivastava M, Broemer M, Meier P, Bergmann A. 2011. *Drosophila* IAP1-mediated ubiquitylation controls activation of the initiator caspase DRONC independent of protein degradation. *PLoS Genet* **7:** e1002261.

Leulier F, Rodriguez A, Khush RS, Abrams JM, Lemaitre B. 2000. The *Drosophila* caspase Dredd is required to resist gram-negative bacterial infection. *EMBO Rep* **1:** 353–358.

Lohmann I, McGinnis N, Bodmer M, McGinnis W. 2002. The *Drosophila* Hox gene *deformed* sculpts head morphology via direct regulation of the apoptosis activator *reaper*. *Cell* **110:** 457–466.

Mills K, Daish T, Harvey KF, Pfleger CM, Hariharan IK, Kumar S. 2006. The *Drosophila melanogaster* Apaf-1 homologue ARK is required for most, but not all, programmed cell death. *J Cell Biol* **172:** 809–815.

Muro I, Berry DL, Huh JR, Chen CH, Huang H, Yoo SJ, Guo M, Baehrecke EH, Hay BA. 2006. The *Drosophila* caspase Ice is important for many apoptotic cell deaths and for spermatid individualization, a nonapoptotic process. *Development* **133:** 3305–3315.

Pritchett TL, Tanner EA, McCall K. 2009. Cracking open cell death in the *Drosophila* ovary. *Apoptosis* **148:** 969–979.

Quinn L, Coombe M, Mills K, Daish T, Colussi P, Kumar S, Richardson H. 2003. Buffy, a *Drosophila* Bcl-2 protein, has anti-apoptotic and cell cycle inhibitory functions. *EMBO J* **22:** 3568–3579.

Rodriguez A, Oliver H, Zou H, Chen P, Wang X, Abrams JM. 1999. Dark is a *Drosophila* homologue of Apaf-1/CED-4 and functions in an evolutionarily conserved death pathway. *Nat Cell Biol* **1:** 272–279.

Sevrioukov EA, Burr J, Huang EW, Assi HH, Monserrate JP, Purves DC, Wu JN, Song EJ, Brachmann CB. 2007. *Drosophila* Bcl-2 proteins participate in stress-induced apoptosis, but are not required for normal development. *Genesis* **45:** 184–193.

Shalini S, Dorstyn L, Dawar S, Kumar S. 2014. Old, new and emerging functions of caspases. *Cell Death Differ* **22:** doi: 10.1038/cdd.2014.216.

Song Z, McCall K, Steller H. 1997. DCP-1, a *Drosophila* cell death protease essential for development. *Science* **275:** 536–540.

Srinivasula SM, Datta P, Kobayashi M, Wu, JW, Fujioka M, Hegde R, Zhang Z, Mukattash R, Fernandes-Alnemri T, Shi Y, Jaynes JB, Alnemri ES. 2002. *sickle*, a novel *Drosophila* death gene in the *reaper/hid/grim* region, encodes an IAP-inhibitory protein. *Curr Biol* **12:** 125–130.

Srivastava M, Scherr H, Lackey M, Xu D, Chen Z, Lu J, Bergmann A. 2007. ARK, the Apaf-1 related killer in *Drosophila*, requires diverse domains for its apoptotic activity. *Cell Death Differ* **14:** 92–102.

Steller H. 2008. Regulation of apoptosis in *Drosophila*. *Cell Death Differ* **15:** 1132–1138.

Tenev T, Zachariou A, Wilson R, Paul A, Meier P. 2002. Jafrac2 is an IAP antagonist that promotes cell death by liberating Dronc from DIAP1. *EMBO J* **21:** 5118–5129.

White K, Grether ME, Abrams JM, Young L, Farrell K, Steller H. 1994. Genetic control of programmed cell death in *Drosophila*. *Science* **264:** 677–683.

Wolff T, Ready DF. 1991. Cell death in normal and rough eye mutants of *Drosophila*. *Development* **113:** 825–839.

Xu D, Li Y, Arcaro M, Lackey M, Bergmann A. 2005. The CARD-carrying caspase Dronc is essential for most, but not all, developmental cell death in *Drosophila*. *Development* **132:** 2125–2134.

Xu D, Wang Y, Willecke R, Chen Z, Ding T, Bergmann A. 2006. The effector caspases drICE and dcp-1 have partially overlapping functions in the apoptotic pathway in *Drosophila*. *Cell Death Differ* **13:** 1697–1706.

Yu X, Wang L, Acehan D, Wang X, Akey CW. 2006. Three-dimensional structure of a double apoptosome formed by the *Drosophila* Apaf-1 related killer. *J Mol Biol* **355:** 577–589.

Yuan S, Yu X, Topf M, Dorstyn L, Kumar S, Ludtke SJ, Akey CW. 2011. Structure of the *Drosophila* apoptosome at 6.9 Å resolution. *Structure* **19:** 128–140.

Terminal Deoxynucleotidyl Transferase (TdT)-Mediated dUTP Nick-End Labeling (TUNEL) for Detection of Apoptotic Cells in *Drosophila*

Donna Denton[1] and Sharad Kumar[1]

Centre for Cancer Biology, University of South Australia, Adelaide, South Australia 5001, Australia

A characteristic feature of apoptosis is DNA fragmentation. This fragmentation can be detected by terminal deoxynucleotidyl transferase (TdT)-mediated dUTP nick-end labeling (TUNEL) of DNA in dying cells. Here, we present a protocol for TUNEL detection of apoptosis in *Drosophila* larval tissue, but these techniques can be adapted for other tissues and developmental stages.

MATERIALS

It is essential that you consult the appropriate Material Safety Data Sheets and your institution's Environmental Health and Safety Office for proper handling of equipment and hazardous material used in this protocol.

RECIPE: Please see the end of this protocol for recipes indicated by <R>. Additional recipes can be found online at http://cshprotocols.cshlp.org/site/recipes.

Reagents

DNase

Fixative (freshly prepared 4% paraformaldehyde in PBS)

Alternatively, a 20% stock can be made and stored at −20°C.

Glycerol (80% in PBS)

Alternatively, Vectashield can be used.

Hoechst 33258 (2 mg/mL stock solution)

PBTx (0.1% Triton X-100 in 1× PBS)

PBTx5 (0.5% Triton X-100 in 1× PBS)

Phosphate-buffered saline (PBS) (1×, pH 7.4; without Ca^{+2} or Mg^{+2}) <R>

Sodium citrate (100 mM in PBTx [1.47 g in 50 mL of PBTx], freshly prepared)

TUNEL kit (e.g., TMR-red; Roche)

There are several commercially available kits with either fluorescent or color detection methods, and these should be used according to the manufacturer's instructions.

Equipment

Confocal or fluorescence microscope, fitted with an appropriate camera

Dissecting microscope (e.g., Olympus)

---

[1]Correspondence: Donna.Denton@unisa.edu.au; Sharad.Kumar@unisa.edu.au

Cite this protocol as *Cold Spring Harb Protoc*; doi:10.1101/pdb.prot086199

Dissection dish

Prepare a dissection dish according to the manufacturer's instructions for the Sylgard 184 Silicone Elastomer kit (Dow Corning). The soft silicone elastomer set in the dish base is useful for preventing damage to dissection equipment.

Dissection equipment

Forceps (Dumont #5 or #55; World Precision Instruments)

Needle (26-gauge hypodermic or tungsten)

Double-sided sticky tape (optional; see Step 11)

Microcentrifuge tubes (1.5-mL)

Microscope coverslips (22 × 22 or 22 × 50 mm) and glass slides (76 × 26 mm)

Water bath (or hybridization oven, or similar) at 65°C

METHOD

1. With forceps, dissect larvae or pupae immersed in 1× PBS on a Sylgard dissection dish under a dissecting microscope at room temperature.

   Dissected tissues can be processed for up to 20 min at room temperature or on ice for up to 1 h.

2. Fix tissue of interest by transferring into a 1.5-mL microcentrifuge tube containing 4% paraformaldehyde in PBS and incubate for 20 min at room temperature (22°C–25°C).

3. Wash three times with PBTx. For each wash, use 1 mL of PBTx and incubate for 5–10 min at room temperature.

4. Wash with 1 mL of PBTx5 for 5 min at room temperature.

5. Permeabilize in 100 mM sodium citrate (freshly prepared in PBTx) for 30 min at 65°C.

6. Wash three times with 1 mL each of PBTx for 5–10 min at room temperature.

7. Perform the TUNEL reaction according to the manufacturer's instructions.

   For each sample, we use 100 µL of Roche TMR-red kit reagents (10 µL of the TdT enzyme and 90 µL of the labeling mix) and incubate for 3 h at 37°C.

   Although TUNEL generally yields reproducible results, to avoid false positives and false negatives, appropriate controls should be included. Consider including a negative control sample that contains only labeling mix and a positive control sample that was treated with DNase. For DNase I treatment, incubate for 10 min at room temperature with DNase I at 3000 U/mL.

   See Troubleshooting.

8. Remove the TUNEL solution. Wash three times with PBTx. For each wash, use 1 mL of PBTx and incubate for 5–10 min at room temperature.

9. (Optional) To detect nuclei, costain samples with 4 µg/mL Hoechst 33258 in PBS for 1 min at room temperature.

10. Wash three times with PBTx. For each wash, use 1 mL of PBTx and incubate for 5–10 min at room temperature.

11. Mount in 80% glycerol in PBS (or in Vectashield) by transferring the sample to a microscope glass slide, dissecting the tissue of interest away from other tissue if required, and then placing a microscope coverslip over tissue.

    Double-sided sticky tape can be used on each end of the microscope slide to prevent the tissue from being squashed by the coverslip.

12. Examine the samples by confocal or fluorescence microscopy.

    Figure 1 shows an example of a TUNEL salivary gland.

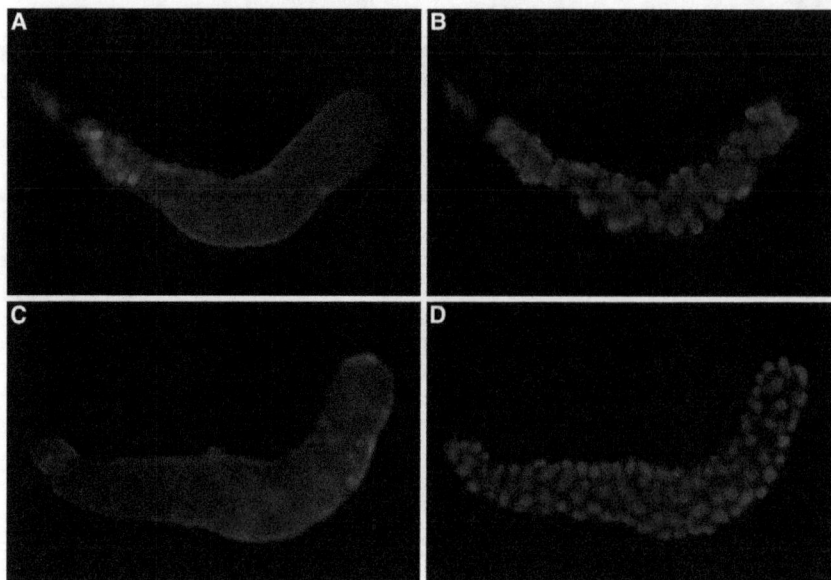

FIGURE 1. Visualization of *Drosophila* DNA fragmentation by terminal deoxynucleotidyl transfer-mediated dUTP nick-end labeling (TUNEL). Salivary glands at 14 h relative to puparium formation from (*A*) control larva, showing TUNEL-positive nuclei, and (*C*) mutant larva lacking the initiator caspase Dronc (Nc), indicating the absence of TUNEL-positive nuclei. DNA stained with Hoechst shows a similar pattern of nuclei in wild-type (*B*) and mutant larvae (*D*).

TROUBLESHOOTING

Problem (Step 7): Difficulties are experienced in labeling.

Solution: Consider using an alternative permeabilization procedure such as proteinase K treatment at Step 5 (incubate the sample at room temperature for 3–5 min in a 10 µg/mL solution of proteinase K that was prepared in PBTx) (Arama and Steller 2006; McCall et al. 2009).

RELATED TECHNIQUES

Slight modifications may be required for TUNEL in other *Drosophila* tissues. For example, TUNEL protocols for embryos include an additional postfixation step (in 4% paraformaldehyde in PBS for 20 min) following permeabilization and preceding Step 6. For additional information, see Arama and Steller (2006) and McCall et al. (2009). TUNEL can also be used for the analysis of cell death in sections from paraffin-embedded samples, necessitating the use of a color detection kit according to the manufacturer's instructions.

Because TUNEL detects DNA strand breaks, the examination of cell death after treatment with inducers of DNA damage, such as γ-irradiation, might require alternative assays, such as those employing acridine orange (see Protocol 2: Use of the Vital Dye Acridine Orange to Observe Dying Cells in *Drosophila* Tissue [Denton and Kumar 2015a]) or immunostaining for active caspase-3 (see Protocol 3: Immunostaining Using an Antibody against Active Caspase-3 to Detect Apoptotic Cells in *Drosophila* [Denton and Kumar 2015b]). This is because TUNEL can potentially detect the DNA damage induced by the treatment itself.

General advice for handling *Drosophila* in the laboratory is available in Sullivan et al. (2000).

Cite this protocol as *Cold Spring Harb Protoc*; doi:10.1101/pdb.prot086199

Studying Apoptosis in *Drosophila*

RECIPE

Phosphate-Buffered Saline (PBS)

Reagent	Amount to add (for 1× solution)	Final concentration (1×)	Amount to add (for 10× stock)	Final concentration (10×)
NaCl	8 g	137 mM	80 g	1.37 M
KCl	0.2 g	2.7 mM	2 g	27 mM
Na$_2$HPO$_4$	1.44 g	10 mM	14.4 g	100 mM
KH$_2$PO$_4$	0.24 g	1.8 mM	2.4 g	18 mM
If necessary, PBS may be supplemented with the following:				
CaCl$_2$·2H$_2$O	0.133 g	1 mM	1.33 g	10 mM
MgCl$_2$·6H$_2$O	0.10 g	0.5 mM	1.0 g	5 mM

PBS can be made as a 1× solution or as a 10× stock. To prepare 1 L of either 1× or 10× PBS, dissolve the reagents listed above in 800 mL of H$_2$O. Adjust the pH to 7.4 (or 7.2, if required) with HCl, and then add H$_2$O to 1 L. Dispense the solution into aliquots and sterilize them by autoclaving for 20 min at 15 psi (1.05 kg/cm^2) on liquid cycle or by filter sterilization. Store PBS at room temperature.

ACKNOWLEDGMENTS

We thank the members of our laboratory for their comments on this protocol. The *Drosophila* cell death research in our laboratory is supported by the National Health and Medical Research Project Grant (1041807) and a Senior Principal Research Fellowship (1002863) to S.K.

REFERENCES

Arama E, Steller H. 2006. Detection of apoptosis by terminal deoxynucleotidyl transferase-mediated dUTP nick-end labeling and acridine orange in *Drosophila* embryos and adult male gonads. *Nat Protoc* 1: 1725–1731.

Denton D, Kumar S. 2015a. Use of the vital dye acridine orange to observe dying cells in *Drosophila* tissue. *Cold Spring Harb Protoc* doi: 10.1101/pdb.prot086207.

Denton D, Kumar S. 2015b. Immunostaining using an antibody against active caspase-3 to detect apoptotic cells in *Drosophila*. *Cold Spring Harb Protoc* doi: 10.1101/pdb.prot086215.

McCall K, Peterson JS, Pritchett TL. 2009. Detection of cell death in *Drosophila*. *Methods Mol Biol* 559: 343–356.

Sullivan W, Ashburner M, Hawley RS. 2000. *Drosophila Protocols*. Cold Spring Harbor Laboratory Press, Cold Spring Harbor, NY.

Using the Vital Dye Acridine Orange to Detect Dying Cells in *Drosophila*

Donna Denton[1] and Sharad Kumar[1]

Centre for Cancer Biology, University of South Australia, Adelaide, South Australia 5001, Australia

Acridine orange is a cell-permeable fluorescent dye that binds to nucleic acids, resulting in an altered spectral emission. Acridine orange staining has been shown to be highly selective for apoptotic cells in *Drosophila*; however, the precise mechanism underlying this effect is not known. Advantages of acridine orange staining include the speed and ease of the staining. But there are disadvantages: It should be performed on unfixed tissue that therefore must be examined immediately, and multiple labeling cannot be performed. Slightly different protocols for the uptake of acridine orange are required for different developmental stages. Here, we present protocols for use of acridine orange to detect apoptosis in *Drosophila* embryos and in larval tissue. Slight modifications might be required for other *Drosophila* tissues.

MATERIALS

It is essential that you consult the appropriate Material Safety Data Sheets and your institution's Environmental Health and Safety Office for proper handling of equipment and hazardous material used in this protocol.

RECIPE: Please see the end of this protocol for recipes indicated by <R>. Additional recipes can be found online at http://cshprotocols.cshlp.org/site/recipes.

Reagents

Acridine orange stock solution (0.5 mg/mL, prepared in ethanol)

The acridine orange stock solution can be stored in the dark at room temperature (22°C–25°C) for several months and should be diluted to the appropriate final concentration in PBS immediately before use.

Bleach (domestic brand) (diluted to 50% in water)
Grape juice agar plates <R>
Heptane
Phosphate-buffered saline (PBS) (1×, pH 7.4; without Ca^{+2} or Mg^{+2}) <R>
Yeast paste

Prepare by mixing dry granular yeast and water into a smooth paste.

Equipment

Artist's paintbrush, small
Cell strainer (BD Falcon) or collection basket

The collection basket can be made by fixing fine wire mesh over a cut-off 1.5-mL microcentrifuge tube.

---

[1]Correspondence: Donna.Denton@unisa.edu.au; Sharad.Kumar@unisa.edu.au

Cite this protocol as *Cold Spring Harb Protoc*; doi:10.1101/pdb.prot086207

Confocal or fluorescence microscope, fitted with an appropriate camera

Dissecting microscope (e.g., Olympus)

Dissection dish

Prepare a dissection dish according to the manufacturer's instructions for the Sylgard 184 Silicone Elastomer kit (Dow Corning). The soft silicone elastomer set in the dish base is useful for preventing damage to dissection equipment.

Dissection equipment

Forceps (Dumont #5 or #55; World Precision Instruments)

Needle (26-gauge hypodermic or tungsten)

Double-sided sticky tape (optional; see Step 11)

Egg-laying chambers

These comprise a tube with mesh over one end with a diameter smaller than that of the lay-dish.

Microscope coverslips (22 × 22 or 22 × 50 mm) and glass slides (76 × 26 mm)

Microcentrifuge tubes (1.5 mL)

METHOD

Procedure for Embryos

1. Set flies to lay for the desired time in egg-laying chambers on top of grape juice agar plates with a small amount of yeast paste.

2. Harvest the embryos from the agar plate by adding water to the plate and transferring the embryos into a collection basket/cell strainer with a small artist's paintbrush. Continue washing the embryos with water in the basket to remove any yeast.

3. Dechorionate the embryos by continually pipetting them in a 50% bleach solution for 2–5 min.

   Because the formulations of bleach stock solutions can vary, dechorionation should be monitored by using a dissection microscope. Continue dechorionation until 80% of the dorsal appendages have dissolved.

4. Wash the dechorionated embryos thoroughly with water to remove bleach.

5. Using a small paintbrush, transfer the embryos to a tube containing equal volumes of heptane and 5 µg/mL acridine orange in PBS.

   For example, use a 1.5-mL tube to which 500 µL of heptane and 500 µL of 5 µg/mL acridine orange in PBS have been added.

6. Shake vigorously for 5 min at room temperature to generate a fine emulsion.

7. Allow the liquid phases to separate, and pipette the embryos from the interface onto a glass microscope slide in a drop of PBS.

   Ensure that the embryos do not dry out.

8. View the samples immediately using confocal or fluorescence microscopy.

   The green channel (522 nm) gives the best sensitivity; however, if the background is high, the red channel (568 nm) can be used (Arama and Steller 2006).

   See Troubleshooting.

Procedure for Larval Tissue

9. Immerse the larvae or pupae in PBS on a Sylgard dissection dish under a dissection microscope. Dissect the tissue of interest away from other tissue with forceps and needle.

   Work in batches of only a few samples at a time to allow for imaging the live tissue.

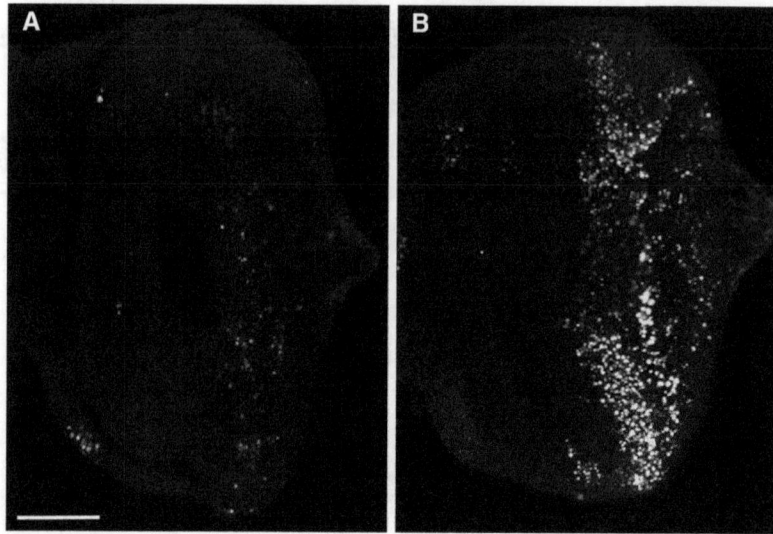

FIGURE 1. Detection of apoptosis in the *Drosophila* eye imaginal disc by using acridine orange. Eye imaginal discs of third-instar larvae stained with acridine orange from (*A*) wild-type and (*B*) mutant larvae expressing the apoptosis activator GMR-hid, revealing strong staining in the posterior half of the eye discs. Posterior is to the *right*. Scale bar, 50 μm. GMR, glass multimer reporter (enhancer).

10. Incubate the dissected tissue in a drop (20–30 μL) of 0.5 μg/mL acridine orange solution (freshly prepared in PBS) for 5–15 min.

    Alternatively, the tissues can be dissected directly in the acridine orange solution on a microscope slide to minimize movement and mechanical damage to the unfixed tissue.

11. Wash the tissue briefly by transferring it to a fresh drop of PBS. Mount in a drop of PBS on a glass microscope slide.

    Alternatively, substitute the acridine orange solution on the slide for PBS.

    Double-sided tape can be used on each end of the microscope slide to prevent the tissue from being squashed by the microscope coverslip.

12. View the samples immediately by using confocal or fluorescence microscopy.

    Figure 1 shows an example of an eye disc stained with acridine orange.

    See Troubleshooting.

TROUBLESHOOTING

Problem (Steps 8 or 12): The PBS mounting solution evaporates too quickly.
Solution: Consider using halocarbon oil as an alternative mounting solution.

Problem (Steps 8 or 12): There is only a poor, or no, fluorescent signal.
Solution: Ensure that there is no detergent present in any of the solutions as this will eliminate staining.

RELATED TECHNIQUES

Additional protocols for acridine orange staining in *Drosophila* embryos, testes, and ovaries can be found in Arama and Steller (2006) and in McCall et al. (2009). General advice for handling *Drosophila* in the laboratory is available in Sullivan et al. (2000).

Cite this protocol as *Cold Spring Harb Protoc*; doi:10.1101/pdb.prot086207

RECIPES

Grape Juice Agar Plates

1. Combine agar, grape juice, sucrose, and water together so that the final concentrations are as follows.

   3% agar
   25% grape juice
   0.3% sucrose

2. Boil by heating in a microwave. Be careful; the solution will be very hot!

3. Mix the solution, and allow it to cool to 60°C.

4. After cooling, add tegosept (*para*-hydroxybenzoate, usually prepared as a 10% stock solution in ethanol) to a final concentration of 0.03%. Mix the solution well.

5. Pour the solution into Petri dishes that are appropriately sized to fit the *Drosophila* egg-laying chambers. Once the plates have set, store them at 4°C (they will keep for several months).

Phosphate-Buffered Saline (PBS)

Reagent	Amount to add (for 1× solution)	Final concentration (1×)	Amount to add (for 10× stock)	Final concentration (10×)
NaCl	8 g	137 mM	80 g	1.37 M
KCl	0.2 g	2.7 mM	2 g	27 mM
Na$_2$HPO$_4$	1.44 g	10 mM	14.4 g	100 mM
KH$_2$PO$_4$	0.24 g	1.8 mM	2.4 g	18 mM
If necessary, PBS may be supplemented with the following:				
CaCl$_2$·2H$_2$O	0.133 g	1 mM	1.33 g	10 mM
MgCl$_2$·6H$_2$O	0.10 g	0.5 mM	1.0 g	5 mM

PBS can be made as a 1× solution or as a 10× stock. To prepare 1 L of either 1× or 10× PBS, dissolve the reagents listed above in 800 mL of H$_2$O. Adjust the pH to 7.4 (or 7.2, if required) with HCl, and then add H$_2$O to 1 L. Dispense the solution into aliquots and sterilize them by autoclaving for 20 min at 15 psi (1.05 kg/cm^2) on liquid cycle or by filter sterilization. Store PBS at room temperature.

ACKNOWLEDGMENTS

We thank the members of our laboratory for their comments on this protocol. The *Drosophila* cell death research in our laboratory is supported by the National Health and Medical Research Project Grant (1041807) and a Senior Principal Research Fellowship (1002863) to S.K.

REFERENCES

Arama E, Steller H. 2006. Detection of apoptosis by terminal deoxynucleotidyl transferase-mediated dUTP nick-end labeling and acridine orange in *Drosophila* embryos and adult male gonads. *Nat Protoc* 1: 1725–1731.

McCall K, Peterson JS, Pritchett TL. 2009. Detection of cell death in *Drosophila*. *Methods Mol Biol* 559: 343–356.
Sullivan W, Ashburner M, Hawley RS. 2000. Drosophila *Protocols*. Cold Spring Harbor Laboratory Press, Cold Spring Harbor, NY.

Immunostaining Using an Antibody against Active Caspase-3 to Detect Apoptotic Cells in *Drosophila*

Donna Denton[1] and Sharad Kumar[1]

Centre for Cancer Biology, University of South Australia, Adelaide, South Australia 5001, Australia

The activation of mammalian caspase-3 after proteolytic cleavage adjacent to residue Asp175 produces a large (17/19-kDa), active subunit. A commercially available antibody recognizes the large, active subunit of caspase-3 but not the full-length inactive caspase-3. This antibody has also been shown to detect active *Drosophila* effector caspases. Here, we present a protocol showing how this antibody can be used to detect apoptotic cells in various *Drosophila* tissues and developmental stages and discuss the specificity of the antibody.

MATERIALS

It is essential that you consult the appropriate Material Safety Data Sheets and your institution's Environmental Health and Safety Office for proper handling of equipment and hazardous material used in this protocol.

RECIPE: Please see the end of this protocol for recipes indicated by <R>. Additional recipes can be found online at http://cshprotocols.cshlp.org/site/recipes.

Reagents

Antibodies

Cleaved caspase-3 (Asp175) antibody (Cell Signaling Technology 9661)

Secondary antibody, fluorescently labeled (e.g., Anti-rabbit Alexa 488 from Molecular Probes)

Alexa-Fluor-conjugated secondary antibodies (Molecular Probes) are available in a wide range of fluorescence-emission wavelengths. We commonly use Alexa Fluor 488 and Alexa Fluor 568, which possess good intensity and photostability.

Blocking solution (1% BSA in PBTx)

Fixative (freshly prepared 4% paraformaldehyde in 1× PBS)

Glycerol (80% in 1× PBS)

Alternatively, Vectashield can be used.

Hoechst 33258 (2 mg/mL stock solution) (optional; see Step 10)

PBTx (0.1% Triton X-100 in 1× PBS)

---

[1]Correspondence: Sharad.Kumar@unisa.edu.au; Donna.Denton@unisa.edu.au

Cite this protocol as *Cold Spring Harb Protoc*; doi:10.1101/pdb.prot086215

Phosphate-buffered saline (PBS) (1×, pH 7.4; without Ca^{+2} or Mg^{+2}) <R>

Equipment

Confocal or fluorescence microscope, fitted with an appropriate camera

Dissecting microscope (e.g., Olympus)

Dissection dish

Prepare a dissection dish according to the manufacturer's instructions for the Sylgard 184 Silicone Elastomer kit (Dow Corning). The soft silicone elastomer set in the dish base is useful for preventing damage to dissection equipment.

Dissection equipment

Forceps (Dumont #5 or #55; World Precision Instruments)

Needle (26-gauge hypodermic or tungsten)

Double-sided sticky tape

Microcentrifuge tubes (1.5-mL)

Microscope coverslips (22 × 22 or 22 × 50 mm) and glass slides (76 × 26 mm)

METHOD

1. With forceps and needle, dissect larvae or pupae immersed in 1× PBS on a Sylgard dissection dish under a dissection microscope.

   Dissected tissues can be processed for up to 20 min at room temperature or on ice for up to 1 h.

2. Fix tissue of interest by transferring into a 1.5-mL microcentrifuge tube containing 4% paraformaldehyde in 1× PBS and incubate for 20 min at room temperature (22°C–25°C).

3. Rinse the tissue three times, each with 1 mL of PBTx. Then wash the tissue three times, each with 1 mL of PBTx for 15 min at room temperature.

4. Block the tissue in 1 mL of blocking solution for 1 h at room temperature.

5. Incubate with the primary antibody (cleaved caspase-3 [Asp175] antibody) diluted 1:100 in blocking solution overnight at 4°C with gentle shaking.

   The primary antibody against cleaved caspase-3 can be used at a higher dilution—this should be determined empirically for each tissue type.

6. Remove the primary antibody solution. Wash the tissue three times, each with 1 mL of PBTx for 10 min at room temperature.

7. Add the fluorescently labeled secondary antibody diluted in PBTx or blocking solution.

   We use an Alexa-Fluor-488-conjugated secondary antibody (Molecular Probes) at a 1:200 dilution.

8. Once the fluorescent secondary antibody has been added, keep the samples in the dark and incubate at room temperature for 1–2 h.

9. Remove the secondary antibody and wash in PBTx as described in Step 6.

10. (Optional) To detect nuclei, costain the samples using Hoechst 33258 at 4 µg/mL in PBS for 1 min.

11. Mount the sample in 80% glycerol in PBS (or in Vectashield) by transferring the sample to a microscope slide, dissecting the tissue of interest away from other tissue (if needed), and placing a microscope coverslip over the tissue sample.

    Double-sided sticky tape can be used on each end of the microscope slide to prevent the tissue from being squashed by the coverslip.

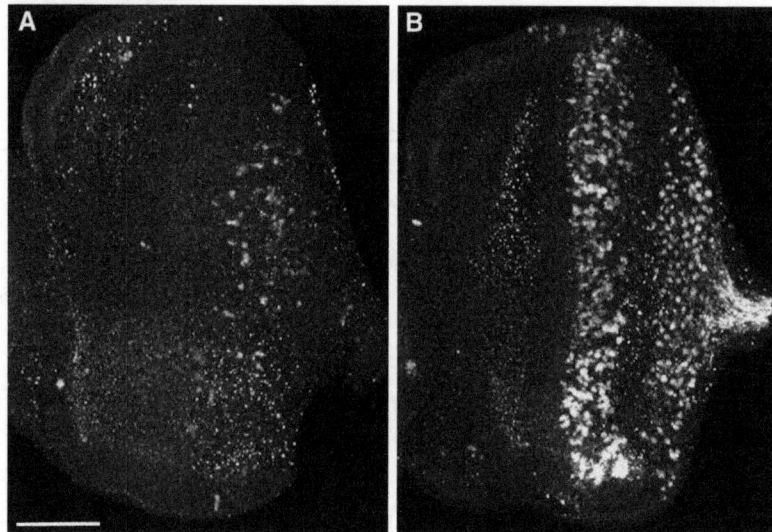

FIGURE 1. Detection of apoptosis in the eye imaginal disc by using an antibody against cleaved caspase-3. Eye imaginal discs of third-instar larvae stained with antibody against cleaved caspase-3, followed by the Alexa-Fluor-488-conjugated secondary antibody detecting active caspases. (*A*) wild-type and (*B*) mutant larvae expressing the apoptosis activator GMR-hid, showing the strong active caspase staining in the posterior half of the eye discs. Posterior is to the *right*. Scale bar, 50 μm. GMR, glass multimer reporter (enhancer).

12. Examine the samples by confocal or fluorescence microscopy.

> *Figure 1 shows an example of an eye disk stained using the antibody against cleaved caspase-3. See Troubleshooting.*

TROUBLESHOOTING

Problem (Step 12): The level of nonspecific background staining is found to be too high.
Solution: Use longer washes and more rinses at Step 6.

Problem (Step 12): The fluorescent signal is lost through photobleaching.
Solution: Use a commercially available antifade mounting medium, such as Vectashield, at Step 11.

DISCUSSION

Here, we have presented a protocol for using a commercially available antibody against the cleaved form of human caspase-3 to detect active, not inactive, effector caspases in various *Drosophila* tissue and developmental stages. Note that the antibody against cleaved caspase-3 not only detects cleaved *Drosophila* effector caspases Drice (*Drosophila* ICE), Decay, and Dcp-1 (caspase-1), but it also detects at least one additional potential Dronc (caspase Nc) substrate (Fan and Bergmann 2010). As the identity of this additional substrate is unknown, it is plausible that this substrate might be involved in some apoptosis-independent aspects of Dronc activity. Given this, Fan and Bergmann suggest that the cleaved-caspase-3 antibody immunoreactivity reflects Dronc, but not necessarily apoptosis-specific, activity (Fan and Bergmann 2010).

 Cite this protocol as *Cold Spring Harb Protoc*; doi:10.1101/pdb.prot086215

RECIPE

Phosphate-Buffered Saline (PBS)

Reagent	Amount to add (for 1× solution)	Final concentration (1×)	Amount to add (for 10× stock)	Final concentration (10×)
NaCl	8 g	137 mM	80 g	1.37 M
KCl	0.2 g	2.7 mM	2 g	27 mM
Na_2HPO_4	1.44 g	10 mM	14.4 g	100 mM
KH_2PO_4	0.24 g	1.8 mM	2.4 g	18 mM
If necessary, PBS may be supplemented with the following:				
$CaCl_2 \cdot 2H_2O$	0.133 g	1 mM	1.33 g	10 mM
$MgCl_2 \cdot 6H_2O$	0.10 g	0.5 mM	1.0 g	5 mM

PBS can be made as a 1× solution or as a 10× stock. To prepare 1 L of either 1× or 10× PBS, dissolve the reagents listed above in 800 mL of H_2O. Adjust the pH to 7.4 (or 7.2, if required) with HCl, and then add H_2O to 1 L. Dispense the solution into aliquots and sterilize them by autoclaving for 20 min at 15 psi (1.05 kg/cm^2) on liquid cycle or by filter sterilization. Store PBS at room temperature.

ACKNOWLEDGMENTS

We thank the members of our laboratory for their comments on this protocol. The *Drosophila* cell death research in our laboratory is supported by the National Health and Medical Research Project Grant (1041807) and a Senior Principal Research Fellowship (1002863) to S.K.

REFERENCES

Fan Y, Bergmann A. 2010. The cleaved-Caspase-3 antibody is a marker of Caspase-9-like DRONC activity in *Drosophila*. *Cell Death Differ* 17: 534–539.

Analyzing the Response of RNAi-Treated *Drosophila* Cells to Death Stimuli by Quantitative Real-Time Polymerase Chain Reaction

Donna Denton[1] and Sharad Kumar[1]

Centre for Cancer Biology, University of South Australia, Adelaide, South Australia 5001, Australia

A useful complement to animal studies is the use of *Drosophila* cell lines to analyze cell-death responses. There are numerous *Drosophila* cell lines available, such as S2 cells, which possess the advantages of being semi-adherent, fast growing, relatively robust, and useful for transfection and knockdown studies, whereas other lines, such as mbn2, are more suitable for analyzing hormone-induced cell death and gene expression. *Drosophila* cell lines are very amenable to knockdown studies as the cells take up double-stranded RNA (dsRNA) from the medium, initiating gene silencing and resulting in a high level of gene knockdown. This means that the cell lines are useful for investigating the response to death stimuli, following gene knockdown, by examining the expression of cell-death genes. This protocol describes the synthesis of dsRNA for treatment of *Drosophila* cells and the subsequent analysis of cell-death gene expression by quantitative real-time polymerase chain reaction (qPCR).

MATERIALS

It is essential that you consult the appropriate Material Safety Data Sheets and your institution's Environmental Health and Safety Office for proper handling of equipment and hazardous material used in this protocol.

RECIPE: Please see the end of this protocol for recipes indicated by <R>. Additional recipes can be found online at http://cshprotocols.cshlp.org/site/recipes.

Reagents

4′,6-diamidino-2-phenylindole (DAPI) (2 µg/mL, prepared in methanol)
Cell-death inducer (e.g., ecdysone, UV, cycloheximide, or etoposide; see Step 19)
Chloroform
Diethylpyrocarbonate (DEPC)-treated water (DNase/RNase-free)
Ethanol (75%)
Fetal bovine serum (FBS), heat-inactivated
Insulin (if required; see Step 9)
Isopropanol
Megascript SP6 Transcription Kit (Ambion/Life Technologies)
Megascript T3 Transcription Kit (Ambion/Life Technologies)
Megascript T7 Transcription Kit (Ambion/Life Technologies)
Methanol

---

[1]Correspondence: sharad.kumar@unisa.edu.au; Donna.Denton@unisa.edu.au

Cite this protocol as *Cold Spring Harb Protoc*; doi:10.1101/pdb.prot086223

PCR products (400–600 bp) to be used for RNAi (see Step 1)
Penicillin–streptomycin (if required; see Step 9)
pGEM-T Easy Vector (Promega)
Phenol:chloroform mixture (1:1)
Phosphate-buffered saline (PBS) ($1\times$, pH 7.4; without Ca^{+2} or Mg^{+2}) <R>
Schneider's *Drosophila* medium (complete medium) (Gibco 21720-024)
TRIzol reagent (Invitrogen/Life Technologies)
Trypan blue

Equipment

Agarose gel electrophoresis materials
Benchtop centrifuge
Conical centrifuge tubes (50-mL, sterile)
Flasks (e.g., T-25 cm^2, T-75 cm^2)
Heat block
Hemocytometer
Incubator (humidified, at 27°C)
Pipettes (5-, 10-, and 25-mL; sterile)
Plates (e.g., six-well; 35-mm)
Primers and kit for real-time PCR

> We use the RT^2 Real-Time SYBR green/ROX PCR MasterMix (QIAGEN). Alternative SYBR green master mixes are available. Ensure that each primer pair is tested for efficiency.

Restriction enzymes and buffers for linearizing the template used for in vitro transcription (see Step 2)
Reverse-transcription kit

> We utilize oligo dT_{19} primers and the High Capacity cDNA Reverse Transcription Kit (Applied Biosciences).

Thermal cycler for qPCR (e.g., Rotor-Gene 6000; Corbett Research)
Tubes (RNase/DNase-free)

METHOD

Synthesis of Double-Stranded RNA

1. Generate template DNA for RNA synthesis by cloning a 400–600-bp region of a gene of interest into a plasmid containing flanking T3, T7, and SP6 promoters, such as the pGEM-T Easy Vector.

   > The 400- to 600-bp regions can be amplified by PCR from plasmids, total cDNA, or genomic DNA.
   > It is necessary also to use a negative-control gene, such as the gene encoding green fluorescent protein (GFP).

2. Linearize the template by using a unique restriction enzyme site on one side of the insert for transcription initiation from the T3 or T7 promoter and a second restriction enzyme site on the other side of the insert for transcription initiation from the SP6 promoter.

   > Use of ~20 µg of DNA for each restriction digest will provide sufficient template for multiple transcription reactions.

3. Generate sense and antisense RNA transcripts by in vitro transcription (we use the Ambion Megascript Transcription Kits), following the kit manufacturer's instructions.

   > To obtain sufficient RNA, set up multiple transcription reactions and pool the reactions together.

4. Extract the RNA with an equal volume of phenol:chloroform mixture and precipitate it by adding an equal volume of isopropanol.

5. Briefly air-dry the pellet and dissolve the RNA in DEPC-treated water at 65°C (use a heat block and incubate tubes for 10 min).

6. Measure the concentration of RNA by spectrometry, and then adjust to 1 mg/mL.

7. Create dsRNAs.

    i. Add equal amounts (usually ~100 µg) of 1 mg/mL sense RNA and 1 mg/mL antisense RNA to an RNase-free tube.

    ii. Anneal the RNAs by heating the tube for 10 min at 70°C in a heat block before cooling slowly to room temperature.

        To achieve slow cooling, the reactions can be removed from the heat block and left on the bench to cool down.

8. Check the size and purity of the dsRNA by running 3–5 µg on a native agarose gel. Use the remaining dsRNA or store it at −20°C.

    The dsRNA is stable for several months when stored at −20°C and can be reannealed before use.

Cell Culture

9. Grow the *Drosophila* cell lines in Schneider's *Drosophila* Medium containing 10% heat-inactivated FBS, supplemented with penicillin–streptomycin if required, in flasks in a humidified incubator at 27°C.

    Additional growth factors might be required for some cell lines—for example, BG2 cells should be supplemented with insulin.

10. Passage the cells when the cell density exceeds 6×10^6 cells/mL.

    i. To dislodge the cells, use a 5-mL pipette to squirt the medium over the cells (this will make a cloudy suspension and will break up cell clumps).

        Drosophila *cells do not require treatment with trypsin to be removed from flasks, but they might require use of a cell scraper to be dislodged. At high cell densities, the cells might pile up or detach from the plastic and grow as clumps in suspension.*

    ii. Determine cell counts by using a hemocytometer and seed at a density $>5 \times 10^5$ cells/mL into fresh complete medium in a final volume of 10–15 mL in a T-75 cm^2 flask.

        As cells can be sensitive to low cell density, a higher cell number per milliliter is better than a low cell density; cell viability can be determined by Trypan blue staining (see Step 12).

    iii. Place flask in an incubator set at 27°C, with the lid of the flask loosened for aeration.

11. Once the cell density has exceeded 6×10^6 cells/mL, repeat Step 10 as necessary to obtain sufficient cells for the intended RNAi treatments.

Viability Assays

Employ one or both of the following strategies.

12. Use the Trypan blue dye-exclusion assay to determine the number of viable cells.

    Dying cells take up Trypan blue. The percentage of viable cells can be estimated by counting the number of dying cells (i.e., those that stain positive for Trypan blue) and viable cells (i.e., those that are unstained) using a hemocytometer.

13. Stain cells with DAPI (2 µg/mL in methanol) and score for cell death, based on nuclear morphology, by microscopy.

    This method allows viability to be assessed based on nuclear morphology—the nuclei of apoptotic cells appear condensed, and, at later stages, apoptotic blebbing of the nuclear membrane is visible.

Treatment of Cells with dsRNA

14. Dislodge exponentially growing cells from a T-75 cm^2 flask by using a 5-mL pipette to wash the medium over the cells. Transfer the cells into a 50-mL conical centrifuge tube, and pellet them at 1000 rpm for 5 min in a benchtop centrifuge.

15. Discard the medium, and resuspend the cells in 10 mL of serum-free medium. Count the cells using a hemocytometer, and then dilute them to a concentration of 2×10^6 cells/mL.

Cite this protocol as *Cold Spring Harb Protoc*; doi:10.1101/pdb.prot086223

16. Seed cells at 2×10^6 cells/mL per well into a six-well plate in 1 mL of serum-free Schneider's *Drosophila* medium.

17. Add dsRNA to give a concentration of 37 nM, mix by pipetting vigorously eight to 10 times, and then incubate for 1 h at 27°C.

    As dsRNA can be slightly toxic to cells, a control dsRNA (such as one targeting expression of GFP) should be included in the analysis.

18. Add 2 mL of medium with serum and incubate for 24–48 h at 27°C.

    The level of knockdown achieved might improve with longer incubation times.

19. Treat cells with appropriate cell-death stimulus.

    There is a range of death inducers available, including application of the molting hormone ecdysone (10 μM), the DNA-damaging agent UV (100 J/m², the protein synthesis inhibitor cycloheximide (20 μg/ mL) or the topoisomerase II poison etoposide (40 μM) (Dorstyn et al. 2002).

    Proceed to Steps 20–30 below to analyze the level of knockdown and cell-death gene expression by qPCR.

Preparation of Total RNA

20. Harvest the experimentally treated cells and pellet by centrifugation at 1000 rpm (200*g*) for 5 min in a large benchtop centrifuge.

    Tissues can also be used for RNA extraction to examine expression of cell-death genes during development.

21. Discard the medium, homogenize the cell pellet in ~100 μL of TRIzol reagent, bring to a final volume of 1 mL with TRIzol, and incubate for 5 min at room temperature.

22. Add 200 μL of chloroform per 1 mL of TRIzol, shake vigorously for 15 sec, and then allow the sample to stand for 2–3 min at room temperature.

23. Centrifuge at 13,000 rpm (16,000*g*) in a small benchtop centrifuge for 15 min at 4°C, and then collect the upper aqueous phase, which contains the RNA.

24. Precipitate the RNA in 500 μL of isopropanol per 1 mL of TRIzol for 10 min at room temperature.

25. Pellet the RNA at 13,000 rpm (16,000*g*) in a small benchtop centrifuge for 10 min at 4°C and remove the supernatant.

26. Wash the pellet with 1 mL of 75% ethanol and then centrifuge at 13,000 rpm (16,000*g*) in a small benchtop centrifuge for 5 min at 4°C.

27. Aspirate the ethanol and then air-dry the pellet (this can be performed in a 60°C heat block for 1–2 min). Then resuspend the pelleted RNA in 20 μL of DEPC-treated water by heating for 10 min at 60°C.

cDNA Synthesis and qPCR

28. Use 1 μg of total RNA for cDNA synthesis with a reverse-transcription kit used per the manufacturer's protocol.

29. Perform real-time PCR on a thermal cycler using the RT² Real-Time SYBR green/ROX PCR MasterMix (QIAGEN) according to the manufacturer's instructions.

30. Perform the reactions in triplicate and normalize the mRNA expression levels against the internal control gene *RpL32* (also known as *rp49*, encoding the *Drosophila* 60S ribosomal protein L32) using the comparative C_T ($2_T^{-\Delta\Delta C}$) method (see Schmittgen and Livak 2008).

DISCUSSION

This protocol has described the use of *Drosophila* S2 cells to knock down the expression of specific genes and subsequently analyze cell-death transcriptional responses (see Denton et al. 2013).

Although S2 cells have many advantages for transfection and knockdown studies, other cell lines that are suitable for particular applications are available. For example, mbn2 cells, although more sensitive than S2 cells to growth conditions, display a robust response to the hormone ecdysone that is useful for analyzing hormone-induced cell death and gene expression. For information on further alternative cell lines, consider consulting the *Drosophila* Genomics Resource Center (DGRC; https://dgrc.cgb.indiana.edu/), which maintains a collection of over 100 *Drosophila* cell lines.

The methodology outlined above has assumed prior technical knowledge and experience of cloning to generate the template for dsRNA synthesis. Template sequences of 400–600 bp dsRNA molecules seem to give the highest level of knockdown. To reduce off-target effects, design the template to gene-specific exon sequences, including the 3′ untranslated region UTR (they can span an intron). Web-based sequence-homology searches, such as the Basic Local Alignment Search Tool (BLAST; NCBI), can be used to ensure that the template sequence does not contain long regions of identity.

RECIPE

Phosphate-Buffered Saline (PBS)

Reagent	Amount to add (for 1× solution)	Final concentration (1×)	Amount to add (for 10× stock)	Final concentration (10×)
NaCl	8 g	137 mM	80 g	1.37 M
KCl	0.2 g	2.7 mM	2 g	27 mM
Na_2HPO_4	1.44 g	10 mM	14.4 g	100 mM
KH_2PO_4	0.24 g	1.8 mM	2.4 g	18 mM

If necessary, PBS may be supplemented with the following:

$CaCl_2 \cdot 2H_2O$	0.133 g	1 mM	1.33 g	10 mM
$MgCl_2 \cdot 6H_2O$	0.10 g	0.5 mM	1.0 g	5 mM

PBS can be made as a 1× solution or as a 10× stock. To prepare 1 L of either 1× or 10× PBS, dissolve the reagents listed above in 800 mL of H_2O. Adjust the pH to 7.4 (or 7.2, if required) with HCl, and then add H_2O to 1 L. Dispense the solution into aliquots and sterilize them by autoclaving for 20 min at 15 psi (1.05 kg/cm^2) on liquid cycle or by filter sterilization. Store PBS at room temperature.

ACKNOWLEDGMENTS

We thank the members of our laboratory for their comments on this protocol. The *Drosophila* cell death research in our laboratory is supported by the National Health and Medical Research Project Grant (1041807) and a Senior Principal Research Fellowship (1002863) to S.K.

REFERENCES

Denton D, Aung-Htut MT, Lorensuhewa N, Nicolson S, Zhu W, Mills K, Cakouros D, Bergmann A, Kumar S. 2013. UTX coordinates steroid hormone-mediated autophagy and cell death. *Nat Commun* **4:** 2916.

Dorstyn L, Read S, Cakouros D, Huh JR, Hay BA, Kumar S. 2002. The role of cytochrome *c* in caspase activation in *Drosophila melanogaster* cells. *J Cell Biol* **156:** 1089–1098.

Schmittgen TD, Livak KJ. 2008. Analyzing real-time PCR data by the comparative C(T) method. *Nat Protoc* **3**(6): 1101–1108.

Using Synthetic Peptide Substrates to Measure *Drosophila* Caspase Activity

Donna Denton[1] and Sharad Kumar[1]

Centre for Cancer Biology, University of South Australia, Adelaide, South Australia 5001, Australia

Central to the apoptotic pathway is the activation of caspases that are members of a highly conserved family of cysteine proteases. Caspases are synthesized as inactive zymogens and are generally activated by proteolytic cleavage to form the catalytically active enzyme. Caspase activity in apoptotic cells can be measured by assessing the cleavage of commercially available synthetic caspase substrates. The synthetic substrates contain a caspase cleavage site conjugated to a fluorochrome, such as 7-amino-4-methylcoumarin (AMC), or a chromophore, such as *p*-nitroaniline (pNA), for colorimetric detection. Here, we present a protocol for the measurement of caspase activity in *Drosophila* cell extracts by cleavage of the target peptide in the synthetic substrate that releases a fluorochrome or color-producing agent. The signal is measured by a spectrophotometer, with the intensity of the signal being proportional to the amount of substrate cleaved.

MATERIALS

It is essential that you consult the appropriate Material Safety Data Sheets and your institution's Environmental Health and Safety Office for proper handling of equipment and hazardous material used in this protocol.

RECIPE: Please see the end of this protocol for recipes indicated by <R>. Additional recipes can be found online at http://cshprotocols.cshlp.org/site/recipes.

Reagents

BCA Protein Assay Kit (Pierce)
Caspase assay lysis buffer <R>
Caspase substrate (e.g., VDVAD-AMC or DEVD-AMC fluorogenic substrates from MP Biomedicals)

Equipment

Benchtop centrifuge
Liquid nitrogen
Microcentrifuge tubes (1.5-mL)
Microtiter plates (96-well)
Pestle for 1.5-mL microcentrifuge tube
Spectrophotometer (e.g., FluoStar BD Biosciences)

[1]Correspondence: Donna.Denton@unisa.edu.au; Sharad.Kumar@unisa.edu.au

Cite this protocol as *Cold Spring Harb Protoc*; doi:10.1101/pdb.prot086231

METHOD

Preparation of Protein Lysates

1. Collect the samples (cell cultures, suitably staged whole animals, or dissected tissue) in 1.5-mL microcentrifuge tubes on ice in an appropriate volume of caspase assay lysis buffer.

    The volume of lysis buffer can be adjusted depending on the protein yield of the sample and the abundance of active caspases—we use 10–20 µL per whole animal, 5 µL per dissected midgut tissue, and 20 µL to resuspend cells originally grown in a well of a six-well plate.

    The samples can be snap-frozen in liquid nitrogen and stored at −70°C before the preparation of protein lysates.

2. Using a handheld pestle for a 1.5 mL microcentrifuge tube, homogenize the sample.

    Approximately five strokes should be sufficient to disrupt cells/tissue and the sample will appear cloudy. If using whole animals, additional strokes may be required; the cuticle will remain and will be pelleted in the debris in Step 4.

3. Lyse the cells by freezing in liquid nitrogen and rapid thawing to room temperature three times.

4. Pellet the cell debris by spinning the extracts at 13,000 rpm (~16,000g) in a benchtop centrifuge for 20 min at 4°C and then transfer the supernatant to a new tube.

5. Determine the protein concentration of the lysates.

    We use the BCA Protein Assay Kit from Pierce (according to the manufacturer's instructions) as it is compatible with components of the caspase assay lysis buffer and works well with small amounts of lysate (dilute 1 µL in 50 µL water and use 10 µL for the assay).

    If necessary, store the protein lysates at −70°C in aliquots.

Measuring Caspase Activity

The assays should be set up in triplicate and include the blank and calibration standards in a 96-well microtiter plate suitable for the spectrophotometer; a positive control, such as purified recombinant caspase, can also be included.

6. Set up the caspase assay on ice by combining 20–50 µg of protein lysate, caspase substrate VDVAD-AMC or DEVD-AMC at a concentration of 100 µM, and caspase assay lysis buffer for a final volume of 100 µL.

7. Prepare a blank control sample with substrate and no protein lysate.

8. Make serial dilutions of the substrate (5, 10, 25, 50, and 100 µM) in caspase assay lysis buffer.

    These will be used to determine the fluorescence calibration standard.

9. Quantify the fluorescence for AMC using a spectrophotometer, with excitation at 385 nm and emission at 460 nm, over a time course (usually 3 h) at 37°C.

10. To obtain the conversion factor, plot the fluorescence calibration standards and obtain the slope of the standard curve.

    The conversion factor is the inverse of the slope (1/slope of standard curve), and, for colorimetric substrates, is the standard concentration divided by the substrate blank reading.

11. Plot the change in fluorescence against time for each sample, adjusting for background for the sample by subtracting the zero time-point value from each time-point for each sample.

12. Determine the linear region of the slope of the line (m) from the formula

$$y = mx + c.$$

13. Calculate the caspase activity, expressed as pmol substrate/min, by multiplying the value of m by the conversion factor and then by the reaction volume (Denton et al. 2008; Kumar and Dorstyn 2009):

Caspase activity = slope of sample (m) × conversion factor × volume (μL).

Cite this protocol as *Cold Spring Harb Protoc*; doi:10.1101/pdb.prot086231

DISCUSSION

Caspase activity can vary greatly between different tissues and developmental stages, depending on the number of cells undergoing death at a given time. Caspases also play some nonapoptotic functions, so low levels of caspase activity can be present in many tissues. The amount of cell or tissue lysates to be used will need to be adjusted depending on these considerations. The protocol and reagents presented here are based on what we currently use in our laboratory. The assay is based on the measurement of active caspase proteolytic activity by cleavage of the target peptide in the synthetic substrate to release the fluorochrome or color-producing agent. There are many caspase peptide substrates available. We use Ac-Asp-Glu-Val-Asp-7-amino-4-methylcoumarin (Ac-DEVD-AMC) and Ac-Val-Asp-Val-Ala-Asp-7-amino-4-methylcoumarin (Ac-VDVAD-AMC) as they show the highest level of activity for *Drosophila* caspases. Other newer and more-sensitive caspase substrates are available, but, as we have not used them, we have refrained from discussing them here.

RECIPE

Caspase Assay Lysis Buffer

Reagent	Final concentration
HEPES (pH 7.5)	50 mM
NaCl	100 mM
EDTA	1 mM
CHAPS	0.1%
Sucrose	10%
DTT	5 mM
Triton X-100	0.5%
Glycerol	4%
Protease inhibitor cocktail (cOmplete, Roche)	1×

After adding the DTT and protease inhibitor cocktail, the lysis buffer can be stored at −20°C in 10-mL aliquots for up to 1 yr.

ACKNOWLEDGMENTS

We thank the members of our laboratory for their comments on this protocol. The *Drosophila* cell death research in our laboratory is supported by the National Health and Medical Research Project Grant (1041807) and a Senior Principal Research Fellowship (1002863) to S.K.

REFERENCES

Denton D, Mills K, Kumar S. 2008. Methods and protocols for studying cell death in *Drosophila*. *Methods Enzymol* **446:** 17–37.

Kumar S, Dorstyn L. 2009. Analyzing caspase activation and caspase activity in apoptotic cells. *Methods Mol Biol* **559:** 3–17.

CHAPTER 14

Monitoring Autophagy in *Drosophila*

Lindsay DeVorkin[1,2] and Sharon M. Gorski[1,2,3]

[1]*The Genome Sciences Centre, BC Cancer Agency, Vancouver, British Columbia V5Z 1L3, Canada;* [2]*Department of Molecular Biology and Biochemistry, Simon Fraser University, Burnaby, British Columbia V5A 1S6, Canada*

Drosophila melanogaster is a well-characterized model system used to study a diverse array of biological processes and their underlying genetic, biochemical, and molecular mechanisms. *Drosophila* research in the area of autophagy has proven to be fruitful, as many of the core autophagy-related (Atg) genes and upstream signaling molecules are evolutionarily conserved. Furthermore, single orthologs of many of the Atg genes exist in *Drosophila*, allowing for an examination of nonredundant components of the core autophagy machinery. Here we provide a brief introduction to autophagy in *Drosophila*, describe why the ovary represents a useful model in which to examine this process, and introduce tools and techniques available for the detection and monitoring of autophagy in the *Drosophila* ovary. These methods can be easily adapted for other *Drosophila* tissues.

INTRODUCTION

Drosophila melanogaster is a powerful model system in which to study diverse biological processes including the dynamic cellular process of autophagy. Autophagy was first observed more than 40 years ago (De Duve 1963), and many of the Atg (autophagy-related) genes first identified in genetic screens in yeast are conserved in *Drosophila* (reviewed in Meléndez and Neufeld [2008] and Zirin and Perrimon [2010]). Several advances in genetic and molecular tools in *Drosophila* have made this system amenable to the detection and manipulation of autophagy in vivo. Studies in *Drosophila* have also helped to elucidate the roles of autophagy in cell survival, cell death, development, immunity, aging, and cellular homeostasis, aiding our understanding of the links between autophagy and human diseases (for review, see McPhee and Baehrecke [2009]).

Autophagy contributes to multiple processes in *Drosophila*, including degradation of larval tissues such as the salivary gland (Berry and Baehrecke 2007) and midgut (Denton et al. 2009), and elimination of the amnioserosa during embryogenesis (Mohseni et al. 2009). It also plays critical roles in innate immunity (Shelly et al. 2009), protein aggregate clearance, neuronal homeostasis, and longevity (Simonsen et al. 2008; Juhasz and Neufeld 2008). Autophagy is regulated by hormones and various signaling pathways during *Drosophila* development (Lee and Baehrecke 2001; Rusten et al. 2004; Calamita and Fanto 2011), and also by exogenous factors such as nutrients. Starvation-induced autophagy has been described in the larval fat body (Scott et al. 2004), the midgut (Wu et al. 2009), and the ovary (Hou et al. 2008; Barth et al. 2011). During nutrient deprivation, autophagy functions, at least in part, to remobilize nutrients to promote cell survival, and thus is considered an adaptive survival response to cell stress. In addition to starvation-induced autophagy in the ovary, developmental autophagy also occurs during *Drosophila* oogenesis (Nezis et al. 2010), making the ovary a particularly attractive model in which to study autophagy.

[3]Correspondence: sgorski@bcgsc.ca

Cite this introduction as *Cold Spring Harb Protoc*; doi:10.1101/pdb.top070441

AUTOPHAGY IN THE *DROSOPHILA* OVARY

Drosophila ovaries are made up of a series of developing egg chambers that arise from the germarium at the anterior region of the ovary. As the egg chambers proceed during development, they move toward the posterior of the ovary in a linear array of 14 well-characterized stages until they reach maturity. Each egg chamber is made up of 15 germline nurse cells and one oocyte surrounded by a layer of somatically derived follicle cells (Spradling 1993). In response to starvation, cell death occurs in the germarium and mid-stage egg chambers (Drummond-Barbosa and Spradling 2001) and is characterized by chromatin condensation and fragmentation, and uptake of the nurse cell cytoplasm by the surrounding follicle cells (Giorgi and Deri 1976). Autophagy also occurs in response to starvation in the germline and somatic follicle cells (Barth et al. 2011) as well as in degenerating mid-stage egg chambers undergoing cell death (Hou et al. 2008). Before completion of oogenesis, nurse cells transfer their cytoplasmic content to the oocyte and undergo developmental cell death (Foley and Cooley 1998), a process that also involves autophagy (Nezis et al. 2010). Therefore, the ovary is a useful system to study both developmentally regulated autophagy as well as starvation-induced autophagy.

In the associated protocols we describe reagents and methods used to assay autophagy in *Drosophila*. The aim of these protocols is not to describe how to decipher the role or function of autophagy (e.g., in cell death vs. survival), but rather aims to describe how to monitor the autophagy process itself. The use of the acidotropic dye LysoTracker (see Protocol 1: LysoTracker Staining to Aid in Monitoring Autophagy in *Drosophila* [DeVorkin and Gorski 2014a]) is a facile assay that is useful as an initial indicator of potential autophagy-associated lysosomal activity in *Drosophila* cell culture or tissues. However, it is not a specific marker for autophagy and should be used in combination with other methods such as monitoring autophagy flux using the *Drosophila* p62 ortholog, Ref(2)P (see Protocol 2: Monitoring Autophagic Flux Using Ref(2)P, the *Drosophila* p62 Ortholog [DeVorkin and Gorski 2014b]) and/or using the UAS-GAL4 system to drive expression of the fluorescent reporter GFP-mCherry-DrAtg8a (see Protocol 3: Monitoring Autophagy in *Drosophila* Using Fluorescent Reporters in the UAS-GAL4 System [DeVorkin and Gorski 2014c]). *Drosophila* is an excellent model system for genetic analyses of Atg genes, and examination of lethal Atg genes is facilitated by techniques for clonal analysis and RNAi (see Protocol 4: Genetic Manipulation of Autophagy in the *Drosophila* Ovary [DeVorkin and Gorski 2014d]). The methods described focus on starvation-induced autophagy in the *Drosophila* germline as a model system, but can be easily adapted for analysis of developmental autophagy in late oogenesis as well as in other tissues including somatic follicle cells.

ACKNOWLEDGMENTS

The authors thank S. Gaumer, T.E. Rusten, N. Katheder, G. Juhasz, and T. Neufeld for reagents and/or fly lines used and J. Hodgson for the fly food recipe. The authors are grateful for a Natural Sciences and Engineering Research Council Discovery grant (RGPIN/371368-2009) and Canadian Institutes of Health Research Operating Grant (MOP-78882) for support; S.M.G. is supported in part by a Canadian Institutes of Health Research New Investigator Award.

REFERENCES

Barth JMI, Szabad J, Hafen E, Kohler K. 2011. Autophagy in *Drosophila* ovaries is induced by starvation and is required for oogenesis. *Cell Death Differ* 18: 915–924.

Berry DL, Baehrecke EH. 2007. Growth arrest and autophagy are required for salivary gland cell degradation in *Drosophila*. *Cell* 131: 1137–1148.

Calamita P, Fanto M. 2011. Slimming down fat makes neuropathic hippo: the Fat/Hippo tumor suppressor pathway protects adult neurons through regulation of autophagy. *Autophagy* 7: 907–909.

Denton D, Shravage B, Simin R, Mills K, Berry DL, Baehrecke EH, Kumar S. 2009. Autophagy, not apoptosis, is essential for midgut cell death in *Drosophila*. *Curr Biol* 19: 1741–1746.

Drummond-Barbosa D, Spradling AC. 2001. Stem cells and their progeny respond to nutritional changes during *Drosophila* oogenesis. *Dev Biol* 231: 265–278.

De Duve C. 1963. The lysosome. *Sci Am* 208: 64–72.

DeVorkin L, Gorski SM. 2014a. LysoTracker staining to aid in monitoring autophagy in *Drosophila*. *Cold Spring Harb Protoc* doi: 10.1101/pdb.prot080325.

DeVorkin L, Gorski SM. 2014b. Monitoring autophagic flux using Ref(2)P, the *Drosophila* p62 ortholog. *Cold Spring Harb Protoc* doi: 10.1101/pdb.prot080333.

Cite this introduction as *Cold Spring Harb Protoc*; doi:10.1101/pdb.top070441

DeVorkin L, Gorski SM. 2014c. Monitoring autophagy in *Drosophila* using fluorescent reporters in the UAS-GAL4 system. *Cold Spring Harb Protoc* doi: 10.1101/pdb.prot080341.

DeVorkin L, Gorski SM. 2014d. Genetic manipulation of autophagy in the *Drosophila* ovary. *Cold Spring Harb Protoc* doi: 10.1101/pdb.prot080358.

Foley K, Cooley L. 1998. Apoptosis in late stage *Drosophila* nurse cells does not require genes within the H99 deficiency. *Development* **125**: 1075–1082.

Giorgi F, Deri P. 1976. Cell death in ovarian chambers of *Drosophila melanogaster*. *J Embryol Exp Morphol* **35**: 521–533.

Hou Y-CC, Chittaranjan S, Barbosa SG, McCall K, Gorski SM. 2008. Effector caspase Dcp-1 and IAP protein Bruce regulate starvation-induced autophagy during *Drosophila melanogaster* oogenesis. *J Cell Biol* **182**: 1127–1139.

Juhasz G, Neufeld TP. 2008. Experimental control and characterization of autophagy in *Drosophila*. In *Autophagosome and phagosome* (ed. Deretic V), Vol. 445, pp. 125–133. Humana Press, Totowa, NJ. http://www.springerprotocols.com/Abstract/doi/10.1007/978-1-59745-157-4_8 (Accessed May 23, 2012).

Lee CY, Baehrecke EH. 2001. Steroid regulation of autophagic programmed cell death during development. *Development* **128**: 1443–1455.

McPhee CK, Baehrecke EH. 2009. Autophagy in *Drosophila melanogaster*. *Biochim Biophys Acta* **1793**: 1452–1460.

Meléndez A, Neufeld TP. 2008. The cell biology of autophagy in metazoans: a developing story. *Development* **135**: 2347–2360.

Mohseni N, McMillan SC, Chaudhary R, Mok J, Reed BH. 2009. Autophagy promotes caspase-dependent cell death during *Drosophila* development. *Autophagy* **5**: 329–338.

Nezis IP, Shravage BV, Sagona AP, Lamark T, Bjørkøy G, Johansen T, Rusten TE, Brech A, Baehrecke EH, Stenmark H. 2010. Autophagic degradation of dBruce controls DNA fragmentation in nurse cells during late *Drosophila melanogaster* oogenesis. *J Cell Biol* **190**: 523–531.

Rusten TE, Lindmo K, Juhász G, Sass M, Seglen PO, Brech A, Stenmark H. 2004. Programmed autophagy in the *Drosophila* fat body is induced by ecdysone through regulation of the PI3K pathway. *Dev Cell* **7**: 179–192.

Scott RC, Schuldiner O, Neufeld TP. 2004. Role and regulation of starvation-induced autophagy in the *Drosophila* fat body. *Dev Cell* **7**: 167–178.

Shelly S, Lukinova N, Bambina S, Berman A, Cherry S. 2009. Autophagy plays an essential anti-viral role in *Drosophila* against vesicular stomatitis virus. *Immunity* **30**: 588–598.

Simonsen A, Cumming RC, Brech A, Isakson P, Schubert DR, Finley KD. 2008. Promoting basal levels of autophagy in the nervous system enhances longevity and oxidant resistance in adult *Drosophila*. *Autophagy* **4**: 176–184.

Spradling AC. 1993. Developmental genetics of oogenesis. In Drosophila *development*. (ed. Bate M, Martinez-Aria A), pp. 1–70. Cold Spring Harbor Laboratory Press, Cold Spring Harbor, NY.

Wu H, Wang MC, Bohmann D. 2009. JNK protects *Drosophila* from oxidative stress by trancriptionally activating autophagy. *Mech Dev* **126**: 624–637.

Zirin J, Perrimon N. 2010. *Drosophila* as a model system to study autophagy. *Semin Immunopathol* **32**: 363–372.

LysoTracker Staining to Aid in Monitoring Autophagy in *Drosophila*

Lindsay DeVorkin[1,2] and Sharon M. Gorski[1,2,3]

[1]*The Genome Sciences Centre, BC Cancer Agency, Vancouver, British Columbia V5Z 1L3, Canada;* [2]*Department of Molecular Biology and Biochemistry, Simon Fraser University, Burnaby, British Columbia V5A 1S6, Canada*

LysoTracker is an acidotropic dye that stains cellular acidic compartments, including lysosomes and autolysosomes. LysoTracker has been used to detect autophagy-associated lysosomal activity in *Drosophila* tissues including the fat body, midgut, salivary gland and ovary, as well as in *Drosophila* cell culture. A low level of LysoTracker staining can be observed under resting or well-fed conditions, and is increased following autophagic stimuli such as starvation. Here we provide a protocol for examining LysoTracker levels in *Drosophila* cultured cells in vitro using standard cell culture methods and flow cytometry. We also describe how to examine LysoTracker in fixed and nonfixed *Drosophila* tissues using fluorescence microscopy. Ovary tissue is used as an example. Dissections of ovaries are relatively easy to perform, given their large size.

MATERIALS

It is essential that you consult the appropriate Material Safety Data Sheets and your institution's Environmental Health and Safety Office for proper handling of equipment and hazardous materials used in this protocol.

RECIPES: Please see the end of this protocol for recipes indicated by <R>. Additional recipes can be found online at http://cshprotocols.cshlp.org/site/recipes.

Reagents

Phosphate-buffered saline (PBS; 1×, pH 7.4) <R>
Reagents needed for LysoTracker staining of cultured cells (see Steps 1–6):
  Cell culture medium appropriate for the cell line (e.g., Schneider's *Drosophila* medium containing 10% fetal bovine serum)
  Drosophila cells of interest grown in appropriate culture flasks
  Glucose (2 mg/mL in 1× PBS, pH 7.4)
  LysoTracker Green (LTG) DND-26 (1 mM; Invitrogen)
  Propidium iodide (PI; 1.0 mg/mL)

Reagents needed for LysoTracker staining of tissues (see Steps 7–23):
  CO_2 gas
  DAPI (0.1 mg/mL)

---

[3]Correspondence: sgorski@bcgsc.ca

Cite this protocol as *Cold Spring Harb Protoc*; doi:10.1101/pdb.prot080325

DRAQ5 DNA stain (optional; see Step 20) <R>

Fly food <R>

> *To supplement fly food with yeast paste, mix 1 g of active dry yeast with 1 mL of distilled H₂O and apply a pea-sized amount of the paste on top of fly food in a vial.*

Fly stocks of interest (1–5 d old)

LysoTracker Red (LTR) DND-99 (Invitrogen)

> *Either LTR or LTG can be used on* Drosophila *tissue to monitor lysosome and autolysosome formation (see Marshall et al. 2012 for LTG example). Because GFP is often used as a marker for clonal analysis, LTR has been used more extensively (Juhasz et al. 2008).*

Nail polish

Paraformaldehyde (16%)

Slowfade Gold Reagent, with or without DAPI (Invitrogen)

Sucrose (10% in 1× PBS, pH 7.4)

Triton X-100

Equipment

Equipment needed for LysoTracker staining of cultured cells (see Steps 1–6):

Centrifuge (benchtop)

Flow cytometer, equipped with a 488 nm laser, and accompanying flow cytometry analysis software

Flow cytometry tubes

Incubator, preset for appropriate cell culture conditions

Equipment needed for LysoTracker staining of tissues (see Steps 7–23):

Coverslips

Diffuser pad for CO₂ anesthesia

Dissecting dish

Dissecting microscope

Fine forceps (e.g., Dumont #5 tweezers, Electron Microscopy Sciences)

Fluorescence microscope (preferably an upright fluorescence microscope)

Fly vials

Image analysis software program (e.g., ImageJ) (optional; see Step 23)

Microscope slides

> *We use three-well, 14-mm hydrophobic printed slides from Electron Microscopy Services.*

Multi-well glass plate for staining

METHOD

> *For analysis of cultured cells, follow Steps 1–6. For analysis of tissues, follow Steps 7–23. Figure 1 summarizes these techniques.*

LysoTracker Analysis of Autophagy in Cultured Cells

Autophagy Induction by Amino Acid Starvation

1. Collect 1×10^6 *Drosophila* cells by centrifugation at 800*g* for 5 min at room temperature. Discard the medium. Resuspend the cell pellet in 500 µL of 2 mg/mL glucose and return the cells to the incubator for an appropriate length of time.

   > *An increase in LTG staining can be seen as early as 30 min following addition of starvation medium (i.e., glucose solution). We observe a robust increase in LTG following 2–4 h of starvation. There is no need to shake the cells while they are incubating in the glucose solution.*

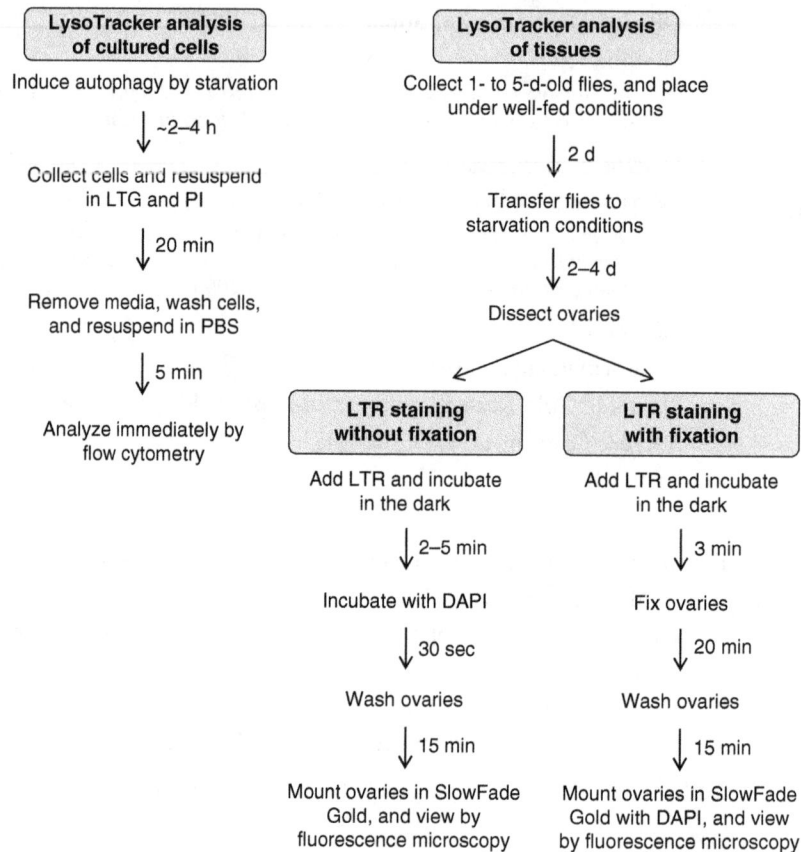

FIGURE 1. Flow chart outlining the procedures described in this protocol.

Staining with LTG

2. Dilute the 1 mM stock LTG solution to a working concentration of 50–75 nM in prewarmed culture medium just before use. Dilute the PI in LTG working solution to a concentration of 2 µg/mL.

   LysoTracker is very sensitive to photobleaching and must be protected from light in all subsequent steps.

3. After incubation, collect the cells from Step 1 by centrifugation at 800g for 5 min at room temperature. Discard the supernatant.

4. Resuspend the cells in 500 µL–1 mL of LTG/PI solution from Step 2 and incubate for 20 min in the incubator protected from light.

5. Centrifuge the cells at 800g for 5 min at 4°C, and then discard the supernatant. Add 500 µL–1 mL of ice-cold 1× PBS to the cell pellet, wash the cells by centrifuging at 800g for 5 min at 4°C, and carefully discard the supernatant. Resuspend the cell pellet with an appropriate volume of ice-cold 1× PBS in tubes suitable for flow cytometry analysis. Place the sample on ice protected from light. Analyze immediately by flow cytometry.

   It is important to analyze samples immediately as LysoTracker is highly sensitive to photobleaching and autophagy can be further induced when cells are in 1× PBS. If cells tend to clump, they should be filtered before flow cytometry analysis.

Flow Cytometry Analysis

6. Obtain fluorescence intensities using the FL1 channel to measure LTG, and the FL3 channel to measure PI. Experimentally determine appropriate gates and compensation. Analyze data obtained from flow cytometry with software such as FlowJo and CellQuest.

   See Troubleshooting.

Cite this protocol as *Cold Spring Harb Protoc*; doi:10.1101/pdb.prot080325

LysoTracker Analysis of Autophagy in Tissues (Ovaries)

Fly Preparation

7. To ensure proper development of ovarian tissue, place approximately ten to twenty 1- to 5-d-old male and female flies on fresh fly food supplemented with yeast paste in a vial for 2 d in uncrowded conditions.

Autophagy Induction by Amino Acid Starvation

8. Place a Kimwipe soaked in 10% sucrose at the bottom of an empty vial. Transfer flies from the vial containing yeast paste to the vial containing the sucrose-soaked Kimwipe. Leave the flies in this vial for 2–4 d.

    Depending on the genetic background of the fly strain, starvation times will have to be experimentally determined to observe changes in autophagy.

Ovary Dissection

9. Anesthetize flies on a CO_2 diffuser pad.

10. Dissect female flies in 1× PBS in a dissecting dish. Grasp each fly between the abdomen and the lower thorax with one set of forceps, and pull at the posterior end of the abdomen to remove the ovaries.

    Handle ovaries gently from the posterior end to ensure midstage egg chambers are not damaged. We typically recommend examining at least 100 ovarioles from at least eight different animals.

11. Remove debris away from the ovaries and place immediately on ice-cold 1× PBS in a multi-well glass plate while other ovaries are dissected.

    Do not let the ovaries dry while dissecting.

12. Using forceps, tease ovarioles apart from each other as carefully as possible to ensure uniform antibody staining.

    We recommend staining five to 10 pairs of ovaries per well. For LysoTracker staining without fixation, proceed to Step 13. For LysoTracker staining with fixation, proceed to Step 19. Fixation may introduce artifacts in the cellular staining pattern of antibodies and fluorescent labels. However, fixed tissue is not as delicate and is much easier to tease apart when compared to nonfixed tissue. In addition, certain antibodies require fixation and permeabilization for penetration into the cell. Nonfixed tissues are advantageous because there are no fixation artifacts and a reliable staining pattern can be observed.

LTR Staining without Fixation

13. Carefully remove PBS from ovaries using a fine pipette tip. Add 200 µL of 0.8 µM LTR in 1× PBS to each well, and incubate in the dark for 2–5 min at room temperature.

14. Remove LTR and add 200 µL of 0.1 mg/mL DAPI. Incubate for 30 sec protected from light.

15. Wash three times (5 min each) with 200 µL of 1× PBS protected from light.

    Proceed immediately to Step 20.

LTR Staining with Fixation

16. Remove PBS from ovaries. Add 200 µL of 50 µM LTR in 1× PBS to each well, and incubate in the dark for 3 min at room temperature.

17. Remove LTR and wash three times (5 min each) with 200 µL of 1× PBS protected from light.

18. Remove PBS and add 200 µL of 4% paraformaldehyde (diluted in 1× PBS). Incubate for 20 min at room temperature, protected from light.

    Continue to protect from light when performing subsequent steps.

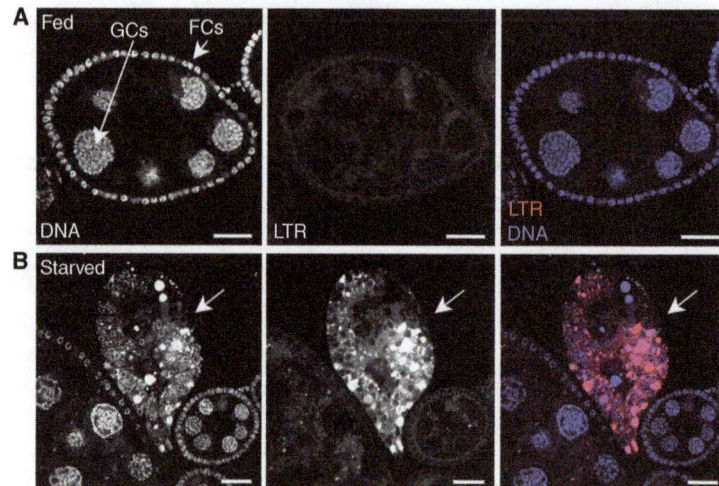

FIGURE 2. LTR staining is increased following starvation in *Drosophila* mid-stage egg chambers. Ovaries were dissected from (*A*) well-fed *w¹¹¹⁸* control flies or (*B*) *w¹¹¹⁸* flies subjected to 4 d of amino acid starvation and were stained with LTR followed by parafomaldehyde fixation. Well-fed flies show low LTR staining, whereas starved flies show a dramatic increase in LTR staining evident in degenerating midstage egg chambers (arrow). Follicle cells (FCs) and germ cells (GCs) are indicated. Scale bar: 25 µm.

19. Wash three times (5 min each) with 200 µL of 0.1% Triton X-100 (diluted in 1× PBS) protected from light.

    Proceed immediately to Step 20.

Fluorescence Microscopy Analysis

20. Place ovaries on a microscope slide in a drop of SlowFade Gold Reagent and carefully tease apart. If tissues were fixed, SlowFade Gold Reagent with DAPI can be used.

    Alternatively, DRAQ5 DNA stain can be used to visualize DNA in fixed tissues. To stain using DRAQ5, remove Triton X-100 solution from tissue after Step 19 and then incubate with diluted DRAQ5 for 10 min in the dark. Mount tissue in Slowfade Gold Reagent (without DAPI). DRAQ5 has a far red excitation and emission.

21. Add coverslip and seal with nail polish.

22. View immediately by fluorescence microscopy.

23. Compare LTR levels between control and experimental samples (or between clones of cells within the same sample; see Protocol 4: Genetic Manipulation of Autophagy in the *Drosophila* Ovary [DeVorkin and Gorski 2014a]). Various image analysis software programs (e.g., ImageJ) can be used to aid in quantification of fluoresecent puncta and/or fluorescence intensities.

    Sample results are shown in Figure 2. See Troubleshooting.

TROUBLESHOOTING

Problem (Steps 6 and 23): There is variability in LysoTracker staining among similar samples.
Solution: The slightest variation in treatments of the samples may alter LysoTracker fluorescence. Check the following.

- For in vitro assays, ensure that equal volumes are added to and removed from cells, and that all solutions are removed during washing steps and when adding LTG/PI.

- Protect samples from light to prevent photobleaching.

- Tissues from animals with different genetic backgrounds may react differently to LTR. When possible, compare LTR staining of tissues that are from the same genetic background.

Cite this protocol as *Cold Spring Harb Protoc*; doi:10.1101/pdb.prot080325

Problem (Step 6): LTG cannot be detected.

Solution: Adjust the concentration of LTG and/or the incubation time. LTG is sensitive to photobleaching, so ensure that samples are protected from light.

Problem (Step 23): LTR cannot be detected.

Solution: Consider the following:

- LTR photobleaches very rapidly. Ensure that samples are protected from light and analyze samples immediately.

- LTR may be sensitive to repeated freeze/thaw cycles. Aliquot LTR and do not refreeze after use.

- The fluorescence signal of LTR is reduced following fixation. Therefore, the fixation time and concentration of LTR may have to be adjusted to observe staining.

DISCUSSION

The use of LysoTracker has been beneficial for examining autophagy-associated lysosomal activity in *Drosophila* cell culture, as well as in several tissues including the fat body (Scott et al. 2004), midgut (Ren et al. 2009), salivary gland (Berry and Baehrecke 2007), and ovary (Hou et al. 2008; Barth et al. 2011). In the ovary, LysoTracker staining helped to reveal that both the germline and the somatic follicle cells undergo autophagy (Hou et al. 2008; Barth et al. 2011). Low levels of LysoTracker can be detected under resting or well-fed conditions in the ovary (Fig. 2A), but are increased on starvation, which is particularly evident in degenerating mid-stage egg chambers (Fig. 2B). Furthermore, inhibition of autophagy in the ovary reduced LysoTracker levels following starvation (Hou et al. 2008; Barth et al. 2011) indicating that it can be a reliable marker of autophagy-associated lysosomal activity.

As a cautionary note, LysoTracker and other acidotropic dyes are not specific markers for autophagy nor are they markers of autophagic flux (Klionsky et al. 2012). Although useful as an initial indicator of potential autophagy-associated lysosomal activity, additional autophagy assays must be used to further substantiate findings.

RELATED TECHNIQUES

Monodansylcadaverine (MDC), an autofluorescent dye that accumulates in lysosomes and autolysosomes, is an alternative to LysoTracker. Complementary techniques include evaluation of autophagosome accumulation using GFP-DrAtg8a (see Protocol 3: Monitoring Autophagy in *Drosophila* Using Fluorescent Reporters in the UAS-GAL4 System [DeVorkin and Gorski 2014b]) and monitoring autophagic flux using *Drosophila* Ref(2)P (see Protocol 2: Monitoring Autophagic Flux Using Ref(2)P, the *Drosophila* p62 Ortholog [DeVorkin and Gorski 2014c]) or GFP-mCherry-DrAtg8a (see Protocol 3: Monitoring Autophagy in *Drosophila* Using Fluorescent Reporters in the UAS-GAL4 System [DeVorkin and Gorski 2014b]).

RECIPES

DRAQ5 DNA Stain

Reagent	Final concentration
DRAQ5 (Biostatus)	1:500
RNase A	100 μg/mL

Prepare 1 mL of DRAQ5 DNA stain in 1× phosphate-buffered saline (pH 7.4).

Fly Food

Ingredients	Amounts (for 300-bottle batch)
Water	17 L
Agar	93 g
Cornmeal	1,716 g
Brewer's yeast	310 g
Sucrose	517 g
Dextrose	1033 g
Antifungal reagents	as appropriate[a]
Antibiotics	as appropriate[b]

[a]Use 8.69 g/L sodium potassium tartrate and 26.1 mL/L of a Tegosept (methyl-*p*-hydroxy benzoate) stock solution. For the Tegosept stock solution, dissolve the Tegosept in 95% ethanol at 100 g/L.

[b]For each batch of food, add 15 mg/L tetracycline plus one of the following two antibiotics (alternate with each batch of food): 50 mg/L ampicillin, or 17 mg/L streptomycin. Make up the antibiotics as fresh stock solutions, with tetracycline dissolved in ethanol at 10 mg/mL and amplicillin or streptomycin dissolved in water at 100 mg/mL.

Mix the agar into 13 L of water; use the other 4 L of water to wet the remaining ingredients (excluding antifungals and antibiotics). Cook the medium in a steam kettle at 15 pounds per square inch (psi) for 20 min with constant stirring. Next, switch off the steam and cover the kettle with a lid. Allow the medium to cool to exactly 55°C and then add both the antibiotics and antifungals, stirring well. Dispense medium into vials and bottles, cover with cheesecloth, and allow it to solidify overnight at room temperature before capping.

This recipe is adapted from that of Lewis (1960) and from recipe 4 from the Bloomington *Drosophila* Stock Centre website (flystocks.bio.indiana.edu/Fly_Work/media-recipes/media-recipes.htm).

Phosphate-Buffered Saline (PBS; 1×, pH 7.4)

Reagent	Quantity (g)	Final concentration (mM)
NaCl	8.0	137.0
KCl	0.20	2.7
Na_2HPO_4	1.14	8.0
KH_2PO_4	0.20	1.5

Combine all ingredients in ~800 mL of H_2O and stir to dissolve. Adjust pH to 7.4 and bring final volume to 1 L. Sterilize by autoclaving.

ACKNOWLEDGMENTS

The authors thank T. Neufeld and Y.C. Hou for helpful advice and J. Hodgson for the fly food recipe. The authors are grateful for a Natural Sciences and Engineering Research Council Discovery grant (RGPIN/371368-2009) and Canadian Institutes of Health Research Operating Grant (MOP-78882) for support; S.M.G. is supported in part by a Canadian Institutes of Health Research New Investigator Award.

REFERENCES

Barth JMI, Szabad J, Hafen E, Kohler K. 2011. Autophagy in *Drosophila* ovaries is induced by starvation and is required for oogenesis. *Cell Death Differ* 18: 915–924.

Berry DL, Baehrecke EH. 2007. Growth arrest and autophagy are required for salivary gland cell degradation in *Drosophila. Cell* 131: 1137–1148.

DeVorkin L, Gorski SM. 2014a. Genetic manipulation of autophagy in the *Drosophila* ovary. *Cold Spring Harb Protoc* doi: 10.1101/pdb.prot080358.

DeVorkin L, Gorski SM. 2014b. Monitoring autophagy in *Drosophila* using fluorescent reporters in the UAS-GAL4 system. *Cold Spring Harb Protoc* doi: 10.1101/pdb.prot080341.

Cite this protocol as *Cold Spring Harb Protoc*; doi:10.1101/pdb.prot080325

DeVorkin L, Gorski SM. 2014c. Monitoring autophagic flux using Ref(2)P, the *Drosophila* p62 ortholog. *Cold Spring Harb Protoc* doi: 10.1101/pdb.prot080333.

Hou Y-CC, Chittaranjan S, Barbosa SG, McCall K, Gorski SM. 2008. Effector caspase Dcp-1 and IAP protein Bruce regulate starvation-induced autophagy during *Drosophila melanogaster* oogenesis. *J Cell Biol* **182**: 1127–1139.

Juhasz G, Hill JH, Yan Y, Sass M, Baehrecke EH, Backer JM, Neufeld TP. 2008. The class III PI(3)K Vps34 promotes autophagy and endocytosis but not TOR signaling in *Drosophila*. *JCB* **181**: 655–666.

Klionsky DJ, Abdalla FC, Abeliovich H, Abraham RT, Acevedo-Arozena A, Adeli K, Agholme L, Agnello M, Agostinis P, Aguirre-Ghiso JA, et al. 2012. Guidelines for the use and interpretation of assays for monitoring autophagy. *Autophagy* **8**: 445–544.

Lewis EB. 1960. A new standard food medium. *Drosophila Information Service* **34**: 117–118.

Marshall L, Rideout EJ, Grewal SS. 2012. Nutrient/TOR-dependent regulation of RNA polymerase III controls tissue and organismal growth in *Drosophila*. *EMBO* **31**: 1916–1930.

Ren C, Finkel SE, Tower J. 2009. Conditional inhibition of autophagy genes in adult *Drosophila* impairs immunity without compromising longevity. *Exp Gerontol* **44**: 228–235.

Scott RC, Schuldiner O, Neufeld TP. 2004. Role and regulation of starvation-induced autophagy in the *Drosophila* fat body. *Develop Cell* **7**: 167–178.

Monitoring Autophagic Flux Using Ref(2)P, the *Drosophila* p62 Ortholog

Lindsay DeVorkin[1,2] and Sharon M. Gorski[1,2,3]

[1]*The Genome Sciences Centre, BC Cancer Agency, Vancouver, British Columbia V5Z 1L3, Canada;* [2]*Department of Molecular Biology and Biochemistry, Simon Fraser University, Burnaby, British Columbia V5A 1S6, Canada*

Human p62, also known as Sequestome-1 (SQSTM1), is a multifunctional scaffold protein that contains many domains, including a Phox/Bem1P (PB1) multimerization domain, an ubiquitin-associated (UBA) domain, and a light chain 3 (LC3) recognition sequence. p62 binds ubiquitinated proteins and targets them for degradation by the proteasome. In addition, p62 directly binds LC3; this may serve as a mechanism to deliver ubiquitinated proteins for degradation by autophagy. During this process, p62 itself is degraded. The inhibition of autophagy leads to the accumulation of p62, indicating that it can be used as a marker of autophagic flux. Ref(2)P (refractory to sigma P), the *Drosophila* ortholog of p62, is also required for the formation of ubiquitinated protein aggregates. Ref(2)P contains a putative LC3-interacting region, and genetic inhibition of autophagy in *Drosophila* leads to the accumulation of Ref(2)P protein levels. Thus, like p62, Ref(2)P may serve as a marker of autophagic flux. Here we provide two procedures to examine Ref(2)P protein levels in *Drosophila* ovaries.

MATERIALS

It is essential that you consult the appropriate Material Safety Data Sheets and your institution's Environmental Health and Safety Office for proper handling of equipment and hazardous materials used in this protocol.

RECIPES: Please see the end of this protocol for recipes indicated by <R>. Additional recipes can be found online at http://cshprotocols.cshlp.org/site/recipes.

Reagents

CO_2 gas

Fly food <R>

> *To supplement fly food with yeast paste, mix 1 g of active dry yeast with 1 mL of distilled H_2O and apply a pea-sized amount of the paste on top of fly food in a vial.*

Fly stocks of interest (1–5 d old)

Phosphate-buffered saline (PBS; 1×, pH 7.4) <R>

Primary antibody to detect Ref(2)P (Nezis et al. 2008; A. Hannigan et al. in revision)

> *For immunofluorescence analysis, we find it is helpful to use Ref(2)P antibody in combination with a membrane stain (e.g., anti-Armadillo [N2 7A1, Developmental Studies Hybridoma Bank] or anti-DE-Cadherin [DCAD2, Developmental Studies Hybridoma Bank]) to help visualize the cell structure of the tissue.*

---

[3]Correspondence: sgorski@bcgsc.ca

Cite this protocol as *Cold Spring Harb Protoc*; doi:10.1101/pdb.prot080333

Reagents needed for immunofluorescence analysis of Ref(2)P only (see Steps 6–18):

    Bovine serum albumin (BSA)

    DRAQ5 DNA stain (optional; see Step 16) <R>

    Nail polish

    Paraformaldehyde (16%)

    Secondary antibody conjugated to, e.g., Alexa 488 or 546

    Slowfade Gold Reagent with DAPI (Invitrogen)

    Sucrose (10% in 1× PBS, pH 7.4)

    Triton X-100

Reagents needed for western blot analysis of Ref(2)P only (see Steps 19–29):

    Chemiluminescent detection kit

    Dry ice

    Horseradish peroxidase-conjugated secondary antibody

    Lysis buffer for protein extraction <R>

    Nonfat powdered milk

    Protein quantitation kit (e.g., BCA Protein Assay Kit, Pierce)

    SDS–PAGE gel and running buffer

        We use NuPAGE Novex 4%–12% Bis–Tris Gels and NuPAGE MES or MOPS SDS running buffer (Life Technologies).

    Tween-20

Equipment

    Diffuser pad for CO_2 anesthesia

    Dissecting dish

    Dissecting microscope

    Fine forceps (e.g., Dumont #5 tweezers, Electron Microscopy Sciences)

    Fly vials

    Kimwipes

    Microscopy equipment needed for Immunofluorescence analysis of Ref(2)P only (see Steps 6–18):

        Confocal microscope

        Coverslips

        Microscope slides

            We use three-well, 14-mm hydrophobic printed slides from Electron Microscopy Services.

        Multi-well glass plate for staining

    Specific equipment needed for western blot analysis of Ref(2)P only (see Steps 19–29):

        Centrifuge (benchtop)

        Heating block

        Homogenizing pestles

        Imaging film

        Imaging software

        Microcentrifuge tubes

        Odyssey system (Licor) (optional)

        Polyvinylidene difluoride (PVDF) membrane

        Rocker platform

        SDS–PAGE and western blot equipment

METHOD

Preparation of Fly Ovaries

Fly Preparation

1. To ensure proper development of ovarian tissue, place ten to twenty 1- to 5-d-old female and male flies on fresh fly food supplemented with yeast paste in a vial for 2 d in uncrowded conditions.

Autophagy Induction by Starvation

2. Place a Kimwipe soaked in 10% sucrose at the bottom of an empty vial. Transfer flies from the vial containing yeast paste to the vial containing the sucrose-soaked Kimwipe. Leave flies in this vial for 2–4 d.

   Depending on the genetic background of the fly strain, starvation times will have to be experimentally determined to observe changes in autophagy.

Ovary Dissection

3. Anesthetize flies on a CO_2 diffuser pad.

4. Dissect female flies in 1× PBS in a dissecting dish. Grasp each fly between the abdomen and the lower thorax with one set of forceps, and pull at the posterior end of the abdomen to remove the ovaries.

   Handle ovaries gently from the posterior end to ensure midstage egg chambers are not damaged.

5. Remove the debris away from the ovaries, and prepare them for analysis as follows.

   - For immunofluorescence analysis, place the ovaries immediately in ice-cold 1× PBS in a multi-well glass plate while other ovaries are dissected. Using forceps, tease ovarioles apart from each other as carefully as possible to ensure uniform antibody staining.

     Do not let the ovaries dry while dissecting. We typically recommend examining at least five to 10 pairs of ovaries. Proceed immediately to Step 6.

   - For western blot analysis, place the ovaries immediately into a sterile 1.5-mL microcentrifuge tube on dry ice while other ovaries are being dissected.

     The number of ovaries required will depend on the genetic background of the strain and the treatment, but in general 5–10 pairs should suffice. Proceed immediately to Step 19.

Immunofluorescence Analysis of Ref(2)P

For a flow chart that summarizes Steps 6–18, see Figure 1.

6. Carefully remove 1× PBS from ovaries using a fine pipette tip. Immediately fix with 200 µL of 4% paraformaldehyde (diluted in 1× PBS) for 20 min at room temperature.

7. Wash three times (5 min each) with 200 µL of 1× PBS containing 0.3% Triton X-100 (PBS-T) at room temperature.

8. Add 200 µL of 1× PBS containing 0.5% Triton X-100 for exactly 5 min at room temperature.

9. Wash two times (5 min each) with 200 µL of PBS-T at room temperature.

10. Block in 200 µL of PBS-T containing 2% BSA for 1 h at room temperature.

11. Wash once for 5 min with 200 µL of PBS-T at room temperature.

12. Incubate in 200 µL of primary antibody diluted 1:1000 (Nezis et al. 2008) or 1:5000 (A. Hannigan et al. in revision) in PBS-T containing 0.5% BSA overnight at 4°C.

13. Wash three times (5 min each) with PBS-T at room temperature.

14. Dilute secondary antibody in PBS-T containing 0.5% BSA and incubate for 1–2 h at room temperature.

15. Wash three times (5 min each) with PBS-T at room temperature.

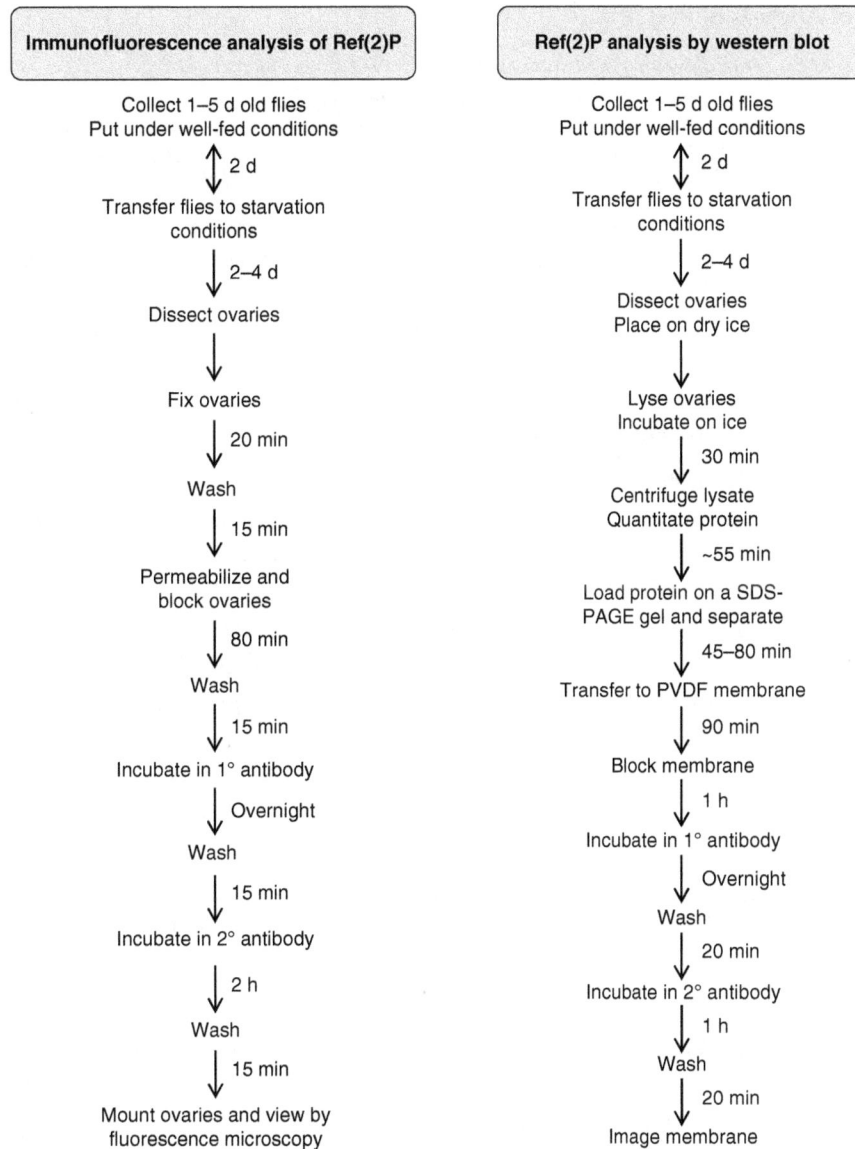

FIGURE 1. Flow chart outlining the procedures described in this protocol.

16. Transfer ovaries to a drop of SlowFade Gold Reagent containing DAPI on a microscope slide and carefully tease apart.

    Alternatively, DRAQ5 DNA stain can be used to visualize DNA. To stain using DRAQ5, remove PBS-T from tissue after Step 15 and then incubate with diluted DRAQ5 for 10 min in the dark. Mount tissue in SlowFade Gold Reagent (without DAPI). DRAQ5 has a far red excitation and emission.

17. Add coverslip and seal with nail polish.

18. Visualize Ref(2)P immunostaining levels by confocal fluorescence microscopy. Compare images to appropriate controls including a no-primary antibody staining control and mosaic clones where possible (see Protocol 4: Genetic Manipulation of Autophagy in the *Drosophila* Ovary [DeVorkin and Gorski 2014a]). View multiple sections within a tissue to ensure there are no artifacts following fixation or antibody staining. When comparing different samples, examine similar regions within the cell (e.g., focal plane, next to nucleus, etc.).

    Fluorescence immunostaining can be quantitated using image analysis software programs (e.g., ImageJ), but we usually quantitate Ref(2)P levels using western blot analysis.

    See Troubleshooting.

Western Blot Analysis of Ref(2)P

For a flow chart that summarizes Steps 19–29, see Figure 1.

As an alternative to chemiluminescence, the Odyssey system (Licor) may be used in this procedure with the following modifications: Block in Odyssey blocking buffer (Licor) rather than milk for 1 h at room temperature. Dilute primary antibodies 1:1000–1:10,000 as experimentally determined in Odyssey blocking buffer. Dilute infrared (IR)-conjugated secondary antibodies 1:10,000 in Odyssey blocking buffer and protect from light.

19. Lyse ovaries in chilled lysis buffer for protein extraction. Homogenize with a pestle until the ovaries appear to be lysed and the lysate is homogenous, and then incubate on ice for 30 min.

    The number and size of ovaries will determine the amount of lysis buffer to add. For 10 pairs of ovaries from flies fed with yeast paste, we suggest starting with 100 µL of lysis buffer.

20. Centrifuge lysate at 15,000 g for 15 min at 4°C. Discard the pellet and keep the supernatant containing the protein.

21. Quantitate amount of protein using a kit or assay of choice.

22. Separate 20–50 µg of protein by SDS–PAGE and transfer to a PVDF membrane.

    The amount of protein to load will depend on the individual cell line. To get a good separation of Ref(2)P and actin, we use NuPAGE Novex 4%–12% Bis–Tris Gels with MES or MOPS SDS running buffer.

23. Block nonspecific binding by incubating the membrane in 5% (w/v) nonfat milk plus 0.1% Tween-20 for 1 h at room temperature with gentle rocking.

24. Incubate in primary antibody diluted 1:1000 (Nezis et al. 2008) or 1:10,000 (Hannigan et al. in revision) in 5% nonfat milk plus 0.1% Tween-20 or in 1× PBS containing 0.1% Tween-20 overnight at 4°C with gentle rocking.

25. Wash membrane four times (5 min each) in 1× PBS containing 0.1% Tween-20 with gentle rocking at room temperature.

26. Dilute horseradish peroxidase-conjugated secondary antibody in 5% nonfat milk plus 0.1% Tween-20. Incubate with secondary antibody for 1 h at room temperature with gentle rocking.

27. Wash membrane four times (15 min each) with 1× PBS containing 0.1% Tween-20 with gentle rocking.

28. Rinse once with 1× PBS. Expose membrane to film and develop per manufacturer's instructions.
    See Troubleshooting.

29. Analyze using imaging software (e.g., ImageQuant or PhotoShop).

    The basic strategy is to determine the intensity of Ref(2)P and actin levels, and normalize the Ref(2)P levels to actin for each sample.

TROUBLESHOOTING

Problem (Step 18): There is nonspecific antibody staining, staining accumulates on the outside of the ovary, and/or no antibody staining is visualized internally.
Solution: The antibody concentration may be too high. Dilute antibody and repeat.

Problem (Steps 18 and 28): High background signal.
Solution: Consider the following.

• Make sure all antibodies are used at the right concentration.

• Ensure that samples are washed thoroughly.

• Antibody incubation times and temperatures (i.e., 4°C vs. room temperature) may vary and therefore may need to be optimized.

Cite this protocol as *Cold Spring Harb Protoc*; doi:10.1101/pdb.prot080333

DISCUSSION

Drosophila Ref(2)P was first characterized for its role in sigma rhabdovirus multiplication (Dezelee et al. 1989). More recently, it was determined that Ref(2)P is the *Drosophila* ortholog of mammalian p62, and requires the PB1 and UBA domains for protein aggregate formation (Nezis et al. 2008). Reduction of autophagy activity causes an accumulation of Ref(2)P-positive protein aggregates (Nezis et al. 2008; Bartlett et al. 2011; Cumming et al. 2008), suggesting that it can be used as a marker of autophagic activity. In mid-stage egg chambers of well-fed flies, Ref(2)P can be observed in the nurse cells as well as in the follicle cells (Fig. 2A). Following induction of autophagy by starvation, mid-stage degenerating egg chambers show a reduction in Ref(2)P detectable by immunofluorescence (Fig. 2B) or by western blot analysis (Fig. 2C). Flies mutant for the autophagy gene *Atg7* show an accumulation of Ref(2)P under both well-fed and starvation conditions (Fig. 2C). These results show that Ref(2)P can be a useful indicator of autophagic flux in *Drosophila* ovaries. Although useful as a marker of autophagic flux, caution must be used when examining Ref(2)P. Ref(2)P was shown to accumulate following genetic inhibition of the proteasome (Nezis et al. 2008), suggesting that Ref(2)P has a dual role in degrading ubiquitinated proteins by autophagy and the proteasome. In addition, Ref(2)P was transcriptionally up-regulated in the larval fat body following starvation (Pircs et al. 2012), and treatment with the proteasomal inhibitor MG132 (Nakaso et al. 2004), or stimulation with LPS

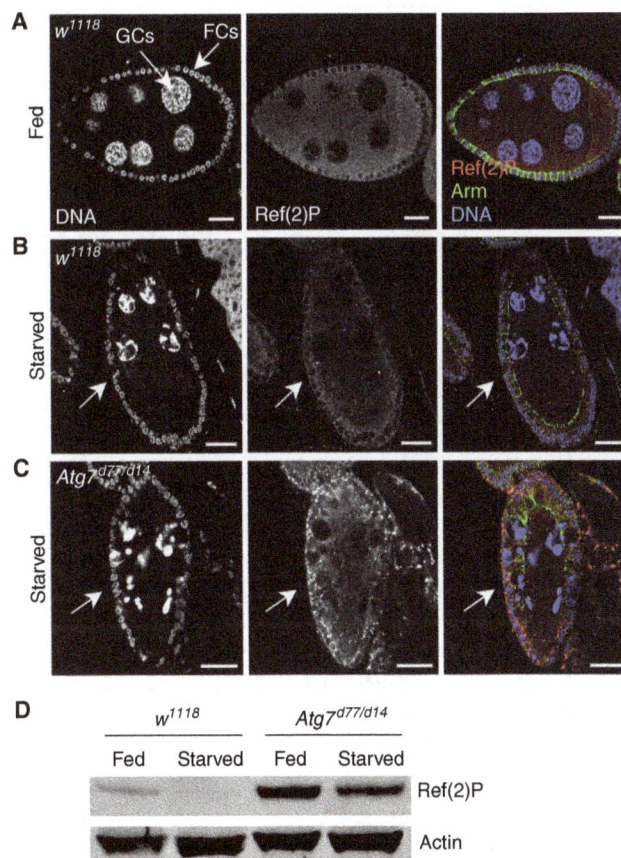

FIGURE 2. Monitoring autophagic flux using *Drosophila* Ref(2)P. Ovaries from well-fed w^{1118} flies (A) or w^{1118} (B) and $Atg7^{d77/d14}$ (C) flies subjected to amino acid starvation for 4 d were dissected and incubated with antibodies to Armadillo (green) and Ref(2)P (red). DNA was examined using DRAQ5 (blue). Far *right* panels are the merged images. Germline nurse cells (GCs) and follicle cells (FCs) are indicated. Arrows in (B) and (C) indicate degenerating mid-stage egg chambers. (D) Lysate was extracted from ovaries of w^{1118} or $Atg7^{d77/d14}$ flies subjected to well-fed or starvation conditions and analyzed by immunoblotting using an antibody to Ref(2)P. Actin served as a loading control. Scale bars: 25 μm.

(Fujita et al. 2011) led to the transcriptional activation of human p62. Therefore, when measuring autophagic activity, it may be useful to include the evaluation of Ref(2)P mRNA levels, and it is best to use Ref(2)P analysis in combination with other flux assays.

RELATED TECHNIQUES

Complementary techniques include evaluation of autophagosome accumulation using fluorescent reporters such as GFP-DrAtg8a or antibodies to *Drosophila* Atg8a (Barth et al. 2011). Moreover, autophagic flux can be monitored using GFP-mCherry-DrAtg8a (see Protocol 3: Monitoring Autophagy in *Drosophila* Using Fluorescent Reporters in the UAS-GAL4 System [DeVorkin and Gorski 2014b]), and autophagy-associated lysosomal activity can be quantified using LysoTracker staining (see Protocol 1: LysoTracker Staining to Aid in Monitoring Autophagy in *Drosophila* [DeVorkin and Gorski 2014c]).

RECIPES

DRAQ5 DNA Stain

Reagent	Final concentration
DRAQ5 (Biostatus)	1:500
RNase A	100 µg/mL

Prepare 1 mL of DRAQ5 DNA stain in 1× phosphate-buffered saline (pH 7.4).

Fly Food

Ingredients	Amounts (for 300-bottle batch)
Water	17 L
Agar	93 g
Cornmeal	1,716 g
Brewer's yeast	310 g
Sucrose	517 g
Dextrose	1033 g
Antifungal reagents	as appropriate[a]
Antibiotics	as appropriate[b]

[a]Use 8.69 g/L sodium potassium tartrate and 26.1 mL/L of a Tegosept (methyl-*p*-hydroxy benzoate) Stock solution. (For the Tegosept stock solution, dissolve the Tegosept in 95% ethanol at 100 g/L.)

[b]For each batch of food, add 15 mg/L tetracycline plus one of the following two antibiotics (alternate with each batch of food): 50 mg/L ampicillin, or 17 mg/L streptomycin. Make up the antibiotics as fresh stock solutions, with tetracycline dissolved in ethanol at 10 mg/mL and amplicillin or streptomycin dissolved in water at 100 mg/mL.

Mix the agar into 13 L of water; use the other 4 L of water to wet the remaining ingredients (excluding antifungals and antibiotics). Cook the medium in a steam kettle at 15 pounds per square inch (psi) for 20 min with constant stirring. Next, switch off the steam and cover the kettle with a lid. Allow the medium to cool to exactly 55°C and then add both the antibiotics and antifungals, stirring well. Dispense medium into vials and bottles, cover with cheese-cloth, and allow it to solidify overnight at room temperature before capping.

This recipe is adapted from that of Lewis (1960) and from recipe 4 from the Bloomington *Drosophila* Stock Centre website (flystocks.bio.indiana.edu/Fly_Work/media-recipes/media-recipes.htm).

Cite this protocol as *Cold Spring Harb Protoc*; doi:10.1101/pdb.prot080333

Lysis Buffer for Protein Extraction

Reagent	Amount	Final concentration
Tris-buffered saline (TBS; 10×, pH 7.5) <R>	5 mL	1×
EDTA (0.5 M)	100 µL	0.05 M
NP-40 (10%)	5 mL	1%
H$_2$O	39.9 mL	

Add one EDTA-free complete protease inhibitor cocktail tablet (Roche)

Phosphate-Buffered Saline (PBS; 1×, pH 7.4)

Reagent	Quantity (g)	Final concentration (mM)
NaCl	8.0	137.0
KCl	0.20	2.7
Na$_2$HPO$_4$	1.14	8.0
KH$_2$PO$_4$	0.20	1.5

Combine all ingredients in ∼800 mL of H$_2$O and stir to dissolve. Adjust pH to 7.4 and bring final volume to 1 L. Sterilize by autoclaving.

Tris-Buffered Saline (TBS; 10×, pH 7.5)

Reagent	Amount	Final concentration
Tris	24.2 g	200 mM
NaCl	87.7 g	1.5 M

Combine ingredients in ∼800 mL of H$_2$O. Adjust pH to 7.5 and bring final volume to 1 L. Sterilize by autoclaving.

ACKNOWLEDGMENTS

The authors thank S. Gaumer, T.E. Rusten, N. Katheder, G. Juhasz and T. Neufeld for reagents and/or fly lines used and J. Hodgson for the fly food recipe. The authors are grateful for a Natural Sciences and Engineering Research Council Discovery grant (RGPIN/371368-2009) for support; S.M.G. is supported in part by a Canadian Institutes of Health Research New Investigator Award.

REFERENCES

Barth JMI, Szabad J, Hafen E, Kohler K. 2011. Autophagy in *Drosophila* ovaries is induced by starvation and is required for oogenesis. *Cell Death Differ* 18: 915–924.

Bartlett BJ, Isakson P, Lewerenz J, Sanchez H, Kotzebue RW, Cumming RC, Harris GL, Nezis IP, Schubert DR, Simonsen A, et al. 2011. p62, Ref(2)P and ubiquitinated proteins are conserved markers of neuronal aging, aggregate formation and progressive autophagic defects. *Autophagy* 7: 572–583.

Cumming RC, Simonsen A, Finley KD. 2008. Quantitative analysis of autophagic activity in *Drosophila* neural tissues by measuring the turnover rates of pathway substrates. *Meth Enzymol* 451: 639–651.

DeVorkin L, Gorski SM. 2014a. Genetic manipulation of autophagy in the *Drosophila* ovary. *Cold Spring Harb Protoc* doi: 10.1101/pdb.prot080358.

DeVorkin L, Gorski SM. 2014b. Monitoring autophagy in *Drosophila* using fluorescent reporters in the UAS-GAL4 system. *Cold Spring Harb Protoc* doi: 10.1101/pdb.prot080341.

DeVorkin L, Gorski SM. 2014c. LysoTracker staining to aid in monitoring autophagy in *Drosophila*. *Cold Spring Harb Protoc* doi: 10.1101/pdb.prot080325.

Dezelee S, Bras F, Contamine D, Lopez-Ferber M, Segretain D, Teninges D. 1989. Molecular analysis of ref(2)P, a *Drosophila* gene implicated in sigma rhabdovirus multiplication and necessary for male fertility. *EMBO J* 8: 3437–3446.

Fujita K-I, Maeda D, Xiao Q, Srinivasula SM. 2011. Nrf2-mediated induction of P62 controls Toll-like receptor-4–driven aggresome-like induced structure formation and autophagic degradation. *Proc Natl Acad Sci* 108: 1427–1432.

Lewis EB. 1960. A new standard food medium. *Drosophila Information Service* 34: 117–118.

Nakaso K, Yoshimoto Y, Nakano T, Takeshima T, Fukuhara Y, Yasui K, Araga S, Yanagawa T, Ishii T, Nakashima K. 2004. Transcriptional activation of p62/A170/ZIP during the formation of the aggregates: Possible mechanisms and the role in Lewy body formation in Parkinson's disease. *Brain Res* 1012: 42–51.

Nezis IP, Simonsen A, Sagona AP, Finley K, Gaumer S, Contamine D, Rusten TE, Stenmark H, Brech A. 2008. Ref(2)P, the *Drosophila melanogaster* homologue of mammalian p62, is required for the formation of protein aggregates in adult brain. *J Cell Biol* 180: 1065–1071.

Pircs K, Nagy P, Varga A, Venkei Z, Erdi B, Hegedus K, Juhasz G. 2012. Advantages and limitations of different p62–based assays for estimating autophagic activity in *Drosophila*. *PLoS ONE* 7: e44214.

Monitoring Autophagy in *Drosophila* Using Fluorescent Reporters in the UAS-GAL4 System

Lindsay DeVorkin[1,2] and Sharon M. Gorski[1,2,3]

[1]*The Genome Sciences Centre, BC Cancer Agency, Vancouver, British Columbia V52 1L3, Canada;* [2]*Department of Molecular Biology and Biochemistry, Simon Fraser University, Burnaby, British Columbia V5A 1S6, Canada*

Following autophagy induction, the autophagy-related protein Atg8 undergoes ubiquitin-like conjugation to phosphatidylethanolamine and inserts into the autophagosome membrane. Transgenic *Drosophila* lines expressing *Drosophila* Atg8 (DrAtg8a) fused to green fluorescent protein (GFP), mCherry, or dual-tagged GFP-mCherry have been constructed and are extremely useful for monitoring autophagy. The use of GFP-mCherry-Atg8a is particularly advantageous because it allows for the assessment of autophagy induction as well as autophagy flux. GFP-mCherry-Atg8a fluoresces yellow in nonacidic structures including the autophagosome. When autophagosomes fuse with lysosomes to form autolysosomes, GFP fluorescence is quenched by the acidic hydrolases and the resulting autolysosome will fluoresce red. The upstream activating sequence (UAS)-GAL4 system allows for the ectopic expression of the gene of interest (in this case, DrAtg8a) in a tissue-specific manner. Here we provide a protocol for monitoring autophagic flux by fluorescence microscopy by expressing UASp-GFP-mCherry-DrAtg8a in the *Drosophila* germline.

MATERIALS

It is essential that you consult the appropriate Material Safety Data Sheets and your institution's Environmental Health and Safety Office for proper handling of equipment and hazardous materials used in this protocol.

RECIPES: Please see the end of this protocol for recipes indicated by <R>. Additional recipes can be found online at http://cshprotocols.cshlp.org/site/recipes.

Reagents

CO_2 gas
DRAQ5 DNA stain (optional; see Step 9) <R>
Drosophila stock of choice
    nosGAL4::VP16 (Rørth 1998)
    UASp-GFP-mCherry-DrAtg8a (Nezis et al. 2010)
    UASp-mCherry-DrAtg8a (Nezis et al. 2009)
    UASp-GFP-DrAtg8a (Barth et al. 2011)
    UASp-GFP-LC3 (Rusten et al. 2004)

[3]Correspondence: sgorski@bcgsc.ca

Cite this protocol as *Cold Spring Harb Protoc*; doi:10.1101/pdb.prot080341

UASt-GFP-Atg8a (Juhász et al. 2008)

UASt-mCherry-Atg8a (Chang and Neufeld 2009)

UASp-GFP-mCherry-Atg8a and UASp-mCherry-Atg8a are available from Bloomington Stock Center. UASp vectors were designed for GAL4-mediated gene expression in the female germline, whereas UASt vectors are suitable for somatic tissue (Rorth, 1998). See Discussion.

GAL4 driver fly stock of choice (see Discussion)

Fly food \<R\>

To supplement fly food with yeast paste, mix 1 g of active dry yeast with 1 mL of distilled H_2O and apply a pea-sized amount of the paste on top of fly food in a vial.

Nail polish

Paraformaldehyde (16%)

Phosphate-buffered saline (PBS; 1×, pH 7.4) \<R\>

SlowFade Gold Reagent with DAPI (Invitrogen)

Sucrose (10% in 1× PBS, pH 7.4)

Equipment

Confocal microscope

Coverslips

Diffuser pad for CO_2 anesthesia

Dissecting dish

Dissecting microscope

Fine forceps (e.g., Dumont #5 tweezers, Electron Microscopy Sciences)

Fly vials

Kimwipe

Microscope slides

We use three-well, 14-mm hydrophobic printed slides from Electron Microscopy Services.

Multi-well glass plate for staining

METHOD

For a flow chart of this protocol, see Figure 1.

Creation of Transgenic Lines Expressing a Fluorescent Tagged Atg8a in the Germline

1. Cross the UAS transgenic line of choice to the appropriate GAL4 driver line.

   See Discussion.

Fly Preparation

2. To ensure proper development of ovarian tissue, place approximately ten to twenty 1- to 5-d-old male and female progeny on fresh fly food supplemented with yeast paste in a vial for 2 d in uncrowded conditions.

Autophagy Induction by Starvation

3. Place a Kimwipe soaked in 10% sucrose at the bottom of an empty vial. Transfer flies from the vial containing yeast paste to the vial containing the sucrose-soaked Kimwipe. Leave the flies in this vial for 2–4 d.

   Depending on the genetic background of the fly strain, starvation times will have to be experimentally determined to observe changes in autophagy.

Monitoring autophagy using
the UAS-GAL4 system

Cross virgin female flies to males

↓ ~7–14 d

Collect 1- to 5-d-old relevant progeny
Put under well fed conditions

↓ 2 d

Transfer flies to starvation
conditions

↓ 2–4 d

Dissect ovaries

↓

Fix ovaries

↓ 20 min

Wash

↓ 15 min

Mount ovaries and view by
fluorescence microscopy

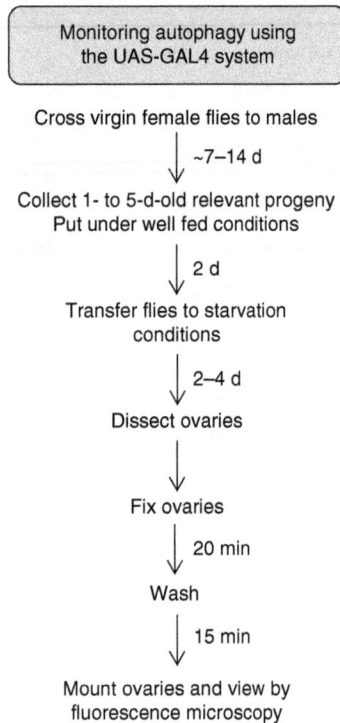

FIGURE 1. Flow chart outlining the procedures described in this protocol.

Ovary Dissection

4. Anesthetize flies on a CO_2 diffuser pad.

5. Dissect female flies in 1× PBS in a dissecting dish. Grasp each fly between the abdomen and the lower thorax with one set of forceps, and pull at the posterior end of the abdomen to remove the ovaries.

    Handle ovaries gently from the posterior end to ensure midstage egg chambers are not damaged. We typically recommend examining at least five to 10 pairs of ovaries.

6. Remove the debris away from the ovaries and place immediately on ice cold 1× PBS in a multi-well glass plate while other ovaries are dissected. Using forceps, tease ovarioles apart from each other as carefully as possible to ensure uniform fixation and staining.

    Do not let the ovaries dry while dissecting.

Ovary Fixation and DNA Staining

7. Carefully remove PBS from ovaries using a fine pipette tip. Immediately fix with 200 µL of 4% paraformaldehyde (diluted in 1× PBS) for 20 min at room temperature.

8. Wash three times (5 min each) with 200 µL of 1× PBS.

9. Transfer ovaries to a drop of SlowFade Gold Reagent with DAPI on a microscope slide and carefully tease apart.

    Alternatively, DRAQ5 DNA stain can be used to visualize DNA. To stain using DRAQ5, remove PBS from tissue after Step 8 and then incubate with DRAQ5 DNA stain for 10 min in the dark. Mount tissue in SlowFade Gold Reagent (without DAPI). DRAQ5 has a far red excitation and emission.

10. Add coverslip and seal with nail polish.

11. View by fluorescence microscopy.

    See Troubleshooting and Discussion.

    Cite this protocol as *Cold Spring Harb Protoc*; doi:10.1101/pdb.prot080341

TROUBLESHOOTING

Problem (Step 11): Few autolysosomes are detected.

Solution: Fixation solutions are often neutral or weak bases and will raise the pH of the cell. Therefore, a reemergence of the GFP signal may occur (Klionsky et al. 2012). It is important to incorporate appropriate controls to ensure the proper detection of autolysosomes.

DISCUSSION

We used this protocol to produce the sample results shown in Figure 2. We crossed the *UASp-GFP-mCherry-DrAtg8a* strain and the *nosGAL4::VP16* driver strain (both homozygous viable) to monitor autophagic flux in the *Drosophila* ovary. The cross was ♀ *UASp-GFP-mCherry-DrAtg8a* × ♂ *nosGAL4::VP16*, and the relevant progeny were *UASp-GFP-mCherry-DrAtg8a/+*; *nosGAL4:: VP16/+*.

Expression of GFP-mCherry-DrAtg8a in the germline of *Drosophila* using the nosGAL4 driver resulted in diffuse yellow cytoplasmic staining throughout the germline under basal conditions (overlap of green and red fluorescence, Fig. 2A). When autophagy was induced by starvation, GFP-mCherry-DrAtg8a inserted into autophagosomes, producing yellow puncta indicative of autophago-somes (Fig. 2B, arrow heads) and red puncta indicative of autolysosomes (or amphisomes) (Fig. 2B, arrows) (Klionsky et al. 2012). The number of red puncta and/or the ratio of yellow to red puncta could be quantitated manually or with the aid of image analysis software such as ImageJ. To better distinguish between follicle cells and germ cells in healthy and degenerating mid-stage egg chambers, membrane markers including DE-Cadherin (DCAD2, DSHB), F-Actin (Phalloidin, Invitrogen) or Armadillo (N2 7A1, DSHB) can be used.

The UAS-GAL4 system can be used for other studies of autophagy in *Drosophila*. It has been used to ectopically express Atg genes (Scott et al. 2004) and upstream autophagy regulators, including RheB and Tsc1 (Scott et al. 2004), to assess the biological effects of their overexpression in a tissue- and/or stage-specific manner (del Valle Rodríguez et al. 2012). Of note, ectopic expression of *Drosophila* Atg1 is a useful tool for inducing autophagy (Scott et al. 2004).

A modified UAS vector (UASp) containing additional GAL4 binding sites, a P-transposase pro-moter and a K10 terminator must be used for expression in the female germline. UASp combined with the nosGAL4::VP16 driver allows for expression throughout oogenesis (Rorth 1998). However, for expression of genes in *Drosophila* somatic tissues, a UASt vector containing a GAL4-responsive promoter is suitable (Brand and Perrimon 1993).

UASp-GFP-mCherry-DrAtg8a/+; nosGAL4::VP16/+

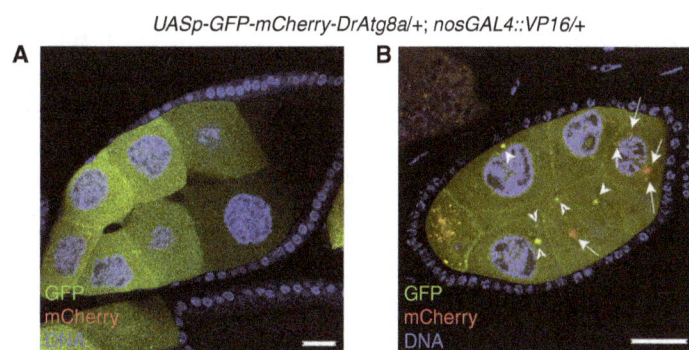

FIGURE 2. Monitoring autophagic flux using the UASp-mCherry-GFP-DrAtg8a transgene. Flies of the genotype *UASp-mCherry-GFP-DrAtg8a/+; nosGAL4/+* were conditioned on yeast paste for 2 d (*A*) or subjected to amino acid starvation for 4 d (*B*) before ovary dissection. Yellow puncta indicated by arrow heads show autophagosomes, and red puncta indicated by arrows show autolysosomes. Scale bars: 25 μm.

Several *Drosophila* transgenic lines have been created that express Atg8a (DrAtg8a) fused to GFP (Rusten et al. 2007), mCherry (Nezis et al. 2009) or dual-tagged GFP-mCherry (Nezis et al. 2010). The dual-tagged GFP-mCherry-Atg8a reporter is advantageous because it enables both the detection of autophagy induction and flux through the system. GFP is sensitive to acidic conditions and therefore its fluorescence is quenched in the autolysosome, whereas mCherry (or mRFP) is more stable (Kimura et al. 2007). In addition to the dual-tagged reporter, the use of GFP-Atg8 in combination with LysoTracker Red has been used to help assess the delivery of autophagosomes to autolysosomes (Rusten et al. 2004).

RELATED TECHNIQUES

Complementary techniques include monitoring autophagic flux using *Drosophila* Ref(2)P (see Protocol 2: Monitoring Autophagic Flux Using Ref(2)P, the *Drosophila* p62 Ortholog [DeVorkin and Gorski 2014a]). Autophagy-associated lysosomal activity can be quantified using LysoTracker staining (see Protocol 1: LysoTracker Staining to Aid in Monitoring Autophagy in *Drosophila* [DeVorkin and Gorski 2014b]).

RECIPES

DRAQ5 DNA Stain

Reagent	Final concentration
DRAQ5 (Biostatus)	1:500
RNase A	100 µg/mL

Prepare 1 mL of DRAQ5 DNA stain in 1× phosphate-buffered saline (pH 7.4).

Fly Food

Ingredients	Amounts (for 300-bottle batch)
Water	17 L
Agar	93 g
Cornmeal	1716 g
Brewer's yeast	310 g
Sucrose	517 g
Dextrose	1033 g
Antifungal reagents	as appropriate[a]
Antibiotics	as appropriate[b]

[a]Use 8.69 g/L sodium potassium tartrate and 26.1 mL/L of a Tegosept (methyl-*p*-hydroxy benzoate) stock solution. (For the Tegosept stock solution, dissolve the Tegosept in 95% ethanol at 100 g/L.)

[b]For each batch of food, add 15 mg/L tetracycline plus one of the following two antibiotics (alternate with each batch of food): 50 mg/L ampicillin, and 17 mg/L streptomycin. Make up the antibiotics as fresh stock solutions, with tetracycline dissolved in ethanol at 10 mg/mL and ampicillin or streptomycin dissolved in water at 100 mg/mL.

Mix the agar into 13 L of water; use the other 4 L of water to wet the remaining ingredients (excluding antifungals and antibiotics). Cook the medium in a steam kettle at 15 pounds per square inch (psi) for 20 min with constant stirring. Next, switch off the steam and cover the kettle with a lid. Allow the medium to cool to exactly 55°C and then add both the antibiotics and antifungals, stirring well. Dispense medium into vials and bottles, cover with cheesecloth, and allow it to solidify overnight at room temperature before capping.

Cite this protocol as *Cold Spring Harb Protoc*; doi:10.1101/pdb.prot080341

This recipe is adapted from that of Lewis (1960) and from recipe 4 from the Bloomington *Drosophila* Stock Centre website (flystocks.bio.indiana.edu/Fly_Work/media-recipes/media-recipes.htm).

Phosphate-Buffered Saline (PBS; 1×, pH 7.4)

Reagent	Quantity (g)	Final concentration (mM)
NaCl	8.0	137.0
KCl	0.20	2.7
Na_2HPO_4	1.14	8.0
KH_2PO_4	0.20	1.5

Combine all ingredients in ~800 mL of H_2O and stir to dissolve. Adjust pH to 7.4 and bring final volume to 1 L. Sterilize by autoclaving.

ACKNOWLEDGMENTS

The authors thank T.E. Rusten, and K. McCall for fly lines used, and J. Jodgson for the fly food recipe. The authors are grateful for a Natural Sciences and Engineering Research Council Discovery grant (RGPIN/371368-2009) for support; S.M.G. is supported in part by a Canadian Institutes of Health Research New Investigator Award.

REFERENCES

Barth JMI, Szabad J, Hafen E, Kohler K. 2011. Autophagy in *Drosophila* ovaries is induced by starvation and is required for oogenesis. *Cell Death Differ* 18: 915–924.

Brand AH, Perrimon N. 1993. Targeted gene expression as a means of altering cell fates and generating dominant phenotypes. *Development* 118: 401–415.

Chang YY, Neufeld TP. 2009. An Atg1/Atg13 complex with multiple roles in TOR-mediated autophagy regulation. *Mol Biol Cell* 20: 2004–2014.

DeVorkin L, Gorski SM. 2014a. Monitoring autophagic flux using Ref(2)P, the *Drosophila* p62 ortholog. *Cold Spring Harb Protoc* doi: 10.1101/pdb.prot080333.

DeVorkin L, Gorski SM. 2014b. LysoTracker staining to aid in monitoring autophagy in *Drosophila*. *Cold Spring Harb Protoc* doi: 10.1101/pdb.prot080325.

Juhász G, Hill JH, Yan Y, Sass M, Baehrecke EH, Backer JM, Neufeld TP. 2008. The class III PI(3)K Vps34 promotes autophagy and endocytosis but not TOR signaling in *Drosophila*. *JCB* 181: 655–666.

Kimura S, Noda T, Yoshimori T. 2007. Dissection of the autophagosome maturation process by a novel reporter protein, tandem fluorescent-tagged LC3. *Autophagy* 3: 452–460.

Klionsky DJ, Abdalla FC, Abeliovich H, Abraham RT, Acevedo-Arozena A, Adeli K, Agholme L, Agnello M, Agostinis P, Aguirre-Ghiso JA, et al. 2012. Guidelines for the use and interpretation of assays for monitoring autophagy. *Autophagy* 8: 445–544.

Lewis EB. 1960. A new standard food medium. *Drosophila Information Service* 34: 117–118.

Nezis IP, Lamark T, Velentzas AD, Rusten TE, Bjørkøy G, Johansen T, Papassideri IS, Stravopodis DJ, Margaritis LH, Stenmark H, et al. 2009. Cell death during *Drosophila melanogaster* early oogenesis is mediated through autophagy. *Autophagy* 5: 298–303.

Nezis IP, Shravage BV, Sagona AP, Lamark T, Bjørkøy G, Johansen T, Rusten TE, Brech A, Baehrecke EH, Stenmark H. 2010. Autophagic degradation of dBruce controls DNA fragmentation in nurse cells during late *Drosophila melanogaster* oogenesis. *The Journal of Cell Biology* 190: 523–531.

Rorth P. 1998. Gal4 in the *Drosophila* female germline. *Mechanisms of Development* 78: 113–118.

Rusten TE, Vaccari T, Lindmo K, Rodahl LMW, Nezis IP, Sem-Jacobsen C, Wendler F, Vincent JP, Brech A, Bilder D, et al. 2007. ESCRTs and Fab1 regulate distinct steps of autophagy. *Curr Biol* 17: 1817–1825.

Rusten TE, Lindmo K, Juhász G, Sass M, Seglen PO, Brech A, Stenmark H. 2004. Programmed autophagy in the *Drosophila* fat body is induced by ecdysone through regulation of the PI3K pathway. *Developmental Cell* 7: 179–192.

Scott RC, Schuldiner O, Neufeld TP. 2004. Role and regulation of starvation-induced autophagy in the *Drosophila* fat body. *Developmental Cell* 7: 167–178.

del Valle Rodríguez A, Didiano D, Desplan C. 2012. Power tools for gene expression and clonal analysis in *Drosophila*. *Nat Methods* 9: 47–55.

Genetic Manipulation of Autophagy in the *Drosophila* Ovary

Lindsay DeVorkin[1,2] and Sharon M. Gorski[1,2,3]

[1]*The Genome Sciences Centre, BC Cancer Agency, Vancouver, British Columbia V5Z 1L3, Canada;* [2]*Department of Molecular Biology and Biochemistry, Simon Fraser University, Burnaby, British Columbia V5A 1S6, Canada*

Flies with mutations in *Atg7* or *Atg8a* are homozygous viable and develop to adulthood, whereas mutations in other autophagy genes, including *Atg1*, are homozygous lethal. Clonal analysis has been instrumental in examining the role and regulation of lethal Atg genes in many aspects of *Drosophila* development and survival. The generation of homozygous mutant clones in an otherwise heterozygous mutant background is possible in mitotically active tissues, and is highly beneficial in that the control cells and experimental cells are subjected to the same developmental and nutritional cues allowing for a side-by-side comparison. Several methods are now available to examine the contribution of lethal autophagy genes during *Drosophila* oogenesis. Here we describe how to generate a homozygous mutant germline using the FLP-DFS (dominant female sterile) technique, how to generate somatic clones, and how to induce targeted gene knockdown in the germline using RNAi.

MATERIALS

It is essential that you consult the appropriate Material Safety Data Sheets and your institution's Environmental Health and Safety Office for proper handling of equipment and hazardous materials used in this protocol.

RECIPES: Please see the end of this protocol for recipes indicated by <R>. Additional recipes can be found online at http://cshprotocols.cshlp.org/site/recipes.

Reagents

Fly food <R>
> *To supplement fly food with yeast paste, mix 1 g of active dry yeast with 1 mL of distilled H_2O and apply a pea-sized amount of the paste on top of fly food in a vial.*

Fly stocks of interest
Phosphate-buffered saline (PBS; 1×, pH 7.4) <R>
Sucrose (10% in 1× PBS, pH 7.4)

Equipment

Dissecting microscope
Diffuser pad for CO_2 anesthesia
Fly vials
Kimwipes
Water bath, 37°C

[3]Correspondence: sgorski@bcgsc.ca

Cite this protocol as *Cold Spring Harb Protoc*; doi:10.1101/pdb.prot080358

METHOD

For generation of germline clones, follow Steps 1-4. For generation of somatic clones, follow Steps 5–8. For analysis of RNAi knockdown in the germline, follow Steps 9–11. For a flow chart of these techniques, see Figure 1.

Generation of Germline Clones

Here we provide an example of a genetic cross used to generate Atg1 mutant germline clones. In the absence of recombination between two FRT sites, the presence of the ovo^D transgene causes female germ cells to arrest during early oogenesis. As a result, only homozygous mutant germline clones will survive.

Fly Crosses

1. Anesthetize flies on a CO_2 diffuser pad and collect virgin females that are heterozygous for the mutation of interest and contain a source of FLP recombinase. Cross with males that possess the *ovo^D* transgene that is distal to the FRT site and has a source of FLP recombinase. For example, cross

$$♀yw, hsflp; Atg1^{\Delta 3D}FRT80B/ \ TM6B \ x♂yw, hsflp; \ ovo^D \ FRT80B/ \ TM6B$$

to produce the following relevant progeny:

$$♀yw, hsflp; Atg1^{\Delta 3D}FRT80B/ \ ovo^D FRT80B.$$

Please refer to protocols for generating flies required for germline clone analysis (Selva and Stronach 2007).

Heat Shock

2. On days 4 and 5, heat shock larval progeny by placing vials containing larvae in a water bath for 1 h at 37°C.

   This induces the generation of mitotic germline clones (GLCs) in the developing ovaries.

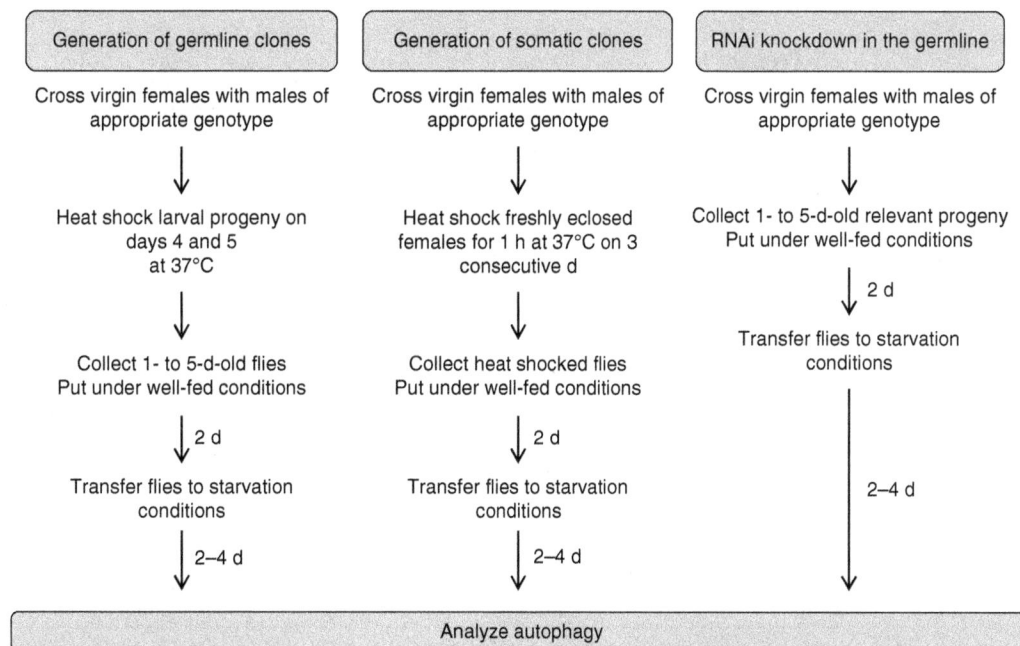

FIGURE 1. Flow chart outlining the procedures described here. Autophagy can then be analyzed using Protocol 1: LysoTracker Staining to Aid in Monitoring Autophagy in *Drosophila* [DeVorkin and Gorski 2014a], Protocol 2: Monitoring Autophagic Flux Using Ref(2)P, the *Drosophila* p62 Ortholog [DeVorkin and Gorski 2014b], or Protocol 3: Monitoring Autophagy in *Drosophila* Using Fluorescent Reporters in the UAS-GAL4 System [DeVorkin and Gorski 2014c].

Fly Preparation

3. To ensure proper development of ovarian tissue, place 1- to 5-d-old progeny on fresh fly food supplemented with yeast paste in a vial for 2 d in uncrowded conditions.

4. Proceed to Step 12.

Generation of Somatic Clones

Here we provide an example of a genetic cross that will yield progeny that can be subjected to heat shock-induced mitotic recombination to generate mosaic clones of somatic follicle cells.

Fly Crosses

5. Anesthetize and collect virgin females that are heterozygous for the mutation of interest (e.g., *Ter94*), contain FRT sites, and have a source of FLP recombinase. Cross with males that possess a marker (e.g., GFP), FRT sites and have a source of FLP recombinase. For example, cross

$$♀w,\ hsflp;\ FRT42B\ Ter94/\ CyO\ x♂w,\ hsflp;\ FRT42B\ UbiGFPnls/\ CyO$$

to produce the following relevant progeny:

$$w,\ hsflp;\ FRT42B\ Ter94/\ FRT42B\ UbiGFPnls.$$

Heat Shock

6. Collect freshly eclosed female progeny into a new vial containing fly food. Heat shock the females by placing the vial into a water bath at 37°C for 1 h. After the heat shock, transfer flies to a fresh vial that also contains wild-type males to help stimulate oocyte maturation (Ravi Ram and Wolfner 2007). Repeat the heat shock and transfer procedure on the following two consecutive days.

Fly Preparation

7. To ensure proper development of ovarian tissue, place flies on fresh fly food supplemented with yeast paste in a vial for 2 d in uncrowded conditions.

8. Proceed to Step 12.

Generation of Female Flies Expressing RNAi in the Germline

For efficient targeted gene knockdown in the germline, two different methods can be used. First, long double-stranded hairpin constructs overexpressed in the germline in combination with Dicer using the nosGAL4 allows for efficient gene knockdown (Handler et al. 2011; Wang and Elgin 2011). In addition, knockdown of target genes using Valium-based vectors available through the Transgenic RNAi Project (TRiP) in combination with the maternal-triple-driver (MTD-GAL4) (Mazzalupo and Cooley 2006) also induces efficient target gene knockdown in the germline. Here we provide an example of a genetic cross for the RNAi knockdown of Atg1 in the germline using the MTD-GAL4 driver. See Table 1 for Drosophila *Atg gene stocks available through TRiP.*

TABLE 1. *Drosophila* Atg gene stocks currently available from TRiP

Gene	Vector	Gene	Vector
Atg1	VALIUM10, 22	*Atg8a*	VALIUM10, 20
Atg2	VALIUM10, 20	*Atg8b*	VALIUM10, 20
Atg4	VALIUM10, 20, 22	*Atg9*	VALIUM10, 20
Atg5	VALIUM10, 20	*Atg12*	VALIUM10, 20
Atg6	VALIUM10, 20	*Atg18*	VALIUM10, 20
Atg7	VALIUM10, 20	*Atg13*	VALIUM20

Cite this protocol as *Cold Spring Harb Protoc*; doi:10.1101/pdb.prot080358

Fly Crosses

9. Anesthetize flies and collect virgin females and cross with males. For example, cross

$$♀ UAS-shRNA-Atg1 \quad x♂ \quad MTD-GAL4$$

to produce the following relevant progeny:

$$♀ UAS-shRNA-Atg1/MTD-GAL4.$$

Fly Preparation

10. Collect 1- to 5-d-old eclosed progeny. To ensure proper development of ovarian tissue, place progeny on fresh fly food supplemented with yeast paste in a vial for 2 d in uncrowded conditions.

11. Proceed to Step 12.

Induction and Analysis of Autophagy

12. Place a Kimwipe soaked in 10% sucrose at the bottom of an empty vial. Transfer flies from the vial containing yeast paste to the vial containing the sucrose-soaked Kimwipe. Leave flies in this vial for 2–4 d.

13. Analyze autophagy in *Drosophila* ovaries. See Protocol 1: LysoTracker Staining to Aid in Monitoring Autophagy in *Drosophila* [DeVorkin and Gorski 2014a] or Protocol 2: Monitoring Autophagic Flux Using Ref(2)P, the *Drosophila* p62 Ortholog [DeVorkin and Gorski 2014b].

    The UAS-GAL4 system using GFP, mCherry or dual-tagged DrAtg8a can also be used (see Protocol 3: Monitoring Autophagy in Drosophila *Using Fluorescent Reporters In the UAS-GAL4 System [DeVorkin and Gorski 2014c]), but the appropriate transgenic reporters and drivers must be incorporated into the genetic background of interest (e.g., Fig. 2A). When visualizing the ovary, we find it is helpful to use a membrane stain (for example anti-Armadillo (N2 7A1, DSHB) or anti-DE-Cadherin (DCAD2, DSHB)) to examine the cell structure of the tissue.*

    See Troubleshooting.

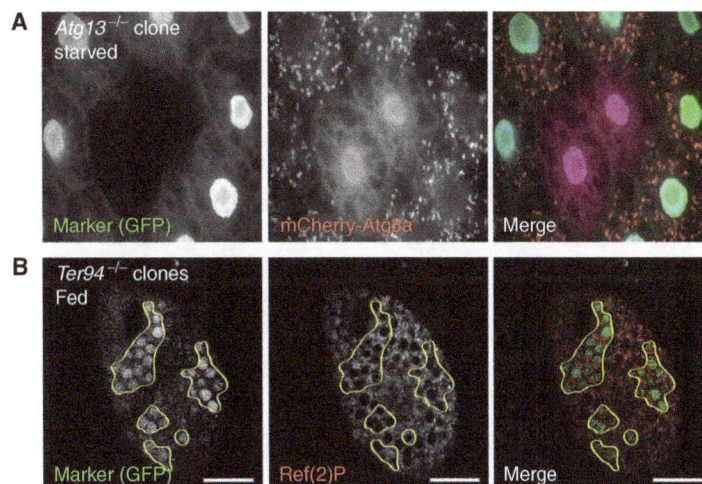

FIGURE 2. Creating somatic cell clones using the heat shock FLP-FRT method. (*A*) Mosaic fat body cell clones from heat-shocked animals of the genotype *hsFLP; Cg-GAL4/+; UAS-mCherryAtg8a FRT82B UAS-GFPnls/FRT82B Atg13$^{\Delta74}$*. Fat body cells lacking nuclear GFP indicate the *Atg13* homozygous mutant clones. In response to 4 h starvation, the *Atg13* homozygous mutant cells fail to induce formation of punctate mCherry-Atg8a. [Images courtesy of Thomas Neufeld, University of Minnesota.] (*B*) Mosaic follicle cell clones from heat-shocked animals of the genotype *hsflp; FRT42B Ter94/ FRT42B UbiGFPnls*. Follicle cells lacking nuclear localized GFP indicate the *Ter94* homozygous mutant clones. In this experiment, we found that loss of Ter94 does not affect Ref(2)P staining in follicle cells. Scale bars: 25 µm.

14. Confirm that the protein of interest has been knocked out or knocked down by immunostaining.

In the case of Atg, this can be performed if and when Atg-specific antibodies suitable for use in Drosophila *tissues are developed.*

TROUBLESHOOTING

Problem (Step 13): A very low number of somatic or germline clones are present.
Solution: The heat shock may not be timed properly. The duration of the heat shock may need to be optimized (start by increasing the duration). Check the cross to verify the correct genotype is present.

Problem (Step 13): Somatic clones are not of the expected size.
Solution: The age at which the progeny are heat shocked will determine the size of the clone. A reduced number of large clones will be present when embryos are heat shocked, whereas there will be many small clones following heat shock 48–72 h after egg laying.

Problem (Step 13): There are unexpected or inconsistent phenotypes following RNAi.
Solutions: Consider the following.

- Functional redundancy of proteins that function within the same pathway may make it difficult to examine the role of a particular gene in autophagy.

- Gene knockdown may lead to agametic ovaries. Try another tissue (e.g., follicle cells) to examine the role of that gene in autophagy.

- Off target effects may lead to changes in the expression of one or more genes that may result in an unexpected or inconsistent phenotype. It is best to use multiple lines with nonoverlapping RNAi to confirm results.

DISCUSSION

To examine the contribution of lethal autophagy genes during *Drosophila* oogenesis, the FLP-DFS system can be used. This system combines the yeast FLP mediated recombinase and the "dominant female sterile" (DFS) mutation in the ovo^D gene to examine recessive lethal mutations during oogenesis. The presence of the dominant ovo^D mutation leads to death of the nonrecombinant germline early in oogenesis, so that the only egg chambers that survive will be the homozygous mutant germline clones (Chou and Perrimon 1996). Several ovo^D lines have been developed for each chromosome and are available at Bloomington *Drosophila* Stock Center.

For creation and analysis of mosaic clones in a somatic tissue, please refer to descriptions of the GAL4 system and descriptions of monitoring autophagy in the larval fat body (Juhasz and Neufeld 2008; Neufeld 2008; Southall et al. 2008) and Fig. 2A. Similarly, in the ovary, the generation of mosaic clones in the somatic follicle cells can be mediated by the heat shock inducible yeast FLP recombinase system (*hsflp*) at FRT sites (Golic 1991) (Fig. 2B). It should be noted that the use of heat shock-flp can induce both germline and follicle cell clones, and in the case of ovo^D, follicle cell clones will be *unmarked*.

RNA interference (RNAi) in somatic tissues in *Drosophila* is achieved by cloning a gene fragment as an inverted repeat and expressing it using the UAS-GAL4 system. The resulting long double-stranded hairpin RNA is processed by Dicer into siRNAs which induce degradation of target mRNA and subsequent gene silencing (Bellés 2010). Several RNAi lines targeting the core autophagy genes are available from Vienna Drosophila Stock Center (VDRC). Until recently, efficient knockdown of target genes in the *Drosophila* germline was not achievable. Now, long double-stranded hairpin constructs overexpressed with Dicer in the germline using the nosGAL4 driver result

Cite this protocol as *Cold Spring Harb Protoc*; doi:10.1101/pdb.prot080358

in efficient target gene knockdown (Handler et al. 2011; Wang and Elgin 2011). In addition, knockdown of target genes in the germline using the Valium22 vector, or the germline and somatic tissues using the Valium20 vector, have been described recently (Ni et al. 2011). Valium22 is a UASp-based vector combining short-hairpin constructs, and when driven by the germline-specific MTD-GAL4 driver (Mazzalupo and Cooley 2006), it induces efficient targeted gene knockdown in the germline (Ni et al. 2011). Current efforts are underway to create a genome wide shRNA collection using Valium-based vectors and will be available through the Transgenic RNAi Project (TRiP). Currently, there are 11 *Drosophila* autophagy gene Valium-based stocks available (Table 1).

RELATED TECHNIQUES

Complementary techniques for clonal analysis in *Drosophila* include the temporal control of GAL4 using the GAL80 repressor, hormone inducible GAL4 and FLP-Out GAL4. Please refer to The GAL4 System: A Versatile Toolkit for Gene Expression in *Drosophila* (Southall et al. 2008) for information on the various systems. For example, mosaic analysis with a repressible cell marker (MARCM) takes advantage of the GAL80 repressor and is used to label homozygous mutant cells in an otherwise heterozygous background (Lee and Luo 2001); see Generation and Staining of MARCM Clones in *Drosophila* (Shrestha and Grueber 2011). In addition, follicle cell clones can be efficiently induced using follicle cell-specific drivers including e22c-GAL4 (Duffy et al. 1998).

RECIPES

Fly Food

Ingredients	Amounts (for 300-bottle batch)
Water	17 L
Agar	93 g
Cornmeal	1,716 g
Brewer's yeast	310 g
Sucrose	517 g
Dextrose	1033 g
Antifungal reagents	as appropriate[a]
Antibiotics	as appropriate[b]

[a]Use 8.69 g/L sodium potassium tartrate and 26.1 mL of a Tegosept (methyl-*p*-hydroxy benzoate) Stock solution. (For the Tegosept stock solution, dissolve the Tegosept in 95% ethanol at 100 g/L.)

[b]For each batch of food, add 15 mg/L tetracycline plus one of the following two antibiotics (alternate with each batch of food): 50 mg/L ampicillin, or 17 mg/L streptomycin. Make up the antibiotics as fresh stock solutions, with tetracycline dissolved in ethanol at 10 mg/mL and ampicillin or streptomycin dissolved in water at 100 mg/mL.

Mix the agar into 13 L of water; use the other 4 L of water to wet the remaining ingredients (excluding antifungals and antibiotics). Cook the medium in a steam kettle at 15 pounds per square inch (psi) for 20 min with constant stirring. Next, switch off the steam and cover the kettle with a lid. Allow the medium to cool to exactly 55°C and then add both the antibiotics and antifungals, stirring well. Dispense medium into vials and bottles, cover with cheesecloth, and allow it to solidify overnight at room temperature before capping.

This recipe is adapted from that of Lewis (1960) and from recipe 4 from the Bloomington *Drosophila* Stock Centre website (flystocks.bio.indiana.edu/Fly_Work/media-recipes/media-recipes.htm).

Phosphate-Buffered Saline (PBS; 1×, pH 7.4)

Reagent	Quantity (g)	Final concentration (mM)
NaCl	8.0	137.0
KCl	0.20	2.7
Na_2HPO_4	1.14	8.0
KH_2PO_4	0.20	1.5

Combine all ingredients in ~800 mL of H_2O and stir to dissolve. Adjust pH to 7.4 and bring final volume to 1 L. Sterilize by autoclaving.

ACKNOWLEDGMENTS

The authors thank T. Neufeld for images, K. McCall for helpful advice, and J. Hodgson for the fly food recipe. The authors are grateful for a Natural Sciences and Engineering Research Council Discovery grant (RGPIN/371368-2009) for support; S.M.G. is supported in part by a Canadian Institutes of Health Research New Investigator Award.

REFERENCES

Bellés X. 2010. Beyond *Drosophila*: RNAi in vivo and functional genomics in insects. *Annu Rev Entomol* 55: 111–128.

Chou TB, Perrimon N. 1996. The autosomal Flp-Dfs technique for generating germline mosaics in *Drosophila Melanogaster*. *Genetics* 144: 1673–1679.

DeVorkin L, Gorski SM. 2014a. LysoTracker staining to aid in monitoring autophagy in *Drosophila*. *Cold Spring Harb Protoc* doi: 10.1101/pdb.prot080325.

DeVorkin L, Gorski SM. 2014b. Monitoring autophagic flux using Ref(2)P, the *Drosophila* p62 ortholog. *Cold Spring Harb Protoc* doi: 10.1101/pdb.prot080333.

DeVorkin L, Gorski SM. 2014c. Monitoring autophagy in *Drosophila* using fluorescent reporters in the UAS-GAL4 system. *Cold Spring Harb Protoc* doi: 10.1101/pdb.prot080341.

Duffy JB, Harrison DA, Perrimon N. 1998. Identifying loci required for follicular patterning using directed mosaics. *Development* 125: 2263–2271.

Golic KG. 1991. Site-specific recombination between homologous chromosomes in *Drosophila*. *Science* 252: 958–961.

Handler D, Olivieri D, Novatchkova M, Gruber FS, Meixner K, Mechtler K, Stark A, Sachidanandam R, Brennecke J. 2011. A systematic analysis of *Drosophila* TUDOR domain-containing proteins identifies Vreteno and the Tdrd12 family as essential primary piRNA pathway factors. *EMBO J* 30: 3977–3993.

Juhasz G, Neufeld TP. 2008. Experimental control and characterization of autophagy in *Drosophila*. In *Autophagosome and phagosome* (ed. Deretic V), Vol. 445, pp. 125–133, Humana Press. Totowa, NJ. http://www.springerprotocols.com/Abstract/doi/10.1007/978-1-59745-157-4_8 (Accessed May 23, 2012).

Lee T, Luo L. 2001. Mosaic analysis with a repressible cell marker (MARCM) for *Drosophila* neural development. *Trends Neurosci* 24: 251–254.

Lewis EB. 1960. A new standard food medium. *Drosophila Information Service* 34: 117–118.

Mazzalupo S, Cooley L. 2006. Illuminating the role of caspases during *Drosophila* oogenesis. *Cell Death Differ* 13: 1950–1959.

Neufeld TP. 2008. Genetic manipulation and monitoring of autophagy in *Drosophila*. *Meth Enzymol* 451: 653–667.

Ni J-Q, Zhou R, Czech B, Liu L-P, Holderbaum L, Yang-Zhou D, Shim H-S, Tao R, Handler D, Karpowicz P, et al. 2011. A genome-scale shRNA resource for transgenic RNAi in *Drosophila*. *Nat Methods* 8: 405–407.

Ravi Ram K, Wolfner MF. 2007. Seminal influences: *Drosophila* Acps and the molecular interplay between males and females during reproduction. *Integr Comp Biol* 47: 427–445.

Selva EM, Stronach BE. 2007. Germline clone analysis for maternally acting *Drosophila* hedgehog components. In *Hedgehog signaling protocols* (ed. Horabin JI), Vol. 397, pp 129–144. Humana Press, Totowa, NJ. http://www.springerprotocols.com/Abstract/doi/10.1007/978-1-59745-516-9_11 (Accessed October 26, 2012).

Southall TD, Elliott DA, Brand AH. 2008. The GAL4 system: A versatile toolkit for gene expression in *Drosophila*. *Cold Spring Harb Protoc* doi: 10.1101/pdb.top49.

Shrestha BR, Grueber WB. 2011. Generation and staining of MARCM clones in *Drosophila*. *Cold Spring Harb Protoc* 2011: pdb.prot5659.

Wang SH, Elgin SCR. 2011. *Drosophila* Piwi functions downstream of piRNA production mediating a chromatin-based transposon silencing mechanism in female germ line. *Proc Natl Acad Sci* 108: 21164–21169.

Cite this protocol as *Cold Spring Harb Protoc*; doi:10.1101/pdb.prot080358

Analysis of Apoptosis in *Caenorhabditis elegans*

Benjamin Lant[1] and W. Brent Derry[2,3]

[1]*Developmental and Stem Cell Biology Program, The Hospital for Sick Children, Toronto, Ontario M5G 1L7, Canada;* [2]*Department of Molecular Genetics, University of Toronto, Toronto, Ontario M5S 1A1, Canada*

The nematode worm *Caenorhabditis elegans* has provided researchers with a wealth of information on the molecular mechanisms controlling programmed cell death (apoptosis). Its genetic tractability, optical clarity, and relatively short lifespan are key advantages for rapid assessment of apoptosis in vivo. The use of forward and reverse genetics methodology, coupled with in vivo imaging, has provided deep insights into how a multicellular organism orchestrates the self-destruction of specific cells during development and in response to exogenous stresses. Strains of *C. elegans* carrying mutations in the core elements of the apoptotic pathway, or in tissue-specific regulators of apoptosis, can be used for genetic analyses to reveal conserved mechanisms by which apoptosis is regulated in the somatic and reproductive (germline) tissue. Here we present an introduction to the study of apoptosis in *C. elegans*, including current techniques for visualization, analysis, and screening.

BACKGROUND

Somatic Apoptosis

During embryonic development, 113 cells die by apoptosis in the *Caenorhabditis elegans* hermaphrodite, and an additional 18 cells undergo apoptosis during the larval stages (Sulston and Horvitz 1981; Sulston et al. 1983). Apoptotic corpses persist for ~40 min and are easily distinguished from healthy cells by their refractile "button-like" appearance under differential interference contrast (DIC) optics—the gold standard for quantification of apoptosis in *C. elegans*.

Forward genetics screens have identified four genes that comprise the core apoptosis pathway in *C. elegans* (Fig. 1). This pathway is required for the execution of almost all cells fated to die during development and in response to stress (Ellis and Horvitz 1986; Hengartner et al. 1992; Yuan and Horvitz 1992; Yuan et al. 1993; Conradt and Horvitz 1998; Gumienny et al. 1999; Gartner et al. 2000; Horvitz 2003). The cell death abnormal gene 3, *ced-3*, encodes a proteolytic caspase protein that is activated by the protein product of the *ced-4* gene, the worm homolog of mammalian apoptotic protease activation factor 1 (Apaf-1) (Yuan and Horvitz 1992; Yuan et al. 1993). The *ced-9* gene encodes a protein that is homologous to the anti-apoptotic B-cell lymphoma 2 (Bcl-2) family of proteins. Ablation of *ced-9* results in massive levels of ectopic apoptosis that are completely suppressed by loss-of-function mutations in *ced-4* or *ced-3*, which was the first genetic evidence that *ced-9*/ Bcl-2 functions upstream of *ced-4*/Apaf-1 to prevent activation of the CED-3 caspase (Hengartner et al. 1992; Hengartner and Horvitz 1994b). Upstream of *ced-9* is the egg laying defective 1 (*egl-1*) gene, which encodes a proapoptotic BH3-only protein that antagonizes the CED-9 protein (Conradt and Horvitz 1998).

---

[3]Correspondence: brent.derry@sickkids.ca

Cite this introduction as *Cold Spring Harb Protoc*; doi:10.1101/pdb.top070458

BH3-only	Bcl2	Apaf1	Caspase

$$egl\text{-}1 \dashv ced\text{-}9 \dashv ced\text{-}4 \rightarrow ced\text{-}3 \rightarrow \text{Apoptosis}$$

FIGURE 1. The core apoptosis pathway of *C. elegans*. Mammalian counterparts are indicated *above* the worm gene names.

The invariant lineage of *C. elegans* has been a major advantage for identifying different regulators of the core apoptosis pathway (see the review by Conradt and Xue [2005]). Activation generally involves increased transcription of *egl-1* by tissue-specific transcription factors. (One notable exception is the male linker cell, which undergoes cell death by a mechanism independent of *egl-1* and the core apoptosis pathway [Abraham et al. 2007; Blum et al. 2012].) Over-expression of apoptosis genes using heat-inducible or lineage-specific promoters can activate (or inhibit) apoptosis in various mutant backgrounds to help establish epistatic relationships (see, for example, Hengartner and Horvitz 1994b; Shaham and Horvitz 1996a,b; Conradt and Horvitz 1998). The ability to perform genetic epistasis experiments with relative ease is one of the key advantages of using *C. elegans* to delineate cell death signaling pathways.

Once committed to apoptosis, the fated cells are recognized by exposed phosphatidylserine (PS) on their surface (the "eat-me" signal), then swiftly engulfed by healthy neighboring cells (Ellis et al. 1991; Grimsley and Ravichandran 2003; Kinchen and Ravichandran 2007). Critical to engulfment is the CED-1 protein, which recognizes corpses following PS exposure; CED-1 can be fused with fluorescent tags and used as an in vivo marker of apoptosis (Zhou et al. 2001; Kinchen et al. 2008). Loss-of-function mutations in engulfment genes such as *ced-1* cause cell corpses to persist, making detection of subtle changes in apoptosis more sensitive when counting corpses in embryos at various stages of development. Because the somatic lineage is invariant, and a number of apoptotic deaths during development occur in the anterior pharynx, it is easy to detect extra "undead" cells in this organ using DIC optics (Fig. 2). (For a comprehensive description of the method for counting cells in the pharynx, we refer readers to the protocol by Schwartz [2007].)

Germline Apoptosis

Physiological Apoptosis

The germline of *C. elegans* is comprised of two symmetric bilateral tubes that contain a population of mitotically proliferating stem cells at their distal ends. As these cells continue to proliferate, they move in a proximal direction and enter the stages of meiosis I. As meiotic cells exit the pachytene stage (around the bends of the gonad arms) they become competent to undergo apoptosis, and it has been estimated that ∼50% of germ cells undergo physiological apoptosis (Gumienny et al. 1999). Similar to developmental cell death in embryos and larvae, apoptotic germ cell corpses display a refractile "button-like" morphology that makes them easy to distinguish by DIC optics (Fig. 3). Mutations in engulfment genes, such as *ced-1*/CD91 or *ced-6*/GULP, cause germ cell corpses to persist because they cannot be engulfed by the surrounding gonadal sheath cells (Gumienny et al. 1999). While the core execution genes *ced-4* and *ced-3* are required for physiological germ cell apoptosis, *egl-1* is dispensable, and a gain-of-function mutation in *ced-9* which suppresses developmental apoptosis in the soma and stress-induced apoptosis in the germline does not affect physiological germ cell apoptosis (Hengartner

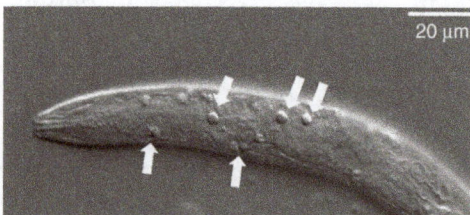

FIGURE 2. Pharyngeal apoptosis in *C. elegans*. The distinct "button-like" morphology is seen in the anterior pharynx at the L1 stage of development. Several corpses arising from developmental apoptosis persist in the head region of animals carrying a mutation in the engulfment gene *ced-5*. Worm viewed by DIC optics using a 63× oil immersion lens.

Cite this introduction as *Cold Spring Harb Protoc*; doi:10.1101/pdb.top070458

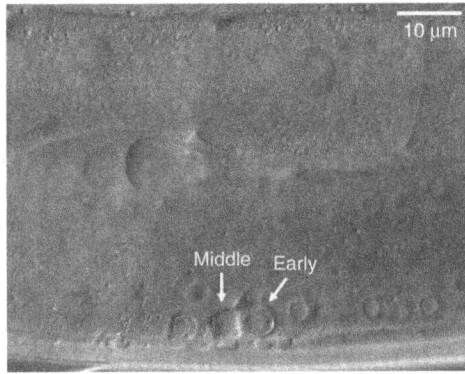

FIGURE 3. Germline apoptosis in *C. elegans*. Corpses appear as "buttons" after they mature and cellularize out of the germline syncytium. Different stages of apoptosis are shown, as the corpse begins to cellularize (early) and form the distinct "button-like" morphology (middle/late) before undergoing engulfment by the surrounding gonadal sheath cells. The process from budding to engulfment takes ~40 min. Germline viewed by DIC optics using a 63× oil immersion lens.

and Horvitz 1994a; Gumienny et al. 1999; Gartner et al. 2000). Thus, there are some interesting differences in how the core cell death pathway regulates apoptosis in the germline versus the soma.

Physiological germ cell apoptosis requires *let-60*/Ras signaling, but because this pathway is also required for germ cells to exit the pachytene stage of meiosis I (where they become competent to undergo apoptosis), it has been unclear whether Ras signaling imparts a direct or indirect effect on germline apoptosis (Gumienny et al. 1999). Recent evidence suggests a more direct role for Ras signaling in promoting germ cell apoptosis, possibly through a combination of *C. elegans* p53-like (*cep-1*)-dependent and -independent mechanisms (Kritikou et al. 2006; Rutkowski et al. 2011; Perrin et al. 2013). Ras pathway activity can be monitored by staining dissected germlines with commercial antibodies to phosphorylated mammalian ERK, which cross-reacts with *C. elegans* phospho-MPK-1 (Arur et al. 2009; Rutkowski et al. 2011; Perrin et al. 2013) (see Protocol 5: Immunostaining for Markers of Apoptosis in the *Caenorhabditis elegans* Germline [Lant and Derry 2014a]). This method provides a reliable readout for assessing Ras activity in the germline, which can be examined in various mutant backgrounds and under different conditions.

Stress-Induced Apoptosis

Germ cells of the *C. elegans* hermaphrodite also undergo apoptosis in response to a variety of environmental stresses, such as DNA damaging agents. (Germ cells of males, however, do not undergo physiological or stress-induced apoptosis.) Stress-induced germ cell apoptosis requires a conserved DNA damage checkpoint and the *cep-1* gene, which is not required for physiological germ cell apoptosis (Gartner et al. 2000; Derry et al. 2001; Schumacher et al. 2001; Stergiou and Hengartner 2004; Gartner et al. 2008). Unlike physiological germ cell apoptosis, stress-induced germ cell apoptosis requires *egl-1,* and a gain-of-function allele in *ced-9* suppresses apoptosis in response to DNA damaging agents (Gartner et al. 2000). In response to DNA damaging agents, such as ionizing radiation (IR), alkylating agents or ultraviolet (UV) irradiation, endogenous CEP-1 protein becomes phosphorylated and activates the transcription of *egl-1* (Hofmann et al. 2002; Quevedo et al. 2007; Gao et al. 2008). CEP-1 activity is negatively regulated by the Akt/PKB kinase AKT-1, the GLD-1 RNA-binding protein, and the E3 ubiquitin ligase SCF^{FSN-1}, which is evident by elevated levels of phosphorylated CEP-1 and *egl-1* transcript in strains carrying loss-of-function mutations in these genes after treatment with DNA damaging agents (Schumacher et al. 2005; Quevedo et al. 2007; Gao et al. 2008). The human equivalent of the SCF^{FSN-1} E3 ubiquitin ligase complex, SCF^{FBXO45}, negatively regulates the p53 family member p73, revealing conservation from worm to human (Peschiaroli et al. 2009). The relative level and subcellular localization of endogenous CEP-1 (as well as core apoptosis proteins) can be monitored by staining dissected germlines with antibodies (see Fig. 1 in Protocol 5: Immunostaining for Markers of Apoptosis in the *Caenorhabditis elegans* Germline [Lant and Derry 2014a]), and CEP-1 phosphorylation status can be analyzed by western blotting (Quevedo et al. 2007; Gao et al. 2008; Sendoel et al. 2010). For an accurate and reproducible estimate of CEP-1 transcriptional activity, *egl-1* transcript levels can be measured by quantitative real-time PCR (Hofmann et al. 2002).

GENETIC ANALYSIS OF APOPTOSIS IN *C. ELEGANS*

C. elegans is an ideal organism for analyzing the genetic control of apoptosis and establishing order in relevant signaling pathways. A vast array of strains is available and encompasses mutations in genes required at all stages of the apoptotic process, including lineage-specific regulators, core pathway components, and corpse engulfment genes. Strains can be ordered by mail from the *Caenorhabditis* Genetics Center (http://www.cbs.umn.edu/CGC/) for a small fee. When analyzing mutants that have abnormally high numbers of refractile corpses, it is important to confirm that these are indeed apoptotic corpses by generating double-mutant strains containing loss-of-function mutations in the core executioner genes *ced-4* or *ced-3*. Because these genes are required for all apoptotic cell death in *C. elegans*, the appearance of abnormal numbers of corpses should be suppressed by *ced-3* and *ced-4* loss-of-function mutations. Likewise, mutants that cause reduced numbers of apoptotic corpses should be crossed into a strain carrying a loss-of-function mutation in the anti-apoptotic gene *ced-9*. If apoptosis is restored by ablation of *ced-9*, then it can be concluded that the core apoptotic machinery is functional in these mutants. Alternatively, ablation of these genes by RNAi can be performed to help establish epistatic relationships in mutants with abnormal numbers of apoptotic corpses (see Protocol 1: Induction of Germline Apoptosis in *Caenorhabditis elegans* [Lant and Derry 2014b]). However, because the penetrance of RNAi can be variable, we recommend that epistatic relationships be confirmed using genetic mutants. Similar to immunostaining for CEP-1 protein, the germlines of mutants with altered levels of apoptosis (physiological or stress-induced) can be stained with antibodies to core apoptosis proteins, such as CED-9 and CED-4, to determine if their levels or localization are altered (Greiss et al. 2008; Perrin et al. 2013; Pourkarimi et al. 2012). For example, using a combination of genetic epistasis and immunocytochemistry, the DAF-2 insulin-like receptor was shown to promote IR-induced germ cell apoptosis by attenuating Ras signaling without detectably affecting the levels or localization of core apoptosis proteins (Perrin et al. 2013). The combination of genetics and cell biology reagents provides a basic toolkit for defining the mechanisms by which various proteins control apoptosis.

Another advantage of *C. elegans* is the availability of tissue-specific promoters to determine the site of focus for apoptosis regulators. For example, physiological and DNA damage-induced germline apoptosis is under regulatory control of both autonomous and nonautonomous signaling molecules (Ito et al. 2010; Sendoel et al. 2010; Li et al. 2012). In addition, *C. elegans* is ideally suited for performing high-throughput reverse genetics screens for apoptosis regulators. Using liquid RNAi cultures in a 96-well plate format, along with the vital dye acridine orange, the entire genome of the nematode can be screened for apoptosis genes with unprecedented efficiency (Lettre et al. 2004) (see Protocol 4: High-Throughput RNAi Screening for Germline Apoptosis Genes in *Caenorhabditis elegans* [Lant and Derry 2014c]).

SUGGESTED TECHNIQUES

We have prepared a series of protocols which cover the basics of assessing apoptosis in *C. elegans*, from basic techniques for visualizing corpses in embryos and the adult germline to the use of vital dyes, fluorescent strains and immunostaining for visualizing the localization and levels of endogenous apoptosis proteins (see Protocol 1: Induction of Germline Apoptosis in *Caenorhabditis elegans* [Lant and Derry 2014b]; Protocol 2: Visualizing Apoptosis in Embryos and the Germline of *Caenorhabditis elegans* [Lant and Derry 2014d]; Protocol 3: Fluorescent Visualization of Germline Apoptosis in Living *Caenorhabditis elegans* (Lant and Derry 2014e); Protocol 4: High-Throughput RNAi Screening for Germline Apoptosis Genes in *Caenorhabditis elegans* [Lant and Derry 2014c]; Protocol 5: Immunostaining for Markers of Apoptosis in the *Caenorhabditis elegans* Germline [Lant and Derry 2014a]). The use of RNAi, either for individual or large groups of genes, enables assessment of apoptotic phenotypes within a few days by simply cultivating worms on bacterial strains expressing double-stranded RNA targeting the gene(s) of interest. Taken as a whole, these protocols allow a

Cite this introduction as *Cold Spring Harb Protoc*; doi:10.1101/pdb.top070458

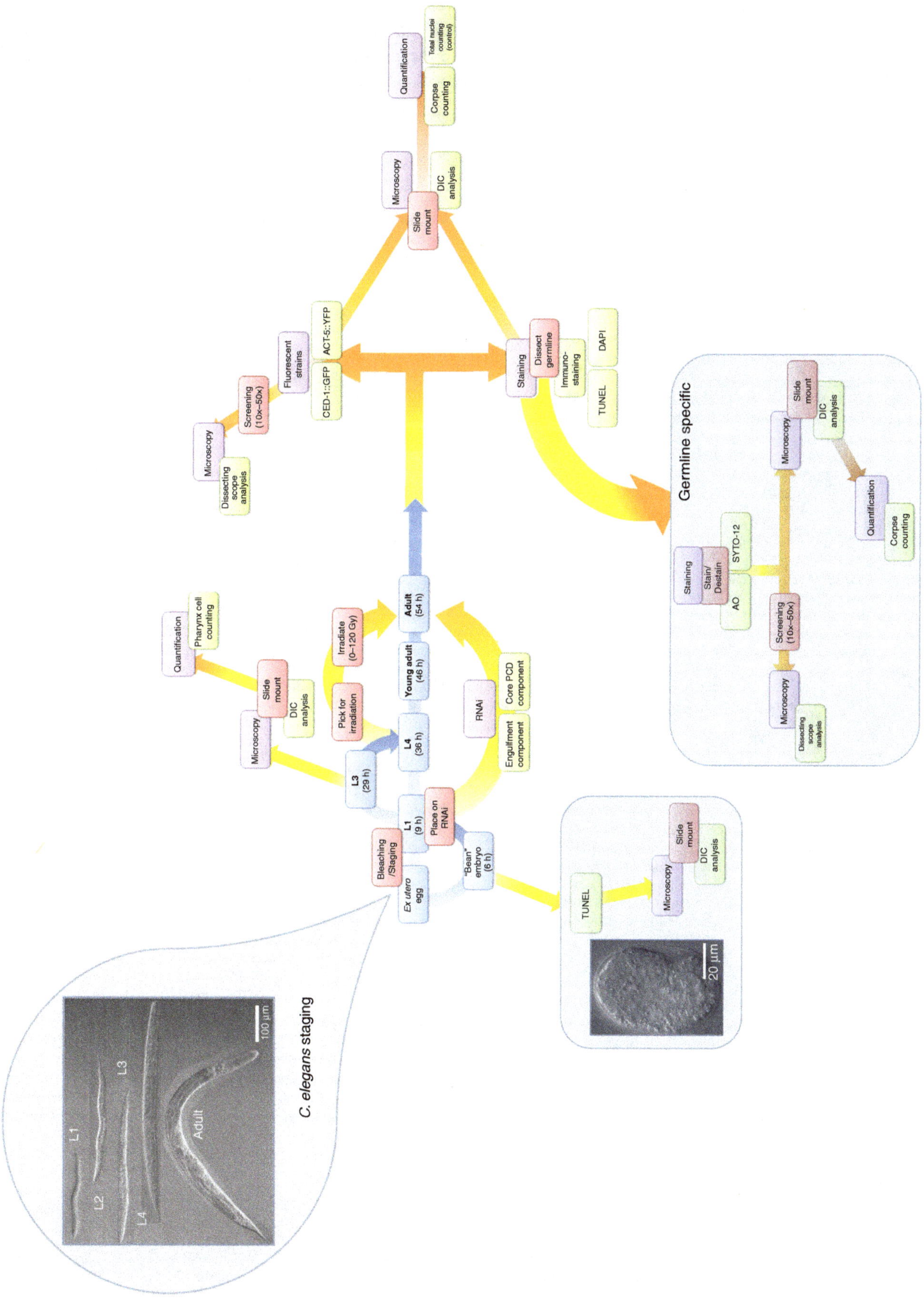

FIGURE 4. An overview of methods used to assess apoptosis in the soma and germline of *C. elegans*. As indicated, some procedures are stage-specific, or only valid in germline analysis.

novice *C. elegans* researcher to quickly develop expertise in assessing apoptosis. They also facilitate methods for deeper analysis, such as high-resolution imaging, genetic epistasis analysis, and genome-wide modifier screens. A flowchart guide to the analysis of apoptosis in *C. elegans* (Fig. 4) places the skills and techniques outlined in our protocols in chronological order according to developmental stages of the worm.

REFERENCES

Abraham MC, Lu Y, Shaham S. 2007. A morphologically conserved non-apoptotic program promotes linker cell death in *Caenorhabditis elegans*. *Dev Cell* 12: 73–86.

Arur S, Ohmachi M, Nayak S, Hayes M, Miranda A, Hay A, Golden A, Schedl T. 2009. Multiple ERK substrates execute single biological processes in *Caenorhabditis elegans* germ-line development. *Proc Natl Acad Sci* 106: 4776–4781.

Blum ES, Abraham MC, Yoshimura S, Lu Y, Shaham S. 2012. Control of nonapoptotic developmental cell death in *Caenorhabditis elegans* by a polyglutamine-repeat protein. *Science* 335: 970–973.

Conradt B, Horvitz HR. 1998. The *C. elegans* protein EGL-1 is required for programmed cell death and interacts with the Bcl-2-like protein CED-9. *Cell* 93: 519–529.

Conradt B, Xue D. Programmed cell death. (October 6, 2005), *WormBook*, ed. The *C. elegans* Research Community, WormBook, doi/10.1895/wormbook.1.32.1, http://www.wormbook.org.

Derry WB, Putzke AP, Rothman JH. 2001. *Caenorhabditis elegans* p53: Role in apoptosis, meiosis, and stress resistance. *Science* 294: 591–595.

Ellis HM, Horvitz HR. 1986. Genetic control of programmed cell death in the nematode *C. elegans*. *Cell* 44: 817–829.

Ellis RE, Jacobson DM, Horvitz HR. 1991. Genes required for the engulfment of cell corpses during programmed cell death in *Caenorhabditis elegans*. *Genetics* 129: 79–94.

Gao MX, Liao EH, Yu B, Wang Y, Zhen M, Derry WB. 2008. The SCF[FSN-1] ubiquitin ligase controls germline apoptosis through CEP-1/p53 in *C. elegans*. *Cell Death Differ* 15: 1054–1062.

Gartner A, Milstein S, Ahmed S, Hodgkin J, Hengartner MO. 2000. A conserved checkpoint pathway mediates DNA damage-induced apoptosis and cell cycle arrest in *C. elegans*. *Mol Cell* 5: 435–443.

Gartner A, Boag PR, Blackwell TK. Germline survival and apoptosis. (September 4, 2008), *WormBook*, ed. The *C. elegans* Research Community, WormBook, doi/10.1895/wormbook.1.145.1, http://www.wormbook.org.

Greiss S, Hall J, Ahmed S, Gartner A. 2008. *C. elegans* SIR-2.1 translocation is linked to a proapoptotic pathway parallel to *cep-1*/p53 during DNA damage-induced apoptosis. *Genes Dev* 22: 2831–2842.

Grimsley C, Ravichandran KS. 2003. Cues for apoptotic cell engulfment: Eat-me, don't eat-me and come-get-me signals. *Trends Cell Biol* 13: 648–656.

Gumienny TL, Lambie E, Hartwieg E, Horvitz HR, Hengartner MO. 1999. Genetic control of programmed cell death in the *Caenorhabditis elegans* hermaphrodite germline. *Development* 126: 1011–1022.

Hengartner MO, Horvitz HR. 1994a. Activation of *C. elegans* cell death protein CED-9 by an amino-acid substitution in a domain conserved in Bcl-2. *Nature* 369: 318–320.

Hengartner MO, Horvitz HR. 1994b. *C elegans* cell survival gene *ced-9* encodes a functional homolog of the mammalian proto-oncogene *bcl-2*. *Cell* 76: 665–676.

Hengartner MO, Ellis RE, Horvitz HR. 1992. *Caenorhabditis elegans* gene *ced-9* protects cells from programmed cell death. *Nature* 356: 494–499.

Hofmann ER, Milstein S, Boulton SJ, Ye M, Hofmann JJ, Stergiou L, Gartner A, Vidal M, Hengartner MO. 2002. *Caenorhabditis elegans* HUS-1 is a DNA damage checkpoint protein required for genome stability and EGL-1-mediated Apoptosis. *Curr Biol* 12: 1908–1918.

Horvitz HR. 2003. Nobel lecture. Worms, life and death. *Biosci Rep* 23: 239–303.

Ito S, Greiss S, Gartner A, Derry WB. 2010. Cell-nonautonomous regulation of *C. elegans* germ cell death by *kri-1*. *Curr Biol* 20: 333–338.

Kinchen JM, Ravichandran KS. 2007. Journey to the grave: Signaling events regulating removal of apoptotic cells. *J Cell Sci* 120: 2143–2149.

Kinchen JM, Doukoumetzidis K, Almendinger J, Stergiou L, Tosello-Trampont A, Sifri CD, Hengartner MO, Ravichandran KS. 2008. A pathway for phagosome maturation during engulfment of apoptotic cells. *Nat Cell Biol* 10: 556–566.

Kritikou EA, Milstein S, Vidalain PO, Lettre G, Bogan E, Doukoumetzidis K, Gray P, Chappell TG, Vidal M, Hengartner MO. 2006. *C. elegans* GLA-3 is a novel component of the MAP kinase MPK-1 signaling pathway required for germ cell survival. *Genes Dev* 20: 2279–2292.

Lant B, Derry WB. 2014a. Immunostaining for markers of apoptosis in the *Caenorhabditis elegans* germline. *Cold Spring Harb Protoc* doi: 10.1101/pdb.prot080242.

Lant B, Derry WB. 2014b. Induction of germline apoptosis in *Caenorhabditis elegans*. *Cold Spring Harb Protoc* doi: 10.1101/pdb. prot080192.

Lant B, Derry WB. 2014c. High-throughput RNAi screening for germline apoptosis genes in *Caenorhabditis elegans*. *Cold Spring Harb Protoc* doi: 10.1101/pdb.prot080234.

Lant B, Derry WB. 2014d. Visualizing apoptosis in embryos and the germline of *Caenorhabditis elegans*. *Cold Spring Harb Protoc* doi: 10.1101/pdb.prot080218.

Lant B, Derry WB. 2014e. Fluorescent visualization of germline apoptosis in living *Caenorhabditis elegans*. *Cold Spring Harb Protoc* doi: 10.1101/pdb. prot080226.

Lettre G, Kritikou EA, Jaeggi M, Calixto A, Fraser AG, Kamath RS, Ahringer J, Hengartner MO. 2004. Genome-wide RNAi identifies p53-dependent and -independent regulators of germ cell apoptosis in *C. elegans*. *Cell Death Differ* 11: 1198–1203.

Li X, Johnson RW, Park D, Chin-Sang I, Chamberlin HM. 2012. Somatic gonad sheath cells and Eph receptor signaling promote germ-cell death in *C. elegans*. *Cell Death Differ* 19: 1080–1089.

Perrin AJ, Gunda M, Yu B, Yen K, Ito S, Forster S, Tissenbaum HA, Derry WB. 2013. Noncanonical control of *C. elegans* germline apoptosis by the insulin/IGF-1 and Ras/MAPK signaling pathways. *Cell Death Differ* 20: 97–107.

Peschiaroli A, Scialpi F, Bernassola F, Pagano M, Melino G. 2009. The F-box protein FBXO45 promotes the proteasome-dependent degradation of p73. *Oncogene* 28: 3157–3166.

Pourkarimi E, Greiss S, Gartner A. 2012. Evidence that CED-9/Bcl2 and CED-4/Apaf-1 localization is not consistent with the current model for *C. elegans* apoptosis induction. *Cell Death Differ* 19: 406–415.

Quevedo C, Kaplan DR, Derry WB. 2007. AKT-1 regulates DNA-damage-induced germline apoptosis in *C. elegans*. *Curr Biol* 17: 286–292.

Rutkowski R, Dickinson R, Stewart G, Craig A, Schimpl M, Keyse SM, Gartner A. 2011. Regulation of *Caenorhabditis elegans* p53/CEP-1-dependent germ cell apoptosis by Ras/MAPK signaling. *PLoS Genet* 7: e1002238.

Schumacher B, Hofmann K, Boulton S, Gartner A. 2001. The *C. elegans* homolog of the p53 tumor suppressor is required for DNA damage-induced apoptosis. *Curr Biol* 11: 1722–1727.

Schumacher B, Hanazawa M, Lee MH, Nayak S, Volkmann K, Hofmann R, Hengartner M, Schedl T, Gartner A. 2005. Translational repression of *C. elegans* p53 by GLD-1 regulates DNA damage-induced apoptosis. *Cell* 120: 357–368.

Schwartz HT. 2007. A protocol describing pharynx counts and a review of other assays of apoptotic cell death in the nematode worm *Caenorhabditis elegans*. *Nat Protoc* 2: 705–714.

Sendoel A, Kohler I, Fellmann C, Lowe SW, Hengartner MO. 2010. HIF-1 antagonizes p53-mediated apoptosis through a secreted neuronal tyrosinase. *Nature* 465: 577–583.

Shaham S, Horvitz HR. 1996a. An alternatively spliced *C. elegans ced-4* RNA encodes a novel cell death inhibitor. *Cell* 86: 201–208.

Shaham S, Horvitz HR. 1996b. Developing *Caenorhabditis elegans* neurons may contain both cell-death protective and killer activities. *Genes Dev* **10:** 578–591.

Stergiou L, Hengartner MO. 2004. Death and more: DNA damage response pathways in the nematode *C. elegans. Cell Death Differ* **11:** 21–28.

Sulston JE, Horvitz HR. 1981. Abnormal cell lineages in mutants of the nematode *Caenorhabditis elegans. Dev Biol* **82:** 41–55.

Sulston JE, Schierenberg E, White JG, Thomson JN. 1983. The embryonic cell lineage of the nematode *Caenorhabditis elegans. Dev Biol* **100:** 64–119.

Yuan J, Horvitz HR. 1992. The *Caenorhabditis elegans* cell death gene *ced-4* encodes a novel protein and is expressed during the period of extensive programmed cell death. *Development* **116:** 309–320.

Yuan J, Shaham S, Ledoux S, Ellis HM, Horvitz HR. 1993. The *C. elegans* cell death gene *ced-3* encodes a protein similar to mammalian interleukin-1 beta-converting enzyme. *Cell* **75:** 641–652.

Zhou Z, Hartwieg E, Horvitz HR. 2001. CED-1 is a transmembrane receptor that mediates cell corpse engulfment in *C. elegans. Cell* **104:** 43–56.

Induction of Germline Apoptosis in *Caenorhabditis elegans*

Benjamin Lant[1] and W. Brent Derry[2,3]

[1]*Developmental and Stem Cell Biology Program, The Hospital for Sick Children, Toronto, Ontario M5G 1L7,*
Canada; [2]*Department of Molecular Genetics, University of Toronto, Toronto, Ontario M5S 1A1, Canada*

RNA interference (RNAi) is an incredibly powerful tool for rapid and efficient knockdown of gene
expression. This technology can be used to induce apoptosis in the germline of *Caenorhabditis elegans*.
Genotoxic stressors such as ionizing radiation (IR), ultraviolet light, chemical mutagens (e.g., *N*-ethyl-
N-nitrosourea [ENU]), and DNA cross-linking reagents can also be used to stimulate apoptosis. These
approaches, described here, combined with the powers of in vivo imaging methods, should keep *C.
elegans* apoptosis researchers busy for several years, sorting out how various signaling pathways influence life and death decisions in this organism.

MATERIALS

It is essential that you consult the appropriate Material Safety Data Sheets and your institution's Environmental
Health and Safety Office for proper handling of equipment and hazardous material used in this protocol.

RECIPES: Please see the end of this protocol for recipes indicated by <R>. Additional recipes can be found online at
http://cshprotocols.cshlp.org/site/recipes.

Reagents

Caenorhabditis elegans strain of interest (e.g., *lin-35[n745]*, *rrf-1[pk1417]*, or *rrf-3[pk1426]*)
*L1 larval stage worms are typically used for the RNAi procedure (see Step 5), whereas L4 worms are used for IR or
ENU treatments (see Steps 6 and 8). The hypochlorite method is a standard technique that is used to synchronize
worms to the L1 larval stage, as well as to remove contamination. This procedure, along with many other basic
worm handling techniques, can be found in Wormbook (Stiernagle).*

ENU (Sigma-Aldrich N3385)
Freshly dilute the ENU stock solution to 5 mM in M9 buffer <R> each time it is used.

HT115 bacteria (see Step 1)
Isopropylthio-β-galactoside (IPTG) (0.1 M)
Luria Bertani (LB) + amp and tet <R>
Prepare both liquid medium and plates containing solid medium.

OP50 bacteria (optional; see Steps 6–9)
Plates containing solid nematode growth medium (NGM) <R>
Plates containing solid RNAi medium <R>
RNAi clones from the *C. elegans* RNAi feeding library (Source BioScience LifeSciences; see Discussion)

[3]Correspondence: brent.derry@sickkids.ca

Cite this protocol as *Cold Spring Harb Protoc*; doi:10.1101/pdb.prot080192

Bacterial RNAi lawns plated on NGM plates remain useful for experiments when stored at 4°C for ~1 mo. However, in our experience, some RNAi bacterial strains do not last this long when stored in the cold. If you plan to use a particular RNAi feeding strain repeatedly, we recommend preparing your own glycerol stock and storing it at −80°C to avoid repeated access of the stock library.

Equipment

Gamma (^{137}Cs source), X, or UV irradiator

Imaging setup (see Protocol 2: Visualizing Apoptosis in Embryos and the Germline of *Caenorhabditis elegans* [Lant and Derry 2014])

For high-resolution imaging and quantification of apoptotic bodies, we recommend a compound microscope that is equipped with differential interference contrast (DIC) optics (63× oil immersion lens).

Incubators at 20°C and 37°C

Microcentrifuge tubes (0.5 mL)

Orbital shaker at 37°C

P200 tips (sterile)

Tubes (15-mL conical)

METHOD

Examining Apoptosis Using RNAi

1. Select a targeted RNAi clone of interest from the *C. elegans* RNAi feeding library. Using a sterile P200 tip, streak from the glycerol stock onto a plate containing LB + amp and tet. Incubate the plates overnight at 37°C.

   Streaking the plate in a cross hatch pattern is optimal for creating single colonies.

   A negative control consisting of an empty RNAi plasmid (or a gene with no strong RNAi phenotype) in the HT115 bacterial strain should be cultured concurrently.

2. On the next day, harvest individual colonies with a sterile P200 tip and transfer to 15-mL conical tubes containing 4 mL of liquid LB + amp and tet. Grow overnight on an orbital shaker at 37°C.

   Liquid RNAi cultures can be stored at 4°C for up to 2 wk for repeated use.

   See Troubleshooting.

3. On the next day, add 20 µL of 0.1 M IPTG to each culture and incubate an additional 4 h on the shaker at 37°C.

   IPTG induces T7 RNA polymerase in the RNAi bacteria.

4. Spread 70 µL of each bacterial culture on plates containing solid RNAi medium, and allow the bacteria to grow overnight at room temperature.

   Ensure that lids are kept on the plates during the overnight growth period to prevent contamination. See Troubleshooting.

5. Add L1-stage worms to the bacterial lawn and allow them to grow until they are at the correct stage to observe apoptosis phenotypes (either young adult for germline apoptosis or after they have begun to lay eggs in the next generation for embryonic apoptosis). Proceed to Step 10.

   Typically worms are placed onto RNAi-expressing bacterial lawns at the L1 stage, but if the targeted gene is essential, plating at the L3 or L4 stage may be necessary to observe a strong loss-of-function phenotype without affecting viability of the organism.

   See Troubleshooting.

Examining Apoptosis with Genotoxic Agents

IR is an efficient genotoxic stressor and causes no obvious morphological changes to somatic tissues. It activates the DNA damage response, thereby stimulating germline apoptosis. As an alternative, alkylating agents such as ENU can be used.

Inducing Apoptosis Using Irradiation

6. One day before irradiation, pick 25 to 40 worms from a plate at the L4 stage. Place the worms on a new NGM plate (seeded with OP50 bacteria if using a mutant strain or the desired RNAi bacterial strain if assessing the apoptosis phenotype of a particular gene after knockdown). Allow the plate to incubate overnight at 20°C.

    See Troubleshooting.

7. On the next day, expose the worms (now at the young adult stage) to a dose of IR (0–120 Gy). Allow them to recover overnight, and proceed to Step 10.

    If there is uncertainty as to what duration of radiation will be most effective to institute a quantifiable effect, the standard starting point should be 60 Gy. If the strain is suspected to have apoptotic resistance, 120 Gy should be applied initially. The irradiated strain should be run alongside equally staged controls (N2) to compare the effects of the treatment.

Inducing Apoptosis Using ENU

8. Pipette 1 μL of 5 mM ENU into the lid of a 0.5-mL microcentrifuge tube anchored on the surface of a small, seeded NGM plate (i.e., in the bacteria). Add 20 worms at the L4 or young adult stage to the ENU solution and incubate in the dark for 4 h at 20°C, with the plate inverted to prevent evaporation.

    Because of the surface tension, using this small volume will prevent the dissolved ENU and worms from falling out of the lid when inverted. Just be careful not to bump the plate too hard when moving.

9. Following incubation, carefully pipette the worms onto a new NGM plate (seeded with OP50 bacteria or target RNAi) and allow them to recover from the ENU treatment. On the next day, proceed to Step 10.

Analyzing Apoptosis

10. Observe and quantify corpses using DIC microscopy with a 63× oil immersion lens.

    Quantification of apoptosis and morphological analysis is described in Protocol 2: Visualizing Apoptosis in Embryos and the Germline of Caenorhabditis elegans *(Lant and Derry 2014). It is imperative to perform at least three independent biological replicates when quantifying germline apoptosis, as this tissue shows more variability compared with developmentally programmed cell death in the invariant somatic lineages. For initial characterization of putative apoptosis-defective mutants, we recommend performing a time course analysis (e.g., 6–48 h post-IR) to define the kinetics of corpse formation.*

TROUBLESHOOTING

Problem (Step 2): The bacterial streak is too thick to pick individual colonies.
Solution: This is the result of too many bacteria on the pipette tip or incubating for too long, causing colony overgrowth. Restreak from the stock and incubate for less time (18 h). Make sure the bacterial culture is not grown for longer than 24 h.

Problem (Step 4): Contamination is present on the bacterial plates.
Solution: Make sure the plate lids are closed to prevent foreign particles from contaminating the bacterial plate. If the glycerol stock is contaminated, pick several individual colonies, verify the presence of the plasmid containing the desired RNAi clone by sequencing or restriction mapping, and grow a new liquid culture. It is also possible that the stock solutions used to prepare the plates or liquid RNAi cultures are contaminated. This can be tested by plating solutions on bacterial growth plates to see if contaminating bacteria grow after 24 h at 37°C. Finally, the worms plated may contain contaminating bacteria. Some worm strains have a greater capacity for contamination (because of storage of bacteria in their gut), and as such the bleaching protocol (Stiernagle) can be modified by adding an extra two bleach wash steps followed by three sterile H_2O washes.

Cite this protocol as *Cold Spring Harb Protoc*; doi:10.1101/pdb.prot080192

Problem (Step 5): RNAi does not provide the expected phenotype, if a phenotype is expected.

Solution: The particular RNAi may not have full penetrance and may require a second generation to show conserved effects. Alternatively, the single colony picked to produce the liquid culture may not have been effective; sometimes, the HT115 strain will mutate plasmids. Pick multiple colonies (making sure to stick to the "lines" of inoculation) and check individually isolated clones for the desired phenotype. However, it may be necessary to reclone the worm gene fragment into a fresh L4440 vector and retransform HT115 bacteria. This has worked well for the *ced-9* RNAi bacteria strain, which we have observed to frequently become ineffective for gene knockdown by the feeding method.

Problem (Step 5): RNAi does not provide a strong phenotype.

Solution: As mentioned above, cultivating the worms on the target RNAi for a further generation may increase the consistency of any phenotypes seen. Alternatively, if the knockdown phenotype is not penetrant we recommend using strains carrying mutations in genes that cause hypersensitivity to RNAi, such as *rrf-3(pk1426)* or *lin-35(n745)* (Simmer et al. 2002; Lehner et al. 2006).

Problem (Step 5): RNAi is too toxic or causes morphological defects that make it difficult to observe apoptosis phenotypes.

Solution: Strains that are resistant to RNAi in the soma, such as the *rrf-1(pk1417)* mutant (Sijen et al. 2001), can be used to selectively ablate genes in the germline. These strains should be used when ablation in the soma causes morphological defects that make assessment of the germline difficult. However, caution should be exercised when using *rrf-1* alleles, as it has been recently reported that these mutants are not resistant to RNAi in all somatic tissues (Kumsta and Hansen 2012).

Problem (Step 6): The L4 stage has been missed, and there is an imperative to irradiate as soon as possible.

Solution: Worms can be irradiated on the same day, provided that they are distinctly at the young adult stage. Hence, the operator must be able to distinguish the stages of adulthood efficiently. Apoptotic nuclei in the germline are difficult to see as the worm ages because the intestine accumulates fat that can occlude the bends of the gonad arms, where apoptosis is detectable.

DISCUSSION

RNAi is not always 100% efficacious, as it causes a reduction in gene function that can be variable in penetrance—compared to the complete loss of function in null mutants or consistent reduction in function of hypomorphic alleles. The main advantages of using RNAi are the incredible ease and speed with which the function of a particular gene can be ascertained. The results from RNAi gene ablation, while not always as penetrant and consistent as those obtained using mutant strains, can be used to systematically screen multiple gene targets concurrently and rapidly. But care must be taken in both culturing RNAi clones and assessing phenotypes. We recommend confirming the identity of RNAi clones by sequencing or restriction mapping and using mutant strains to confirm RNAi phenotypes and epistatic interactions with other genes. Observing a phenotype by RNAi may not be effective or reproducible in the first generation, depending on the nature of the gene or the RNAi clone. This will require troubleshooting, either of the developmental stage when the worms are placed on the RNAi feeding strains or of the number of generations on RNAi necessary to obtain a penetrant phenotype.

The Source BioScience LifeSciences RNAi feeding library provides an extensive array of RNAi clones stored as frozen glycerol stocks in a 384-well plate format. This library represents over 86% of the predicted genes in the *C. elegans* genome, including almost all known apoptosis genes. For control experiments, RNAi ablation of core apoptosis pathway genes (Fig. 1) can be used to ensure that assays

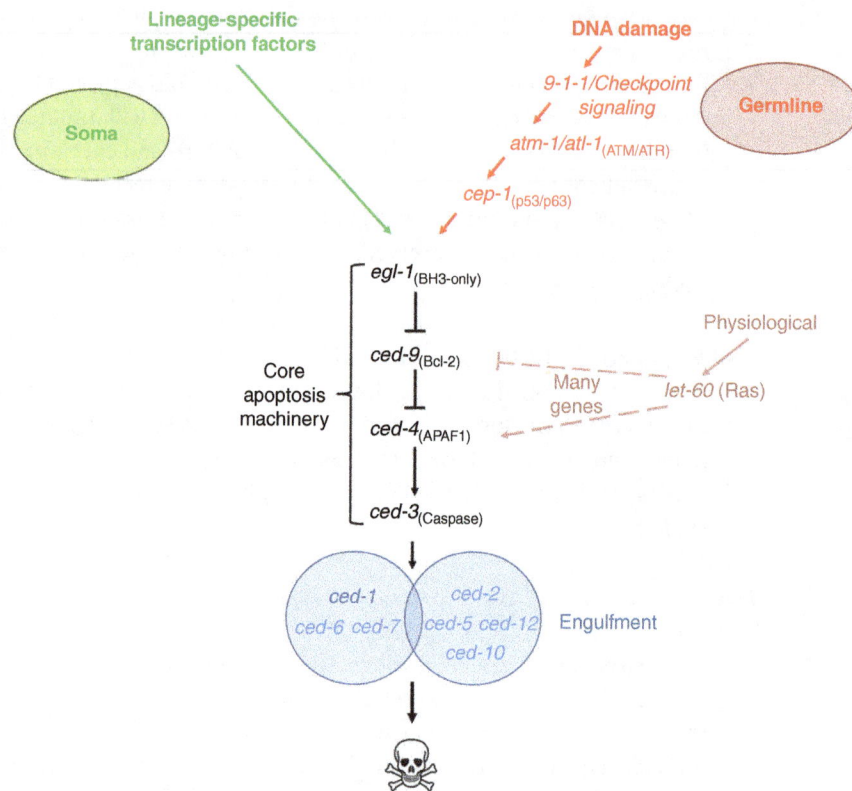

FIGURE 1. Apoptosis signaling in *C. elegans*. Somatic apoptosis uses the core machinery and is regulated by controlling the levels of *egl-1* through lineage-specific transcription factors. Germline apoptosis can be broken into two categories: physiological and stress-induced. Stress-induced apoptosis (indicated as DNA damage) requires the activities of both *cep-1* and *egl-1*, whereas physiological apoptosis is controlled by independent pathways that activate the core apoptotic proteins CED-4 and CED-3 (Gumienny et al. 1999). Following core apoptosis machinery activation and culminating in the activation of the CED-3 executioner caspase, DNA degradation and engulfment machinery is also activated (full components not shown), resulting in the complete death and removal of the cell.

and reagents are working properly. In addition, targeting known apoptosis genes that stimulate or repress apoptosis when knocked down is useful in establishing epistatic order for new apoptosis genes for which mutant strains are available. Whole-genome profiling studies are revealing previously undetected variations in the genomes of various *C. elegans* strains, many of which are predicted to disrupt gene function (Flibotte et al. 2010). Thus, it is also important to confirm that the apoptotic phenotype of a given mutant strain is actually caused by the suspected gene by either rescuing the phenotype with a transgene or validating with different mutant alleles. While rescue of a mutant phenotype by expression of a wild-type transgene is the desirable method to confirm that genotype corresponds to phenotype, it is technically challenging to express genes in the *C. elegans* germline. As an alternative, the RNAi feeding method can be used to confirm that ablation of a suspect gene in wild-type animals can recapitulate the apoptosis phenotype. Whenever possible, apoptosis phenotypes detected using RNAi should be confirmed using strains that contain loss-of-function alleles to the corresponding gene.

Stimulation of apoptosis with genotoxic stress can be achieved through a number of strategies. The source of toxicity should be selected appropriately. We outlined IR treatment here, as it is a common activator of the DNA damage response in the *C. elegans* germline and requires much less optimization than a drug dissolved in buffer, such as ENU. However, RNAi to the *rad-51* gene causes double-strand breaks in DNA that can be used as an alternative to irradiation to activate the *cep-1*-dependent germline apoptosis signaling pathway (Lettre et al. 2004). By stimulating germline apoptosis via irradiation, the effects of a mutation (or gene ablation by RNAi) are exaggerated in proportion with the radiation dose. Of course, using an appropriate control for comparison is imperative, and

Cite this protocol as *Cold Spring Harb Protoc*; doi:10.1101/pdb.prot080192

provides verifiable and consistent baselines for stress-induced apoptosis (for both qualitative and quantitative metrics).

RECIPES

Luria Bertani (LB) + Amp and Tet

1. To prepare solid medium, mix 10 g of tryptone, 5 g of yeast extract, 10 g of NaCl, and 15 g of agar, and bring to 1 L with H_2O. (To prepare liquid medium, combine all ingredients except agar.)
2. Autoclave for 30 min.
3. Let cool for 1 h in a 55°C water bath.
4. Add 2 mL of 50 mg/mL ampicillin and 2 mL of 5 mg/mL tetracycline.
5. While the agar-containing medium is still warm, pour a thin layer (~0.5 cm) into each Petri dish and let cool to solidify.
6. Store dishes or liquid medium at 4°C for up to 1 mo.

M9 Buffer for Worms

1. Dissolve 3 g of KH_2PO_4, 6 g of Na_2HPO_4, and 5 g of NaCl in 1 L of H_2O.
2. Autoclave for 20 min.
3. Add 1 mL of 1 M $MgSO_4$.
4. Store at room temperature. After 2 wk, check for visible contamination before use.

Nematode Growth Medium (NGM)

1. For solid NGM, mix 3 g of NaCl, 2.5 g of peptone, and 20 g of agar and bring to 1 L with H_2O. (For liquid NGM, prepare without agar.)
2. Autoclave for 1 h.
3. Let cool for 1 h in a 55°C water bath.
4. Add 1 mL of cholesterol (5 mg/mL in ethanol), 1 mL of 1 M $CaCl_2$, 1 mL of 1 M $MgSO_4$, and 25 mL of 1 M (pH 6.0) KPO_4, mixing after each addition.
5. Using an automated plate pourer (peristaltic pump), pour the medium into sterile plastic plates to set. During pouring, keep the NGM on a hot plate with stirrer to prevent the medium from solidifying.
6. Store NGM plates at 4°C until use. Seed NGM plates with bacteria the day after pouring. Warm the plates to room temperature before adding the worms. (For liquid NGM, store at room temperature for up to 1 mo. Regularly check for excess cloudiness to ensure bacterial contamination is absent; this is particularly important for liquid RNAi cultures.)

RNAi Medium

1. Prepare 1 L of nematode growth medium (NGM) by following Steps 1–4 of the NGM recipe. <R>
2. Add 2.5 mL of 0.1 M isopropylthio-β-galactoside (IPTG) and 500 μL of 50 mg/mL carbenicillin.
3. Using an automated plate pourer (peristaltic pump), pour the medium into sterile plastic plates to set. During pouring, keep the medium on a hot plate with stirrer to prevent the medium from solidifying.
4. Store RNAi plates for up to 1 mo at 4°C.

REFERENCES

Flibotte S, Edgley ML, Chaudhry I, Taylor J, Neil SE, Rogula A, Zapf R, Hirst M, Butterfield Y, Jones SJ, et al. 2010. Whole-genome profiling of mutagenesis in *Caenorhabditis elegans*. *Genetics* **185**: 431–441.

Gumienny TL, Lambie E, Hartwieg E, Horvitz HR, Hengartner MO. 1999. Genetic control of programmed cell death in the *Caenorhabditis elegans* hermaphrodite germline. *Development* **126**: 1011–1022.

Kumsta C, Hansen M. 2012. *C elegans rrf-1* mutations maintain RNAi efficiency in the soma in addition to the germline. *PloS ONE* **7**: e35428.

Lant B, Derry WB. 2014. Visualizing apoptosis in embryos and the germline of *Caenorhabditis elegans*. *Cold Spring Harb Protoc* doi: 10.1101/pdb.prot080218.

Lehner B, Calixto A, Crombie C, Tischler J, Fortunato A, Chalfie M, Fraser AG. 2006. Loss of LIN-35, the *Caenorhabditis elegans* ortholog of the tumor suppressor p105Rb, results in enhanced RNA interference. *Genome Biol* **7**: R4.

Lettre G, Kritikou EA, Jaeggi M, Calixto A, Fraser AG, Kamath RS, Ahringer J, Hengartner MO. 2004. Genome-wide RNAi identifies p53-dependent and -independent regulators of germ cell apoptosis in *C. elegans*. *Cell Death Differ* **11**: 1198–1203.

Sijen T, Fleenor J, Simmer F, Thijssen KL, Parrish S, Timmons L, Plasterk RH, Fire A. 2001. On the role of RNA amplification in dsRNA-triggered gene silencing. *Cell* **107**: 465–476.

Simmer F, Tijsterman M, Parrish S, Koushika SP, Nonet ML, Fire A, Ahringer J, Plasterk RH. 2002. Loss of the putative RNA-directed RNA polymerase RRF-3 makes *C. elegans* hypersensitive to RNAi. *Curr Biol* **12**: 1317–1319.

Stiernagle T. Maintenance of *C. elegans*. in *WormBook: The online review of* C. elegans *biology* (ed. The *C. elegans* Research Community). WormBook.

Visualizing Apoptosis in Embryos and the Germline of *Caenorhabditis elegans*

Benjamin Lant[1] and W. Brent Derry[2,3]

[1]*Developmental and Stem Cell Biology Program, The Hospital for Sick Children, Toronto, Ontario M5G 1L7, Canada;* [2]*Department of Molecular Genetics, University of Toronto, Toronto, Ontario M5S 1A1, Canada*

Visualizing apoptosis in developing embryos or the germline of *Caenorhabditis elegans* is remarkably easy because of the transparency of the organism. The invariant pattern of cell division and programmed cell death during development makes it possible to quantify small but reproducible changes in apoptosis, which are easy to detect by light microscopy because of the refractile properties of dying cells. Although apoptotic death is easy to visualize and quantify in the germline of adult hermaphrodites, the pattern of cell death is variable, especially when triggered by stress. The most convenient method for visualization of apoptosis in vivo is light microscopy, which requires immobilizing live embryos or adult animals on slides. This protocol describes the basic methods for visualizing and analyzing apoptosis in living animals.

MATERIALS

It is essential that you consult the appropriate Material Safety Data Sheets and your institution's Environmental Health and Safety Office for proper handling of equipment and hazardous material used in this protocol.

RECIPE: Please see the end of this protocol for recipes indicated by <R>. Additional recipes can be found online at http://cshprotocols.cshlp.org/site/recipes.

Reagents

Agar solution (4% [w/v], maintained in the liquid phase in a 55°C water bath)

> C. elegans *at the appropriate stage of development, in which apoptosis has been stimulated.*
> *Activate apoptosis in the worms of interest by perturbing gene function (e.g., with RNAi) or by exposing to various stressors (e.g., DNA-damaging agents). (See Protocol 1: Induction of Germline Apoptosis in* Caenorhabditis elegans *[Lant and Derry 2014].) Careful staging of the worms, particularly with embryos, will be needed, and typically will arise only from repeated viewings of worms/embryos at different stages.*

M9 buffer for worms <R>
Nail polish and/or petroleum jelly
Tetramisole (20 mM)

Equipment

Compound microscope with differential interference contrast (DIC) optics

> *Numerous transgenic strains of* C. elegans *carrying fluorescent markers are available; these can be crossed into any mutant background to assess various stages of apoptosis in living animals. Thus, for high-resolution imaging*

[3]Correspondence: brent.derry@sickkids.ca

Cite this protocol as *Cold Spring Harb Protoc*; doi:10.1101/pdb.prot080218

and quantification of apoptotic bodies, we recommend a compound microscope that is equipped with both DIC and fluorescence optics (e.g., Leica DMRA).

Coverslips
Eyelash glued to a toothpick
Glass slides
Image capturing software (e.g., Volocity [PerkinElmer])
Immersion oil
Incubator at 37°C
Wire (platinum)

METHOD

Preparing Slides

Note that slide mounting is necessary to assess apoptosis under DIC optical resolution, but depending on the strain, apoptotic response may be visualized using fluorescent markers with a dissecting microscope, which does not require slide mounting of the worms.

1. Add a small drop (~25 μL) of warm 4% agar to a slide. Making sure that the agar does not set, quickly lay another slide on top, perpendicular to the lower slide. Press firmly to flatten the warm agar into a thin pad (~1 mm thick). Aim to maintain a level pad so the worms remain on a level plane when viewed under the microscope.

2. Before mounting the embryos or worms, carefully remove the top slide by sliding so the thin agar pad remains on the bottom slide.

   Be careful; sometimes the agar pad will slip off or buckle as the top slide is removed.

3. If desired, dry the slide/agar pad overnight in a 37°C incubator.

   For experiments of long duration (e.g., determining corpse persistence times) the agar pad on which the worms or embryos are placed will desiccate and shrink, making it difficult to record 4D images without going out of focus. Thus, we recommend performing this step for all time-course studies.

   Proceed to Steps 4–5 (for embryos) or Steps 6–9 (for worms).

Visualizing Apoptosis in Developing Embryos

4. Place a 2.5-μL drop of M9 buffer on a thin agar pad on a microscope slide. Pick embryos from a plate that is well fed (i.e., still has plenty of bacteria) and place them into the M9 buffer.

   When visualizing embryos, it is not necessary to use tetramisole, as embryos do not require anesthesia. Instead, we recommend mounting embryos in M9 buffer. It is important to ensure that the samples remain hydrated as they are picked and placed onto the agar pad. If the liquid evaporates, add a few microliters of M9 buffer to the embryos on the agar pad to keep them hydrated while transferring to the pad.

5. Using an eyelash glued to a toothpick, carefully drag the embryos into straight lines near one side of the M9 drop. Carefully place a coverslip over the top, sandwiching the embryos between the agar pad and coverslip. If necessary, add ~18 μL of M9 buffer under the edge of the coverslip to prevent dehydration. Seal the coverslip with nail polish or warm petroleum jelly to prevent evaporation of buffer during the observation period.

   Take care that the M9 does not evaporate during this time, which will cause embryo morphology to deteriorate and make it difficult to visualize apoptotic corpses.

   Proceed to Step 10.

Visualizing Germ Cell Apoptosis In Vivo

Because the germline continues to proliferate and expand throughout development, it is important to carefully stage worms used for this procedure so they are all at the same stage (e.g., 24 h post-L4). This ensures that their germlines are uniform in size.

Cite this protocol as *Cold Spring Harb Protoc*; doi:10.1101/pdb.prot080218

6. Place a drop (~5 µL) of 20 mM tetramisole on the agar pad. Carefully pick worms with a platinum wire coated with a small amount of bacteria from the NGM plate and transfer them into the liquid.

   It is important to ensure that the worms remain hydrated as they are picked and placed onto the agar pad. If the liquid evaporates, the integrity of the animals will degrade and make it very difficult to accurately visualize apoptotic corpse morphology.

7. Once the appropriate number of worms has been added to the agar pad, carefully place a coverslip over the top, sandwiching the samples between the agar pad and coverslip. If necessary, add ~18 µL of 20 mM tetramisole under the edge of the coverslip to prevent dehydration. Seal the coverslip with nail polish or warm petroleum jelly to prevent evaporation of tetramisole (and subsequent desiccation of the sample) during the observation period.

   Embryos develop for several hours after mounting, thereby permitting time-course analysis of apoptosis during development. In contrast, the observation of the adult germline should be limited to a few hours after mounting, as germline morphology tends to degrade after prolonged periods in tetramisole.

8. Using the 10× lens on a compound microscope, focus on the target worms.

   Often, this magnification, in conjunction with high signal labeling (such as acridine orange [AO] staining), permits the identification of the most visible/unobstructed germlines; this essentially acts as prescreening for higher magnification visualization.

9. Move to the 63× oil immersion lens and focus on nuclei in the bend region of the germline. Confirm focus on the image capturing software, as there is often a differential between the ocular lens and the camera capture.

   We find that the posterior end of the germline is often easier to visualize, as the anterior bend tends to be obstructed by the intestine.

   Proceed to Step 10.

Analyzing Apoptosis

10. Quantify corpse numbers and corpse persistence time as desired.

    Embryos will develop to hatching under the coverslip, which allows visualization of developmental apoptosis at various times over several hours. Using morphological criteria to stage the embryos, the best stages to quantify apoptosis are at the "lima bean" stage (6 h post fertilization; Fig. 1), the comma stage, the 1.5-fold stage, and the 2-fold stage (see IntroFIG7 at http://www.wormatlas.org/ver1/handbook/anatomyintro/anatomyintro.htm). Apoptosis peaks during the stages of development where embryos begin to undergo elongation—the comma and 1.5-fold stages.

    With 4D microscopy, it is easy to both quantify the number of corpses and their persistence time to determine whether abnormal numbers of corpses at specific stages of development are caused by engulfment defects. Corpses generally persist in embryos for ~30 min before getting engulfed.

    As in embryos, apoptosis can be visualized in germlines of anesthetized adult worms for several hours. Thus, it is possible to quantify the persistence time of germline corpses to determine whether abnormal numbers are the result of engulfment defects. Germline corpses generally persist for ~45 min before getting engulfed by the surrounding gonadal sheath cells.

    See Troubleshooting and Discussion.

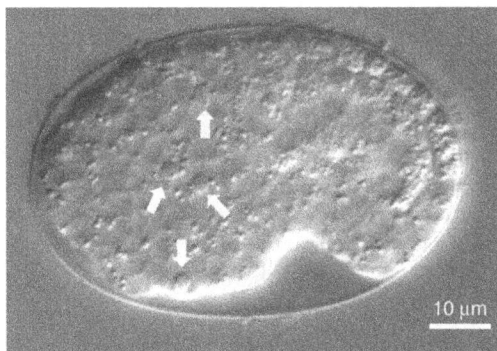

FIGURE 1. Corpses in the *C. elegans* embryo (indicated by arrows) appear as cellularized bodies or "buttons" and are spatially distinct. The above embryo is in the "bean" stage, which is ~6 h into ex utero embryonic development. Embryo viewed by DIC using 630× oil immersion magnification.

TROUBLESHOOTING

Problem (Step 10): Embryos are dead on viewing.

Solutions: There are a number of reasons that you may be observing dead embryos on the slide.

- One possibility (if you are employing RNAi, or observing a mutant strain or products of a cross) is that the eggs may be dead before plating as part of a mutation phenotype. In some cases, a mutant phenotype may result in embryonic lethality after much developmental apoptosis has occurred, for example in the case of large chromosomal deficiencies (Sugimoto et al. 2001). For such studies, it is important to obtain embryos at very early stages of development (2- to 8-cell stage) by cutting open gravid hermaphrodites to follow their lineages to precisely score apoptotic events. We also refer the reader to the less labor-intensive method of scoring surviving nuclei in the pharynx if the animals survive to hatching, where many extra cells are located when apoptosis is suppressed (Schwartz 2007).

- A second possible cause may be transfer time from plate to slide. Often if this is not done rapidly (and the embryos desiccate on the pick) you will be unable to observe developmental apoptosis. Finally, embryos will cease dividing and eventually die if they are in an anoxic environment. Thus, mounting embryos with minimal amounts of bacteria and ensuring that there are plenty of air bubbles in the agar pad help prevent hypoxic conditions under the coverslip.

Problem (Step 10): Air bubbles are present (in the agar pad or tetramisole) that obscure worm physiology.

Solution: Before the agar setting, if air bubbles have formed, use the edge of a coverslip to puncture the bubbles. Similarly, with the application of tetramisole, if air bubbles are noticeable on the pad, use the edge of a coverslip to puncture them before putting worms in the anesthetic. Note that air bubbles can be beneficial in preventing hypoxic conditions, when making long duration (i.e., time course) observations. In these cases, make sure to manipulate the worms in the tetramisole so that they do not lie directly on the bubbles (causing imaging issues).

Problem (Step 10): Worms are dead/severely damaged on viewing.

Solution: There are a number of reasons that this may occur. (Making sure the worms you mount are not already dead or damaged on the plate is a good habit!) In the transfer between plate and slide (and tetramisole), do not leave the worm on the pick in excess of 30 sec, as desiccation will begin and can result in death by the time the worm is mounted. After the worms have been placed on the slide, do not be tempted to apply pressure on the coverslip to "secure" its hold. This will cause the worms to burst and prevent you from accurately quantifying apoptosis. The capillary force between the coverslip and the tetramisole should keep it in place (unless there is too much solution on the slide—in which case the coverslip may slide off).

Problem (Step 10): Germlines appear damaged or obstructed.

Solution: Damaged germlines can be the result of a specific mutation or RNAi treatment, or damage can be incurred by desiccation from prolonged time on the pick. As mentioned above, damaged germlines or unexpected apoptotic phenotypes (particularly in terms of engulfment rates) can result when worms are dying on the slide. As in embryos, the hypoxic conditions imposed by slide mounting can affect germline quality in adult worms.

Cases of germline obstruction from intestinal "blocking" are common, and are more prominent in the anterior of the worm, particularly in older animals which tend to retain their eggs. We recommend mounting more than 50 worms per condition and only quantifying corpses in animals with clearly visible germlines in all focal planes. In our experience, the posterior germline is consistently less obstructed by the intestine and therefore better for quantifying apoptosis.

Problem (Step 10): Germline corpses are persisting in excess of "standard" engulfment time.

Cite this protocol as *Cold Spring Harb Protoc*; doi:10.1101/pdb.prot080218

Solution: You may have discovered a gene that regulates corpse engulfment, or there is insufficient oxygen for the worms to remain viable. If you suspect that defective engulfment is not linked to the particular condition or gene that you are studying, the persistence may be caused by lack of oxygen in the slide. Avoid hypoxic conditions as referred to above. Bacterial consumption of oxygen is usually not a problem, because in most cases, it is not necessary to have adults under coverslips for more than a few hours.

Problem (Step 10): Despite the induction of DNA damage (via genetic or genotoxic means), the apoptotic effect appears too small to quantify.

Solution: An alternative method would be to quantify germline apoptosis in an engulfment-defective background, such as the *ced-1* mutant. By preventing engulfment, corpses will persist (giving the impression of increased corpse numbers), which will shift the distribution of corpses to a normal curve (see, e.g., Ross et al. 2011).

DISCUSSION

The methods described here are the most simple, linear preparative means to assess apoptosis. Corpse counting and quantification of the cellularized "buttons" (Fig. 2) in the germline requires practice. For quantification, it is critical to have experience with discriminating apoptotic corpse morphology. Thus, it is recommended to practice double-blind quantification of apoptotic corpses in various mutant strains that have high and low levels of apoptosis to gain expertise. Attuning your eyes to these morphological features, and differentiating early from late, is simply a matter of patience and repetition.

Assessing corpse numbers throughout the stacks (Z-plane) of the germline is critical to accurately obtaining corpse numbers, which occasionally show only small variation between strains/conditions. It is important to ensure that the animals, whether embryos or adults, are carefully staged so data are reproducible. A steady hand is also required when moving embryos, so make sure to monitor your caffeine intake before the experiment! In addition, we highly recommend performing double-blind apoptosis counts to prevent observer bias. By quantifying corpse persistence times it is possible to determine whether abnormal corpse numbers are caused by defects in the execution or engulfment phases of the apoptotic program. Whenever possible, we also recommend confirming results using multiple alleles, transgene rescue (when working with mutant strains), and RNAi. This is particularly important for analyzing germline apoptosis in strains/conditions where the differences in average corpse number are subtle. We therefore recommend using engulfment defective mutants such as *ced-1* or *ced-5* to increase the sensitivity of detecting subtle changes in apoptosis.

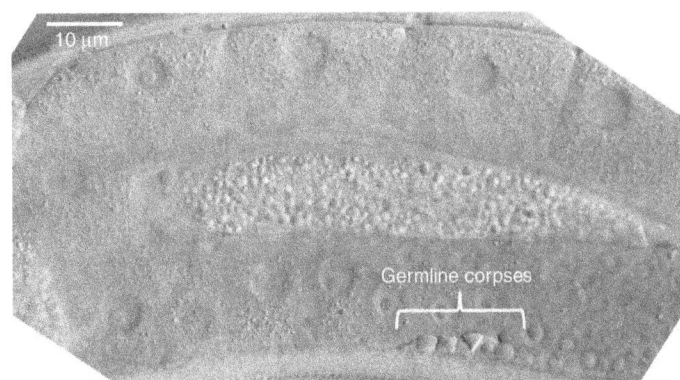

FIGURE 2. *C. elegans* germline indicating corpses as a result of irradiation (60 Gy). Corpses appear, as in the embryo, as cellularized bodies or "buttons" in (and just before) the bend of the pachytene region of the germline. Germline viewed under 630× oil immersion magnification.

Data Analysis

Proper statistical methods are necessary for comparing apoptotic corpse numbers. This is particularly important when comparing corpse numbers in undamaged germlines with those that have been subject to stress (Fig. 2). In our experience, physiological apoptosis tends to follow a Poisson distribution in wild type and most mutant animals, whereas after stress, corpse numbers follow a normal (bell curve) distribution. Comparing these types of data distributions cannot be achieved with a Student's *t*-test, for example. Rather, we recommend the use of statistical methods such as the Mann–Whitney U test, which does not require different data sets to follow the same distribution. For a good refresher on the proper use of error bars and statistics in biology, we recommend the reviews by Cumming et al. (2007) and Fay and Gerow (2013).

It is important to consider the potential impact of proliferative changes when analyzing corpse numbers in the germlines of various mutant strains. Because the somatic lineages of *C. elegans* follow invariant patterns of cell division and cell death, analysis of apoptotic events in developing embryos is very precise. In the germline, cell deaths are more stochastic and can be influenced by changes in proliferation. To ensure that decreased numbers of corpses observed in the germline are not simply the result of reduced proliferation and therefore fewer cells capable of undergoing apoptosis, it is important to compare corpse numbers with the total number of germ cells. This can be achieved using fluorescent reporters such as mCherry-tagged histone H2B (i.e., strain OD141) that mark all nuclei in the germline. Alternatively, germlines can be fixed and stained with DAPI and the nuclei quantified. In the case of somatic apoptosis, the invariant lineage of *C. elegans* makes it possible to visualize extra (undead) cells in vivo without the use of fluorescent markers, but this usually takes a very patient and seasoned pro.

RELATED TECHNIQUES

Visualizing germline apoptosis can be enhanced through the use of strains that express fluorescent apoptosis markers. Many markers of apoptosis routinely used in cultured cells, such as terminal deoxynucleotidyl transferase (TdT)-mediated dUTP nick end labeling (TUNEL), AO staining, and immunohistochemistry to a growing list of apoptotic proteins, can also be applied to *C. elegans*.

RECIPE

M9 Buffer for Worms

1. Dissolve 3 g of KH_2PO_4, 6 g of Na_2HPO_4, and 5 g of NaCl in 1 L of H_2O.
2. Autoclave for 20 min.
3. Add 1 mL of 1 M $MgSO_4$.
4. Store at room temperature. After 2 wk, check for visible contamination before use.

REFERENCES

Cumming G, Fidler F, Vaux DL. 2007. Error bars in experimental biology. *J Cell Biol* **177**: 7–11.

Fay DS, Gerow K. 2013. A biologist's guide to statistical thinking and analysis. *WormBook: The online review of* C. elegans *biology*. (ed. The *C. elegans* Research Community). WormBook. 1–54.

Lant B, Derry WB. 2014. Induction of germline apoptosis in *C. elegans*. *Cold Spring Harb Protoc* doi: 10.1101/pdb.prot080192.

Ross AJ, Li M, Yu B, Gao MX, Derry WB. 2011. The EEL-1 ubiquitin ligase promotes DNA damage-induced germ cell apoptosis in *C. elegans*. *Cell Death Differ* **18**: 1140–1149.

Schwartz HT. 2007. A protocol describing pharynx counts and a review of other assays of apoptotic cell death in the nematode worm *Caenorhabditis elegans*. *Nat Protoc* **2**: 705–714.

Stiernagle T. Maintenance of *C. elegans*. in *WormBook: The online review of* C. elegans *biology* (ed. The *C. elegans* Research Community). WormBook.

Sugimoto A, Kusano A, Hozak RR, Derry WB, Zhu J, Rothman JH. 2001. Many genomic regions are required for normal embryonic programmed cell death in *Caenorhabditis elegans*. *Genetics* **158**: 237–252.

Cite this protocol as *Cold Spring Harb Protoc*; doi:10.1101/pdb.prot080218

Fluorescent Visualization of Germline Apoptosis in Living *Caenorhabditis elegans*

Benjamin Lant[1] and W. Brent Derry[2,3]

[1]*Developmental and Stem Cell Biology Program, The Hospital for Sick Children, Toronto, Ontario M5G 1L7, Canada;* [2]*Department of Molecular Genetics, University of Toronto, Toronto, Ontario M5S 1A1, Canada*

Visualization of apoptosis using fluorescent tools is quite straightforward in living *Caenorhabditis elegans*. Several transgenic lines are available that mark dying cells with fluorescent proteins, making it possible to quantify kinetics at various stages of the apoptotic process. Proteins required for the engulfment of cell corpses are particularly useful for detecting early apoptotic stages using this approach. For example, expression of the engulfment protein CED-1 fused to green fluorescent protein (CED-1::GFP) creates a halo around cells during early apoptosis, before their refractile morphology can be detected by differential interference contrast (DIC) optics. In addition, vital dyes such as acridine orange (AO) and SYTO-12 are selectively retained in apoptotic cells and can be used to visualize apoptosis in the germlines of living animals. It is also possible to use vital dyes in combination with transgenic strains expressing fluorescent markers of cell corpses to examine, in detail, multiple stages of apoptosis in vivo. Because of the high optical contrast of fluorescent reagents, apoptosis can be visualized even at low magnification, facilitating the use of screening platforms to identify apoptosis regulators. This protocol describes multiple uses of fluorescent reagents for visualization of germline apoptosis in living *C. elegans*, including AO staining, time-course studies using fluorescent proteins, and low-throughput screening of cell death genes using RNA interference (RNAi).

MATERIALS

It is essential that you consult the appropriate Material Safety Data Sheets and your institution's Environmental Health and Safety Office for proper handling of equipment and hazardous material used in this protocol.

RECIPES: Please see the end of this protocol for recipes indicated by <R>. Additional recipes can be found online at http://cshprotocols.cshlp.org/site/recipes.

Reagents

Acridine orange (AO) (10-mg/mL stock solution) (Molecular Probes, Inc.)
Immediately before use, dilute 7.5 µL of AO stock solution into 1 mL of M9 buffer for worms <R>.

C. elegans strain of interest, in which apoptosis has been stimulated.
We recommend the following fluorescent reporter strains because of their broad applications for quantifying apoptosis and utility for screening; see Discussion.

- *MD701 (CED-1::GFP fusion protein)*
  Genotype: bcIs39 (Plim-7::CED-1::GFP + lin-15[+])V

[3]Correspondence: brent.derry@sickkids.ca

Cite this protocol as *Cold Spring Harb Protoc*; doi:10.1101/pdb.prot080226

- WS2170 (YFP::ACT-5 fusion protein)

  Genotype: opIs110 (Plim-7::YFP::ACT-5 + unc-119[+])IV

Both are available from the Caenorhabditis Genetics Center (CGC) and can be used as germline apoptosis indicators in conjunction with RNAi treatment, application of stressors, or after crossing into mutant strains. Alternatively, there are other fluorescent reporter strains available from the CGC that mark various stages of the engulfment process.

Young adult worms are used for AO staining (see Step 1), while both embryos and adults can be used for time-course studies (see Step 5). The RNAi procedure begins with worms at the L1 larval stage (see Step 9). Activate apoptosis in the worms of interest by perturbing gene function (e.g., with RNAi) or by exposing the germline to various stressors (e.g., DNA-damaging agents). (See Protocol 1: Induction of Germline Apoptosis in Caenorhabditis elegans [Lant and Derry 2014a].)

Plates (35-mm or 24-well) containing solid nematode growth medium (NGM) <R>

These plates may be seeded with bacteria (e.g., HT115 or OP50) as appropriate.

RNAi clones from the *C. elegans* RNAi feeding library (Source BioScience LifeSciences)

Tetramisole (20 mM)

Equipment

Compound microscope with differential interference contrast (DIC) and fluorescence optics (e.g., Leica DMRA)

Fluorescence dissecting microscope (with optional camera attachment)

Immersion oil

Slide-mounting setup, as described in Protocol 2: Visualizing Apoptosis in Embryos and the Germline of *Caenorhabditis elegans* (Lant and Derry 2014b)

METHOD

Staining Apoptotic Corpses in the Adult Germline with AO

AO is used for staining of the young adult germline. Vital dyes such as AO are not effective for detecting apoptosis in developing embryos because they cannot penetrate the egg shell. In addition, vital dyes rely on engulfment signals; thus, engulfment-defective mutants cannot be labeled with AO (Gartner et al. 2008; Neukomm et al. 2011).

1. Pipette 200 µL of freshly diluted AO solution (75 µg/mL) onto a plate containing at least 25 target young adult worms on a bacterial lawn. Distribute the solution evenly by carefully rotating the plate. Allow the plate to dry with its lid slightly ajar for ~20 min in the dark.

   We recommend using at least 25 target worms to ensure that germlines can be visualized in a minimum of 15 animals per condition.

2. Replace the lids so that plates are completely covered and allow the worms to continue feeding on the AO-soaked bacterial lawn in the dark for 1 h.

   The ingested AO will dissipate throughout the worm's body and concentrate in apoptotic nuclei.

3. Transfer the worms to a clean NGM plate containing the appropriate bacterial lawn. Incubate the plates in the dark for 3 h to clear excess AO from the intestines. Observe the worms under a dissecting microscope before slide mounting to be sure background has been reduced appropriately (Fig. 1).

   For RNAi experiments, plates used for this step can be seeded with the target RNAi bacteria or the HT115 strain, as the RNAi effect will last for several hours.

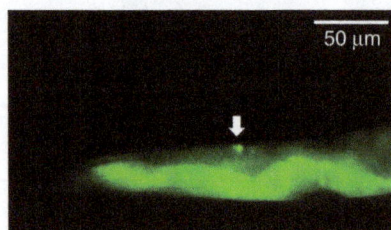

FIGURE 1. AO-stained germline viewed at 5× magnification (in the GFP channel) using a fluorescence dissecting microscope. The apoptotic corpse, indicated by the arrow, is clearly identifiable by the strong AO signal. The intensity of AO staining in the corpse relative to the surrounding cells is a good indicator (or prescreen) of how effective the staining/destaining process has been.

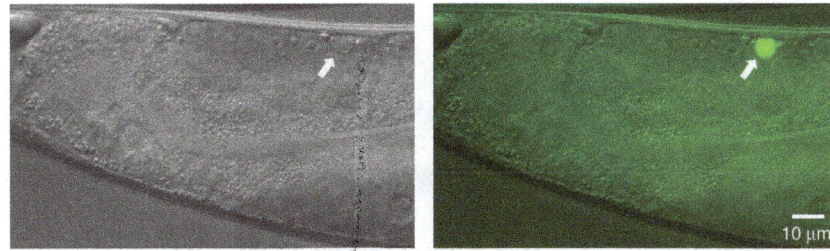

FIGURE 2. AO staining of a germline corpse. The vital dye enters the corpse as membrane permeability increases (mid- to late-stage apoptosis) and intercalates DNA. The *left* panel shows a DIC image of a germ cell corpse (arrow) among healthy germ cells in the pachytene bend region of the posterior germline. Shown in the *right* panel is the same corpse stained with AO (arrow). Images taken at 630× magnification.

> *The time required for destaining may require some optimization, based on the eating habits of your worms. You will rarely obtain a signal that is too low, but too much background is common. A general ratio of 1:3 staining:destaining time should be appropriate.*
>
> *See Troubleshooting.*

4. Slide-mount the worms on agar pads as described in Protocol 2: Visualizing Apoptosis in Embryos and the Germline of *Caenorhabditis elegans* (Lant and Derry 2014b). Observe on a compound microscope with fluorescence optics.

> *When viewing AO-stained corpses, be careful not to mistake germ cell corpses with spermatocytes, which will be closer to the uterus, past the bend region of the gonad, and will also take up AO (and SYTO-12) dyes. Stained corpses are found in the pachytene region of the distal germline (Fig. 2).*
>
> *See Troubleshooting.*

Time-Course Studies Using Fluorescent Proteins

> *CED-1::GFP can be used to visualize apoptotic corpses in both developing embryos and the germline (Fig. 3) (Kinchen et al. 2008; Yu et al. 2008); however, we find ACT-5::YFP to be a superior marker for the fluorescent detection of apoptotic cells in the germline (Fig. 4) (Kinchen et al. 2008). (See Discussion.)*
>
> *When using either of the recommended strains in conjunction with AO staining (as shown in Fig. 5), destain more rigorously to optimize detection of both fluorescent signals; we recommend a 1:4 ratio of staining:destaining time.*

5. Slide-mount embryos or worms of a fluorescent strain as described in Protocol 2: Visualizing Apoptosis in Embryos and the Germline of *Caenorhabditis elegans* (Lant and Derry 2014b).

6. Using the 10× lens on a compound microscope, identify the samples on the slide. Move to the 63× oil immersion lens and focus on a point of the embryo or germline with readily visible corpses (the more, the better).

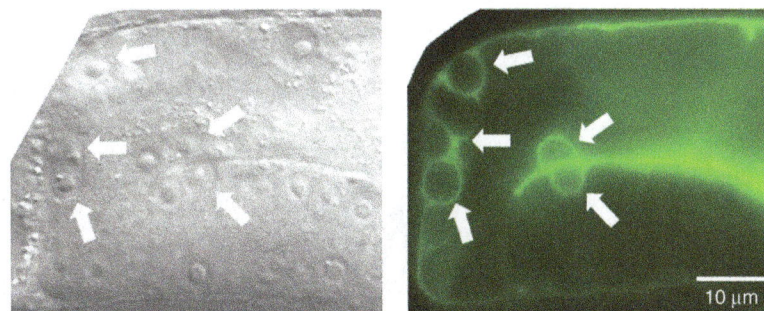

FIGURE 3. Germline corpses in a worm expressing the CED-1 protein fused to GFP in the gonadal sheath cells using the *lim-7* promoter following irradiation (60 Gy). The CED-1::GFP marker indicates early-stage corpses, as gonadal sheath cells surround the apoptotic bodies. The *left* panel shows the corpses under DIC optics, while the *right* panel shows the corpses in the GFP channel. Images taken at 630× magnification.

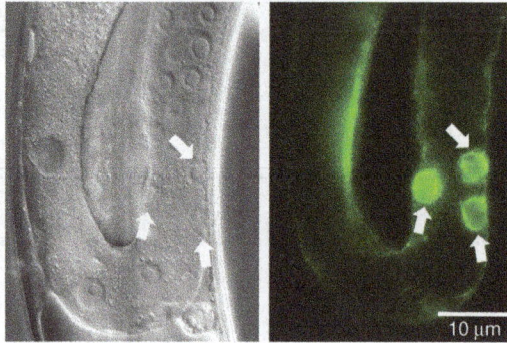

FIGURE 4. Germline corpses in worms expressing the actin protein ACT-5 fused to the yellow fluorescent protein (YFP) following irradiation (60 Gy). As gonadal sheath cells surround the apoptotic bodies, actin cables are remodeled, pushing the corpse up (giving the "button"-like appearance). The *left* panel shows the germ cell corpses under DIC optics (bright field), and the *right* panel shows the same germline in the YFP channel. ACT-5::YFP, like CED-1::GFP, forms halos around early stage corpses. White arrows indicate corpses. Images taken at 630× magnification.

Corpses detected by CED-1::GFP and ACT-5::YFP mark early apoptotic stages, so you may not always observe refractile bodies coinciding with strong fluorescence signals. We and others have also noticed CED-1::GFP signal surrounding some cells that do not undergo apoptosis (Craig et al. 2012), so exercise caution when quantifying corpses with this marker.

7. Capture a sequence of images over the span of ~1 h; one image in several Z planes every 3–5 min is ideal. Shutter the fluorescent light source in between images to minimize photobleaching. If possible, note the depth (in the Z plane) of the starting image because the agar pad will dry over time and offset the image stacks. Compensate for shrinkage of the agar pad and/or stage drift by adjusting the stage periodically to maintain the bottom plane of the worm or embryo where the first image was taken.

   Alongside the fluorescent image, an image in the corresponding plane in white light should also be taken to allow for comparison of changes in corpse morphology in both spectra—specifically the formation of button-like corpses, which follows loss of fluorescent signals. The difference between fluorescence and white light imaging is shown in Figure 5.

   See Troubleshooting.

Low-Throughput RNAi Screening Using Fluorescent Reagents

The following method describes low-throughput screening of RNAi-induced apoptosis using low magnification (i.e., a dissecting microscope). This is appropriate for screening <100 genes. When screening very low numbers, you may instead wish to use Protocol 1: Induction of Germline Apoptosis in Caenorhabditis elegans *(Lant and Derry 2014a). To*

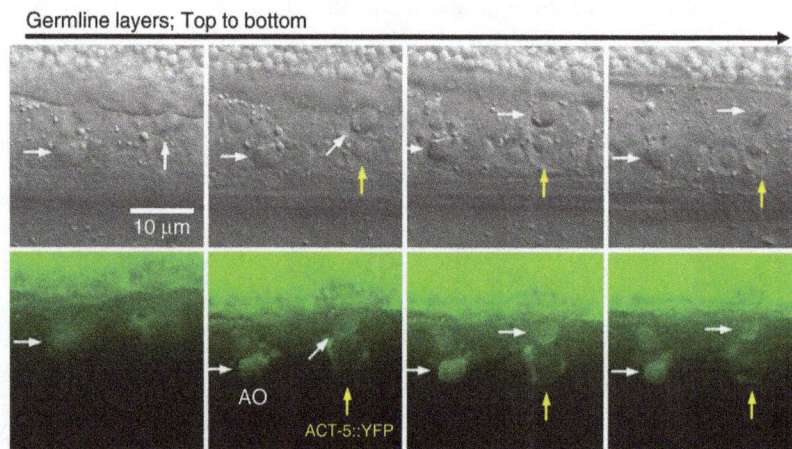

FIGURE 5. Concurrent imaging of early- and late-stage apoptotic corpses in the *C. elegans* germline. The images from *left* to *right* represent layers (from *top* to *bottom*) of the germline. White arrows indicate AO-stained late-stage corpses, while the yellow arrows indicate an ACT-5::GFP-labeled early-stage corpse. Images taken at 630× magnification. (Scale is indicated in the first panel.)

   Cite this protocol as *Cold Spring Harb Protoc*; doi:10.1101/pdb.prot080226

screen 100 or more genes, see Protocol: **High-Throughput RNAi Screening for Germline Apoptosis Genes in** **Caenorhabditis elegans** *(Lant and Derry 2014c). Low-magnification prescreening of apoptosis induced by various stressors can also be performed on single plates by following Steps 11–13.*

8. Add 30 μL of RNAi liquid culture to each well of a 24-well plate containing solid NGM.

   Plates should be thoroughly labeled, or a template created for reference, to avoid confusing RNAi samples between wells.

9. Using the hypochlorite method to synchronize populations of worms at the L1 (Stiernagle 2006), add 20–30 L1 worms to each well and grow to the young adult stage.

10. (Optional) Stain the worms with AO.

    i. Transfer the worms to a fresh 24-well plate containing solid NGM.

    ii. Pipette 20 μL of freshly diluted AO solution (75 μg/mL) into each well. Allow the worms to stain in the dark for 1 h.

    iii. Transfer the worms to a fresh 24-well plate containing solid NGM seeded with OP50 bacteria. Destain for 3 h.

       As mentioned above, a 1:3 stain:destain ratio is a starting point, but may require optimization.

11. Observe the worms under the lowest magnification lens of a dissecting microscope. Move to the desired well and select the appropriate fluorescence channel.

12. Focus on worms that show fluorescence at a low magnification. Apply a drop of tetramisole to the desired well. Wait ∼30 sec for the tetramisole to immobilize the worms.

    This process—maintaining your eye on the moving worm through the eyepiece, while attempting to drop tetramisole onto the plate in the correct location—can be tricky at first.

    Note that CED-1::GFP and ACT-5::YFP should both produce visible signals under low magnification with a dissecting microscope, but the resolution of single cells (i.e., with visible individual halos) is clear only at higher magnifications. Thus, these markers are largely inappropriate for quantification of apoptosis during low-magnification screening; see Discussion.

13. Increase the magnification to visualize germlines in more detail. For high resolution imaging, carefully pick and transfer the worms to slides for mounting as described in Protocol 2: Visualizing Apoptosis in Embryos and the Germline of *Caenorhabditis elegans* (Lant and Derry 2014b).

    As with DIC optics, the intestine can sometimes obscure the signal and make it difficult to resolve individual corpses.

    If your dissecting microscope has a camera attachment, it may be possible to quantify apoptosis using a snapshot of the image; see Discussion.

TROUBLESHOOTING

Problem (Step 3): The AO background signal is too strong.

Solution: While a 1:3 staining:destaining ratio is suggested, it may be necessary to increase the destaining time, especially when using the aforementioned fluorescent strains in conjunction with AO. Check the worms in 20- to 30-min intervals until the background signal is minimal. Luckily, using a dissecting microscope with a fluorescence attachment, you can prescreen the signal ratio before slide mounting.

Problem (Step 4): The AO corpse signal appears faded.

Solution: Fading can be caused by prolonged exposure to fluorescent lighting, which eventually will cause photobleaching of the image. Alternatively, there may have been insufficient stain uptake, in which case prescreening for corpses should be used (Fig. 1). Note that ablation of genes required for the engulfment of corpses prevents vital dyes from entering apoptotic cells (Neukomm et al. 2011).

Problem (Step 7): More corpses are visible by DIC optics than with fluorescence in the CED-1:: GFP strain.

Solution: This is a fairly common caveat of using the CED-1::GFP strain, which shows a dissipation of signal at high levels of apoptosis.

Problem (Step 7): Corpses labeled with CED-1::GFP or ACT-5::YFP persist for longer than expected periods of time, sometimes not completing engulfment.

Solution: Alas, this also seems to be a caveat of using these strains. Given the role of CED-1 and actin in the engulfment process, the labeling of these proteins can prevent some corpses from completing engulfment. We have also noticed mildly reduced levels of overall apoptosis in strains expressing ACT-5::YFP.

DISCUSSION

Markers for apoptosis and apoptotic corpses themselves can be detected through physically staining the *C. elegans* adult germline, because one of the hallmark features of apoptotic cells is the condensation of chromatin into pycnotic bodies. DNA intercalating dyes (such as 4′,6-diamidino-2-phenyl-indole [DAPI]) and covalent labeling of fragmented DNA using terminal deoxynucleotidal transferase (TdT)-mediated dUTP nick end labeling (TUNEL) have long been used for visualizing apoptosis, but these reagents are limited to use in fixed specimens.

More accurate imaging of early and late stages of apoptosis in the germline of living animals can be achieved using vital dyes (such as AO and SYTO-12; Gumienny et al. 1999; Derry et al. 2001; Lettre et al. 2004) as well as transgenic lines that express fluorescent markers. Visualization of cell corpses by fluorescence depends on the resolution and intensity of the markers used. Many transgenic strains expressing fluorescent markers specific for various stages of apoptosis are bright enough to create detectable changes using a low-resolution fluorescence dissection microscope. However, some markers are less bright and are only detectable when visualized using a higher resolution compound fluorescence microscope (see below).

Fluorescently tagged proteins such as CED-1 are regularly used to visualize apoptosis in both embryos and the germline of living animals. CED-1 is required for the engulfment of apoptotic cells, and labels corpses before and during engulfment (Zhou et al. 2001). Strains carrying mutations in the engulfment genes (i.e., *ced-1[e1754]*) are useful for determining whether subtle changes in corpse numbers are statistically significant, because subtle differences are greatly enhanced when they accumulate in the absence of engulfment. The CED-1::GFP marker is also a good tool for quantifying apoptosis, as it forms distinct rings around early-stage apoptotic corpses in the soma and germline (Fig. 3). One caveat is that it is difficult to distinguish individual corpses at high levels of apoptosis because the signal becomes distorted. Conditions that cause higher numbers of apoptotic corpses—judged by morphological criteria in the bright field or with fluorescent reagents—should be validated in apoptosis-resistant mutants, such as *cep-1(gk138)* for genotoxic stress and *ced-3(n717)* or *ced-4(n1162)* to inhibit the core apoptotic pathway. Another potential problem is that a fraction of CED-1::GFP-labeled cells do not undergo apoptosis (Craig et al. 2012), which will skew quantification if not carefully addressed.

An alternative germline apoptosis marker expresses actin (ACT-5) fused to yellow fluorescent protein (YFP) in the gonadal sheath cells (Fig. 4). Like CED-1::GFP, ACT-5::YFP forms a fluorescent halo around cells beginning engulfment, but in our experience there is less background compared with CED-1::GFP, making it easier to distinguish individual corpses. However, the ACT-5::YFP strain has a modest resistance to DNA damage-induced apoptosis in the germline (B Lant and WB Derry, unpubl.), so exercise caution if using it for this purpose. Regardless, both strains can be used to indicate which cells should undergo the stereotypical morphological changes of apoptosis. As such, time-course experiments can be performed in which both GFP/YFP and white light channels are

Cite this protocol as *Cold Spring Harb Protoc*; doi:10.1101/pdb.prot080226

photographed over a period of ~45 min (the standard time for corpse development and engulfment). It is also possible to use the combination of a vital dye on a fluorescent strain to visualize distinct steps of early, mid- and late-stage corpses in real time (Fig. 5). The main caveat with this approach is that the fluorescence of the dye will often overwhelm the signal of the integrated fluorescent marker. Thus, extensive optimization of dye signal is necessary.

QUANTIFICATION OF APOPTOSIS USING FLUORESCENT REAGENTS

If multiple genes are being assessed for apoptosis defects in fluorescent strains (or using vital dyes), we recommend prescreening samples using a fluorescence dissecting microscope. This method is rapid, relatively light on handling, and does not kill the worms during observation. Corpses stained with vital dyes or marked with fluorescently tagged proteins can be detected at 10× magnification, but are ideally viewed at higher magnifications (i.e., 50×), especially in the case of weaker reporters. Both CED-1::GFP and ACT-5::YFP can be used to differentiate high and low levels of apoptosis using a fluorescence dissecting microscope. However, the signals of both are low compared with AO staining, and are therefore not ideal for quantification at this level of magnification. For quantification it is best to visualize corpses in these strains at 630× using a high-resolution compound microscope.

From experience, only AO staining facilitates consistent quantification potential for screening at lower magnification (10×–50×; magnifications typically found on standard dissecting microscopes). The ability to quantify AO-stained corpses is also dependent on magnification capacity; quantification is possible with the 2× lens, but a 5× lens is preferable. Because worms move out of view quickly at higher magnification, it is best to anesthetize them before attempting quantification. Using AO-stained or fluorescently labeled corpses to quantify apoptosis is much like standard corpse counting (see Protocol 2: Visualizing Apoptosis in Embryos and the Germline of *Caenorhabditis elegans* [Lant and Derry 2014b]). Images taken in a fluorescent channel should be confirmed for corpse morphology in the bright field. If you are unsure about the different stages of corpse morphology, practice identifying early-stage corpses (Figs. 2–4). Again, it should be noted that while screening can give an approximation of corpse numbers (a qualitative measure more than anything), quantification in strains which have been stained or contain fluorescent markers should only be used on images taken at high magnification (>400×).

RELATED TECHNIQUES

The TUNEL assay can be used to identify apoptotic corpses in *C. elegans*, and protocols are readily available elsewhere (Wu et al. 2000; Parusel et al. 2006; Craig et al. 2012).

RECIPES

M9 Buffer for Worms

1. Dissolve 3 g of KH_2PO_4, 6 g of Na_2HPO_4, and 5 g of NaCl in 1 L of H_2O.
2. Autoclave for 20 min.
3. Add 1 mL of 1 M $MgSO_4$.
4. Store at room temperature. After 2 wk, check for visible contamination before use.

Nematode Growth Medium (NGM)

1. For solid NGM, mix 3 g of NaCl, 2.5 g of peptone, and 20 g of agar and bring to 1 L with H_2O. (For liquid NGM, prepare without agar.)

2. Autoclave for 1 h.

3. Let cool for 1 h in a 55°C water bath.

4. Add 1 mL of cholesterol (5 mg/mL in ethanol), 1 mL of 1 M $CaCl_2$, 1 mL of 1 M $MgSO_4$, and 25 mL of 1 M (pH 6.0) KPO_4, mixing after each addition.

5. Using an automated plate pourer (peristaltic pump), pour the medium into sterile plastic plates to set. During pouring, keep the NGM on a hot plate with stirrer to prevent the medium from solidifying.

6. Store NGM plates at 4°C until use. Seed NGM plates with bacteria the day after pouring. Warm the plates to room temperature before adding the worms. (For liquid NGM, store at room temperature for up to 1 mo. Regularly check for excess cloudiness to ensure bacterial contamination is absent.)

REFERENCES

Craig AL, Moser SC, Bailly AP, Gartner A. 2012. Methods for studying the DNA damage response in the *Caenorhabditis elegans* germ line. *Methods Cell Biol* 107: 321–352.

Derry WB, Putzke AP, Rothman JH. 2001. *Caenorhabditis elegans* p53: Role in apoptosis, meiosis, and stress resistance. *Science* 294: 591–595.

Gartner A, Boag PR, Blackwell TK. Germline survival and apoptosis. (September 4, 2008), *WormBook*, ed. The *C. elegans* Research Community, WormBook, doi/10.1895/wormbook.1.145.1, http://www.wormbook.org.

Gumienny TL, Lambie E, Hartwieg E, Horvitz HR, Hengartner MO. 1999. Genetic control of programmed cell death in the *Caenorhabditis elegans* hermaphrodite germline. *Development* 126: 1011–1022.

Kinchen JM, Doukoumetzidis K, Almendinger J, Stergiou L, Tosello-Trampont A, Sifri CD, Hengartner MO, Ravichandran KS. 2008. A pathway for phagosome maturation during engulfment of apoptotic cells. *Nat Cell Biol* 10: 556–566.

Lant B, Derry WB. 2014a. Induction of germline apoptosis in *Caenorhabditis elegans*. *Cold Spring Harb Protoc* doi: 10.1101/pdb.prot080192.

Lant B, Derry WB. 2014b. Visualizing apoptosis in embryos and the germline of *Caenorhabditis elegans*. *Cold Spring Harb Protoc* doi: 10.1101/pdb.prot080218.

Lant B, Derry WB. 2014c. High-throughput RNAi screening for germline apoptosis genes in *Caenorhabditis elegans*. *Cold Spring Harb Protoc* doi: 10.1101/pdb.prot080234.

Lettre G, Kritikou EA, Jaeggi M, Calixto A, Fraser AG, Kamath RS, Ahringer J, Hengartner MO. 2004. Genome-wide RNAi identifies p53-dependent and -independent regulators of germ cell apoptosis in *C. elegans*. *Cell Death Differ* 11: 1198–1203.

Neukomm LJ, Frei AP, Cabello J, Kinchen JM, Zaidel-Bar R, Ma Z, Haney LB, Hardin J, Ravichandran KS, Moreno S, et al. 2011. Loss of the RhoGAP SRGP-1 promotes the clearance of dead and injured cells in *Caenorhabditis elegans*. *Nat Cell Biol* 13: 79–86.

Parusel CT, Kritikou EA, Hengartner MO, Krek W, Gotta M. 2006. URI-1 is required for DNA stability in *C. elegans* (February 11, 2006), *WormBook*, ed. The *C. elegans* Research Community, WormBook, doi/10.1895/wormbook.1.101.1, http://www.wormbook.org.

Stiernagle T. Maintenance of *C. elegans* (February 11, 2006), *WormBook*, ed. The *C. elegans* Research Community, WormBook, doi:10.1895/wormbook.1.101.1, http://www.wormbook.org.

Wu YC, Stanfield GM, Horvitz HR. 2000. NUC-1, a *Caenorhabditis elegans* DNase II homolog, functions in an intermediate step of DNA degradation during apoptosis. *Genes Dev* 14: 536–548.

Yu X, Lu N, Zhou Z. 2008. Phagocytic receptor CED-1 initiates a signaling pathway for degrading engulfed apoptotic cells. *PLoS Biol* 6: e61.

Zhou Z, Hartwieg E, Horvitz HR. 2001. CED-1 is a transmembrane receptor that mediates cell corpse engulfment in *C. elegans*. *Cell* 104: 43–56.

Cite this protocol as *Cold Spring Harb Protoc*; doi:10.1101/pdb.prot080226

High-Throughput RNAi Screening for Germline Apoptosis Genes in *Caenorhabditis elegans*

Benjamin Lant[1] and W. Brent Derry[2,3]

[1]*Developmental and Stem Cell Biology Program, The Hospital for Sick Children, Toronto, Ontario M5G 1L7, Canada;* [2]*Department of Molecular Genetics, University of Toronto, Toronto, Ontario M5S 1A1, Canada*

Among the greatest tools that *Caenorhabditis elegans* can provide researchers are the capabilities to perform high-throughput, genome-wide screens. Using bacterial RNAi libraries, which cover the majority (>85%) of the worm genome, genes can be rapidly and systematically evaluated for apoptosis phenotypes in the germline. Screens can be designed to directly assess the levels of apoptotic corpses under normal physiological conditions using transgenic strains expressing fluorescent reporters that mark apoptotic bodies. Vital dyes that are selectively retained in apoptotic cells, such as acridine orange (AO), can also be used to screen for genes that regulate germline apoptosis. Using these reagents, screens can be performed in wild-type worms or mutant backgrounds that suppress or enhance apoptosis phenotypes. This protocol describes methods for designing and carrying out high-throughput or targeted RNAi screens for germline apoptosis regulators.

MATERIALS

It is essential that you consult the appropriate Material Safety Data Sheets and your institution's Environmental Health and Safety Office for proper handling of equipment and hazardous material used in this protocol.

RECIPES: Please see the end of this protocol for recipes indicated by <R>. Additional recipes can be found online at http://cshprotocols.cshlp.org/site/recipes.

Reagents

Acridine orange (AO) (10-mg/mL stock solution) (Molecular Probes A3568) (optional)

Immediately before use, dilute 7.5 µL of AO stock solution into 1 mL of M9 buffer for worms <R>.

C. elegans strain of interest (e.g., fluorescent reporter or mutant strain)

Before screening, synchronize tens of thousands of worms at the L1 stage using the hypochlorite method (see WormBook [Stiernagle]). Worms should be prepared no more than 3 d before use (Step 8); see Discussion.
Recommended fluorescent strains are described in Protocol 3: Fluorescent Visualization of Germline Apoptosis in Living Caenorhabditis elegans *(Lant and Derry 2014a).*
If you are screening a mutant strain, you should simultaneously run controls (typically of the N2 strain) to assess the individual effects of gene ablation on apoptosis. If your purpose is to identify genes that restore apoptosis to a resistant strain, this may not be necessary.

Isopropylthio-β-galactoside (IPTG) (0.1 M)

IPTG induces expression of the T7 RNA polymerase in the RNAi bacteria (Timmons et al. 2001).

Luria Bertani (LB) + amp and tet (liquid) <R>

---

[3]Correspondence: brent.derry@sickkids.ca

Cite this protocol as *Cold Spring Harb Protoc*; doi:10.1101/pdb.prot080234

M9 buffer for worms (for use with AO staining) <R>

Nematode growth medium (NGM; liquid) <R>

> *The number of genes you plan to screen will determine how best to culture the bacteria. The method provided here assumes that a large number of genes (>1000) will be screened; hence, liquid culture in 96-well plates is appropriate. Alternatively, if you are screening <500 genes, solid medium in 24-well plates can be substituted (but can be costly in terms of time and money). To screen a small number of genes (<100), see Protocol 3: Fluorescent Visualization of Germline Apoptosis in Living Caenorhabditis elegans (Lant and Derry 2014a).*

Plates containing solid NGM (see Step 13) <R>

RNAi clones from the *C. elegans* RNAi feeding library (Source BioScience LifeSciences)

Tetramisole (20 mM)

Equipment

Automated liquid media dispenser

Centrifuge

Flask (large)

Fluorescence dissecting microscope

Incubator at 37°C

Opaque bag or tin foil (for optional AO staining)

Orbital shaker (lateral only) at 37°C and room temperature

P200 tips (sterile)

Parafilm

Plastic bags (sealable)

Plates (conical- and flat-bottom), 96-well

Replicator, 96-pin

> *We recommend the disposable 96 LONG-PIN REPAD from Singer Instruments, which can be cleaned with 70% ethanol and reused many times.*

Slides

METHOD

> *Careful handling of liquid cultures must be used at all times. Although it seems obvious, the bacteria in plates can very easily be spilled and enter surrounding wells. Take special care in keeping the plates horizontal, especially in transition to and from the orbital shaker.*

1. Using an automated liquid medium dispenser, aliquot ~150 μL of LB + amp and tet into each well of a conical-bottom 96-well plate.

2. Thaw the RNAi library glycerol stock for 5–10 min at room temperature and inoculate the liquid cultures as follows.
   - If you are targeting a specific class of genes (i.e., the "kinome"), use individual sterile P200 tips to take small samples of bacteria from the target wells. Transfer each tip containing bacteria directly into one well of the 96-well plate containing medium. Incubate at room temperature for 5 min and discard the tip.
   - If you are conducting a full genome screen, use a 96-pin replicator to inoculate the liquid cultures.
     > *The 96-pin replicator will create a pattern of even and odd numbers and letters from the 384-well library plate. A → C → E …; 1 → 3 → 5 …etc. Hence, rigorous labeling and documentation will be needed, lest you end up in a puddle of misplaced genes!*

3. Place the lid on the plate and seal the wells with Parafilm to prevent evaporation. Wrap the plate in a wet paper towel, place in a sealed plastic bag and incubate on an orbital shaker at 37°C.

4. After 24 h of incubation, confirm bacterial growth by observing cloudiness of the medium. Add 6 μL of IPTG (0.1 M) per well. Rewrap the plate and place it back in the orbital shaker for 1 h at 37°C.

Cite this protocol as *Cold Spring Harb Protoc*; doi:10.1101/pdb.prot080234

5. Following incubation, centrifuge the plates at 3500 rpm to pellet the bacteria. Shake the plates sharply into a large flask to discard the supernatant.

   The pellet should remain in the wells.

6. Add 150 µL of liquid NGM to each well and resuspend the bacteria.

   To delay the screen at this step, plates can be stored at 4°C for up to 7 d.

   Depending on your experimental design, a resuspension volume of 150 µL should allow for creation of both an experimental and a control plate, and provide extra liquid culture for subsequent confirmations on solid plates.

7. Transfer 50 µL of each liquid RNAi culture into individual wells of a flat-bottom 96-well plate.

8. Determine the number of L1 worms per microliter in your sample by placing a small volume (2, 5, or 10 µL) of worms on a slide in tetramisole. Count the worms using a dissecting microscope. Repeat at least five times to obtain an accurate estimate. Dilute the worms to a concentration of 1 worm/µL.

9. Pipette 10 µL of worms into each well of the 96-well plate from Step 7. Seal the plate as above and place on an orbital shaker at room temperature. Grow worms to the young adult stage (∼2–3 d).

   The ideal stage for observation of the worms is the young adult stage; exact timing will depend on the growth rate of the strain. Thus, we highly recommend determining the rate of growth of the strain of interest before screening. If you are unsure on the timing of larval stages for your particular worm strains, it may be prudent to quantify growth earlier in the incubation period. It is important to note that as worms age, their intestines will increasingly obstruct the germline, so avoid trying to assess apoptosis >48 h after the young adult stage. Should you wish to determine the effects of stresses such as DNA damage on apoptosis it is important to treat worms with stress 24 h after the L4 stage, then assess apoptosis 24 h later (i.e., 48 h after L4).

   See Troubleshooting.

10. (Optional) Stain the worms with AO.

    i. Add 20 µL of freshly diluted AO solution (75 µg/mL) to each well of the 96-well plate containing worms. Wrap the plate in an opaque bag or tin foil to prevent light from breaking down the dye. Place the plate on an orbital shaker at room temperature for 1 h.

    ii. After 1 h, unpackage the plates and place them at 4°C for 10 min.

       The worms will stop moving at the colder temperature and descend to the bottoms of the wells.

    iii. Carefully remove the liquid from the wells, leaving the worms intact at the bottom of each well. Add 50 µL of M9 buffer to the worms. Place the plates at 4°C for 10 min, remove the liquid, and repeat the M9 wash.

    iv. Repackage the plates (including light resistant covering) and place them on the orbital shaker at room temperature for 3 h to destain. Observe the worms under the fluorescence channel of a dissecting microscope to be sure background has been reduced appropriately.

       The critical factor in apoptosis detection by vital dyes is the stain:destain ratio; see Protocol 3: Fluorescent Visualization of Germline Apoptosis in Living Caenorhabditis elegans *(Lant and Derry 2014a). A 1:3 ratio is suggested here, but may not be exact for a given experiment or strain. It is important to check staining levels at this step to ascertain whether more destaining time is necessary before proceeding.*

       See Troubleshooting.

11. Before analysis on the dissecting microscope, add 5 µL of tetramisole to each well.

    Addition of tetramisole will straighten and immobilize the worms, thereby increasing visibility of the germline.

12. Under the dissecting microscope, locate and focus on the worms in the bright field then switch to the green fluorescent protein (GFP) channel to observe the fluorescent corpses. Use the 10× magnification initially to localize the fluorescence and then move to a higher magnification (20×– 50×) to look specifically at the corpses in the germline. Look at germlines of multiple worms to avoid recording anomalies.

    The average number of apoptotic corpses is generally equivalent in the anterior and posterior germline regions of the same animals, but we find the posterior gonad easier to visualize because it is less obstructed by the intestine.

    See Troubleshooting.

13. Following the screen, or concurrent to running it, confirm any positive "hits" (i.e., candidate apoptosis genes) by reculturing the candidate RNAi strain (or using leftover liquid stock from Step 6, if this is within a week of initial growth) as outlined in Protocol 1: Induction of Germline Apoptosis in *Caenorhabditis elegans* (Lant and Derry 2014b).

> *If you have a large number of candidates to test, it may be more convenient to rescreen on 24-well plates containing solid NGM rather than on individual plates. It may also be necessary to cultivate the worms on the RNAi feeding strains for a second generation if the first generation produces inconsistent results.*

TROUBLESHOOTING

Problem (Step 9): Contamination is observed after incubation.

Solution: Contamination can be seen as a distinct discoloration in the liquid medium (often brown/yellow, compared to a much lighter cloudy control). In these cases, compare the coloration with surrounding wells to confirm. Often confusion can occur if there are fewer worms in certain wells (because the worms grow slower, or the RNAi induces synthetic lethality), causing bacteria in the well to be consumed more slowly than in surrounding lanes. If indeed contamination is occurring, the well will need to be replicated. There should be excess liquid culture from Step 6, which can be used on solid medium to redo any number of samples. If the stocks are contaminated, it may be necessary to streak out the bacterial strain on selective medium, then pick and culture individual colonies to remove the contamination.

Problem (Step 9): The worms appear to be dead.

Solution: They may very well be! An RNAi screen will produce a number of phenotypes, many of which are lethal. As you will be attempting to view only adult worms, embryonic lethal phenotypes will not be an issue, but synthetic lethality, sterility, and a number of growth defects should be expected. Refer to Wormbase (http://www.wormbase.org/#012-3-6) for detailed information on specific genes and results from previous RNAi screens to confirm an observed phenotype.

Problem (Step 10): There is too much fluorescent background signal in the intestine or too little apoptotic fluorescence after AO staining.

Solution: The stain:destain ratio is critical to successful visualization of corpses. Unfortunately, it requires a level of trial and error (especially if conducted solely in liquid culture), and can vary between samples. Of course, low apoptotic fluorescence may be a symptom of the target RNAi. Hence, it is important to exercise judgment as to whether low germline apoptosis is relative to the background staining or is indeed a real phenotype.

Problem (Step 12): The germline region is too difficult to assess because there are excessive numbers of worms in the well.

Solution: If there are worms settling on top of each other in a well, it may be beneficial to transfer them onto solid plates for visualization. Transferring to solid medium (in 24-well plates) will allow the worms to separate themselves and make detection of apoptotic cells easier, but this requires material and can be labor intensive. Transfer can be done during destaining (in which case the plates must be seeded with OP50 bacteria), or following destaining during visualization (in which case the plates do not need to be seeded with bacteria).

DISCUSSION

Using RNAi for High-Throughput Screening

Screening the entire genome for apoptosis phenotypes at high resolution would be a very labor-intensive endeavor, requiring tens of thousands of worms to be mounted on slides to quantify

Cite this protocol as *Cold Spring Harb Protoc*; doi:10.1101/pdb.prot080234

corpses in each germline. Screening by lower resolution greatly accelerates throughput and reduces handling time significantly, and prevents excess use of vital dyes. The utility of high-throughput screening for apoptosis regulators by fluorescence dissecting microscopy is, of course, subject to the designs of the project, but is generally advantageous because it limits the number of genes to be screened at high resolution. For example, such an approach was used to identify the SCF^{FSN-1} E3 ubiquitin ligase (*fsn-1*) (Gao et al. 2008). Initially, regulators of the Skp-Cullin-F-box (SCF) class of E3 ubiquitin ligases (i.e., *ned-8*) were found to cause lower levels of lethality in *cep-1* mutants compared with wild-type controls in a genome-wide RNAi screen. This suggested that CEP-1 might be under negative regulatory control by a cullin-based E3 ligase, which in turn reduced the number of candidate genes to be screened for apoptosis phenotypes at high resolution. Systematic analysis of ∼300 substrate-specific adaptors of SCF ligases (F-box genes) using DIC microscopy led to the identification of *fsn-1* (Gao et al. 2008). Thus, a first pass screen at low resolution to identify genes that cause selective lethality in wild-type copies but not mutants of proapoptotic genes (such as *cep-1*) narrowed down the search to specific classes of genes (i.e., F-box genes). Targeting high-resolution screens to a limited number of genes has also been successfully used to identify apoptosis regulators from other ubiquitin ligases, such as the HECT domain E3 ligase EEL-1 (Ross et al. 2011).

The high-throughput method described here facilitates the rapid assessment of potentially thousands of RNAi targets, and once the techniques have been learned, the procedure is fairly simple and largely revolves around volume of work rather than specific skills. Data analysis (see below) requires simultaneous microscopy and recognition of apoptotic corpses. In terms of timing, experience and comfort with the techniques are the major determinants of throughput rate. It is a process that requires numerous elements of preparation, so we highly recommend starting slowly then scaling up after the basics have been established. Timing the culturing of the worms and the bacteria must be done carefully, so that neither the worms nor bacteria are compromised by being ready too early. This is particularly pertinent in the case of the worm bleaching procedure, because arrested L1 worms will only survive a limited amount of time in M9 buffer! Not having enough worms to adequately populate hundreds or thousands of wells can seriously delay your screen. Hence, planning both bleaching days and RNAi inoculation days will be critical before starting the project. Subsequent calculations of the number of worm plates, volumes of buffer and tips (and you will go through them at an astounding rate!) that you will need are equally important.

It should be cautioned that the strategy described in this protocol will not always catch all relevant genes because of the inherent variability of RNAi, but exploiting more general phenotypes such as synthetic lethality (or buffering of lethality) or body morphology defects can effectively identify a subset of genes for more tedious high-resolution analysis. In addition, follow-up analysis of reproducible apoptotic phenotypes identified by RNAi should be performed using available mutant strains. Confirmation of results and quantitative analysis should be conducted using high-resolution optics (described in Protocol 3: Fluorescent Visualization of Germline Apoptosis in Living *Caenorhabditis elegans* [Lant and Derry 2014a]). As an alternative to RNAi-induced apoptosis, screening can be performed after exposing worms to various stressors, such as DNA damage via gamma irradiation, hypoxia, or small bioactive compounds (see Protocol 1: Induction of Germline Apoptosis in *Caenorhabditis elegans* [Lant and Derry 2014b]).

Data Analysis

High-throughput screens produce extensive quantities of data, which can be challenging to manage. The details of your screen will determine how much or little you decide to filter when setting thresholds for strong, intermediate, and weak phenotypes. A simple system of qualitative assessment (i.e., increased by twofold or more, decreased by twofold or more, or undetectable differences) of apoptosis is advised. This allows you to pick the so-called low hanging fruit, which enables analysis of a select number of candidate genes for deeper mechanistic studies. For example, AO staining was successfully used to conduct a genome-wide RNAi screen for genes that cause increased levels of apoptosis in the germline, which were then quantified by secondary analysis at higher resolution (Lettre et al. 2004). Alternatively, a screen can use a more global parameter, such as sterility, as a read out for massive apoptosis to reduce the number of hits, with apoptosis being directly assessed on a secondary/confirmatory run. Statistical analysis of apoptosis levels is important if you detect genes that cause intermediate effects on germline apoptosis. Thus, at least three independent replicates must be performed and, as previously mentioned, any gene that is identified as a candidate apoptosis regulator by RNAi should be confirmed in mutant strains, if available.

While the process may seem daunting, the ability to cover an entire genome in a matter of a few months can provide feedstock for several years of future studies.

RECIPES

Luria Bertani (LB) + Amp and Tet

1. To prepare solid medium, mix 10 g of tryptone, 5 g of yeast extract, 10 g of NaCl, and 15 g of agar, and bring to 1 L with H_2O. (To prepare liquid medium, combine all ingredients except agar.)
2. Autoclave for 30 min.
3. Let cool for 1 h in a 55°C water bath.
4. Add 2 mL of 50 mg/mL ampicillin and 2 mL of 5 mg/mL tetracycline.
5. While the agar-containing medium is still warm, pour a thin layer into each Petri dish and let cool to solidify.
6. Store dishes or liquid medium at 4°C for up to 1 mo.

M9 Buffer for Worms

1. Dissolve 3 g of KH_2PO_4, 6 g of Na_2HPO_4, and 5 g of NaCl in 1 L of H_2O.
2. Autoclave for 20 min.
3. Add 1 mL of 1 M $MgSO_4$.
4. Store at room temperature. After 2 wk, check for visible contamination before use.

Nematode Growth Medium (NGM)

1. For solid NGM, mix 3 g of NaCl, 2.5 g of peptone, and 20 g of agar and bring to 1 L with H_2O. (For liquid NGM, prepare without agar.)
2. Autoclave for 1 h.
3. Let cool for 1 h in a 55°C water bath.
4. Add 1 mL of cholesterol (5 mg/mL in ethanol), 1 mL of 1 M $CaCl_2$, 1 mL of 1 M $MgSO_4$, and 25 mL of 1 M (pH 6.0) KPO_4, mixing after each addition.
5. Using an automated plate pourer (peristaltic pump), pour the medium into sterile plastic plates to set. During pouring, keep the NGM on a hot plate with stirrer to prevent the medium from solidifying.
6. Store NGM plates at 4°C until use. Seed NGM plates with bacteria the day after pouring. Warm the plates to room temperature before adding the worms. (For liquid NGM, store at room temperature for up to 1 mo. Regularly check for excess cloudiness to ensure bacterial contamination is absent.)

 Cite this protocol as *Cold Spring Harb Protoc*; doi:10.1101/pdb.prot080234

ACKNOWLEDGMENTS

We are grateful to B. Yu and M. Gunda for their insight and help with procedures. This work was funded by grants from the Excellence in Radiation Research for the 21st Century (EIRR21) Radiation Medicine Research Program (B.L.), operating grants from the Canadian Institutes of Health Research (CIHR) (MOP79495) (W.B.D.), and infrastructure grants from the Canada Foundation for Innovation and Ontario Research Fund (Project No. 9782) (W.B.D.).

REFERENCES

Gao MX, Liao EH, Yu B, Wang Y, Zhen M, Derry WB. 2008. The SCFFSN-1 ubiquitin ligase controls germline apoptosis through CEP-1/p53 in *C. elegans*. *Cell Death Differ* **15**: 1054–1062.

Lant B, Derry WB. 2014a. Fluorescent visualization of germline apoptosis in living *Caenorhabditis elegans*. *Cold Spring Harb Protoc* doi: 10.1101/pdb. prot080226.

Lant B, Derry WB. 2014b. Induction of germline apoptosis in *Caenorhabditis elegans*. *Cold Spring Harb Protoc* doi: 10.1101/pdb.prot080192.

Lettre G, Kritikou EA, Jaeggi M, Calixto A, Fraser AG, Kamath RS, Ahringer J, Hengartner MO. 2004. Genome-wide RNAi identifies p53-dependent and -independent regulators of germ cell apoptosis in *C. elegans*. *Cell Death Differ* **11**: 1198–1203.

Ross AJ, Li M, Yu B, Gao MX, Derry WB. 2011. The EEL-1 ubiquitin ligase promotes DNA damage-induced germ cell apoptosis in *C. elegans*. *Cell Death Differ* **18**: 1140–1149.

Stiernagle T. Maintenance of *C. elegans*. (February 11, 2006), *WormBook*, ed. The *C. elegans* Research Community, wormbook. 1.0101.1, http://www. wormbook.org.

Timmons L, Court DL, Fire A. 2001. Ingestion of bacterially expressed dsRNAs can produce specific and potent genetic interference in *Caenorhabditis elegans*. *Gene* **263**: 103–112.

Immunostaining for Markers of Apoptosis in the *Caenorhabditis elegans* Germline

Benjamin Lant and W. Brent Derry[1]

Developmental and Stem Cell Biology Program, The Hospital for Sick Children, Toronto, Ontario M5G 1L7, Canada; Department of Molecular Genetics, University of Toronto, Toronto, Ontario M5S 1A1, Canada

The transparency of *Caenorhabditis elegans* makes it an ideal organism for visualizing proteins by immunofluorescence microscopy; however, the tough cuticle of worms and the egg shell surrounding embryos pose challenges in achieving effective fixation so that antibodies can diffuse into cells. In this protocol, we describe immunostaining of apoptosis-related proteins in the *C. elegans* adult germline using fluorescent reagents. Protein localization and abundance can be determined in various mutant backgrounds and under a variety of conditions, such as exposure to genotoxic stress. The number of antibodies specific to *C. elegans* proteins is quite limited compared with other organisms, but there is a growing list of immunological reagents directed against proteins in other organisms that cross-react with the homologous *C. elegans* proteins.

MATERIALS

It is essential that you consult the appropriate Material Safety Data Sheets and your institution's Environmental Health and Safety Office for proper handling of equipment and hazardous material used in this protocol.

RECIPES: Please see the end of this protocol for recipes indicated by <R>. Additional recipes can be found online at http://cshprotocols.cshlp.org/site/recipes.

Reagents

4′,6-Diamidino-2-phenylindole (DAPI) (optional; see Step 18)

DNA can be stained with DAPI to assess nuclear morphology and/or quantify nuclei in the germline. Prepare a solution of 1 µg/mL DAPI in PBST.

Acetone (chilled to −20°C)

Antibody (primary)

Antibody (secondary, fluorescently labeled)

Blocking solution

Prepare a blocking solution of 0.1% BSA in PBST.

Bovine serum albumin (BSA)

C. elegans strain of interest, at the L4 stage

Always use strains with a null mutation in the gene corresponding to the protein of interest as negative controls to ensure antibody specificity. If mutants are not available, RNA interference (RNAi) can be used to knockdown the

[1]Correspondence: brent.derry@sickkids.ca

Cite this protocol as *Cold Spring Harb Protoc*; doi:10.1101/pdb.prot080242

gene; see Protocol 1: Induction of Germline Apoptosis in Caenorhabditis elegans *(Lant and Derry 2014). The method described here can also be used to determine if RNAi clones are working as expected.*

Image-iT FX signal enhancer (Life Technologies)

Methanol (chilled to −20°C)

Paraformaldehyde solution (4%) <R>

On the day of staining, prepare a solution of 2% paraformaldehyde by diluting the 4% solution in an equal volume of 1× PBS <R>.

PBST

Prepare a PBST solution of 0.1% Tween-20 in 1× PBS.

Phosphate-buffered saline (PBS) (1×) (pH 7.4) <R>

Plates containing solid nematode growth medium (NGM) <R>, seeded with OP50 or RNAi bacteria

ProLong Gold antifade reagent (Invitrogen P36930)

Tetramisole (20 mM)

Triton X-100

Prepare a solution of 1% Triton X-100 in 1× PBS <R>.

Unique reagents required for preparing acetone blocking powder (optional; see Steps 22–31):

Acetone (chilled to −20°C)

C. elegans

We recommend using at least six large (10-cm diameter) plates worth of young adult worms for preparation of acetone blocking powder.

NaCl (0.1 M, chilled)

Sucrose (60% in 0.1 M NaCl)

Equipment

Centrifuge

Compound microscope with differential interference contrast (DIC) and fluorescence optics

Coverslips

Dry ice

Forceps

Glass transfer pipettes

Glass pipettes are recommended because worms are prone to stick to plastic tips.

Humidified chamber at room temperature

Imaging software

Metal heating block

Nail polish

Needles (25-gauge)

Razor blade

Sealing wax

Slides coated with poly-L-lysine

Tubes (2-mL)

Unique equipment required for preparing acetone blocking powder (optional; see Steps 22–31):

Incubator at 37°C

Liquid nitrogen

Microcentrifuge tubes

Mortar and pestle

Petri dishes (sterile plastic)

Rotating platform

Tubes (15-mL conical)

METHOD

Steps 1–21 describe the standard immunostaining procedure. If you notice high background signal after staining, we recommend repeating the procedure and blocking the specimens before staining using an acetone worm powder, prepared as in Steps 22–31.

Immunostaining Dissected Gonads

Gonad dissection (Step 4) can be a difficult skill to master. It requires a steady hand and should be practiced before excising the gonads of target worms for fixation and staining. Finding the ideal pressure for fixing excised gonads to slides (Step 7) also may take some practice.

If you wish to compare staining between different genotypes or assay conditions, all samples should be prepared on the same day with the same solutions and antibody dilutions. Small variations in antibody binding can have large effects on signal intensity. In addition, for all antibody treatments, staining, and subsequent washes, it is crucial that the durations are equivalent between controls and replicates. Variation in incubation times can affect signal strength.

1. One day before staining, select at least 50 L4-stage worms per condition and incubate them on an NGM plate containing bacteria overnight at room temperature.

2. On the following day, wash the target worms off the plate with 1 mL of 1× PBS. Transfer the worms, in PBS, into a 2-mL tube and centrifuge at 2000 rpm for 1 min. Decant the supernatant and discard.

   A worm pellet of ~70 µL should remain.

3. Using a glass transfer pipette, transfer the pellet of worms onto a coated slide. Apply a 1-µL drop of 20 mM tetramisole on top of the worms to immobilize them.

4. Isolate a single immobilized worm. Using two 25-gauge needles like scissors, make cuts below the pharynx and above the tip of the tail to remove the head and tail from the worm so that the gonad extrudes from the body.

   It is important to work quickly, as gonads will not pop out of the decapitated bodies after ~10 min in solution.

5. After ~50 gonads have been extruded, apply 30 µL of 2% paraformaldehyde solution to the worms. Incubate the slide at room temperature for 10 min.

   If gonads detach from slide at this step, see Troubleshooting.

6. Remove the excess fluid carefully with a pipette or capillary and place a coverslip diagonally over the sample, such that the corners of the coverslip hang off the edge of the slide. Press down gently on the coverslip with the back of a pair of forceps.

   Be sure to leave enough of the coverslip overhanging such that it can be used for leverage, otherwise the glass will break during coverslip removal in Step 9. Be careful not to exert too much pressure and disintegrate the germlines.

7. Place the slide on a cold metal block supported on a bed of dry ice for at least 15 min until the dissected worms and germlines freeze to the slide.

8. Repeat Steps 2–7 for any additional plates of worms.

9. Use a razor blade to carefully lever the coverslips off the slides and immediately immerse the slides in ice-cold 1:1 methanol:acetone at −20°C for 1 min.

10. Remove each slide from the solution and allow it to dry.

11. To permeabilize the samples, wash the slides with 1% Triton X-100 for 10 min at room temperature.

12. Apply 200 µL of PBST to each slide, cover with a piece of Parafilm to prevent evaporation, and incubate for 10 min in a humidified chamber. Remove the excess fluid. Repeat this step two more times.

13. Apply two drops of Image-iT FX signal enhancer to each slide and incubate for 20 min in a humidified chamber.

Cite this protocol as *Cold Spring Harb Protoc*; doi:10.1101/pdb.prot080242

14. Remove the excess fluid and add 200 µL of blocking solution to each slide. Incubate for 30 min in a humidified chamber.

15. Remove the excess fluid and add 50 µL of primary antibody (dilute antibody to appropriate concentration in PBST + 1% BSA) to each slide. Cover the slides with sealing wax to prevent evaporation and incubate overnight in a humidified chamber.

    Alternatively, primary antibody can be combined with acetone blocking powder to reduce nonspecific staining. Preparation and use of acetone blocking powder is outlined in Steps 22–31.

16. On the following day, wash the slides three times with 200 µL of PBST for 10 min per wash in a humidified chamber.

17. Add 50 µL of secondary antibody (diluted to 1:500 in PBST + 1% BSA) to each slide. Incubate the slides for 1 h in a humidified chamber.

18. (Optional) Perform DAPI staining as follows.

    i. Wash the slides twice with 200 µL of PBST for 10 min per wash in a humidified chamber.

    ii. Incubate slides in DAPI solution for 10 min.

19. Following the addition of secondary antibody (and DAPI, if applicable) staining, wash the slides three times with 200 µL of PBST for 10 min per wash in a humidified chamber.

20. Mount the samples in ProLong Gold, using less than one drop to prevent coverslip movement. Apply a coverslip to each slide and incubate in the dark overnight at 4°C.

21. On the following day, seal the coverslips to the slides using a light coating of nail polish and visualize samples by fluorescence microscopy.

    Although it seems counterintuitive not to seal the slide overnight, this will allow excess PBST to evaporate which in turn improves visibility.

    Immunostaining for a given protein target or phosphorylation site can be assessed qualitatively (as the presence or absence of signal) or by quantitative analysis, which is a standard capability of most imaging software. If you are looking at staining in the nuclei, it is important that intensity values are normalized to the intensity values of the surrounding cytosol. It is also critical to use the same exposure and gain settings on the imaging software when comparing intensities between samples.

    See Troubleshooting.

Preparing Acetone Blocking Powder

22. Grow worms to adulthood on 5 × 10 mm NGM plates then wash the worms from the plates into a 15-mL conical tube using 1× PBS. Centrifuge the worms at 1000 rpm for 2 min to pellet, and carefully remove the supernatant.

    When centrifuging worms for this procedure, you will often notice a "tornado" of worms following a spin. Resist the urge to crank up the RPMs! (Although technically worms will survive spins of 5000 rpm.) Allow the samples to settle after each spin and take extra care removing the supernatant, as the top layer of worms will not be overly compact.

    It is important to obtain a sufficient quantity of adult worms for preparation of the acetone blocking powder. We recommend predetermining the number of L1 stage worms for each mutant strain that will nearly starve out a 100-mm plate by the time they develop to adults. We also recommend growing dense lawns of OP50 bacteria to maximize the number of worms for this procedure.

23. Wash the pellet three times with cold 0.1 M NaCl. Centrifuge at 1000 rpm for 2 min and remove the supernatant after each wash.

24. Fill the tube with 5 mL of cold 0.1 M NaCl and 5 mL of 60% sucrose and mix. Centrifuge at 2000 rpm for 3 min. Use a glass transfer pipette to collect and transfer the worms to a fresh tube.

25. Resuspend the worms in 10 mL of cold dH$_2$O, centrifuge at 1000 rpm for 2 min and remove the supernatant. Repeat for a second wash with cold dH$_2$O.

26. Fix the worms with 5 mL of 4% paraformaldehyde at 4°C for 4 h on a rotating platform. Once the fixation is complete, centrifuge the worms at 1000 rpm for 2 min and remove the supernatant.

The volume of paraformaldehyde is dependent on how many worms you are using; 5 mL corresponds to six large plates.

27. Wash the worms with 10 mL of PBST, centrifuge at 1000 rpm for 2 min and remove the supernatant.

28. Incubate the worms with 5 mL of cold acetone for 20 min. Centrifuge at 1000 rpm for 2 min and remove the supernatant.

29. Wash the worms twice with 10 mL of PBST followed by once with 10 mL of PBS. Centrifuge at 1000 rpm for 2 min and remove the supernatant after each wash.

30. Transfer the worm pellet into a mortar, cover the sample with liquid nitrogen, and grind it into a fine powder with a pestle. Place this homogenate on a sterile plastic Petri dish and dry overnight in a 37°C incubator.

    The worm homogenate powder can be stored in microcentrifuge tubes at 4°C for subsequent experiments.

31. To use the powder during immunostaining, mix the powder with just enough blocking solution to form a paste. Add the target antibody to a 1:1 ratio (i.e., 50 µL of paste to 50 µL of primary antibody) and proceed with Step 15.

TROUBLESHOOTING

Problem (Step 5): The extracted gonad will not stick to the slide with the addition of liquids.

Solution: Once you have extracted the gonads onto the poly-L-lysine-coated slide, the subsequent application of liquids can loosen the samples from the slide. If you experience this, you may need to apply more coats of poly-L-lysine solution to the slides to ensure the samples stick. Be gentle when handling slides with immobilized germlines!

Problem (Step 21): The fluorescent signal of the target protein is too low.

Solution: There are a number of possible explanations. Most importantly, you may be seeing a significant result (i.e., a biologically relevant variation in protein expression). Alternatively, there may have been an error in antibody dilution or the antibodies may have denatured. Consider changing either the primary or secondary antibody dilution to increase signal. If you have a worm strain that overexpresses the target protein, you can use it to ensure the antibodies are working well.

Problem (Step 21): Background fluorescence is too high.

Solution: If you have not tried it already, use the acetone blocking powder. The use of Image-iT FX Signal Enhancer should also reduce the level of nonspecific fluorescence, but it may be necessary to dilute the antibodies to reduce background signal. Finally, increasing the number of washes following either primary or secondary antibody incubation will also help optimize the signal-to-background ratio.

Problem (Step 21): Germlines appear damaged.

Solution: This may be an issue with fixation. Reduction of fixation time to 5 min may eliminate germlines that appear altered/damaged in the visualization step. Alternatively, germlines can appear to have exploded if the coverslip is pressed on with too much force. Practice, practice, practice!

DISCUSSION

Immunostaining allows for in situ visualization of numerous protein features, such as subcellular localization or relative levels in different genetic backgrounds or environmental conditions. Germline immunostaining is a relatively straightforward technique but requires some practice and optimization to achieve the best results. The largest issue to consider is reproducibility of staining. A second

Cite this protocol as *Cold Spring Harb Protoc*; doi:10.1101/pdb.prot080242

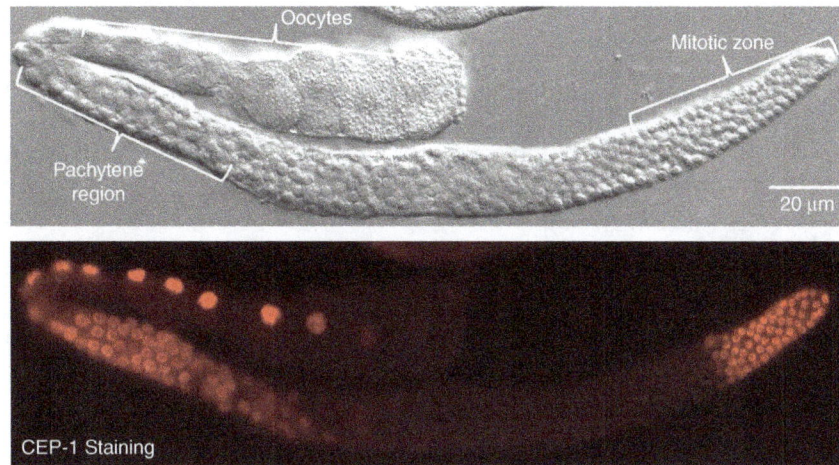

FIGURE 1. CEP-1 immunostaining of an excised germline. The *top* panel shows the white light image, clearly indicating the regions of the germline (moving *right* to *left*, from the distal [mitotic] tip through the meiotic transition zone to the bend of the pachytene region and into the developing region of the oocytes). The *bottom* panel shows CEP-1 staining, indicating strong signal in the mitotic zone, pachytene region, and oocytes. Germlines visualized at 400× magnification.

common problem is the abundance of background signal. However, using acetone blocking powder in conjunction with an antifade reagent (i.e., ProLong Gold) should ensure that images are fairly clean.

Unfortunately, there are a limited number of antibodies available which are specific to *C. elegans* proteins, but the list is growing. Antibodies to *C. elegans* core apoptosis proteins have proven to be very useful for a number of studies in the germline and soma (Greiss et al. 2008; Ito et al. 2010; Sendoel et al. 2010; Pourkarimi et al. 2012). However, most of these reagents are not commercially available and therefore require making polite requests from the *C. elegans* researchers in whose laboratories they were generated. For the intrepid, *C. elegans* antigens can be used to develop antibodies; this process is described in the review by Shakes et al. (2012).

Despite the availability of worm-specific antibodies as a limiting factor, many antibodies to well-conserved proteins or epitopes work quite well in *C. elegans*. For example, antibodies raised to the diphosphorylated form of mammalian Erk cross-react with phosphorylated MPK-1 in the *C. elegans* germline (Rutkowski et al. 2011; Perrin et al. 2013). These antibodies are emerging as important reagents for analyzing apoptotic signaling in the germline, because the MAPK pathway promotes both physiological and stress-induced apoptosis in this tissue (Kritikou et al. 2006; Rutkowski et al. 2011; Perrin et al. 2013). Antibodies to the p53-like protein CEP-1 are also commercially available (Santa Cruz Biotechnology) (Fig. 1), and antibodies that recognize ATM consensus phosphorylation sites (pS/TQ) can be used to detect these epitopes in *C. elegans* as a readout for DNA damage checkpoint activation (Vermezovic et al. 2012).

RELATED TECHNIQUES

Immunostaining can be performed using embryos as well as whole worms (both larva and adults); see the review by Shakes et al. [2012] or refer to *WormBook* (Duerr).

Individual proteins can be isolated from worm tissue and detected by immune (western) blotting for quantification (via densitometry) of protein levels in response to mutation or stress. This procedure is described in *WormBook* (Duerr). The approach holds the caveat of requiring high protein concentrations, such that typically the whole worm (rather than a specific tissue) is used for preparation.

RECIPES

Nematode Growth Medium (NGM)

1. For solid NGM, mix 3 g of NaCl, 2.5 g of peptone, and 20 g of agar and bring to 1 L with H_2O. (For liquid NGM, prepare without agar.)
2. Autoclave for 1 h.
3. Let cool for 1 h in a 55°C water bath.
4. Add 1 mL of cholesterol (5 mg/mL in ethanol), 1 mL of 1 M $CaCl_2$, 1 mL of 1 M $MgSO_4$, and 25 mL of 1 M (pH 6.0) KPO_4, mixing after each addition.
5. Using an automated plate pourer (peristaltic pump), pour the medium into sterile plastic plates to set. During pouring, keep the NGM on a hot plate with stirrer to prevent the medium from solidifying.
6. Store NGM plates at 4°C until use. Seed NGM plates with bacteria the day after pouring. Warm the plates to room temperature before adding the worms. (For liquid NGM, store at room temperature for up to 1 mo. Regularly check for excess cloudiness to ensure bacterial contamination is absent.)

Paraformaldehyde Solution (4%)

1. Add 30 mg of paraformaldehyde to 1 µL of 1 M NaOH in 386 µL of H_2O. Incubate at 60°C in a heat block and vortex intermittently until the paraformaldehyde enters solution. (If this process takes longer than 30 min, an additional 1 µL of 1 M NaOH can be added.)
2. Place the heat block containing the paraformaldehyde solution on dry ice, cover, and allow the solution to cool.
3. Add 386 µL of phosphate buffer for worms.
4. Store on ice until use.

   This solution must be prepared fresh every time.

Phosphate-Buffered Saline (PBS)

Reagent	Amount to add (for 1× solution)	Final concentration (1×)	Amount to add (for 10× stock)	Final concentration (10×)
NaCl	8 g	137 mM	80 g	1.37 M
KCl	0.2 g	2.7 mM	2 g	27 mM
Na_2HPO_4	1.44 g	10 mM	14.4 g	100 mM
KH_2PO_4	0.24 g	1.8 mM	2.4 g	18 mM

If necessary, PBS may be supplemented with the following:

$CaCl_2 \cdot 2H_2O$	0.133 g	1 mM	1.33 g	10 mM
$MgCl_2 \cdot 6H_2O$	0.10 g	0.5 mM	1.0 g	5 mM

PBS can be made as a 1× solution or as a 10× stock. To prepare 1 L of either 1× or 10× PBS, dissolve the reagents listed above in 800 mL of H_2O. Adjust the pH to 7.4 (or 7.2, if required) with HCl, and then add H_2O to 1 L. Dispense the solution into aliquots and sterilize them by autoclaving for 20 min at 15 psi (1.05 kg/cm^2) on liquid cycle or by filter sterilization. Store PBS at room temperature.

Phosphate Buffer for Worms

Add 200 mg of KH_2PO_4 to 940 mg of Na_2HPO_4 in 25 mL of H_2O. Adjust pH to 7.2. Store at 4°C or room temperature for up to 1 mo.

ACKNOWLEDGMENTS

We are grateful to B. Yu and M. Gunda for their insight and help with procedures. This work was funded by grants from the Excellence in Radiation Research for the 21st century (EIRR21) Radiation Medicine Research Program (B.L.), operating grants from the Canadian Institutes of Health Research (CIHR) (MOP79495) (W.B.D.), and infrastructure grants from the Canada Foundation for Innovation and Ontario Research Fund (Project No. 9782) (W.B.D.).

REFERENCES

Duerr JS. Immunohistochemistry. (June 19, 2006), *WormBook*, ed. The *C. elegans* Research Community, WormBook, doi:10.1895/wormbook.1.105.1, http://www.wormbook.org.

Greiss S, Hall J, Ahmed S, Gartner A. 2008. *C. elegans* SIR-2.1 translocation is linked to a proapoptotic pathway parallel to cep-1/p53 during DNA damage-induced apoptosis. *Genes Dev* 22: 2831–2842.

Ito S, Greiss S, Gartner A, Derry WB. 2010. Cell-nonautonomous regulation of *C. elegans* germ cell death by kri-1. *Curr Biol* 20: 333–338.

Kritikou EA, Milstein S, Vidalain PO, Lettre G, Bogan E, Doukoumetzidis K, Gray P, Chappell TG, Vidal M, Hengartner MO. 2006. *C. elegans* GLA-3 is a novel component of the MAP kinase MPK-1 signaling pathway required for germ cell survival. *Genes Dev* 20: 2279–2292.

Lant B, Derry WB. 2014. Induction of germline apoptosis in *Caenorhabditis elegans. Cold Spring Harb Protoc* doi:10.1101/pdb.prot080192.

Perrin AJ, Gunda M, Yu B, Yen K, Ito S, Forster S, Tissenbaum HA, Derry WB. 2013. Noncanonical control of *C. elegans* germline apoptosis by the insulin/IGF-1 and Ras/MAPK signaling pathways. *Cell Death Differ* 20: 97–107.

Pourkarimi E, Greiss S, Gartner A. 2012. Evidence that CED-9/Bcl2 and CED-4/Apaf-1 localization is not consistent with the current model for *C. elegans* apoptosis induction. *Cell Death Differ* 19: 406–415.

Rutkowski R, Dickinson R, Stewart G, Craig A, Schimpl M, Keyse SM, Gartner A. 2011. Regulation of *Caenorhabditis elegans* p53/CEP-1-dependent germ cell apoptosis by Ras/MAPK signaling. *PLoS Genet* 7: e1002238.

Sendoel A, Kohler I, Fellmann C, Lowe SW, Hengartner MO. 2010. HIF-1 antagonizes p53-mediated apoptosis through a secreted neuronal tyrosinase. *Nature* 465: 577–583.

Shakes DC, Miller DM 3rd, Nonet ML. 2012. Immunofluorescence microscopy. *Methods Cell Biol* 107: 35–66.

Vermezovic J, Stergiou L, Hengartner MO, d'Adda di Fagagna F. 2012. Differential regulation of DNA damage response activation between somatic and germline cells in *Caenorhabditis elegans. Cell Death Differ* 19: 1847–1855.

CHAPTER 16

Detection of Autophagy in *Caenorhabditis elegans*

Nicholas J. Palmisano[1,2] and Alicia Meléndez[1,2,3]

[1]*Department of Biology, Queens College-CUNY, Flushing, New York 11367;* [2]*The Graduate Center, The City University of New York, New York 10016*

Autophagy is a dynamic and catabolic process that results in the breakdown and recycling of cellular components through the autophagosomal–lysosomal pathway. Many autophagy genes identified in yeasts and mammals have orthologs in the nematode *Caenorhabditis elegans*. In recent years, gene inactivation by RNA interference (RNAi) and chromosomal mutations has been useful to probe the functions of autophagy in *C. elegans*, and a role for autophagy has been shown to contribute to multiple processes, such as the adaptation to stress, longevity, cell death, cell growth control, clearance of aggregation-prone proteins, degradation of P granules during embryogenesis, and apoptotic cell clearance. Here, we discuss some of these roles and describe methods that can be used to study autophagy in *C. elegans*. Specifically, we summarize how to visualize autophagy in embryos, larva, or adults, how to detect the lipidation of the ubiquitin-like modifier LGG-1 by western blot, and how to inactivate autophagy genes by RNAi.

AUTOPHAGY IN *C. ELEGANS*

Autophagy is a lysosome-mediated pathway resulting in the degradation and recycling of long-lived proteins, protein aggregates, as well as damaged and old organelles (Levine and Klionsky 2004). It is highly conserved and has been shown to be a fundamental catabolic process in eukaryotes that is required for key developmental and pathological events. Autophagy was first described in mammals, through morphological studies of rat liver cells (Deter et al. 1967). However, it was in the budding yeast *Saccharomyces cerevisiae* where many autophagy genes (*ATG* genes) were discovered by screening for mutations that decreased the survival of yeast cells under starvation, as well as mutations that disrupted the cytoplasm-to-vacuole targeting (Cvt) process (Tsukada and Ohsumi 1993; Thumm et al. 1994; Harding et al. 1995, 1996; Hutchins and Klionsky 2001; Klionsky et al. 2003).

The process of autophagy comprises several distinct steps: formation of a phagophore (also referred to as an isolation membrane or preautophagosomal structure); elongation and closure of the phagophore to form the double-membrane autophagosome; transport and fusion of the autophagosome with a lysosome; and finally degradation of the autophagosomal contents and recycling of degraded material (Fig. 1) (Mizushima 2007; Xie and Klionsky 2007; Nakatogawa et al. 2009). In addition to fusing with a lysosome, an autophagosome can also fuse with an endosome to form a hybrid organelle called the amphisome (Liou et al. 1997; Jing and Tang 1999). When an amphisome or autophagosome fuses with a lysosome, it is referred to as an autophagolysosome (or an autolysosome).

The evolutionary conservation of autophagy genes between budding yeast and *Caenorhabditis elegans* allowed for the identification of genes that encode core components of the autophagic ma-

[3]Correspondence: alicia.melendez@qc.cuny.edu

Cite this introduction as *Cold Spring Harb Protoc*; doi:10.1101/pdb.top070466

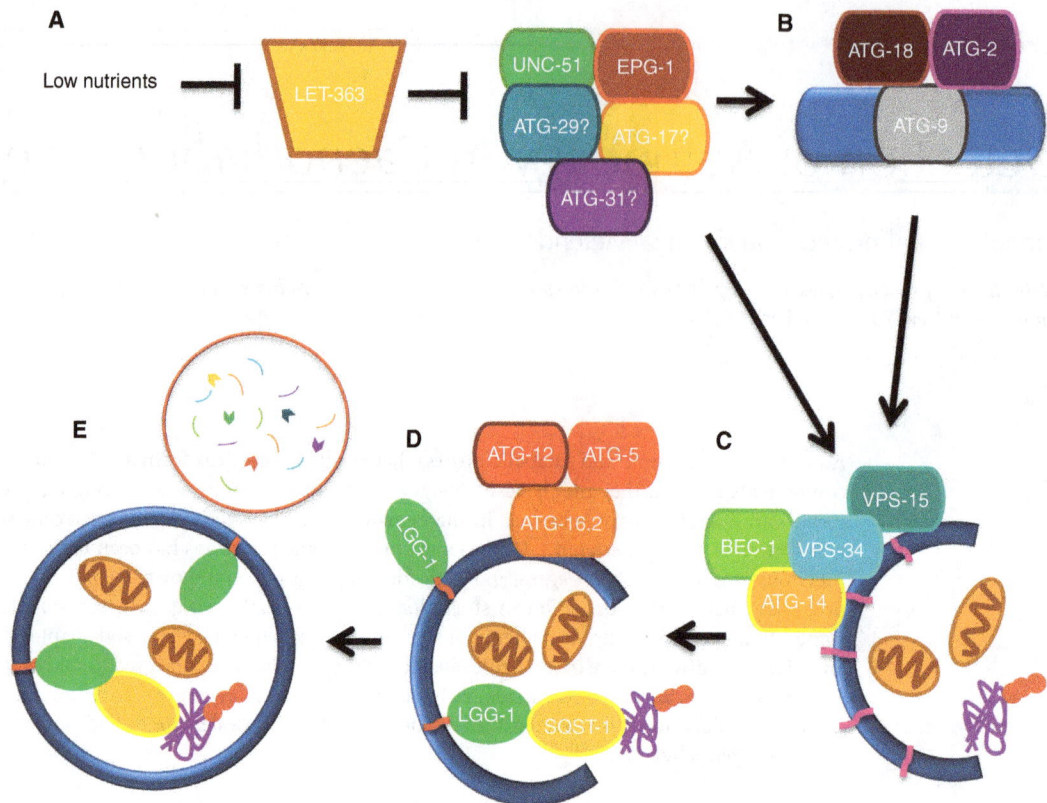

FIGURE 1. Autophagy in the nematode *Caenorhabditis elegans*. (*A*) The process of autophagy has been delineated by studies in yeast and mammalian cells. It is presumed that induction of autophagy begins with the activation of UNC-51, via loss of signaling from the *C. elegans* ortholog of the target of rapamycin (TOR), LET-363. (*B*) Autophagosome formation requires the integral protein ATG-9, thought to contribute membrane to the developing autophagosome. (*C*) Nucleation requires the class III phosphoinositide 3-kinase (PI3K) complex (BEC-1–VPS-34–ATG-14), which recruits downstream autophagy proteins to the isolation membranes (IMs) in mammals or preautophagosomal structure (PAS) in yeast, through the production of phosphatidylinositol 3-phosphate (PI3P; pale purple). (*D*) Two conjugation complexes (LGG-1 and ATG-12) are required for elongation of the IMs and completion of the developing autophagosome. LGG-1 conjugated to phosphatidylethanolamine (PE, red) binds to both the inner and outer membranes of the autophagosome. LGG-1 also has the ability to bind to autophagic adaptor proteins, such as SQST-1 (encoded by *T12G3.1*), which bind to polyubiquitylated aggregates. (*E*) The complete autophagosome eventually fuses with the lysosome (depicted in a red circle), leading to the degradation of cargo within the autophagosome.

chinery in *C. elegans* on the basis of genomic sequence homology (Table 1) (Meléndez et al. 2003; Meléndez and Levine 2009). Genetic screens for mutations that disrupt the degradation of P granules led to the discovery of autophagy genes not previously identified in *C. elegans* on the basis of sequence homology, including: *epg-1*, the ortholog of yeast *ATG13*, and *epg-8*, the ortholog of yeast *ATG14* (Table 1) (Tian et al. 2009; Yang and Zhang 2011). Although, the similarities between *S. cerevisiae*, mammals and *C. elegans* autophagy proteins suggest that the molecular mechanisms of autophagosome formation might be conserved (Fig. 1; Table 1) (Meléndez and Levine 2009), the presence of genes recently identified in *C. elegans* that do not have a yeast ortholog could indicate that autophagy involves more-complex membrane dynamics in higher eukaryotes. Hence, it is important to uncover further details about the roles of autophagy genes in autophagosome formation and maturation in *C. elegans*, and uncover further details about the function of these genes in the different settings where autophagy is required.

The role of autophagy genes in the development of *C. elegans* has emerged from studies using chromosomal mutations or RNA interference (RNAi) against autophagy genes. Chromosomal mutations exist for many of the autophagy genes found in *C. elegans*, and many RNAi clones are available (Table 1).

Cite this introduction as *Cold Spring Harb Protoc*; doi:10.1101/pdb.top070466

TABLE 1. Autophagy genes in the nematode *Caenorhabditis elegans*

C. elegans Atg gene	Allele	Yeast/mammalian ortholog	Phenotype in C. elegans	References
let-363[a]	h98	TOR1,2/mTOR	Let, LL	Brown et al. (1994); Noda and Ohsumi (1998); Vellai et al. (2003); Jia et al. (2004); Hansen et al. (2007, 2008)
unc-51[a]	e369	ATG1/Ulk1/2	Unc, AbD, Pg, Egl	Hedgecock et al. (1985); Ogura et al. (1994); Matsuura et al. (1997); Kuroyanagi et al. (1998); Meléndez et al. (2003); Zhang et al. (2009)
epg-1[a]	bp414	ATG13/Atg13	Dv, Pg	Funakoshi et al. (1997); Chan et al. (2009); Tian et al. (2009)
bec-1[a]	ok691	ATG6,VPS30/beclin 1	Let, AbD, St, SL, Pg pQ	Seaman et al. (1997); Kametaka et al. (1998); Kihara et al. (2001); Meléndez et al. (2003); Takacs-Vellai et al. (2005); Jia et al. (2007); Hansen et al. (2008); Zhao et al. (2009); Ruck et al. (2011)
let-512/vps-34[a]	h797	VPS34/Vps34	Let, SL, Pg	Seglen and Gordon (1982); Volinia et al. (1995); Roggo et al. (2002); Zhao et al. (2009); Ruck et al. (2011)
ZK930.1[a]	ok3132	VPS15/p150	ND	Panaretou et al. (1997); Kihara et al. (2001); Kovács et al. (2003)
epg-8[a]	bp251	ATG14/Atg14L, Barkor	Dv, Pg	Kihara et al. (2001); Obara et al. (2006); Sun et al. (2008); Fan et al. (2011); Yang and Zhang (2011)
epg-6	bp242	–/WIPI4	Dv, Pg	Lu et al. (2011)
epg-3[a]	bp405	–/VMP1	Dv, Pg	Tian et al. (2010)
epg-4[a]	bp425	–/EI24,PIG8	Dv, Pg	Tian et al. (2010)
atg-3	bp412	ATG3/Atg3	Pg	Tanida et al. (2002); Zhang et al. (2009)
atg-4.1[a,b]	tm3949	ATG4/Atg4	Pg	Kirisako et al. (2000); Tanida et al. (2004); Zhang et al. (2009)
atg-4.2[a,b]	tm3948	ATG4/Atg4	ND	Kirisako et al. (2000); Tanida et al. (2004); Zhang et al. (2009)
atg-5	bp545	ATG5/Atg5	ND	Mizushima et al. (1998, 2001); Tian et al. (2010)
atg7/M7.5[a]	tm831	ATG7/Atg7	AbD, SL, Pg, pQ	Kim et al. (1999); Tanida et al. (2001); Meléndez et al. (2003)
lgg-1[a]	bp500, tm3489	ATG8/LC3	Let, Dv, AbD, SL, Pg	Kirisako et al. (2000); He et al. (2003); Meléndez et al. (2003); Zhang et al. (2009); Alberti et al. (2010)
lgg-2[a]	–	ATG8/LC3	Let, Dv, AbD, SL, Pg	Kirisako et al. (2000); He et al. (2003); Meléndez et al. (2003); Zhang et al. (2009); Alberti et al. (2010)
atg-10[a]	bp588	ATG10/Atg10	ND	Shintani et al. (1999); Mizushima et al. (2002); Meléndez et al. (2003); Tian et al. (2010)
lgg-3R[a]	gk1857	ATG12/Atg12	SL, Pg	Mizushima et al. (1998); Meléndez et al. (2003); Hars et al. (2007)
atg-16.1[a,b]	–	ATG16/Atg16L1	ND	Kuma et al. (2002); Mizushima et al. (2003); Tian et al. (2010)
atg-16.2[a,b]	ok3224	ATG16/Atg16L1	ND	Kuma et al. (2002); Mizushima et al. (2003); Tian et al. (2010)
atg-2	bp576	ATG2/Atg2		Shintani et al. (2001); Wang et al. (2001); Lu et al. (2011)
atg-9[b]	bp564	ATG9/Atg9		Noda et al. (2000); Yamada et al. (2005); Reggiori and Klionoky 2006)
atg-18[b]	gk378	ATG18/WIPI1/2	Let, AbD, Pg, pQ	Barth et al. (2001); Meléndez et al. (2003); Jia et al. (2007); Polson et al. (2010); Tian et al. (2010)
epg-2	bp287	–/–	Pg	Tian et al. (2010)
epg-5[b]	bp450	–/KIAA1632	Dv, Pg	Tian et al. (2010)
sepa-1	bp402	–/–	Pg	Zhang et al. (2009)
sqst-1	ok2892	–/p62(SQSTM1)		Tian et al. (2010); Lu et al. (2011)

Let, lethal; Unc, uncoordinated; Dv, decreased viability of L1s during starvation; AbD, abnormal dauer; St, sterile; LL, long lifespan; SL, short lifespan; Pg, P granule accumulation; Egl, egg laying defective; pQ, polyQ expansion susceptibility; ND, not determined;

[a]RNAi clone available; [b]Paralogs in *C. elegans*.

AUTOPHAGY IN *C. ELEGANS* DEVELOPMENT AND AGING

L1 Arrest after Starvation

Autophagy plays a role mediating the developmental changes associated with survival during extracellular and/or intracellular stress, such as starvation (Levine and Klionsky 2004). In the absence of food, L1 larvae undergo a reversible developmental arrest and can survive for 1–2 wk (Johnson et al. 1984). The insulin-like/IGF-1 signaling (IIS) pathway, which comprises the insulin-like/IGF-1 receptor *daf-2* and the Forkhead box protein O transcription factor *daf-16*, is involved in regulating L1 arrest triggered by starvation (Gems et al. 1998; Baugh and Sternberg 2006; Fukuyama et al. 2006). Interestingly, reduced levels of autophagy have been shown to greatly decrease the survival of starved L1 larvae, emphasizing the importance of autophagy during early stages of development (Kang et al. 2007; Tian et al. 2009, 2010; Alberti et al. 2010; Lu et al. 2011; Yang and Zhang 2011).

Dauer Development

During the first larval stages, animals that are exposed to a limited food supply develop into an alternative L3 larval stage, termed dauer (Albert et al. 1981). Dauer development is associated with morphological and behavioral changes that allow for survival under harsh conditions and stress (Cassada and Russell 1975; Golden and Riddle 1984). The regulation of dauer development has been well characterized and requires the insulin-like/IGF-1, guanylyl cyclase, and transforming growth factor-β signaling pathways, as mutations in any of these pathways can result in a dauer-constitutive phenotype (Daf-c) or a dauer-defective phenotype (Daf-d) (Estevez et al. 1993; Thomas et al. 1993; Gottlieb and Ruvkun 1994; Ren et al. 1996; Schackwitz et al. 1996; Patterson et al. 1997; Birnby et al. 2000; Inoue and Thomas 2000; da Graca et al. 2004). Dauer development is associated with an increase in autophagy, which appears to be required for the cell remodeling associated with proper dauer formation (Meléndez et al. 2003).

Longevity Pathways

In *C. elegans*, aging is controlled by multiple longevity pathways, such as insulin-like growth factor signaling, target of rapamycin (TOR) signaling, dietary restriction, mitochondrial activity, and germline signaling (Antebi 2007). Recent genetic studies suggest that autophagy interacts with many of these longevity signals to regulate *C. elegans* aging (Meléndez et al. 2003; Hansen et al. 2008; Toth et al. 2008; Lapierre et al. 2011). Mutants in the insulin-like/IGF-1 receptor (IIR) gene, *daf-2*, display an increase in autophagy, as detected by an increase in the number of punctate structures labeled with the autophagy marker, GFP::LGG-1, in hypodermal seam cells, a cell type commonly used to visualize autophagy in *C. elegans* (Meléndez et al. 2003; Hansen et al. 2008). A reduction in autophagy during development, or only during adulthood, shortens the long life span of *daf-2* mutants (Meléndez et al. 2003; Hars et al. 2007; Hansen et al. 2008). Reduced food intake without malnutrition, otherwise referred to as dietary restriction, occurs in *eat-2* mutants (Avery 1993). These animals lack a nicotinic acetylcholine receptor specific to the pharynx, thereby showing reduced pharyngeal pumping, and have a phenotype of extended life span (Raizen et al. 1995; Lakowski and Hekimi 1998). Consistent with a role for TOR in dietary restriction, *eat-2* mutants have reduced TOR signaling, display an increase in autophagy, and require autophagy for their phenotype of long life span (Jia and Levine 2007; Hansen et al. 2008; Toth et al. 2008). The reduction in mitochondrial respiration in *isp-1* mutants extends life span (Dillin et al. 2002; Lee et al. 2003), and this phenotype is also dependent on autophagy (Toth et al. 2008). Finally, *glp-1/Notch* germline-less mutants induce autophagy, and require autophagy for life span extension (Lapierre et al. 2011). Interestingly, protein HLH-30, the ortholog of the mammalian TFEB transcription factor, is required for the life span extension associated with the longevity pathways described above and also regulates autophagy (Lapierre et al. 2013). In conclusion, autophagy is required as part of most longevity pathways in *C. elegans*, the only

Cite this introduction as *Cold Spring Harb Protoc*; doi:10.1101/pdb.top070466

exception thus far being the longevity associated with a reduction in protein translation (Pan et al. 2007; Hansen et al. 2008).

Degradation of Paternal Mitochondria

Directly after fertilization, autophagy is induced, resulting in the elimination of spermatozoon-specific organelles, including paternal mitochondria (Al Rawi et al. 2011; Sato and Sato 2011). Whether autophagy also acts in higher eukaryotes to degrade paternal mitochondria is not known; however, an increase in ubiquitylation and the localization of the Atg8-family protein LC3 (*Map1lc3a*) near the sperm mid-piece at the point of entry are suggestive that this is the case in fertilized mouse zygotes (Al Rawi et al. 2011).

Autophagy in Apoptosis, Necrosis, and Cell Clearance

Although autophagy has a role in homeostasis as an important prosurvival mechanism in response to stress, an excess in autophagy can result in cell death (Kang et al. 2007). Autophagy is also required for necrotic cell death, a type of cell death characterized by the loss of integrity of the cell plasma membrane (Toth et al. 2007; Samara et al. 2008). Additionally, similar to the situation in mammalian cells, BEC-1, a component of the class III phosphoinositide 3-kinase complex (Fig. 1), interacts with an antiapoptotic ortholog of Bcl-2, CED-9, suggesting the existence of cross talk between autophagy and apoptosis (Takacs-Vellai et al. 2005; Erdelyi et al. 2011). Autophagy proteins have been shown to play a role in the proper degradation of apoptotic cell corpses in *C. elegans* as, in autophagy-deficient animals, apoptotic cells are internalized but not properly degraded (Ruck et al. 2011; Li et al. 2012). Interestingly, rescue experiments indicate that autophagy genes are required within the engulfing cell to promote apoptotic cell degradation (Li et al. 2012).

DETECTING AUTOPHAGY IN *C. ELEGANS*

Autophagy can be monitored by transmission electron microscopy, fluorescent image analysis of the GFP::LGG-1 reporter or other autophagy reporters (Table 2), and by western blotting. The latter technique takes advantage of changes in the level of lipidation of LGG-1 that occur during autophagy and result in the protein running faster on a gel in its lipidated form. It should be noted that an increase in the number of autophagosomes does not necessarily reflect an induction of autophagy (Klionsky 2012), and it is therefore important to distinguish between induction of autophagy, an increase in autophagic flux, and the accumulation of autophagosomes due to inefficient or blocked autophagy (Klionsky 2012). Usually, it is useful to infer the turnover of autophagosomes in the presence and absence of lysosomal degradation. In *C. elegans*, this can be achieved by RNAi knockdown of genes with lysosomal function, such as *cup-5* (Kostich et al. 2000; Fares and Greenwald 2001; Sun et al. 2011), or the addition of inhibitors such as bafilomycin A1, or chloroquine, routinely used in mammalian cells to inhibit fusion between autophagosomes and lysosomes, which have also been successful in *C. elegans* (Oka and Futai 2000; Ji et al. 2006; Pivtoraiko et al. 2010). Clearly, the use of multiple assays to verify an increase in functional autophagy is recommended. A comprehensive list of guidelines was recently reported (Klionsky 2012). Here, we describe four protocols for the basic study of autophagy in *C. elegans*: first, we present a method for the detection of autophagy by using LGG-1 tagged with the fluorescent marker GFP (see Protocol 1: Detection of Autophagy in *Caenorhabditis elegans* Using *GFP::LGG-1* as an Autophagy Marker [Palmisano and Meléndez 2015a]); second, we give a methodology for monitoring the degradation of P granules, which is dependent on autophagy in embryos (see Protocol 2: Detection of Autophagy in *Caenorhabditis elegans* Embryos Using Markers of P Granule Degradation [Palmisano and Meléndez 2015b]); third, we show how western blotting can be used to evaluate the lipidation of LGG-1 and discuss how caution has to be applied to the interpretation of the results of such an analysis (see Protocol 3: Detection of Autophagy in *Caenorhabditis elegans* by Western Blotting Analysis of LGG-1 [Palmisano and Meléndez 2015c]); and

TABLE 2. Fluorescent reporters for monitoring autophagy in the nematode *Caenorhabditis elegans*

Protein	Function	Tissue expression[a]	Transgenes[b]	References
LGG-1	Microtubule-associated protein-1/ubiquitin-like protein	Intestine, hypodermis, muscle, pharynx, neurons, vulva, somatic gonad, germline	adIs2122[Plgg-1::GFP::LGG-1; rol-6(su1006)]izEx1[Plgg-1::GFP::LGG-1; rol-6(su1006)]izEx5[Plgg-1::GFP::LGG-1; Podr-1::RFP]vkEx1093[Pnhx-2::mCherry::LGG-1]dkIs399[Ppie-1::GFP::lgg-1, unc-119 (+)]Is[Ppie-1::GFP::mCherry::LGG-1; unc-119(+))Ex[Plgg-1::DsRED::LGG-1; Pmyo-2::GFP]	Meléndez et al. (2003); Kang et al. (2007); Samara et al. (2008); Gosai et al. (2010); Manil-Segalen et al. (2014)
LGG-2	Ubiquitin-like protein	Hypodermis, intestine, vulva, pharynx, neurons, muscle	RD108 Ex[Plgg-2::GFP::LGG-2; rol-6(su1006)]RD217 unc119(ed3)III; Ex[unc-119(+); Ppie-1::gfp::mcherry::lgg-1]VIG9 unc119(ed3)III; Is[unc-119(+); Plgg-2::gfp::lgg-2]	Alberti et al. (2010); Manil-Segalen et al. (2014)
DFCP1	Double FYVE-containing protein	Head, tail, vulva, neurons	bpIs168[Pnfya-1::DFCP1::GFP; unc-76(+)]	Derubeis et al. (2000); Cheung et al. (2001); Axe et al. (2008); Tian et al. (2010)
ATG-16.1	WD repeat-containing protein	Intestine, head, pharynx, muscle, neurons	[Patg-16.1::ATG-16.1::GFP; rol-6(su10006)]	Zhang et al. (2013)
ATG-16.2	WD repeat-containing protein	Intestine, head, pharynx, muscle, neurons	[Patg-16.2::ATG-16.2::GFP; rol-6(su10006)]	Zhang et al. (2013)
ATG-9	Integral membrane protein	Head, tail, vulva, neurons	bpIs211[Pnfya-1::ATG-9::GFP; unc-76(+)]	Noda et al. (2000); Lu et al. (2011); Liang et al. (2012); Lin et al. (2013)
EPG-1	Atg13 homolog	Neurons, pharynx, muscle	bpIs175[Pepg-1::EPG-1::GFP; rol-6(su1006)]	Tian et al. (2009)
EPG-9	Atg101 homolog	Intestine, pharynx, neurons	bpIs214[Pepg-9::EPG-9::GFP; unc-76(+)]	Liang et al. (2012)
BEC-1	Coiled-coil protein	Intestine, hypodermis, vulva, neurons, somatic gonad	swEx520 [Pbec-1::BEC-1::GFP; rol-6(su1006)]grEx129 [Pbec-1::BEC-1::mRFP; lin-15(+)]Ex[Pced-1::mCherry::BEC-1; rol-6(su1006)]Ex[Pegl1::mCherry::BEC-1; rol-6(su1006)]	Takacs-Vellai et al. (2005); Rowland et al. (2006); Ruck et al. (2011); Huang et al. (2012)
SQST-1	p62/autophagy adaptor protein	Hypodermis, neurons, intestine, vulva, muscle	bpIs151[Psqst-1::SQST-1::GFP; unc-76(+)]	Hunt-Newbury et a. (2007); Pankiv et al. (2007); Tian et al. (2010)
SEPA-1	Autophagy adaptor protein	Intestine, head, tail	bpIs131[Psepa-1::SEPA-1::GFP; unc-76(+)]	Zhang et al. (2009); Tian et al. (2010)
PGL-1	RNA-binding protein/P-granule component	P-granules, intestine	bnIs1[Ppie-1::GFP::PGL-1; unc-119(+)]bnIs26 [Pelt-2::PGL-1::GFP; Pmyo-2::mCherry]	Cheeks et al. (2004); Zhang et al. (2009); Updike et al. (2011)

[a]Tissue expression can vary depending on the specific promoter used.

[b]The transgenes shown are those found in autophagy studies or those that might be beneficial in autophagy studies; additional transgenes might be available for each gene.

Cite this introduction as *Cold Spring Harb Protoc*; doi:10.1101/pdb.top070466

finally, we present a protocol for investigating the function of autophagy genes by RNAi, a technique that has the advantage of circumventing some of the drawbacks associated with strong loss-of-function mutations (see Protocol 4: RNAi-Mediated Inactivation of Autophagy Genes in *Caenorhabditis elegans* [Palmisano and Meléndez 2015d]).

ACKNOWLEDGMENTS

We thank all members of the Meléndez laboratory for helpful discussions and comments on the manuscript. We extend special thanks to Melissa Silvestrini and Kristina Ames for comments on an earlier draft. The work in the Meléndez laboratory is supported by a National Science Foundation Research Initiation Award (0818802) and a National Institutes of Health 1 R15 GM102846-01 award; A.M. is an Ellison Medical Foundation New Scholar in Aging and N.J.P. is supported by a CUNY Graduate Center Dissertation award.

REFERENCES

Al Rawi S, Louvet-Vallee S, Djeddi A, Sachse M, Culetto E, Hajjar C, Boyd L, Legouis R, Galy V. 2011. Postfertilization autophagy of sperm organelles prevents paternal mitochondrial DNA transmission. *Science* **334**: 1144–1147.

Albert PS, Brown SJ, Riddle DL. 1981. Sensory control of dauer larva formation in *Caenorhabditis elegans*. *J Comp Neurol* **198**: 435–451.

Alberti A, Michelet X, Djeddi A, Legouis R. 2010. The autophagosomal protein LGG-2 acts synergistically with LGG-1 in dauer formation and longevity in *C. elegans*. *Autophagy* **6**: 622–633.

Antebi A. 2007. Genetics of aging in *Caenorhabditis elegans*. *PLoS Genet* **3**: 1565–1571.

Avery L. 1993. The genetics of feeding in *Caenorhabditis elegans*. *Genetics* **133**: 897–917.

Axe EL, Walker SA, Manifava M, Chandra P, Roderick HL, Habermann A, Griffiths G, Ktistakis NT. 2008. Autophagosome formation from membrane compartments enriched in phosphatidylinositol 3-phosphate and dynamically connected to the endoplasmic reticulum. *J Cell Biol* **182**: 685–701.

Barth H, Meiling-Wesse K, Epple UD, Thumm M. 2001. Autophagy and the cytoplasm to vacuole targeting pathway both require Aut10p. *FEBS Lett* **508**: 23–28.

Baugh LR, Sternberg PW. 2006. DAF-16/FOXO regulates transcription of cki-1/Cip/Kip and repression of lin-4 during *C. elegans* L1 arrest. *Curr Biol* **16**: 780–785.

Birnby DA, Link EM, Vowels JJ, Tian H, Colacurcio PL, Thomas JH. 2000. A Transmembrane guanylyl cyclase (DAF-11) and Hsp90 (DAF-21) regulate a common set of chemosensory behaviors in *Caenorhabditis elegans*. *Genetics* **155**: 85–104.

Brown EJ, Albers MW, Bum Shin T, Ichikawa K, Keith CT, Lane WS, Schreiber SL. 1994. A mammalian protein targeted by G1-arresting rapamycin–receptor complex. *Nature* **369**: 756–758.

Cassada RC, Russell RL. 1975. The dauerlarva, a post-embryonic developmental variant of the nematode *Caenorhabditis elegans*. *Dev Biol* **46**: 326–342.

Chan EY, Longatti A, McKnight NC, Tooze SA. 2009. Kinase-inactivated ULK proteins inhibit autophagy via their conserved C-terminal domains using an Atg13-independent mechanism. *Mol Cell Biol* **29**: 157–171.

Cheeks RJ, Canman JC, Gabriel WN, Meyer N, Strome S, Goldstein B. 2004. *C. elegans* PAR proteins function by mobilizing and stabilizing asymmetrically localized protein complexes. *Curr Biol* **14**: 851–862.

Cheung PC, Trinkle-Mulcahy L, Cohen P, Lucocq JM. 2001. Characterization of a novel phosphatidylinositol 3-phosphate-binding protein containing two FYVE fingers in tandem that is targeted to the Golgi. *Biochem J* **355**: 113–121.

da Graca LS, Zimmerman KK, Mitchell MC, Kozhan-Gorodetska M, Sekiewicz K, Morales Y, Patterson GI. 2004. DAF-5 is a Ski oncoprotein homolog that functions in a neuronal TGF β pathway to regulate *C. elegans* dauer development. *Development* **131**: 435–446.

Derubeis AR, Young MF, Jia L, Robey PG, Fisher LW. 2000. Double FYVE-containing protein 1 (DFCP1): Isolation, cloning and characterization of a novel FYVE finger protein from a human bone marrow cDNA library. *Gene* **255**: 195–203.

Deter RL, Baudhuin P, De Duve C. 1967. Participation of lysosomes in cellular autophagy induced in rat liver by glucagon. *J Cell Biol* **35**: C11–C16.

Dillin A, Hsu AL, Arantes-Oliveira N, Lehrer-Graiwer J, Hsin H, Fraser AG, Kamath RS, Ahringer J, Kenyon C. 2002. Rates of behavior and aging specified by mitochondrial function during development. *Science* **298**: 2398–2401.

Erdelyi P, Borsos E, Takacs-Vellai K, Kovacs T, Kovacs AL, Sigmond T, Hargitai B, Pasztor L, Sengupta T, Dengg M, et al. 2011. Shared developmental roles and transcriptional control of autophagy and apoptosis in *Caenorhabditis elegans*. *J Cell Sci* **124**: 1510–1518.

Estevez M, Attisano L, Wrana JL, Albert PS, Massague J, Riddle DL. 1993. The *daf-4* gene encodes a bone morphogenetic protein receptor controlling *C. elegans* dauer larva development. *Nature* **365**: 644–649.

Fan W, Nassiri A, Zhong Q. 2011. Autophagosome targeting and membrane curvature sensing by Barkor/Atg14(L). *Proc Natl Acad Sci* **108**: 7769–7774.

Fares H, Greenwald I. 2001. Regulation of endocytosis by CUP-5, the *Caenorhabditis elegans* mucolipin-1 homolog. *Nat Genet* **28**: 64–68.

Fukuyama M, Rougvie AE, Rothman JH. 2006. *C elegans* DAF-18/PTEN mediates nutrient-dependent arrest of cell cycle and growth in the germline. *Curr Biol* **16**: 773–779.

Funakoshi T, Matsuura A, Noda T, Ohsumi Y. 1997. Analyses of APG13 gene involved in autophagy in yeast, *Saccharomyces cerevisiae*. *Gene* **192**: 207–213.

Gems D, Sutton AJ, Sundermeyer ML, Albert PS, King KV, Edgley ML, Larsen PL, Riddle DL. 1998. Two pleiotropic classes of daf-2 mutation affect larval arrest, adult behavior, reproduction and longevity in *Caenorhabditis elegans*. *Genetics* **150**: 129–155.

Golden JW, Riddle DL. 1984. The *Caenorhabditis elegans* dauer larva: Developmental effects of pheromone, food, and temperature. *Dev Biol* **102**: 368–378.

Gosai SJ, Kwak JH, Luke CJ, Long OS, King DE, Kovatch KJ, Johnston PA, Shun TY, Lazo JS, Perlmutter DH, et al. 2010. Automated high-content live animal drug screening using *C. elegans* expressing the aggregation prone serpin alpha1-antitrypsin Z. *PLoS ONE* **5**: e15460.

Gottlieb S, Ruvkun G. 1994. daf-2, daf-16 and daf-23: Genetically interacting genes controlling Dauer formation in *Caenorhabditis elegans*. *Genetics* **137**: 107–120.

Hansen M, Taubert S, Crawford D, Libina N, Lee SJ, Kenyon C. 2007. Lifespan extension by conditions that inhibit translation in *Caenorhabditis elegans*. *Aging Cell* **6**: 95–110.

Hansen M, Chandra A, Mitic LL, Onken B, Driscoll M, Kenyon C. 2008. A role for autophagy in the extension of lifespan by dietary restriction in *C. elegans*. *PLoS Genet* **4**: e24.

Harding TM, Morano KA, Scott SV, Klionsky DJ. 1995. Isolation and characterization of yeast mutants in the cytoplasm to vacuole protein targeting pathway. *J Cell Biol* **131:** 591–602.

Harding TM, Hefner-Gravink A, Thumm M, Klionsky DJ. 1996. Genetic and phenotypic overlap between autophagy and the cytoplasm to vacuole protein targeting pathway. *J Biol Chem* **271:** 17621–17624.

Hars ES, Qi H, Ryazanov AG, Jin S, Cai L, Hu C, Liu LF. 2007. Autophagy regulates ageing in *C. elegans*. *Autophagy* **3:** 93–95.

He H, Dang Y, Dai F, Guo Z, Wu J, She X, Pei Y, Chen Y, Ling W, Wu C, et al. 2003. Post-translational modifications of three members of the human MAP1LC3 family and detection of a novel type of modification for MAP1LC3B. *J Biol Chem* **278:** 29278–29287.

Hedgecock EM, Culotti JG, Thomson JN, Perkins LA. 1985. Axonal guidance mutants of *Caenorhabditis elegans* identified by filling sensory neurons with fluorescein dyes. *Developmental Biol* **111:** 158–170.

Huang S, Jia K, Wang Y, Zhou Z, Levine B. 2012. Autophagy genes function in apoptotic cell corpse clearance during *C. elegans* embryonic development. *Autophagy* **9:** 138–149.

Hunt-Newbury R, Viveiros R, Johnsen R, Mah A, Anastas D, Fang L, Halfnight E, Lee D, Lin J, Lorch A, et al. 2007. High-throughput in vivo analysis of gene expression in *Caenorhabditis elegans*. *PLoS Biol* **5:** e237.

Hutchins MU, Klionsky DJ. 2001. Vacuolar localization of oligomeric alpha-mannosidase requires the cytoplasm to vacuole targeting and autophagy pathway components in Saccharomyces cerevisiae. *J Biol Chem* **276:** 20491–20498.

Inoue T, Thomas JH. 2000. Suppressors of transforming growth factor-β pathway mutants in the *Caenorhabditis elegans* dauer formation pathway. *Genetics* **156:** 1035–1046.

Ji YJ, Choi KY, Song HO, Park BJ, Yu JR, Kagawa H, Song WK, Ahnn J. 2006. VHA-8, the E subunit of V-ATPase, is essential for pH homeostasis and larval development in *C. elegans*. *FEBS Lett* **580:** 3161–3166.

Jia K, Levine B. 2007. Autophagy is required for dietary restriction-mediated life span extension in *C. elegans*. *Autophagy* **3:** 597–599.

Jia K, Chen D, Riddle DL. 2004. The TOR pathway interacts with the insulin signaling pathway to regulate *C. elegans* larval development, metabolism and life span. *Development* **131:** 3897–3906.

Jia K, Hart AC, Levine B. 2007. Autophagy genes protect against disease caused by polyglutamine expansion proteins in *Caenorhabditis elegans*. *Autophagy* **3:** 21–25.

Jing Y, Tang XM. 1999. The convergent point of the endocytic and autophagic pathways in leydig cells. *Cell Res* **9:** 243–253.

Johnson TE, Mitchell DH, Kline S, Kemal R, Foy J. 1984. Arresting development arrests aging in the nematode *Caenorhabditis elegans*. *Mech Ageing Dev* **28:** 23–40.

Kametaka S, Okano T, Ohsumi M, Ohsumi Y. 1998. Apg14p and Apg6/Vps30p form a protein complex essential for autophagy in the yeast, *Saccharomyces cerevisiae*. *J Biol Chem* **273:** 22284–22291.

Kang C, You YJ, Avery L. 2007. Dual roles of autophagy in the survival of *Caenorhabditis elegans* during starvation. *Genes Dev* **21:** 2161–2171.

Kihara A, Noda T, Ishihara N, Ohsumi Y. 2001. Two distinct Vps34 phosphatidylinositol 3-kinase complexes function in autophagy and carboxypeptidase Y sorting in *Saccharomyces cerevisiae*. *J Cell Biol* **152:** 519–530.

Kim J, Dalton VM, Eggerton KP, Scott SV, Klionsky DJ. 1999. Apg7p/Cvt2p is required for the cytoplasm-to-vacuole targeting, macroautophagy, and peroxisome degradation pathways. *Mol Biol Cell* **10:** 1337–1351.

Kirisako T, Ichimura Y, Okada H, Kabeya Y, Mizushima N, Yoshimori T, Ohsumi M, Takao T, Noda T, Ohsumi Y. 2000. The reversible modification regulates the membrane-binding state of Apg8/Aut7 essential for autophagy and the cytoplasm to vacuole targeting pathway. *J Cell Biol* **151:** 263–276.

Klionsky DJ. 2012. Guidelines for the use and interpretation of assays for monitoring autophagy. *Autophagy* **8:** 1–100.

Klionsky DJ, Cregg JM, Dunn WA Jr, Emr SD, Sakai Y, Sandoval IV, Sibirny A, Subramani S, Thumm M, Veenhuis M, et al. 2003. A unified nomenclature for yeast autophagy-related genes. *Dev Cell* **5:** 539–545.

Kostich M, Fire A, Fambrough DM. 2000. Identification and molecular-genetic characterization of a LAMP/CD68-like protein from *Caenorhabditis elegans*. *J Cell Sci* **113:** 2595–2606.

Kovács AL, Vellai T, Müller F. 2003. Autophagy in *Caenorhabditis elegans*. In *Autophagy* (Klionsky D), pp. 219–225.

Kuma A, Mizushima N, Ishihara N, Ohsumi Y. 2002. Formation of the approximately 350-kDa Apg12-Apg5.Apg16 multimeric complex, mediated by Apg16 oligomerization, is essential for autophagy in yeast. *J Biol Chem* **277:** 18619–18625.

Kuroyanagi H, Yan J, Seki N, Yamanouchi Y, Suzuki Y, Takano T, Muramatsu M, Shirasawa T. 1998. Human ULK1, a novel serine/threonine kinase related to UNC-51 kinase of *Caenorhabditis elegans*: cDNA cloning, expression, and chromosomal assignment. *Genomics* **51:** 76–85.

Lakowski B, Hekimi S. 1998. The genetics of caloric restriction in *Caenorhabditis elegans*. *Proc Natl Acad Sci* **95:** 13091–13096.

Lapierre LR, Gelino S, Melendez A, Hansen M. 2011. Autophagy and lipid metabolism coordinately modulate life span in germline-less *C. elegans*. *Curr Biol* **21:** 1507–1514.

Lapierre LR, De Magalhaes Filho CD, McQuary PR, Chu CC, Visvikis O, Chang JT, Gelino S, Ong B, Davis AE, Irazoqui JE, et al. 2013. The TFEB orthologue HLH-30 regulates autophagy and modulates longevity in *Caenorhabditis elegans*. *Nat Commun* **4:** 2267.

Lee CY, Clough EA, Yellon P, Teslovich TM, Stephan DA, Baehrecke EH. 2003. Genome-wide analyses of steroid- and radiation-triggered programmed cell death in *Drosophila*. *Curr Biol* **13:** 350–357.

Levine B, Klionsky DJ. 2004. Development by self-digestion: Molecular mechanisms and biological functions of autophagy. *Dev Cell* **6:** 463–477.

Li W, Zou W, Yang Y, Chai Y, Chen B, Cheng S, Tian D, Wang X, Vale RD, Ou G. 2012. Autophagy genes function sequentially to promote apoptotic cell corpse degradation in the engulfing cell. *J Cell Biol* **197:** 27–35.

Liang Q, Yang P, Tian E, Han J, Zhang H. 2012. The *C. elegans* ATG101 homolog EPG-9 directly interacts with EPG-1/Atg13 and is essential for autophagy. *Autophagy* **8:** 1426–1433.

Lin L, Yang P, Huang X, Zhang H, Lu Q, Zhang H. 2013. The scaffold protein EPG-7 links cargo-receptor complexes with the autophagic assembly machinery. *J Cell Biol* **201:** 113–129.

Liou W, Geuze HJ, Geelen MJ, Slot JW. 1997. The autophagic and endocytic pathways converge at the nascent autophagic vacuoles. *J Cell Biol* **136:** 61–70.

Lu Q, Yang P, Huang X, Hu W, Guo B, Wu F, Lin L, Kovacs AL, Yu L, Zhang H. 2011. The WD40 repeat PtdIns(3)P-binding protein EPG-6 regulates progression of omegasomes to autophagosomes. *Dev Cell* **21:** 343–357.

Manil-Segalen M, Lefebvre C, Jenzer C, Trichet M, Boulogne C, Satiat-Jeunemaitre B, Legouis R. 2014. The *C. elegans* LC3 acts downstream of GABARAP to degrade autophagosomes by interacting with the HOPS subunit VPS39. *Dev Cell* **28:** 43–55.

Matsuura A, Miki T, Wada Y, Ohsumi Y. 1997. Apg1p, a novel protein kinase required for the autophagic process in *Saccharomyces cerevisiae*. *Gene* **192:** 245–250.

Meléndez A, Levine B. 2009. Autophagy in *C. elegans*. *WormBook* doi: 10.1895/wormbook.1.147.1.

Meléndez A, Talloczy Z, Seaman M, Eskelinen EL, Hall DH, Levine B. 2003. Autophagy genes are essential for dauer development and life-span extension in *C. elegans*. *Science* **301:** 1387–1391.

Mizushima N. 2007. Autophagy: Process and function. *Genes Dev* **21:** 2861–2873.

Mizushima N, Sugita H, Yoshimori T, Ohsumi Y. 1998. A new protein conjugation system in human the counterpart of the yeast Apg12p conjugation system essential for autophagy. *J Biol Chem* **273:** 33889–33892.

Mizushima N, Yamamoto A, Hatano M, Kobayashi Y, Kabeya Y, Suzuki K, Tokuhisa T, Ohsumi Y, Yoshimori T. 2001. Dissection of autophagosome formation using Apg5-deficient mouse embryonic stem cells. *J Cell Biol* **152:** 657–667.

Mizushima N, Yoshimori T, Ohsumi Y. 2002. Mouse Apg10 as an Apg12-conjugating enzyme: Analysis by the conjugation-mediated yeast two-hybrid method. *FEBS Lett* **532:** 450–454.

Mizushima N, Kuma A, Kobayashi Y, Yamamoto A, Matsubae M, Takao T, Natsume T, Ohsumi Y, Yoshimori T. 2003. Mouse Apg16L, a novel WD-repeat protein, targets to the autophagic isolation membrane with the Apg12-Apg5 conjugate. *J Cell Sci* **116:** 1679–1688.

Nakatogawa H, Suzuki K, Kamada Y, Ohsumi Y. 2009. Dynamics and diversity in autophagy mechanisms: Lessons from yeast. *Nat Rev Mol Cell Biol* **10:** 458–467.

Cite this introduction as *Cold Spring Harb Protoc*; doi:10.1101/pdb.top070466

Noda T, Ohsumi Y. 1998. Tor, a phosphatidylinositol kinase homologue, controls autophagy in yeast. *J Biol Chem* 273: 3963–3966.

Noda T, Kim J, Huang WP, Baba M, Tokunaga C, Ohsumi Y, Klionsky DJ. 2000. Apg9p/Cvt7p is an integral membrane protein required for transport vesicle formation in the Cvt and autophagy pathways. *J Cell Biol* 148: 465–480.

Obara K, Sekito T, Ohsumi Y. 2006. Assortment of phosphatidylinositol 3-kinase complexes—Atg14p directs association of complex I to the preautophagosomal structure in *Saccharomyces cerevisiae*. *Mol Biol Cell* 17: 1527–1539.

Ogura K, Wicky C, Magnenat L, Tobler H, Mori I, Muller F, Ohshima Y. 1994. *Caenorhabditis elegans* unc-51 gene required for axonal elongation encodes a novel serine/threonine kinase. *Genes Dev* 8: 2389–2400.

Oka T, Futai M. 2000. Requirement of V-ATPase for ovulation and embryogenesis in *Caenorhabditis elegans*. *J Biol Chem* 275: 29556–29561.

Palmisano NJ, Meléndez A. 2015a. Detection of autophagy in *Caenorhabditis elegans* using GFP::LGG-1 as an autophagy marker. *Cold Spring Harb Protoc* doi: 10.1101/pdb.prot086496.

Palmisano NJ, Meléndez A. 2015b. Visualization of autophagy in *Caenorhabditis elegans* embryos using markers of P granule degradation. *Cold Spring Harb Protoc* doi: 10.1101/pdb.prot086504.

Palmisano NJ, Meléndez A. 2015c. Detection of autophagy in *Caenorhabditis elegans* by western blotting analysis of LGG-1. *Cold Spring Harb Protoc* doi: 10.1101/pdb.prot086512.

Palmisano NJ, Meléndez A. 2015d. RNAi-mediated inactivation of autophagy genes in *Caenorhabditis elegans*. *Cold Spring Harb Protoc* doi: 10.1101/pdb.prot086520.

Pan KZ, Palter JE, Rogers AN, Olsen A, Chen D, Lithgow GJ, Kapahi P. 2007. Inhibition of mRNA translation extends lifespan in *Caenorhabditis elegans*. *Aging Cell* 6: 111–119.

Panaretou C, Domin J, Cockcroft S, Waterfield MD. 1997. Characterization of p150, an adaptor protein for the human phosphatidylinositol (PtdIns) 3-kinase. *J Biol Chem* 272: 2477–2485.

Pankiv S, Clausen TH, Lamark T, Brech A, Bruun JA, Outzen H, Øvervatn A, Bjørkoy G, Johansen T. 2007. p62/SQSTM1 binds directly to Atg8/LC3 to facilitate degradation of ubiquitinated protein aggregates by autophagy. *J Biol Chem* 282: 24131–24145.

Patterson GI, Koweek A, Wong A, Liu Y, Ruvkun G. 1997. The DAF-3 Smad protein antagonizes TGF-β-related receptor signaling in the *Caenorhabditis elegans* dauer pathway. *Genes Dev* 11: 2679–2690.

Pivtoraiko VN, Harrington AJ, Mader BJ, Luker AM, Caldwell GA, Caldwell KA, Roth KA, Shacka JJ. 2010. Low-dose bafilomycin attenuates neuronal cell death associated with autophagy-lysosome pathway dysfunction. *J Neurochem* 114: 1193–1204.

Polson HE, de Lartigue J, Rigden D, Reedijk M, Urbé S, Clague M, Tooze S. 2010. Mammalian Atg18 (WIPI2) localizes to omegasome-anchored phagophores and positively regulates LC3 lipidation. *Autophagy* 6: 506–522.

Raizen DM, Lee RY, Avery L. 1995. Interacting genes required for pharyngeal excitation by motor neuron MC in *Caenorhabditis elegans*. *Genetics* 141: 1365–1382.

Reggiori F, Klionsky DJ. 2006. Atg9 sorting from mitochondria is impaired in early secretion and VFT-complex mutants in *Saccharomyces cerevisiae*. *J Cell Sci* 119: 2903–2911.

Ren P, Lim CS, Johnsen R, Albert PS, Pilgrim D, Riddle DL. 1996. Control of *C. elegans* larval development by neuronal expression of a TGF-β homolog. *Science* 274: 1389–1391.

Roggo L, Bernard V, Kovacs AL, Rose AM, Savoy F, Zetka M, Wymann MP, Muller F. 2002. Membrane transport in *Caenorhabditis elegans*: An essential role for VPS34 at the nuclear membrane. *EMBO J* 21: 1673–1683.

Rowland AM, Richmond JE, Olsen JG, Hall DH, Bamber BA. 2006. Presynaptic terminals independently regulate synaptic clustering and autophagy of GABAA receptors in *Caenorhabditis elegans*. *J Neurosci* 26: 1711–1720.

Ruck A, Attonito J, Garces KT, Nunez L, Palmisano NJ, Rubel Z, Bai Z, Nguyen KC, Sun L, Grant BD, et al. 2011. The Atg6/Vps30/Beclin1 ortholog BEC-1 mediates endocytic retrograde transport in addition to autophagy in *C. elegans*. *Autophagy* 7: 386–400.

Samara C, Syntichaki P, Tavernarakis N. 2008. Autophagy is required for necrotic cell death in *Caenorhabditis elegans*. *Cell Death Differ* 15: 105–112.

Sato M, Sato K. 2011. Degradation of paternal mitochondria by fertilization-triggered autophagy in *C. elegans* embryos. *Science* 334: 1141–1144.

Schackwitz WS, Inoue T, Thomas JH. 1996. Chemosensory neurons function in parallel to mediate a pheromone response in *C. elegans*. *Neuron* 17: 719–728.

Seaman MNJ, Marcusson EG, Cereghino JL, Emr SD. 1997. Endosome to Golgi retrieval of the vacuolar protein sorting receptor, Vps10p, requires the function of the VPS29, VPS30, and VPS35 gene products. *J Cell Biol* 137: 79–92.

Seglen PO, Gordon PB. 1982. 3-Methyladenine: Specific inhibitor of autophagic/lysosomal protein degradation in isolated rat hepatocytes. *Proc Natl Acad Sci* 79: 1889–1892.

Shintani T, Mizushima N, Ogawa Y, Matsuura A, Noda T, Ohsumi Y. 1999. Apg10p, a novel protein-conjugating enzyme essential for autophagy in yeast. *EMBO J* 18: 5234–5241.

Shintani T, Suzuki K, Kamada Y, Noda T, Ohsumi Y. 2001. Apg2p functions in autophagosome formation on the perivacuolar structure. *J Biol Chem* 276: 30452–30460.

Sun Q, Fan W, Chen K, Ding X, Chen S, Zhong Q. 2008. Identification of Barkor as a mammalian autophagy-specific factor for Beclin 1 and class III phosphatidylinositol 3-kinase. *Proc Natl Acad Sci* 105: 19211–19216.

Sun T, Wang X, Lu Q, Ren H, Zhang H. 2011. CUP-5, the *C. elegans* ortholog of the mammalian lysosomal channel protein MLN1/TRPML1, is required for proteolytic degradation in autolysosomes. *Autophagy* 7: 1308–1315.

Takacs-Vellai K, Vellai T, Puoti A, Passannante M, Wicky C, Streit A, Kovacs AL, Muller F. 2005. Inactivation of the autophagy gene bec-1 triggers apoptotic cell death in *C. elegans*. *Curr Biol* 15: 1513–1517.

Tanida I, Tanida-Miyake E, Ueno T, Kominami E. 2001. The human homolog of *Saccharomyces cerevisiae* Apg7p is a protein-activating enzyme for multiple substrates including human Apg12p, GATE-16, GABARAP, and MAP-LC3. *J Biol Chem* 276: 1701–1706.

Tanida I, Tanida-Miyake E, Komatsu M, Ueno T, Kominami E. 2002. Human Apg3p/Aut1p homologue is an authentic E2 enzyme for multiple substrates, GATE-16, GABARAP, and MAP-LC3, and facilitates the conjugation of hApg12p to hApg5p. *J Biol Chem* 277: 13739–13744.

Tanida I, Sou YS, Ezaki J, Minematsu-Ikeguchi N, Ueno T, Kominami E. 2004. HsAtg4B/HsApg4B/autophagin-1 cleaves the carboxyl termini of three human Atg8 homologues and delipidates microtubule-associated protein light chain 3- and GABAA receptor-associated protein-phospholipid conjugates. *J Biol Chem* 279: 36268–36276.

Thomas JH, Birnby DA, Vowels JJ. 1993. Evidence for parallel processing of sensory information controlling dauer formation in *Caenorhabditis elegans*. *Genetics* 134: 1105–1117.

Thumm M, Egner R, Koch B, Schlumpberger M, Straub M, Veenhuis M, Wolf DH. 1994. Isolation of autophagocytosis mutants of *Saccharomyces cerevisiae*. *FEBS Lett* 349: 275–280.

Tian E, Wang F, Han Ja, Zhang H. 2009. epg-1 functions in autophagy-regulated processes and may encode a highly divergent Atg13 homolog in *C elegans*. *Autophagy* 5: 608–615.

Tian Y, Li Z, Hu W, Ren H, Tian E, Zhao Y, Lu Q, Huang X, Yang P, Li X, et al. 2010. *C elegans* screen identifies autophagy genes specific to multicellular organisms. *Cell* 141: 1042–1055.

Toth ML, Simon P, Kovacs AL, Vellai T. 2007. Influence of autophagy genes on ion-channel-dependent neuronal degeneration in *Caenorhabditis elegans*. *J Cell Sci* 120: 1134–1141.

Toth ML, Sigmond T, Borsos E, Barna J, Erdelyi P, Takacs-Vellai K, Orosz L, Kovacs AL, Csikos G, Sass M, et al. 2008. Longevity pathways converge on autophagy genes to regulate life span in *Caenorhabditis elegans*. *Autophagy* 4: 330–338.

Tsukada M, Ohsumi Y. 1993. Isolation and characterization of autophagy-defective mutants of *Saccharomyces cerevisiae*. *FEBS Lett* 333: 169–174.

Updike DL, Hachey SJ, Kreher J, Strome S. 2011. P granules extend the nuclear pore complex environment in the *C. elegans* germ line. *J Cell Biol* 192: 939–948.

Vellai T, Takacs-Vellai K, Zhang Y, Kovacs AL, Orosz L, Muller F. 2003. Genetics: Influence of TOR kinase on lifespan in *C. elegans*. *Nature* 426: 620.

Volinia S, Dhand R, Vanhaesebroeck B, MacDougall LK, Stein R, Zvelebil MJ, Domin J, Panaretou C, Waterfield MD. 1995. A human phosphatidylinositol 3-kinase complex related to the yeast Vps34p-Vps15p protein sorting system. *EMBO J* **14:** 3339–3348.

Wang CW, Kim J, Huang WP, Abeliovich H, Stromhaug PE, Dunn WA Jr, Klionsky DJ. 2001. Apg2 is a novel protein required for the cytoplasm to vacuole targeting, autophagy, and pexophagy pathways. *J Biol Chem* **276:** 30442–30451.

Xie Z, Klionsky DJ. 2007. Autophagosome formation: Core machinery and adaptations. *Nat Cell Biol* **9:** 1102–1109.

Yamada T, Carson AR, Caniggia I, Umebayashi K, Yoshimori T, Nakabayashi K, Scherer SW. 2005. Endothelial nitric-oxide synthase antisense (NOS3AS) gene encodes an autophagy-related protein (APG9-like2) highly expressed in trophoblast. *J Biol Chem* **280:** 18283–18290.

Yang P, Zhang H. 2011. The coiled-coil domain protein EPG-8 plays an essential role in the autophagy pathway in *C. elegans*. *Autophagy* **7:** 159–165.

Zhang Y, Yan L, Zhou Z, Yang P, Tian E, Zhang K, Zhao Y, Li Z, Song B, Han J, et al. 2009. SEPA-1 mediates the specific recognition and degradation of P granule components by autophagy in *C. elegans*. *Cell* **136:** 308–321.

Zhang H, Wu F, Wang X, Du H, Wang X, Zhang H. 2013. The two *C. elegans* ATG-16 homologs have partially redundant functions in the basal autophagy pathway. *Autophagy* **9:** 1965–1974.

Zhao Y, Tian E, Zhang H. 2009. Selective autophagic degradation of maternally-loaded germline P granule components in somatic cells during *C. elegans* embryogenesis. *Autophagy* **5:** 717–719.

Cite this introduction as *Cold Spring Harb Protoc;* doi:10.1101/pdb.top070466

Detection of Autophagy in *Caenorhabditis elegans* Using *GFP::LGG-1* as an Autophagy Marker

Nicholas J. Palmisano[1,2] and Alicia Meléndez[1,2,3]

[1]*Department of Biology, Queens College-CUNY, Flushing, New York 11367;* [2]*The Graduate Center, The City University of New York, New York 10016*

In yeast and mammalian cells, the autophagy protein Atg8/LC3 (microtubule-associated proteins 1A/1B light chain 3B encoded by *MAP1LC3B*) has been the marker of choice to detect double-membraned autophagosomes that are produced during the process of autophagy. A lipid-conjugated form of Atg8/LC3B is localized to the inner and outer membrane of the early-forming structure known as the phagophore. During maturation of autophagosomes, Atg8/LC3 bound to the inner autophagosome membrane remains in situ as the autophagosomes fuse with lysosomes. The nematode *Caenorhabditis elegans* is thought to conduct a similar process, meaning that tagging the nematode ortholog of Atg8/LC3—known as LGG-1—with a fluorophore has become a widely accepted method to visualize autophagosomes. Under normal growth conditions, GFP-modified LGG-1 displays a diffuse expression pattern throughout a variety of tissues, whereas, when under conditions that induce autophagy, the GFP::LGG-1 tag labels positive punctate structures, and its overall level of expression increases. Here, we present a protocol for using fluorescent reporters of LGG-1 coupled to GFP to monitor autophagosomes in vivo. We also discuss the use of alternative fluorescent markers and the possible utility of the LGG-1 paralog LGG-2.

MATERIALS

It is essential that you consult the appropriate Material Safety Data Sheets and your institution's Environmental Health and Safety Office for proper handling of equipment and hazardous material used in this protocol.

RECIPES: Please see the end of this protocol for recipes indicated by <R>. Additional recipes can be found online at http://cshprotocols.cshlp.org/site/recipes.

Reagents

Agarose (2%–3%)
M9 minimal medium buffer <R>
NGM agar plates <R>
Sodium azide (NaN$_3$; 25 mM)
Suitable strain(s) of *Caenorhabditis elegans*

> QU1:*izEx1 [P$_{lgg-1}$::GFP::LGG-1]* (Meléndez et al. 2003)
> DA2123:*adIs2122[Plgg-1::GFP::LGG-1; rol-6(su1006)]* (Kang et al. 2007)
> *bpIs168[Pnfya-1::DFCP1::GFP; unc-76(+)]* (Tian et al. 2010)
> FR758: *swEx520 [Pbec-1::BEC-1::GFP + rol-6(su1006)]* (Takacs-Vellai et al. 2005)
> *bpIs211[Pnfya-1::ATG-9::GFP; unc-76(+)]* (Lu et al. 2011)

[3]Correspondence: Alicia.Meléndez@qc.cuny.edu

Cite this protocol as *Cold Spring Harb Protoc*; doi:10.1101/pdb.prot086496

Equipment

Fluorescence microscope (e.g., Zeiss AxioImager A1)
Image-analysis programs (e.g., ImageJ)
Incubator
Microscope coverslips (1.22 × 13 mm)
Microscope slides (75 × 25 × 0.96 mm)
OP50 *Escherichia coli*
Platinum wire pick (standard)

METHOD

For a summary of the whole procedure (Steps 1–8), see Figure 1.

Animal Handling and Larval Preparation

1. Incubate ∼30 gravid adults carrying the *Is[P_{lgg-1}::GFP::LGG-1]* integrated transgene, or other transgene of interest, on standard OP50 *E. coli*-seeded nematode growth medium (NGM) agar plates overnight in an incubator set at the desired temperature.

   This overnight incubation should allow for many eggs to be laid.

2. On the next day, perform a wash-off/hatch-off protocol to remove all adults and larvae from each plate using ∼1 mL of sterile M9 minimal medium buffer at room temperature.

   At this point, only eggs will remain on the plates.

3. Return the plates to the incubator and incubate for 1–2 h at the desired temperature.

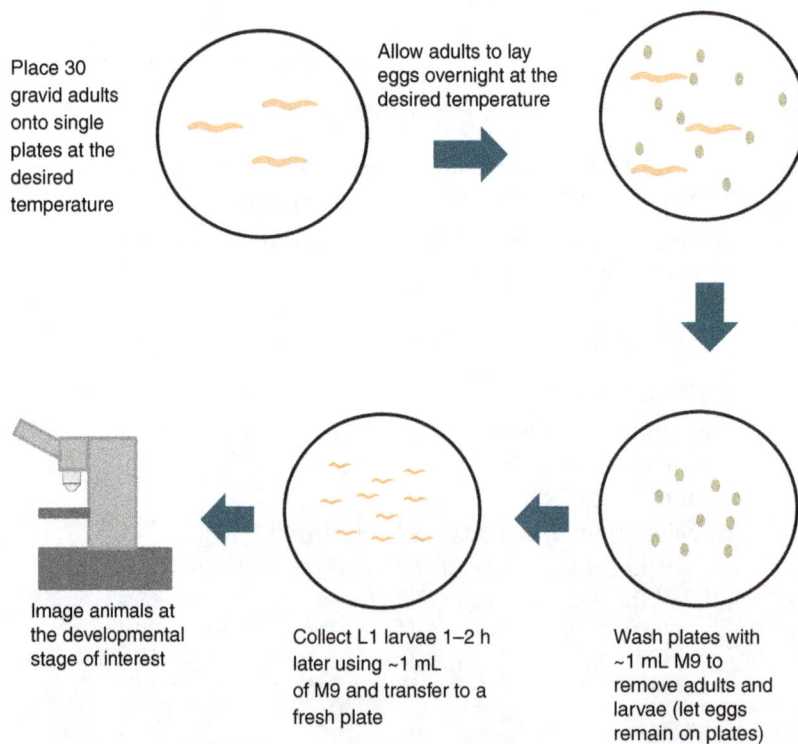

Place 30 gravid adults onto single plates at the desired temperature

Allow adults to lay eggs overnight at the desired temperature

Wash plates with ∼1 mL M9 to remove adults and larvae (let eggs remain on plates)

Collect L1 larvae 1–2 h later using ∼1 mL of M9 and transfer to a fresh plate

Image animals at the developmental stage of interest

FIGURE 1. Flowchart summarizing a protocol for using GFP::LGG-1 as an autophagy marker in larvae of the nematode *C. elegans*.

Cite this protocol as *Cold Spring Harb Protoc*; doi:10.1101/pdb.prot086496

4. Using ∼300–500 µL of sterile M9 buffer and a platinum wire pick, collect L1 progeny that have hatched in this 1- to 2-h timeframe. Transfer ∼50–100 µL of these synchronized L1 larvae to freshly seeded OP50 NGM agar plates and incubate at the desired temperature. Commence imaging of the animals (see Steps 5–8) as soon as the desired strain reaches the stage of interest.

> *The volume of M9 and number of worms will vary depending on the concentration of the animals grown. The collection period can vary depending on how tightly synchronized the investigator needs the animals to be; also, if using a dauer-constitutive strain such as daf-2(e1370), incubate the L1 animals for 48 h at 25°C to ensure that all animals are able to enter the dauer stage.*

Imaging

5. Apply a single drop of molten agarose (2%–3%) onto a microscope slide. Immediately, place another microscope slide on top of the agarose drop to form a thin agarose pad. After the agarose has solidified, remove the top slide.

6. Place ∼2 µL of 25 mM sodium azide in the center of the agarose pad and submerge the desired number of animals into the solution. Immediately, apply a microscope coverslip to prevent evaporation of the sodium azide solution. Wait until the animals are completely anesthetized before imaging with a fluorescence microscope.

> *It is important not to keep the animals in sodium azide for longer than 15 min total.*
> *See Troubleshooting.*

7. View under a fluorescence microscope using 63× or 100× magnification.

8. Evaluate an increase or decrease in the number of autophagosomes present by quantifying the GFP::LGG-1 punctate structures either by manually counting fluorescence-positive dots within a cell or by using image-analysis programs such as ImageJ (Manil-Segalen et al. 2014a). Be careful to note the size and frequency of GFP::LGG-1-positive punctate structures. If planning to obtain quantifiable data, keep the exposure time constant for all of the images acquired.

> *These punctate structures have been counted in hypodermal seam cells (lateral rows of cells arranged on each side of the body), as well as in 1–500-cell stage embryos (Manil-Segalen et al. 2014b); other tissues where GFP::LGG-1-positive vesicles have been analyzed are in the pharynx and intestine (Kang et al. 2007; Lapierre et al. 2011).*
>
> *See Troubleshooting and Discussion.*

TROUBLESHOOTING

Problem (Step 6): The animals, while submerged in sodium azide, do not become anesthetized.
Solution: The concentration of sodium azide might be too low; remake the solution or increase the concentration until the desired effect is attained.

Problem (Step 6): The sodium azide evaporates before animals are anesthetized.
Solution: Use additional sodium azide or prepare the slide using fewer animals—the issue is important as, if the animals are dehydrated, their anatomy will be affected.

Problem (Step 8): L3 larval hypodermal seam cells are not clearly visible at low or high magnifications during image acquisition.
Solution: Increasing the exposure time during acquisition might be required to image seam cells that are difficult to visualize clearly.

DISCUSSION

Establishing the GFP::LGG-1 Reporter for Autophagosomes

Atg8/LC3 has been the marker of choice to detect double-membraned autophagosomes in yeast and mammalian cells (Mizushima et al. 2004; Klionsky 2012). The phosphatidylethanolamine-conjugated form of Atg8/LC3B is localized to the inner and outer membrane of the phagophore. During maturation of the autophagosome, Atg8/LC3 on the outer membrane is cleaved and recycled, whereas Atg8/LC3 bound to the inner membrane remains in situ as the autophagosome fuses with the lysosome (Huang et al. 2000; Tanida et al. 2005). As this is thought to occur also in the nematode *C. elegans*, tagging the nematode ortholog of Atg8/LC3, LGG-1, with a fluorophore has become an accepted method to visualize autophagosomes. Reporters comprising LGG-1 coupled to fluorescent GFP, DsRED, and mCherry have been used to monitor autophagosomes in vivo (Meléndez et al. 2003; Maiuri et al. 2007; Gosai et al. 2010). In *C. elegans*, there are two orthologs of Atg8/LC3—LGG-1 and LGG-2. It has been shown that GFP::LGG-1 is expressed throughout development in multiple tissues, such as the nervous system, muscle, intestine, pharynx, vulva, hypodermis, and somatic gonad (Meléndez et al. 2003). Under normal conditions of growth, GFP::LGG-1 has a diffuse expression pattern throughout these tissues, whereas, under conditions that induce autophagy, GFP::LGG-1 labels positive punctate structures, and its overall expression increases (Meléndez et al. 2003; Kang et al. 2007; Lapierre et al. 2011).

An increase in GFP::LGG-1-positive punctate structures has been observed in loss-of-function mutations in the insulin-like/IGF-1 pathway using *daf-2(e1370)* mutants (Meléndez et al. 2003) and also in germline-less animals using *glp-1(e2141)* mutants (Meléndez et al. 2003; Lapierre et al. 2011). In addition, dietary-restricted animals such as *eat-2(ad1116)* mutants also display an increase in the number of structures positive for GFP::LGG-1 (Hansen et al. 2008). The GFP::LGG-1-positive vesicles were shown, by electron-microscopy analysis, to correlate with the formation of autophagosomes in *daf-2* mutants (Meléndez et al. 2003), making GFP::LGG-1 an acceptable marker for autophagosomes. Thus, one can test whether a gene under investigation is required for autophagy by monitoring changes in the expression and localization of GFP::LGG-1 in animals that carry any of the above loss-of-function mutations (Meléndez et al. 2003; Hansen et al. 2008; Lapierre et al. 2011). It can also be tested whether the gene of interest affects any of the phenotypes that have been shown to require autophagy, such as survival of L1 larvae during starvation, dauer development, longevity, and degradation of P granules.

In addition to using the GFP::LGG-1 reporter, an antibody against LGG-1 has also been used to monitor the levels of endogenous LGG-1 and the presence of autophagosomes in embryos (Chen et al. 2010; Tian et al. 2010; Lu et al. 2011; Yang and Zhang 2011). Wild-type embryos display a number of punctate structures that are positive for the LGG-1 antibody, starting at the 50-cell stage and lasting throughout embryogenesis. Formation of prominent LGG-1-positive puncta occurs at the 100-cell stage (Tian et al. 2010). Loss of autophagy activity can result in diffuse staining or aggregation of endogenous LGG-1, as in *atg-3* and *epg-1/atg-13* mutants, respectively (Tian et al. 2010). Theoretically, an antibody against GFP could also be used to monitor autophagy in fixed embryos or cells expressing the GFP::LGG-1 transgene, as has been achieved in mammalian cells expressing GFP–LC3 (Tian et al. 2010; Klionsky 2012).

Detection of LGG-2, a paralog of LGG-1 important for the acidification and degradation of autophagosomes, has also been used to monitor autophagy in *C. elegans* (Alberti et al. 2010; Manil-Segalen et al. 2014b). GFP::LGG-2 is expressed in a range of tissues similar to those for GFP::LGG-1, and both *lgg-1* and *lgg-2* have been shown to act synergistically in mediating dauer formation and longevity (Alberti et al. 2010). Although LGG-1 and LGG-2 proteins are similar in structure, under conditions that induce autophagy, they appear to localize differently. For example, L1 larvae display a punctate expression pattern for GFP::LGG-2, and yet a more ubiquitous expression pattern for GFP::LGG-1, under starvation conditions (Alberti et al. 2010). However, *lgg-1*, but not *lgg-2*, rescues the decreased survival associated with *atg8*-depleted yeast (Alberti et al. 2010), and therefore *lgg-1* appears to be more functionally related to *ATG8*.

Counting and Data-Acquisition Advice

Manually counting GFP-positive dots can be more accurate than using computer software, as imaging programs can inappropriately interpret background fluorescence as positive punctate structures. However, manually counting GFP-positive punctate structures can provide challenges of its own, especially when large quantities of puncta are present in a cell and/or if the fluorescence intensity of GFP is variable, making it difficult to distinguish each individual punctate structure. In such cases, it might be advisable to use both methods together to acquire more accurate results, although this might be somewhat tedious.

Whether deciding to use computer software programs or to manually quantify fluorescently labeled punctate structures, the parameters used during the quantification process should be clearly stated to ensure repeatability. For example, Alberti et al. used ImageJ to analyze punctate structures between 0.3 and 4 μm^2 in size and a circularity of 0 to 1 (Alberti et al. 2010). Additionally, if GFP::LGG-1-positive puncta of different sizes are observed within a sample, it might be best to group them according to their size. This is especially useful when dealing with the presence of GFP::LGG-1-positive aggregates, which can be associated with defective vesicles.

Finally, because cells display basal levels of autophagy, quantification of GFP::LGG-1-positive punctate structures should be noted on a per-cell basis, as opposed to the total number of cells containing puncta (Klionsky 2012). It is important to note, however, that the size of puncta should not be the only parameter used to measure autophagic activity (Klionsky 2012).

Caveats and Options for Additional Markers

Although GFP::LGG-1 is a widely accepted marker for visualizing autophagy in *C. elegans*, a cautious approach should be adopted when evaluating changes in the expression level of GFP::LGG-1. For example, increased numbers of GFP::LGG-1 puncta and/or increased puncta size can represent increased autophagic flux or an inhibitory event such as decreased lysosomal activity (Klionsky 2012). Additionally, in mammalian cells, the orthologous marker GFP-LC3 has been documented to associate with polyubiquitylated proteins both during autophagy and also independent of autophagy (Kuma et al. 2007; Pankiv et al. 2007; Shvets and Elazar 2008), which might also be true for GFP::LGG-1. Therefore, additional reporters, alone or in combination with GFP::LGG-1, might be required in *C. elegans* to distinguish between functional autophagosomes and cytosolic GFP::LGG-1 aggregates, the latter being indicative of defective autophagosomes.

Other protein markers that can be used in *C. elegans* to monitor the various steps of autophagy include: BEC-1::GFP, DFCP1::GFP, and ATG-9::GFP (Takacs-Vellai et al. 2005; Tian et al. 2010; Lu et al. 2011). BEC-1 (Atg6/Vps30) is a component of the class III phosphoinositide 3-kinase complex. DFCP1 is a double-FYVE-domain-containing protein that binds to phosphatidylinositol 3-phosphate localized to endoplasmic reticulum-derived early autophagosomes called omegasomes (Derubeis et al. 2000; Cheung et al. 2001; Axe et al. 2008). ATG-9 is the only integral membrane protein to date that is speculated to contribute membrane to the developing autophagosome (Noda et al. 2000). In conclusion, these additional markers might be particularly useful for visualizing and possibly quantifying early and late autophagosomes.

As an alternative to using the GFP::LGG-1 for monitoring autophagosomes, investigators can use the tandem GFP::mCherry::LGG-1 reporter, similar to a prior approach by others with LC3 in mammalian cells (Kimura et al. 2007; Pankiv et al. 2007; Manil-Segalen et al. 2014b). When analyzing a tandem reporter, colocalization of GFP and mCherry fluorescence can indicate a compartment that has not fused with the lysosome. As autophagosomes fuse with lysosomes, the GFP signal becomes sensitized to the acidic conditions of lysosomes, whereas the mCherry signal remains stable. Thus, the mCherry signal, when detected alone, corresponds to amphisomes (hybrid organelles derived from fusion of autophagosomes with endosomes) or autolysosomes (from fusion of amphisomes or autophagosomes with lysosomes) (Kimura et al. 2007; Manil-Segalen et al. 2014b).

RELATED TECHNIQUES

Complementary methodologies available include the use of GFP::LC3 fluorescence microscopy (Mizushima 2009) and Tandem mCherry-GFP–LC3 fluorescence microscopy (Kimura et al. 2007).

RECIPES

M9 Minimal Medium Buffer

Reagent	Concentration
KH_2PO_4	22 mM
Na_2HPO_4	22 mM
NaCl	85 mM
$MgSO_4$	1 mM

Autoclave for 15 min on liquid cycle. Allow medium to cool and store at room temperature.

NGM Agar Plates

68 g Bacto agar powder
12 g NaCl
10 g Bacto peptone
4 mL cholesterol (5 mg/mL in ethanol)
3.9 L deionized water
4 mL $CaCl_2$ (1 M)
4 mL $MgSO_4$ (1 M)
100 mL KPO_4 (1 M)

To prepare 4 L of nematode growth medium (NGM), mix the first five ingredients, and autoclave for 70 min on liquid cycle. Ensure that the agar is dissolved, let cool, and then add the last three ingredients. Pour the plates (these can be stored for up to 6 wk at 4°C).

REFERENCES

Alberti A, Michelet X, Djeddi A, Legouis R. 2010. The autophagosomal protein LGG-2 acts synergistically with LGG-1 in dauer formation and longevity in C. elegans. Autophagy 6: 622–633.

Axe EL, Walker SA, Manifava M, Chandra P, Roderick HL, Habermann A, Griffiths G, Ktistakis NT. 2008. Autophagosome formation from membrane compartments enriched in phosphatidylinositol 3-phosphate and dynamically connected to the endoplasmic reticulum. J Cell Biol 182: 685–701.

Chen B, Jiang Y, Zeng S, Yan J, Li X, Zhang Y, Zou W, Wang X. 2010. Endocytic sorting and recycling require membrane phosphatidylserine asymmetry maintained by TAT-1/CHAT-1. PLoS Genet 6: e1001235.

Cheung PC, Trinkle-Mulcahy L, Cohen P, Lucocq JM. 2001. Characterization of a novel phosphatidylinositol 3-phosphate-binding protein containing two FYVE fingers in tandem that is targeted to the Golgi. Biochem J 355: 113–121.

Derubeis AR, Young MF, Jia L, Robey PG, Fisher LW. 2000. Double FYVE-containing protein 1 (DFCP1): Isolation, cloning and characterization of a novel FYVE finger protein from a human bone marrow cDNA library. Gene 255: 195–203.

Gosai SJ, Kwak JH, Luke CJ, Long OS, King DE, Kovatch KJ, Johnston PA, Shun TY, Lazo JS, Perlmutter DH, et al. 2010. Automated high-content live animal drug screening using C. elegans expressing the aggregation prone serpin alpha1-antitrypsin Z. PLoS ONE 5: e15460.

Hansen M, Chandra A, Mitic LL, Onken B, Driscoll M, Kenyon C. 2008. A role for autophagy in the extension of lifespan by dietary restriction in C. elegans. PLoS Genet 4: e24.

Huang WP, Scott SV, Kim J, Klionsky DJ. 2000. The itinerary of a vesicle component, Aut7p/Cvt5p, terminates in the yeast vacuole via the autophagy/Cvt pathways. J Biol Chem 275: 5845–5851.

Kang C, You YJ, Avery L. 2007. Dual roles of autophagy in the survival of Caenorhabditis elegans during starvation. Genes Dev 21: 2161–2171.

Kimura S, Noda T, Yoshimori T. 2007. Dissection of the autophagosome maturation process by a novel reporter protein, tandem fluorescent-tagged LC3. Autophagy 3: 452–460.

Klionsky DJ. 2012. Guidelines for the use and interpretation of assays for monitoring autophagy. Autophagy 8: 1–100.

Kuma A, Matsui M, Mizushima N. 2007. LC3, an autophagosome marker, can be incorporated into protein aggregates independent of autophagy: Caution in the interpretation of LC3 localization. Autophagy 3: 323–328.

Lapierre LR, Gelino S, Meléndez A, Hansen M. 2011. Autophagy and lipid metabolism coordinately modulate life span in germline-less C. elegans. Curr Biol 21: 1507–1514.

Lu Q, Yang P, Huang X, Hu W, Guo B, Wu F, Lin L, Kovacs AL, Yu L, Zhang H. 2011. The WD40 repeat PtdIns(3)P-binding protein EPG-6 regulates progression of omegasomes to autophagosomes. Dev Cell 21: 343–357.

Maiuri MC, Le Toumelin G, Criollo A, Rain JC, Gautier F, Juin P, Tasdemir E, Pierron G, Troulinaki K, Tavernarakis N, et al. 2007. Functional and physical interaction between Bcl-X(L) and a BH3-like domain in Beclin-1. EMBO J 26: 2527–2539.

Manil-Segalen M, Culetto E, Legouis R, Lefebvre C. 2014a. Interactions between endosomal maturation and autophagy: Analysis of ESCRT machinery during Caenorhabditis elegans development. Methods Enzymol 534: 93–118.

Manil-Segalen M, Lefebvre C, Jenzer C, Trichet M, Boulogne C, Satiat-Jeunemaitre B, Legouis R. 2014b. The C. elegans LC3 acts downstream of GABARAP to degrade autophagosomes by interacting with the HOPS subunit VPS39. Dev Cell 28: 43–55.

Meléndez A, Talloczy Z, Seaman M, Eskelinen EL, Hall DH, Levine B. 2003. Autophagy genes are essential for dauer development and life-span extension in C. elegans. Science 301: 1387–1391.

Cite this protocol as Cold Spring Harb Protoc; doi:10.1101/pdb.prot086496

Mizushima N. 2009. Methods for monitoring autopagy using GFP-LC3 transgenic mice. *Methods Enzymol* **452:** 13–23.

Mizushima N, Yamamoto A, Matsui M, Yoshimori T, Ohsumi Y. 2004. In vivo analysis of autophagy in response to nutrient starvation using transgenic mice expressing a fluorescent autophagosome marker. *Mol Biol Cell* **15:** 1101–1111.

Noda T, Kim J, Huang WP, Baba M, Tokunaga C, Ohsumi Y, Klionsky DJ. 2000. Apg9p/Cvt7p is an integral membrane protein required for transport vesicle formation in the Cvt and autophagy pathways. *J Cell Biol* **148:** 465–480.

Pankiv S, Clausen TH, Lamark T, Brech A, Bruun JA, Outzen H, Overvatn A, Bjorkoy G, Johansen T. 2007. p62/SQSTM1 binds directly to Atg8/LC3 to facilitate degradation of ubiquitinated protein aggregates by autophagy. *J Biol Chem* **282:** 24131–24145.

Shvets E, Elazar Z. 2008. Autophagy-independent incorporation of GFP-LC3 into protein aggregates is dependent on its interaction with pp62/SQSTM1. *Autophagy* **4:** 1054–1056.

Takacs-Vellai K, Vellai T, Puoti A, Passannante M, Wicky C, Streit A, Kovacs AL, Muller F. 2005. Inactivation of the autophagy gene *bec-1* triggers apoptotic cell death in *C. elegans*. *Curr Biol* **15:** 1513–1517.

Tanida I, Minematsu-Ikeguchi N, Ueno T, Kominami E. 2005. Lysosomal turnover, but not a cellular level, of endogenous LC3 is a marker for autophagy. *Autophagy* **1:** 84–91.

Tian Y, Li Z, Hu W, Ren H, Tian E, Zhao Y, Lu Q, Huang X, Yang P, Li X, et al. 2010. *C. elegans* screen identifies autophagy genes specific to multicellular organisms. *Cell* **141:** 1042–1055.

Yang P, Zhang H. 2011. The coiled-coil domain protein EPG-8 plays an essential role in the autophagy pathway in *C. elegans*. *Autophagy* **7:** 159–165.

Detecting Autophagy in *Caenorhabditis elegans* Embryos Using Markers of P Granule Degradation

Nicholas J. Palmisano[1,2] and Alicia Meléndez[1,2,3]

[1]Department of Biology, Queens College-CUNY, Flushing, New York 11367; [2]The Graduate Center, The City University of New York, New York 10016

Autophagy plays an active role during the early stages of embryogenesis in the nematode *Caenorhabditis elegans*. Although their exact function is unknown, P granules are ribonucleoprotein particles that play a role in germ cell specification. The localization of P granules is restricted to the germline precursor cells in wild-type embryos, as a result of their degradation in the somatic cell lineage. Autophagy is known to be required for the degradation of P granules, as mutations in various autophagy genes, including those encoding the adaptor SEPA-1 and the p62-like adaptor SQST-1, result in the accumulation of the P granule components PGL-1 and PGL-3 (termed PGL granules) in the somatic cells of *C. elegans* embryos. In this protocol, we present a methodology for using fusion reporters of SEPA-1, SQST-1, and PGL-1 that have aided in the identification of new genes for normal autophagy activity by screening for mutant animals that lack the degradation of these autophagy substrates.

MATERIALS

It is essential that you consult the appropriate Material Safety Data Sheets and your institution's Environmental Health and Safety Office for proper handling of equipment and hazardous material used in this protocol.

RECIPES: Please see the end of this protocol for recipes indicated by <R>. Additional recipes can be found online at http://cshprotocols.cshlp.org/site/recipes.

Reagents

Agarose (2%–3%)

Escherichia coli OP50 or equivalent strain (nematode food source)

M9 minimal medium buffer <R>

NGM agar plates <R>

Suitable strains of *Caenorhabditis elegans*

    bpIs151[Psqst-1::SQST-1::GFP, unc-76(+)] (Tian et al. 2010)
    bpIs131[Psepa-1::SEPA-1::GFP, unc-76(+)] (Tian et al. 2010)
    bnIs1[Ppie-1::GFP::PGL-1, unc-119(+)] (Tian et al. 2010)
    RD210: unc-119(ed3); Is[Ppie-1::GFP::mCherry::LGG-1; unc-119(+)] (Manil-Ségalen et al. 2014)

Equipment

Fluorescence microscope (e.g., Zeiss AxioImager A1)

Microscope coverslips (1.22 × 13 mm)

[3]Correspondence: Alicia.Meléndez@qc.cuny.edu

Cite this protocol as *Cold Spring Harb Protoc*; doi:10.1101/pdb.prot086504

Microscope slides (75 × 25 × 0.96 mm)
Platinum wire pick (standard)

METHOD

For a summary of the whole procedure, see Figure 1.

1. Allow about 50 gravid adults of the chosen nematode strain to lay eggs on a single nematode growth media (NGM) agar plate containing *E. coli* for 10–15 min at the desired temperature, and then transfer the gravid adults to fresh plates.

    Repeat as desired. This will result in multiple plates containing tightly synchronized embryos.

2. Wait until the recently laid embryos reach the desired stage of embryogenesis (e.g., for ~100-cell-stage embryos, wait an additional ~200 min at 20°C). (See Lewis and Fleming 1995.)

    The amount of time required to reach the desired stage of embryogenesis will vary according to temperature. Commence imaging (as described below) as soon as the stage of interest is reached.

    See Troubleshooting.

3. Apply a single drop of molten agarose (2%–3%) onto a microscope slide. Immediately, place another microscope slide on top of the agarose drop to form a thin agarose pad. After the agarose has solidified, remove the top slide.

4. Pipette ~2.5 μL of M9 minimal medium buffer onto the center of the agarose pad.

5. Dab a platinum wire pick into *E. coli* bacteria, and then *gently* pick embryos and submerge them into the M9 minimal medium buffer.

    See Troubleshooting.

6. Use 63× or 100× magnifications to quantify the puncta and/or take images of the fluorescent positive puncta found within the embryos.

    Do not keep the embryos on the slides for >15 min.

    See Troubleshooting and Discussion.

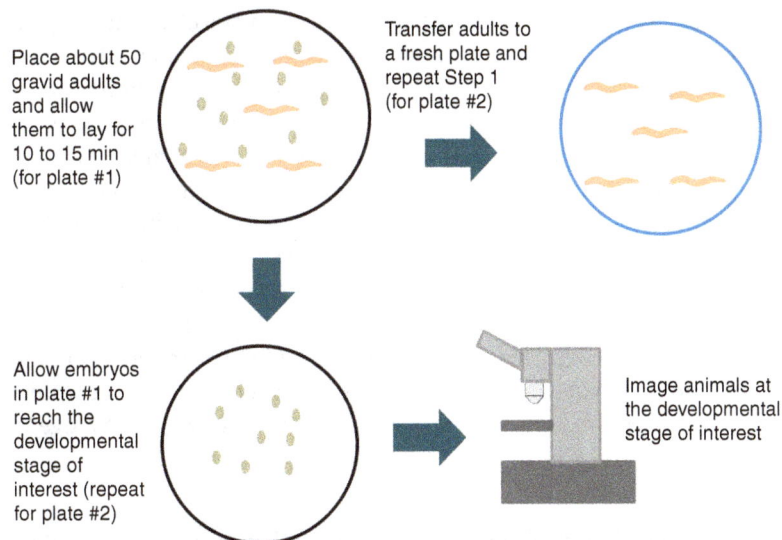

Place about 50 gravid adults and allow them to lay for 10 to 15 min (for plate #1)

Transfer adults to a fresh plate and repeat Step 1 (for plate #2)

Allow embryos in plate #1 to reach the developmental stage of interest (repeat for plate #2)

Image animals at the developmental stage of interest

FIGURE 1. Flowchart summarizing a protocol for using fluorescent marker proteins in screens using mutant animals that lack the degradation of autophagy substrates in larvae of the nematode *Caenorhabditis elegans*.

TROUBLESHOOTING

Problem (Step 2): Embryos have not reached the expected stage from the time of egg laying.
Solution: It is possible that the particular strain being used has a delay in development compared with that of wild-type animals. Reevaluate the time required to reach the developmental stage of interest.

Problem (Step 5): Difficulties arise when attempting to transfer embryos from the platinum wire pick to the agarose pad.
Solution: Carefully, press the platinum wire against the agarose pad several times, without tearing it, to release embryos.

Problem (Step 6): Positive puncta found within embryos are too large to quantify.
Solution: Image processing and analysis software, such as ImageJ or MetaMorph, can be used to quantify the number of puncta and/or fluorescence intensity by following the instructions provided by the software supplier.

DISCUSSION

Useful Reporters of Autophagy

Autophagy plays an active role during the early stages of embryogenesis in *C. elegans* (Meléndez and Neufeld 2008; Zhang et al. 2009; Kovacs and Zhang 2010; Al Rawi et al. 2011; Sato and Sato 2011). Although their exact function is unknown, P granules are ribonucleoprotein particles thought to play a part in germ cell specification (Strome and Lehmann 2007). The localization of P granules is restricted to the germline precursor cells in wild-type embryos as a result of the degradation of the P granules in the somatic cell lineage (Hird et al. 1996). Autophagy is required for the degradation of P granules as mutations in various autophagy genes result in the accumulation of the P granule components, PGL-1 and PGL-3 (termed PGL granules), in the somatic cells of *C. elegans* embryos (Zhang et al. 2009; Zhao et al. 2009; Tian et al. 2010).

SEPA-1 has been shown to be an adaptor protein important for the degradation of PGL granules in somatic cells through its interaction with LGG-1 (the nematode ortholog of Atg8/LC3), linking PGL granules to the autophagosome for granule removal (Zhang et al. 2009). In wild-type animals, SEPA-1 protein aggregates begin to form at the 16-cell-stage embryo and peak at the 100-cell stage, but disappear significantly by the comma stage (Zhang et al. 2009). However, in autophagy mutants, SEPA-1 aggregates persist past the comma stage, confirming that SEPA-1 is targeted to the autophagosome for removal (Zhang et al. 2009; Tian et al. 2010; Lu et al. 2011).

The gene *sqst-1* (*T12G3.1*) encodes the *C. elegans* homolog of the adaptor protein p62/SQSTM1, which has also been investigated during embryogenesis (Tian et al. 2010; Lu et al. 2011). In mammals, p62 binds to both LC3 and polyubiquitylated proteins, linking them to the autolysosomal pathway (Kihara et al. 2001; Bjorkoy et al. 2005; Pankiv et al. 2007; Tian et al. 2010). The expression of the marker SQST-1::GFP, in which SQST-1 is tagged with green fluorescent protein (GFP), is diffuse in the cytoplasm throughout embryogenesis; however, in autophagy mutants, the SQST-1::GFP protein localizes to positive punctate structures, similar to those of SEPA-1 (Tian et al. 2010; Lu et al. 2011).

Overall, the use of SEPA-1, SQST-1, and PGL-1 fusion reporters has facilitated the identification of new autophagy genes, by enabling screening for mutant animals that lack the degradation of these autophagy substrates (Tian et al. 2010). Therefore, such reporters can be used to identify additional genes required for normal autophagy activity during embryogenesis.

Data Analysis and Caveats

Changes in autophagic activity during embryogenesis are usually visualized by monitoring changes in the expression pattern of a fluorescent reporter relative to the control. Such changes can be easily observed as, under wild-type conditions, some reporters can have a diffuse cytoplasmic expression,

Cite this protocol as *Cold Spring Harb Protoc*; doi:10.1101/pdb.prot086504

whereas, under conditions of defective autophagy, the reporter can accumulate into aggregate-like structures or positive punctate structures. An example of this can be found by comparing wild-type animals and autophagy-defective embryos expressing proteins tagged with GFP—for example, SQST-1::GFP or GFP::LGG-1 (Tian et al. 2010; Lu et al. 2011). Other changes might not be so prominent and can include an increase in the number or size of preexisting protein aggregates, as is found when comparing wild-type animals and autophagy-mutant embryos treated with primary antibodies against LGG-1 (Tian et al. 2010; Lu et al. 2011).

To determine whether subtle changes in the expression of a reporter protein are significant will require additional methods. One method is to quantify the positive punctate structures; however, manual quantification of aggregates within embryos can be extremely difficult as autofluorescence is usually too high and the number of positive puncta too great. Image-analysis programs can provide better quantification abilities; however, even with image-analysis programs, one might not distinguish true punctate structures from autofluorescence. One option for measuring changes in the number of autophagosomes, or active autophagy, during embryogenesis is to measure the fluorescence intensity of the autophagy reporter used (Morselli et al. 2011). Such changes in fluorescence intensity can be used to determine whether there is elevated or decreased expression of the reporter, which can be indicative of increased or decreased autophagy.

When evaluating reporter expression, to determine whether a gene of interest plays a role in autophagy during embryogenesis, changes in the expression pattern of the reporter might not be apparent compared with that of controls. As a result, additional methods, such as image-processing software, will be required to ensure that such changes are significant. If choosing to monitor the expression of GFP::LGG-1 or SQST-1::GFP, care should also be taken. In mammals, p62 can form cytoplasmic inclusions unrelated to autophagosomes (Zatloukal et al. 2002; Bjorkoy et al. 2005). Furthermore, in mammals, LC3 can become incorporated into p62 cytosolic aggregates through its direct interaction with p62 (Shvets et al. 2008). Although this has not been formally shown in *C. elegans*, if visualizing SQST-1::GFP- or GFP::LGG-1-positive structures, additional reporters might be required to determine whether such aggregates are true autophagosomes or membrane-free inclusion bodies. Thus, in summary, when analyzing the activity of autophagy in embryos, a careful approach and successful interpretation of results can require additional methods and/or techniques.

RELATED TECHNIQUES

Complementary methodologies available include the use of antibodies against LGG-1, SQST-1, and SEPA-1 (Tian et al. 2010; Lu et al. 2011) rather than using strains expressing the GFP and mCherry fluorescent tags described in this protocol.

RECIPES

M9 Minimal Medium Buffer

Reagent	Concentration (mM)
KH_2PO_4	22 mM
Na_2HPO_4	22 mM
NaCl	85 mM
$MgSO_4$	1 mM

Autoclave for 15 min on liquid cycle. Store for up to 2 wk at room temperature. Prewarm to the appropriate experimental temperature before use.

NGM Agar Plates

68 g Bacto agar powder
12 g NaCl
10 g Bacto peptone
4 mL cholesterol (5 mg/mL in ethanol)
3.9 L deionized water
4 mL $CaCl_2$ (1 M)
4 mL $MgSO_4$ (1 M)
100 mL KPO_4 (1 M)

To prepare 4 L of normal growth media (NGM), mix the first five ingredients, and autoclave for 70 min on liquid cycle. Ensure that the agar is dissolved, let cool, and then add the last three ingredients. Pour the plates (these can be stored for up to 6 wk at 4°C.).

REFERENCES

Al Rawi S, Louvet-Vallee S, Djeddi A, Sachse M, Culetto E, Hajjar C, Boyd L, Legouis R, Galy V. 2011. Postfertilization autophagy of sperm organelles prevents paternal mitochondrial DNA transmission. *Science* **334**: 1144–1147.

Bjorkoy G, Lamark T, Brech A, Outzen H, Perander M, Overvatn A, Stenmark H, Johansen T. 2005. p62/SQSTM1 forms protein aggregates degraded by autophagy and has a protective effect on huntingtin-induced cell death. *J Cell Biol* **171**: 603–614.

Hird SN, Paulsen JE, Strome S. 1996. Segregation of germ granules in living *Caenorhabditis elegans* embryos: Celltype-specific mechanisms for cytoplasmic localisation. *Development* **122**: 1303–1312.

Kihara A, Noda T, Ishihara N, Ohsumi Y. 2001. Two distinct Vps34 phosphatidylinositol 3-kinase complexes function in autophagy and carboxypeptidase Y sorting in *Saccharomyces cerevisiae*. *J Cell Biol* **152**: 519–530.

Kovacs AL, Zhang H. 2010. Role of autophagy in *Caenorhabditis elegans*. *FEBS Lett* **584**: 1335–1341.

Lewis JA, Fleming JT. 1995. Basic culture methods. *Methods Cell Biol* **48**: 3–29.

Lu Q, Yang P, Huang X, Hu W, Guo B, Wu F, Lin L, Kovacs AL, Yu L, Zhang H. 2011. The WD40 repeat PtdIns(3)P-binding protein EPG-6 regulates progression of omegasomes to autophagosomes. *Dev Cell* **21**: 343–357.

Manil-Ségalen M, Culetto E, Legouis R, Lefebvre C. 2014. Interactions between endosomal maturation and autophagy: Analysis of ESCRT machinery during *Caenorhabditis elegans* development. *Methods Enzymol* **534**: 93–118.

Meléndez A, Neufeld TP. 2008. The cell biology of autophagy in metazoans: A developing story. *Development* **135**: 2347–2360.

Morselli E, Marino G, Bennetzen MV, Eisenberg T, Megalou E, Schroeder S, Cabrera S, Benit P, Rustin P, Criollo A, et al. 2011. Spermidine and resveratrol induce autophagy by distinct pathways converging on the acetylproteome. *J Cell Biol* **192**: 615–629.

Pankiv S, Clausen TH, Lamark T, Brech A, Bruun JA, Outzen H, Overvatn A, Bjorkoy G, Johansen T. 2007. p62/SQSTM1 binds directly to Atg8/LC3 to facilitate degradation of ubiquitinated protein aggregates by autophagy. *J Biol Chem* **282**: 24131–24145.

Sato M, Sato K. 2011. Degradation of paternal mitochondria by fertilization-triggered autophagy in *C. elegans* embryos. *Science* **334**: 1141–1144.

Shvets E, Fass E, Scherz-Shouval R, Elazar Z. 2008. The N-terminus and Phe52 residue of LC3 recruit p62/SQSTM1 into autophagosomes. *J Cell Sci* **121**: 2685–2695.

Strome S, Lehmann R. 2007. Germ versus soma decisions: Lessons from flies and worms. *Science* **316**: 392–393.

Tian Y, Li Z, Hu W, Ren H, Tian E, Zhao Y, Lu Q, Huang X, Yang P, Li X, et al. 2010. *C. elegans* screen identifies autophagy genes specific to multicellular organisms. *Cell* **141**: 1042–1055.

Zatloukal K, Stumptner C, Fuchsbichler A, Heid H, Schnoelzer M, Kenner L, Kleinert R, Prinz M, Aguzzi A, Denk H. 2002. p62 is a common component of cytoplasmic inclusions in protein aggregation diseases. *Am J Pathol* **160**: 255–263.

Zhang Y, Yan L, Zhou Z, Yang P, Tian E, Zhang K, Zhao Y, Li Z, Song B, Han J, et al. 2009. SEPA-1 mediates the specific recognition and degradation of P granule components by autophagy in *C. elegans*. *Cell* **136**: 308–321.

Zhao Y, Tian E, Zhang H. 2009. Selective autophagic degradation of maternally-loaded germline P granule components in somatic cells during *C. elegans* embryogenesis. *Autophagy* **5**: 717–719.

Cite this protocol as *Cold Spring Harb Protoc*; doi:10.1101/pdb.prot086504

Detection of Autophagy in *Caenorhabditis elegans* by Western Blotting Analysis of LGG-1

Nicholas J. Palmisano[1,2] and Alicia Meléndez[1,2,3]

[1]*Department of Biology, Queens College-CUNY, Flushing, New York 11367;* [2]*The Graduate Center, The City University of New York, New York 10016*

A common way to measure the induction of autophagy in yeast and mammalian cells is to compare the amount of Atg8/LC3-I with that of Atg8-PE/LC3-II by using western blot analysis. This is because changes in the amount of LC3-II correlate closely with changes in the number of autophagosomes present in cells. Atg8/LC3 is initially synthesized as an unprocessed form, which is proteolytically processed to form Atg8/LC3-I, and then this is modified into the phosphatidylethanolamine (PE)-conjugated Atg8-PE/LC3-II form. Atg8/LC3-II is membrane bound, whereas Atg8-PE/LC3-I is cytosolic. By associating with both the inner and outer membranes of the autophagosome, Atg8-PE/LC3-II is the only autophagy reporter that is reliably associated with completed autophagosomes. In the nematode *Caenorhabditis elegans*, the ortholog of Atg8/LC3 is LGG-1. Here, we discuss how changes in the levels of LGG-1-II (and the paralog LGG-2) protein can, with appropriate controls, be used to monitor autophagy activity in nematodes and present a protocol for monitoring changes in the protein levels of different forms of LGG-1 by western blotting.

MATERIALS

It is essential that you consult the appropriate Material Safety Data Sheets and your institution's Environmental Health and Safety Office for proper handling of equipment and hazardous material used in this protocol.

RECIPES: Please see the end of this protocol for recipes indicated by <R>. Additional recipes can be found online at http://cshprotocols.cshlp.org/site/recipes.

Reagents

Bleach solution <R>
Enhanced chemiluminescence (ECL) detection kit
Feeding plates containing *Caenorhabditis elegans*
Isopropanol
M9 minimal medium buffer <R>
Ponceau S solution (diluted to 1× with distilled water before use) <R>
Primary antibodies against green fluorescent protein (GFP) and/or LGG-1
Protease inhibitor cocktail
SDS gel-loading buffer (2×; diluted to 1× with distilled water before use) <R>
SDS-polyacrylamide gel <R>

[3]Correspondence: alicia.melendez@qc.cuny.edu

Cite this protocol as *Cold Spring Harb Protoc*; doi:10.1101/pdb.prot086512

Secondary antibody
Stacking gel (5%) <R>
TBST blocking buffer (TBST containing 5% w/v nonfat dry milk)
Tris-buffered saline (TBS; 10×, pH 7.5) <R>
Tris-buffered saline containing 0.1% Tween-20 (TBST)
Tris-glycine buffer (diluted to 1× with distilled water before use) <R>
Western transfer buffer (1×) <R>

Equipment

Eppendorf 5417C table top centrifuge (or equivalat)
Erlenmeyer flask
Fiber pads
Filter paper
Gel electrophoresis cassette and power supply
Gel transfer cell and sandwich cassette
Microcentrifuge tubes
Nitrocellulose (or PVDF) membrane
Plastic bag (or Tupperware dish)
Rocker
Sponge pad
Whatman filter paper

METHOD

For a summary of the whole procedure, see Figure 1.

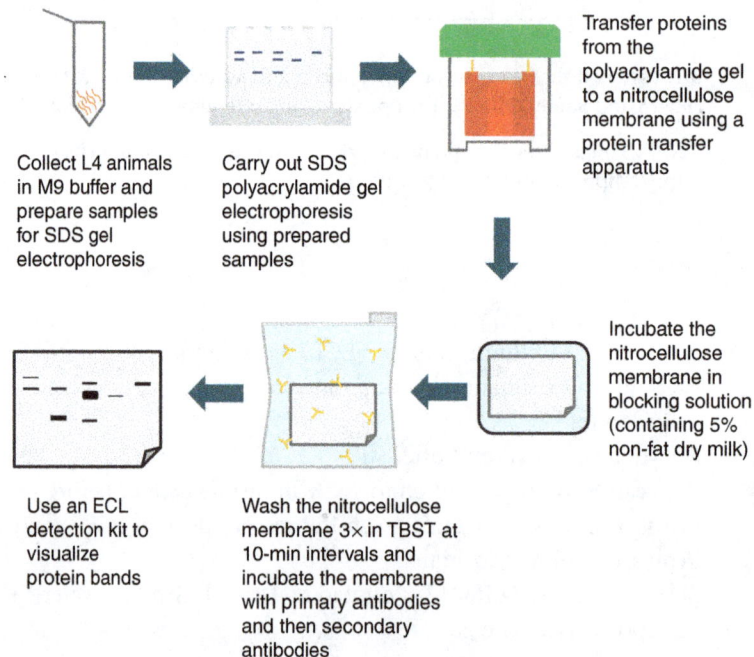

FIGURE 1. Flowchart summarizing a protocol for quantitating the levels of autophagy marker proteins by western blotting in the nematode *Caenorhabditis elegans*. ECL, enhanced chemiluminescence; L4, larval stage 4; TBST, Tris-buffered saline containing 0.1% Tween-20.

Cite this protocol as *Cold Spring Harb Protoc*; doi:10.1101/pdb.prot086512

Sample Preparation

1. Collect worms of the appropriate stage as follows:

   For larvae preparation

   i. Wash all L4 larvae or young adult hermaphrodites from feeding plates using 1 mL of M9 minimal medium buffer, and transfer animals to a 1.5-mL microcentrifuge tube.

   ii. Rinse worms several times by adding 100–300 µL of M9 buffer and centrifuging for 1 min at 2000 rpm. Between washes, discard the supernatants, leaving a pellet of animals.
   Proceed to Step 2.

   For embryo preparation

   i. Collect all gravid adults by washing plates with 1 mL of sterile water. Transfer worms to a 15 mL conical tube containing 500 µL of freshly prepared bleach solution.

   ii. Shake well or vortex tube for a few seconds, and repeat a few times, for no longer than 10 min.
   The adults dissolve, whereas the embryos remain intact.

   iii. Remove the bleach solution by centrifuging the 15 mL conical tube at 2000 rpm and carefully pipette out the supernatant.

   iv. Rinse worms several times by adding 1 mL of M9 buffer and centrifuging for 1 min at 2000 rpm. Between washes, discard the supernatants, leaving a pellet of embryos.
   Proceed to Step 2.

2. Snap-freeze the animals in liquid nitrogen and add an equal volume of 1× SDS gel-loading buffer containing a protease inhibitor cocktail to the sample and boil for 3–10 min at 100°C.
   It is important to also prepare marker proteins of known molecular mass for control purposes.

3. After boiling, centrifuge the samples at 2000 rpm for 1 min at room temperature, cool on ice for 5 min, and then load the samples onto an SDS-PAGE gel (see Step 7).

SDS-PAGE

This procedure was modified from Sambrook and Russell (2001).

4. Assemble the polyacrylamide gel apparatus by inserting two glass plates into the gel caster, as described by the manufacturer.

5. Prepare a 12%–20% polyacrylamide resolving gel.
   i. Immediately after preparing the gel mixture in an Erlenmeyer flask, swirl it and rapidly pour it into the gap between the glass plates.
   Be sure to leave enough space for the 5% stacking gel.

   ii. Overlay the polyacrylamide gel with isopropanol.

   iii. Allow the gel to polymerize for ∼30 min.

   iv. Once polymerization is complete, pour off the isopropanol and rinse the top of the resolving gel a few times with deionized water. Remove all residual isopropanol and water.

6. Prepare a 5% stacking gel.
   i. Immediately after preparing the gel mixture, rapidly pour it onto the polymerized resolving gel.

   ii. Carefully insert a clean gel comb into the stacking gel solution by avoiding air bubbles.

   iii. Allow the stacking gel to polymerize at room temperature.

   iv. After polymerization is complete, mount the gel in the electrophoresis apparatus and add the Tris-glycine buffer to the top and bottom reservoirs of the apparatus. Remove the gel comb and wash all wells with Tris-glycine buffer to remove any unpolymerized acrylamide.

7. Load ~15 µL of sample into each well.

8. Attach the electrophoresis apparatus to the power supply, as described by the manufacturer's instructions. Run the gel at ~180 to 200 V or until the bromophenol blue dye in the sample buffer reaches the bottom of the gel.

9. Once electrophoresis is complete, gently remove the glass plates from the gel apparatus and carefully release the polyacrylamide gel.

Membrane Transfer

This procedure was modified from Bio-Rad Laboratories Mini-Trans-Blot instruction manual.

10. Cool the 1× western transfer buffer to 4°C. Cut the nitrocellulose membrane (or PVDF membrane) and filter paper to the dimensions of the polyacrylamide gel. Soak the membrane, filter paper, and fiber pad in western transfer buffer until ready to use.

11. Prepare the western blot sandwich in the following order: start with the clear side of the case, followed by the sponge pad, Whatman filter paper, membrane, gel, Whatman filter paper, sponge pad, and the black side of the cassette. Close the cassette firmly and ensure that all bubbles are removed from the sandwich.

12. Place the transfer cassette in the transfer cell with the black side facing black, and fill the cell with cooled western transfer buffer. Transfer for 1 h at 4°C and ~90 V.

    Make sure that the cassette is positioned in the correct direction so that the proteins in the gel transfer to the membrane. The voltage required may vary according to the manufacturer's instructions. The membrane can be stored at 4°C for a few weeks after proteins have been transferred.

Immunoblotting

This procedure was modified from You et al. (2006).

13. (Optional) Check the membrane for successful transfer of proteins before blocking (Step 14) using Ponceau S stain.

    i. Incubate the blot for 1–5 min in 1× Ponceau S solution (1×) on a rocker.

    ii. Rinse the blot with distilled water to rid the blot of stain, until protein bands are clearly visible.

    iii. Wrap blot with plastic wrap and take a picture or a photocopy before proceeding to Step 14.

    Keep the membrane from drying out.

14. Incubate the membrane in TBST blocking buffer for ~1 h at room temperature on a rocker.

15. Dilute the primary antibody (i.e., anti-LGG-1 [Tian et al. 2010] or anti-GFP [Kang et al. 2007; Alberti et al. 2010]), in 2 mL of TBST blocking buffer at the appropriate concentration and incubate the membrane overnight at 4°C on a rocker.

    For primary antibodies, use the dilution factor suggested by the manufacturer; anti-GFP from Roche has been commonly used at a 1:500 dilution (Alberti et al. 2010; Djeddi et al. 2012). As antibodies vary, and the protocols can vary, determine the concentration of the primary antibody empirically before the start of the experiment.

16. Wash the membrane three times with ~50–100 mL of TBST for 5 min each on a rocker.

17. Incubate the membrane with the secondary antibody at the correct concentration (as directed by the antibody manufacturer, or as previously described (You et al. 2006; Alberti et al. 2010), for 1 h at room temperature on a rocker.

18. Rinse the membrane three times with ~50–100 mL of TBST for 10 min each on a rocker.

19. Rinse the membrane once with distilled water.

20. Use an ECL detection kit, as instructed by the manufacturer, to visualize the protein bands.

    For valuable troubleshooting advice, see the Abcam website (http://www.abcam.com/protocols/western-blot-troubleshooting-tips).

Cite this protocol as *Cold Spring Harb Protoc*; doi:10.1101/pdb.prot086512

DISCUSSION

Establishing LGG-1 as a Marker of Autophagy

A common way to measure the induction of autophagy is to compare the amount of Atg8/LC3-I with that of Atg8-PE/LC3-II using western blot analysis (Kabeya et al. 2000; Mizushima and Yoshimori 2007). This is because changes in the amount of LC3-II are closely associated with changes in the number of autophagosomes present in a cell (Kabeya et al. 2000). Atg8/LC3 is initially synthesized as an unprocessed form, which is proteolytically processed by Atg4 to form Atg8/LC3-I, and then modified into the phosphatidylethanolamine (PE)-conjugated Atg8-PE/LC3-II form (Kabeya et al. 2000; Kirisako et al. 2000). Atg8/LC3-II is the membrane-bound form of Atg8/LC3, whereas Atg8-PE/LC3-I is cytosolic (Kabeya et al. 2000). Atg8-PE/LC3-II associates with both the inner and outer membrane of the autophagosome, thus Atg8-PE/LC3-II is the only autophagy reporter that is reliably associated with completed autophagosomes (Klionsky 2012).

As with mammalian LC3 and yeast Atg8, the carboxyl terminus of LGG-1 appears to be conjugated to PE, with LGG-1-I and LGG-1-II being visible on a western blot as one major band and minor band, respectively (Kang et al. 2007; Alberti et al. 2010; Tian et al. 2010). In contrast, LGG-2 contains two minor bands as opposed to one; yet it is not clearly defined which minor band represents the lipidated form of LGG-2 (Alberti et al. 2010). Under normal nonstress conditions, the protein levels of LGG-1-I are higher than that of LGG-1-II. Under conditions that induce autophagy, LGG-1-I levels still appear higher than that of LGG-1-II; however, an overall increase in the amount of LGG-1-II is apparent (Kang et al. 2007; Alberti et al. 2010). A similar increase is also found for LGG-2 protein levels (Alberti et al. 2010). Therefore, changes in LGG-1-II and LGG-2 protein levels can be used to monitor autophagy activity in *C. elegans*.

Although Atg8-PE/LC3-II is reliably associated with the autophagosome, its protein levels might not change in a predictable manner upon autophagy induction (Mizushima and Yoshimori 2007; Klionsky 2012). For example, upon induction of autophagy in mammalian cells, the total levels of LC3 might not change; instead an increase in the conversion of LC3-I to LC3-II, or a decrease in the level of LC3-II relative to that of LC3-I, might result. The decrease in LC3-II can be due to the rapid lysosomal degradation of LC3-II (Huang et al. 2000; Klionsky 2012). Therefore, although changes in the protein levels of LGG-1 can be used to monitor autophagy, caution should be used when evaluating such changes.

Issues Arising from LGG-1 Western Data Analysis

Western blot analysis provides a convenient way to measure any changes in the levels of the different forms of LGG-1; however, caution is advised when interpreting the quantity of LGG-1-II using western blot analysis. A direct way to measure changes in overall LGG-1-II levels is through quantification—by comparing the protein levels of LGG-1-II with the protein levels of a housekeeping gene product (i.e., tubulin) or with those of LGG-1-I (Kang et al. 2007; Michelet et al. 2009; Alberti et al. 2010; Barth et al. 2010). In addition, it is important to use appropriate standardization controls to ensure equal loading between samples as this can change the amount of LGG-1-I and LGG-1-II protein between samples (Klionsky 2012). Furthermore, the stress condition of the animals before experimental manipulation should be at a minimum to ensure unaltered levels of LGG-1-I and LGG-1-II protein at the start of an experiment.

Increased levels of LGG-1-II relative to LGG-1-I can reflect autophagosome accumulation, owing to increased autophagy, or an accumulation of autophagosomes, as a result of defective lysosomal degradation (Michelet et al. 2009; Alberti et al. 2010; Barth et al. 2010; Lu et al. 2011; Klionsky 2012). Alternatively, based on mammalian studies, lower levels of LGG-1-II compared with LGG-1-I can represent defective autophagy, as a result of poor LGG-1-I to LGG-1-II conversion, or increased autophagic flux, resulting in the rapid degradation of LGG-1-II (Mizushima and Yoshimori 2007). The use of lysosomal inhibitors is one way to distinguish between all these possibilities (Oka and Futai 2000; Ji et al. 2006; Mizushima and Yoshimori 2007; Pivtoraiko et al. 2010).

Alternatively, protein extracts isolated from mutant animals previously shown to alter the lipidation of LGG-1 can also be useful as positive and/or negative controls in blots that measure changes in LGG-1 protein levels. Guanine nucleotide-binding protein subunit beta-2 (GPB-2) is a G-protein β subunit involved in the muscarinic signaling pathway, and *gpb-2* mutants, following starvation, have elevated levels of autophagy in pharyngeal muscles, visualized by the expression of GFP::LGG-1, and also have a higher ratio of lipidated LGG-1 to nonlipidated LGG-1, when compared with wild-type controls (Kang et al. 2007). *lgg-2* mutants, which have defects in the acidification and degradation of autophagosomes, produce elevated levels of both lipidated and nonlipidated forms of LGG-1 (Manil-Segalen et al. 2014). In addition, protein extracts isolated from animals fed double-stranded RNA (dsRNA) against the *C. elegans* ortholog of the target of rapamycin (TOR), *let-363* (Long et al. 2002), and *rab-7*, the small GTPase involved in endosome–lysosomal fusion events (Bucci et al. 2000), can also be used as controls to monitor changes in the levels of the different forms of LGG-1 protein (Alberti et al. 2010). RNA interference (RNAi) against *let-363* induces autophagy, observed by elevated levels of GFP::LGG-1 in the hypodermis and intestine, and has been reported to cause a decrease in the levels of nonlipidated LGG-1, but an increase in the levels of lipidated LGG-1, when compared with empty-vector controls. In contrast, RNAi against *rab-7*, which leads to increased levels of GFP::LGG-1 as a result of defective lysosomal fusion, results in elevated levels of both lipidated and nonlipidated LGG-1, compared with controls (Alberti et al. 2010).

It is important to note that the loss of certain autophagy genes can inhibit autophagy without affecting LC3-II/Atg8-PE formation (Klionsky 2012), which appears to be true also for LGG-1-II in *C. elegans* (Tian et al. 2010; Lu et al. 2011; Liang et al. 2012). Therefore, additional methods might be required to determine whether autophagy is functional when observing changes in the protein levels of LGG-1.

Additionally, LGG-1-II levels can be influenced by the type of antibodies against LGG-1 used, as well as the type of membrane used during protein transfer (Barth et al. 2010; Klionsky 2012). In *C. elegans*, experiments that visualize the lipidated and nonlipidated forms of LGG-1 by western blotting can use primary antibodies against LGG-1, such as antibodies specific to GFP, for strains expressing GFP::LGG-1 (Kang et al. 2007; Alberti et al. 2010; Tian et al. 2010).

In mammalian cells, during the initial periods of starvation, the amount of LC3-I can be inversely proportional to that of LC3-II; however, as the starvation period is prolonged, the levels of both LC3-I and LC3-II have been shown to decrease (Mizushima and Yoshimori 2007). Although this has not been fully examined for LGG-1 in *C. elegans*, one should consider the appropriate length of time that animals are exposed to starvation. Finally, LGG-1 has been shown to localize to phagosomes in cells that engulf and degrade apoptotic cells in *C. elegans* (Li et al. 2012). Thus, conditions that induce apoptosis should be at a minimum when evaluating the levels of LGG-1-II during autophagy.

In summary, a way to measure autophagy induction is to analyze the levels of LGG-1-II and compare them with the levels of LGG-1-I; however, several considerations should be made to ensure the proper interpretation of results.

Concluding Remarks

This protocol has discussed how western blot analysis can be used to monitor autophagy by evaluating the levels of lipidated and nonlipidated Atg8/LC3/LGG-1. However, an increase in the lipidated form of Atg8/LC3/LGG-1-II can reflect the induction of autophagy and/or inhibition of autophagy, and is therefore not a direct measure of autophagic flux without the use of additional methods (Klionsky 2012). In mammalian cells, the accumulation of LC3-II can result from an increase in autophagy activity or defective lysosomal degradation (Mizushima and Yoshimori 2007). The addition of lysosomal protease inhibitors, the introduction of a mutation, or RNAi treatment that results in defective lysosomal degradation, can differentiate between an increase in autophagy or the reduction of autophagic function. As described by Mizushima and Yoshimori, an additional increase in LGG-1-II levels, observed under conditions that block the fusion between autophagosomes and lysosomes, would be indicative of autophagy induction (Mizushima and Yoshimori 2007). In contrast, no change in LGG-1-II levels after treatment with agents or RNAi that decrease lysosomal degradation is indicative of a block in the autophagic pathway. Thus, in summary, with careful interpretation of results, changes in the protein levels of LGG-1 can provide a way to measure changes in autophagy activity.

RECIPES

Bleach Solution

3.5 mL sterile H_2O
0.5 mL NaOH (5 N)
1 mL household bleach (sodium hypochlorite)

Make the solution fresh each time it is needed.

M9 Minimal Medium Buffer

Reagent	Concentration
KH_2PO_4	22 mM
Na_2HPO_4	22 mM
NaCl	85 mM
$MgSO_4$	1 mM

Autoclave for 15 min on liquid cycle. Allow medium to cool and store at room temperature.

Ponceau S Solution

Ponceau S (0.2%)
Trichloroacetic acid (3%)
Sulfosalicylic acid (3%)

For a 10× stock, make a solution of 2% Ponceau S, 30% Trichloroacetic acid, and 30% Sulfosalicylic acid. Bring up to volume with H_2O. The solution is stable at room temperature for over 1 year.

SDS Gel-Loading Buffer (2×)

100 mM Tris-Cl (pH 6.8)
4% (w/v) SDS (sodium dodecyl sulfate; electrophoresis grade)
0.2% (w/v) bromophenol blue
20% (v/v) glycerol
200 mM DTT (dithiothreitol)

Store the SDS gel-loading buffer without DTT at room temperature. Add DTT from a 1 M stock just before the buffer is used.
200 mM β-mercaptoethanol can be used instead of DTT.

SDS-Polyacrylamide Gel

30% Acrylamide mix (Sigma-Aldrich A3574)
1.5 M Tris (pH 8.8)
10% SDS
10% ammonium sulfate
0.04% TEMED (*N,N,N′,N′*-tetramethylethylenediamine)

Combine ingredients immediately before use.

Stacking Gel (5%)

To prepare 5% stacking gel mixture, combine in the following order:
2 mL of 30% acrylamide mix
3 mL of 0.5 M Tris-HCl (pH 6.8)
0.12 mL of 10% (w/v) SDS
6.76 mL of H_2O
0.12 mL of 10% ammonium persulfate
0.006 mL of N,N,N',N'-tetramethylelthylenediamine (TEMED)

Tris-Buffered Saline (TBS; 10×, pH 7.5)

Reagent	Amount	Final concentration
Tris	24.2 g	200 mM
NaCl	87.7 g	1.5 M

Combine ingredients in ~800 mL of H_2O. Adjust pH to 7.5 and bring final volume to 1 L. Sterilize by autoclaving.

Tris-Glycine Buffer

Prepare a 5× stock solution in 1 liter of H_2O.
15.1 g Tris base
94 g glycine (electrophoresis grade)
50 mL 10% SDS (electrophoresis grade)

The 1× working solution is 25 mM Tris-Cl/250 mM glycine/0.1% SDS. Use Tris-glycine buffers for SDS-polyacrylamide gels.

Western Transfer Buffer

25 mM Tris-Cl
192 mM Glycine
20% (v/v) methanol, pH 8.3

REFERENCES

Alberti A, Michelet X, Djeddi A, Legouis R. 2010. The autophagosomal protein LGG-2 acts synergistically with LGG-1 in dauer formation and longevity in C. elegans. Autophagy 6: 622–633.

Barth S, Glick D, Macleod KF. 2010. Autophagy: Assays and artifacts. J Pathol 221: 117–124.

Bio-Rad Laboratories. Mini-Trans Blot Electrophoretic Transfer Cell Instruction Manual, Rev K. Bio-Rad Laboratories, Hercules, CA.

Bucci C, Thomsen P, Nicoziani P, McCarthy J, van Deurs B. 2000. Rab7: A key to lysosome biogenesis. Mol Biol Cell 11: 467–480.

Djeddi A, Michelet X, Culetto E, Alberti A, Barois N, Legouis R. 2012. Induction of autophagy in ESCRT mutants is an adaptive response for cell survival in C. elegans. J Cell Sci 125: 685–694.

Huang WP, Scott SV, Kim J, Klionsky DJ. 2000. The itinerary of a vesicle component, Aut7p/Cvt5p, terminates in the yeast vacuole via the autophagy/Cvt pathways. J Biol Chem 275: 5845–5851.

Ji YJ, Choi KY, Song HO, Park BJ, Yu JR, Kagawa H, Song WK, Ahnn J. 2006. VHA-8, the E subunit of V-ATPase, is essential for pH homeostasis and larval development in C. elegans. FEBS Lett 580: 3161–3166.

Kabeya Y, Mizushima N, Ueno T, Yamamoto A, Kirisako T, Noda T, Kominami E, Ohsumi Y, Yoshimori T. 2000. LC3, a mammalian homologue of yeast Apg8p, is localized in autophagosome membranes after processing. EMBO J 19: 5720–5728.

Kang C, You YJ, Avery L. 2007. Dual roles of autophagy in the survival of Caenorhabditis elegans during starvation. Genes Dev 21: 2161–2171.

Kirisako T, Ichimura Y, Okada H, Kabeya Y, Mizushima N, Yoshimori T, Ohsumi M, Takao T, Noda T, Ohsumi Y. 2000. The reversible modification regulates the membrane-binding state of Apg8/Aut7 essential for autophagy and the cytoplasm to vacuole targeting pathway. J Cell Biol 151: 263–276.

Klionsky DJ. 2012. Guidelines for the use and interpretation of assays for monitoring autophagy. Autophagy 8: 1–100.

Liang Q, Yang P, Tian E, Han J, Zhang H. 2012. The C. elegans ATG101 homolog EPG-9 directly interacts with EPG-1/Atg13 and is essential for autophagy. Autophagy 8: 1426–1433.

Long X, Spycher C, Han ZS, Rose AM, Muller F, Avruch J. 2002. TOR deficiency in C. elegans causes developmental arrest and intestinal atrophy by inhibition of mRNA translation. Curr Biol 12: 1448–1461.

Lu Q, Yang P, Huang X, Hu W, Guo B, Wu F, Lin L, Kovacs AL, Yu L, Zhang H. 2011. The WD40 repeat PtdIns(3)P-binding protein EPG-6 regulates progression of omegasomes to autophagosomes. Dev Cell 21: 343–357.

Manil-Segalen M, Lefebvre C, Jenzer C, Trichet M, Boulogne C, Satiat-Jeunemaitre B, Legouis R. 2014. The C. elegans LC3 acts downstream of GABARAP to degrade autophagosomes by interacting with the HOPS subunit VPS39. Dev Cell 28: 43–55.

Michelet X, Alberti A, Benkemoun L, Roudier N, Lefebvre C, Legouis R. 2009. The ESCRT-III protein CeVPS-32 is enriched in domains distinct from CeVPS-27 and CeVPS-23 at the endosomal membrane of epithelial cells. Biol Cell 101: 599–615.

Mizushima N, Yoshimori T. 2007. How to interpret LC3 immunoblotting. *Autophagy* **3:** 542–545.

Oka T, Futai M. 2000. Requirement of V-ATPase for ovulation and embryogenesis in *Caenorhabditis elegans*. *J Biol Chem* **275:** 29556–29561.

Pivtoraiko VN, Harrington AJ, Mader BJ, Luker AM, Caldwell GA, Caldwell KA, Roth KA, Shacka JJ. 2010. Low-dose bafilomycin attenuates neuronal cell death associated with autophagy-lysosome pathway dysfunction. *J Neurochem* **114:** 1193–1204.

Sambrook J, Russell DW. 2001. *Molecular cloning: A laboratory manual,* 3rd ed. Cold Spring Harbor Laboratory Press, Cold Spring Harbor, NY.

Tian Y, Li Z, Hu W, Ren H, Tian E, Zhao Y, Lu Q, Huang X, Yang P, Li X, et al. 2010. *C elegans* screen identifies autophagy genes specific to multicellular organisms. *Cell* **141:** 1042–1055.

You YJ, Kim J, Cobb M, Avery L. 2006. Starvation activates MAP kinase through the muscarinic acetylcholine pathway in *Caenorhabditis elegans* pharynx. *Cell Metab* **3:** 237–245.

RNAi-Mediated Inactivation of Autophagy Genes in *Caenorhabditis elegans*

Nicholas J. Palmisano[1,2] and Alicia Meléndez[1,2,3]

[1]*Department of Biology, Queens College-CUNY, Flushing, New York 11367;* [2]*The Graduate Center, The City University of New York, New York 10016*

RNA interference (RNAi) is a process that results in the sequence-specific silencing of endogenous mRNA through the introduction of double-stranded RNA (dsRNA). In the nematode *Caenorhabditis elegans*, RNA inactivation can be used at any specific developmental stage or during adulthood to inhibit a given target gene. Investigators can take advantage of the fact that, in *C. elegans*, RNAi is unusual in that it is systemic, meaning that dsRNA can spread throughout the animal and can affect virtually all tissues except neurons. Here, we describe a protocol for the most common method to achieve RNAi in *C. elegans*, which is to feed them bacteria that express dsRNA complementary to a specific target gene. This method has various advantages, including the availability of libraries that essentially cover the whole genome, the ability to treat animals at any developmental stage, and that it is relatively cost effective. We also discuss how RNAi specific to autophagy genes has proven to be an excellent method to study the role of these genes in autophagy, as well as other cellular and developmental processes, while also highlighting the caveats that must be applied.

MATERIALS

It is essential that you consult the appropriate Material Safety Data Sheets and your institution's Environmental Health and Safety Office for proper handling of equipment and hazardous material used in this protocol.

RECIPES: Please see the end of this protocol for recipes indicated by <R>. Additional recipes can be found online at http://cshprotocols.cshlp.org/site/recipes.

Reagents

Agarose (2%–3%)

Ampicillin

Caenorhabditis elegans strains of interest (larval stage L4 animals)

> *The strains to be analyzed should be well fed on nematode growth media (NGM) agar plates seeded with* Escherichia coli *bacteria and cultured for at least two generations before the experiment.*

Escherichia coli (HT115 expressing double-stranded RNA against target gene under the isopropylthio-β-galactoside [IPTG]-inducible T7 RNA polymerase promoter)

LB (Luria-Bertani) liquid medium <R>

M9 minimal medium buffer <R>

NGM RNA interference (RNAi) agar plates <R>

Sodium azide (NaN$_3$; 25 mM)

---

[3]Correspondence: alicia.melendez@qc.cuny.edu

Cite this protocol as *Cold Spring Harb Protoc*; doi:10.1101/pdb.prot086520

Equipment

Fluorescence microscope (Zeiss AxioImager A1)
Microscope coverslips (1.22 × 13 mm)
Microscope slides (75 × 25 × 0.96 mm)
Platinum wire pick (standard)
Shaker

METHOD

For a summary of the whole procedure, see Figure 1.

1. Inoculate each bacterial RNAi clone of interest in 5 mL of liquid medium (LB containing 100 mg/mL ampicillin) overnight at 37°C in a shaker at 250 rpm.

   Be aware of the danger of contamination or incorrect clones. A bacterial clone that is isolated from an RNAi library for the first time should be verified to ensure the presence of the correct insert in the plasmid. Glycerol stocks can then be made from the verified colony.

2. Seed experimental RNAi plates with 500 μL of the bacterial liquid culture, and allow them to dry overnight.

   Allowing plates to dry overnight before adding worms ensures sufficient induction of T7 RNA polymerase activity.

3. Place ~30 L4 larval stage animals on RNAi plates and allow animals to ingest bacteria overnight at the desired temperature.

   Some C. elegans *strains are temperature sensitive, and therefore care should be taken when deciding on the temperature at which the experiment should be performed.*

4. On the following day, transfer the animals (now adults) to new RNAi plates using ~50 to 100 μL of M9 minimal medium buffer and allow them to lay eggs overnight at the desired temperature. Discard the first set of plates (from Step 3).

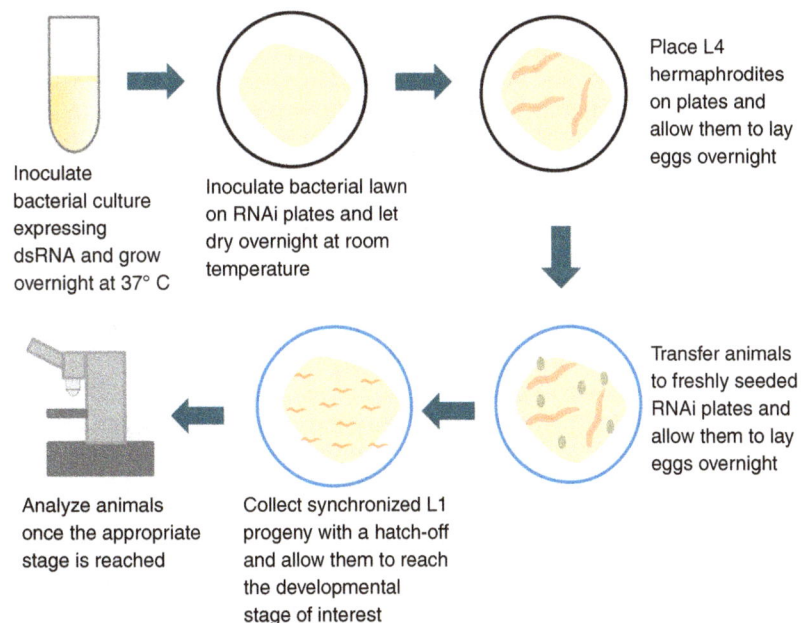

FIGURE 1. Flowchart summarizing a protocol for achieving RNAi-mediated inactivation of autophagy genes in the nematode *Caenorhabditis elegans*. L1, larval stage 1; L4, larval stage 4; RNAi, RNA interference.

*The adults can be transferred to fresh RNAi plates for two more days (as a third set or a fourth set), to obtain more F1 animals with knockdown of the specific target gene. If desired, F1 progeny treated with RNAi can be tightly synchronized using the wash-off/hatch-off method with M9 minimal medium buffer (see Protocol: **Detection of Autophagy Using GFP::LGG-1 as an Autophagy Marker** [Palmisano and Meléndez 2015a]).*

5. Incubate the second set of plates at the appropriate temperature until the F1 progeny reach the developmental stage of interest.

   Only analyzing progeny from the second set of plates ensures that the F1 progeny to be analyzed are derived from mothers that ingested the double-stranded RNA (dsRNA).

   See Troubleshooting.

6. Apply a single drop of molten agarose (2%–3%) onto a microscope slide. Immediately, place another microscope slide on top of the agarose drop to form a thin agarose pad. After the agarose has solidified, remove the top slide.

7. Pipette ~3 µL of 25 mM sodium azide onto the center of the agarose pad. Transfer 10–20 animals into the sodium azide drop and immediately apply a coverslip to prevent evaporation of the sodium azide solution. Wait ~30 sec until the animals are completely anesthetized, and then perform imaging (see Protocol 1: Detection of Autophagy in *Caenorhabditis elegans* Using *GFP:: LGG-1* as an Autophagy Marker [Palmisano and Meléndez 2015a] and Protocol 2: Detection of Autophagy in *Caenorhabditis elegans* Embryos Using Markers of P Granule Degradation [Palmisano and Meléndez 2015b]).

   For sample results, see Figure 2.

   See Troubleshooting.

TROUBLESHOOTING

Problem (Step 5): Exposure to a specific RNAi clone results in embryonic lethality.

Solution: It might be that the particular gene targeted by RNAi is required during embryogenesis. To bypass the requirement for the essential gene, treat L1 animals with the specific RNAi clone and analyze animals once the appropriate stage has been reached.

Problem (Step 7): Animals submerged in sodium azide do not become anesthetized.

Solution: The concentration of sodium azide might be too low. Remake the solution or increase the concentration, as needed.

Problem (Step 7): Knockdown of a particular gene does not give a phenotype.

Solution: Consider the following:

- The specific RNAi bacterial clone might be contaminated with another clone. It is important to ensure that the plasmid contains the correct DNA sequence corresponding to the gene of interest. If contamination is suspected, streak a fresh LB plate (containing 100 mg/mL ampicillin and 10 mg/mL tetracycline) to isolate new bacterial colonies and inoculate a new colony into 5 mL of LB medium, and then proceed with Step 1.

- The animals might not have been exposed to the RNAi clone for an adequate amount of time. Allow the worms to ingest the bacteria expressing the dsRNA for a longer period of time.

- To strengthen the effect of the RNAi, seed the plates with the bacteria expressing the dsRNA, allow the bacterial seed to dry, and then add 0.5–1 mM IPTG directly to the bacteria to induce expression. For this change in the experimental method, the NGM RNAi agar plates should be made only with carbenicillin and should not contain IPTG.

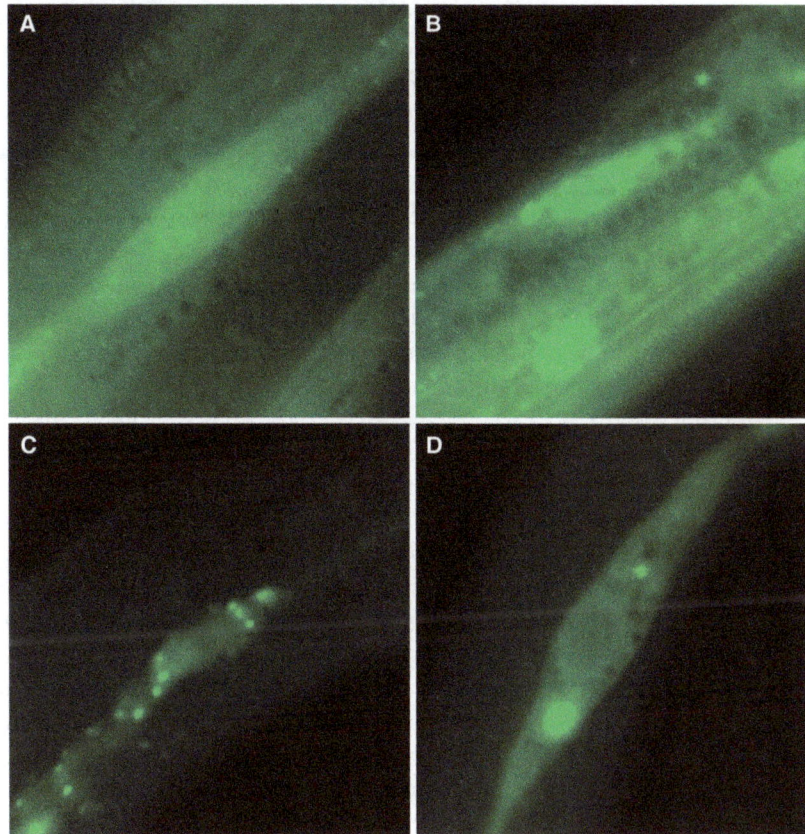

FIGURE 2. Expression of the fluorescent marker GFP::LGG-1 in hypodermal seam cells of *daf-2(e1370)* mutants of the nematode *Caenorhabditis elegans*. (*A*) *daf-2(e1370)* mutants grown on OP50 *Escherichia coli*, at 15°C, display a diffuse localization of GFP::LGG-1. (*B*) *daf-2(e1370)* mutants grown on OP50 *E. coli*, at 25°C, display an increase in GFP::LGG-1-positive puncta (up to 12 GFP::LGG-1-positive puncta per seam cell) that represent early autophagic structures or autophagosomes. (*C*) *daf-2(e1370)* mutants grown on control RNAi *E. coli* (transformed with empty vector, L4440), at 25°C, display the characteristic GFP::LGG-1-positive punctate structures. (*D*) *daf-2(e1370)* mutants fed *bec-1* RNAi, and raised at 25°C, display an increase in GFP::LGG-1 expression and large GFP::LGG-1-positive aggregates.

DISCUSSION

RNAi Enables the Study of Autophagy in Worms

RNAi is a process that results in the sequence-specific silencing of endogenous mRNA through the introduction of dsRNA (Fire et al. 1998). In *C. elegans*, RNA inactivation can be applied at any specific developmental stage, or only during adulthood to avoid developmental requirements for a given target gene. *Caenorhabditis elegans* is unusual in that RNAi is systemic, meaning that the dsRNA can spread throughout the animal and affect virtually all tissues except neurons (Grishok et al. 2000; Winston et al. 2002). There are multiple ways to deliver dsRNA; however, the most common method used in *C. elegans* is to feed the animals bacteria that express dsRNA complementary to a specific target gene (Timmons and Fire 1998; Kamath et al. 2001). This method is advantageous because libraries are available that represent most genes in *C. elegans* (Kamath et al. 2003; Rual et al. 2004), and we have the ability to knock down a given target gene beginning at any developmental stage.

The dsRNA can also be administered through injection or soaking; however, these methods require the in vitro preparation of dsRNA and are more time consuming (Ahringer 2006). One problem with the soaking method is that it can also introduce a level of stress as *C. elegans* do not feed well in liquid media and can undergo starvation (Klass 1977). However, the injection method has been shown to be more efficient in the inactivation of certain target genes (Ahringer 2006).

Autophagy genes were shown to be required for the changes associated with dauer morphogenesis in *daf-2* dauer constitutive mutants. *daf-2* encodes the *C. elegans* Insulin-like/IGF-1 receptor (IIR). RNAi has been greatly advantageous for the study of autophagy genes in the development of *C. elegans*. RNAi treatment against *unc-51*, *bec-1*, *atg-7*, *lgg-1*, and *atg-18* (by injection of dsRNA) has been shown to block dauer morphogenesis of *daf-2/IIR* dauer constitutive mutants, and to result in the appearance of GFP::LGG-1-positive protein aggregates in hypodermal seam cells (Meléndez et al. 2003). Defects in dauer development and GFP::LGG-1 expression are also observed after feeding animals RNAi against *bec-1* or against several other autophagy genes. However, we find that the phenotypes observed after feeding the RNAi bacteria may not be as penetrant as those after the dsRNA is delivered by injection. Although the exact nature of the GFP::LGG-1 aggregates is not clear, they have been suggested to be polyubiquitylated proteins that accumulate as a result of ineffective lysosomal degradation (Szeto et al. 2006). Therefore, GFP::LGG-1 localization should be carefully interpreted when using RNAi.

Knockdown of *bec-1* gene activity (by injection or feeding), and knockdown of *atg-12* and *atg-7* gene activity (by feeding only), has shown that these autophagy genes are also required for the longevity phenotype of *daf-2/IIR* mutants (Meléndez et al. 2003; Hars et al. 2007; Hansen et al. 2008). RNAi, by feeding during adulthood, against *vps-34*, *atg-7*, *unc-51*, *atg-18*, *lgg-1*, or *bec-1* has also been shown to decrease the life span extension of dietary-restricted *eat-2* mutants (Hansen et al. 2008) and, more recently, of germline-less *glp-1/Notch* mutants (Lapierre et al. 2011). The decrease in life span is not due to the effects of RNAi treatment during development, which can lead to decreased longevity as RNAi treatment was conducted during adulthood (Hansen et al. 2008; Meléndez et al. 2008). In conclusion, RNAi specific to autophagy genes has proven to be an excellent method to study the role of these genes in autophagy, as well as other cellular and developmental processes. For a description of the methods applicable to longevity assays, the reader is referred elsewhere (Meléndez et al. 2008).

Data Analysis

Confirmation that gene knockdown has occurred can be easily determined by measuring a reduction in mRNA levels of the target gene by reverse transcription polymerase chain reaction (PCR), or by measuring a reduction in protein levels by western blot analysis (Ahringer 2006). In addition, the success of an RNAi experiment can be confirmed through the use of appropriate positive and negative controls. For example, when using strains that contain a green-fluorescent protein (GFP)-tagged fluorescent reporter, a positive control that can be used is to feed animals with RNAi-expressing bacteria targeting GFP (Timmons et al. 2001). After treatment with GFP RNAi, the expression of any GFP reporter should decrease significantly. A more formal negative control for RNAi is to feed animals with bacteria transformed with the empty vector control *L4440* (Timmons and Fire 1998). Results obtained after RNAi treatment against a gene of interest should be compared with the results obtained after RNAi against the negative control with empty vector to determine whether the RNAi clone tested is the direct cause for any observed phenotype.

The effects of RNAi can be further enhanced by using certain mutations that have been shown to increase the sensitivity of animals to RNAi, such as mutations in the *lin-35*, *lin-15b*, *eri-1*, and *rrf-3* loci (Simmer et al. 2002; Kennedy et al. 2004; Wang et al. 2005; Lehner et al. 2006; Schmitz et al. 2007). To investigate the specific tissue that requires the activity of a the target gene, by RNAi, several strains can be used that carry mutations that confer RNAi resistance in particular tissues (Smardon et al. 2000; Sijen et al. 2001; Kumsta and Hansen 2012). Furthermore, as neurons are commonly known to be refractory to RNAi treatment, to treat neurons by RNAi requires a strain with neuronal expression of SID-1, a transmembrane protein required for systemic RNAi (Calixto et al. 2010). When expressed in neurons, SID-1 increases the response of those neurons to RNAi; however, neuronal expression of *sid-1* has been shown to decrease the effects of nonneuronal RNAi (Calixto et al. 2010), which has to be taken into consideration when evaluating gene knockdown in both neuronal and nonneuronal tissues. Additionally, expression of *sid-1(+)* from a cell-specific promoter in a *sid-1(−)* mutant can be used as a

Cite this protocol as *Cold Spring Harb Protoc*; doi:10.1101/pdb.prot086520

cell-specific method of feeding RNAi and limiting the effect of gene knockdown to specific cell types (i.e., neurons) (Calixto et al. 2010).

In *C. elegans*, *ego-1*, and *rrf-1* encode RNA-dependent RNA polymerases, with tissue-specific RNAi-processing function—*ego-1* partially targets the germline and *rrf-1* targets the somatic tissues (Smardon et al. 2000; Sijen et al. 2001; Suzuki et al. 2004; Qadota et al. 2007; Kumsta and Hansen 2012). Thus, to enquire whether a particular gene acts in somatic tissues, or in the germline, one can use a strain containing a mutation in the *rrf-1* locus (Sijen et al. 2001). Two mutations of *rrf-1*, *pk1417*, and *ok589*, are available and have been shown to have no effect on life span or thermotolerance (Kumsta and Hansen 2012). However, care should be taken when using *rrf-1* mutants as these mutants have been found capable of processing RNAi in somatic tissues, particularly in the intestine and a subset of hypodermal cells (Kumsta and Hansen 2012). Another consideration is whether *rrf-1* exerts any phenotype in the process being studied as it has been shown that the *rrf-1* mutation induces the expression of several transgenes and increases the expression of *sod-2*, a DAF-16/FOXO transcription target (Kumsta and Hansen 2012).

Another method that has been previously used to achieve RNAi in specific tissues is to use an *rde-1* mutant strain that is resistant to RNAi (Tabara et al. 1999) and to express the wild-type *rde-1* cDNA under the control of tissue-specific promoters to achieve RNAi sensitivity in the tissues that express *rde-1* (Suzuki et al. 2004; Qadota et al. 2007). However, several considerations have to be made when using the tissue-specific promoters as they might not rescue completely and they can sometimes be expressed in other tissues. For genes that express and/or function in multiple tissues, the assay of tissue-specific RNAi can be very helpful in distinguishing the activity of that gene in a particular tissue or cell type.

There are multiple phenotypes that have been shown to depend on autophagy gene activity and thus can be used to determine whether a particular gene of interest functions in autophagy. One can treat *daf-2/IIR* mutants that express the GFP::LGG-1 reporter shown to label structures corresponding to autophagosomes, with RNAi against the gene of interest and evaluate whether the RNAi treatment affects dauer formation or the induction of GFP::LGG-1-positive punctate structures in seam cells. When conducting such an experiment, RNAi specific to autophagy genes, such as *bec-1*, *unc-51*, or *atg-18*, should be used as positive controls. RNAi against *bec-1*, *unc-51*, or *atg-18* has been shown to increase the diffuse expression of GFP::LGG-1 and also results in the formation of GFP::LGG-1 protein aggregates (see Fig. 2) (Meléndez et al. 2003). However, the nature of the GFP::LGG-1 aggregates that arise because of inhibition, or knockdown of autophagy gene activity, is not fully understood. Thus, the formation of GFP::LGG-1 protein aggregates that arise as a result of knocking down the gene of interest should be further examined to determine where the gene of interest acts in the autophagy pathway. A next step would be to investigate whether the specific RNAi treatment affects the formation of autophagosomes, lysosomal degradation, or the accumulation of polyubiquitylated proteins in such a way as to exceed the degradative capacity of the lysosome.

Concluding Remarks

RNAi is sometimes more advantageous to study gene function than mutations, as pleiotropic effects during development can be avoided by treating animals as adults or at a specific developmental stage. For example, in the case of *bec-1*, deletion mutants that lack both maternal and zygotic *bec-1* activity are embryonic lethal; however, *bec-1* RNAi animals live until young adulthood and display severe vacuolization and incoordination (Ruck et al. 2011). Therefore, RNAi provides a means to study gene function by circumventing some of the drawbacks associated with strong loss-of-function mutations.

When investigating whether a particular gene functions in autophagy, *daf-2* mutants expressing GFP::LGG-1 can be treated with RNAi specific for the gene of interest and evaluated for changes in the expression pattern of GFP::LGG-1. Other mutant backgrounds, such as *eat-2* dietary-restricted or *glp-1* germline-less mutants, can also be used for investigating whether a particular gene functions in autophagy; however, as for *daf-2* mutants, appropriate controls are required to make accurate conclusions and comparisons from the results obtained.

RECIPES

LB (Luria-Bertani) Liquid Medium

Reagent	Amount to add
H_2O	950 mL
Tryptone	10 g
NaCl	10 g
Yeast extract	5 g

Combine the reagents and shake until the solutes have dissolved. Adjust the pH to 7.0 with 5 N NaOH (~0.2 mL). Adjust the final volume of the solution to 1 L with H_2O. Sterilize by autoclaving for 20 min at 15 psi (1.05 kg/cm^2) on liquid cycle.

M9 Minimal Medium Buffer

Reagent	Concentration (mM)
KH_2PO_4	22 mM
Na_2HPO_4	22 mM
NaCl	85 mM
$MgSO_4$	1 mM

Autoclave for 15 min on liquid cycle. Allow medium to cool and store at room temperature.

NGM RNAi Agar Plates

68 g Bacto agar powder
12 g NaCl
10 g Bacto peptone
4 mL cholesterol (5 mg/mL in ethanol)
3.9 L deionized water
4 mL $CaCl_2$ (1 M)
4 mL $MgSO_4$ (1 M)
100 mL KPO_4 (1 M)
50 mg/mL carbenicilin
1 M Isopropyl B-D-1-thiogalactopyranoside (IPTG)

To prepare 4 L of normal growth media (NGM), mix the first five ingredients, and autoclave for 70 min on liquid cycle. Ensure that the agar is dissolved, let cool, and then add the last five ingredients. Pour the plates (these can be stored for up to 6 wk at 4°C).

REFERENCES

Ahringer J. 2006. Reverse genetics. *WormBook: The online review of C. elegans biology.*

Calixto A, Chelur D, Topalidou I, Chen X, Chalfie M. 2010. Enhanced neuronal RNAi in *C. elegans* using SID-1. *Nat Methods* 7: 554–559.

Fire A, Xu S, Montgomery MK, Kostas SA, Driver SE, Mello CC. 1998. Potent and specific genetic interference by double-stranded RNA in *Caenorhabditis elegans. Nature* 391: 806–811.

Grishok A, Tabara H, Mello CC. 2000. Genetic requirements for inheritance of RNAi in *C. elegans. Science* 287: 2494–2497.

Hansen M, Chandra A, Mitic LL, Onken B, Driscoll M, Kenyon C. 2008. A role for autophagy in the extension of lifespan by dietary restriction in *C. elegans. PLoS Genet* 4: e24.

Hars ES, Qi H, Ryazanov AG, Jin S, Cai L, Hu C, Liu LF. 2007. Autophagy regulates ageing in *C. elegans. Autophagy* 3: 93–95.

Kamath RS, Martinez-Campos M, Zipperlen P, Fraser AG, Ahringer J. 2001. Effectiveness of specific RNA-mediated interference through ingested double-stranded RNA in *Caenorhabditis elegans. Genome Biol* 2: RESEARCH0002.

Kamath RS, Fraser AG, Dong Y, Poulin G, Durbin R, Gotta M, Kanapin A, Le Bot N, Moreno S, Sohrmann M, et al. 2003. Systematic functional analysis of the *Caenorhabditis elegans* genome using RNAi. *Nature* **421**: 231–237.

Kennedy S, Wang D, Ruvkun G. 2004. A conserved siRNA-degrading RNase negatively regulates RNA interference in *C. elegans*. *Nature* **427**: 645–649.

Klass MR. 1977. Aging in the nematode *Caenorhabditis elegans*: Major biological and environmental factors influencing life span. *Mech Ageing Dev* **6**: 413–429.

Kumsta C, Hansen M. 2012. *C. elegans* rrf-1 mutations maintain RNAi efficiency in the soma in addition to the germline. *PLoS ONE* **7**: e35428.

Lapierre LR, Gelino S, Meléndez A, Hansen M. 2011. Autophagy and lipid metabolism coordinately modulate life span in germline-less *C. elegans*. *Curr Biol* **21**: 1507–1514.

Lehner B, Calixto A, Crombie C, Tischler J, Fortunato A, Chalfie M, Fraser AG. 2006. Loss of LIN-35, the *Caenorhabditis elegans* ortholog of the tumor suppressor p105Rb, results in enhanced RNA interference. *Genome Biol* **7**: R4.

Meléndez A, Talloczy Z, Seaman M, Eskelinen EL, Hall DH, Levine B. 2003. Autophagy genes are essential for dauer development and life-span extension in *C. elegans*. *Science* **301**: 1387–1391.

Meléndez A, Hall DH, Hansen M. 2008. Monitoring the role of autophagy in *C. elegans* aging. *Methods Enzymol* **451**: 493–520.

Palmisano NJ, Meléndez A. 2015a. Detection of autophagy in *Caenorhabditis elegans* using *GFP::LGG-1* as an autophagy marker. *Cold Spring Harb Protoc* doi: 10.1101/pdb.prot086496.

Palmisano NJ, Meléndez A. 2015b. Visualization of autophagy in *Caenorhabditis elegans* embryos. *Cold Spring Harb Protoc* doi: 10.1101/pdb.prot086504.

Qadota H, Inoue M, Hikita T, Koppen M, Hardin JD, Amano M, Moerman DG, Kaibuchi K. 2007. Establishment of a tissue-specific RNAi system in *C. elegans*. *Gene* **400**: 166–173.

Rual JF, Ceron J, Koreth J, Hao T, Nicot AS, Hirozane-Kishikawa T, Vandenhaute J, Orkin SH, Hill DE, van den Heuvel S, et al. 2004. Toward improving *Caenorhabditis elegans* phenome mapping with an ORFeome-based RNAi library. *Genome Res* **14**: 2162–2168.

Ruck A, Attonito J, Garces KT, Nunez L, Palmisano NJ, Rubel Z, Bai Z, Nguyen KC, Sun L, Grant BD, et al. 2011. The Atg6/Vps30/Beclin1 ortholog BEC-1 mediates endocytic retrograde transport in addition to autophagy in *C. elegans*. *Autophagy* **7**: 386–400.

Schmitz C, Kinge P, Hutter H. 2007. Axon guidance genes identified in a large-scale RNAi screen using the RNAi-hypersensitive *Caenorhabditis elegans* strain nre-1(hd20) lin-15b(hd126). *Proc Natl Acad Sci* **104**: 834–839.

Sijen T, Fleenor J, Simmer F, Thijssen KL, Parrish S, Timmons L, Plasterk RH, Fire A. 2001. On the role of RNA amplification in dsRNA triggered gene silencing. *Cell* **107**: 465–476.

Simmer F, Tijsterman M, Parrish S, Koushika SP, Nonet ML, Fire A, Ahringer J, Plasterk RH. 2002. Loss of the putative RNA-directed RNA polymerase RRF-3 makes *C. elegans* hypersensitive to RNAi. *Curr Biol* **12**: 1317–1319.

Smardon A, Spoerke JM, Stacey SC, Klein ME, Mackin N, Maine EM. 2000. EGO-1 is related to RNA-directed RNA polymerase and functions in germ-line development and RNA interference in *C. elegans*. *Curr Biol* **10**: 169–178.

Suzuki M, Sagoh N, Iwasaki H, Inoue H, Takahashi K. 2004. Metalloproteases with EGF, CUB, and thrombospondin-1 domains function in molting of *Caenorhabditis elegans*. *Biol Chem* **385**: 565–568.

Szeto J, Kaniuk NA, Canadien V, Nisman R, Mizushima N, Yoshimori T, Bazett-Jones DP, Brumell JH. 2006. ALIS are stress-induced protein storage compartments for substrates of the proteasome and autophagy. *Autophagy* **2**: 189–199.

Tabara H, Sarkissian M, Kelly WG, Fleenor J, Grishok A, Timmons L, Fire A, Mello CC. 1999. The *rde-1* gene, RNA interference, and transposon silencing in *C. elegans*. *Cell* **99**: 123–132.

Timmons L, Court DL, Fire A. 2001. Ingestion of bacterially expressed dsRNAs can produce specific and potent genetic interference in *Caenorhabditis elegans*. *Gene* **263**: 103–112.

Timmons L, Fire A. 1998. Specific interference by ingested dsRNA. *Nature* **395**: 854.

Wang D, Kennedy S, Conte D Jr, Kim JK, Gabel HW, Kamath RS, Mello CC, Ruvkun G. 2005. Somatic misexpression of germline P granules and enhanced RNA interference in retinoblastoma pathway mutants. *Nature* **436**: 593–597.

Winston WM, Molodowitch C, Hunter CP. 2002. Systemic RNAi in *C. elegans* requires the putative transmembrane protein SID-1. *Science* **295**: 2456–2459.

High-Throughput Approaches to Measuring Cell Death

Darren N. Saunders,[1,2] Katrina J. Falkenberg,[3,4] and Kaylene J. Simpson[3,4,5,6]

[1]*Cancer Research Program, The Kinghorn Cancer Centre, Garvan Institute of Medical Research, Darlinghurst, New South Wales 2010, Australia;* [2]*St. Vincent's Clinical School, University of New South Wales Medicine, Sydney, New South Wales 2000, Australia;* [3]*Victorian Centre for Functional Genomics, Peter MacCallum Cancer Centre, East Melbourne, Victoria 3002, Australia;* [4]*Department of Pathology, The University of Melbourne, Parkville, Victoria 3052, Australia;* [5]*Sir Peter MacCallum Department of Oncology, The University of Melbourne, Parkville, Victoria 3052, Australia*

Cell death is integral to developmental and disease processes, and high-throughput screening (HTS) has been instrumental both for understanding biological mechanisms underlying cell death and for discovering novel therapeutic agents targeting these pathways. The various cell death modalities and their distinctive morphological and biochemical features have led to the development of a staggering variety of assays to measure these features, many of which have been adapted to HTS format. Although not all cell death assays are readily amenable to a high-throughput format, the potential power of HTS assays and increasing accessibility to associated technology make it likely that new approaches will continue to emerge. In particular, many recent studies in this field have used multiplex assays and high-content imaging to measure several features concurrently. Here, we discuss a broad array of considerations for designing HTS cell death assays, including some common challenges and pitfalls. We aim to provide a framework for deciding the most appropriate biological readouts, assay strategy and mode, workflow, controls, validation, and bioinformatics.

INTRODUCTION

Our understanding of the molecular pathways controlling cell death is rapidly improving, and with it our ability to modulate these pathways for therapeutic benefit. High-throughput screening (HTS) has been instrumental both for understanding the biological mechanisms underlying cell death and for discovering novel cytoprotective and cytotoxic agents targeting these pathways. In particular, HTS approaches have been used for drug discovery and pharmacogenomics, as well as for identifying genetic regulation of cell death (Dompe et al. 2011) and understanding mechanisms of drug resistance (Potratz et al. 2010) (Table 1).

Cell death can occur via multiple pathways, each showing distinct morphological and biochemical features. Accordingly, a staggering variety of assays have been developed to measure these features, many of which have been adapted to HTS format and are available commercially (Table 1). HTS assays for cell death can take many forms, depending on the specific question(s) being asked, the parameters being measured, and the nature of the library being used to modulate cell behavior (i.e., chemical or genetic). They are generally linked to a few common reporter systems that are easily adapted for use in the various high-throughput instruments available, including morphology, fluorescence, bioluminescence, and phenotypic selection (e.g., survival/resistance or synthetic lethality). Fluorescent protein reporters used in HTS to measure cell death or cell death signaling include Bax-GFP (Galluzzi et al.

[6]Correspondence: kaylene.simpson@petermac.org

Cite this introduction as *Cold Spring Harb Protoc*; doi:10.1101/pdb.top072561

TABLE 1. Cell death assays amenable to HTS

Biological readout	Format	Commercial assay	Notes	Suppliers	Reference
Caspase activity	HC or PR	Caspase-Glo, ApoLive Glo (multiplex with cell titer Fluor) Caspase-3 activation kit	Various reagents available for specific caspases; multiplex capable; potential for nonspecific activation of substrates; end point	Promega Cellomics (Thermo)	Stec et al. 2012
Cell viability—ATP	PR	Cell Titer-Glo	End point	Promega	Boutros et al. 2004; Nguyen et al. 2006
Cell viability—redox status	PR	Rezazurin (alamar blue) Cell Titer-Blue	Nontoxic, can be multiplexed with HC	Life Technologies Promega	Bauer et al. 2010
Cell viability—live cell protease	PR	Cell titer Fluor	Nontoxic, can be multiplexed with HC	Promega	
Cell viability—esterase	HC or PR	Calcein-AM	Live cell only	Life Technologies, Sigma, BD	Gilbert et al. 2011
Protease activity	PR	CytoTox-glo	Dead cell protease detection; reflects membrane integrity; end point	Promega	Niles et al. 2009
DNA cleavage	HC	TUNEL	Measures DNA strand nicks associated with apoptotic nuclease activity; end point	Life Technologies	Parrish and Xue 2003
Nuclear morphology	HC	DAPI, DRAQ-5, BOBO-3, etc.	DNA binding dyes; end point and live cell; often multiplexed	Various (numerous)	De Pasquale et al. 1990
Mitochondrial function (membrane potential)	PR	MitoProbe JC-1CMXRos (mitotracker red)	Measures mitochondrial membrane integrity; end point and live cell	Life Technologies	Qiu et al. 2010
DNA synthesis	HC	EdU/BrdU	Usually coupled with measurement of DNA content	Life Technologies	Poon et al. 2008
Plasma membrane integrity (LDH release, protease release, DNA binding)	HC or PR	CytoTox-glo, CytoTox-ONE, CellTox Green Image-iT DEAD, Ethidium	Often coupled with other markers in multiplex assays particularly suited to kinetic assays, nontoxic	Promega Roche Life Technologies	Buenz et al. 2007

HC, high content; PR, plate reader

2010), GFP-H2B (Neumann et al. 2010), and protein fragment complementation (PCA/BiFC) (Mac-Donald et al. 2006).

Common challenges when measuring cell death and apoptosis by HTS include the highly dynamic and multifaceted nature of cell death signaling pathways, a diversity of responses to specific stimuli, the heterogeneity of cell lines, and defining the most appropriate timing for the assay readout. Avoiding reliance on a single morphological or biochemical feature by deploying two or more complementary biological readouts can help avoid many of these potential pitfalls. Recent advances in single-cell analysis may also be a powerful means to overcome some of these issues (Snijder et al. 2012).

Although not all cell death assays are readily amenable to a high-throughput format, increasing interest in HTS assays and their potential power as well as increasing accessibility to technology make it likely that new approaches will emerge. We discuss here a broad array of considerations when designing HTS cell death assays. Our intention is to provide a framework for deciding the most appropriate biological readouts, assay strategy and mode, workflow, controls, validation, and bioinformatics. We also provide a protocol that encompasses high-throughput imaging and plate reader-based assays for cell viability and caspase 3/7 activity (see Protocol 1: A High-Throughput, Multiplex Cell Death Assay Using an RNAi Screening Approach [Falkenberg et al. 2014]).

INITIAL CONSIDERATIONS

In designing an HTS assay for cell death, it is critical to first determine the measurement most relevant to the particular signaling pathway or phenotype of interest. For a particular assay to be adaptable to high-throughout format, it should satisfy the normal criteria related to reproducibility, sensitivity, and

Cite this introduction as *Cold Spring Harb Protoc*; doi:10.1101/pdb.top072561

specificity. However, there are three key additional considerations: (1) compatibility with multiwell format readouts (e.g., a plate reader or high-content imager), (2) compatibility with a high-throughput workflow (i.e., the assay should not be too complex or time consuming), and (3) expense of substrates and specialized reagents, particularly when being used in large quantities.

From a biological perspective, it is important to consider the various cell death modalities when deciding which assay(s) may be most appropriate. For example, apoptosis and necrosis are distinguishable by very specific biochemical and morphological features. Many assays that measure cell viability as a surrogate of cell death are unable to distinguish between these modalities or discern the effects of cytostasis.

A selection of cell death assays with demonstrated utility in the HTS format is shown in Table 1. These assays represent a reasonably broad spectrum of the biological mechanisms relevant to the various cell death modalities, including proliferation (cell number), protease activity (e.g., caspases), cell cycle (GFP-cyclins, DNA content and synthesis), cell viability (e.g., membrane permeability, metabolic substrates), cellular energetic or reductive potential (e.g., ATP or NADH levels), mitochondrial function (cytochrome c release, membrane potential), and DNA cleavage (e.g., TUNEL). More recently, multiplex assays combining multiple readouts in a single workflow have been used, as described in the accompanying protocol (Protocol 1: A High-Throughput, Multiplex Cell Death Assay Using an RNAi Screening Approach [Falkenberg et al. 2014]). Further, in vitro and in vivo selection-based screens are also possible in high-throughput format (e.g., using arrayed or pooled shRNA/ORFeome screens or survival of mixed populations).

ASSAY MODE

As noted in Table 1, there are many cell death assays to choose from. However, there are limited tools for performing these assays. Principally, high-throughput cell death assays can be measured using high-content imaging (via a dedicated microscope that is fully automated with computer-learned algorithms for analysis), an intermediate low-resolution imager for simple cell counting, or a microplate reader that quantifies fluorescence or luminescence. The distinction to note is that high-content imaging generally reflects a specific number of cells in a well, whereas plate reader quantitation represents the entire contents of the well.

High-Content Imaging Mode

High-content imaging enables quantitation of specific cellular features on a per cell basis (Snijder et al. 2012). It requires robust assays and protocols and can be highly specific (e.g., measuring a particular molecular process in a cell death pathway) or more general (e.g., counting cells and distinguishing live from dead). Imaging assays typically require a lot of dedicated time and effort to develop because of the inherent nature of "teaching" the computational algorithm the specific intricacies of the phenotype, followed by statistical validation. Image analysis relies on quantifying a statistically representative number of cells and fields per well. Some of the more straightforward assays include quantitation of either healthy or dying cells using DAPI or Hoechst staining and defining the size of a healthy versus condensed nucleus. More detailed analysis includes detection of caspase activity and/or cytochrome C release from mitochondria. One particularly challenging screen used GFP:H2B reporter cell lines to follow cytokinesis via live imaging (Tsui et al. 2009), whereas another quantified an exhaustive array of basic cell features such as cell division, proliferation, survival, and migration (Neumann et al. 2010). The cell death features that can be measured by high-content imaging are extensive (see Table 1 for selected examples) and can often be modified from low-throughput approaches. Other considerations for assay development and design are outlined below in "Assay Design and Strategy."

The following hardware is needed for high-throughput imaging:

- A specialized microscope with multiple objectives and multiple fluorophore capabilities that is capable of reading at least 96- to 384-well format plates.

- Automation to deliver plates to the instrument (depending on the number of plates, automation improves throughput for plates that take many hours to read and allows images to be acquired overnight and through the weekend).

- Barcode read-enabled instrumentation for plate label incorporation into file names of all data acquisition.

- Software that is intuitive, allows visualization of all well data, and can export cell features in a manageable data format (e.g., .xls, .csv, or .txt).

- Brightfield imaging. It is not often used in high-content screening, but it can be useful for visualizing cells and can be used in conjunction with, for example, β-galactosidase senescence assays.

- Sufficient data processing and storage capacity, and IT support. Image size depends on many parameters such as the image resolution (2×2 or 1×1 binning), the number of fluorophores, the number of fields imaged, the number of cells counted, and the number of parameters measured. Together, the amount of data from a single 384-well plate can range from several Gb to ~25 Gb. Factor those figures into a large compound screen or genome-scale RNAi screen, in the order of 60 or more library plates, screened in duplicate and multiple terabytes are easily required. The amount of data collected and ultimately stored for the purpose of probity should not be underestimated; a large server and storage capacity is absolutely required, particularly for screening facilities.

Because experience and technical knowledge can be a significant barrier to accessing this technology, a local field application specialist or dedicated microscopy specialist can be invaluable. Consider the following when you plan for high-content imaging:

- The appropriate cell density must be empirically determined; for example, overconfluent cells will not be easy to segment and identify.

- The complexity of the assay should be considered in terms of minimizing wash steps and the feasibility of performing all steps in a relatively defined period.

- The number of channels to be imaged will vastly alter the imaging time.

- The number of cell features collected must also be considered. Capturing information that you do not intend to use increases data storage dramatically, so think carefully about the information required. It is always possible to reanalyze images after they have been captured.

Plate Reader Mode

Most plate reader assays measure the response of the entire cell population within the well, although some instruments are capable of multiple sampling across individual wells. Assay development is generally much faster using a plate reader-based assay compared with a microscopy-based assay but will be limited to only one or two key parameters. It is important to note that in comparison with image analysis, in which information can be obtained on a single-cell level, detail can be lost when taking an aggregate measurement. Often, a plate reader-based viability assay will be multiplexed into an imaging-based screen in which cell health is not easily measured (e.g., wound healing [Simpson et al. 2008]). Because there is no permanent record of the well contents as there is for imaging, it is imperative to confirm during assay development that the specific assay readout accurately depicts what is visible. For example, when measuring cell health using an ATP-based readout, a parallel plate must be evaluated at least visually but if possible, by accurately counting nuclei-stained cells on a high-throughput microscope. This step is essential both to validate the assay and to ensure there is a significant correlation between cell number and the relative viability reading. Once confirmed, a screen can continue with only the plate reader quantitation, but we recommend a quick visual inspection of randomly selected plates before addition of fluorescent or luminescent reagents. If quantifying a phenotype that is not easily interpreted visually (e.g., caspase activity), then benchmarking the readout against a more traditional low-throughput approach (FACS-based annexinV/PI or active caspase-stained cells) is essential. Regardless of the assay, a comparable dynamic range and signal-to-noise ratio should be detected between the positive and negative controls in high-throughput versus traditional low-throughput methods.

Plate reader assays involve the following instrumentation and software:

- Plate management automation. Automation provides timed accuracy and allows the researcher to continue working while plates are reading.

- Software to batch export files, so all the plates read at a single time point can be incorporated in one output file (individual files work fine, but it is better to keep all data in one file).

- Barcode read-enabled instrumentation to incorporate the barcode into the output file to ensure correct plate ID, thus eliminating any concerns if plates are accidentally stacked out of sequence.

- A plate reader that permits multi-sampling injection. These readers can be very useful for kinetic studies and luminescence assays.

Optimize gain and exposure times to ensure high sensitivity and dynamic range without "overflowing" the sample (when the gain has reached maximum capacity for readout value).

Adherent Versus Suspension Cells

One of the most critical considerations in defining the assay type and mode of measurement is the choice of cell type, which ideally reflects the biological question you are asking. The most suitable cells from a biological standpoint, however, may not always be the best choice for the assay. For example, suspension cells in particular are not easily amenable to high-throughput assays from the perspective of handling and assay readouts, but they can perform sufficiently well in plate reader assays if media changes are possible (or can be avoided) during the workflow and if staining is optimal. If suspension cells can settle or be centrifuged to the bottom of the plate, imaging might be possible but will still be challenging.

ASSAY DESIGN AND STRATEGY

Many different elements must be considered collectively when planning and executing an HTS approach.

The Model System

Understanding the limitations and advantages of your model system, and in particular the cell line(s) being used, is critical to a successful high-throughput screen. Are you investigating cell death in a global fashion with no mechanistic insight? Do you need to distinguish growth inhibition versus cell death? Cytostasis is a common drug response that is difficult to detect with assays that measure ATP content, in part because of the rate of proliferation during the course of the assay (especially if only 2–3 d). However, to understand the mechanism driving the death response, the mode of action of the drug must be considered (see "Drug Action"). Answers to these questions will help establish the assays most suitable for evaluating a particular emergent phenotype (Table 1). Importantly, the cell line(s) chosen should reflect the biological question being asked. For example, if you are performing an siRNA screen to identify novel ovarian cancer oncogenes, an appropriate ovarian cell line should be chosen (such as a tumorigenic line) in which loss of a gene target will result in death or a less aggressive phenotype. Powerful insights have been gained using synthetic lethal genetic approaches with isogenic cell lines (with genetically altered lines created, for example, by gene ablation or overexpression in a parental line) or matched sensitive-versus-resistant cell lines derived from the same genetic background (Luo et al. 2009). We highly recommended that cell lines be routinely assessed for mycoplasma contamination. You must be "in tune" with the cells being used and understand their morphology, passaging and seeding densities, transfection conditions, and response to any treatment. This is critical in being assured that the cells are responding as expected throughout the course of an assay. Importantly, pilot assays should be performed to statistically evaluate cell responses, because differences that appear satisfactory on first glance may often not be statistically robust.

Drug Action

Many chemotherapeutics kill by inducing apoptosis via a caspase activation cascade, but their mechanisms for doing so vary. Targets include intrinsic and extrinsic pathways, mitochondria, and death/ growth factor receptor pathways (Brunelle and Zhang 2010). Understanding the mechanism of action for the drug you are studying will help you devise sophisticated analyses, which may not be amenable for a primary screen but will often be suitable for triaging hits in a lower-throughput, secondary screen format (i.e., one plate versus 20 plates). The mechanism of action also informs the type of screening approach to be used, particularly with respect to the timeframe and duration of the screen (e.g., compound versus siRNA screen; synthetic lethal or drug resistance).

An understanding of drug action also directly affects the choice of controls to include on each plate. Positive and negative controls become vital for establishing the dynamic range of the death response. For example, the extent of cell survival in control wells determines the concentration of drug required to induce cell death or sensitize cells to death. If screening for drug resistance, the drug concentration used should be no greater than that needed to produce 90% cell death in control wells. This will avoid drug concentrations so toxic that even resistant cells struggle to survive and assay sensitivity is compromised.

Maximizing Data Density

In general, plate reader-based assays produce relatively limited amounts of data, as only one or two parameters can be measured in an assay. In contrast, high-content imaging is capable of producing seemingly endless amounts of multidimensional cell feature data. This can be a distinct advantage, but identifying the most meaningful data and the best way to extract it is critical. Recent outstanding publications that capture exhaustive numbers of cell features are clearly the result of many years of dedicated effort, from acquisition of data to advanced computational analysis (Neumann et al. 2010; Snijder et al. 2012). These huge, multidimensional approaches can be particularly daunting and in reality are not currently achievable for most researchers outside specialist facilities. Keeping such publications in mind, however, it is important to be aware of the main questions being asked, the key data to be collected (specific cell features, timing of response, live versus fixed cells), and what additional information can be collected that could add value—not necessarily in the present but possibly in the future. If cell-based imaging is set up thoughtfully, encompassing key elements like nuclei staining, a cell mask, and possibly a mitochondrial marker, and a significant number of cells and fields are imaged, archived images can always be reanalyzed with new algorithms to answer different or related questions.

Live Versus Fixed Analysis—Temporal Context

There are clear advantages to quantifying cell death over a time course to establish the rate and mechanism of cell death, particularly when looking for the most effective drug or gene target. Live imaging, however, is not significantly high throughput if one is trying to keep the timing between treatment and imaging consistent, given the time it takes to capture multiple sites in a 384-well plate. There are several dyes that can be used to track live cells, but the production of a cell line harboring a fluorescent reporter system (Tsui et al. 2009) is a very efficient and consistent tool for tracking cells. Another consideration is that live imaging can generate an enormous amount of data, depending on the time course. For example, compare imaging five fields every 30 sec for 3 h with imaging five fields once every hour for 24 h. An exciting new product, the CellTox Green Express Cytotoxicity Assay (Promega), involves a nontoxic dye that binds to DNA in cells with compromised membrane integ-

Cite this introduction as *Cold Spring Harb Protoc*; doi:10.1101/pdb.top072561

rity; the dye can be presented in culture media over the time course of an experiment, emitting fluorescence quantified by a plate reader or microscope.

Fixing cells and staining with fluorescently labeled antibodies remains the most common method for image-based analysis, and the range of different fluorophores matching the read capabilities of a high-content microscope allows multiplexing. A stable fluorescent reporter cell line also performs very well for fixed end point readouts and can significantly reduce screening costs by not requiring any antibodies. With fixed cell staining, it is possible to accurately conclude an experiment within a defined window of time; the method is generally robust enough that imaging can occur over days, if necessary, or the samples can be stored at 4°C and imaged later.

Cell Density

Cell density is critical for assay success and must be considered at multiple points during assay development. The appropriate density at which to start and end the assay is affected by the number of population doublings the cells must undergo and the timeframe of efficacy of the drug or other treatment used (e.g., siRNA knockdown). Optimal transfection density, for example, is tied to cell number and lipid concentration, and the final cell density is generally based on controls.

In the context of a high-throughput chemical screen, compounds are generally serially diluted over a broad dose–response range and applied to cells plated at a density that results in controls reaching ~90% confluence at the assay end point (cells can be seeded before or at the same time as drug addition). When performing a synthetic lethal functional genomics screen, cell density at transfection is optimized based on the desired end point density and transfection/infection efficiency. The end point density is a combination of the level of death in the positive and negative controls, the length of drug treatment, and the response time after treatment. Drug activity and response can be acute (hours), or cells can be cultured in drug for days. Overconfluent cells at an end point can cause misleading viability data and affect the measurement of other variables, because such cells can effectively alter their physiological state and responsiveness. For siRNA screens, the knockdown window is normally 72–96 h. This timeframe can be extended effectively as far as 120–144 h, but you must verify these times in your cell line. Starting density can then be extrapolated by knowing the end point cell number and estimating the doubling time over the course of the assay. The optimal transfectable density must also be considered, and is usually ~30% confluence. If the density is too sparse, cells would not proliferate well and the lipid may be toxic, and if the density is too high, the transfection efficiency will be dramatically reduced and the effective length of time in culture (especially in a 384-well format) will be reduced.

Instrument-specific aspects of imaging also are affected by cell density. Microscopes can have difficulty defining and counting confluent cells, and imaging very sparse wells can take extraordinary lengths of time as the microscope struggles to find cells and subsequently a focal plane (i.e., increasing sampling time). In general, if sparse wells is not the phenotype you are studying, we recommend defining a minimum threshold cell number to be obtained in three to five fields, which if not met, will cause the microscope to move to the next well; there is clearly no need to waste time imaging nothing.

Controls

Appropriate and robust controls, including negative (or no effect), positive (assay-specific biological controls), and technical controls, are essential to any high-throughput approach, especially for screens that might take weeks to complete. These controls must be defined by your biological question and the technical considerations inherent in the specific platform used. Assay-specific controls must be both informative and statistically robust. It is not unusual that a suitable positive control cannot be found for a specific RNAi screen, in some ways reflecting the discovery-based nature of this approach. In practice, however, it is often possible to set up viability screens using known potent death genes identified from numerous screening efforts (such as PLK1, COPB2, WEE1, and AURKA). Controls not only report the technical elements of a screen on a per-plate basis, but are also essential for normalization during data analysis, not just on a per-plate basis but also for comparison of all plates.

The same controls must be included in every plate, with sufficient replicates to ensure statistical significance, and their location on the microplate depends on the format of the screen's library. Where possible, edge columns and rows should be avoided to minimize potential edge effects. Statistical analysis of controls is outlined briefly below and in detail in the accompanying protocol (see Protocol 1: A High-Throughput, Multiplex Cell Death Assay Using an RNAi Screening Approach [Falkenberg et al. 2014]).

Dynamic Range and Magnitude of Effect

The dynamic range of an assay represents the breadth between positive and negative control values, such that they are statistically significant from each other, allowing "hits" to be identified from all the samples. The signal-to-noise ratio reflects the level of signal above background, which should also be significantly higher. Luminescence assays tend to give very low background (i.e., in media-only conditions) and very robust, highly sensitive readings that result in large numbers. Signal-to-noise issues can be more pervasive in imaging assays, depending on how clean the background staining is and how strong the signal is. However, postprocessing correction can significantly improve signal-to-noise ratio and segmentation performance (Poon et al. 2008).

Timing of Effect

Drug mechanism of action and concentration can directly affect the rate of cell death. During assay development, it is important to establish the window of activity for treating cells such that the subsequent effects can be measured. If possible, we recommend evaluating multiple time points (if working with an end point assay, set up as many replicate plates as required). For example, in a synthetic lethal RNAi screen, stimulation with drug for 4 h results in control cells that are largely still viable but primed and sensitized for death. The time it takes for the maximum number of control cells to die must then be evaluated and the window for the assay based on that. The best option for some screens may be a series of kinetic measurements rather than a single end point. If that approach is too daunting on a genome scale, then consider an end point screen followed by a kinetic screen with a smaller high-confidence gene/compound list.

Workflow Considerations

When working in high-throughput mode, the sheer number of plates necessitates a trade-off between complexity and throughput, whereby workflows are often simplified to increase throughput. This does not mean compromising on experimental rigor. Thus, all assays must be reviewed in light of the question, "Can I make all of these manipulations within a defined window of time so each plate is treated equally?" For example, traditional low-throughput immunohistochemistry might require fixing, blocking, and staining, with three washes between each step. This process is very difficult to automate, because of the time the steps require and the forceful effect of the dispensing and media change apparatus on cells (which, although determined to be acceptable at the outset of a screen, can dislodge the cells after days of culture and high repetition). The use of commercial assays must be reviewed in the same manner; incubations for very short time frames (e.g., 10 sec followed by 1 min) are not possible with large numbers of plates. To avoid compromising the outcome of the experiment, the design of the assay must take into account the trade-offs between quick and relatively easy procedures versus those that require extensive processing.

Understanding Assay Limitations

There is no "one size fits all" assay that is suitable for every high-throughput cell death approach. It is, therefore, critical to be aware of exactly what is being measured and what, if any, limitations there might be on those measurements. For example, the stability of reagents after they are added to cells determines when assay outputs must be measured, and data quality can deteriorate quickly outside of that time frame. Many reagents can have diminished activity when freeze–thawed, and plate readers

Cite this introduction as *Cold Spring Harb Protoc*; doi:10.1101/pdb.top072561

and microscopes can take many minutes or hours to read a plate—conditions that might not be compatible with the assay or throughput requirements.

Assay Reproducibility

The use of liquid-handling automation for high-throughput approaches is essential for reducing variability. Reproducibility can be controlled by maintaining consistency on the timing of all aspects of the experiment, including cell dispensing, media changes, stimulation or drug treatment, and the assay itself, performed at the same time and the same way each time. In other words, a 2-min shake and 10-min incubation must be exactly that for each plate, each time it is done. The coefficient of variation (for the same sample treatment over multiple wells) is the key measurement to consider when assessing assay reproducibility (defined as the standard deviation divided by the mean, expressed as a percentage). Ideally, this number is ~10%, with values higher than 25% being unacceptable. The vast majority of high-throughput screens are performed with multiple technical replicates (at least duplicates), and depending on the assay, multiple biological replicates. Variability can also arise from the materials used in screens. We recommend banking a large number of vials of cells at the same passage before starting a screen and to using a fresh passage for each set of transfections or compound addition. Assay variability can be introduced by changes in lot number of essential reagents, including serum, compound or dilution batch, transfection lipid, media and additives, primary and secondary antibodies, and death assay reagents. We recommend that when possible, you purchase enough resources for the entire primary screen and any subsequent validation. All lots must be validated before embarking on a screen.

All liquid-handling instruments, including those that do nanoliter or microliter dispensing, cell dispensing, media changes, and siRNA transfection mixing (the actual mixing step is very important to ensure homogeneity in a very small volume), must be subject to routine quality control for volumetric dispensing, using a fluorescent reagent such as tartrazine. Microscopes should be routinely evaluated for consistent light intensity, even illumination field, and filter integrity to ensure that the exposure level is constant (these issues are of special concern if exposure times start to alter over experiments that last weeks). Bulbs should be changed well within their expected life span, although light source deterioration generally occurs long before the bulb goes out. The new LED light system microscopes have addressed these major limitations by having an extremely long life and more even illumination.

Validation Screens

The discussion points above are a broad overview centering on the concepts for a primary, first-pass screen that aims to identify and quantify the occurrence of cell death but generally not provide specific mechanistic insights. In general, once a hit list of primary candidates has been obtained, the researcher needs to be prepared to evaluate orthogonal assays to refine this list. These validation screens may come in the form of alternative high-throughput assays, or a shift back to a lower-throughput system that relies on more standard assays, which in this field often involve flow cytometry-based measurements.

INFORMATICS

No high-throughput approach is possible without strong computational and biostatistical support. Bioinformatics approaches range from determining assay robustness to collating a comprehensive compound library or genome-scale RNAi data. Robust statistical methods must be used to identify the most potent and consistent phenotypes from the complete screen data. The Z'-factor, a statistical measure of the dynamic range between positive and negative controls, encompasses the average and standard deviation of each control, which must be at least two standard deviations from each other. To ensure statistical significance, this method requires a large number of wells for each control reagent;

we recommend at least 10 of any control in a 384-well plate. For RNAi screening, a value of >0.3 is acceptable, whereas compound screening requires a value of >0.7. Great emphasis is placed on the Z'-factor during assay development and thereafter during the screen to ensure controls are consistent from week to week. The Z'-factor calculation is as follows:

$$Z'\text{-factor} = 1 - (3 * sd_{hc} + 3 * sd_{lc})/|mean_{hc} - mean_{lc}|,$$

where $*$ indicates "multiplied by," "mean" indicates the average, "sd" indicates the standard deviation, "hc" indicates the high-value control, and "lc" indicates the low-value control.

For an siRNA/miRNA screen, the most common approach for defining screen hits is to use a z-score or robust z-score (sample-based normalization) (Birmingham et al. 2009). This approach standardizes the strength of an siRNA phenotype relative to the rest of the sample distribution, reported as the number of standard deviations away from the mean. The z-score can be calculated as follows. A z-score approach can be applied to a plate of data or an entire genome's worth of data to identify hits:

$$z\text{-score} = (\text{sample value} - \text{sample mean})/\text{standard deviation}.$$

An alternative to the z-score is the robust z-score, which takes into account the plate median and median absolute deviation to eliminate the effects of outliers:

$$\text{robust } z\text{-score} = (\text{sample value} - \text{sample median})/\text{sample median absolute deviation}.$$

For a compound screen, an IC_{50} (half maximal inhibitory concentration; usually growth inhibition or cell death) measurement is made for the entire population assayed and used to compare the different treatments.

The number of "hits," or targets that are significantly identified in the screen, will vary for many reasons, including the cell line responsiveness to the assay, the actual assay itself, and the robustness of the experimental design and its controls. The number of hits selected will depend on the end user and his or her goals for the screen. Many researchers choose to integrate published interaction data with their hit list to triage the candidates to a manageable number.

ACKNOWLEDGMENTS

D.N.S. is supported by the Australian National Health and Medical Research Council (NHMRC), New South Wales Office of Science and Medical Research, Cancer Institute New South Wales, and the Mostyn Family Foundation. K.J.F. is a recipient of an Australian Postgraduate award. The VCFG (K.J.S.) is funded by the Australian Cancer Research Foundation (ACRF), the Victorian Department of Industry, Innovation and Regional Development (DIIRD), the Australian Phenomics Network (APN) supported by funding from the Australian Government's Education Investment Fund through the Super Science Initiative, the Australasian Genomics Technologies Association (AMATA), the Brockhoff Foundation, and the Peter MacCallum Cancer Centre Foundation.

REFERENCES

Bauer JA, Ye F, Marshall CB, Lehmann BD, Pendleton CS, Shyr Y, Arteaga CL, Pietenpol JA. 2010. RNA interference (RNAi) screening approach identifies agents that enhance paclitaxel activity in breast cancer cells. *Breast Cancer Res* **12**: R41.

Birmingham A, Selfors LM, Forster T, Wrobel D, Kennedy CJ, Shanks E, Santoyo-Lopez J, Dunican DJ, Long A, Kelleher D, et al. 2009. Statistical methods for analysis of high-throughput RNA interference screens. *Nat Methods* **6**: 569–575.

Boutros M, Kiger AA, Armknecht S, Kerr K, Hild M, Koch B, Haas SA, Paro R, Perrimon N. 2004. Genome-wide RNAi analysis of growth and viability in *Drosophila* cells. *Science* **303**: 832–835.

Brunelle JK, Zhang B. 2010. Apoptosis assays for quantifying the bioactivity of anticancer drug products. *Drug Resist Updat* **13**: 172–179.

Buenz EJ, Limburg PJ, Howe CL. 2007. A high-throughput 3-parameter flow cytometry-based cell death assay. *Cytometry A* **71**: 170–173.

Cite this introduction as *Cold Spring Harb Protoc*; doi:10.1101/pdb.top072561

De Pasquale B, Righetti G, Menotti A. 1990. L-carnitine for the treatment of acute myocardial infarct. *Cardiologia* **35**: 591–596.

Dompe N, Rivers CS, Li L, Cordes S, Schwickart M, Punnoose EA, Amler L, Seshagiri S, Tang J, Modrusan Z, et al. 2011. A whole-genome RNAi screen identifies an 8q22 gene cluster that inhibits death receptor-mediated apoptosis. *Proc Natl Acad Sci* **108**: E943–E951.

Falkenberg KJ, Saunders DN, Simpson KJ. 2014. A high-throughput multiplex cell death assay using an RNAi screening approach. *Cold Spring Harb Protoc* doi: 10.1101/pdb.prot080267.

Galluzzi L, Morselli E, Vitale I, Kepp O, Senovilla L, Criollo A, Servant N, Paccard C, Hupe P, Robert T, et al. 2010. miR-181a and miR-630 regulate cisplatin-induced cancer cell death. *Cancer Res* **70**: 1793–1803.

Gilbert DF, Erdmann G, Zhang X, Fritzsche A, Demir K, Jaedicke A, Muehlenberg K, Wanker EE, Boutros M. 2011. A novel multiplex cell viability assay for high-throughput RNAi screening. *PLoS ONE* **6**: e28338.

Luo J, Emanuele MJ, Li D, Creighton CJ, Schlabach MR, Westbrook TF, Wong KK, Elledge SJ. 2009. A genome-wide RNAi screen identifies multiple synthetic lethal interactions with the Ras oncogene. *Cell* **137**: 835–848.

MacDonald ML, Lamerdin J, Owens S, Keon BH, Bilter GK, Shang Z, Huang Z, Yu H, Dias J, Minami T, et al. 2006. Identifying off-target effects and hidden phenotypes of drugs in human cells. *Nat Chem Biol* **2**: 329–337.

Neumann B, Walter T, Heriche JK, Bulkescher J, Erfle H, Conrad C, Rogers P, Poser I, Held M, Liebel U, et al. 2010. Phenotypic profiling of the human genome by time-lapse microscopy reveals cell division genes. *Nature* **464**: 721–727.

Nguyen DG, Wolff KC, Yin H, Caldwell JS, Kuhen KL. 2006. "UnPAKing" human immunodeficiency virus (HIV) replication: Using small interfering RNA screening to identify novel cofactors and elucidate the role of group I PAKs in HIV infection. *J Virol* **80**: 130–137.

Niles AL, Moravec RA, Riss TL. 2009. In vitro viability and cytotoxicity testing and same-well multi-parametric combinations for high throughput screening. *Curr Chem Genomics* **3**: 33–41.

Parrish JZ, Xue D. 2003. Functional genomic analysis of apoptotic DNA degradation in *C. elegans*. *Mol Cell* **11**: 987–996.

Poon SS, Wong JT, Saunders DN, Ma QC, McKinney S, Fee J, Aparicio SA. 2008. Intensity calibration and automated cell cycle gating for high-throughput image-based siRNA screens of mammalian cells. *Cytometry A* **73**: 904–917.

Potratz JC, Saunders DN, Wai DH, Ng TL, McKinney SE, Carboni JM, Gottardis MM, Triche TJ, Jurgens H, Pollak MN, et al. 2010. Synthetic lethality screens reveal RPS6 and MST1R as modifiers of insulin-like growth factor-1 receptor inhibitor activity in childhood sarcomas. *Cancer Res* **70**: 8770–8781.

Qiu BY, Turner N, Li YY, Gu M, Huang MW, Wu F, Pang T, Nan FJ, Ye JM, Li JY, et al. 2010. High-throughput assay for modulators of mitochondrial membrane potential identifies a novel compound with beneficial effects on db/db mice. *Diabetes* **59**: 256–265.

Simpson KJ, Selfors LM, Bui J, Reynolds A, Leake D, Khvorova A, Brugge JS. 2008. Identification of genes that regulate epithelial cell migration using an siRNA screening approach. *Nat Cell Biol* **10**: 1027–1038.

Snijder B, Sacher R, Ramo P, Liberali P, Mench K, Wolfrum N, Burleigh L, Scott CC, Verheije MH, Mercer J, et al. 2012. Single-cell analysis of population context advances RNAi screening at multiple levels. *Mol Syst Biol* **8**: 579.

Stec E, Locco L, Szymanski S, Bartz SR, Toniatti C, Needham RH, Palmieri A, Carleton M, Cleary MA, Jackson AL, et al. 2012. A multiplexed siRNA screening strategy to identify genes in the PARP pathway. *J Biomol Screen* **17**: 1316–1328.

Tsui M, Xie T, Orth JD, Carpenter AE, Rudnicki S, Kim S, Shamu CE, Mitchison TJ. 2009. An intermittent live cell imaging screen for siRNA enhancers and suppressors of a kinesin-5 inhibitor. *PLoS ONE* **4**: e7339.

A High-Throughput, Multiplex Cell Death Assay Using an RNAi Screening Approach

Katrina J. Falkenberg,[1,2] Darren N. Saunders,[4,5] and Kaylene J. Simpson[1,2,3,6]

[1]Victorian Centre for Functional Genomics, Peter MacCallum Cancer Centre, East Melbourne, Victoria 3002, Australia; [2]Department of Pathology, The University of Melbourne, Parkville, Victoria 3052, Australia; [3]Sir Peter MacCallum Department of Oncology, The University of Melbourne, Parkville, Victoria 3052, Australia; [4]Cancer Research Program, The Kinghorn Cancer Centre, Garvan Institute of Medical Research, Darlinghurst, New South Wales 2010, Australia; [5]St. Vincent's Clinical School, University of New South Wales Medicine, Sydney, New South Wales 2000, Australia

This protocol outlines a high-throughput, multiplex cell death assay and its use in conjunction with a genome-scale siRNA screen to identify genes that cooperate with a drug to induce apoptosis. The assay, ApoLive-Glo (Promega), measures viability of drug-treated, reverse-transfected cells via the fluorescent CellTiter-Fluor reagent, which includes a substrate that is cleaved by a live cell protease. ApoLive-Glo also quantitates cell death by the amount of cleaved caspases 3 and 7 using a luminescent Caspase-Glo 3/7 caspase activation assay. The advantage of the multiplex assay is that it distinguishes rapid cell death from the slower activation of caspase activity, permitting measurement of different stages of cell death in the same sample at a single time point. In parallel, a high-content imaging protocol involving 4′,6-diamidino-2-phenylindole-stained nuclei is used as a cost-effective way to quantitate viability of vehicle-treated control cells. Automation and robotic liquid handling are built into the protocol to increase speed of workflow and improve reproducibility. A screen using these assays will identify gene targets that are essential for viability irrespective of drug treatment and gene targets that cause a synergistic enhancement of cell death in the presence of drug. Candidate target activity can then be validated by conventional flow cytometry-based assays.

MATERIALS

It is essential that you consult the appropriate Material Safety Data Sheets and your institution's Environmental Health and Safety Office for proper handling of equipment and hazardous materials used in this protocol.

Reagents

ApoLive-Glo multiplexed assay (includes Caspase-Glo 3/7 substrate, Caspase-Glo 3/7 buffer, GF-AFC substrate, and assay buffer; Promega G6411)

For consistency in a long-term screening project spanning weeks or months, it is important to prepare fresh Caspase-Glo 3/7 reagent every time, so order the appropriate pack size. The combined Caspase-Glo 3/7 substrate and buffer can be stored at −20°C for later use, but storage decreases signal intensity by ~25%, so it should be used in independent experiments and not combined with fresh reagent. When calculating reagent quantities for a screen as well as preparing reagents for use, always include an extra 10% to account

[6]Correspondence: kaylene.simpson@petermac.org

Cite this protocol as *Cold Spring Harb Protoc*; doi:10.1101/pdb.prot080267

for pipetting error. This is especially important when working with large numbers of plates, where a small volumetric error can become large across many plates. When using liquid-handling automation, you must also prepare a dead volume. This is the extra volume required for priming (filling all the tubes), avoiding bubbles in the dispenser tubing, and accounting for reagent remaining in the reservoirs. Calculate how many plates you can process in one batch. The more plates you can handle at once, the less reagent is lost as dead volume (and the shorter the total screening time), although the additional 10% volume needed for pipetting error is unaffected.

Cells

The ApoLive-Glo assay is most amenable to adherent cells. Suspension cells are very difficult to work with in high throughput format, particularly from the perspective of performing media changes with automation. In addition, suspension cells cannot be transfected with siRNAs. However, the ApoLive-Glo assay can be used with suspension cells in high throughput if no media change is required (e.g., compound screens). However, to be cost effective, cells must be plated in a minimum volume, which is feasible with adherent cells but not as feasible for suspension cells. Fixing and staining with DAPI is not feasible for suspension cell lines in high throughput. In low throughput, we recommend a cytospin.

CellTiter-Fluor (CTF) Cell Viability Assay (includes GF-AFC substrate and assay buffer; Promega G6082)

This reagent is needed to supplement that which is supplied with the ApoLive-Glo assay, because of the dilution of the caspase component. GF-AFC substrate and buffer can be freeze-thawed as separate reagents. The combined substrate and buffer, called CTF reagent, can be stored at 4°C for up to 1 wk with minimal loss of activity but should not be frozen. For consistency, we prepare fresh CTF reagent immediately before use.

DAPI solution (2×)

To prepare 1 mL of 2× DAPI solution, combine 2 μL of 5 mg/mL 4',6-diamidino-2-phenylindole (DAPI) with 40 μL of 10% Triton X-100 and 958 μL of 50 mM Tris (pH 7.5). Minimize freeze/thaw cycles of DAPI.

Microplates with RNAi library, 384-well, opaque-walled, clear-bottomed (see Step 1)

Use white-walled plates for ApoLive-Glo and black-walled plates for DAPI staining. The choice of brand is determined by the optical abilities of the preferred imager and cost. If you are not performing high end, very detailed microscopy, a lower quality plate can be used, but be aware that use of such plates often correlates with problems such as uneven plate surface, which can impact imaging. All plates should be evaluated on a per instrument basis.

Paraformaldehyde (PFA) (4% in Tris)

To prepare a 4% working solution of PFA, dilute one part of, 16%, PFA (e.g., Electron Microscopy Sciences 15710) with three parts of Tris (50 mM, pH 7.5).

PBS
Reverse transfection reagents
Tris (50 mM, pH 7.5), filtered

To make 1 L of this reagent, dissolve 6.057 g of Tris base in 800 mL of deionized H_2O. Adjust the solution to pH 7.5 with concentrated HCl and bring the volume to 1 L with deionized H_2O. Immediately before use, filter through a 0.45-μm filter.

Equipment

The high-end instrumentation required for this protocol is generally installed in a core facility that is managed by automation specialists. These specialists, who are well versed in many assays, develop robust protocols for instrument usage and train users. They also assist with assay development and optimization as required.

Automation for liquid handling, capable of dispensing 5–50 μL (e.g., EL406 microplate washer dispenser; BioTek)

Centrifuge, with microplate buckets

Foil, aluminum (optional)

Freezer, preset to −20°C

Incubator, humidified, preset to 37°C

Microplate barcode labeler (alphanumeric) and readers for high-throughput-enabled instrumentation (optional)

Barcode instruments allow the user to track plates and avoid errors caused by stacking plates in the wrong order or orientation.

Microplate heat sealer, automated (optional)

This highly recommended item saves the user from sealing plates by hand. The seals are cut by the instrument to fit entirely within plate boundaries, preventing the plate-stacking instrumentation "gripper fingers" from becoming stuck on overhanging edges. If you instead choose to manually adhere seals, pay particular attention to overhangs.

Microplate reader, multiwell, capable of reading luminescence, with filters for CTF assay (excitation: 380–400 nm, emission: 505 nm) and a dichroic mirror (e.g., Synergy H4 Hybrid Multi-Mode Microplate Reader; BioTek)

Microplate seals, foil

Microplate stacker (optional)

Stackers automatically deliver microplates to the reader and microscope. Their even timing improves screening accuracy.

Microscope, high-throughput (i.e., high-content imager)

Orbital plate shaker, small, with variable speed option

Refrigerator, preset to 4°C

Tissue culture equipment, including a biosafety cabinet

Uncouple the light from the operation of the cabinet so you have independent control of it (for the purposes of fixing and staining) when the unit is running. Few biosafety cabinets operate this way out of the box, but they are relatively easy to modify.

Water bath, preset to 37°C

METHOD

The high-throughput screen described here consists of two parallel arms: One for cells treated with drug and one for cells treated with drug-free vehicle. An RNAi screen for identifying synthetic lethal targets is depicted in Figure 1A. The drug-treated cells are evaluated for general viability using CTF and for apoptosis using Caspase-Glo 3/7. Together, these assays constitute the ApoLive-Glo multiplex assay. (For more information, see the ApoLive-Glo Multiplex Assay Technical Manual from Promega at http://www.promega.com/resources/protocols/technical-manuals/101/apolive-glo-multi-plex-assay-protocol/.) The control, vehicle-treated cells are subjected to nuclear staining by DAPI followed by high-content cell counting as a surrogate readout of cell death (it is a true readout of the combination of cell death and proliferation). Figure 1B depicts the steps required to perform these assays and the time required for each step.

RNAi Screen: Reverse Transfection and Treatment

1. Reverse-transfect the cells with the siRNA library and lipid-based transfection reagent in 384-well, opaque-walled microplates according to the chosen transfection protocol.

   i. Prepare duplicate plates for each treatment (control and drug-treated).

   ii. Use clear-bottomed plates. For multiplexed CTF and Caspase-Glo 3/7 assays, use white-walled plates; for CTF alone, use either black- or white-walled plates; and for fluorescence microscopy (e.g., DAPI), use black-walled plates.

   Although white-bottom plates are often recommended for luminescence assays to eliminate cross-well luminescence bleed-through, the opaque bottoms prevent visualizing the cells, which is important for evaluating cell density, efficiency of positive controls, and general health of the cells, as well as for identifying contamination. We have found that the bleed-through of a high-responding well to an adjacent nonresponding well is minimal (~1%–2% of the high-luminescence value).

   iii. Plate cells in column 1 for mock transfection (lipid only) to evaluate the cell dispensing automation (during analysis, determine variance in that column to confirm that all rows of the plate were consistently dispensed).

   iv. Include positive controls, both technical (e.g., PLK1 for cell death) and assay-specific (see Introduction: High-Throughput Approaches to Measuring Cell Death [Saunders et al. 2014]). We use columns 2 and 23 for our positive and negative controls.

   See Troubleshooting.

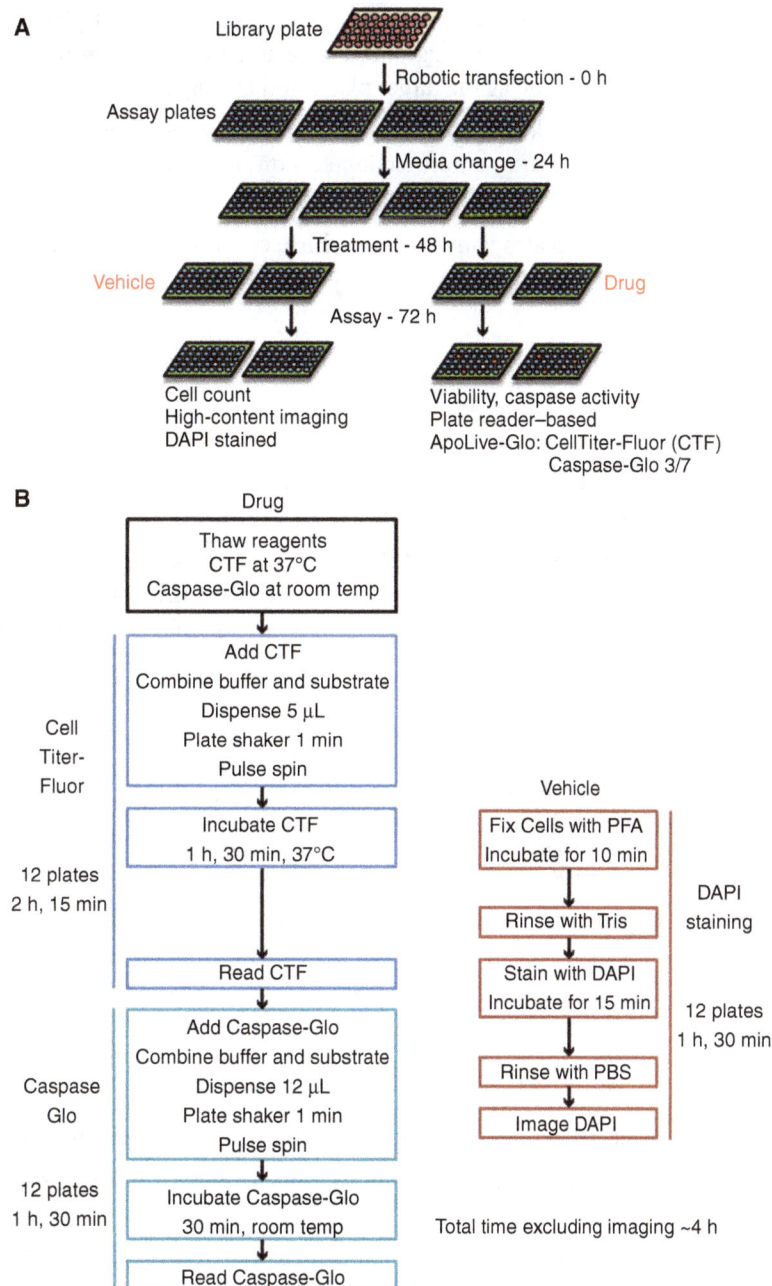

FIGURE 1. Experimental workflow. (*A*) The experimental steps of a genome-scale siRNA screen. A single 384-well siRNA library plate was used to reverse transfect four assay plates. Final siRNA concentration was 40 nM, and each well received lipid and cells. Plates were incubated for 24 h before changing the medium and a further 24 h before drug treatment. Duplicate plates received drug or the vehicle control. After 24 h of treatment, the vehicle plates were assayed for cell count using high-content imaging with DAPI nuclear staining. Drug-treated plates were assayed for viability and caspase activity using the plate reader–based ApoLive-Glo assay. (*B*) The experimental steps of the ApoLive-Glo assay coupled with DAPI nuclear staining for quantitation of cell number on the vehicle control. The time to complete each segment of the assay is described. To decrease overall hands-on time, DAPI staining is performed in parallel with the ApoLive-Glo incubations and quantitation.

    v. Include a nontargeting siRNA as a negative control.

   vi. Leave one column free of cells as a "medium only" control. This important control is used for subtracting a background reading for plate reader-based assays. We use column 24 of the 384-well plate for this purpose, but whichever column you choose, stay consistent once the choice made. If you choose to work in 96-well plates there is far less room for controls; therefore, we suggest to create an additional "sentinel plate" that is medium alone, treated the same as the target plates and averaged to provide a medium-only background value.

vii. Consider the total number of plates and workflow. We are able to process six siRNA library plates in one transfection session of 20 min, resulting in 24 experimental assay plates (two for the vehicle arm and two for the drug-treated arm for each library plate).

2. Incubate plates with drug or vehicle control. Each well should contain 20 μL.

The Assays

All assay steps are conducted at room temperature unless otherwise stated. When using multiple plates, always run the plates in the same order for each experimental step and for each assay when multiplexing. This ensures that the workflow timing remains consistent and provides a fail-safe if the barcode reader is unable to detect a barcode. Once work has begun on a set of plates, it must be completed to the end of the procedure.

3. Thaw the GF-AFC substrate and assay buffer at 37°C (a 10-mL bottle of buffer takes ~10 min in a water bath).

4. Thaw the Caspase-Glo 3/7 lyophilized substrate and buffer at room temperature while the CTF assay is being performed (be prepared; a 10-mL bottle of buffer takes ~2 h to thaw on the bench!).

    Alternatively, the Caspase-Glo 3/7 reagents can be thawed overnight at 4°C, but remember to equilibrate them to room temperature before use.

CTF Assay

CTF is a viability assay that makes use of a fluorogenic, cell-permeable peptide substrate. Live cells take up the substrate, and constitutive protease activity cleaves it to the fluorescent form, which generates a fluorescent signal proportional to the number of live cells. This protease activity is only retained in live cells with intact plasma membranes. When membrane integrity is lost, the protease is inactive and the fluorescent signal reduced. The reduction in fluorescent signal corresponds to the reduction in numbers of live cells.

5. Prepare the CTF reagent by combining 10 μL of GF-AFC substrate with 2.5 mL of assay buffer. Mix well.

6. With an automated dispenser, add 5 μL of CTF reagent to each well of the plates of cells that have been treated with drug.

    The volume added in this step is CRITICAL, because it is the amount that must be added to the 20 μL of medium already in each well of the 384-well plate.

7. Shake the plates on an orbital plate shaker for 1 min at ~1/4 of maximum speed.

8. Centrifuge the plates briefly (pulse-spin them) to 60g at room temperature to remove reagents from the sides of the wells.

9. Incubate the plates at 37°C in a humidified incubator for 1.5 h.

    To accelerate the workflow, the control plates that were not treated with drug can be fixed and DAPI-stained during the CTF incubation (Steps 17–28) (see Fig. 1B).

10. Read the plates with a plate reader: measure the fluorescence at 380–400 nm_{Ex}/505 nm_{Em}. Settings for each instrument must be determined empirically, and once determined, must remain constant throughout the screen. Increase the gain, sensitivity, or integration time (luminescence) to increase signal intensity and dynamic range (the difference between the positive and negative controls).

    While the CTF plates are being assayed on the plate reader, high-throughput imaging of the DAPI-stained plates can be initiated (Step 29).

    See Troubleshooting.

Caspase-Glo 3/7 Assay

Caspase-Glo 3/7 is a luminescent end point assay measuring the combined activity of caspases 3 and 7, the executioner caspases and central mediators of the intrinsic and extrinsic apoptosis pathways. The assay contains a lumino-

genic caspase 3/7 substrate, containing the tetrapeptide sequence DEVD. Addition of the combined Caspase-Glo 3/7 reagent results in immediate lysis of cells, allowing cleavage of the caspase substrate into its mature form, which is a luciferase substrate. The luciferase reaction releases light, the amount of which is proportional to the amount of caspase activation in the well.

11. Add the thawed bottle of Caspase-Glo 3/7 buffer from Step 4 to the substrate and mix well. Keep the mix protected from light by using it in a biosafety cabinet with the light off.

12. With an automated dispenser, dispense 12 µL of substrate mix to each well of the 384-well plates that were taken through the CTF assay.

    This addition creates a 1:2 reagent dilution with the 25 µL of medium already in each well. Refreeze any remaining reagent at −20°C.

13. Shake the plates on an orbital plate shaker for 1 min at ∼1/4 of maximum speed.

14. Centrifuge the plates briefly (pulse-spin them) to 60g at room temperature to remove reagents from the sides of the wells.

15. Incubate the plates at room temperature for exactly 30 min.

16. Measure luminescence with a plate reader.

    See Troubleshooting.

Nuclear Staining for Cell Counting

Nuclear staining should be performed on the plates of cells that were not treated with drug (i.e., the vehicle-treated cells). Evaluate all liquid-handling steps to ensure that cell attachment to the plate is not disturbed during the fixing and staining steps.

17. Aspirate the contents of each well using a high aspirate setting. We leave ∼15 µL/well, but this volume depends on the adherence capacity of the cells being used.

18. Add a volume of 4% PFA in Tris that is equal to the volume left behind after aspiration (in our case, 15 µL), for a final concentration of 2% PFA.

    For good coverage in a 384-well plate, aim for ∼25 µL final volume. You can optimize for a lower volume if you want to conserve reagents. Alternatively, if you have good, adherent cells and a low volume (∼5 µL) remaining after aspirating, you can dilute the PFA to 2% in advance and use that directly.

19. Incubate the plates for 10 min at room temperature.

20. Aspirate the contents of each well.

21. Rinse each well with 50 µL of 50 mM Tris (pH 7.5).

    This volume removes most of the PFA, which can interfere with the clarity of DAPI images.

22. Aspirate the contents of each well using a high aspirate setting; leave ∼10 µL/well.

23. Add 10 µL of 2× DAPI solution, resulting in a final concentration of 5 µg/mL DAPI.

    As in Step 18, if the volume remaining after aspiration is low, a 1× DAPI solution can be added.

    The 2× DAPI solution contains Triton X-100 to permeabilize cells in conjunction with staining, effectively combining these steps.

24. Incubate the plates for 15 min with the lights off in a biosafety hood.

    Alternatively, to reduce the exposure to light, cover the top plate in the microplate stack with foil or another plate.

25. Aspirate the contents of each well.

26. Add 50 µL of PBS to each well.

    This volume prevents evaporation leading to dried wells and allows subsequent restaining with other fluorophores.

27. Centrifuge the plates briefly (pulse-spin them) to 60g at room temperature to remove reagents from the sides of the wells.

28. Seal the plates with foil seals, ideally by using an automated heat sealer.

29. Image the plates on a high-content imager immediately, or store the plates at 4°C wrapped in foil before imaging.

> *DAPI is very robust, so the plates can be stored at least a week before imaging, but this timing depends on which other fluorophores are used.*
> *See Troubleshooting.*

30. Store the plates wrapped in foil at 4°C in case they need to be reimaged.

Data Analysis

The methods below describe how to analyze the CTF, Caspase-Glo 3/7, and cell counting data to generate a list of screen hits (see Fig. 2). Briefly, the cell counting data from DAPI nuclear staining is analyzed using cell number and field count values to reveal the wells that are dead in the vehicle arm. These wells can be removed from downstream analysis in the drug arm. Next, CTF and Caspase 3/7 data from the drug-treated arm undergo normalization, followed by identification of the most-reduced CTF wells (synthetic lethal cell death hits) and the most-increased Caspase-Glo 3/7 wells (sensitizer hits).

Quantitation of Cell Number

Perform cell counting on the plates of cells that were treated with vehicle and then stained with DAPI.

31. Average the replicate plates for cell count and field number.

32. To evaluate the health of the well, define a binning strategy for cell counts. If the predetermined cell count (based on control healthy wells) cannot be reached within the predetermined field number, the well conditions have a toxic effect. As long as the predetermined cell count is reached, field number gives an indication of the health of the well. In the example shown in Figure 3, we set a count of 1500 cells in up to 25 fields, but these numbers will vary among screens and depend on cell line, sensitivity to death stimulus, etc.:

- Toxic bin: <1500 cells counted in 25 fields

- Very healthy bin: ≥1500 cells counted in ≤11 fields (note: the negative controls consistently count ≥1500 cells in 7–8 fields)

- Moderately healthy bin: all other wells (≥1500 cells counted in 12–25 fields, inclusive)

  > *See Discussion.*

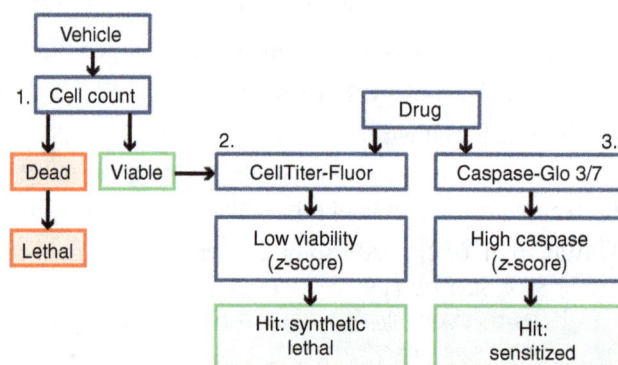

FIGURE 2. Analysis pipeline, combining screen arms (vehicle and drug) and assays (cell counting, CTF, and Caspase-Glo 3/7). (1) DAPI cell counting on the vehicle-treated plates determines viability on gene knockdown alone. In an siRNA screen, a well is considered dead, and the siRNA lethal, if the cell number threshold is not reached. These gene targets are excluded from further analysis in the drug-treated arm. A well is viable if the cell number threshold is reached. The combined effect of gene knockdown and drug treatment on viable wells is evaluated with (2) CTF and (3) Caspase-Glo 3/7. CTF hits are those wells with a low-viability z-score, where a synthetic lethal interaction has taken place between gene knockdown and drug treatment. Caspase-Glo 3/7 hits are those wells with a high-caspase activation z-score (see Fig. 4). These wells are sensitized to drug treatment, because they show substantial caspase activity but are not yet dead. The same target may fall into both the reduced viability and caspase activation hit bins because they are not mutually exclusive, but in practice we find this situation uncommon. In the first instance, following convention, z-score cutoffs of $z \leq -2$ or $z \geq 2$ are a starting point for defining hits, but cutoffs may be determined empirically as in the case of the screen described, based on the range and spread of the data and subsequent validation considerations (Birmingham et al. 2009; see also Fig. 4).

Cite this protocol as *Cold Spring Harb Protoc*; doi:10.1101/pdb.prot080267

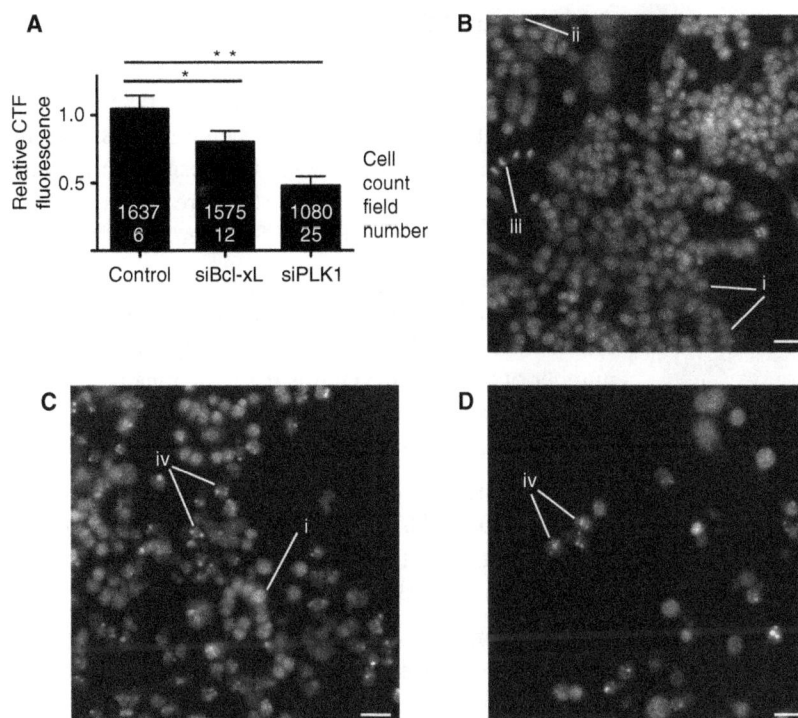

FIGURE 3. Concordance between CTF and DAPI cell counting. (*A*) Results for CTF and DAPI cell counting showing excellent correlation between the two assays. Cells underwent lipid-based reverse transfection with 40 nM siRNA in duplicate 384-well plates and were incubated for 72 h before being subjected to CTF or DAPI staining. PLK1 knockdown greatly reduces cell viability as measured by CTF. Bcl-xL knockdown reduces viability to a smaller extent. Data are expressed relative to lipid transfection (mock), mean, and standard deviation from a single experiment plotted. Lipid transfection gives a very healthy well, counting 1500 cells in only six fields. Bcl-xL knockdown reduces viability such that it takes six more fields to achieve this cell number (total of 12 fields). PLK1 knockdown is toxic to the cells. Even in 25 fields, we are unable to count 1500 cells. *$p \leq 0.01$, **$p \leq 0.001$. Representative high-content images of DAPI-stained cells (*B*) mock, (*C*) siBcl-xL, and (*D*) siPLK1 showing different levels of cell death within the wells. Specific features are illustrated: (i) healthy nuclei, (ii) an object falling on the edge of the field which is thereby excluded from the analysis, (iii) a cell undergoing division, and (iv) apoptotic nuclei included in the cell count. Scale bar, 30 μm.

33. Evaluate the readouts for each control type on every plate to ensure that they all reach the threshold value within the same field counts. The coefficient of variation (CV) measurement can be used here: For a given number of replicates of the same treatment, the CV, expressed as a percentage, is the standard deviation divided by the average (see Introduction: High-Throughput Approaches to Measuring Cell Death [Saunders et al. 2014]). If the variability between cell number and field count for each type of control is >25%, the assay is too variable to statistically interpret any hits.

34. If the screen is performed over many weeks, compare the raw data on a "per plate" and "per week" basis to ensure that the cells are performing as expected. A fold change ratio of positive to negative controls can also be compared weekly to ensure consistency.

CTF and Caspase-Glo 3/7 Analysis

Perform these analyses on the plates of drug-treated cells.

35. Average the luminescence signal from the media-only column, and subtract this background from all other values.

36. Similar to the analysis of the DAPI-stained cells, ensure that the positive and negative controls on each plate have reported raw and normalized values that are consistent with optimization and

data on a weekly basis. Calculate the CV (in %) to ensure that variability is within an acceptable range.

37. Calculate the Z'-factor for each positive control relative to the same technical negative control (e.g., the PLK1 death control versus the nontargeting siRNA negative control). The Z'-factor is a measure of the dynamic range between the positive and negative control and should, at the very least, be >0. If a plate fails, it must be rescreened (see Introduction: High-Throughput Approaches to Measuring Cell Death [Saunders et al. 2014]).

38. Perform one of the following analysis options.
    - Perform sample-based normalization: In an unbiased screen, a z-score (or robust z-score) can be calculated for all samples (Birmingham et al. 2009). The z-score is a measure of the number of standard deviations from the mean (or, in the case of robust z-scores, absolute deviations from the median). This normalization is appropriate only when most of the samples are assumed to be negative.

    - Perform control-based normalization: An alternative analysis is to generate fold change to a negative control. This should be done on a per plate basis and can be used for biased data sets. Either this method or the z-score method can be performed when working with a limited number of plates.

39. When analyzing data from an entire RNAi screen, it is necessary to reduce any potential bias from plate to plate or week to week, in addition to factoring in the layout of the library (e.g., if an entire genome is plated in gene families, a kinome plate may yield many more hits than a plate of uncharacterized targets from the remaining genome). The sample data from each library plate is normalized to the average of the technical negative control samples on each plate, and then the replicate plates are averaged. The averaged value for each individual siRNA target is then collated and the robust z-score is calculated across the library collection (use the same approach for a genome or small custom library).

40. Define a cutoff for z-score or fold change based on positive and negative controls. A z-score cutoff of $z \geq 2$ or $z \leq -2$ is commonly used (Birmingham et al. 2009). The cutoff can be made more stringent for screens with a large dynamic range and depending on the number of hits desired for further analysis. Our laboratory routinely takes the top 400–500 candidates through for secondary analysis, therefore a decision is made on the actual z-score cutoff for selection based on the number of candidates that reflect this. In the example shown in Figure 4A, there is a skewed distribution that might be expected from a screen set up to identify death. If we were to take all the Caspase-Glo targets from a z-score cutoff of 2, there would have been several thousand targets, an unmanageable amount. By defining the number of targets to pursue in a validation screen, you can then specify the z-score cutoff for caspase activation and for cell viability. For comparison, Figure 4B shows an example of a more normal, less-skewed data set by which targets that decrease and increase viability can be identified from either end of the z-score distribution.

TROUBLESHOOTING

Problem: Cells detach from the plate during media changes.
Solution: Automated liquid handlers can subject cells to more powerful forces during aspiration and dispensing than hand pipetting does. Keep in mind that when hand pipetting, we place the tip down the side of the well to avoid disturbing cells, whereas most liquid handling robots aspirate and dispense with vertical tips. Minimize the number of washing steps where possible, and test for z heights (vertical distance above the plate) that allow cells to remain adhered to the plate. Optimize liquid aspiration steps to remove as much liquid as possible without disturbing cell adhesion (the volume remaining must be factored in with all subsequent calculations for volumes required for each assay). Plating density is also crucial; if cells are too sparse, they are more likely to come off. The rate of liquid dispensing is a compromise between high accuracy

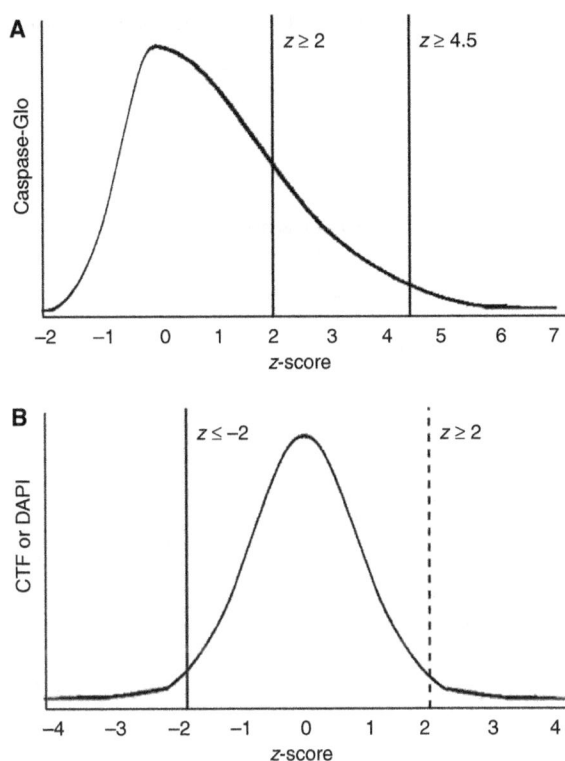

FIGURE 4. A z-score analysis of CTF and Caspase-Glo 3/7 assays. Representative cartoon plots illustrate the types of z-score distribution curves and subsequent cutoff values that could be achieved for a genome-scale screen for (A) Caspase-Glo 3/7 and (B) CTF or DAPI. The schematic in (A) shows a skewed distribution of response to caspase-mediated cell death (Caspase-Glo 3/7) where a long tail has thousands of targets with a z-score cutoff >2. Moving to a z-score cutoff of 4.5 clearly changes the number of targets that will be studied. The schematic in B shows a more normal data distribution for cell viability based on CTF or counting of DAPI-stained nuclei, where there are targets outside >2 and <−2 that increase and decrease viability, respectively. In interpreting these figures, the values on the y-axis are not important; emphasis is placed on the profile of the distribution of the z-score across the entire data set (the x-axis).

(fast flow rate) and low mechanical force to the cells (low flow rate). We dispense at a high flow rate to improve accuracy, but to reduce the impact of media dispensing on cells, we installed an angled deflector shield on the dispense head of the Biotek 406 dispenser (our liquid handler of choice). The angled shield reduces the downward force by shifting the dispensed media to off-center.

If cell adhesion issues persist, a cellular substrate such as poly-L-lysine or an extracellular matrix such as collagen or fibronectin can be used. These substrates do not interfere with the Caspase-Glo 3/7 and CTF assays, but their presence can lead to optical issues during microscopy, such as changes in the focal plane across the well. Before using them, evaluate substrates for your specific cell line and assay. We do not recommend firmly adhering cells if you are investigating the cytoskeleton and migration.

Problem (Step 1): Assay-specific positive controls cannot be identified.
Solution: To accurately determine the dynamic range of your assay and the potential of the screen to identify hits, you need an assay-specific positive control. This should be the same entity that you are screening: If you are conducting an siRNA screen, you need an siRNA, and for a compound screen, you need a drug. Although a drug can be used in assay development, it will not reflect any variability that might arise with weekly transfections. If a positive control for your biological system is unknown (which is common), an in-depth literature search can help. If you cannot identify a positive control, you can still complete a screen, but be aware that gain settings, for example, might need to be altered to accommodate higher readings than calibrated when the screen was developed. If all else fails, technical controls such as known genes or drugs that cause cell death can be used to generate at least some expectation of the extent of data values.

Problem: A plate's barcode is not being read.
Solution: Apply the following procedures to ensure that barcodes are always read.

- Make sure that the barcodes are correctly placed on the plate (the barcode should be level and placed in the middle of the side of the plate).

- Avoid scratching the barcode with a fingernail or other apparatus (gloves or not, this is a common problem when applying barcodes).

- If there is a mirror involved in the barcode reader, clean it weekly with 70% ethanol to prevent dust from causing barcode reader errors.

Problem (Steps 10 and 16): There is too much noise in the assay, the background is too high, or the dynamic range of the plate reader-based assay is too low.

Solution: Improving your assay's signal-to-noise ratio increases dynamic range and reduces noise in the assay. As much as possible, reduce background. For the plate reader–based assays, keep control wells healthy by allowing the cells sufficient nutrients (make sure to change the medium during the assay) and space (do not grow the cells to overconfluence). Reagents do not always need to be used at the manufacturer's specified concentration or dilution, so you can experiment with amounts and concentrations to get cleaner, more robust data.

Problem (Steps 10 and 16): The assay's dynamic range changes over time.

Solution: The luminescent readout of the Caspase-Glo 3/7 assay is very sensitive to changes in incubation time, so the incubations for all plates must be equal. You might have to scale down the number of plates done in one batch if, for example, the time it takes to read them means the plates at the end are sitting around for longer.

Problem (Steps 10 and 16): Fluorescence or luminescence readings are saturated or out of range.

Solution: Maxed-out readings can happen when you have optimized the instrument gain based on control wells that respond to a lesser degree than some of the screen hits. To avoid maxing out with very high readings, take care not to set the gain too high. Plates with saturated signals can be read again with a lower gain. If this modification is not applied to all of the plates, plate-based normalization must be used.

Problem (Step 29): High-content imaging shows high background or low clarity.

Solution: Residual PFA remaining in the wells after fixation can make images fuzzy, so it is important to remove as much of it as possible. In addition, the three-dimensional nature of matrices compared to the plastic they are coating can make it difficult to find a focal plane. Try to reduce the amount of matrix/substrate coating the wells, or use a higher *z* height to avoid using any substrate at all. Optimize for a high signal-to-noise ratio, especially when using fluorophore-conjugated antibodies.

DISCUSSION

High-Content Microscopy

Each instrument has its own proprietary tools for software-driven data analysis, but cell counting is very standard. During optimization, pay close attention to defining the average nucleus size for the population, particularly for excluding dead and dying cells. In the example shown in Figure 3, working in 384-well format and at a magnification of 20× on a Cellomics ArrayScan Vti microscope, we imaged 1500 cells or 25 fields, whichever came first. This field number was chosen, because it includes the entire center of the well but excludes fields on the edge of the well, where cell confluence was often different than in the center. The cutoff of 1500 cells was chosen by observing control healthy wells in comparison with moderately healthy and dead wells under a light microscope followed by nuclear staining, counting, and statistical evaluation. Control, healthy wells reached 1500 cells in few fields (seven to eight fields), and moderately healthy wells reached this cutoff with fields to spare (15 to 20 fields), but clearly apoptotic wells were unable to make this cell count threshold in the 25 fields (see Fig. 3). There is no hard-and-fast rule dictating the required number of cells to be imaged. Rather, the number of cells will be based on the phenotype of interest, the size of cells, the percentage of positive cells in control wells (penetrance of the phenotype), and robust statistical outcomes. As a general guideline, it is usually sufficient to image between 800 and 2000 cells, and we recommend this as a ballpark figure when getting started. Cells bordering fields were excluded from the analysis to prevent cells being counted multiple times. Note that for simple cell counting, it is also possible to use lower magnification to significantly reduce the amount of data generated and the processing time required.

Selecting, Configuring, and Validating the Multiplex Assay

We have presented here a strategy for performing a high-throughput siRNA screen with a multiplexed cell death assay readout to identify genes that when knocked down cooperate with a drug to induce apoptosis (i.e., synthetic lethality). For this screen, we required a plate reader-compatible, high-throughput assay that would specifically detect apoptosis, as opposed to total cell number. Assays based on ATP content and mitochondrial metabolism, such as CellTiter-Glo (Promega) and ala-marBlue (Invitrogen), respectively, were unsuitable because they measure a combination of proliferation and death (Fig. 5A,B). Detection of caspase 3/7 activation, however, using either fluorescent- or luminescent-based detection (Promega), provided a robust readout that correlated with traditional flow cytometry-based annexin V staining (Fig. 5C,D). The luminescent assay had greater sensitivity and dynamic range than the fluorescent version of the assay. The end point nature of this assay (requiring lysis of cells) raised the possibility of false negatives arising from wells that had undergone very rapid apoptosis, resulting in a loss of cells and insignificant caspase activity at the assay end point. To counter this problem, we included a viability assay that would distinguish rapid cell death from slower development of caspase activity within the window of our assay. The advantage of this mul-

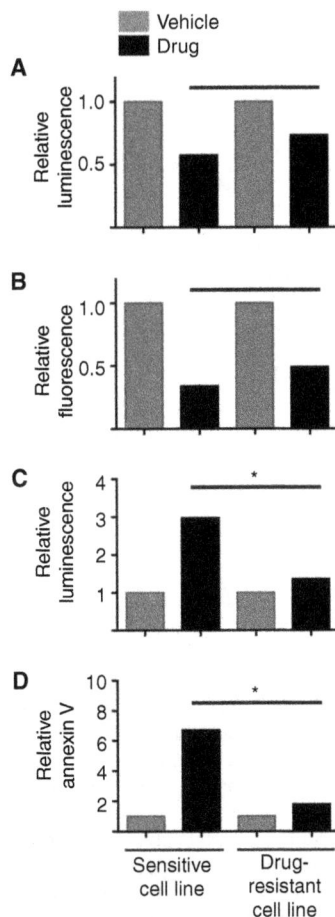

FIGURE 5. Finding an appropriate cell death readout. A schematic comparison modeled on experimental data of different cell viability readouts using cell lines that are sensitive or resistant to drug treatment. (A) CellTiter-Glo measures ATP content. (B) alamarBlue measures mitochondrial metabolism. Both assays show a significant difference between vehicle and drug treatment for both sensitive and drug-resistant cells, deeming them unsuitable for the screen. (C) Caspase-Glo 3/7 correlates significantly with (D) annexin V staining, proving the suitability of the caspase reagent for high-throughput use. Cells were treated with drug or vehicle for 24 h before being subjected to the relevant assay. Data are expressed relative to the vehicle mean. Asterisk represents a theoretical statistical significance in cell death between the sensitive and resistant lines after drug treatment.

tiplex format is that we could identify different stages of cell death in the same sample at a single time point.

In an attempt to reduce reagent cost, we evaluated the CTF and Caspase-Glo 3/7 reagents at 1:1 (recommended by the manufacturer), 1:2, and 1:4 dilutions. We found that signal intensity and dynamic range was maintained for Caspase-Glo 3/7 at a 1:2 dilution but was lost at a 1:4 dilution. No further dilution of CTF was possible without severely abrogating the assay signal. To further reduce assay cost, we chose not to use ApoLive-Glo for the vehicle control arm of the screen and instead used high-content imaging to count cell number after fixation and nuclear staining with DAPI. We imaged 1500 cells in up to 25 fields and determined the relative "health" of the well by the number of fields it took to sample 1500 cells (i.e., we used field number as a measure of viability). We showed that cell count correlated significantly with CTF-determined viability (Fig. 3). Intensity of DAPI fluorescence, as a measure of DNA content, is also a useful gross overall indicator of cell cycle distribution (e.g., 2N, 4N). This analysis allowed us to identify conditions causing strong cell cycle arrest in G1 or G2/M phase but was not sensitive enough to evaluate more subtle differences in cell cycle profile. If cell cycle analysis is central to your screen, we recommend including a dedicated cell cycle marker such as Ki67, phosphorylated histone H2B, or DNA synthesis via BrdU/EdU incorporation (Poon et al. 2008).

We show that three different gene targets that caused synthetic lethality as measured by caspase 3/7 activity were further validated by conventional low-throughput flow cytometry-based assays for phosphatidylserine externalization (annexin V binding), DNA fragmentation (subG1 DNA content with propidium iodide [PI]), and caspase activation (cleaved caspase 3 intracellular staining) (Fig. 6). These assays confirm the utility of the ApoLive-Glo assay for identifying apoptosis in a high-throughput, plate reader-based format.

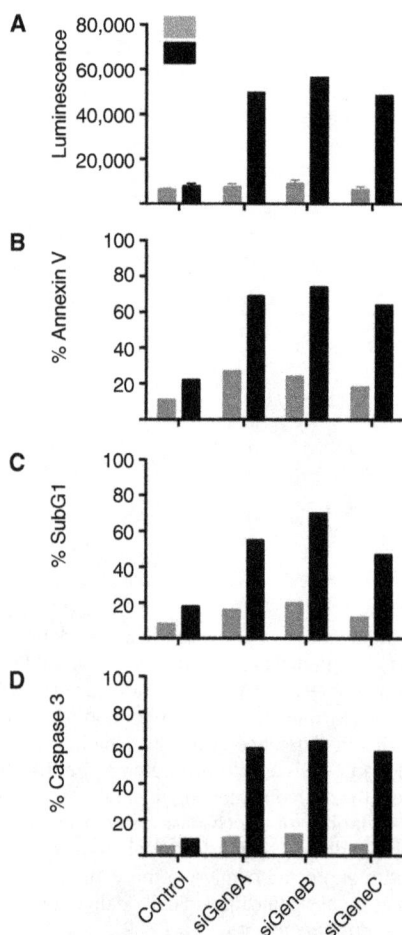

FIGURE 6. Confirmation of plate reader assays using standard FACS-based readouts. A schematic representation modeled on experimental data comparing three individual gene targets between the plate reader-based caspase activation assay and standard FACs readouts for cell death. Effect of gene knockdown with and without drug treatment using (A) Caspase-Glo 3/7 (raw luminescence units), (B) annexin V staining (percent cells), (C) subG1 DNA content (percent cells), and (D) intracellular cleaved caspase 3 staining assays (percent cells). Results reproduce in all assays, confirming the utility of Caspase-Glo 3/7 for identifying apoptotic cells.

ACKNOWLEDGMENTS

We thank members of the Victorian Centre for Functional Genomics (VCFG), Kate Gould for bioinformatics analysis of screen data, and Daniel Thomas and Yanny Handoko for expert technical guidance on all automation and screening support. The assay development was funded by Merck Sharp and Dohme, and an Australian National Health and Medical Research Council (NHMRC) project grant #1028871 (to Professor Ricky Johnstone, Peter MacCallum Cancer Centre). K.J.F. is a recipient of an Australian Postgraduate award. D.N.S. is supported by the NHMRC, New South Wales Office of Science and Medical Research, Cancer Institute New South Wales, and the Mostyn Family Foundation. The VCFG (KJS) is funded by the Australian Cancer Research Foundation (ACRF), the Victorian Department of Industry, Innovation and Regional Development (DIIRD), the Australian Phenomics Network (APN) supported by funding from the Australian Government's Education Investment Fund through the Super Science Initiative, the Australasian Genomics Technologies Association (AMATA), the Brockhoff Foundation and the Peter MacCallum Cancer Centre Foundation.

REFERENCES

Birmingham A, Selfors LM, Forster T, Wrobel D, Kennedy CJ, Shanks E, Santoyo-Lopez J, Dunican DJ, Long A, Kelleher D, et al. 2009. Statistical methods for analysis of high-throughput RNA interference screens. *Nat Methods* 6: 569–575.

Poon SS, Wong JT, Saunders DN, Ma QC, McKinney S, Fee J, Aparicio SA. 2008. Intensity calibration and automated cell cycle gating for high-throughput image-based siRNA screens of mammalian cells. *Cytometry A* 73: 904–917.

Saunders DN, Falkenberg KJ, Simpson KJ. 2014. High-throughput approaches to measuring cell death. *Cold Spring Harb Protoc* doi: 10.1101/pdb.top072561.

General Safety and Hazardous Material Information

This manual should be used by laboratory personnel with experience in laboratory and chemical safety or students under the supervision of such trained personnel. The procedures, chemicals, and equipment referenced in this manual are hazardous and can cause serious injury unless performed, handled, and used with care and in a manner consistent with safe laboratory practices. Students and researchers using the procedures in this manual do so at their own risk. It is essential for your safety that you consult the appropriate Material Safety Data Sheets, the manufacturers' manuals accompanying equipment, and your institution's Environmental Health and Safety Office, as well as the General Safety and Disposal Cautions in this appendix for proper handling of hazardous materials in this manual. Cold Spring Harbor Laboratory makes no representations or warranties with respect to the material set forth in this manual and has no liability in connection with the use of these materials.

All registered trademarks, trade names, and brand names mentioned in this book are the property of the respective owners. Readers should please consult individual manufacturers and other resources for current and specific product information.

Users should always consult individual manufacturers, the manufacturers' safety guidelines and other resources, including local safety offices, for current and specific product information and for guidance regarding the use and disposal of hazardous materials.

PRIMARY SAFETY INFORMATION RESOURCES FOR LABORATORY PERSONNEL

Institutional Safety Office. The best source of toxicity, hazard, storage, and disposal information is your institutional safety office, which maintains and makes available the most current information. Always consult this office for proper use and disposal procedures.

Post the phone numbers for your local safety office, security office, poison control center, and laboratory emergency personnel in an obvious place in your laboratory.

Material Safety Data Sheets (MSDSs). The Occupational Safety and Health Administration (OSHA) requires that MSDSs accompany all hazardous products that are shipped. These data sheets contain detailed safety information. MSDSs should be filed in the laboratory in a central location as a reference guide.

GENERAL SAFETY AND DISPOSAL CAUTIONS

The guidance offered here is intended to be generally applicable. However, proper waste disposal procedures vary among institutions; therefore, always consult your local safety office for specific instructions. All chemically constituted waste must be disposed of in a suitable container clearly labeled with the type of material it contains and the date the waste was initiated.

It is essential for laboratory workers to be familiar with the potential hazards of materials used in laboratory experiments and to follow recommended procedures for their use, handling, storage, and disposal.

The following general cautions should always be observed.

- **Before beginning the procedure,** become completely familiar with the properties of substances to be used.

- **The absence of a warning** does not necessarily mean that the material is safe, because information may not always be complete or available.

- **If exposed** to toxic substances, contact your local safety office immediately for instructions.

- **Use proper disposal procedures** for all chemical, biological, and radioactive waste.

- **For specific guidelines on appropriate gloves to use,** consult your local safety office.

- **Handle concentrated acids and bases** with great care. Wear goggles and appropriate gloves. A face shield should be worn when handling large quantities.

  Do not mix strong acids with organic solvents because they may react. Sulfuric acid and nitric acid especially may react highly exothermically and cause fires and explosions.

  Do not mix strong bases with halogenated solvents because they may form reactive carbenes that can lead to explosions.

- **Handle and store pressurized gas containers** with caution because they may contain flammable, toxic, or corrosive gases; asphyxiants; or oxidizers. For proper procedures, consult the Material Safety Data Sheet that is required to be provided by your vendor.

- **Never pipette** solutions using mouth suction. This method is not sterile and can be dangerous. Always use a pipette aid or bulb.

- **Keep halogenated and nonhalogenated** solvents separately (e.g., mixing chloroform and acetone can cause unexpected reactions in the presence of bases). Halogenated solvents are organic solvents such as chloroform, dichloromethane, trichlorotrifluoroethane, and dichloroethane. Non-halogenated solvents include pentane, heptane, ethanol, methanol, benzene, toluene, *N,N*-dimethylformamide (DMF), dimethylsulfoxide (DMSO), and acetonitrile.

- **Laser radiation,** visible or invisible, can cause severe damage to the eyes and skin. Take proper precautions to prevent exposure to direct and reflected beams. Always follow the manufacturer's safety guidelines and consult your local safety office. See caution below for more detailed information.

- **Flash lamps,** because of their light intensity, can be harmful to the eyes. They also may explode on occasion. Wear appropriate eye protection and follow the manufacturer's guidelines.

- **Photographic fixatives, developers, and photoresists** also contain chemicals that can be harmful. Handle them with care and follow the manufacturer's directions.

- **Power supplies and electrophoresis equipment** pose serious fire hazard and electrical shock hazards if not used properly.

- **Microwave ovens and autoclaves** in the laboratory require certain precautions. Accidents have occurred involving their use (e.g., when melting agar or Bacto Agar stored in bottles or when sterilizing). If the screw top is not completely removed and there is inadequate space for the steam to vent, the bottles can explode and cause severe injury when the containers are removed from the microwave or autoclave. Always completely remove bottle caps before microwaving or autoclaving. An alternative method for routine agarose gels that do not require sterile agar is to weigh out the agar and place the solution in a flask.

- **Ultrasonicators** use high-frequency sound waves (16–100 kHz) for cell disruption and other purposes. This "ultrasound," conducted through air, does not pose a direct hazard to humans, but the associated high volumes of audible sound can cause a variety of effects, including headache, nausea, and tinnitus. Direct contact of the body with high-intensity ultrasound (not medical imaging equipment) should be avoided. Use appropriate ear protection and display signs on the door(s) of laboratories where the units are used.

- **Use extreme caution when handling cutting devices,** such as microtome blades, scalpels, razor blades, or needles. Microtome blades are extremely sharp! Use care when sectioning. If unfamiliar with their use, have an experienced user demonstrate proper procedures. For proper disposal, use the "sharps" disposal container in your laboratory. Discard used needles *unshielded*, with the syringe still attached. This prevents injuries and possible infections when manipulating used needles because many accidents occur while trying to replace the needle shield. Injuries may also be caused by broken Pasteur pipettes, coverslips, or slides.

- **Procedures for the humane treatment of animals** must be observed at all times. Consult your local animal facility for guidelines. Animals, such as rats, are known to induce allergies that can increase in intensity with repeated exposure. Always wear a lab coat and gloves when handling these animals. If allergies to dander or saliva are known, wear a mask.

DISPOSAL OF LABORATORY WASTE

There are specific regulatory requirements for the disposal of all medical waste and biological samples mandated by the U.S. Environmental Protection Agency (see http://www.epa.gov/epa waste/hazard/tsd/index.htm) and regulated by the individual states and territories (see http://www.epa.gov/epawaste/wyl/stateprograms.htm). Medical and biological samples that require special handling and disposal are generally termed Medical Pathological Waste (MPW), and medical, veterinary, and biological facilities will have programs for the collection of MPW and its disposal. Restrictions on how radioactive waste can be disposed of as regulated by the U.S. Nuclear Regulatory Commission can be found in 10 CFR 20.2001, General requirements for waste disposal (see http://www.nrc.gov/reading-rm/doc-collections/cfr/part020/part020-2001.html) or the individual Agreement States. The preferred method for the disposal of radioactively contaminated MPW is decay-in-storage (see http://www.nrc.gov/reading-rm/doc-collections/cfr/part035/part035-0092.html).

Waste and any materials contaminated with biohazardous materials must be decontaminated and disposed of as regulated medical waste. No harmful substances should be released into the environment in an uncontrolled manner. This includes all tissue samples, needles, syringes, scalpels, etc. Be sure to contact your institution's safety office concerning the proper practices associated with the handling and disposal of biohazardous waste.

Some basic rules are outlined below. For treatment of radioactive and biological waste, see sections on Radioactive Safety Procedures and Biological Safety Procedures.

- In practice, only **neutral aqueous solutions** without heavy metal ions and without organic solvents can be poured down the drain (e.g., most buffers). Acid and basic aqueous solutions need to be neutralized cautiously before their disposal by this method.

- For proper disposal of **strong acids and bases,** dilute them by placing the acid or base onto ice and neutralize them. Do not pour water into them. If the solution does not contain any other toxic compound, the salts can be flushed down the drain.

- For disposal of **other liquid waste,** similar chemicals can be collected and disposed of together, whereas chemically different wastes should be collected separately. This avoids chemical reactions between components of the mixture (see above). Collect at least inorganic aqueous waste, non-halogenated solvents, and halogenated solvents separately.

- Waste **from photo processing and automatic developers** should be collected separately to recycle the silver traces found in it.

Appendix

RADIOACTIVE SAFETY PROCEDURES

In the United States and other countries, the access to radioactive substances is strictly controlled. You may be required to become a registered user (e.g., by attending a mandatory seminar and receiving a personal dosimeter). A convenient calculator to perform routine radioactivity calculations can also be found at http://www.graphpad.com/quickcalcs/ChemMenu.cfm.

If you have never worked with radioactivity before, follow the steps below.

- *Try to avoid it!* Many experiments that are traditionally performed with the help of radioactivity can now be done using alternatives based on fluorescence or chemiluminescence and colorimetric assays, including, for example, DNA sequencing, Southern and northern blots, and protein kinase assays. However, in other cases (e.g., metabolic labeling of cells), use of radioactivity cannot be avoided.

- **Be informed.** While planning an experiment that involves the use of radioactivity, include the physicochemical properties of the isotope (half-life, emission type, and energy), the chemical form of the radioactivity, its radioactive concentration (specific activity), total amount, and its chemical concentration. Order and use only as much as is really needed.

- **Familiarize yourself** with the designated working area. Perform a mental and practical dry run (replacing radioactivity with a colored solution) to make sure that all equipment needed is available and to get used to working behind a shield. Handle your samples as if sterility would be required to avoid contamination.

- **Always wear appropriate gloves,** lab coat, and safety goggles when handling radioactive material.

- **Check the work area** for contamination before, during, and after your experiment (including your lab coat, hands, and shoes).

- **Localize your radioactivity.** Avoid formation of aerosols or contamination of large volumes of buffers.

- **Liquid scintillation cocktails** are often used to quantitate radioactivity. They contain organic solvents and small amounts of organic compounds. Try to avoid contact with the skin. After use, they should be regarded as radioactive waste; the filled vials are usually collected in designated containers, separate from other (aqueous) liquid radioactive waste.

- **Dispose of radioactive waste** only into designated, shielded containers (separated by isotope, physical form [dry/liquid], and chemical form [aqueous/organic solvent phase]). Always consult your safety office for further guidance in the appropriate disposal of radioactive materials.

- Among the experiments requiring **special precautions** are those that use $[^{35}S]$methionine and ^{125}I, because of the dangers of airborne radioactivity. $[^{35}S]$methionine decomposes during storage into sulfoxide gases, which are released when the vial is opened. The isotope ^{125}I accumulates in the thyroid and is a potential health hazard. ^{125}I is used for the preparation of Bolton–Hunter reagent to radioiodinate proteins. Consult your local safety office for further guidance in the appropriate use and disposal of these radioactive materials before initiating any experiments. Wear appropriate gloves when handling potentially volatile radioactive substances, and work only in a radioiodine fume hood.

BIOLOGICAL SAFETY PROCEDURES

Biological safety fulfills three purposes: to avoid contamination of your biological sample with other species; to avoid exposure of the researcher to the sample; and to avoid release of living material into the environment. Biological safety begins with the receipt of the living sample; continues with its storage, handling, and propagation; and ends only with the proper disposal of all contaminated materials. A catalog of operations known as "sterile handling" is usually employed in manipulating living matter. However, the actual manner of treatment largely depends on the actual sample, which can be quite diverse: *Escherichia coli* and other bacterial strains, yeasts, tissues of animal or plant origin, cultures of mammalian cells, or even derivatives from human blood are routinely handled in a biological laboratory. Two of these, bacteria and human blood products, are discussed in more detail below.

The Department of Health, Education, and Welfare (HEW) has classified various bacteria into different categories with regard to shipping requirements (see Sanderson and Zeigler 1991). Non-pathogenic strains of *E.coli* (such as K12) and *Bacillus subtilis* are in Class 1 and are considered to present no or minimal hazard under normal shipping conditions. However, *Salmonella*, *Haemophilus*, and certain strains of *Streptomyces* and *Pseudomonas* are in Class 2. Class 2 bacteria are "[a]gents of ordinary potential hazard: agents which produce disease of varying degrees of severity... but which are contained by ordinary laboratory techniques." Contact your institution's safety office concerning shipping biological material.

Human blood, blood products, and tissues may contain occult infectious materials such as hepatitis B virus and human immunodeficiency virus (HIV) that may result in laboratory-acquired infections. Investigators working with lymphoblast cell lines transformed by Epstein–Barr virus (EBV) are also at risk of EBV infection. Any human blood, blood products, or tissues should be considered a biohazard and should be handled accordingly until proved otherwise. Wear appropriate disposable gloves, use mechanical pipetting devices, work in a biological safety cabinet, protect against the possibility of aerosol generation, and disinfect all waste materials before disposal. Autoclave contaminated plasticware before disposal; autoclave contaminated liquids or treat with bleach (10% [v/v] final concentration) for at least 30 minutes before disposal (this is valid also for used bacterial media).

Always consult your local institutional safety officer for specific handling and disposal procedures of your samples. Further information can be found in the Frequently Asked Questions of the ATCC homepage (http://www.atcc.org) and is also available from the National Institute of Environmental Health and Human Services, Biological Safety (http://www.niehs.nih.gov/about/stewardship).

GENERAL PROPERTIES OF COMMON HAZARDOUS CHEMICALS

The hazardous materials list can be summarized in the following categories.

- **Inorganic acids,** such as hydrochloric, sulfuric, nitric, or phosphoric, are colorless liquids with stinging vapors. Avoid spills on skin or clothing. Spills should be diluted with large amounts of water. The concentrated forms of these acids can destroy paper, textiles, and skin and cause serious injury to the eyes.

- **Inorganic bases,** such as sodium hydroxide, are white solids that dissolve in water and under heat development. Concentrated solutions will slowly dissolve skin and even fingernails.

- **Salts of heavy metals** are usually colored, powdered solids that dissolve in water. Many of them are potent enzyme inhibitors and therefore toxic to humans and the environment (e.g., fish and algae).

- Most **organic solvents** are flammable volatile liquids. Avoid breathing the vapors, which can cause nausea or dizziness. Also avoid skin contact.

- **Other organic compounds** including organosulfur compounds, such as mercaptoethanol or organic amines, can have very unpleasant odors. Others are highly reactive and should be handled with appropriate care.

- If improperly handled, **dyes and their solutions** can stain not only your sample but also your skin and clothing. Some are also mutagenic (e.g., ethidium bromide), carcinogenic, and toxic.

- **Nearly all names ending with "ase"** (e.g., catalase, β-glucuronidase, or zymolyase) refer to enzymes. There are also other enzymes with nonsystematic names such as pepsin. Many of them are provided by manufacturers in preparations containing buffering substances, etc. Be aware of the individual properties of materials contained in these substances.

- **Toxic compounds** are often used to manipulate cells. They can be dangerous and should be handled appropriately.

- Be aware that several of the compounds listed have not been thoroughly studied with respect to their toxicological properties. Handle each chemical with appropriate respect. Although the toxic effects of a compound can be quantified (e.g., LD_{50} values), this is not possible for carcinogens or mutagens where one single exposure can have an effect. Also realize that dangers related to a given compound may also depend on its physical state (fine powder vs. large crystals/diethyl ether vs. glycerol/dry ice vs. carbon dioxide under pressure in a gas bomb). Anticipate under which circumstances during an experiment exposure is most likely to occur and how best to protect yourself and your environment.

Cold Spring Harbor Laboratory Press (CSHLP) has used its best efforts in collecting and preparing the material contained herein but does not assume, and hereby disclaims, any liability for any loss or damage caused by errors and omissions in the publication, whether such errors and omissions result from negligence, accident, or any other cause. CSHLP does not assume responsibility for the user's failure to consult more complete information regarding the hazardous substances listed in this publication.

REFERENCE

Sanderson KE, Zeigler DR. 1991. Storing, shipping, and maintaining records on bacterial strains. *Methods Enzymol* **204**: 248–264.

WWW RESOURCES

ATCC Home page http://www.atcc.org

ATCC, for Sample Handling (in Frequently Asked Questions) http://www.atcc.org/CulturesandProducts/TechnicalSupport/FrequentlyAskedQuestions/tabid/469/Default.aspx

GraphPad Software, Radioactivity Calculations http://www.graphpad.com/quickcalcs/ChemMenu.cfm

National Institute of Environmental Health and Human Services, Biological Safety (NIEHS) http://www.niehs.nih.gov/about/stewardship

U.S. Environmental Protection Agency (EPA), Federal waste disposal regulations, Laboratory http://www.epa.gov/epawaste/hazard/tsd/index.htm

U.S. Environmental Protection Agency (EPA), Individual States and Territories http://www.epa.gov/epawaste/wyl/stateprograms.htm

U.S. Nuclear Regulatory Commission (NRC), Medical Pathological Radioactively Contaminated Waste (Decay-in-Storage) http://www.nrc.gov/reading-rm/doc-collections/cfr/part035/part035-0092.html

U.S. Nuclear Regulatory Commission (NRC), Radioactive Waste Disposal Regulations: General Requirements http://www.nrc.gov/reading-rm/doc-collections/cfr/part020/part020-2001.html

Index

509